Advances in Industrial and Hazardous Wastes Treatment
Series Editors: Lawrence K. Wang and Mu-Hao Sung Wang

Control of Heavy Metals in the Environment, Vol. 1
Edited by Lawrence K. Wang, Mu-Hao Sung Wang, Yung-Tse Hung, and Jiaping Paul Chen

Control of Heavy Metals in the Environment, Vol. 2
Edited by Lawrence K. Wang, Mu-Hao Sung Wang, Yung-Tse Hung, and Jiaping Paul Chen

Control of Heavy Metals in the Environment, Vol. 3
Edited by Lawrence K. Wang, Mu-Hao Sung Wang, Yung-Tse Hung, and Jiaping Paul Chen

Industrial Water and Waste Treatment: Fundamentals, Vol. 1
Edited by Lawrence K. Wang, Mu-Hao Sung Wang, and Yung-Tse Hung

Industrial Water and Waste Treatment: Research Developments, Vol. 2
Edited by Lawrence K. Wang, Mu-Hao Sung Wang, and Yung-Tse Hung

Industrial Water and Waste Treatment: Technology Developments, Vol. 3
Edited by Lawrence K. Wang, Mu-Hao Sung Wang, and Yung-Tse Hung

Industrial Water and Waste Treatment: Process Developments, Vol. 4
Edited by Lawrence K. Wang, Mu-Hao Sung Wang, and Yung-Tse Hung

Industrial Water and Waste Treatment: Pollution Control, Vol. 5
Edited by Lawrence K. Wang, Mu-Hao Sung Wang, and Yung-Tse Hung

Industrial Water and Waste Treatment: Resources Conservation, Vol. 6
Edited by Lawrence K. Wang, Mu-Hao Sung Wang, and Yung-Tse Hung

Industrial Water and Waste Treatment: Case Studies, Vol. 7
Edited by Lawrence K. Wang, Mu-Hao Sung Wang, and Yung-Tse Hung

Industrial Water and Waste Treatment: Innovative Solutions, Vol. 8
Edited by Lawrence K. Wang, Mu-Hao Sung Wang, and Yung-Tse Hung

Industrial Water and Waste Treatment

Fundamentals

Edited by

Lawrence K. Wang, Mu-Hao Sung Wang,
and Yung-Tse Hung

CRC Press
Taylor & Francis Group
Boca Raton London New York

CRC Press is an imprint of the
Taylor & Francis Group, an **informa** business

Designed cover image: Shutterstock

First edition published 2026
by CRC Press
2385 NW Executive Center Drive, Suite 320, Boca Raton FL 33431

and by CRC Press
4 Park Square, Milton Park, Abingdon, Oxon, OX14 4RN

CRC Press is an imprint of Taylor & Francis Group, LLC

ISBN: 978-1-032-77295-0 (hbk)
ISBN: 978-1-032-77986-7 (pbk)
ISBN: 978-1-003-53235-4 (ebk)

DOI: 10.1201/9781003532354

Typeset in Times
by codeMantra

Contents

Preface

The industrial sector stands as a cornerstone of modern civilization, driving economic growth and technological advancements. However, with great power comes great responsibility, and the environmental impact of industrial activities cannot be overlooked. The treatment of industrial water and waste is a critical aspect of sustainable development, ensuring that our progress does not come at the expense of our planet's health.

Industrial Water and Waste Treatment: Fundamentals is the first volume in the *Industrial Water and Waste Treatment* series, meticulously edited by Lawrence K. Wang, Mu-Hao Sung Wang, and Yung-Tse Hung. This comprehensive volume, published by CRC Press, Florida, USA, serves as an essential resource for engineers, scientists, and environmental professionals dedicated to the field of industrial water and waste treatment.

The chapters within this book cover a wide array of topics, each contributing to a holistic understanding of the subject. The journey begins with an introduction to wastewater engineering design, followed by in-depth explorations of biological and physicochemical treatment methods. The fundamentals and innovations in controlling iron, manganese, calcium, and magnesium are thoroughly examined, providing readers with the latest advancements in these areas.

Biodegradation and the treatment of tannery solid wastes are discussed, highlighting the importance of sustainable waste management practices. This book also delves into the principles and innovations of waste stabilization ponds, nanomaterials for industrial wastewater treatment, and the use of clay minerals for fluoride wastewater treatment.

One of the most exciting aspects of this volume is the exploration of cutting-edge technologies, such as microbial electrochemical technology-assisted bioremediation and thermochemical processes for polyolefin plastic waste treatment. These chapters underscore the potential for transforming industrial waste into valuable resources, aligning with the principles of resource recycling and conservation.

This book further addresses solvent extraction fundamentals, oil and grease analysis using trichloroethane, and colorimetric technology for the determination of lignin and tannin in industrial waters. The fundamentals of ion exchange engineering and sequencing batch reactors are also covered, providing readers with a comprehensive understanding of these critical processes.

As we navigate the complexities of industrial water and waste treatment, this volume serves as a beacon of knowledge and innovation. It is our hope that the insights and advancements presented in this book will inspire and empower professionals to develop sustainable solutions that protect our environment and ensure a prosperous future for generations to come.

Lawrence K. Wang,
Mu-Hao Sung Wang, and
Yung-Tse Hung

Editors

Lawrence K. Wang has served the society as a professor, inventor, chief engineer, chief editor, and public servant (UN, USEPA, New York State) for over 50 years, with experience in the field of environmental science, technology, engineering, and mathematics (STEM). He is a licensed Professional Engineer, a certified Laboratory Director, a licensed Water Operator, and an OSHA Hazardous Waste Management Instructor. He has expertise in developing various innovative technologies, educational programs, licensing courses, international projects, and humanitarian organizations, all for promoting world peace. He is a retired Acting President/Professor of the Lenox Institute of Water Technology, USA; a Senior Advisor of the United Nations Industrial Development Organization (UNIDO), Vienna, Austria; and a former Professor/Visiting Professor of Rensselaer Polytechnic Institute, Stevens Institute of Technology, University of Illinois, National Cheng Kung University, Zhejiang University, and Tongji University. He is the author of 750+ papers and 50+ books and is credited with 29 invention patents. He holds a BSCE from National Cheng-Kung University, Taiwan, ROC; an MSCE from the University of Missouri-Rolla; an MS from the University of Rhode Island; and a PhD from Rutgers University, USA. Currently, he is the Book Series Editor of CRC Press. He has been a Delegate of the People to People International Foundation, a Diplomate of the American Academy of Environmental Engineers, a member of ASCE, AIChE, ASPE, WEF, AWWA, CIE, and OCEESA, and a recipient of Water Environment Federation Kenneth Research Award (NY), Five-Star Innovative Engineering Award (first dissolved air flotation drinking water plant in Americas), Ukraine Bulava/Mace Award (UNIDO Humanitarian Environmental Engineering Service to Ukraine), Korean Pollution Control Association Award (transfer of flotation technology to South Korea), and Overseas Environmental Engineers and Scientists Association Medal of Honor.

Mu-Hao Sung Wang has been an engineer of the New York State Department of Environmental Conservation, an editor of CRC Press, Springer Nature Switzerland, and Lenox Institute Press, and a university professor at the Stevens Institute of Technology, National Cheng-Kung University, and the Lenox Institute of Water Technology. Totally, she has been a government official and an educator in the USA and Taiwan for over 50 years. Dr. Wang is a licensed Professional Engineer and a Diplomate of the American Academy of Environmental Engineers (AAEE). Her publications have been in the areas of water quality, modeling, environmental sustainability, solid and hazardous waste management, NPDES, flotation technology, industrial waste treatment, and analytical methods. She is the author of over 50 publications and an inventor of 14 US and foreign patents. She holds a BSCE degree from National Cheng-Kung University, Taiwan, ROC; an MS degree from the University of Rhode Island, RI, USA; and a PhD degree from Rutgers University, NJ, USA. She is the Coeditor of the *Advances in Industrial and Hazardous Wastes Treatment* series (CRC Press); and a member of AWWA, NYWWA, NEWWA, WEF, NEWEA, CIE, and OCEESA.

Yung-Tse Hung served as Professor of Civil Engineering at Cleveland State University, Cleveland, Ohio, USA, since 1981 until his retirement in 2024. He is a Fellow of the American Society of Civil Engineers, a licensed Professional Engineer in Ohio and North Dakota, and a Diplomate of the American Academy of Environmental Engineers. He has taught at 16 universities in 8 countries. His research interests and publications have been involved with biological treatment processes, solid wastes, hazardous waste management, and industrial waste treatment. He is credited with over 811 publications and presentations, 40 books, 242 book chapters, and approximately 198 peer-reviewed journal articles in water and wastewater treatment. He received his BSCE and MSCE degrees from National Cheng-Kung University, Taiwan, ROC, and his PhD degree from the University of Texas at Austin, USA. He is the Editor-in-Chief of *International Journal of Environmental Pollution Control and Management, International Journal of Environmental Engineering,* and *International Journal of Environmental Engineering Science,* and the Coeditor of the *Advances in Industrial and Hazardous Wastes Treatment* series (CRC Press). He is also the Permanent Executive Director and the Ex-President of OCEESA (Overseas Chinese Environmental Engineers and Scientists Association).

Contributors

Shaifa Abrarin
Institute of National Analytical Research and Service (INARS)
Bangladesh Council of Scientific and Industrial Research (BCSIR)
Dhanmondi, Dhaka, Bangladesh

Motasem Y. D. Alazaiza
Department of Environmental Engineering
A'Sharqiyah University
Ibra, Oman

Ahmed Albahnasawi
Department of Environmental Engineering
Gebze Technical University
Gebze-Kocaeli, Turkey

Hamidi Abdul Aziz
No 2, Lorong Jawi Jaya 11, Taman Jawi Jaya
14200 Sungai Jawi, Pulau Pinang, Malaysia

Krzysztof Biernat
Fuel and Bioeconomy Research Group
Łukasiewicz Research Network – Automotive Industry Institute
Warsaw, Poland

João Henrique Zimnoch dos Santos
Federal University of Rio Grande do Sul
Porto Alegre-RS, Brazil

Murat Eyvaz
Department of Environmental Engineering
Gebze Technical University
Gebze-Kocaeli, Turkey

Ercan Gürbulak
Department of Environmental Engineering
Gebze Technical University
Gebze-Kocaeli, Turkey

Mariliz Gutterres
Chemical Engineering Department
Federal University of Rio Grande do Sul
Porto Alegre-RS, Brazil

Noureddine Hamdi
Higher Institute of Water Sciences and Techniques of Gabes
University of Gabes
Zrig, Tunisia
and
National Center of Research in Materials Sciences
Soliman, Tunisia

Yung-Tse Hung
Department of Civil and Environmental Engineering
Cleveland State University, Cleveland, Ohio, USA

Abu Reza Md. Towfiqul Islam
Department of Disaster Management
Begum Rokeya University
Rangpur, Bangladesh

Rahat Khan
Institute of Nuclear Science and Technology
Bangladesh Atomic Energy Commission (BAEC)
Savar, Dhaka, Bangladesh

Victória Vieira Kopp
Chemical Engineering Department
Federal University of Rio Grande do Sul
Porto Alegre-RS, Brazil

Mohammed Mahbubul Matin
Department of Chemistry
Faculty of Science
University of Chittagong
Chittagong, Bangladesh

Anna Matuszewska
Fuel and Bioeconomy Research Group
Łukasiewicz Research Network – Automotive Industry Institute
Warsaw, Poland

Kenneth A. Mikkelson
Aqua-Aerobic Systems, Inc.,
Rockford, Illinois, USA

Marlena Owczuk
Fuel and Bioeconomy Research Group
Łukasiewicz Research Network – Automotive Industry Institute
Warsaw, Poland

Nor Habsah Md Sabiani
School of Civil Engineering
Universiti Sains Malaysia
Pulau Pinang, Malaysia

Rakesh Kumar Sharma
Department of Life Sciences
Chhatrapati Shahu Ji Maharaj University
Kanpur, Uttar Pradesh, India

Akansha Shrivastava
Department of Biosciences
Manipal University Jaipur
Jaipur, Rajasthan, India

Md. Abu Bakar Siddique
Institute of National Analytical Research and Service
Bangladesh Council of Scientific and Industrial Research (BCSIR)
Dhanmondi, Dhaka, Bangladesh

Pragati Singh
Department of Biosciences
Manipal University Jaipur
Jaipur, Rajasthan, India

Lawrence K. Wang
Lenox Institute of Water Technology
Auburndale, Massachustees, USA

Mu-Hao Sung Wang
Lenox Institute of Water Technology
Auburndale, Massachustees, USA

Kawthar Yahya
Higher Institute of Water Sciences and Techniques of Gabes
University of Gabes
Zrig, Tunisia
and
National Center of Research in Materials Sciences
Soliman, Tunisia

Ebubekir Yüksel
Department of Environmental Engineering
Gebze Technical University
Gebze-Kocaeli, Turkey

1 Wastewater Engineering Design

Introduction[1]

Hamidi Abdul Aziz

LIST OF ABBREVIATIONS

ANN	Ammoniacal nitrogen
ASP	Activated sludge process
BOD	Biochemical oxygen demand
COD	Chemical oxygen demand
Commission	See SPAN
CST	Communal septic tank
DN	Nominal internal diameter
DWF	Dry weather flow
EA	Extended aeration
HRT	Hydraulic retention time
IST	Individual septic tank
IWK	Indah Water Konsortium
NH_3-N	Ammoniacal nitrogen
NO_3-N	Nitrate-nitrogen
Q_{ave}	Average flow or DWF
Q_{max}	Maximum flow
Q_{min}	Minimum flow
Q_{peak}	Peak flow
RAS	Return activated sludge
RBC	Rotating biological contactor
SBR	Sequential batch reactor
SLR	Solids loading rate
SOR	Surface overflow rate
SPAN	Suruhanjaya Perkhidmatan Air Negara
SS	Suspended solids, also referred as TSS
SVI	Sludge volume index
TOC	Total organic carbon
TSS	Total suspended solids
WAS	Waste activated sludge
WLR	Weir loading rate
WWTP	Wastewater treatment plant

1.1 INTRODUCTION

Wastewater treatment is crucial to prevent environmental pollution. In a broad term, wastewater is used water that has been contaminated due to human use. Sources of wastewater include: domestic from households, municipal from communities (also called sewage), or industrial from industrial

DOI: 10.1201/9781003532354-1

activities. The term sewage is nowadays regarded as an older term and is being more and more replaced by "domestic wastewater" (the term used in this book).

Households may produce wastewater from toilets, sinks, dishwashers, washing machines, bathtubs, and showers. Wastewater from sinks, tubs, showers, dishwashers, and clothes washers is termed greywater. Sanitary wastewater or blackwater is the water used to flush toilets (faeces and urine), combined with the human waste that it flushes away. Wastewater is also rich in soaps and detergents, and toilet paper.

Wastewater usually travels from a building through a pipeline called sewer (combined or separate), which will carry it elsewhere, or into an on-site sewage facility (of which there are many kinds). A sanitary sewer conveys only blackwater. A combined sewer which includes stormwater runoff is not so common in recent days due to its high volume, especially in countries with high rainfall.

The characteristics of wastewater vary depending on the source. Wastewater is normally characterized by volume or rate of flow, physical condition, chemical and toxic constituents, and its bacteriologic status (which organisms it contains and in what quantities). Wastewater that is discharged to the environment without suitable treatment causes water pollution. Hence, the wastewater must be treated before being released into receiving waters. Wastewater treatment normally involves physical, chemical, and biological processes to remove contaminants and produce treated wastewater. In many developing countries, the bulk of domestic and industrial wastewater is discharged without any treatment or with partial treatment only.

A wastewater treatment plant (hereafter referred to as WWTP) treats wastewater from a community, a region, or a particular city. After treatment, the treated effluent is discharged to a receiving water body, normally in drains or rivers, to maintain the water cycle. Figures 1.1 and 1.2 show a typical schematic diagram of the wastewater treatment process of a city.

1.2 MAIN OBJECTIVES IN WASTEWATER TREATMENT

Wastewater treatment is a process used to convert wastewater into treated effluent, which is discharged into receiving waters such as drains or rivers or returned to the water cycle for recovery or reuse. As populations increase by leaps and bounds, it places more pressure on the environment, threatening sources of fresh water supplies; it was recognized that the problem of wastewater needed proper management.

Generally, wastewater is treated for the following reasons:

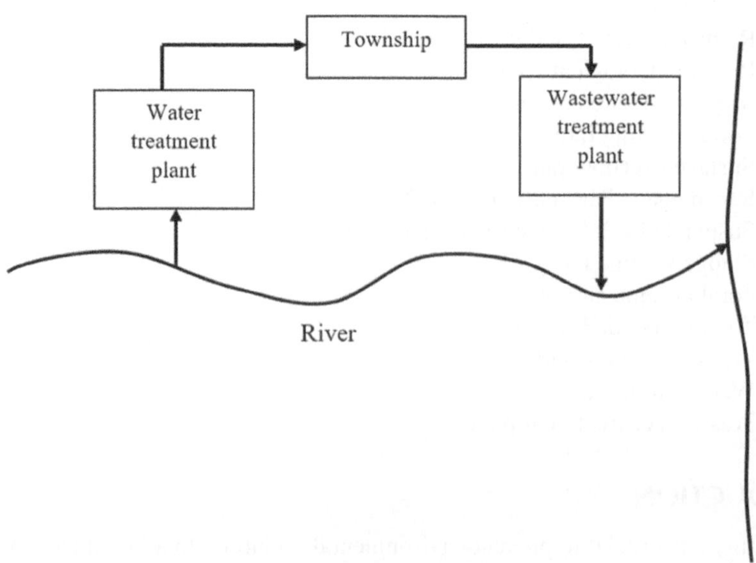

FIGURE 1.1 Example of a wastewater treatment system of a city.

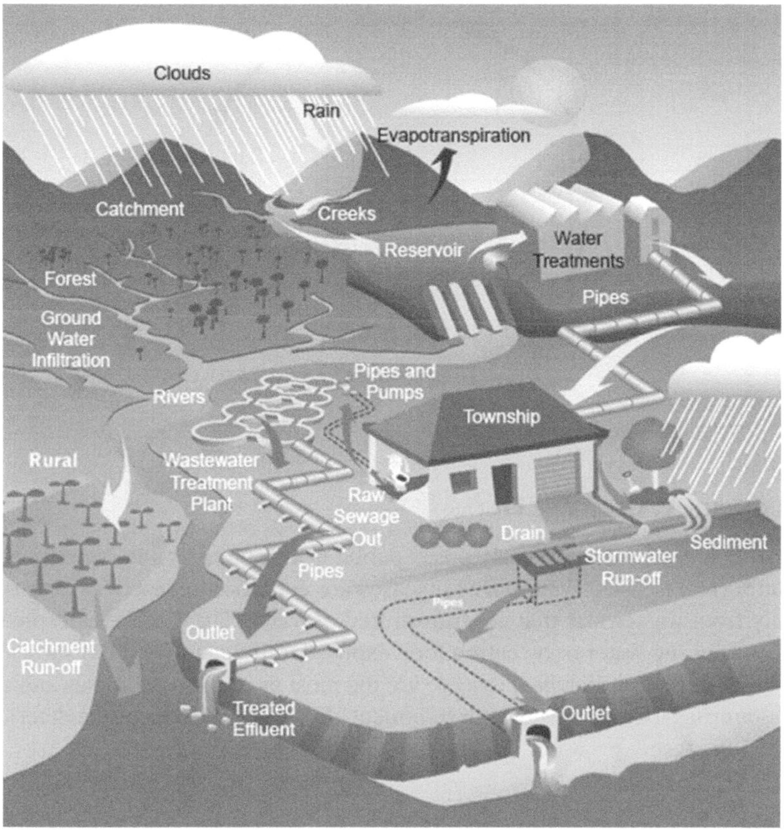

FIGURE 1.2 Basic water cycle.

1.2.1 To Avoid Contamination

As explained earlier, wastewater treatment is crucial to prevent environmental pollution. This is because, as we know, wastewater contains a high level of pollutants such as biochemical oxygen demand (BOD), chemical oxygen demand (COD), and suspended solids (SS) with values around 250, 300, and 350 mg/L, respectively in Malaysia. They cause various pollution issues such as degradation of water quality which impacts aquatic life. The standard discharge limits for BOD, COD, and SS (Standard A) in Malaysia are 20, 100, and 40 mg/L, respectively. Therefore, the wastewater must be treated before discharge to the environment.

1.2.2 To Improve Sanitation

The treatment of wastewater is part of the overarching field of sanitation. Sanitation also includes the management of human waste and solid waste as well as stormwater (drainage) management. By-products from WWTPs, such as screenings, grit, and wastewater sludge, also require proper disposal. The generated wastewater should be properly managed from the source, transported via sewer lines, collected at the treatment plant, and treated. This process is part of the basic infrastructure necessity of a community in the development of human civilization. The importance of this treatment system is in line with the importance of basic amenities such as water and electricity. The development of an urban area would have to take into account aspects of wastewater treatment facilities.

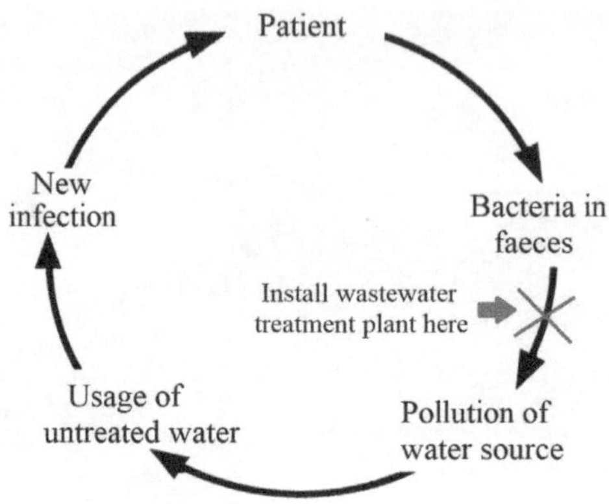

FIGURE 1.3 A cycle of a waterborne disease.

1.2.3 To Control Waterborne Diseases

Wastewater that is not managed properly poses public health hazards and increases the risk of the spread of waterborne diseases. Waterborne diseases are conditions caused by pathogenic microorganisms mainly from wastewater that are transmitted in water. The disease can spread while bathing, washing, or drinking water or by eating food exposed to infected water, as shown in Figure 1.3. Various forms of waterborne diarrheal disease are the most prominent examples and affect children in developing countries. Hence, wastewater treatment eliminates or reduces the deteriorative impact of wastewater on the receiving water quality and its environment.

1.2.4 Others

Clean water is critical to plants and animals that live in water. This is important to the fishing industry, sport fishing enthusiasts, and future generations. Our rivers and ocean waters teem with life that depends on shorelines, beaches, and marshes. They are critical habitats for hundreds of species of fish and other aquatic life. All these must be protected from pollution. Migratory waterbirds use the water for resting and feeding. Water is also a great playground. The scenic and recreational nature of our waters may be lost due to pollution from untreated wastewater.

1.3 EVOLUTION OF WASTEWATER TREATMENT SYSTEMS

This section does not attempt to give a detailed history of wastewater treatment but instead gives a brief and general overview of the wastewater treatment from the old days until now.

Many ancient cities had drainage systems, but they were primarily intended to carry rainwater away from roofs and pavements. A notable example is the drainage system of ancient Rome. It included many surface conduits that were connected to a large vaulted channel called the Cloaca Maxima ("Great Sewer"), which carried drainage water to the Tiber River. Built of stone and on a grand scale, the Cloaca Maxima is one of the oldest existing monuments of Roman engineering.

There was little progress in urban drainage or sewerage during the Middle Ages. Privy vaults and cesspools were used, but most wastes were simply dumped into gutters to be flushed through the drains by floods. Toilets (water closets) were installed in houses in the early 19th century, but they were usually connected to cesspools, not to sewers. In densely populated areas, local conditions soon became intolerable because the cesspools were seldom emptied and frequently overflowed. The threat to public health became apparent. In England in the middle of the 19th century, outbreaks of cholera were traced directly to well-water supplies contaminated with human waste from privy vaults and cesspools. It soon became necessary for all water closets in the larger towns to be connected directly to the storm sewers. This transferred wastewater from the ground near houses to nearby bodies of water. Thus, a new problem emerged: surface water pollution.

It used to be said that "the solution to pollution is dilution." When small amounts of wastewater are discharged into a flowing body of water, a natural process of stream self-purification occurs. Densely populated communities generate such large quantities of wastewater; however, that dilution alone does not prevent pollution. This makes it necessary to treat or purify wastewater to some degree before disposal.

The construction of centralized WWTPs began in the late 19th and early 20th centuries, principally in the United Kingdom and the United States. Instead of discharging wastewater directly into receiving water, it was first passed through a combination of physical, biological, and chemical processes that removed some or most of the pollutants. Also, beginning in the 1900s, new wastewater collection systems were designed to separate stormwater from domestic wastewater so that treatment plants did not become overloaded during periods of wet weather.

After the middle of the 20th century, increasing public concern for environmental quality led to broader and more stringent regulation of wastewater disposal practices. Higher levels of treatment were required. For example, pretreatment of industrial wastewater, with the aim of preventing toxic chemicals from interfering with the biological processes used at WWTPs, often became a necessity. Wastewater treatment technology advanced to the point where it became possible to remove virtually all pollutants from wastewater. This was so expensive, however, that such high levels of treatment were not usually justified.

WWTPs became large, complex facilities that required considerable amounts of energy for their operation. After the rise of oil prices in the 1970s, concern for energy conservation became a more important factor in the design of new pollution control systems. Consequently, land disposal and subsurface disposal of wastewater began to receive increased attention where feasible. Such "low-tech" pollution control methods not only might help to conserve energy but also might serve to recycle nutrients and replenish groundwater supplies.

Today's wastewater treatments are much more technologically advanced than they were in previous years. Many centralized mechanized treatments are run via a computer system, and they are run more efficiently.

In Malaysia, basic sanitation facilities in the late 1960s such as pit latrines, bucket systems, and pour-flush systems were slowly replaced by more modern systems like aerated lagoons, activated sludge systems, package systems, and a variety of mechanical plants, as summarized in Figure 1.4. However, wastewater remains one of the major pollutants of our inland waterways. In the 1900s, the emergence of new treatment technologies was mainly driven by the basic need to treat wastewater to control waterborne diseases. Today, the environmental regulations, including effluent discharge standards, are becoming more stringent with increasing awareness towards sustainable environmental management. The public is also more educated and more alert on the need to preserve the water source and the environment. Hence the introduction of more innovative design in municipal wastewater treatment technologies is needed to meet the stricter regulatory requirements. Many modern centralized WWTPs, such as an activated sludge process (ASP), have been used in Malaysia since the 1970s–1980s. Currently, the Sequential Batch Reactor (SBR) system with various modifications is already in operation at many sites in the country.

1.4 COMMON TERMS IN WASTEWATER ENGINEERING

The following terms are commonly used in wastewater engineering:

Aerobic Treatment (or Oxic Treatment): A process where wastewater or material is broken down by microorganisms in the presence of dissolved oxygen (DO).

Anaerobic Treatment: A process where wastewater or material is broken down by microorganisms without the aid of DO.

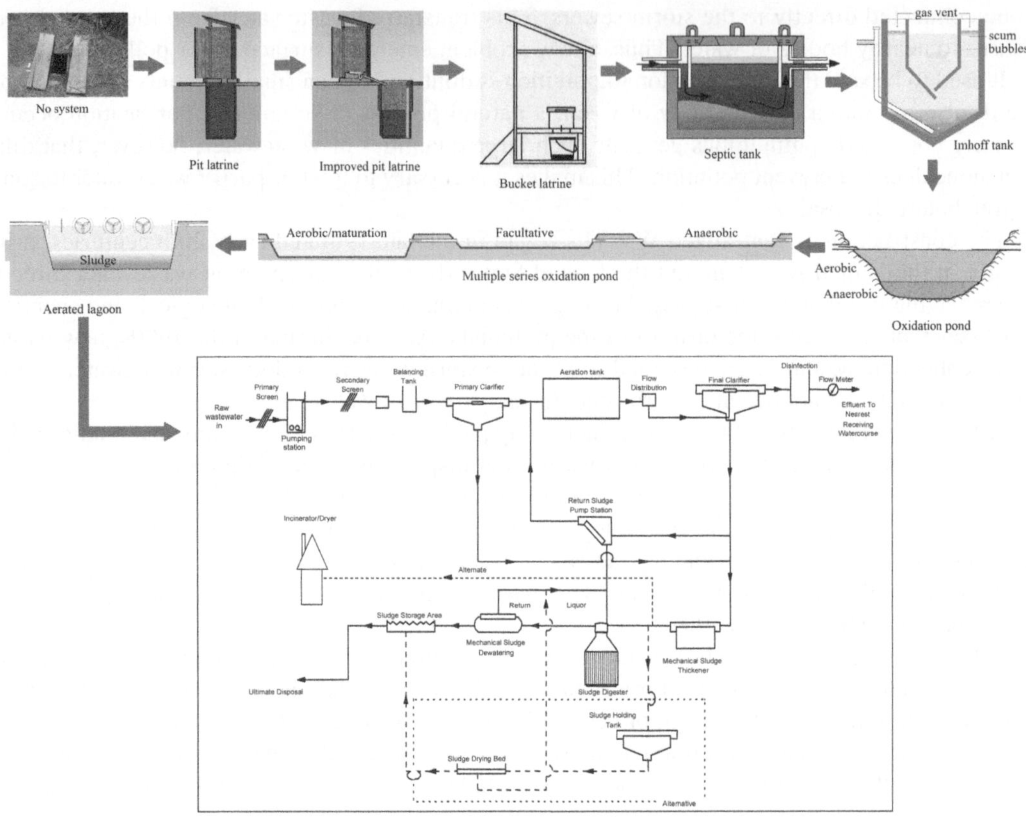

FIGURE 1.4 Evolution of wastewater treatment in Malaysia.

Anoxic: A biological environment that is deficient in molecular oxygen (very low concentration of molecular oxygen (O_2)) but may contain chemically bound oxygen, such as nitrates (NO_3^-) nitrites (NO_2^-), sulphate (SO_4^{2-}), and phosphate (PO_4^{2-}).

Bacteria: Microscopic living organisms. They are a group of universally distributed, rigid, essentially unicellular, microscopic organisms lacking chlorophyll. They are characterized as spheroids, rod-like, or curved entities but occasionally appear as sheets, chains, or branched filaments.

Biomass: The bacteria used to treat the wastewater are collectively called the biomass. The bacteria grow, multiply, and die in the presence of food, nutrients, and air (in aerobic systems) to make the system work.

Biosolid: The sludge after digestion.

Centralized Wastewater Treatment System: A public system in which wastewater is collected by sewer that transports the wastewater to a central treatment facility.

Coliform Organisms: A group of bacteria recognized as indicators of faecal pollution.

Combine Sewer: A sewer that carries wastewater together with rainwater. The system is not practical for countries with high rainfall.

Digestion: Process by which organisms break down sludge, creating as by-products methane gas, carbon dioxide, solid organic material, and water. (Aerobic digestion takes place in the presence of oxygen, and anaerobic digestion takes place in the absence of oxygen.)

Disinfection: Final step in the tertiary wastewater treatment process, when chlorine or sodium hypochlorite is added to the treated wastewater to kill disease-causing organisms. Ultraviolet light is another means of disinfection.

Dry Weather Flow (DWF) Rate: An average discharge of wastewater based on the total population multiplied by the rate of water consumption per capita per day.

Effluent: "Cleaned" wastewater which flows out of a treatment plant. It is also commonly termed treated effluent or final effluent.

***Escherichia coli* (*E. coli*):** A species of bacteria found in large numbers in the intestinal tract of warm-blooded animals, such as a human.

Force Mains: Sewer line with forced pressure or flows with pumping force.

Freeboard: The vertical distance from the normal water surface to the top of the confining wall. It is normally added to the designed depth as a safety factor.

Grit: Heavy, inorganic, and non-biodegradable matter, such as sand or pebbles.

Indah Water Konsortium (IWK): A company owned by Minister of Finance Incorporated, is Malaysia's national sewerage company which has been entrusted with the task of developing and maintaining a modern and efficient sewerage system for all Malaysians.

Individual or a Separate Wastewater Treatment System: An individual wastewater treatment system such as septic and Imhoff tank that works for specific premises and does not connect to public sewer.

Infiltration: Groundwater that seeps into or outside the sewer through leakages of sewer (see Figure 1.5).

Influent: Untreated wastewater when it flows into a treatment plant.

Inorganic Material: A substance that has no element or carbon bond. It will not respond to biological action. In wastewater engineering, it usually refers to grit, salt, and metal.

Organic Material: Material that has the main element which is bound to a carbon-carbon bond. In wastewater engineering, organic materials commonly refer to a material that contains elements carbon (C), hydrogen (H), oxygen (O), nitrogen (N), and sulphur (S) present in the molecule. For example, a sugar molecule is composed of $C_6H_{12}O_6$.

Primary Treatment: First stage of wastewater treatment that uses physical processes such as screening, settlement, and filtration.

Raw Sewage: Untreated wastewater.

Receiving Waters: Drains, rivers, lakes, or other water sources that receive treated or untreated effluent.

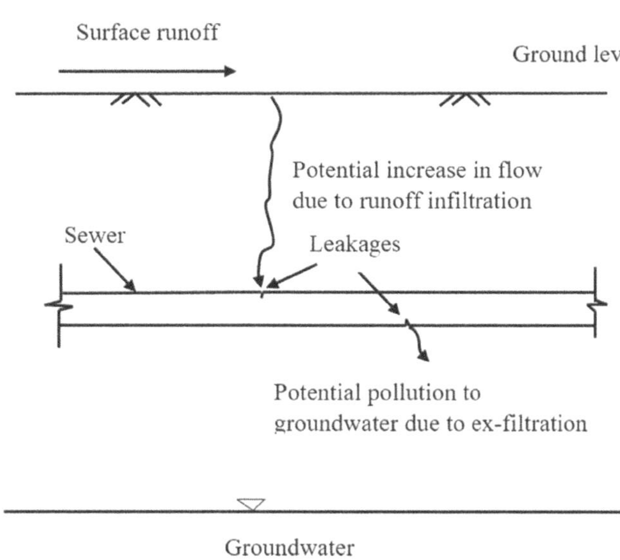

FIGURE 1.5 Infiltration and insertion sewers.

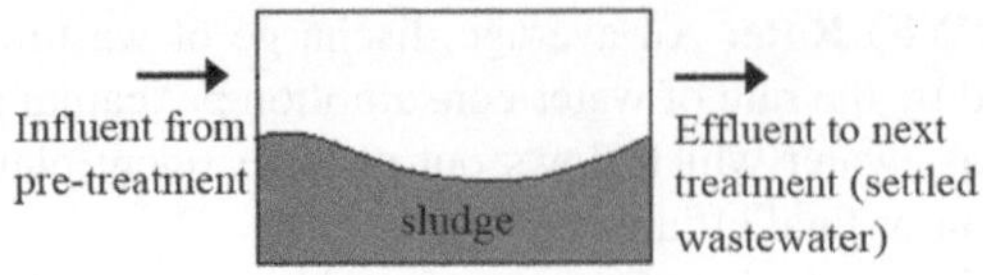

FIGURE 1.6 Sludge.

Secondary Treatment: Second stage of wastewater treatment uses a biological process in which bacteria consume organic matter, then settle out as sludge.

Sedimentation: Process by which solid material settles out of a liquid. Sludge settles out of wastewater in sedimentation tanks.

Separate Sewer: A sewer that only carries wastewater. This system is practised in Malaysia.

Sewerage: The infrastructure that conveys wastewater or stormwater using sewers. It encompasses components such as receiving drains, manholes and connections, pumping stations, storm overflows, and screening chambers of the combined sewer or sanitary sewer.

Sewers: Pipes that carry the wastewater.

Sludge: Deposited solids separated from wastewater during treatment. It usually refers to the discharged wastewater (effluent) from the primary sedimentation tank, as shown in Figure 1.6.

SPAN (Suruhanjaya Perkhidmatan Air Negara): A National Water Services Commission which was established in 2008 to regulate the water services industry in Malaysia (including wastewater). It is also referred to as Commission.

Storm Sewers: A pipeline or channel system that carries surface water and/or runoff to public waters but does not feed into the sewer system.

Storm Water: Surface water and/or runoff to public waters.

Tertiary Treatment: The final stage of wastewater treatment which aims to raise the effluent quality to be domestic and industrial standards or to meet special requirements for discharging into receiving water. It normally involves the treatment of nutrients such as nitrogen and phosphorous.

Wastewater: Wastewater from homes, businesses, or industry that contains organic and inorganic materials in the form of dissolved, suspended, and particulate matter. It usually refers to water from the toilet, sink, and toilet. It can easily be defined as any discharge that enters the sewer line. For example, if the washing machine drain is connected to the bathroom, which channels it into the sewer line, it is considered as wastewater or sewage, which requires full treatment at the treatment plant. If it is placed in the kitchen and the flow enters the drain, it is termed stormwater.

Wastewater Loading: The concentration of a wastewater parameter multiplies with its flow reported in terms of mass per time. For example, BOD load means the BOD concentration in a unit of mass per volume (for example, in mg/L) multiplied with its flow in a unit of volume over time (for example, in m^3/day). Its load is reported as kg/day. The same applies to COD, ammoniacal nitrogen, and others.

Water Pollution: A general term signifying the introduction into the water of microorganisms, chemicals, wastes, or wastewater that renders the water unfit for its intended use.

1.5 WASTEWATER COMPOSITIONS

1.5.1 Definition

Domestic wastewater contains organic and inorganic solids and pathogenic microorganisms with 99.9% water and 0.1% solids. The breakdown of wastewater is shown in Figure 1.7.

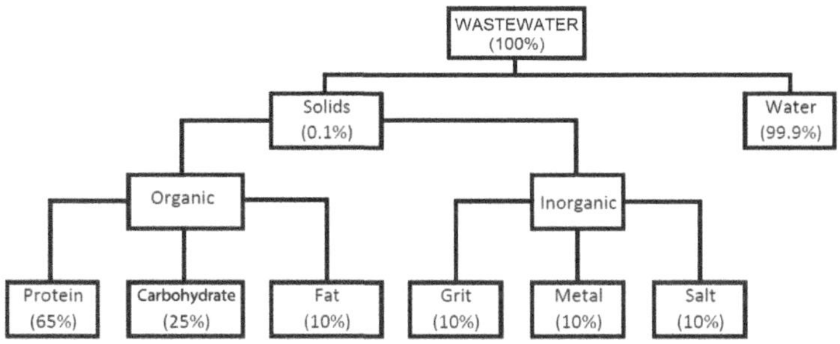

FIGURE 1.7 Classification of wastewater.

Temperature ↑	Viscocity ↓	Settlement ↑	Removal ↑
Temperature ↓	Viscocity ↑	Settlement ↓	Removal ↓

FIGURE 1.8 Settlement rate at different temperatures.

1.5.2 Wastewater Parameters

Wastewater is characterized in terms of its physical, chemical, and biological compositions. The main wastewater characteristics or compositions are discussed below:

1.5.2.1 Physical Characteristics

1.5.2.1.1 Temperature

Generally, wastewater temperature is slightly higher than the water temperature. The solubility of DO in wastewater also depends on the temperature. The higher the temperature, the lower the degree of oxygen solubility as described by Le Chatelier's principle. This principle states that if the equilibrium of a system is disturbed by a change in one or more of the determining factors (such as temperature, pressure, or concentration), the system tends to adjust itself to a new equilibrium by counteracting as far as possible the effect of the change.

Temperature factors also influence the degree of solids settlement. The higher the temperature, the lower the viscosity of the sludge, hence the settlement will be better (higher removal rate of SS). As opposed to the viscous sample (which happens at low temperature), the degree of settlement is lower. Hence the solids do not settle much and may release as partially treated effluent (lower removal rate of SS), as shown in Figure 1.8.

1.5.2.1.2 Colour

Colour is not necessarily a problem in wastewater treatment. Instead, it is an indicator of the condition of the wastewater. Condition refers to the age of the wastewater, which is determined qualitatively by its colour and odour. Fresh wastewater is a light brownish-grey colour. The colour of wastewater changes sequentially from grey to dark grey and ultimately to black as the travel time in the collection system increases (flow becomes increasingly more septic) and more anaerobic conditions develop. Domestic wastewaters will have present, in varying amounts, all kinds of floating matter such as faecal solids, a bit of food, oil, garbage, paper, rags, wood, plastics, and other materials disposed of in the daily life of a community. Under certain conditions, because of biochemical changes caused by bacteria, the colour of the wastewater will gradually change from grey to black. As this happens, foul and unpleasant odours develop, and black solids appear on the surface or throughout the liquid. Wastewater that has undergone such a change is called septic.

1.5.2.1.3 Turbidity

Turbidity is the cloudiness or haziness of a fluid caused by large numbers of individual particles (fine SS) that are generally invisible to the naked eye, similar to smoke in the air.

1.5.2.1.4 Odour

In wastewater, odours are of major concern, especially to those who reside near a WWTP. These odours are generated by gases produced by the decomposition of organic matter or by substances added to the wastewater. The odour from fresh wastewater is less objectionable than the odour from wastewater that has undergone anaerobic decomposition. The most characteristic odour of stale or septic wastewater is that of hydrogen sulphide (H_2S), which is produced by anaerobic microorganisms that reduce sulphate to sulphide. The malodorous compounds responsible for producing objectionable odours in water can be detected by diluting a sample with odour-free water until the least detectable odour level is achieved. This is recorded as TON (Threshold Odour Number). Any commercially available gas monitor can measure the concentration of malodorous gases such as hydrogen sulphide, ammonia, mercaptans, etc., emitted into the air from wastewater.

1.5.2.1.5 Solids

Solids in domestic wastewater contain about 50% organics. They come from the food that we take, which mainly contains organics such as carbohydrates, glucose, protein, and oil and fat. These are substances that contain carbon, hydrogen, and oxygen, some of which may be combined with nitrogen, sulphur, or phosphorous. These compounds are subject to decay or decomposition through the activity of bacteria and other living organisms. They are rich in various forms of solids (dissolve, suspended) when released as wastewater. Once combusted at high temperatures, the organics are volatile.

The inorganic solids (the remaining 50%) are substances that are inert and not subject to decay. Inorganic solids are frequently called mineral substances and include fine sand, gravel, and silt, as well as mineral salts. In general, they are non-combustible.

The amount of these solids, both organic and inorganic, gives wastewater the characteristic frequently termed as its "strength." The amount or concentration of the organic solids and their capacity to undergo decay or decomposition is the most important part of this strength. The greater the concentration of organic or volatile solids, the stronger the wastewater. "Strong" wastewater can be defined as one containing a large number of solids, particularly organic solids. "Weak" wastewater is one containing only a small amount of organic solids.

Solids can also be grouped depending on their physical state as total solids, SS, colloidal solids, and dissolved solids, each of which can include both organic and inorganic solids.

Total solids, as the term implies, includes all the solid constituents of wastewater. Total solids are the total of the organic and inorganic solids or the total of the suspended and dissolved solids. In average domestic wastewater, total solids are about half organic and half inorganic, and about two-thirds in solution (dissolved) and one-third in suspension. The organic solids, which are subject to decay, constitute the main problem in wastewater treatment.

SS are those which are visible and in suspension in the water. They are the solids that can be removed from the wastewater by physical or mechanical means, such as sedimentation or filtration. SS are approximately 70% organic solids and 30% inorganic solids, the latter being principally sand and grit. The SS portion consists of settleable solids and colloidal solids (non-settleable solids).

Settleable solids are that portion of the SS which are of enough size and weight to settle in a given period, usually 1 hour. There are those that will settle in an Imhoff Cone in 1 hour. The results are reported as millilitres of settled solids per litre of wastewater. Settleable solids are approximately 75% organic and 25% inorganic.

Colloidal solids are solid particles of size less than 2 μm equivalent spherical diameter, also identified as clay. It may be organic or inorganic and is non-settleable.

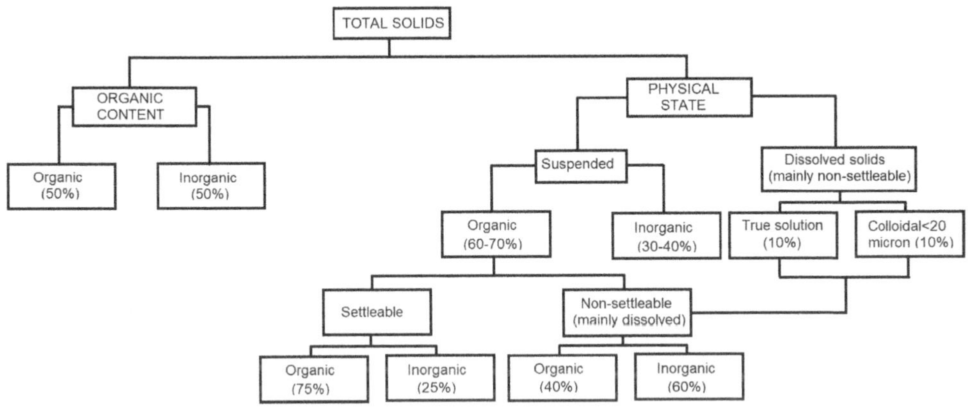

FIGURE 1.9 Breakdown of solids found in wastewater .

Volatile suspended solid (VSS) is a water quality measurement obtained from the loss on ignition of the mass of measured total SS. This ignition generally takes place in an oven at a temperature of 550°C–600°C. It represents the amount of **VOLATILE MATTER (ORGANICS)** present in the solid fraction of the measured solution.

Dissolved solids are smaller in size than suspended and colloidal solids. As used, the term means all the solids which pass through the filter pad of a Gooch Crucible. Of the total dissolved solids, about 90% are in true solution and about 10% colloidal. Dissolved solids are about 40% organic and 60% inorganic.

The breakdown of solids found in wastewater is summarized in Figure 1.9.

1.5.2.2 Chemical Properties

Chemical characteristics of wastewater consist of parameters that could not be observed, tasted, smelt, or felt.

1.5.2.2.1 pH

pH refers to the acidity of the sample. Domestic wastewater before treatment typically has a pH of 6.5–8.5 but a final effluent of 7.0–7.2. Usually, the pH of fresh wastewater was slightly more than 7. However, this value should be slightly acidic wastewater that has undergone the anaerobic process. This is because, in the anaerobic process (Equation 1.1), the gas produced is methane (CH_4), carbon dioxide (CO_2), and hydrogen sulphide (H_2S). These gases will drop the pH.

$$\text{Organic matter (in sewage)} + \text{Bacteria (without oxygen)} \rightarrow CO_2 + CH_4 + H_2S + \text{New cells} \quad (1.1)$$

Product main anaerobic reaction

Most of the WWTP requires the effluent to be maintained at neutrality. Advanced treatment is necessary should the level fall within the neutrality range, especially for the case of the treatment plant that also receives effluent other than the domestic wastewater.

1.5.2.2.2 Nitrogen

The different forms of nitrogen and phosphorus in wastewater are referred to collectively as nutrients.

Most nitrogen excreted by humans is in the form of organic nitrogen (dead cell material, proteins, amino acids) and urea. Through natural biochemical processes, organic nitrogen in wastewater is slowly converted into ammonia, which is the form of nitrogen best able to be utilized as a nutrient by microorganisms in the treatment process. Some wastewaters may be nitrogen deficient and require supplemental ammonia for adequate reproduction.

Ammonium (NH_4^+) is the ionized form of ammonia (NH_3) which is oxidized by nitrifying bacteria (*Nitrosomonas*). Unionized ammonia (NH_3) is the toxic form and will predominate at higher pH. Ammonium (NH_4^+) is relatively non-toxic and predominates at lower pH values. Excessive ammoniacal nitrogen in waterways can cause taste and (pungent) odour problems, apart from introducing a psychological problem to consumers, who will be under the impression that the water source is contaminated with wastewater, even though this is not the case.

Nitrogen (organically bonded nitrogen with C bond) contains mainly proteins (for example, amino acids $-NH_2CHRCOOH$), peptides, nucleic acids, urea.

$$\text{Total Inorganic Nitrogen (TIN)} = \text{nitrate } (NO_3^-) + \text{nitrite } (NO_2^-) + \text{ammonia } (NH_3)$$

$$\begin{aligned}\text{Total Nitrogen} = {} & \text{organic N (proteins, peptides, nucleic acids, urea)} \\ & + \text{TIN}\left[\text{nitrate } (NO_3^-) + \text{nitrite } (NO_2^-) + \text{ammonia } (NH_3)\right]\end{aligned}$$

$$\text{TKN (Total Kjeldahl Nitrogen)} = \text{organic nitrogen (particulate and soluble)} + \text{ammonia}$$

TKN is usually much higher on influent (untreated waste) samples than effluent samples. TKN, ammonia (common term as ammoniacal nitrogen), and nitrate (commonly term as nitrate-nitrogen) are the most common forms of nitrogen in wastewater engineering. For example, fresh wastewater typically contains between 30 and 50 mg/L of ammonia nitrogen [1], while the final effluent contains nitrates which are usually less than 1 mg/L [2]. The total nitrogen (TKN) content amounts to about 15%–20% of the BOD_5 in the domestic effluent.

Nitrification is an aerobic process that involves the conversion of ammonia to nitrite and then nitrate through nitrifying bacteria. *Nitrosomonas*, as well as other nitrifying bacteria, convert ammonia to nitrite. *Nitrobacter*, as well as other nitrifying bacteria, convert nitrite to nitrate in an anaerobic condition.

$$\underset{\text{Ammoniacal nitrogen}}{NH_3\text{-}N} \xrightarrow{\text{Nitrosomonas bacteria}} \underset{\text{Nitrite nitrogen}}{NO_2\text{-}N} \xrightarrow{\text{Nitrobacter bacteria}} \underset{\text{Nitrate nitrogen}}{NO_3\text{-}N} \quad (1.2)$$

The nitrification process (aerobic)

$$\begin{aligned}& NO_3^- \rightarrow NO_2^- \rightarrow NO \rightarrow N_2O \rightarrow N_2 \\ & \text{Nitrate} \rightarrow \text{Nitrite} \rightarrow \text{Nitrogen oxide} \rightarrow \text{Nitrous oxide} \rightarrow \text{Nitrogen gas}\end{aligned} \quad (1.3)$$

The denitrification process (anaerobic)

The typical amount of medium strength nitrogen in raw wastewater is TKN and total nitrogen (35 mg/L), ammoniacal nitrogen (25 mg/L), organic nitrogen (10 mg/L), with nitrite and nitrate (0 mg/L).

1.5.2.2.3 Phosphorus

Phosphorus is essential to the growth of algae and other biological organisms. The amount of phosphorus compounds present in wastewater discharge must be controlled to avoid noxious algal blooms occurring in surface water. The usual forms of phosphorus found in aqueous solutions include orthophosphate, polyphosphate, and organic phosphate. Phosphorus is also a source of nutrients for the wastewater oxidation process. It is typically present at a level of 5–15 mg/L [2]. Phosphorus is a constituent of human wastewater. Organically bound phosphorus originates from the body and food waste and, upon biological decomposition of these solids, is converted to orthophosphates. Polyphosphates are used in synthetic detergents and used to contribute as much as one-half of the total phosphates in wastewater. Most household phosphate inputs now come from human waste and automatic dishwasher detergent. Polyphosphates can be hydrolysed to orthophosphates. Thus, the principal form of phosphorus in wastewater is assumed to be orthophosphates, although the other forms may exist. Orthophosphates consist of the negative ions PO_4^{3-}, HPO_4^{2-}, and $H_2PO_4^-$. These may form chemical combinations with cations (positively charged ions).

1.5.2.2.4 Chloride

Domestic wastewater is a rich source of chlorides because human excreta, mainly urine, is rich in chloride. It does not present a major pollution threat. But, chloride ion concentration is an important factor to be considered if the treated effluent is used for irrigation. The primary salt utilized in the home is sodium chloride (NaCl). It enters the wastewater mainly from urea, which is in the form of sodium chloride in the urine. Values of 50–100 mg/L are common in wastewater [1].

1.5.2.2.5 Sulphate

The sulphate ion occurs naturally in most water supplies and is present in wastewater as well. Sulphate is present in the effluent through some sources, primarily through soap and detergents. Sulphate is reduced biologically under anaerobic conditions to sulphide, which in turn can combine with hydrogen to form hydrogen sulphide (H_2S). The presence of sulphate can cause problems to the sewer because of anaerobic conditions; it can cause the release of hydrogen sulphide gas which can form sulphuric acid when combined with water. The resulting acid can erode the sewer.

1.5.2.2.6 Metals

There are also some metals that may be present in wastewater, particularly through contamination. These include copper, zinc, mercury, and others. They normally present at trace levels.

1.5.2.2.7 Organic Materials

1.5.2.2.7.1 Biochemical Oxygen Demand (BOD_5) BOD is a sum parameter and the amount of oxygen required to oxidize organic matter present in the water biochemically. So, BOD is an indirect measure of the concentration of organic contamination in water. BOD analysis does not oxidize all the organic matter present in the waste; only the organics that are biochemically degradable during n days period at 20°C are oxidized. The day period is given as an index in BOD_n. The standard for usual measurements is a 5-day period. The general range of BOD observed for raw wastewater is 100–400 mg/L. In Malaysia, the average value is 250 mg/L.

1.5.2.2.7.2 Chemical Oxygen Demand (COD) The equivalent amount of oxygen required to oxidize organic matter present in a water sample using a strong chemical oxidizing agent is called COD. COD is also a sum parameter and is used to measure the content of organic matter of wastewater. The COD values include the oxygen demand created by biodegradable as well as non-biodegradable substances. As a result, COD values are greater than BOD. In comparison with BOD_5, COD measurement has an advantage in that it requires a short digestion period of about 3 hours rather than the incubation of 5 days period required for BOD_5 measurement. For many types of wastes, it is possible to correlate COD with BOD. Once the correlation has been established, COD measurements can be used to good advantage for treatment-plant control and operation. In general, the COD of raw wastewater at various places is reported to be in the range of 200–700 mg/L. In Malaysia, the average value is 350 mg/L.

1.5.2.2.7.3 Total Organic Carbon Another means for measuring the organic matter present in water is the total organic carbon (TOC) test, which is especially applicable to small concentrations of organic matter. The wastewater content of carbon bound in organic molecules is the TOC. Organic carbon comprises nearly all carbon compounds except a few carbon species, which are looked at as inorganic (carbon dioxide, hydrogen carbonate, carbonate, cyanide, and some further examples which are not commonly found in wastewaters).

1.5.2.2.7.4 Oil and Grease (O&G) Oils, fats, waxes, and fatty acids are the major constituents included in this category in domestic wastewater. The presence of a significant amount of oil and grease in wastewater hinders the transportation of wastes through pipelines. It causes scum in aeration basins of activated sludge plants, which interferes with the biological oxidation of wastes and produces a low-quality settling sludge.

1.5.2.3 Biological Properties

Biologically, wastewater contains various microorganisms, but the ones that are of concern are those classified as protists, plants, and animals. The category of Protista includes bacteria, fungi, protozoa, and algae. Plants include ferns, mosses, seed plants, and liverworts. Invertebrates and vertebrates are included in the animal category. In terms of wastewater treatment, the most important category is Protista, especially bacteria, algae, and protozoa. Also, wastewater contains many pathogenic organisms, which generally originate from humans who are infected with disease or who are carriers of disease. Typically, the concentration of faecal coliforms found in raw wastewater is about several hundred thousand to tens of million per 100 mL of sample.

The Coliform group of bacteria comprises mainly species of the genera *Citrobacter, Enterobacter, Escherichia,* and *Klebsiella* and includes faecal coliforms, of which *Escherichia coli* (*E. coli*) is the predominant species. Several of the coliforms can grow outside of the intestine, especially in hot climates. Hence their enumeration is unsuitable as a parameter for monitoring wastewater reuse systems. The faecal coliform test may also include some non-faecal organisms which can grow at 44°C, so the *E. coli* count is the most satisfactory indicator parameter for wastewater use in agriculture.

1.6 CONCENTRATION OF WASTEWATER

Characterization of wastes is essential for an effective and economical waste management programme. It helps in the choice of treatment methods deciding the extent of treatment, assessing the beneficial uses of wastes, and utilizing the waste purification capacity of natural bodies of water in a planned and controlled manner. While analysis of wastewater in each case is advisable, data from the other cities may be utilized during the initial stage of planning. Domestic wastewater comprises spent water from the kitchen, bathroom, lavatory, etc. The factors that contribute to variations in characteristics of the domestic wastewater are daily per capita use of water; quality of water supply; the type, condition, and extent of the sewerage system; and habits of the people. Municipal wastewater, which contains both domestic and industrial wastewater, may differ from place to place depending upon the type of industries and industrial establishment. Similarly, other types of wastewater may also differ in volume and concentrations. Typical concentrations of various types of wastewater are given in Tables 1.1 and 1.2.

$$\text{BOD load} = (3) = (1) * (2)$$

$$(1) = \text{BOD}\ (\text{g/L}) \times \text{Flow}\ (\text{L/unit.day})$$

$$(2) = \text{TKN}\ (\text{g/L}) \times \text{Flow}\ (\text{L/unit.day})$$

TABLE 1.1
Concentration, Volume Wastewater Loading [1]

Type of Wastewater	BOD (mg/L) (1)	The Volume of Wastewater (L/Unit) (2)	BOD Load (g/Unit) (3)
Cow	450–4,330	45–136	182
Pig	1,275–13,260	11–23	114
Chicken and duck	8,500–40,000	0.2	7–15
Wastewater	250–400	227	68

TABLE 1.2
Composition of the BOD and TKN Various Wastewater [1]

Type of Wastewater	BOD Load Per Day (g/Unit.Day) (1)	TKN Load Per Day (g/Unit.Day) (2)
Wastewater	55	—
Cow	498	222
Buffalo	347	107
Sheep	102	22
Pig	137	23

Example 1.1

Calculate the BOD of wastewater produced by 1,000 people with a total flow rate of 225 m^3/day. Assume BOD as 55 g/person.day.

Solution:

Population = 1,000 people
Flow rate (Q) = 225 m^3/day
BOD rate = 55 g/person.day
BOD rate for 1,000 persons

$$= (1{,}000) \times (55 \text{ g/person.day})$$

$$= 55{,}000 \text{ g/day}$$

Since Q = 225 m^3/day, then by dividing 55,000 g/day of Q, we will get the BOD:

$$\text{BOD} = \frac{55{,}000 \text{ g/day}}{225 \text{ m}^3\text{/day}}$$

$$\text{BOD} = \frac{55{,}000 \text{ g/day} \times 10^3 \text{ mg/day}}{225 \times 10^3 \text{ L/day}}$$

$$= 244.44 \text{ mg/L}$$

$$= 244 \text{ mg/L (no decimals in BOD value).}$$

1.7 WASTEWATER FLOW AND SEWER DESIGN

1.7.1 Wastewater Flow Rate

Determining the rates of wastewater flow is a fundamental step in the design of wastewater collection and treatment systems. The flow rate will affect the size of the sewer and the size of each treatment plant component. In a situation where wastewater flow rate data are limited or unavailable, wastewater flow rate estimates must be developed from water consumption records and population data. Estimated residential flow rates need to account for both averages and peak flows.

For areas served with the public sewerage system, wastewater flow rates are commonly determined from existing long-term (the best) or short-term records or by direct field measurements. For new developments, wastewater flow rates are derived from an analysis of population data and corresponding water consumption rate or estimates of per capita wastewater flow rates from similar communities.

1.7.2 Factors That Influence Wastewater Flow Rate

Wastewater flow rate is an important factor in the design of wastewater treatment units. Many factors influence the quantity of flow. These are discussed in the following sections.

1.7.2.1 Climate

Climatic effects such as precipitation and temperature can significantly impact water consumption. Water use is at its peak when it is hot and dry, and more wastewaters are produced.

1.7.2.2 Population Size

Population size affects not only the average per capita water use but also the peak rate of use. The rate of use fluctuates depending on the water demand. More wastewater is produced for a bigger community. The amount of water used per capita will also vary from one country to another. The example is given in Table 1.3.

1.7.2.3 Development Intensity

The density of development (i.e. bungalow, semi-detached, terrace house, condominiums, apartments, and flats) affects water consumption. Bungalow and semi-detached homes may have more water-using appliances such as washing machines and dishwashers than apartments. Exterior water use for flats and apartments is generally much less because of reduced needs for landscape watering.

TABLE 1.3
The Rate of Water Use by Category

Community	L/Capita.Day
Highly populated	300
Scotland	250
England	180
Primitive	50
Malaysia	210
	225[a]
Israel	280
WHO	165

[a] Used in the design of WWTP in Malaysia.

1.7.2.4 Socio-economic Status

The affluence or economic capabilities of a community affect water use which results in wastewater flows. Water use and wastewater flow rates are generally higher for the higher income group.

1.7.2.5 Sufficient and Quality Supply

A water supply that is dependable and of good quality encourages use by its customers. Poor pressure and limited quantities during peak or dry periods or that have objectionable taste may have lower water use.

1.7.2.6 Water Conservation Practices

There is also the effect of conservation practices of water usage on the quantity of wastewater. The higher the conservation, the lower the wastewater flow.

1.7.2.7 Public Water Supply

Water authority usually charges their customers based on the water used. Metering the individual consumer's supply and billing at established meter rates indirectly prevents waste of water by users and tends to reduce actual water use, hence minimizing the wastewater flow.

1.7.2.8 Type of Sewer

A combined sewer system (sanitary wastewater combines with rainwater) will have a higher flow than the separate sewer system.

1.7.2.9 Loss of Water

Leakages of sewers will cause the infiltration of wastewater into the environment, hence reducing the flow to the treatment plant. Exterior use of water for car washing or watering the plants will divert the water to storm water which does not go to the treatment plant. Not all supplied water becomes wastewater as some may flow as stormwater, especially due to exterior water usage.

1.7.2.10 Season

Season can also influence the rate of water consumption, which will indirectly affect the quantity of waste produced. For example, during school holidays and the festive season there is an increase in total water consumption and hence increase in the wastewater flow.

1.7.3 Dry Weather Flow

In Malaysia, DWF is the **AVERAGE FLOW** calculated based on the total population multiplied by the water consumption rate per person per day. The design of WWTP units will be based on the DWF and peak flow. The latter is detailed in Section 1.18.

Example 1.2

Calculate the DWF of an area with 10,000 population (PE) with a water consumption rate (q) of 250 L/capita.day.

Solution:

$$\begin{aligned}\text{DWF} &= \text{PE} \times q \\ &= 10{,}000 \text{ capita} \times 250 \text{ L/capita.day} \\ &= 2{,}500{,}000 \text{ L/day} \\ &= 2{,}500 \text{ m}^3\text{/day}\end{aligned}$$

Example 1.3

Calculate the DWF for a 20,000 population with water consumption of 200 L/capita.day, and the ratio of wastewater/supplied water is 0.7.

Solution:

$$\begin{aligned}\text{DWF} &= P \times q \\ &= 20{,}000 \text{ capita} \times 250 \text{ L/capita.day} \times 0.7 \\ &= 3{,}500{,}000 \text{ L/day} \\ &= 3{,}500 \text{ m}^3/\text{day}\end{aligned}$$

1.7.4 Typical Wastewater Concentrations

Wastewater concentration varies from one country to another. It is very much related to water availability and consumption rate. Some of the typical values in the UK are given in Table 1.4. The values in Malaysia are given in Table 1.5.

1.7.5 Per Capita Organic Load

Per capita organic load for the specific parameter of wastewater is a very important parameter in the design stage of the wastewater treatment system. It can be expressed in terms of BOD, COD, SS, etc. It will be used in the calculation of Population Equivalent (PE), which will be discussed in the following section.

TABLE 1.4
Typical Values of Wastewater [3]

	Wastewater (mg/L)		
Parameter	**Raw**	**Settled**	**Final Effluent**
BOD	500	175	20
COD	700	400	90
TOC	200	90	30
SS	400	200	30
NH_3-N	40	40	5
NO_3-N	<1	<1	20

TABLE 1.5
Influent Value of Wastewater in Malaysia in Accordance to Malaysian Sewerage Industry Guidelines (MSIG), Volume 4, Third Edition, 2009 [4]

Parameter	g/Capita.Day	mg/L
Biochemical oxygen demand (BOD_5)	56	250
Suspended solids (SS)	68	300
Chemical oxygen demand (COD)	113	500
Total nitrogen (TN)	11	50
Ammoniacal nitrogen (NH_3-N)	7	30
Total phosphorus (TP)	2	10
Oil and grease (O&G)	11	50

The per capita organic load produced by a person is normally expressed in grams/capita.day unit. For example, for BOD, the per capita organic load is derived as follows:

$$\begin{aligned}\text{Organic Load per person (OL) in terms of BOD} &= Q \times \text{BOD} \\ &= (\text{m}^3/\text{capita.day}) \times (\text{mg/L}) \\ &= (\text{m}^3/\text{capita.day}) \times (\text{g/m}^3) \\ &= (\text{grams/capita.day})\end{aligned} \quad (1.4)$$

Typical values of the per capita organic load in terms of BOD and SS are between 55 and 65 and 68 g/capita.day, respectively (Jackson et al., 1990). In Malaysia, the value is 56 and 68 g/capita.day, respectively, for BOD and SS, as detailed in Table 1.3.

Example 1.4

If the organic load per capita in terms of BOD is 59 g/day and has a flow rate of 182 L/day, calculate:

i. The value of BOD in mg/L.
ii. The new value should the flow reduce to 136 L/day.
iii. Comment on your answer.

Solution:

i. Given, OL = 59 g/capita.day

$$Q = 182 \text{ L/capita.day}$$

From Equation 1.4,

$$\text{Organic Load per person (OL) in terms of BOD} = Q \times \text{BOD}$$

$$59 \text{ g/capita.day} = 182 \text{ L/capita.day} \times \text{BOD}$$

$$\text{BOD} = \frac{59 \times 10^3 (\text{mg/capita.day})}{182 \text{ L/capita.day}}$$

$$= 324 \text{ mg/L}$$

ii. Given, OL = 59 g/capita.day

$$Q = 136 \text{ L/capita.day}$$

$$59 \text{ g/capita.day} = 138 \text{ L/kapita.day} \times \text{BOD}$$

$$\text{BOD} = \frac{59 \times 10^3 (\text{mg/capita.day})}{136 \text{ L/capita.day}}$$

$$= 434 \text{ mg/L}$$

iii. With the reduction of Q, the BOD will be more concentrated.

1.7.6 Population Equivalent

It is often known that the domestic wastewater treatment system also receives other non-toxic wastewater from different premises, such as industries and commercial places. They also contribute to the pollution load. As for domestic wastewater, the design of the treatment unit is based on the served population. However, for commercial places and industries, their loads are compared with the number of a person's load and expressed as PE. In other words, PE or unit per capita loading (PE) in wastewater treatment is the number that expresses the ratio of the sum of the pollution load produced by industrial facilities and services to the individual pollution load in household wastewater produced by one person.

The PE can be expressed as follows:

PE = organic load from the premises in a day/Organic load of 1 person in a day

For example, PE in terms of BOD and SS in Malaysia can be expressed as shown in Equations 1.5 and 1.6, respectively:

$$\text{PE} = \frac{Q \times \text{BOD}}{56\ \text{g/capita.day}} \tag{1.5}$$

and PE in the form of SS:

$$\text{PE} = \frac{Q \times \text{SS}}{68\ \text{g/capita.day}} \tag{1.6}$$

Example 1.5

Calculate the PE for a juice processing plant (BOD =1,000 mg/L) that channels 1,000 m^3/day of its wastewater into a domestic WWTP. Take the organic load per person (OL) in terms of BOD.

Solution:

From Equation 1.5,

$$\text{PE} = \frac{Q \times \text{BOD}}{56\ \text{g/capita.day}}$$

$$= \frac{1{,}000\,\frac{\text{m}^3}{\text{day}} \times 1{,}000\ \text{mg/L}}{56\ \text{g/capita.day}}$$

$$= \frac{1{,}000\,\frac{\text{m}^3}{\text{day}} \times 1\ \text{kg/m}^3}{56\ \text{g/capita.day}}$$

$$= \frac{1{,}000\,\frac{\text{m}^3}{\text{day}} \times 1{,}000\ \text{mg/L}}{56\ \text{g/capita.day}}$$

$$= 17.86$$

$$\text{PE} = 18\ \text{capita (or person)}$$

From the above example, the WWTP should cater for additional 18 PE in terms of BOD.

Example 1.6

An industry has a flow rate of 1,500 L/day and 2,000 mg/L SS. Calculate the PE if the organic load per person (OL) in terms of SS is 68/capita.day.

Solution:

From Equation 1.6,

$$PE = \frac{Q \times SS}{68 \text{ g/capita.day}}$$

$$= \frac{1{,}500\,(\text{L/day}) \times 2{,}000 \times 10^{-3} \text{ g/L}}{60 \text{ g/capita.day}}$$

$$= 50 \text{ capita (or person)}$$

From the above example, the WWTP should cater for additional 50 PE in terms of SS.

Some of the typical values of PE for different industries are given in Table 1.6.

TABLE 1.6
Population Equivalents of Wastewater from Selected Industries [5]

Type of Industry	Type	PE in Terms of BOD
Food	Tomato	50–185
	Potato	215–545
	Carrot	180–390
	Citrus fruit	55
	Beef	20–600
	Fish	300–2,300
	Sugarcane	50
	Margarine	500
	Slaughterhouse	10–100
	Sweets/candies	40–150
Confined animals' breeding	Pigs	35–100
	Cattle	65–150
	Poultry	15–20
Drinks	Brewery	150–350
	Soft drinks	50–100
	Wine	5
Textiles	Cotton	2,800
	Wool	5,600
	Nylon	800
	Polyester	3,700
	Dyeing	2,000–3,500
	Bleaching	250–350
Leather and tanneries	Tanning	1,000–3,500
	Shoes	300
Pulp and paper	Pulp	600
	Paper	100–300
Chemical industries	Paint	20
	Soap	1,000
	Petroleum refinery	1
	PVC	200
Steelworks	Foundry	12–30
	Plating	8–50

TABLE 1.7
Recommended Population Equivalent

Type of Premises/Establishment	Recommended Population Equivalent (PE)
Residential	5 per house
Commercial:	3 per 100 m^2 gross area
Includes offices, shopping complex, entertainment/recreational centres, restaurants, cafeteria, theatres	
Schools/Educational Institutions:	
• Day schools/institutions	0.2 per student
• Fully residential	1 per student
• Partial residential	0.2 per non-residential student 1 per residential student
Hospitals	4 per bed
Hotels with dining and laundry facilities	4 per room
Factories, excluding process water	0.3 per staff
Market (wet type)	3 per stall
Market (dry type)	1 per stall
Petrol kiosks/service stations	15 per toilet
Bus terminal	4 per bus bay
Taxi terminal	4 per taxi bay
Mosque/Church/Temple	0.2 per person
Stadium	0.2 per person
Swimming pool/sports complex	0.5 per person
Public toilet	15 per toilet
Airport	0.2 per passenger 0.3 per employee
Laundry	10 per machine
Prison	1 per person
Golf course	20 per hole

Source: Malaysian Standard 1228 [6].

There are many methods for calculating expected flow rates. One method is to calculate the design parameter called the "Population Equivalent" (PE) of a catchment and convert it to a flow rate. Design calculation for all unit processes shall be done in sequence, starting from inlet works to biological treatments and sludge treatments. Calculation of PE is to be based on the recommended values, as shown in Table 1.7.

The PE may be converted to a flow rate using a simple formula such as set out in Malaysian Standards 1228 (MS1228) [6]. The water consumption rate (q) is 225 L/capita.day.

1.7.7 Storm Sewer

The storm sewer is a pipeline that carries rainwater or stormwater into a stormwater drainage system. In Malaysia, as we receive a high density of rain, we use a separate system to carry sanitary wastewater and stormwater.

The quantity of stormwater depends on many factors, such as topography, rainfall intensity, and permeability of the soil. There are several available methods to determine the quantity of stormwater. The most common is the rational method. The method determines the quantity of flow (Q). By knowing the Q, the size of the storm sewer can be determined, as shown in Equation 1.7.

According to the rational method,

$$Q = 0278AIR \tag{1.7}$$

where

Q = the flow (m^3/s)

A = catchment area (km^2)

I = rainfall intensity (mm/hr)

R = runoff coefficient (no units). R is high for impermeable and low for permeable areas. For an area with a composite permeability, then:

$$R = \frac{a_1 R_1 + a_2 R_2 + \cdots + a_n R_n}{a_1 + a_2 + \cdots + a_n}$$

where a_1 = area 1

R_1 = the runoff coefficient for area 1 (from a Standard table)

The management of stormwater is dealt with in detail in hydraulics and hydrology.

1.7.8 Effluent Standard

In Malaysia, the standard discharge limit of effluent must comply with Standard A and B of the Environmental Quality Act 1974 [7] under the Environmental Quality (Sewage) Regulations 2009, Second Schedule, Regulation 7. This applies to all new WWTPs, as detailed in Table 1.8. For an existing wastewater treatment system (approved before January 1999), the standard is given in Table 1.9. Standard A applies to discharge into any inland waters within catchment areas listed in the Third Schedule of the Regulation, while Standard B applies to any other inland waters or Malaysian waters.

1.7.9 Dilution of Effluent

The effluent of WWTP is normally channelled into nearby drains or rivers. Without proper treatment, this practice may cause pollution to the water body. The question is, what will happen when this effluent is mixed with the receiving water. It is important to appreciate the effluent dilution principle in understanding this phenomenon.

TABLE 1.8
Standard A and B of the Environmental Quality Act 1974 [7] under Environmental Quality (Sewage) Regulations 2009, Second Schedule, Regulation 7 (For New WWTP)

		Standard	
Parameter	**Unit**	**A**	**B**
Temperature	°C	40	40
pH value	—	6.0–9.0	5.5–9.0
BOD at 20°C	mg/L	20	50
COD	mg/L	120	200
Suspended solids	mg/L	50	100
Oil and grease	mg/L	5.0	10.0
Ammoniacal nitrogen (enclosed water body)	mg/L	5.0	5.0
Ammoniacal nitrogen (river)	mg/L	10.0	20.0
Nitrate-nitrogen (river)	mg/L	20.0	50.0
Nitrate-nitrogen (enclosed water body)	mg/L	10.0	10.0
Phosphorous (enclosed water body)	mg/L	5.0	10.0

TABLE 1.9
Standard A and B of the Environmental Quality Act 1974 [7] under Environmental Quality (Sewage) Regulations 2009 (Existing Wastewater Treatment System Approved before January 1999)

	Communal Septic Tank		Imhoff Tank		Aerated Lagoon		Oxidation Pond		Mechanical System	
Parameter	**A**	**B**	**A**	**B**	**A**	**B**	**A**	**B**	**A**	**B**
BOD at 20°C (mg/L)	200	200	175	175	100	100	120	120	60	60
COD (mg/L)					300	300	360	360	180	240
Suspended solids (mg/L)	180	180	150	150	120	120	150	150	100	120
Oil and grease (mg/L)									20	20
Ammoniacal nitrogen (mg/L)			100	100	80	80	70	70	60	60

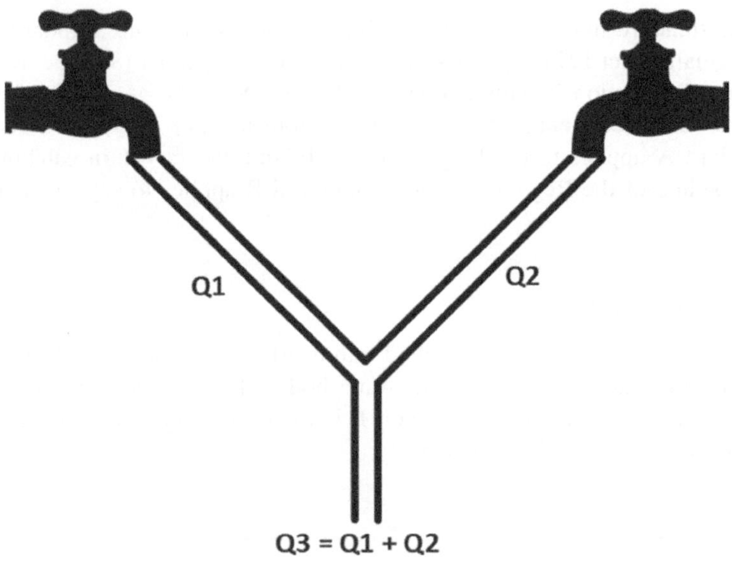

FIGURE 1.10 Dilution principle of effluent.

In this principle, we can estimate what will happen downstream of the effluent discharge point if we know the original data of the river water. This principle is usually used for the main parameters such as BOD, COD, or SS. Water pollution modelling work is based on this principle whereby we calculate the mass balance of the pollution load and the flow as given in Equations 1.15 and 1.16 and Figure 1.10.

For the case of BOD (two pollution loads only converging into a river, Q_3):

$$Q_3 = Q_1 + Q_2 \tag{1.8}$$

$$(Q_1\text{BOD}_1) + (Q_2\text{BOD}_2) = (Q_3\text{BOD}_3) \tag{1.9}$$

Solving these two equations with two unknowns and replacing Equation 1.8 with Equation 1.9:

$$(Q_1\text{BOD}_1) + (Q_2\text{BOD}_2) = (Q_1 + Q_2)(\text{BOD}_3) \tag{1.10}$$

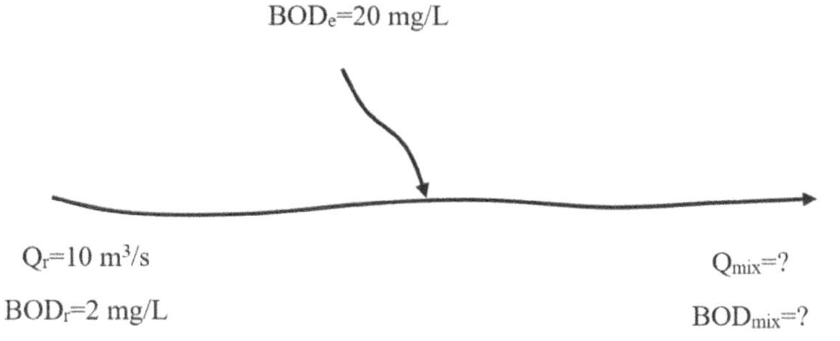

FIGURE 1.11 Dilution of effluent.

Example 1.7

A WWTP releases a BOD and flow of 20 mg/L and 2.0 m³/s, respectively, into a river that has a flow rate of 10 m³/s and BOD 2.0 mg/L. Estimate roughly the BOD downstream of the discharge point.

Solution:

Take:

Effluent = e
Existing river = r
River after mix = "mix" as shown in Figure 1.11.
$Q_{e=}2\ \text{m}^3/\text{day}$
$Q_{r=}10\ \text{m}^3/\text{s}$
$BOD_e = 20$ mg/L
$BOD_r = 2$ mg/L
From Equation 1.8,

$$Q_{\text{mix}} = Q_e + Q_r$$

$$Q_{\text{mix}} = 2\ \text{m}^3/\text{s} + 10\ \text{m}^3/\text{s}$$

$$Q_{\text{mix}} = 12\ \text{m}^3/\text{s}$$

$$(Q_e\text{BOD}_e) + (Q_r\text{BOD}_r) = (Q_1 + Q_2)(\text{BOD}_{\text{mix}}) \tag{1.11}$$

$$\left(2\ \text{m}^3/\text{s}\right)\left(20\ \text{mg/L}\right) + \left(10\ \text{m}^3/\text{s}\right)\left(2\ \text{mg/L}\right) = \left(12\ \text{m}^3/\text{s}\right)\left(\text{BOD}_3\ \text{mg/L}\right)$$

$$\left(2\ \text{m}^3/\text{s}\right)\left(20\ \text{mg/L}\right) + \left(10\ \text{m}^3/\text{s}\right)\left(2\ \text{mg/L}\right) = \left(12\ \text{m}^3/\text{s}\right)\left(\text{BOD}_3\ \text{mg/L}\right)$$

$$\text{BOD}_{\text{mix}} = \frac{40 + 20}{12}\ \text{mg/L}$$

$$\text{BOD}_{\text{mix}} = 5\ \text{mg/L}$$

1.7.10 Pollution of River/Water Course

The release of untreated or partially treated effluent into rivers may cause serious pollution issues. The DO level will drop due to the biodegradation of organic matter in the mix by the microorganism. This disrupts aquatic life, such as fish that needs oxygen to survive. At the same time, the river tends to attain oxygen naturally by photosynthesis and other sources in a process called "self-purification". Therefore, there is a balance between oxygen demand and the oxygen supply in the water body, as detailed in Table 1.10.

TABLE 1.10
Oxygen Balance in the River

Oxygen Supplied	Oxygen Demand
Atmosphere	BOD
Photosynthesis	Aquatic life
Aeration	Nitrification

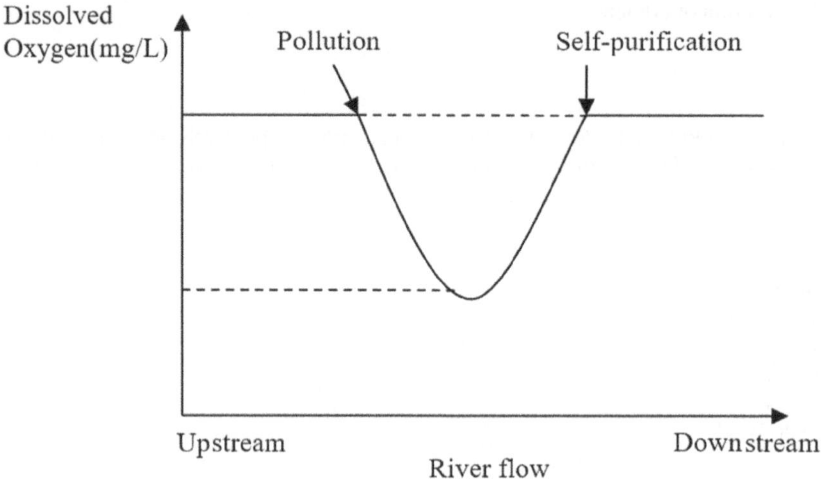

FIGURE 1.12 Self-purification process in a river.

Self-purification can be defined as the ability of the river to recover to its original condition (an increase of DO level before the pollution) when the water flows at some distance downstream of the pollution point load. This principle is shown in Figure 1.12.

1.7.11 Collection System

The purpose of a wastewater collection system is to remove wastewater from points of origin to a treatment facility. The collection system consists of the sewers (pipes and conduits) and plumbing necessary to convey wastewater. It is necessary that the collection system is designed so that the wastewater will reach the treatment system as soon as possible after entering the sewer. If the length of time in the sewers is too long, the wastewater will become anaerobic when it reaches the treatment facilities.

Sanitary wastewater collection systems are designed to remove domestic wastewater only. Surface drainage is excluded to avoid constructing large sewers and treating large volumes of wastewater diluted by rainwater during storms, especially in countries with heavy rainfall.

There are three types of sewer systems, i.e. a combined system (sanitary sewer combined with stormwater), a separate system, and a semi-combined system. In Malaysia, we use a separate system where the sewer channels only sanitary wastewater. A combined and semi-combined system is not practised in Malaysia as we receive a high density of rain yearly.

The design, construction, and testing of sewer networks and network pump stations are detailed in Volume 3 of the Malaysian Sewerage Industry Guidelines, Third Edition, National Water Services Commission (SPAN), 2009 [8].

TABLE 1.11
The Difference between Sewer and Drinking Water Pipes

Sewer	Drinking Water Pipe
Flows under gravity	Flows under pressure
Slope is important	The slope is not important
Has a self-purification velocity	Without a self-purification velocity
Designed without full	Designed as full
Designed based on Manning, Colebrook-White, and Hazen-Williams equations, others	Designed based on Darcy Equation, others.

1.7.11.1 The Difference between Sewer and Drinking Water Pipes

There are some major differences between sewer and drinking water pipes, as summarized in Table 1.11.

A self-cleansing velocity may be defined as an ideal and minimum velocity in a sewer that allows the solid particles to remain in suspension, without settling at the bottom of the sewer, which may cause clogging, at the same time, protecting scouring from happening that may damage the sewer line (if the velocity is too fast). Velocity is important because if the speed is too low, solids can settle and cause blockages. If the velocity is too fast, the rough solids can erode the sewer. Usually, the slope is such that a flow rate of 0.03 m/s or more is maintained when the line is flowing half full to full.

Pumping is necessary where the slope of the sewer does not produce the required minimum velocity of 0.03 m/s, where wastewater must be lifted to a higher elevation. Wastewater can be pumped from pumping stations through pressure lines (force mains) regardless of their slope, or it can be raised to a higher elevation at pumping stations (lift stations), so that gravity flow will again produce the required velocity.

1.7.11.2 Sewer Material

The Malaysian Sewerage Industry Guidelines (MSIG), Volume 3, 2009 [8], provides acceptable sewer materials for gravity flow, force mains (rising mains), and vacuum sewers. The types of sewer which flow under gravity include vitrified clay (VC) and reinforced concrete (for rigid pipe), glass reinforced plastic (GRP), ductile iron (DI), and high-density polyethylene (HDPE) (for flexible pipes). For force mains (rising mains), the types of acceptable sewer are ductile iron, GRP, acrylonitrile butadiene styrene (ABS), HDPE, and steel. ABS (for internal use) and HDPE (for external use) are used for vacuum sewers. Some of the common materials are discussed below:

1.7.11.2.1 Vitrified Clay

i. Only size 150 mm or above shall be used.
ii. The minimum size is 225 mm.
iii. Pipe shall not be used on unstable ground.
iv. Flexible joints are recommended.

VC pipe is manufactured in Malaysia in diameters of 100–600 mm and lengths ranging from 0.91 to 2.50 m.

1.7.11.2.2 Reinforced Concrete

i. Pipe protection linings are required.
ii. Only sizes 600 mm or above are allowed in compliance with the policy.
iii. Flexible joints are recommended.

RC pipe is manufactured in Malaysia in diameters from 150 to 3,600 mm. The standard pipe length is 3.05 m.

1.7.11.2.3 Glass-Reinforced Plastic

i. Pipe shall not be used in ground contaminated with a high concentration of chemicals such as a solvent that can degrade the pipe.
ii. Pipe shall not accept any industrial or other aggressive discharges that may affect the pipe integrity.
iii. Pipe shall be used only when no fittings are required.
iv. Only sizes 600 mm or above are allowed.

GRP pipe is currently required to be imported into Malaysia.

1.7.11.2.4 Ductile Iron

i. The use is only allowed for applications needing high pipe strength.
ii. Pipe protection linings and coatings are required.
iii. Polyethylene (PE) sleeving is required for all buried applications.

DI pipe is manufactured in Malaysia for diameters from 80 to 1,200 mm. The diameter of the imported pipe can be up to 2,000 mm. Standard lengths are 6.0 m.

1.7.11.2.5 High-Density Polyethylene

1.7.11.2.5.1 Solid Wall Polyethylene Pipe PE pipe is resistant to the sulphuric acid concentrations that might be found in septic sewage under the worst conditions.

PE solid wall pipe is available locally in diameters up to 1,000 mm and in standard lengths of 6 and 12 m. This pipe is normally butted fusion jointed. Pipe size of 160 mm or less may be flange jointed or electrofusion jointed. PE pipe is classified by pressure rating with static working pressures up to 1.6 MPa.

1.7.11.2.5.2 Profiled Wall Polyethylene Pipe A profiled wall pipe is a pipe with a plain inside surface and with a ribbed or corrugated outside surface. The ribs or corrugations are normally either aligned circumferentially or helically. These corrugated or ribbed profiles optimize the pipe ring stiffness to weight ratio. The pipe can be designed with a double-wall profile or triple-wall profile.

Corrugated high-density PE pipe is available in Malaysia in a range of sizes from 100 to 3,000 mm nominal diameter and in standard 6 m lengths. The standard joint is a flexible spigot-socket joint with a rubber seal.

1.7.11.3 Factors to Consider in Sewer Selection

Among the factors to be considered in selecting appropriate sewer include the following:

a. Lifespan.
b. Previous experience.
c. Resistant to corrosion (external and internal).
d. The strength of the structure.
e. Costs.
f. Availability.
g. The coefficient of roughness.

1.7.12 Sewer Design

Except for force mains, sewers are laid to permit the gravity flow of their contents. Unlike water in a water distribution system, the contents of a sewer do not flow under pressure. Hence, the design is based on the open channel concept. Sewers should be designed precisely for the conveyance of wastewater produced, especially on the diameter and slope for the appropriate flow.

1.7.12.1 Specific Requirements

1.7.12.1.1 Designed as Not Full

Sewer is not designed for full capacity (usually designed 1/2 to 3/4 full) for the following reasons:

i. To protect its tendency from becoming anaerobic when full.
ii. To provide room for future expansion.
iii. To consider possible infiltration due to leakages.

1.7.12.1.2 Designed at Q_{peak}

The design of the sewer shall be based on the worst-case scenario. It shall be designed at Q_{peak} in accordance with MS1228 (1991) [6]:

$$Q_p = F_p \times \text{DWF}$$
$$Q_p = 4.7p^{-0.11} \times \text{DWF} \tag{1.12}$$

where

Q_p = Peak flow $\left(\text{m}^3/\text{day}\right)$
F_p = Peak factor $= 4.7p^{-0.11}$
p = Population (PE) in thousand; if PE = 10,000, p = 10; if PE < 1,000, take p = 1
DWF = Dry Weather Flow $\left(\text{m}^3/\text{day}\right)$
= Average flow $\left(\text{m}^3/\text{day}\right)$ = PE × water consumption rate

The peak factor, F_p = 3–4, is quite commonly practised in many countries.

1.7.12.1.3 A Self-Cleaning Velocity

Sewer must comply with the cleansing velocities against slime and sediments. Minimum velocity values at the full bore of 0.8 m/s are commonly specified. However, it has been found that larger pipe diameters require higher velocity to cleanse the sediment. This is mainly due to higher sediment depths in large diameter pipes. Refer to Tables 1.12–1.14 for the ranges of cleansing velocities.

The self-cleansing velocity is normally checked at two conditions, i.e. at Q_{maximum} (Q_{max}) and Q_{minimum} (Q_{min}), as detailed in Table 1.10.

$$^*Q_{\text{min}} \text{ is taken at } \frac{\text{DWF}}{F_p} = \frac{1}{4.7p^{-0.11}} \times \text{DWF} \tag{1.13}$$

TABLE 1.12
Requirement of a Self-Cleansing Velocity of Sewer

Criteria of Q	The Range of Velocity in the Sewer (m/s)
At Q_{max}	<0.8–4.0
At Q_{min}*	Should not be less than the values in Table 1.11 (depending on the size of sewer)

Note: * denotes Equation 1.13.

1.7.12.1.4 Other Factors

i. Wherever practical, sewers are laid in straight lines. Corners and sharp bends slow the flow rate, permit clogging, and make line cleaning difficult.
ii. All sewers shall be sited in public road reserve so that access can be gained for maintenance purposes.
iii. Sewer pipes should not be constructed on a slope or within a slope failure envelope.

TABLE 1.13
A Self-Cleansing Velocity of a Sewer According to the Diameter

Diameter of Sewer (mm)	Minimum Velocity (m/s)
150–250	1.00
300–600	0.75
>600	0.60

TABLE 1.14
Sewer Design Criteria Based on MS1228 (1991) [6], MSIG (Volume III, 2009) [8]

Parameter	Design Criteria
Flow (Q)	$Q_p = 4.7p^{-0.11} \times \text{DWF}$
Velocity (self-cleaning)	Q_{min}; $v > 0.6 - 1.0$ m/s[a] Q_{max}; $v < 0.8 - 4.0$ m/s
Minimum diameter	225 mm
Minimum depth	1.2 m from the ground surface
Gradient	Should not be steeper than 1:50
Distance to water supply pipe	Should not be laid on top of drinking water pipe
Distance to building	Should not be constructed under buildings
Economic factor	Should be as shorter as possible

[a] Based on the sewer diameter in Table 1.11.

iv. An overflow pipe shall be provided at the last manhole before the network pump station and/or sewage treatment plant. Otherwise, it should be located at the manhole sited at the lowest ground level.

v. The minimum size of domestic connections to the public sewer shall be 150 mm in diameter.

1.7.12.1.5 Summary

Factors to be considered in sewer design are summarized in Table 1.14.

1.7.12.2 Design Formula

Wastewater may also flow under pressure (forced main) for a specific requirement; however, in most cases, it flows under gravity. Hence, the design will be based on an open channel principle. The common method used includes Manning, Colebrook-White, Hazen-Williams, Chezy, Kutter, Scobey formulae, and others. However, only the Manning approach is considered in this book.

According to Manning,

$$V = \frac{1}{n}(R)^{\frac{2}{3}}(s)^{\frac{1}{2}} \tag{1.14}$$

and

$$Q = \frac{1}{n}(A)(R)^{\frac{2}{3}}(s)^{\frac{1}{2}} \tag{1.15}$$

where

V = velocity of wastewater (m/s) at full flow
Q = flow rate of wastewater (m^3/s) at full flow
R = wetted area = A/P (covered by wastewater)[1]
A = wetted area of sewer (m^2)
P = perimeter (m)
D = diameter (m)
S = gradient of sewer (m/m)
n = the roughness coefficient (Table 1.15)

Wetted area is the area that is covered by water in the sewer. Example: $R = A/P$ is an area divided by the perimeter *(P)* **AT FULL CONDITION**, as shown in Figure 1.13.

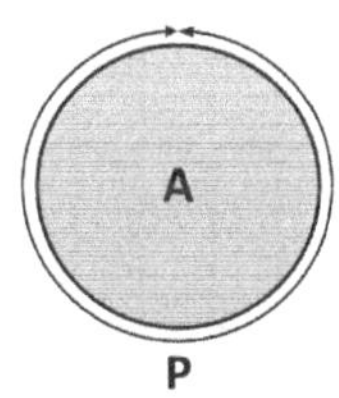

FIGURE 1.13 Wetted area used in Manning equation.

TABLE 1.15
Manning Roughness Coefficient

Type of Sewer	Manning Coefficient, na
Uncoated cast-iron	0.012–0.015
Coated cast-iron	0.011–0.013
Ductile iron (DI)	0.012–0.015
Vitrified clay pipe (VC)	0.010–0.017
Concrete	0.012–0.016
Steel	0.010
Asbestos cement	0.011
Soil and gravel	0.018–0.020
Earth drain	0.025–0.035

Source: MSIG Volume 3 [8], Barnes et al. [9], Wilson [10], and Metcalf and Eddy [11].
[a] Good condition with lower value.

1.7.12.3 Manning Chart

In the design, we usually require determining the diameter of the sewer. For this, the Manning chart, as shown in Figure 1.14, will be used.

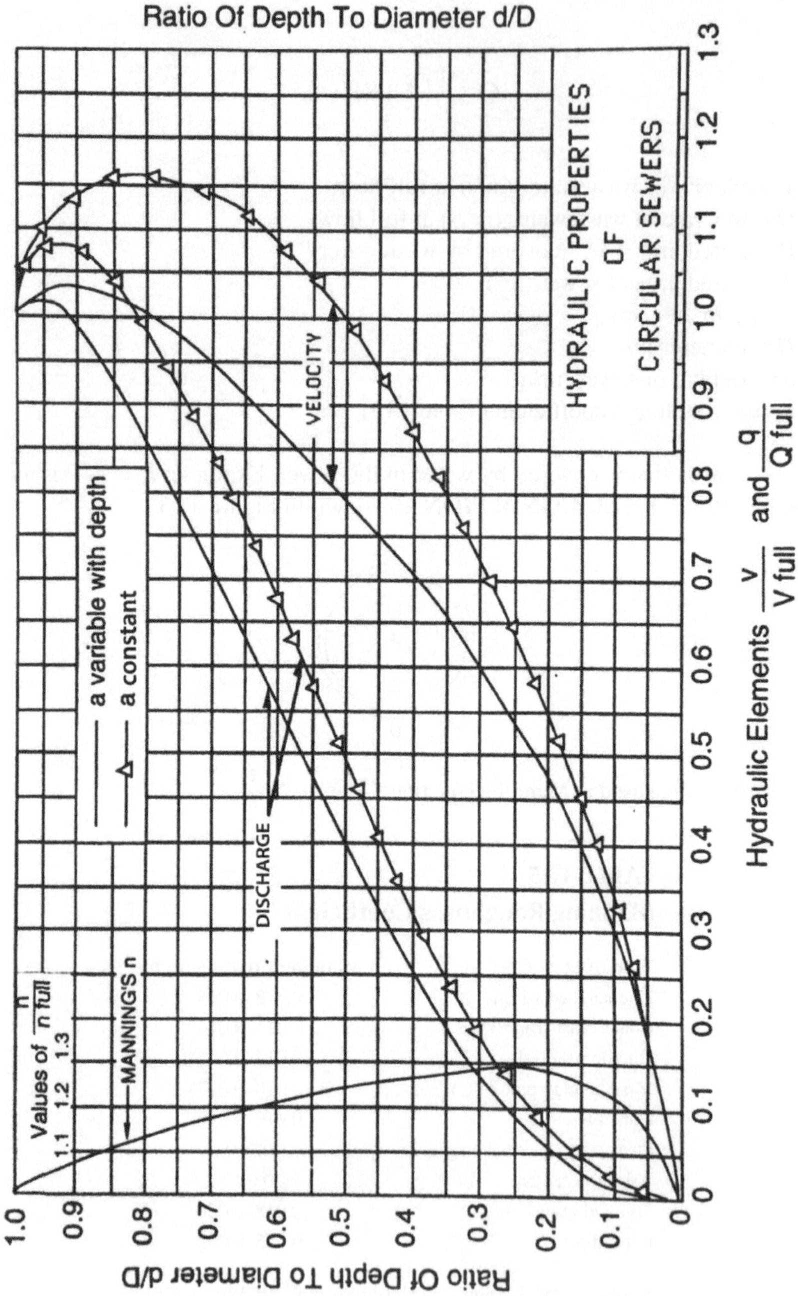

FIGURE 1.14 Manning chart.

1.7.12.4 Design Steps

1. Determine the ratio of the actual depth of wastewater against the full diameter, d/D. For example, if it flows 70% full, it means $d/D=0.7$.
2. Determine the value of q/Q_{full} and v/V_{full} from the Manning chart (Figure 1.14) for d/D in (1). q and v are at the actual flow and velocity at $d/D=0.7$.
3. Determine Q_{full} and V_{full}. Please note that the **Manning formula is used only at conditions Q_{full} and V_{full}** using Equations 1.14 and 1.15. The unit for Q and V is m^3/s and m/s, respectively.

4. Calculate the diameter of sewer (D) from (3).
5. Determine V_{full} from Equation 1.14.
6. Check the self-cleaning velocity at Q_{max} and Q_{min}.

Example 1.8

Determine the actual velocity of an uncoated cast-iron sewer with a diameter 1.8 m that flows 60% full at a gradient 1:600.

Solution:

The given data:

Diameter, D=1.8 m

Gradient, s=1:600

d/D=0.6, where d=the actual depth, D is the diameter of the sewer

Manning coefficient n=0.015 (uncoated cast-iron)

The actual velocity, v=?

Solution:

At d/D=0.6 (see Figure 1.15).

From Manning chart (Figure 1.14), at d/D=0.6,

$$\frac{v}{V_{full}} = 1.08$$

From the Manning formula (Equation 1.14)

$$V = \frac{1}{n}(R)^{\frac{2}{3}}(s)^{\frac{1}{2}}$$

Determine R at full flow:

$$R = \frac{A}{P}$$

$$R = \left(\frac{\pi}{4}\right)D^2$$

$$= \frac{D}{4}$$

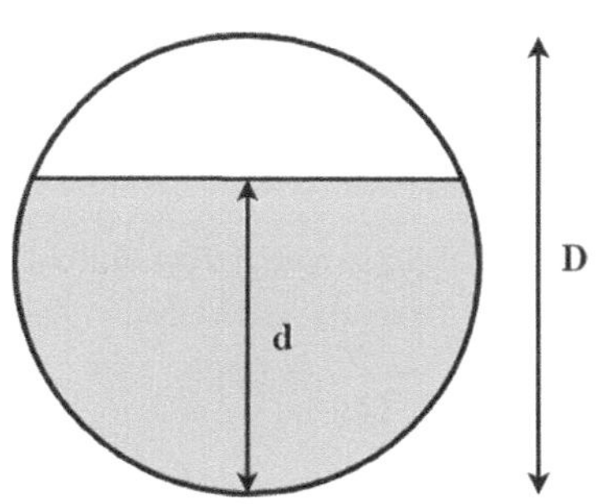

FIGURE 1.15 Depth of wastewater in the sewer.

Then,

$$V = \frac{1}{n}(R)^{\frac{2}{3}}(s)^{\frac{1}{2}}$$

$$V = \frac{1}{0.015}\left(\frac{1.8}{4}\right)^{\frac{2}{3}}\left(\frac{1}{600}\right)^{\frac{1}{2}}$$

$$= 1.598 \text{ m/s (at the full condition)}$$

From above,

$$\frac{v}{V_{\text{full}}} = 1.08$$

$$\frac{v}{1.598 \text{ m}^3/\text{s}} = 1.08$$

$$v = 1.725 \text{ m/s (actual velocity at 60\% full)}$$

Example 1.9

Design a separated, coated cast-iron sewer (n=0.013) flowing at 70% full at a peak flow of 0:42 m^3/s and a minimum flow of 0.047 m^3/s. Take gradient as 1:600.

Solution:

Available data:

Q_{peak}=0.42 m^3/s
Q_{min}=0.047 m^3/s
s=1:600
n=0.013
d/D=0.7 (70% full)
From Equations 1.14 and 1.15:

$$V = \frac{1}{n}(R)^{\frac{2}{3}}(s)^{\frac{1}{2}}$$

$$Q = \frac{1}{n}(A)(R)^{\frac{2}{3}}(s)^{\frac{1}{2}}$$

From Manning chart (Figure 1.14), at d/D=0.7,

$$\frac{q}{Q_{\text{full}}} = 0.85$$

$$\frac{v}{V_{\text{full}}} = 1.14$$

The maximum design speed (actual flow in sewer at 70% full)=0.42 m^3/s

$$\frac{q}{Q_{\text{full}}} = 0.85$$

$$\frac{0.42}{Q_{\text{full}}} = 0.85$$

So, Q Manning (full) = 0.494 m^3/s

(I) DETERMINE THE SIZE OF THE SEWER

Enter in the Manning formula (Equation 1.15),

$$Q = \frac{1}{n}(A)(R)^{\frac{2}{3}}(s)^{\frac{1}{2}}$$

$$R = \frac{A}{P} = \text{the flow rate at FULL condition (Manning)}$$

$$R = \left(\frac{\pi}{4}\right)D^2$$

$$= \frac{D}{4}$$

So,

$$0.494 = \frac{1}{0.013}\left(\frac{\pi}{4}\right)D^2\left(\frac{D}{4}\right)^{\frac{2}{3}}\left(\frac{1}{600}\right)^{\frac{1}{2}}$$

$$D^{8/3} = \frac{0.494}{0.9788}$$

D = 0.77 m (choose the next available size of sewer in the market)

$$\text{Say}, D = 0.8\text{ m}$$

(II) DETERMINE THE VELOCITY AT FULL FLOW (MANNING)

From Equation 1.14,

$$V = \frac{1}{n}(R)^{\frac{2}{3}}(s)^{\frac{1}{2}}$$

$$V = \frac{1}{0.013}\left(\frac{0.8}{4}\right)^{\frac{2}{3}}\frac{1}{600}^{\frac{1}{2}}$$

$$= 1.07\text{ m/s (full flow)}$$

(III) CHECK SELF-CLEANING VELOCITIES AT Q_{MAX} AND Q_{MIN}

(a) v at Q_{max}

$$\text{From } \frac{v}{V_{full}} = 1.14,$$

$$\frac{v}{1.07\text{ m/s}} = 1.14$$

$$v = 1.14\,(1.07)\text{ m/s}$$

$$= 1.22\text{ m/s (Within an acceptable range of } 0.8 < v < 4.0\text{ m/s in Table 1.12)}$$

(b) v at Q_{min}

Given that the minimum flow at 70% full = 0.047 m³/s

Q_{full} (Manning) = 0.494 m³/s

From the above data,

$$\frac{q}{Q_{full}} = \frac{0.047}{0.494}$$

$$= 0.095$$

From Manning chart (Figure 1.14),

$$\text{At } \frac{q}{Q_{full}} = 0.095\ (x \text{ axis}),\ \frac{d}{D} = 0.215$$

$$\text{At } d/D = 0.215, \quad v/V_{full} = 0.575, \quad \text{where } v = \text{the minimum flow}$$

$$\frac{v_{min}}{V_{full}} = 0.575$$

From before, $V_{full} = 1.07$ m/s (from ii).

So, $v_{minimum} = 0.575\,(1.07) = 0.615$ m/s > 0.6 m/s for 0.8 m diameter sewer (Table 1.11), hence acceptable.

1.7.12.5 Structural Design of Sewers

The structural design of a buried sewer can be divided into the following two categories, i.e. rigid pipe and flexible pipe. All two structural designs shall take into account how the sewer is supported to determine the loading that the sewer can safely withstand. There are various categories of bedding (Type A, B, C, D) that shall be applied for the rigid and flexible type of sewer. An example of Type B bedding is shown in Figure 1.16. Other types of bedding with detailed requirements are given in MSIG Volume 3, 2009 [8].

1.7.12.6 Related Guidelines

The following guidelines are useful in designing the sewer with related procedures for planning, approval, and upgrading.

1. Technical Drawing (SPAN DW 0101:2011) – Standard and Typical Drawing for Sewerage System, Part 1: Sewer and Appurtenances – Gravity Sewer, First Edition, Suruhanjaya Perkhidmatan Air Negara (SPAN), February 2012 [12].
2. Malaysian Sewerage Industry Guidelines (MSIG), Volume 2: Sewerage works procedures, 2nd Edition, Suruhanjaya Perkhidmatan Air Negara (SPAN), July 2013 [13]. It covers sewerage planning, fee, application for approval procedures, with related application forms.
3. Malaysian Sewerage Industry Guidelines (MSIG), Volume 3 Sewer Networks and Pump Stations, 3rd Edition, Suruhanjaya Perkhidmatan Air Negara (SPAN), 2009 [8].
4. Upgrading procedures for public sewerage system for the developer (applicant), Suruhanjaya Perkhidmatan Air Negara (SPAN), June 2016 [14].

1.7.13 Manhole

A manhole (sewer hole) is the top opening to an underground sewer used to house an access point for making connections, inspection, or performing maintenance. Detailed requirements are given in MSIG, Volume 3, 2009 [8].

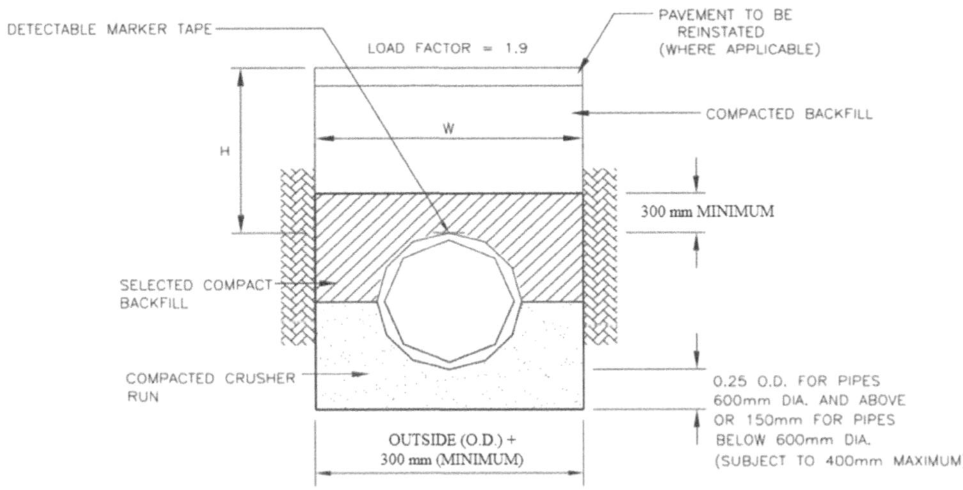

CLASS 'B' BEDDING (COMPACTED CRUSHER RUN)

Pipe embedded in carefully compacted crusher run 5-20 size range extending halfway up to sides of the pipe. The remainder site fills and tops are to be compacted carefully with backfill.

The selected compact backfill for vitrified clay pipe (VC) shall be compacted dry soil (free from organic vegetable matters and excavation/compacted dry red earth/sand/chipping/stones not greater than 25.

FIGURE 1.16 Class B bedding compacted crusher run (MSIG Volume 3, 2009) [8].

1.7.13.1 Specific Requirements

a. The minimum diameter of manhole chamber constructed from pre-cast concrete rings shall be as given in Table 1.16 and Figure 1.17.
b. Provision of back-drop manhole shall be based on the following criteria:
    i. For pipe size equal to 225 mm or less, a back-drop manhole shall be provided when the difference in invert level is equal to 900 mm or more.
    ii. For pipe size more than 225 mm, a back-drop manhole shall be provided when the difference in invert level is equal to 1,000 mm or more.

    An example of the back-drop manhole is shown in Figure 1.18.
c. All manholes shall be sited in public road reserve so that access can be gained for maintenance purposes.
d. Manhole shall be provided for the following locations:
    i. The starting end of all gravity sewers may be replaced by a terminal layout
    ii. Every change in direction or alignment for sewers less than 600 mm in diameter
    iii. Every change in gradient
    iv. Every junction of two or more sewers
    v. Every change in the size of the sewer
e. Generally, the spacing between manholes shall not be more than 100 m for sewers less than 1.0 m in diameter. For sewers with a diameter larger than 1.0 m, the spacing between manholes shall not be more than 150 m.
f. Where site conditions prevent manhole construction on the existing public sewer, a manhole shall be provided on the connection pipe as near to the public sewer as possible.

TABLE 1.16
Minimum Manhole Diameters

Depth to Soffit* from the Cover Level of Manhole (m)	Nominal Internal Diameter (DN)* Pipe in a Manhole (mm)	The Minimum Internal Dimension of the Manhole Chamber (mm)
<1.5	<150	1,000
	225–300	1,200
	375–450	1,350
	525–710	1,500
	820–900	1,800
	>900	Subject to designer's requirements based on site condition
≥1.5	<300	1,200
	375–450	1,350
	525–710	1,500
	820–900	1,800
	>900	Subject to designer's requirements based on site condition

*Refer to Figure 1.17 for soffit and invert levels and DN (Nominal internal diameter).

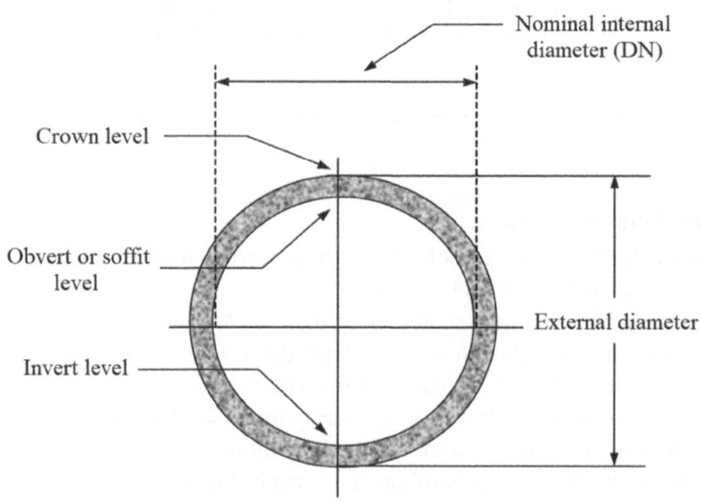

FIGURE 1.17 Various terms used in sewer design (MSIG, Volume 3, 2009) [8].

g. There are eight categories of a manhole (Type A to H), with selected diagrams shown in Figures 1.191.21. Details of other types of manhole and construction are shown in Appendix A of the MSIG, Volume 3, 2009 [8] (Figures 1.20 and 1.21).

   A typical shallow precast concrete manhole (ground level to invert of pipe 1.2 m ≤ depth <2.5 m) is shown in Figure 1.22 (a) plan view, (b) Section A-A, (c) Section B-B, (d) Section AA (free drop < 600 mm).

h. Individual houses will need a manhole to connect the internal wastewater to the public sewer. For a house with spare space, it is normally located just outside the rear door. However, if there is no available space outside, the manhole is normally installed in the kitchen next to the rear door.

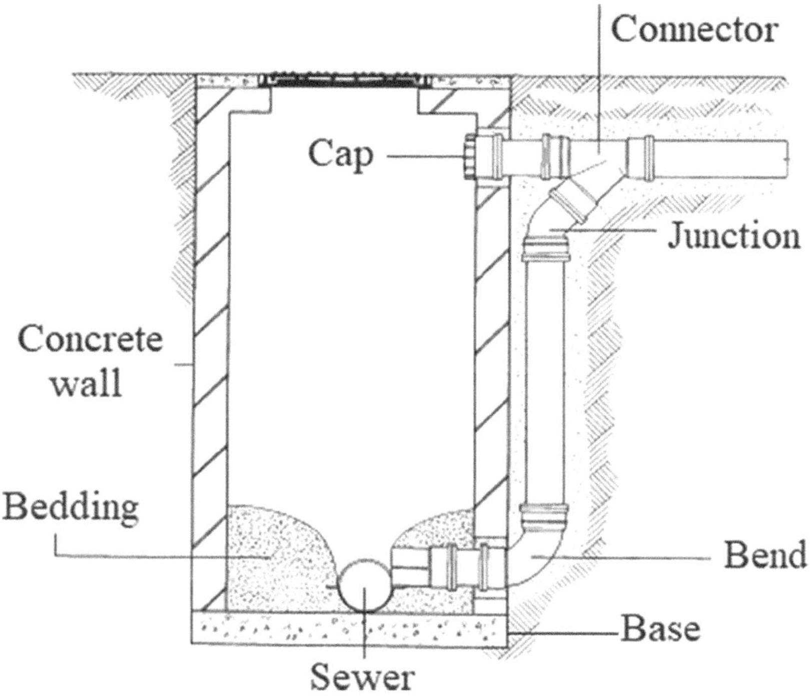

FIGURE 1.18 Backdrop manhole (MSIG, Volume 3, 2009) [8].

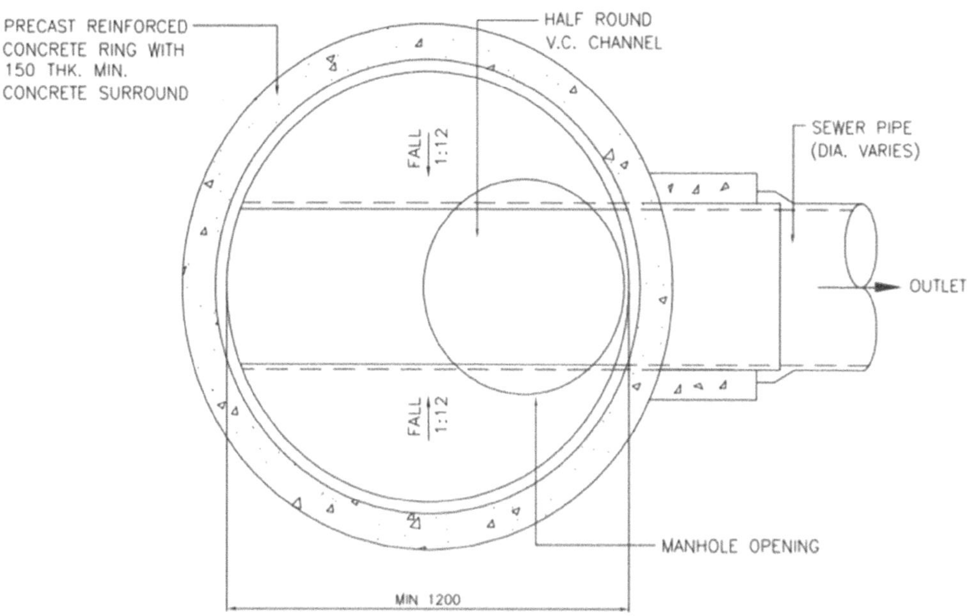

FIGURE 1.19 A plan view of a manhole type A (first manhole) with one outgoing sewer (MSIG, Volume 3, 2009) [8].

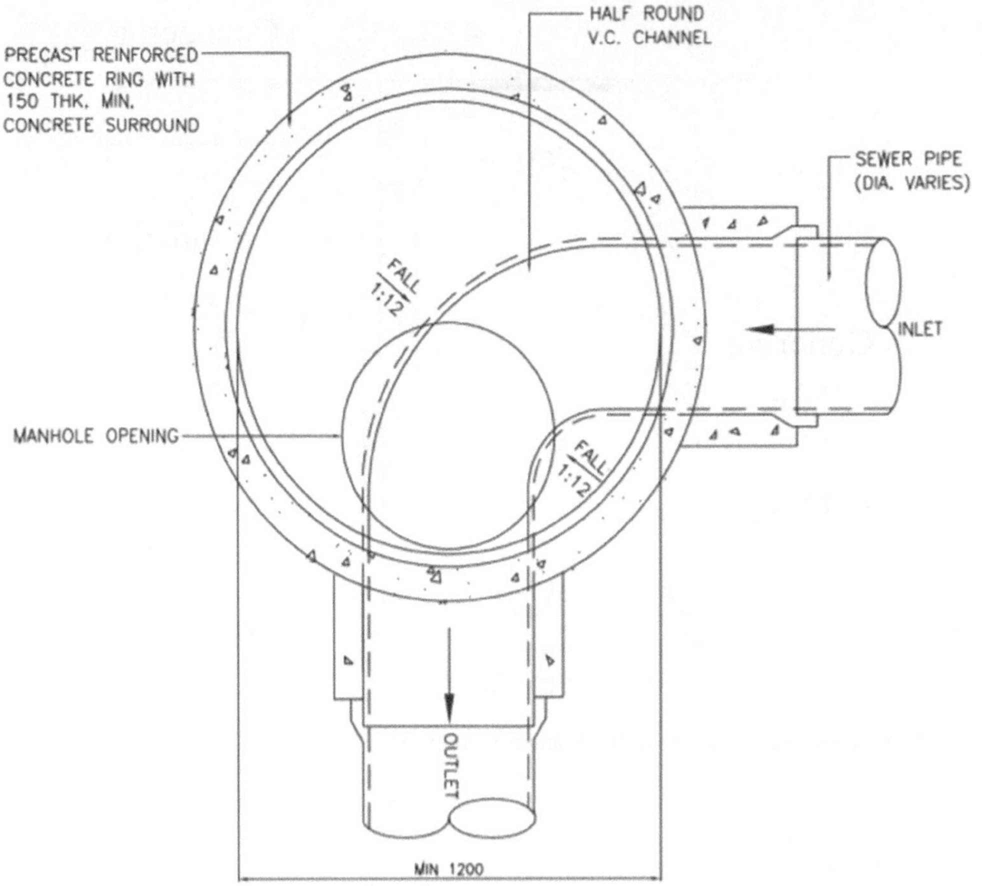

FIGURE 1.20 A plan view of a manhole type C (change of direction 90° bend) with one incoming and one outgoing sewer (MSIG, Volume 3, 2009) [8].

1.7.13.2 Structural Design Considerations for Manhole

The basic structural requirements for a manhole as detailed in MSIG, Volume 3, 2009 [8] are as follows:

a. Concrete used in situ shall be 25 MPa Portland cement unless shown otherwise by the qualified person.
b. A cement blinding with a minimum of 50 mm thickness shall be placed before pouring the concrete manhole base.
c. The base of the manhole shall not be less than 300 mm thickness, which is measured from the channel invert.
d. A minimum of 150 mm thick Grade 25 concrete in situ shall be encased to the precast concrete section. Brick manholes shall not be used.
e. Drop connection pipes and fittings in the manhole shall be of the same diameter and material as the connecting sewer.
f. Manhole covers in roads shall be set to the road profile and shall be flushed with the road surface.

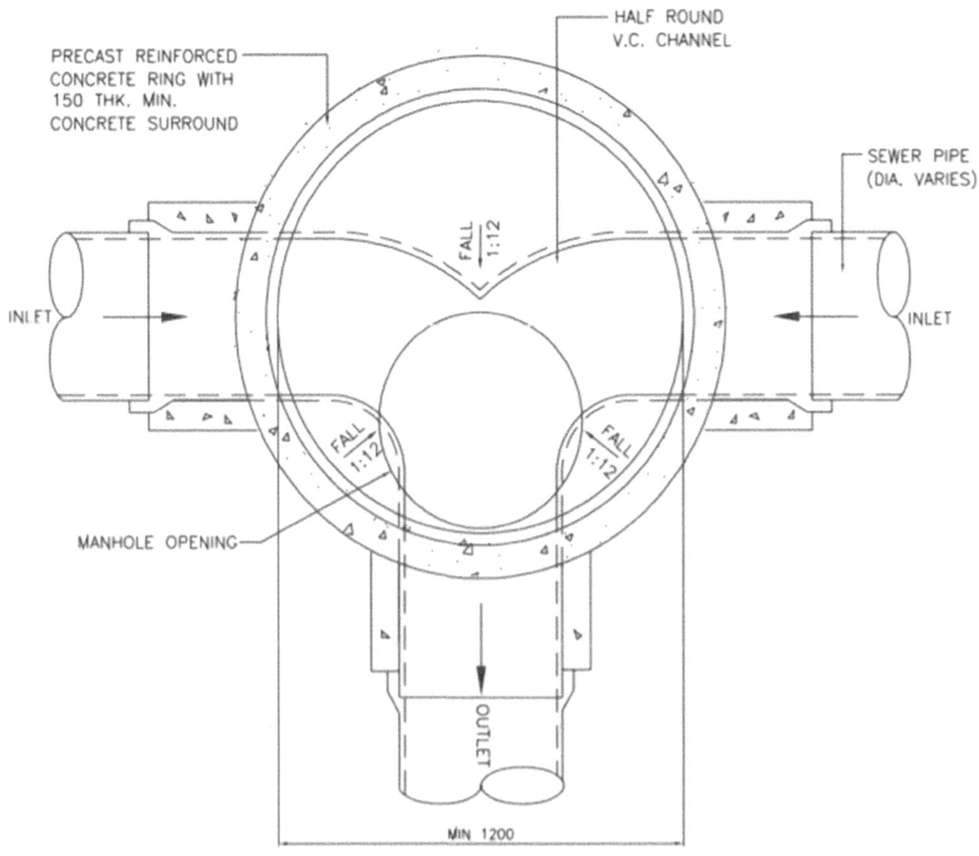

FIGURE 1.21 A plan view of a manhole type F with two incoming and one outgoing sewer (MSIG, Volume 3, 2009) [8].

g. Manhole structures shall not be constructed on a slope or within a slope failure envelope.
h. The maximum depth shall be equal to, or less than 9 m and all manholes deeper than 6 m are subjected to the Commission's prior approval.
i. Manhole covers and frames shall comply with the specifications in this guideline and BS EN124.

Typical examples and dimensions of manhole covers are given in Figure 1.23.

1.7.13.3 Construction and Installation

Construction and installation of manholes involve various aspects, which include protection of affected services, structure, pavement, and vegetation. It also involves excavation procedures and requirements, pipe laying, and also manhole connections. There are specific requirements for pipe laying, which involves pipe bedding, fittings, jacking, concrete pipe support, pipe cutting, and also backfill of the trench. All of these are detailed in MSIG, Volume 3, 2009 [8].

1.7.13.4 Sewer Connection Networking

There are many types of sewer connections. Three of the most common are shown in Figures 1.24–1.27.

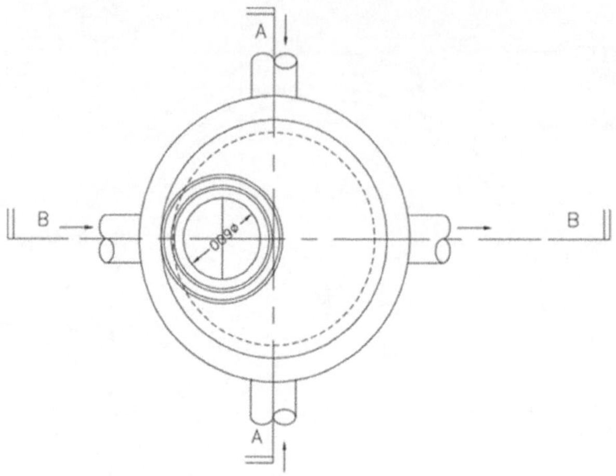

(a) Plan view

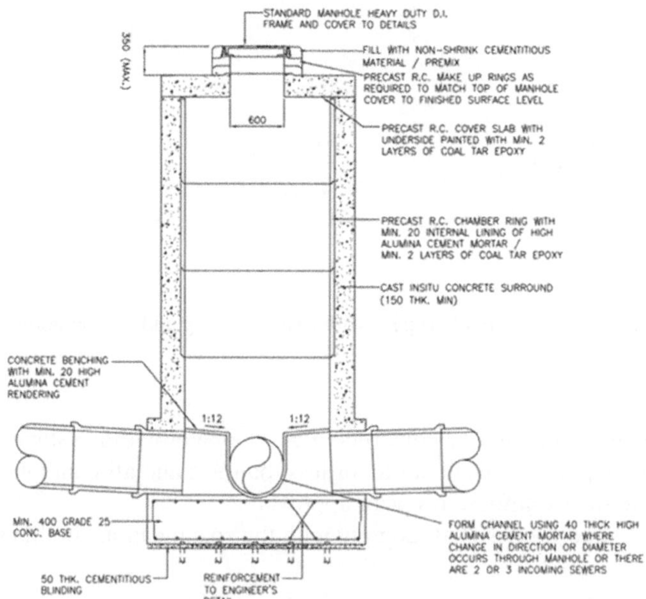

(b) Section A-A

FIGURE 1.22 A typical shallow precast concrete manhole (ground level to invert of pipe $1.2\text{ m} \leq \text{depth} < 2.5\text{ m}$) (MSIG, Volume 3, 2009) [8].

(Continued)

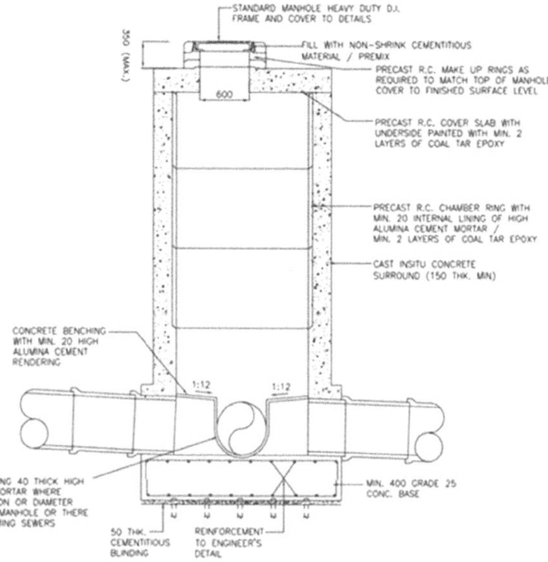

(c) Section B-B

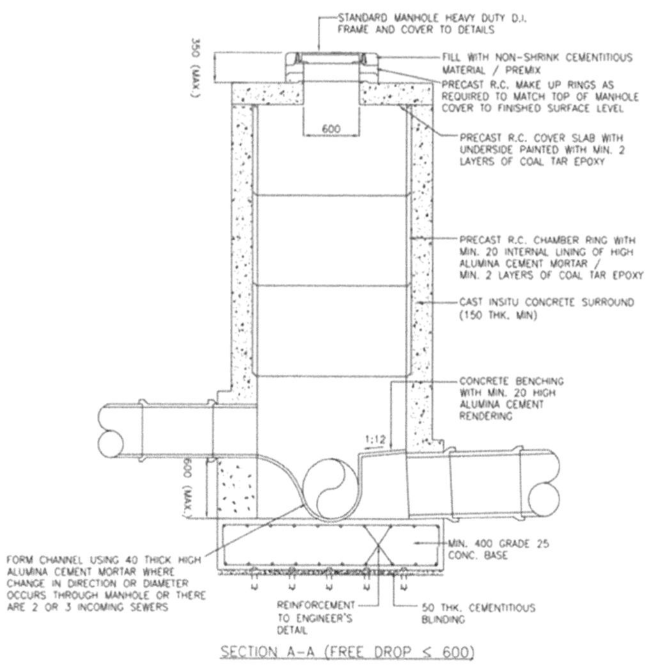

(d) Section A-A (free drop<600 mm)

FIGURE 1.22 (*Continued*) A typical shallow precast concrete manhole (ground level to invert of pipe 1.2 m ≤ depth < 2.5 m) (MSIG, Volume 3, 2009) [8].

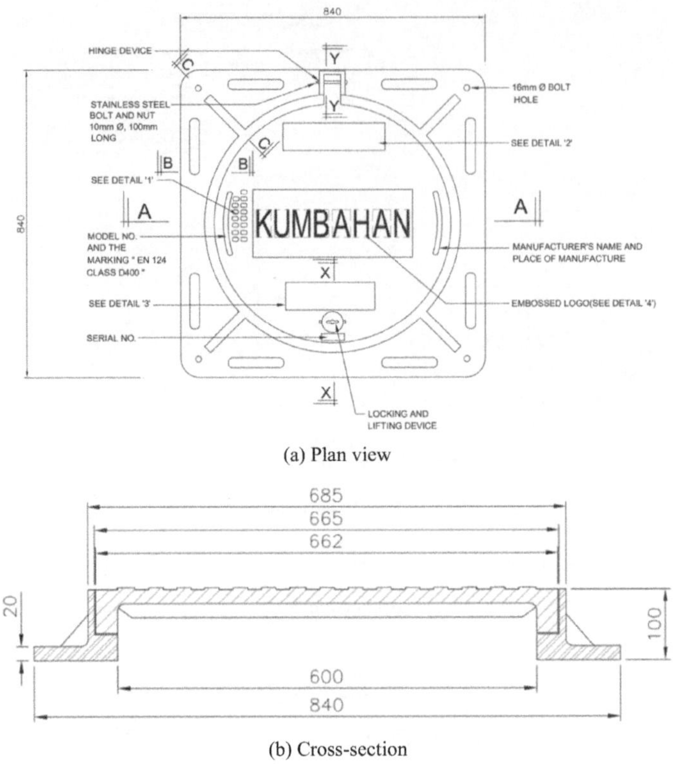

FIGURE 1.23 Typical view of a manhole cover (MSIG, Volume 3, 2009) [8].

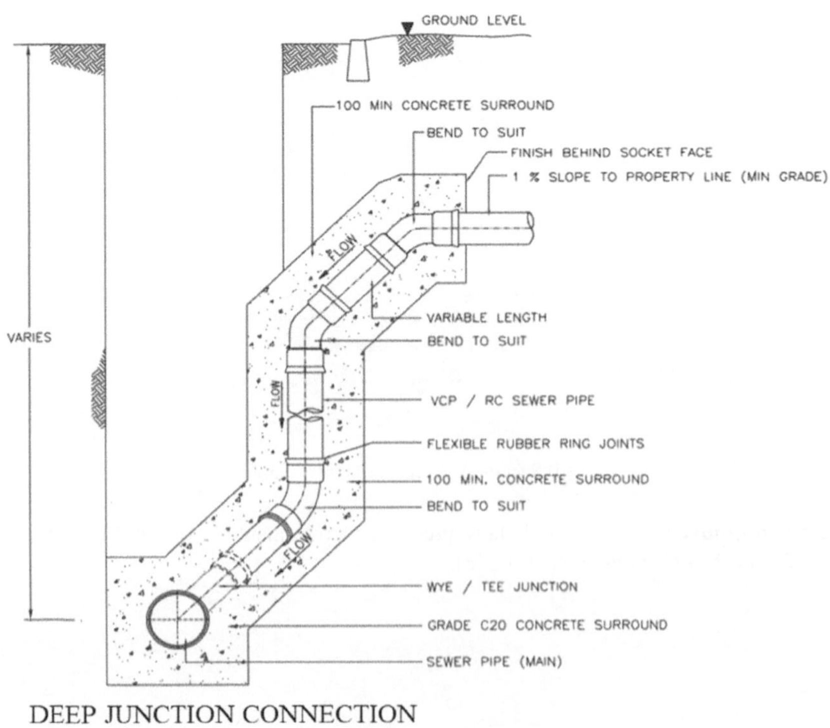

FIGURE 1.24 Deep junction connection (MSIG, Volume 3, 2009) [8].

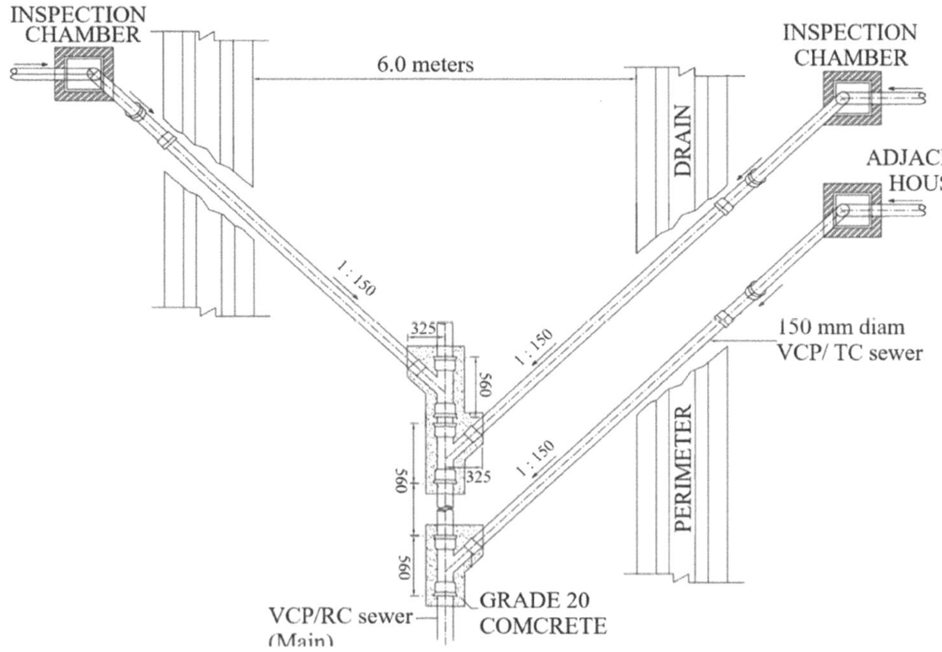

FIGURE 1.25 Lateral connection (plan view) (MSIG, Volume 3, 2009) [8].

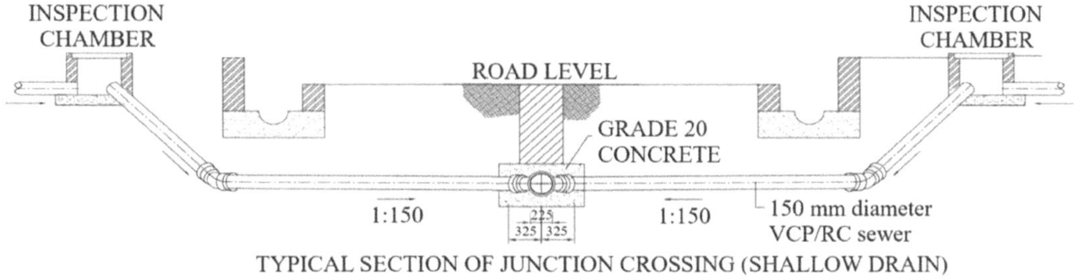

FIGURE 1.26 Typical section of junction connection (for shallow drain crossing) (MSIG, Volume 3, 2009) [8].

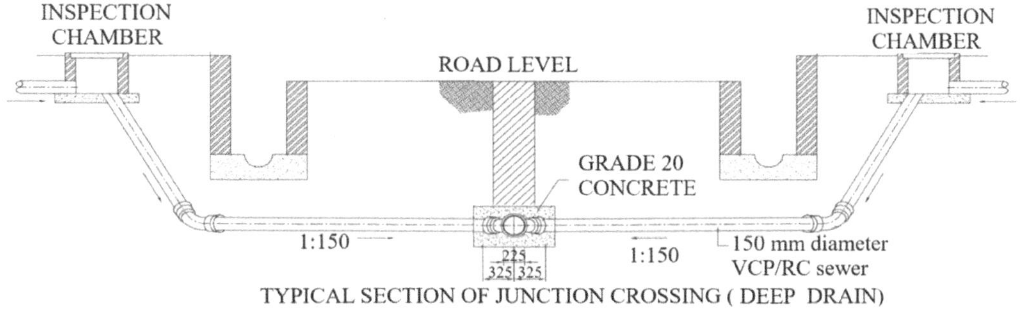

FIGURE 1.27 Typical section of junction connection (for deeper drain crossing) (MSIG, Volume 3, 2009) [8].

1.8 BIODEGRADATION OF ORGANIC MATTER

1.8.1 Introduction

Wastewater treatment is a process used to convert wastewater into a treated effluent that can be safely discharged into receiving water or returned to the water cycle with minimal impact on the environment. The treatment process that takes place in a WWTP often involves multi-stage processes from physical, biological, or sometimes chemical process. The removal of organic matter such as BOD, COD, and some forms of ammoniacal nitrogen happens using a biological process where the pollutants are removed or broken down by microorganisms. Biodegradation is the process by which organic substances are broken down into smaller compounds by enzymes produced by living microbial organisms. Although we can design sophisticated treatment units without knowing the basic principles of how the wastewater is treated, we may face difficulties to troubleshoot operational problems that may occur at any instance.

In wastewater engineering, biodegradation of organic matter is very important after the physical processes. The dissolved organics which cannot be treated physically need to be broken down at the biological stage. Biodegradable matter is generally an organic material that provides nutrient for microorganisms. The breakdown of organic substances (such as BOD) by living microorganisms, particularly bacteria and fungi, plays a major role in biodegradation because of their abundance, species diversity, catabolic versatility, and ability to adapt to a wide variety of environmental conditions. This process can occur aerobically (with oxygen) or anaerobically (without oxygen). The result is often the complete conversion of the organic substance to stable inorganic products (e.g. carbon dioxide and water).

For example, when glucose (which is rich in carbon and exhibits a higher BOD value) is biodegraded in the aeration tank, the carbon is converted into CO_2. In this manner, the carbon content has been reduced, hence reducing the BOD level after the process is complete. Similarly, the nitrogen, phosphorous, and sulphate at influent are also biodegraded into simpler substances such as nitrate (NO_3), phosphate (PO_4), and sulphate (SO_4), respectively. This aerobic process will reduce the amount of N, P, and S at effluent. A good example is the oxidation of glucose (a monosaccharide) in aerobic respiration, as expressed in Equations 1.16–1.18.

$$\underset{\substack{\text{Glucose}\\\text{(High C hence}\\\text{high BOD)}}}{C_6H_{12}O_6} + \underset{\text{Oxygen}}{6O_2} \rightarrow \underset{\substack{\text{Carbon dioxide}\\\text{(Less C, hence}\\\text{low BOD)}}}{6CO_2} + \underset{\text{Water}}{6H_2O} \tag{1.16}$$

Similarly, for Sulphur (S), Phosphorus (P), Nitrogen (N),

$$\underset{\substack{\text{(High S,P,N)}\\\text{(Complexed}\\\text{state)}}}{S+P+N} + O_2 \rightarrow + \underset{\text{Low S}}{\text{Sulphate } (SO_4)} + \underset{\text{Low P}}{\text{Phosphate } (PO_4)} + \underset{\text{Low N}}{\text{Nitrate } (NO_3)} + H_2O \tag{1.17}$$

$$\underset{\text{Unbalanced}}{C_nH_aO_bN_c} + O_2 \rightarrow CO_2 + H_2O + NO_3 \tag{1.18}$$

1.8.2 Types of Organic Matter

Organic matter (or organic material) in wastewater engineering is a substance that has come from carbon-bonded compounds (carbon bonds) that originate from plants and animals. It is capable of being biodegraded aerobically or anaerobically by microorganisms. It enters wastewater from the food and drinks that we eat and flushed in the toilet. It is rich in carbohydrates, proteins, lipids, fat, oil, and others. In compound form, it is combined with other substances such as hydrogen (H), oxygen (O), nitrogen (N), and sulphur (S). Some basic compound structures are as follows: carbohydrates and glucose ($C_6H_6O_{12}$), fat as oleic acid ($C_{17}H_{33}COOH$), protein as creatine ($C_4H_9N_3O_2$).

1.8.3 Aerobic and Anaerobic Reactions

There are two types of reactions, namely, aerobic reaction and anaerobic reaction.

1.8.3.1 Aerobic Reaction

Aerobic biodegradation is the breakdown of organic contaminants by microorganisms in the presence of oxygen. The aerobic reaction involves almost all the biological treatment plants, such as in the aeration tank of an Activated Sludge Process and aerated lagoon. After the physical process (screening, grit and grease removal, and sedimentation tank), the unsettled organics will be biodegraded in the next stage of treatment called biological process or sometimes referred to as a secondary treatment. At this stage, the organic contaminants such as carbohydrate and glucose (which are indicated by BOD and COD) are degraded under aerobic conditions by aerobic bacteria called aerobes. Oxygen is used in this biological process to break down the small molecules into water, carbon dioxide, and other oxides. The reactions also release energy. There are two main processes in this reaction, as shown in Figure 1.28:

i. **Synthesis**: This is a process to produce new cells of microorganisms. This process uses the energy which is released by respiration action.
ii. **Respiration**: This is a process to degrade the wastewater to a simple form, and this produces energy.

1.8.3.2 Anaerobic Reaction

Anaerobic reaction, commonly known as anaerobic digestion, is a series of processes in which microorganisms break down biodegradable material in the absence of oxygen. The digestion process begins with bacterial hydrolysis of the input materials to break down insoluble organic substances such as carbohydrates and make them available for the bacteria. Acetogen then converts the sugars and amino acids into carbon dioxide, hydrogen, ammonia, and organic acid. Acetogenic bacteria then convert these resulting organic acids into acetic acid, along with additional ammonia, hydrogen, and carbon dioxide. Methanogen finally can convert these products to

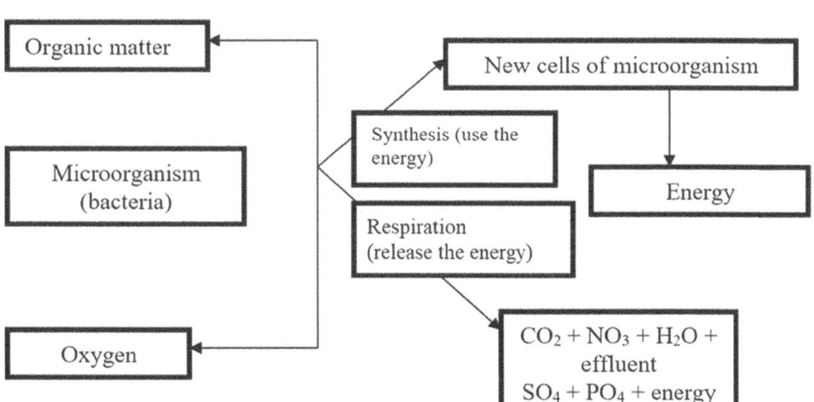

FIGURE 1.28 Aerobic reaction.

methane and carbon dioxide. Several bacteria are involved in the process of anaerobic digestion, including acetic acid-forming bacteria and methane-forming bacteria. These bacteria feed upon the initial organic substance, which undergoes a few different processes converting it to intermediate molecules including sugars, hydrogen, and acetic acid before finally being converted to biogas.

Anaerobic digestion is widely used to treat highly concentrated organics, wastewater sludge, and septic and Imhoff tanks because it provides volume and mass reduction of the input material. It occurs in two stages as follows:

i. **Acid forming (Stage 1)**: Organic matter will be oxidized by a microorganism to form new cells. However, this is just an intermediate process (not stable).
ii. **Methane forming (Stage 2)**: Bacteria react with unstable acids to form new stable components such as methane (CH_4), CO_2, hydrogen sulphide (H_2S), nitrate (NO_3), and water. These stable components will be released into the environment.

1.8.4 Enzyme Reaction

In practice, almost all organic compounds and materials are subject to biodegradation processes. The significance, however, is in the relative rates of such processes, such as days, weeks, years, or centuries. Several factors determine the rate at which this degradation of organic compounds occurs. Salient factors include oxygen, availability of nutrients, light, and temperature. Their biodegradability limits the degradation rate of many organic compounds. Biodegradation is normally in solution.

The reaction rate or rate of reaction is the speed at which reactants are converted into products. There are many influencing factors. Some reactions are naturally faster than others. Other factors include concentration, pressure, and temperature.

The enzyme controls biological response speed. This enzyme is an organic catalyst released by the bacteria. It can accelerate the reaction without themselves undergoing any permanent chemical change. They are neither used up in the reaction nor do they appear as reaction products. The enzyme is very sensitive to the environment, such as temperature, pH, and metal ions. For example, the biological process is rapid at a higher temperature.

1.8.5 Biological Growth

Bacteria in the aeration basin will biodegrade the organic matter (a source of nutrients or food) into simpler substances via synthesis and respiration processes to develop new cells and release energy. This process continues if there is an oxygen supply and until all the organics have been used. The process stops at one stage when the food supply is ceased due to possible blockage or there is no oxygen supply due to possible power failure.

To see this behaviour, it is good to understand this phenomenon by taking into consideration the following situation. Assume we have a group of microorganism cultures provided with a food source in a given time within an aerated basin as a batch experiment (as shown in Figure 1.29). The different growth stages of bacteria over time in relation to the amount of food or nutrient (BOD) available are shown in Figure 1.30.

There are four distinct phases of bacterial growth as shown in Figure 1.31: lag (or delayed) phase, log (or exponent) phase, stationary growth phase, and decline phase (also referred to as endogenous respiration phase).

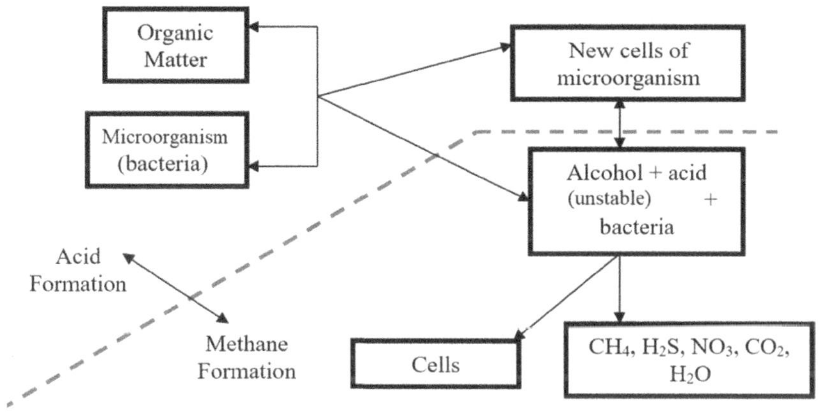

FIGURE 1.29 Anaerobic reaction.

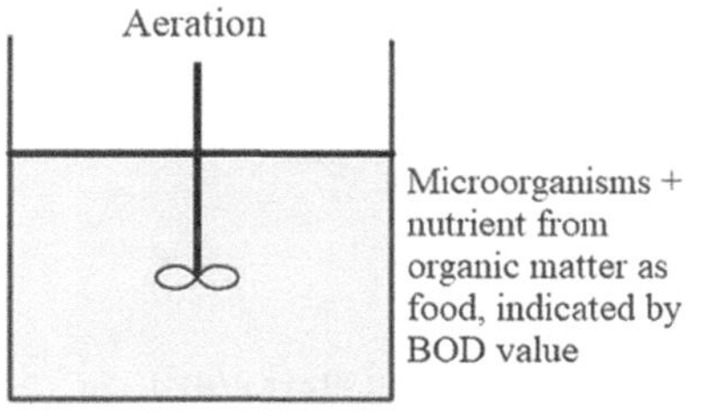

FIGURE 1.30 A growth rate experiment for a given bacterial culture.

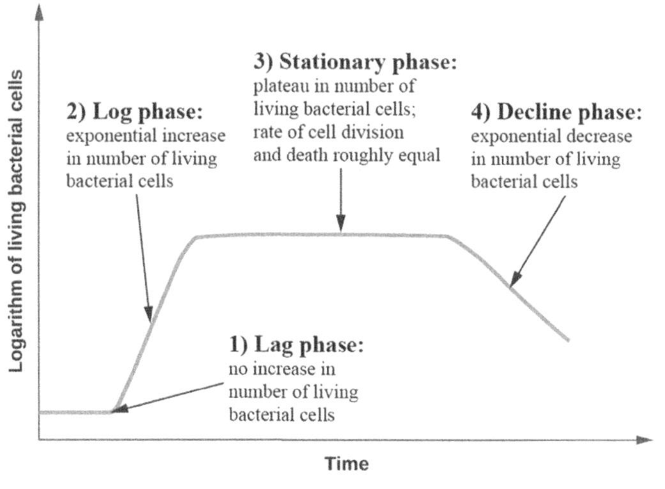

FIGURE 1.31 The bacterial growth curve.

1.8.5.1 Lag (or Delayed) Phase

During this phase, there is a large amount of food available compared to the number of bacteria. It is called the "Lag Phase" because it takes the bacteria a certain amount of time to respond to the available food. There will be an increase in the number of living bacteria cells.

1.8.5.2 Log (or Exponent) Phase

In the log phase, the bacteria have begun consuming the BOD and start replicating exponentially. There will be an exponential increase in the number of living bacteria cells.

1.8.5.3 Stationary Phase

The available food begins to diminish, and bacterial reproduction begins to slow as there is more competition for food. The number of living bacteria cells plateaued at this phase where the rate of cell division and death is roughly equal.

1.8.5.4 Decline or (Endogenous Respiration) Phase

At this stage, there will be an exponential decrease in the number of living bacterial cells. Most of the food (BOD) has been converted, and there is little food available for the bacteria to consume. Most of the bacteria have formed flocs, and some bacterial cells die. Flocs can settle out to the bottom of the basin, which leaves a clearer effluent with little BOD behind.

The above growth rate can be related as follows (Equation 1.19).

Microorganism growth rate = Growth rate in the presence of food – decay rate

$$\frac{dx}{dt} = -y\frac{dc}{dt} - kdx \tag{1.19}$$

where

dx/dt = growth rate (kg/m^3.day)

x = the concentration of microorganism (kg/m^3)

$-y$ = the coefficient of growth (no unit), equivalent to the mass of the microorganism. The negative value indicates a shortage of food over time

dc/dt = the food usage rate (kg/m^3.day)

kd = decay coefficient (time^{-1})

1.8.6 Biochemical Oxygen Demand

1.8.6.1 Principle of BOD

BOD is the amount of DO needed (i.e. demanded) naturally by aerobic biological organisms to break down organic material present in a given water sample at a certain temperature over a specific period. The BOD value is most commonly expressed in milligrams of oxygen consumed per litre of the sample during 5 days of incubation at 20°C. It is the main parameter as a surrogate of the degree of organic pollution of water and wastewater. BOD is the most important parameter used to gauge the effectiveness of WWTPs and enacted as one of the standard discharge effluent parameters under the Environmental Quality Act, 1974 [7]. There are two types of oxidation in wastewater, the oxidation of carbon (C) and oxidation of nitrogen (N). Both types of oxidation require oxygen. Oxidation in the BOD test occurs slowly (biodegraded naturally) and is not always complete in 5 days. The percentage of degradation on the 5th day varies from one sample to another. For example, the percentage glucose breakdown is 100% on day 5 and is 65% for domestic wastewater, but only about 40% of complex organic materials such as fats and oils [3].

1.8.6.2 BOD Test

There are two methods to test the BOD, i.e. the dilution method and the manometric method.

1.8.6.2.1 Dilution Method

This standard method is internationally recognized (Method 5210B in the Standard Methods for the Examination of Water and Wastewater) [15]. The Royal Commission introduced BOD experiments

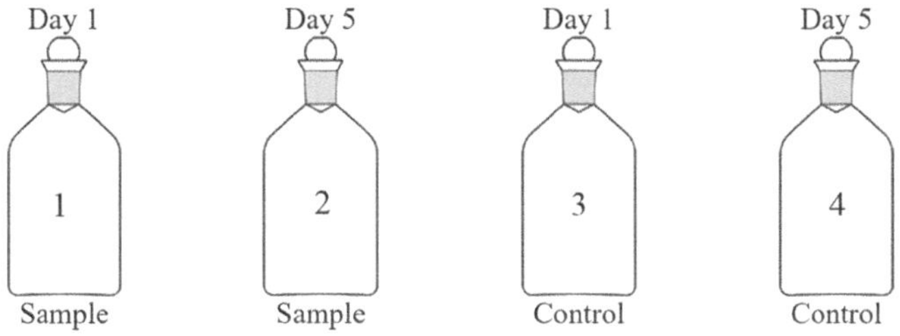

FIGURE 1.32 Standard 300 mL bottles used in BOD test.

on wastewater disposal, UK. To obtain BOD_5, DO concentrations in a sample must be measured before and after the incubation period and appropriately adjusted by the sample's corresponding dilution factor. This analysis is performed using 300 mL incubation bottles (Figure 1.32) in which buffered dilution water is dosed with seed microorganisms and stored for 5 days in the darkroom at 20°C to prevent DO production via photosynthesis. A control sample (dilution water blanks) containing various nutrients is also prepared for the same conditions. This is done to confirm the quality of the dilution water that is used to dilute the other samples that do not take up the DO during the incubation period. For the measurement of carbonaceous BOD, a nitrification inhibitor is required to hinder the oxidation of ammoniacal nitrogen. The percentage of degradation on day 5 depends on the type of organic matter present in the sample. For example, the percentage of glucose is 100% on day 5, 65% for domestic wastewater, and 40% for a complex organic matter such as oil and grease.

The set-up of BOD experiments requires the preparation of sample and control bottles, as shown in Figure 1.32. Two bottles are used for samples (bottles 1 and 2), and two bottles (bottles 3 and 4) as a control (containing water dilution made from distilled water). The DO levels in bottles 1 and 3 are determined on day 1, followed by DO level measurements on day 5 for bottles 2 and 4. The BOD then can be expressed as below (Equation 1.20).

Dilution of the sample is usually necessary. This is because, without dilution, the sample may be too concentrated, and the DO value may be exhausted after 5 days and could not be tested. A good dilution factor may leave the DO balance of 2.0–7.5 mg/L on day 5 [9], which can be tested. The dilution factor also depends on the type of sample. Concentrated samples require higher dilution. For example, the normal dilution factor for domestic wastewater is between 1:100 and 1:1,000.

$$BOD_5 = \frac{(D_1 - D_2) - (S_1 - S_2) f}{P} \tag{1.20}$$

where

BOD_5 = BOD on day 5 (mg/L)
D_1 = Dissolved oxygen of diluted sample immediately after preparation (mg/L)
D_2 = Dissolved oxygen of diluted sample after 5 – day incubation (mg/L)
S_1 = Dissolved oxygen of seed control immediately after preparation (mg/L)
S_2 = Dissolved oxygen of seed control after 5-day incubation (mg/L)
f = ratio of seed in diluted sample to seed in seed control
= % seed in diluted sample : % seed in seed control
P = decimal volumetric fraction of sample used

BOD test accuracy is about 10%–20% (Barnes et al. [9]). In other words, the BOD result has only 80%–90% accuracy (or degree of error of 10%–20%). Therefore, the results of the BOD are not reported in the decimal point.

Example 1.10

Fifteen millilitres of the waste sample was added directly into a 300 mL BOD incubation bottle. Determine the BOD_5 if the following test results are obtained:

Day 1 DO for sample = 8.55 mg/L
Day 5 DO for sample = 2.4 mg/L
Day 1 DO for control = 8.75 mg/L
Day 5 DO for control = 8.53 mg/L

$$D_1 = 8.55 \text{ mg/L}$$

$$D_2 = 2.4 \text{ mg/L}$$

$$S_1 = 8.75 \text{ mg/L}$$

$$S_2 = 8.53 \text{ mg/L}$$

$$f = \frac{300 - 15}{300} = 0.95$$

$$P = 15 \text{ mL} / 300 \text{ mL} = 0.05$$

Solution:

From Equation 1.20,

$$\text{BOD}_5 = \frac{(8.55 - 2.4) - (8.75 - 8.53)0.95}{0.05}$$

$$= 199 \text{ mg/L}$$

1.8.6.2.2 Manometric Method

This method is limited to the measurement of oxygen consumption due only to carbonaceous oxidation. Ammonia oxidation is inhibited. In this test, the sample is kept in a sealed container fitted with a pressure sensor (Figure 1.33). A substance that absorbs carbon dioxide (typically lithium hydroxide) is added to the container above the sample level. The sample is stored in the same conditions as in the dilution method. Oxygen is consumed and, as ammonia oxidation is inhibited, carbon dioxide is released. The total amount of gas, and thus the pressure, decreases because carbon dioxide is absorbed. From the drop of pressure, the sensor electronics computes and displays the consumed quantity of oxygen.

1.8.6.3 The Kinetics of BOD

The kinetics of BOD indicates the speed of the food supply (c) consumption in the BOD test over time. It is assumed that it follows the first-order reaction, where the degree of organic matter lost by time, dc/dt, depends on the concentration of organic matter (c) only.

The loss of organic matter (dc/dt) is directly proportional to the concentration of food (c) as shown in Equation 1.21.

FIGURE 1.33 An example of a manometric BOD apparatus.

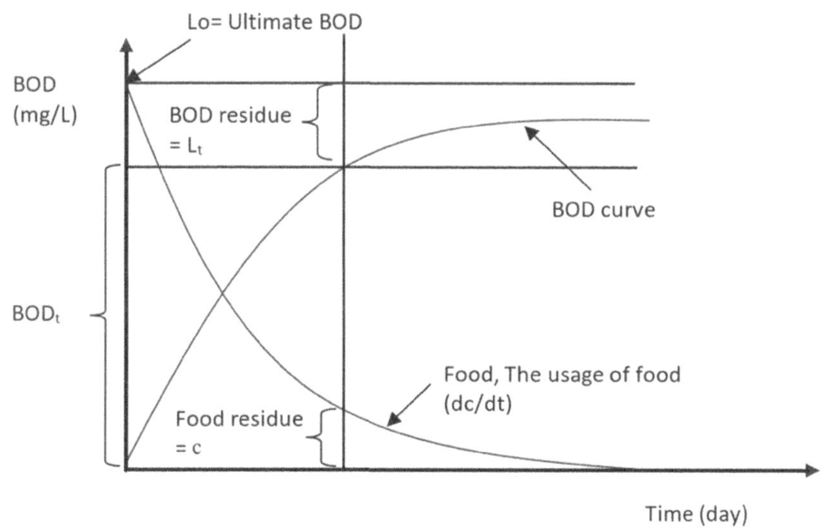

FIGURE 1.34 The balance chart between food consumption and the BOD in the kinetics of BOD.

$$\frac{dc}{dt} \alpha - c \tag{1.21}$$

where

$\frac{dc}{dt}$ = the degree of organic matter lost by time

$-c$ = residue of organic matter at a time, t.

The residue of food after a time t is equivalent to the residual of BOD for the sample, as shown in Figure 1.34. The ultimate BOD (L_o) is the stage whereby 100% BOD degradation has occurred. For example, for domestic wastewater, the BOD_5 represent 65%–68% of the ultimate BOD (32%–38% residual as non-degraded BOD).

From Figure of 1.34:

$$\text{BOD}_t = L_o - L_t \tag{1.22}$$

where

BOD_t = BOD value on day t
L_o = the ultimate BOD
L_t = the residual BOD on day t

The residual food can be expressed as in Equation 1.21,

$$\frac{dc}{dt} \alpha - c$$

$$\frac{dc}{dt} = -kc \tag{1.23}$$

where k is constant.

We equate the remaining BOD ≡ the remaining food

$$L_t \equiv c$$

Hence,

$$\frac{dL_t}{dt} = \frac{dc}{dt}$$

From Equation 1.23,

$$\frac{dL_t}{dt} = -kL_t$$

where k is constant, and a negative value means reducing residual.

Integrating from $t=0$ ($L=L_o$) to $t=t$ ($L=L_t$), then,

$$\int_{L_o}^{L_t} \frac{dL_t}{d_t} = -kdt$$

$$\ln\frac{L_t}{L_o} = -kt \text{ negative value means}$$

$$\ln\frac{L_t}{L_o} = -kt$$

$$\frac{L_t}{L_o} = e^{-kt}$$

$$L_t = (L_o)e^{-kt}$$

Taking log on both sides,

$$\log L_t = \log(L_o)e^{-kt}$$

$$\log\frac{L_t}{L_o}=\log e^{-kt}$$

$$\log\frac{L_t}{L_o}=-k(t)\log e$$

$$\log\frac{L_t}{L_o}=-k_1t$$

where

$k_1=k\ \log_e$, or
$k_1=0.4343\,k$
$k_1=k/2.3$
k_1(base 10), k (base e)
Then,

$$\frac{L_t}{L_o}=10^{-k_1t}$$

$$L_t=L_o10^{-k_1t} \tag{1.24}$$

or

$$L_t=L_oe^{-kt} \tag{1.25}$$

Inserting into Equation 1.22,

$$BOD_t=L_o-L_t$$

$$BOD_t=L_o-L_o10^{-k_1t} \tag{1.26}$$

$$BOD_t=L_o\left(1-10^{-k_1t}\right)$$

or

$$BOD_t=L_o\left(1-e^{-kt}\right) \tag{1.27}$$

k_1=coefficient of BOD (day^{-1}) (k_1 base 10)
k=coefficient of BOD (day^{-1}) (k_1 base e)

Equations 1.26 and 1.27 are true when the test is done at a temperature of 20°C (k_1 at 20°C). Corrections for other temperatures can be done as shown in Equations 1.28 and 1.29.

$$k_T=k_{20}(1.0147)^{T-20} \tag{1.28}$$

$$L_T=L_{20}\left[1+0.02(T-20)\right] \tag{1.29}$$

Please note that the BOD rate of reaction is directly proportional to the BOD coefficient, k_1. The higher the value of k_1 means the reaction is much faster. Different compounds have different k_1 values, as shown in Figure 1.35.

Some observations from Figure 1.35:

1. BOD_5 for k_1 =0.01 day^{-1} equals 10.4% of L_o. The examples are fat, oil, and grease. It may take a few days, months, or years to reach the ultimate BOD.
2. BOD_5 for k_1=0.1 day^{-1} equals 68.4% of L_o. An example is protein compounds.
3. BOD_5 for k_1=0.5 day^{-1} equals 99.7% of L_o. Examples are glucose and carbohydrate.

The typical k_1 value for domestic wastewater is 0.17–0.30 day^{-1}. Depending on the types, the typical k_1 value for industrial wastewater is 0.05 day^{-1}.

Example 1.11

A wastewater sample has its BOD_5^{20} value of 200 mg/L and the BOD rate of the coefficient, k_1=0.17 day^{-1} at 30°C. Calculate the value of the ultimate BOD, L_o.

Solution:

From Equation 1.28,

$$k_T = k_{20}(1.0147)^{T-20}$$

$$k_{30} = 0.17(1.0147)^{30-20}$$

$$= 0.269\ \text{day}^{-1}$$

Inserting into Equation 1.26,

$$\text{BOD}_t = L_o\left(1-10^{-k_1 t}\right)$$

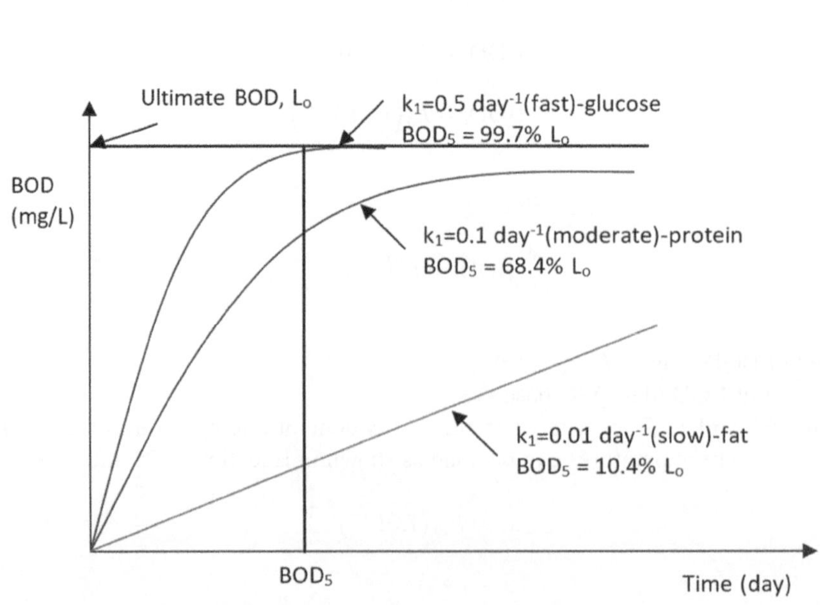

FIGURE 1.35 The BOD rate of reaction at the different values of k_1.

$$\text{BOD}_5 = L_o\left(1-10^{-k_1(5)}\right)$$

$$200 = L_o\left(1-10^{-0.269(5)}\right)$$

$$200 = L_o(0.955)$$

$$L_o = 209.42 \text{ mg/L}$$

$$= 209 \text{ mg/L}$$

Example 1.12

The ultimate BOD value (L_o) for a BOD kinetic experiment conducted on domestic wastewater is 750 mg/L. The experiment was conducted at 28°C, determine:

i. The L_o at 20°C.
ii. The value of BOD_5 if the BOD rate of the coefficient at 20°C is 0.16 day^{-1}.

Solution:

i. From Equation 1.29,

$$L_T = L_{20}\left[1+0.02(T-20)\right]$$

$$L_{28} = L_{20}\left[1+0.02(30-20)\right]$$

$$750 = L_{20}(1.2)$$

$$L_{20} = 625 \text{ mg/L}$$

ii. From Equation 1.26,

$$\text{BOD}_t = L_o\left(1-10^{-k_1 t}\right)$$

$$\text{BOD}_5 = L_o\left(1-10^{-k_1(5)}\right)$$

$$\text{BOD}_5 = 625\left(1-10^{-0.16(5)}\right)$$

$$\text{BOD}_5 = 625(0.842)$$

$$= 526 \text{ mg/L}$$

1.8.6.4 Determination of BOD Rate of the Coefficient, k_1 and the Ultimate BOD, L_o

The BOD rate of the coefficient, k_1 and the ultimate BOD, L_o, can be determined experimentally in the laboratory. However, the scattered plot obtained (Figure 1.36) makes it difficult to predict the L_o, hence k_1 value could not be calculated (from Figure 1.38). Thomas method is the alternative approach to resolve this issue [2].

According to Thomas,

$$\left(1-e^{-kt}\right) = kt\left(1+\frac{kt}{6}\right)^{-3} \qquad (1.30)$$

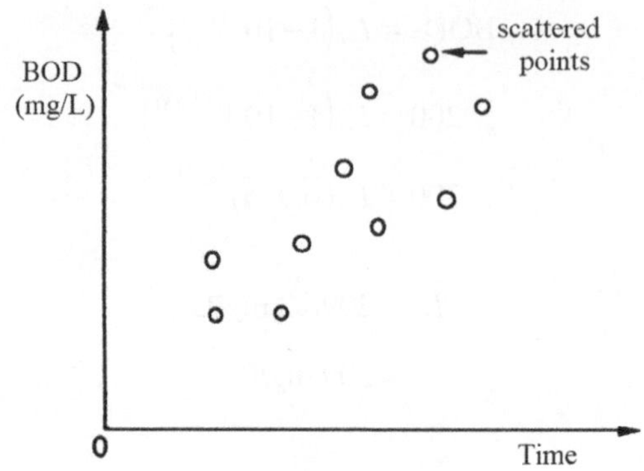

FIGURE 1.36 Scatter points by the direct plot of BOD experiment.

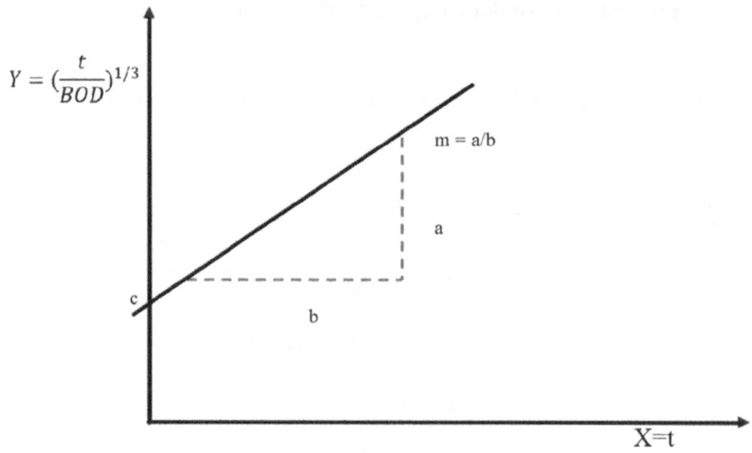

FIGURE 1.37 Thomas plot.

Inserting Equation 1.30 into Equation 1.27,

$$\text{BOD}_t = L_o\left(1-e^{-kt}\right)$$

$$\text{BOD}_t = L_o kt\left(1+\frac{kt}{6}\right)^{-3} \tag{1.27}$$

By dividing both sides with t, then

$$\left(\frac{1}{\text{BOD}}\right)^{1/3} = (kL_o)^{-1/3} + \left(\frac{k^{\frac{2}{3}}}{6L_o^{\frac{1}{3}}}\right)t$$

This is a straight-line equation, $Y = mX + c$, where m is the slope and c is where it cuts the y-axis.

By plotting $\left(\frac{1}{\text{BOD}}\right)^{1/3}$ against t in Figure 1.36, we can determine the values of k(base e) and L_o. From Figure 1.37, we can get

$$y = \left(\frac{t}{\text{BOD}}\right)^{1/3}$$

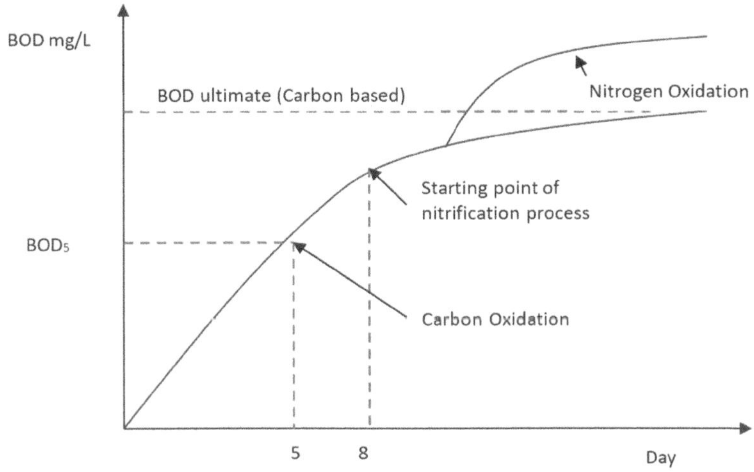

FIGURE 1.38 Oxidation process of carbon and nitrogen in the wastewater sample.

$$c = (KL_o)^{-1/3}$$

$$m = \frac{k^{2/3}}{6L_o^{1/3}}$$

1.8.6.5 Nitrification in the BOD Reaction

There are two stages of decomposition in the BOD test: a carbonaceous stage and a nitrogenous stage (Figure 1.38).

i. The carbonaceous stage, or first stage, represents that portion of oxygen demand involved in the conversion of organic carbon to carbon dioxide.
ii. The nitrogenous stage, or second stage, represents a combined carbonaceous plus nitrogenous demand, when organic nitrogen, ammonia, and nitrite are converted to nitrate.

Domestic nitrogenous oxygen demand generally begins after about 6 days. For some wastewater, especially discharge from WWTPs utilizing biological treatment processes, nitrification can occur in less than 5 days if ammonia, nitrite, and nitrifying bacteria are present. In this case, a chemical compound that prevents nitrification should be added (allyl thiourea, ATU) to the sample if the intent is to measure only the carbonaceous demand so that only the oxidation of carbon is measured. The results are reported as carbonaceous BOD (CBOD) or as $CBOD_5$ when a nitrification inhibitor is used.

NOTE

1 In accordance with Malaysian Sewerage Industry Guidelines, National Water Services Commission (SPAN) and with design examples.

REFERENCES

1. Nicoll, E.H. *Small Water Pollution Control Works: Design and Practice*, Ellis Horwood Ltd, Chichester, 1988.
2. Jackson, M.H., Morris, G.P., Smith, P.G., Crawford, J.F. Environmental Health Reference Book, Butterworth-Heinemann Ltd, London, 1989.
3. Tebbutt, T.H.Y. *Principles of Water Quality Control*, 5th Edition, Butterworth-Heinemann, London, 1997.
4. Suruhanjaya Perkhidmatan Air Negara (SPAN), *Malaysian Sewerage Industry Guidelines (MSIG), Volume 4 Sewage Treatment Plants*, 3rd Edition, SPAN, Kuala Lumpur, 2009.
5. Sperling, M.V., Chernicharo, C.A.D.L. *Biological Wastewater Treatment in Warm Climate Regions, Vol. 1*, IWA Publishing, London, 2005.
6. Standards & Industrial Research Institute of Malaysia (SIRIM). MS 1228. *Code of Practice for Design and Installation of Sewerage Systems*, SIRIM, Kuala Lumpur, 1991.
7. Government of Malaysia, Environmental Quality Act 1974 (Act 127), *Regulations, Rules & Orders*, International Law Book Services, Kuala Lumpur, 2022.
8. Suruhanjaya Perkhidmatan Air Negara (SPAN), *Malaysian Sewerage Industry Guidelines (MSIG), Volume 3 Sewer Networks and Pump Stations*, 3rd Edition, SPAN, Kuala Lumpur, 2009.
9. Barnes, D., Bliss, P.J., Gould, B.W., Vallentine, H.R. *Water and Wastewater Engineering Systems*, Longman Scientific & Technical, Essex, 1981.
10. Wilson, F. *Design Calculations in Wastewater Treatment*, Spon, London, 1981.
11. Matcalf & Eddy. *Wastewater Engineering Treatment Disposal Reuse*, 3rd Edition, McGraw Hill, New York, 1991.
12. Suruhanjaya Perkhidmatan Air Negara (SPAN), Technical Drawing (SPAN DW 0101:2011). *Standard and Typical Drawing for Sewerage System, Part 1: Sewer and Appurtenances-Gravity Sewer*, 1st Edition, SPAN, Kuala Lumpur, February 2012.
13. Suruhanjaya Perkhidmatan Air Negara (SPAN), *Malaysian Sewerage Industry Guidelines (MSIG), Volume 2 Sewerage Works Procedures*, 2nd Edition (revised), SPA), Kuala Lumpur, July 2013.
14. Suruhanjaya Perkhidmatan Air Negara (SPAN), *Upgrading Procedures for a Public Sewerage System for a Developer (Applicant)*, SPAN, Kuala Lumpur, 2016.
15. American Public Health Association, American Water Works Association, Water Environment Federation. 2540 SOLIDS. In: Lipps, W.C., Braun-Howland, E.B., Baxter, T.E., eds. *Standard Methods for the Examination of Water and Wastewater*, 24th Edition, APHA Press, Washington, DC, 2023.

2 Wastewater Engineering Design

Biological Treatment[1]

Hamidi Abdul Aziz

LIST OF ABBREVIATIONS

ANN	Ammoniacal nitrogen
ASP	Activated sludge process
BOD	Biochemical oxygen demand
COD	Chemical oxygen demand
Commission	See SPAN
CST	Communal septic tank
DN	Nominal internal diameter
DWF	Dry weather flow
EA	Extended aeration
HRT	Hydraulic retention time
IST	Individual septic tank
IWK	Indah Water Konsortium
NH_3-N	Ammoniacal nitrogen
NO_3-N	Nitrate nitrogen
Q_{ave}	Average flow or DWF
Q_{max}	Maximum flow
Q_{min}	Minimum flow
Q_{peak}	Peak flow
RAS	Return activated sludge
RBC	Rotating biological contactor
SBR	Sequential batch reactor
SLR	Solids loading rate
SOR	Surface overflow rate
SPAN	Suruhanjaya Perkhidmatan Air Negara
SS	Suspended solids, also referred as TSS
SVI	Sludge volume index
TOC	Total organic carbon
TSS	Total suspended solids
WAS	Waste activated sludge
WLR	Weir loading rate
WWTP	Wastewater treatment plant

DOI: 10.1201/9781003532354-2

2.1 INTRODUCTION

Biological or secondary treatment is the most important stage in the wastewater treatment process. At this stage, the remaining dissolved and non-settleable organics from primary or physical processes are biodegraded or broken down by microorganisms into simpler substances that can be discharged safely into the environment. In this secondary treatment, a wide variety of microorganisms, primarily bacteria, are involved in aerobic and anaerobic processes. "Aerobic" refers to a process in which oxygen is present, while "anaerobic" describes a biological process in which oxygen is absent. The mechanisms have been discussed in Chapter 3.

Treatment processes are divided into the system, which utilizes oxygen to break down organic matter (aerobic) and treatment, which does not utilize oxygen (anaerobic). The breakdown of organic matter can occur while in suspension (suspended growth) or on the surface of some media (attached growth). Also, processes using ponds are also sometimes used where large areas of land are available. It can also be categorized into its oxidation stage, i.e. aerobic and anaerobic, as detailed in Table 2.1.

2.2 SUSPENDED GROWTH/CULTURES SYSTEM

2.2.1 Activated Sludge Process (ASP)

The activated sludge process (ASP) is a multi-chamber reactor unit that makes use of (mostly) aerobic microorganisms to degrade organics in wastewater into stable compounds which produce high-quality effluent. These active bacterial cultures are commonly known as activated sludge.

TABLE 2.1
Category of the Biological Treatment Unit

	Category of Wastewater Treatments				
Details	**Aerobic**	**Anaerobic**	**Suspended Growth/ Culture**	**Fixed Growth/ Film**	**Ponding System**
Mechanism	Requires oxygen	Does not require oxygen	The microorganisms (biomass) treating the wastes are suspended (aerated) in the wastewater being treated	The biomass grows on media (such as rock, plastic, wood, or other natural or synthetic solid material), and the wastewater passes over its surface and within its porous structure	The natural processes of algal and bacteria growth exist in a mutually dependent relationship
Example	a. Conventional activated sludge (CAS) system b. Extended aeration (EA) c. Oxidation ditch (OD) system d. Sequencing batch reactor (SBR) e. Aerated lagoon	a. Anaerobic digestion b. Anaerobic filter c. Septic tank d. Imhoff tank	a. Conventional activated sludge (CAS) system b. Extended aeration (EA) c. Oxidation ditch (OD) system d. Sequencing batch reactor (SBR) e. Aerated lagoon	a. Trickling filters b. Constructed wetlands c. Rotating biological contactors (RBC)	a. Anaerobic b. Facultative c. Maturation

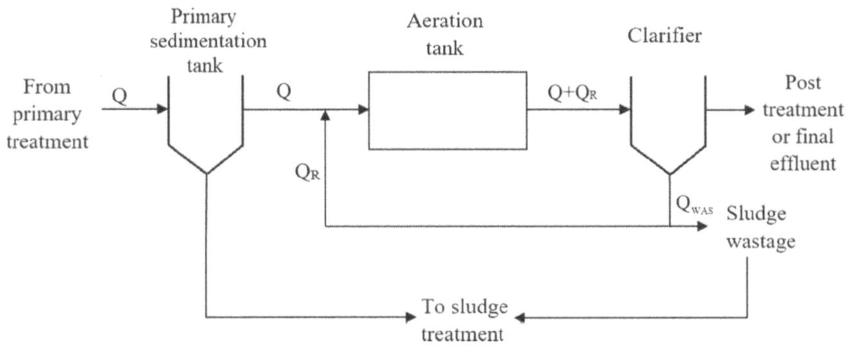

FIGURE 2.1 Typical activated sludge process.

To maintain aerobic conditions and to keep the active biomass suspended, a constant and well-timed supply of oxygen is required. The microorganisms oxidize the organic carbon in the wastewater to produce new cells, carbon dioxide and water. Although aerobic bacteria are the most common organisms, aerobic, anaerobic, and/or nitrifying bacteria along with higher organisms can be present. During aeration and mixing, the bacteria form small clusters or flocs.

The ASP is most suitably used to treat a high volume of wastewater and where land is limited. The typical model of an activated sludge process is shown in Figure 2.1.

The process can be summarized as follows:

1. The effluent from the primary sedimentation tank (the supernatant, or settled wastewater) enters the aeration tank.
2. Aeration is done, and the biomass mixes with the wastewater, and they are active in suspension. Oxidation and biodegradation take place in this tank. The important parameter in this tank is the mixed liquor suspended solids (MLSS), a measure of suspended solids (SS) mixed with biomass. MLSS is maintained at a certain concentration and ratio in this tank to make it ideal in terms of food to microorganisms (F:M).
3. Effluent from the aeration tank flows to the clarifier or secondary sedimentation tank, and the flocs will settle here.
4. Supernatant from clarifier will be removed as effluent or for further advanced treatment (tertiary tank).
5. Some of the settled sludge in the clarifier is recycled back (Q_R) to the aeration tank to maintain the active biomass (MLSS) within 1,500–3,000 mg/L. The remaining sludge is disposed of (Q_w) for further treatment.

There are a few types, depending on how the oxygen is supplied to the system. The most common include conventional, extended aeration (EA), tapered aeration, step aeration, oxidation ditch (OD), sequencing batch reactor (SBR), pure oxygen, high rate, plug flow, completely mixed, contact stabilization, and various modifications of ASP. The widely practised system in Malaysia is discussed here.

2.2.1.1 Conventional ASP

The conventional activated sludge process is one of the many categories of the ASP. All conventional activated sludge systems used at sewage treatment plants (STPs) for Class 3 (5,001–20,000 population equivalent [PE]), Class 4 (>20,000 PE), and where requested by the Commission [1] must be designed with an anoxic zone to achieve a total nitrogen removal to comply with the requirements, as well as to minimize potential rising sludge at secondary sedimentation system. The anoxic zone

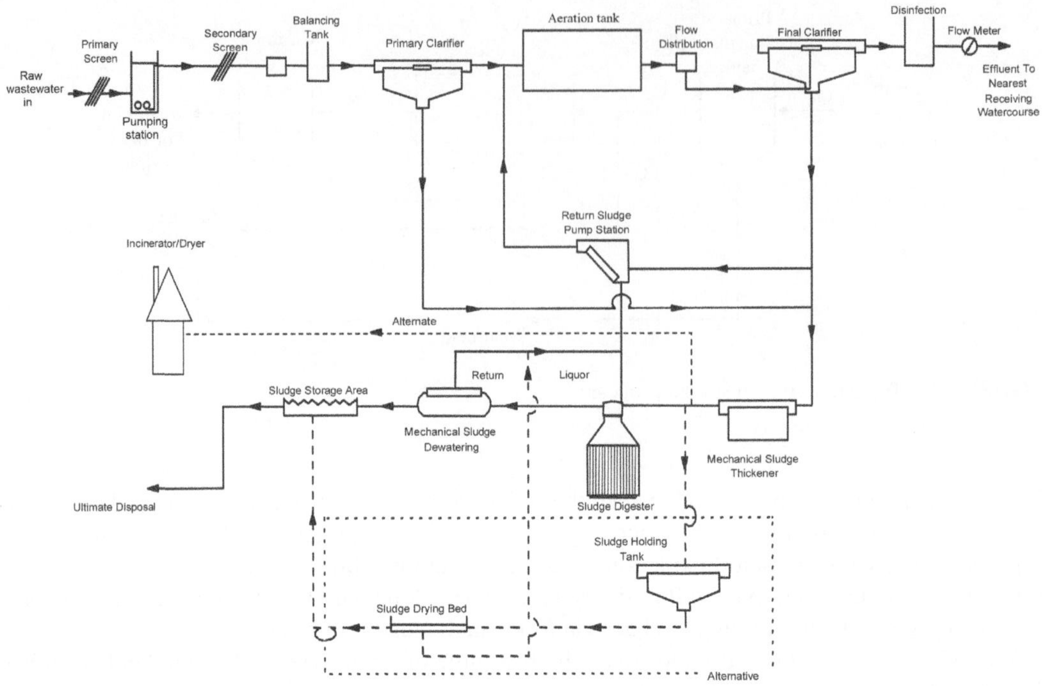

FIGURE 2.2 Conventional activated sludge process (ASP) (MSIG, Volume 4, 2009) [1].

must be mixed without inducing dissolved oxygen (DO). A typical process flow diagram of conventional ASP is shown in Figure 2.2.

The design criteria for conventional ASP in accordance with Malaysian Sewerage Industry Guidelines (MSIG), Volume 4, 2009 [1] is given in Table 2.2.

2.2.1.1.1 Food to Microorganism Ratio (F:M)

$$F:M = \frac{Q_{ave} \times \text{BOD}}{V \times \text{MLSS}} \text{ kg BOD/kg.MLSS.day} \quad (2.1)$$

where

$Q_{ave} = \text{DWF}\left(\text{m}^3/\text{day}\right)$

$V = \text{Volume}\left(\text{m}^3\right)$

$\text{MLSS} = \text{Mixed Liquor Suspended Solids}\left(\text{mg/L}\right)$

2.2.1.1.2 Hydraulic Retention Time (t)

$$t = \frac{V}{Q_{ave}} \text{ (hour)} \quad (2.2)$$

where

$Q_{ave} = \text{DWF}\left(\text{m}^3/\text{day}\right)$

$V = \text{Volume}\left(\text{m}^3\right)$

TABLE 2.2
Design Parameters for Conventional Activated Sludge System

Description	Unit	Design Criteria
Primary sedimentation system		Must be provided
Minimum number of aeration tanks		2
Organic loading (F/M ratio)[a]	kg BOD/kg MLSS.day	0.25–0.50
Hydraulic retention time (HRT)[b]	Hours	6–16 (for a system where only ammonia removal is required) 12–16 (for plants that require total nitrogen removal)
Oxygen requirements (for BOD and ammonia nitrogen removal)	kgO_2/kg substrate	2.0
Mixed liquor suspended solids (MLSS)	mg/L	1,500–3,000 Typical: 2,500
Dissolved oxygen (DO) level in the tank	mg/L	1.0
Aeration device rating		Continuous, 24 hours
Sludge yield	kg sludge produced/kg BOD_5 consumed	0.8–1.0
Sludge age	Day	5–10[d]
Waste activated sludge, Q_{WAS}	m^3/day	Refer equation[e]
Return activated sludge flow, Q_{RAS}	m^3/day	Refer equation[f]
Q_{RAS}/Q_{IN} ratio (R)	Extended aeration	0.75–1.0
	Conventional activated sludge	0.25–0.5
Mixed liquor suspended solids recirculation for denitrification purpose		4–6 of Q_{ave}
RAS pump rating	Hours/day	24
Volumetric loading (VOL)[c]	kg BOD_5/m^3.day	0.3–0.6
Minimum mixing requirement	W/m^3	20
	Tank dimension	
Water depth	m	3–5
Length: Width		3:1
Max width of the joined tank	m	<30

Source: MSIG, Volume 4, 2009 [1].
[a] Refer Equation 2.1.
[b] Refer Equation 2.2
[c] Refer Equation 2.3
[d] Refer Equations 2.4–2.6
[e] Refer Equation 2.7.
[f] Refer Equation 2.8.

2.2.1.1.3 *Volumetric Loading (VOL)*

$$\text{VOL} = \frac{Q_{ave} \times \text{BOD}}{V} \text{ kg BOD/m}^3\text{.day} \tag{2.3}$$

where

$Q_{ave} = \text{DWF} \left(\text{m}^3/\text{day}\right)$

$V = \text{Volume} \left(\text{m}^3\right)$

2.2.1.1.4 Sludge Age (θ_c)

Sludge age is also termed mean cell residence time based on solids in the aeration tank.

There are a few methods for determining the sludge age (Equations 2.4–2.6).

$$\text{Sludge age } (\theta_c) = \frac{\text{MLSS in aeration tank}}{\text{SS wasted+SS in the effluent}} \tag{2.4}$$

$$\frac{1}{\theta_c} = YU - k_d \tag{2.5}$$

where

θ_c = Sludge age (day)

Y = Yield coefficient of growth (g/g)

$U = \frac{dc}{dt}$, Substrate utilization rate (day^{-1})

k_d = Endogenous decay coefficient (day^{-1})

$$\theta_c = \frac{V \times \text{MLSS}}{Q_{\text{WAS}} \times C_u} \tag{2.6}$$

where

θ_c = Sludge age (day)
V = Volume (m^3)
MLSS = Mixed liquor suspended solids (mg/L)
$Q_{\text{WAS}} = Q_{\text{W}}$ = Waste activated sludge (m^3/day)
C_u = Underflow concentration (mg/L)
A good sludge age (θ_c) is normally between 3 and 10 days.
$\theta_c > 10$ indicates sludge bulking due to filamentous bacteria, which hindered settlement.
$\theta_c < 3$ indicates the sludge is too little and takes a long time to settle.

2.2.1.1.5 Waste Activated Sludge, $Q_{WAS} = Q_w$

$$Q_{\text{WAS}} = \frac{\left[\frac{V_T \times \text{MLSS}}{\theta_c}\right] - \left[Q_{\text{ave}} \times \text{SS}_{\text{eff}}\right]}{C_u} \tag{2.7}$$

where

V_T = Volume of reactor (m^3)
MLSS = Mixed liquor suspended solids (kg/m^3)
θ_c = Sludge age (days)
Q_{ave} = Average flow (m^3/day)
SS_{eff} = Effluent suspended solids (kg/m^3)
C_u = Underflow concentration (kg/m^3)

2.2.1.1.6 Return Activated Sludge Flow, Q_{RAS}

$$Q_{\text{RAS}} = \frac{\text{MLSS}}{C_u - \text{MLSS}} \times Q_{\text{ave}} \tag{2.8}$$

where

C_u = Underflow concentration

2.2.1.1.7 *Return Sludge Ratio (R)*

There are two methods for determining the R. These are given in Equations 2.9 and 2.10.

$$R = \frac{Q_{\text{RAS}}}{\text{DWF}} \tag{2.9}$$

$$X_a = X_R \times \frac{R}{(1+R)} \tag{2.10}$$

where

- R=Return sludge ratio
- Q_{RAS}=Return sludge flow rate (m³/day)
- DWF=Average flow entering the aeration tank
- x_a=Concentration of MLSS (mg/L)
- x_R or C_u=Concentration of return sludge (mg/L)

Equation 2.10 can be derived from the mass balance between the flow and concentration, as shown in Figure 2.3.

where

- C_o=Concentration of influent organic matter (mg/L)
- C_1=Concentration of effluent organic matter (mg/L)
- X_a=Concentration of MLSS (mg/L)
- X_R=Concentration of return sludge (m³/day)
- Q_i=Flow rate of influent (m³/day)
- Q_R=Flow rate of return sludge (m³/day)
- Q_e=Flow rate of effluent (m³/day)

Balance the system by assuming:

i. The system is 100% efficient, with effluent concentration zero, C_1=0.
ii. All sludge is returned to the system with no wasted sludge, Q_{WAS}=0.

$$\left[(\text{Concentration})\times(\text{Flow rate})\right]_{\text{Influent}} = \left[(\text{Concentration})\times(\text{Flow rate})\right]_{\text{Effluent}}$$

$$(C \times Q)_{\text{Influent}} = (C \times Q)_{\text{Effluent}}$$

$$\left[(C)\times(Q)\right]_{\text{Total}} = \left[(C)\times(Q)\right]_1 + \left[(C)\times(Q)\right]_2 + \cdots + \left[(C)\times(Q)\right]_n$$

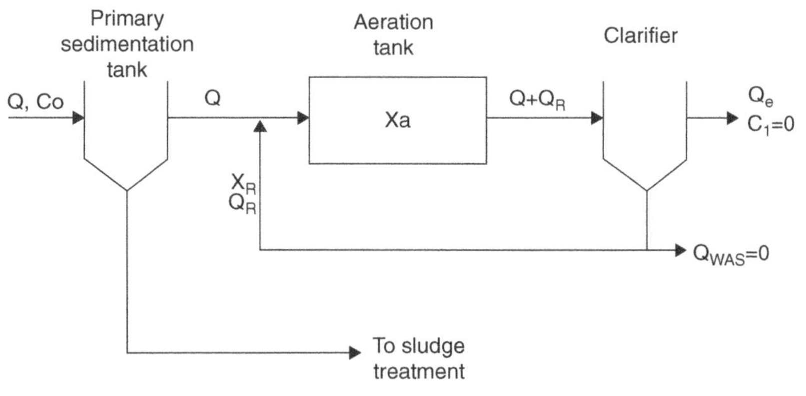

FIGURE 2.3 The mass balance in ASP.

Therefore, we have two equations:

$$(X_a)(Q+Q_R)=(Q_R)(X_R) \tag{2.11}$$

$$R=\frac{Q_R}{Q} \tag{2.12}$$

A solution of these two equations:

$$\frac{X_a}{X_R}=\frac{Q_R}{Q+Q_R}$$

$$\frac{X_R}{X_a}=\frac{Q+Q_R}{Q_R}$$

$$\frac{X_R}{X_a}=\frac{Q}{Q_R}+1$$

$$\frac{X_R}{X_a}=\frac{1}{R}+1$$

$$\frac{X_R}{X_a}=\frac{1+R}{R}$$

$$\frac{X_a}{X_R}=\frac{R}{1+R}$$

$$X_a=X_R\left(\frac{R}{1+R}\right) \tag{2.10}$$

Proven

2.2.1.1.8 Oxygen Requirement

The amount of oxygen required in the aeration tank can be determined using Equation 2.13:

$$O_2=\frac{Q_{\text{ave}}\times \text{BOD}_5}{\dfrac{\text{BOD}_5}{\text{BOD}_L}}-1.42\times P_x \tag{2.13}$$

$$P_x=\frac{Y}{1+(k_d\times\theta_c)}\times(Q\times\text{BOD})$$

where

O_2 = Total oxygen needed (kg/day)

Q = Flow rate of influent to the aeration tank = DWF (m^3/day)

BOD_L = Ultimate BOD

P_x = Sludge yield (kg/day)

Y = Yield coefficient of growth (g/g)

k_d = Endogenous decay coefficient (day^{-1})

θ_c = Sludge age (days)

$\frac{BOD_5}{BOD_L}$ for domestic wastewater is about 0.68.

The normal value of O_2 is 0.8 kg O_2/kg BOD_5 removed.

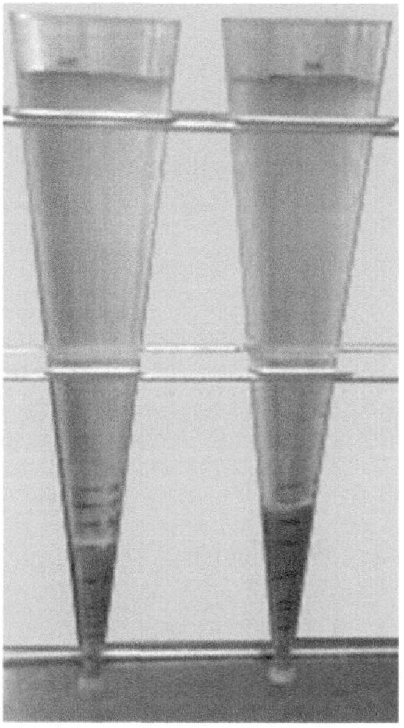

FIGURE 2.4 SVI test using Imhoff cone.

2.2.1.1.9 The Sludge Volume Index

The sludge volume index (SVI) is used to describe the settling characteristics of sludge in the aeration tank in the ASP. It is the volume in millilitres occupied by 1 g of an MLSS after 30 minutes settling in Imhoff cone (Figure 2.4). SVI typically is used to monitor settling characteristics of activated sludge that indicates the tendency of activated sludge solids (aerated solids) to thicken or to become concentrated during the sedimentation/thickening process. SVI can be calculated using Equation 2.14.

$$SVI = \frac{\text{settled MLSS volume in 30 minutes } (\text{mL/L}) \times 1,000}{\text{MLSS } (\text{mg/L})} \tag{2.14}$$

Good SVI is between 50 and 100 mL/g.

SVI > 150 indicates sludge bulking due to filamentous bacteria, which hindered settlement.

SVI < 40 indicates the sludge is too little and is difficult to settle.

2.2.1.1.10 Nutrient Ratio

Appropriate nutrients are needed to ensure suitable environmental conditions for microorganisms. The ratio of the nutrient is BOD:N:P which is 100:5:1.

2.2.1.1.11 Dissolved Oxygen Level

DO is normally maintained between 1 and 2 mg/L in the aeration tank.

Example 2.1

Based on the given data, design an aeration tank of a rectangular conventional activated sludge process:

$Q_{ave} = 1,500\ m^3/day$

MLSS = 2,500 mg/L

$BOD_5 = 250$ mg/L
$Y = 0.5$ mg/mg
$k_d = 0{:}06\ \text{day}^{-1}$

$$\theta_c = 10\ \text{days}$$

$SS_{eff} = 50$ mg/L $= 0.05$ kg/m^3
$C_u = 12{,}500$ mg/L $= 12.5$ kg/m^3

Solution:

Q_{ave} = DWF $= 1{,}500$ m^3/day
MLSS $= 2{,}500$ mg/L $= 2.5$ kg/m^3

(I) VOLUME OF TANK

From Figure 2.2, assume F:M $= 0.3$ kg BOD/kg MLSS.day
From Equation 2.1,

$$\text{Organic load} = F{:}\,M = \frac{Q \times \text{BOD}}{V \times \text{MLSS}}\ \text{kg BOD/kg MLSS.day}$$

$$0.3 = \frac{1{,}500\ \text{m}^3/\text{day} \times 0.25\ \text{kg/m}^3}{V \times 2.5\ \text{kg/m}^3}$$

$$V = 500\ \text{m}^3$$

Dimension the tank accordingly based on the given ratio.
Take $L{:}W = 3{:}1$ and depth $(D) = 3$ m

$L{:}\,W = 3:1$ and depth $(D) = 3$ m, as follows:

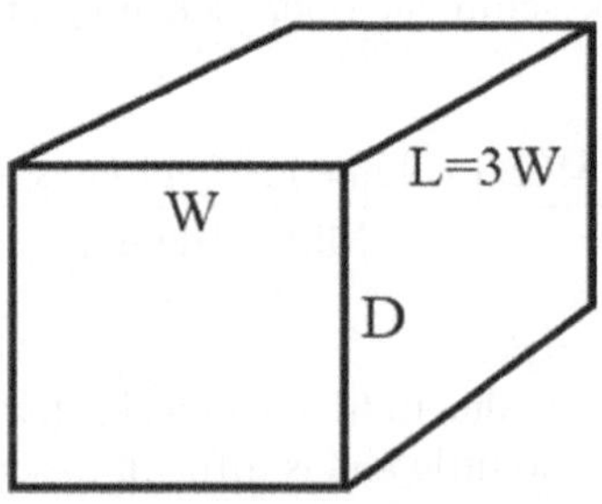

$$V = W \times 3W \times D$$

$$500 = 3W^2 \times 3$$

$W = 7.45$ m, say 7.5 m

$$L = 3W = 3(7.5)$$

$$= 22.5\ \text{m}$$

The new dimension (excluding freeboard) is as follows:

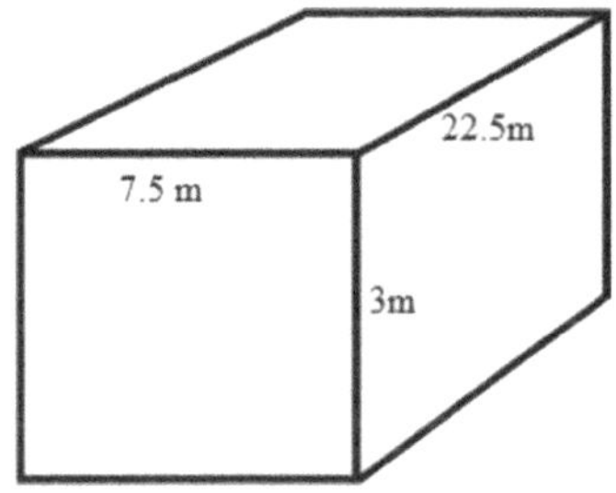

New volume, $V = 7.5\ \text{m} \times 3\ \text{m} \times 22.5\ \text{m}$

$$V = 506.25\ \text{m}^3$$

(II) RETENTION TIME (*t*)

From Equation 2.2,

$$t = \frac{V}{Q_{\text{ave}}}\ (\text{hours})$$

$$t = \frac{506.25\ \text{m}^3}{1{,}500\ \text{m}^3/\text{day}}$$

$$= 0.3375\ \text{day}$$

$$= 8.1\ \text{hours (within 6 – 16 hours, hence acceptable)}$$

(III) VOLUMETRIC ORGANIC LOADING (VOL)

From Equation 2.3,

$$\text{VOL} = \frac{Q_{\text{ave}} \times \text{BOD}}{V}\ \text{kg BOD/m}^3.\text{day}$$

$$= \frac{1{,}500\ \text{m}^3/\text{day} \times 0.25\ \text{kg/m}^3}{506.25\ \text{m}^3}$$

$$= 0.74\ \text{kg BOD/m}^3.\text{day (not within 0.3-0.6, hence not acceptable)}$$

Make the volume of the tank slightly bigger, say $V = 650\ \text{m}^3$

Redimension the shape of the tank with a new volume of $650\ \text{m}^3$

$$\text{New VOL} = \frac{1{,}500\ \text{m}^3/\text{day} \times 0.25\ \text{kg/m}^3}{650\ \text{m}^3}$$

$$= 0.577\ \text{kg BOD/m}^3.\text{day (within 0.3-0.6, hence acceptable)}$$

(IV) WASTE ACTIVATED SLUDGE (Q_{WAS})

From Equation 2.7,

$$Q_{\text{WAS}} = \frac{\left[\dfrac{V_T \times \text{MLSS}}{\theta_c}\right] - [Q_{\text{ave}} \times \text{SS}_{\text{eff}}]}{C_u}$$

$$Q_{\text{WAS}} = \frac{\left[\dfrac{650\ \text{m}^3 \times 2.5\ \text{kg/m}^3}{10\ \text{days}}\right] - \left[1{,}500\ \text{m}^3/\text{day} \times 0.05\ \text{kg/m}^3\right]}{12.5\ \text{m}^3/\text{day}}$$

$$= 7\ \text{m}^3/\text{day}$$

(V) RETURN ACTIVATED SLUDGE FLOW, Q_{RAS}

From Equation 2.8,

$$Q_{\text{RAS}} = \frac{\text{MLSS}}{C_u - \text{MLSS}} \times Q_{\text{ave}}$$

$$Q_{\text{RAS}} = \frac{2.5\ \text{kg/m}^3}{(12.5 - 2.5)\ \text{kg/m}^3} \times 1{,}500\ \text{m}^3/\text{day}$$

$$= 375\ \text{m}^3/\text{day}$$

(VI) Q_{RAS}/Q_{IN} (R) RATIO

$$Q_{\text{IN}} = Q_{\text{ave}} = \text{DWF}$$

$$R = \frac{Q_{\text{RAS}}}{\text{DWF}}$$

$$= \frac{375\ \text{m}^3/\text{day}}{1{,}500\ \text{m}^3/\text{day}}$$

$$= 0.25\ \text{(within an acceptable range of } 0.25 - 0.50\text{)}$$

(VII) OXYGEN REQUIREMENT

From Equation 2.13,

$$O_2 = \frac{Q_{\text{ave}} \times \text{BOD}_5}{\dfrac{\text{BOD}_5}{\text{BOD}_L}} - 1.42 \times P_x$$

Solve P_x first:

$$P_x = \frac{Y}{1 + (k_d \times \theta_c)} \times (Q \times \text{BOD})$$

$$P_x = \frac{0.5\ \text{mg/mg}}{1 + \left(0.06\ \text{day}^{-1} \times 10\ \text{days}\right)} \times \left(1{,}500\ \text{m}^3/\text{day} \times 0.25\ \text{kg/m}^3\right)$$

$$= 117.19\ \text{kg/day}$$

$$O_2 = \frac{1{,}500\ \text{m}^3/\text{day} \times 0.25\ \text{kg/m}^3}{0.68} - 1.42 \times \left(117.19\ \text{kg/day}\right)$$

$$= 385.06\ \text{kg/day}$$

The volume of air required, considering air contains 20.95% oxygen by weight and density of air 1.1839 kg/m^3 at 25°C and 1 atm is

$$= \frac{\text{Amount of oxygen } (\text{kg/day})}{\text{Density of oxygen } (\text{kg/m}^3) \times \% \text{ Oxygen}}$$

$$= \frac{385.06 \ (\text{kg/day})}{1.1839 \ (\text{kg/m}^3) \times 0.2095}$$

$$= 1{,}552.5 \text{ m}^3/\text{day}$$

Considering oxygen transfer efficiency of 8%, the air required is

$$= \frac{1{,}552.5 \text{ m}^3/\text{day}}{0.08}$$

$$= 19{,}406.25 \text{ m}^3/\text{day}$$

$$= 13.48 \text{ m}^3/\text{min}$$

Considering the safety factor of 2, the air requirement is

$$= 2 \times 13.48 \text{ m}^3/\text{min}$$

$$= 26.95 \text{ m}^3/\text{min}$$

2.2.1.2 Extended Aeration System

The EA process is like the conventional activated sludge process except that it operates in the endogenous respiration phase of the growth curve, which requires a low organic loading and long aeration time. The system produces high MLSS concentration, high RAS pumping rate, and low sludge wastage. The advantage of having long hydraulic retention times is that it allows the plant to operate effectively over widely varying flow and waste loadings.

EA plants shall be designed as either plug flow or completely mixed. An anoxic zone at the head of the reactor must be provided for denitrification. The anoxic zone must be mixed without inducing DO. A typical process flow diagram is given in Figure 2.5.

The design criteria for an extended ASP in accordance with MSIG, Volume 4, 2009 [1] is given in Table 2.3.

SPAN has produced specific technical specifications on an EA system for 150–5,000 PE. The Technical Specification of Wastewater Treatment Systems: Standard Wastewater Treatment Plant, Part 1-Extended Aeration (150 PE-5,000 PE) (SPAN TS 1402:2016), SPAN, 2016 [2] covers the following requirements:

a. **Section 1:** Layout and Design
b. **Section 2:** Construction and Installation
c. **Section 3:** Testing and Commissioning

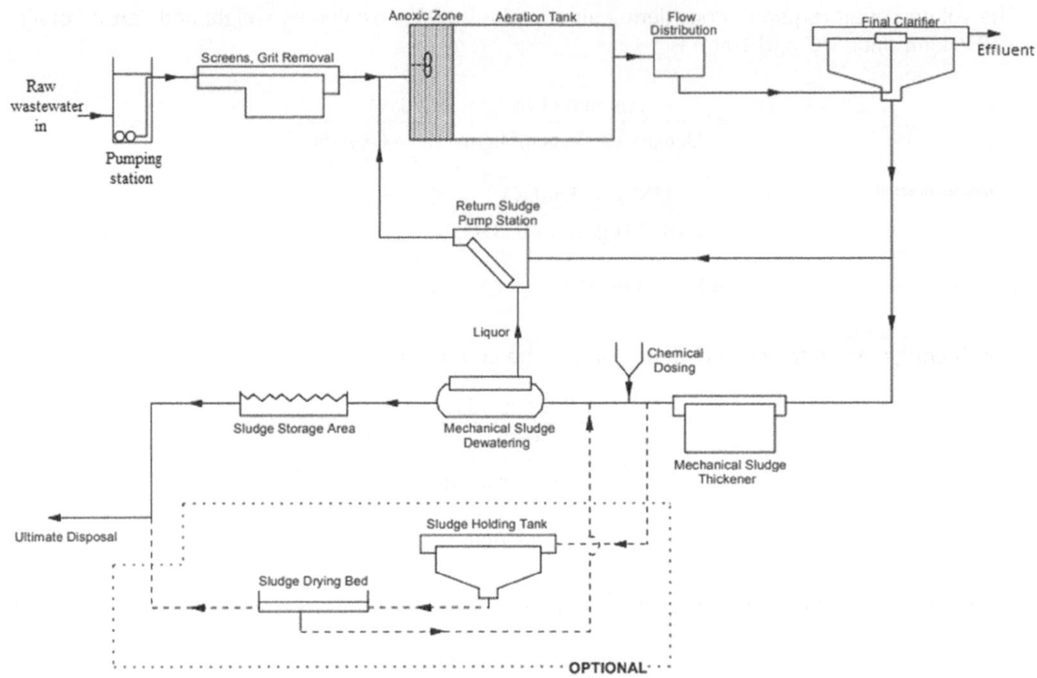

FIGURE 2.5 An extended aeration ASP (MSIG, Volume 4, 2009) [1].

TABLE 2.3
Design Parameters for Extended Aeration

Description	Unit	Design Criteria
Minimum number of aeration tanks		2
Organic loading (F/M ratio)[a]		0.05–0.1
Hydraulic retention time (HRT)[b]	Hours	18–24
Oxygen requirements (for BOD and ammonia nitrogen removal)	$kg\ O_2/kg_{substrate}$	2.0
Mixed liquor suspended solids (MLSS)	mg/L	2,500–5,000 Typical: 3,000
Dissolved oxygen (DO) level in the tank	mg/L	2.0
Sludge yield	kg sludge produced/kg BOD_5 consumed	0.4 (at 24 hours HRT) 0.6 (at 18 hours HRT)
Sludge age	day	>20[d]
Waste activated sludge flow, Q_{WAS}	3	Refer equation[e]
Return activated sludge flow, Q_{RAS}	m^3/day	Refer equation[f]
RAS pump rating	Hours/day	24
Recirculation ratio, Q_{RAS}/Q_{INFLOW}		0.5–1.0
MLSS recycle ratio		4–6 times of Q_{ave}
Volumetric loading[c]	3	0.1–0.4

(*Continued*)

TABLE 2.3 (*Continued*)
Design Parameters for Extended Aeration

Description	Unit	Design Criteria
Minimum mixing requirement	W/m³	20
	Tank dimension	
Water depth	m	3–5
Length: Width	Ratio	3:1
Max width of the joined tank	m	<60

Source: MSIG, Volume 4, 2009 [1].
Cross-reference to Figure 2.2.
[a] Refer Equation 2.1.
[b] Refer Equation 2.2.
[c] Refer Equation 2.3.
[d] Refer Equations 2.4–2.6.
[e] Refer Equation 2.7.
[f] Refer Equation 2.8.

2.2.1.3 Sequencing Batch Reactor System

SBRs or sequential batch reactors are a type of activated sludge process. SBR is a fill-and-draw where wastewater is added to a single "batch" reactor, treated to remove undesirable components, and then discharged. Equalization, aeration, and clarification can all be achieved using single or multiple batch reactors (typically a minimum of two). There are five basic sequences in a cycle, namely 1 – Fill, 2 – React (Aeration), 3 – Settle, 4 – Decant, and 5 – Idle. Typically, all actions in the reactor occur at different sequences of time, where the system is intermittently filled and intermittently decant. The recent development of the SBR system has led to the emergence of various modifications in the operating sequences.

In the reaction stage, oxygen supplied should be in accordance with the load to the system within the time frame of the reaction cycle. This requires higher oxygen capacity per unit time than a continuously aerated system. In the decanting stage, enough time should be given to allow for MLSS to settle before effluent decanting begins. Decanting time is typically much shorter than the fill time. Hence, the effluent flow rate will also be much higher than the influent flow rate.

The design criteria for the SBR system in accordance with MSIG, Volume 4, 2009 [1] is given in Table 2.4.

2.2.2 Aerated Lagoon

Unlike an oxidation pond which functions with no aeration, an aerated lagoon (or aerated pond) is also a simple wastewater treatment system consisting of a pond with aeration to promote the biological oxidation of wastewaters. The pond preceded the pre-treatment processes (Figure 2.6). The aeration provides mixing so that the microorganisms meet the dissolved and suspended organic matter for biodegradation (Figure 2.7). If operated properly, an aerated lagoon can be one of the most efficient, cost-effective, and low-maintenance methods of wastewater treatment. The design is according to the volume load and the retention time. The difference between the aerated lagoon with an activated sludge system is that there will be no sludge return to the aeration system in an aerated lagoon system.

There are two types of an aerated lagoon: partial-mix and completely mixed system. In a partial-mix aerated lagoon, the aeration is designed to provide the oxygen needed for biological treatment and to provide only partial-mix conditions on the top and anaerobic digestion to occur at the bottom. Because of this being a partial-mix environment, there is not enough energy/mixing to keep 100%

TABLE 2.4
Design Requirements for SBR System

Parameter	Unit	Continuous Fill and Intermittently Decant[a]	Intermittently Fill and Intermittently Decant
No. of reactors	Unit	2 (minimum)	2 (minimum)
Hydraulic retention time at Q_{ave} (at average water level)	Hours	18–24	18–24
F/M ratio	Day^{-1}	0.05–0.08	0.05–0.30
Sludge age	Day	20–30	10–30
Sludge yield	kg sludge/kg BOD_5 load	0.75–0.85	0.75–1.10
MLSS (end of decant)	mg/L	3,000–4,500	3,000–4,500
Cycle time	Hours	4–8	4–8
DO (reactor)	mg/L	0 to ~6.5	0 to ~6.5
DO (effluent)		2.0	2.0
Oxygen requirement	kg O_2/kg substrate	$\frac{\text{Cycle time}}{\text{Aeration time}} \times \frac{2.0\text{ kg }O_2}{\text{kg substrate}}$	$\frac{\text{Cycle time}}{\text{Aeration time}} \times \frac{2.0\text{ kg }O_2}{\text{kg substrate}}$
Decant time	Hours	≥1.0	≥1.0
Decant depth	m	Max 0.5	Max 1.0
Decant volume	%	Not more than 25% of the volume of biological reactor at TWL	Not more than 30% of the volume of biological reactor at TWL
Decanting device loading rate[b]	m^3/m/h	≤20 for decant draw-down from TWL	≤20 for decant draw-down from TWL
Minimum number of the decanter		2 nos. independent decanter per tank	2 nos. independent decanter per tank
Max. decanter length	m	4.0	4.0
WAS	kg sludge/day	WAS = Total solids in system/Sludge age	WAS = Total solids in system/Sludge age
Fill volume	m^3	$V_{fill} = (Q_p \times 1.5\text{ hours}) + (T_{fill} - 1.5) \times Q_{ave}$ (if no balancing tank) $V_{fill} = Q_{ave} \times T_{fill}$ (if preceeded by balancing tank)	$V_{fill} = (Q_p \times 1.5\text{ hours}) + (T_{fill} - 1.5) \times Q_{ave}$ (if no balancing tank) $V_{fill} = Q_{ave} \times T_{fill}$ (if preceeded by balancing tank)

Source: MSIG, Volume 4, 2009 [1].
RAS may be necessary where the length to width ratio poses a dilution effect on the inlet.
[a] The length to width ratio is 3:1.
[b] Decanting device loading rate shall be based on V_{fill}/decant time during decanting.

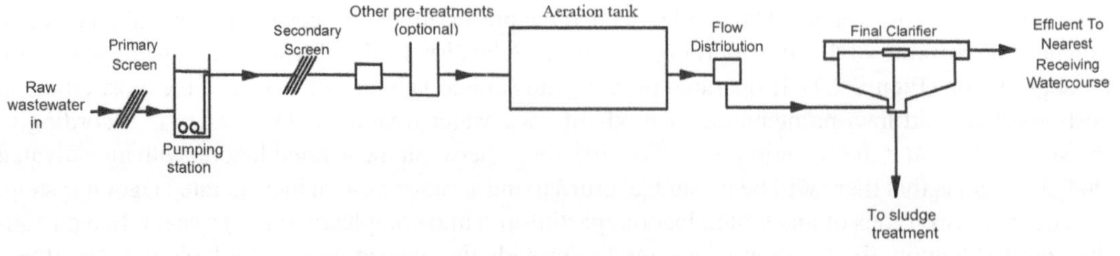

FIGURE 2.6 A typical aerated lagoon in Malaysia.

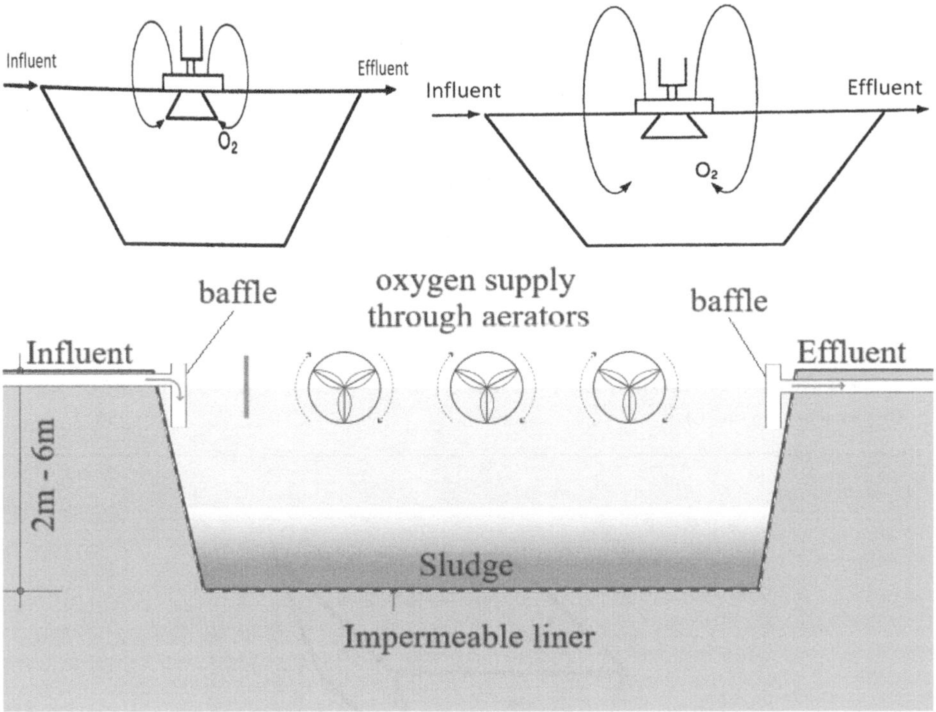

FIGURE 2.7 An aeration in an aerated lagoon system.

of the biological solids from settling. A small percentage of the settled solids will settle at the bottom in an anaerobic zone or sludge storage area, where they biodegrade anaerobically and slowly. Depths generally range from 2 to 7 m, with the shallower lagoons likely to have been facilitative lagoons upgraded with aeration to improve treatment. Most aerated lagoons in the United States are partially mixed and use either surface or diffused aeration.

Completely mixed aerated lagoon is designed to keep the solids in full suspension, as in the case of an activated sludge system. The depth of the pond is normally between 2 and 5 m. It is reported that when the biological solids are always kept in suspension, they biodegrade up to 10 times faster as compared to a partial-mix environment. The main reason for selecting a completely mixed system includes needing to achieve more treatment in the same footprint (limited area). The mechanisms of removal are the same as what has been discussed in the ASP, except there is no sludge recirculation.

The flocs from both partially mixed and completely mixed aerated lagoon will be settled in the clarifier. The land area requirement for an aerated lagoon can be referred to in Figure 2.6. Table 2.5 provides the design criteria of an aerated lagoon system.

The design of an aerated lagoon system is based on three common formulas (Equation 2.15–2.18):

$$\text{Retention time } (t) = \frac{V}{Q_{\text{ave}}} \text{ day} \tag{2.15}$$

$$\text{Aerial Organic Loading (AOL)} = \frac{Q_{\text{ave}}(\text{BOD})}{A_s} \text{ kg BOD/m}^2\text{.day} \tag{2.16}$$

$$\text{Volumetric Organic Loading (VOL)} = \frac{Q_{\text{ave}}(\text{BOD})}{V} \text{ kg BOD/m}^3\text{.day} \tag{2.17}$$

$$F:M = \frac{Q_{\text{ave}}(\text{BOD})}{V(\text{SS})} \text{ kg BOD/kg SS.day} \tag{2.18}$$

TABLE 2.5
Design Parameters for Aerated Lagoon

	Type of Lagoons	
Parameters	**Partially Mixed**	**Completely Mixed**
Design flow	DWF	DWF
Depth (m)	2–7	2–5
Detention time (day)	>2.5	>1
Ratio of L:W	2:1–4:1	2:1–4:1
Oxygen requirement (kg O_2/kg BOD, consumed)	1.5–1.8	0.8–1.1
% BOD_5 removed	60%–70%	50%
Power of oxidation (kW/1,000 m^3)	3	5
Dissolved oxygen (mg/L)	>3	>3

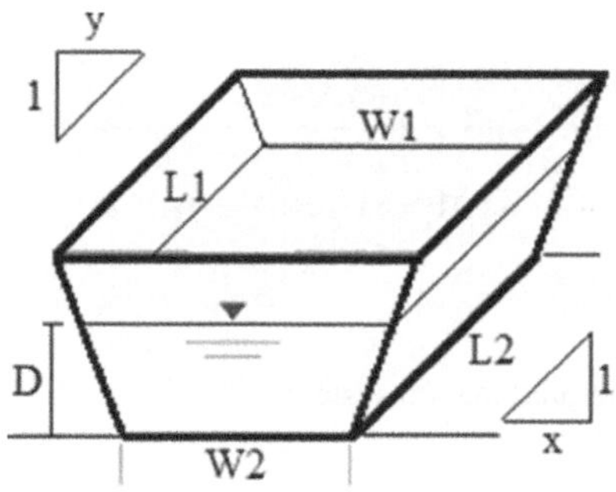

FIGURE 2.8 Volume of a truncated pyramid.

where

$Q_{ave} = \text{DWF}\ (\text{m}^3/\text{day})$

$A_s = \text{Surface area of lagoon (top)}\ (\text{m}^2)$

$V = \text{Volume of lagoon}\ (\text{m}^3)$

$F : M = \text{Food to microorganism ratio}$

$\text{SS} = \text{Suspended solids (mg/L)}$

The shape of the lagoon is normally trapezoidal, as shown in Figure 2.8.

The volume can be determined as follows:

$$\text{Water volume } (V) = \frac{1}{6}(D)\left[(W1 \times L1) + (W2 \times L2) + (W1 + W2)(L1 + L2)\right] \quad (2.19)$$

Example 2.2

A rectangular aerated lagoon with the bottom surface of 50×70 and 3 m depth (side slopes at all sides 2:1) receives wastewater of 2,500 m^3/day with BOD_5 250 mg/L and SS 500 mg/L. Determine the:

i. Retention time (t)
ii. VOL
iii. F:M ratio

Solution:

Using Figure 2.8 and Equation 2.19,

Given the base, $W2 = 50\,\text{m}$, $L2 = 70\,\text{m}$

Compute the top surface area ($W1$ and $L1$). Since side slopes are 2:1 at all sides,

For $W1$:

By ratio,

$$\frac{x}{2} = \frac{3}{1}$$

$$x = 6\ \text{m}$$

Hence

$$W1 = 6\ \text{m} + 50\ \text{m} + 6\ \text{m}$$

$$W1 = 62\ \text{m}$$

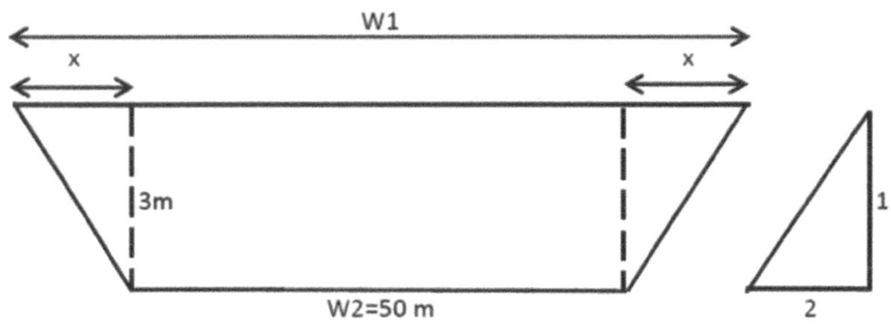

Similarly, for $L1$:

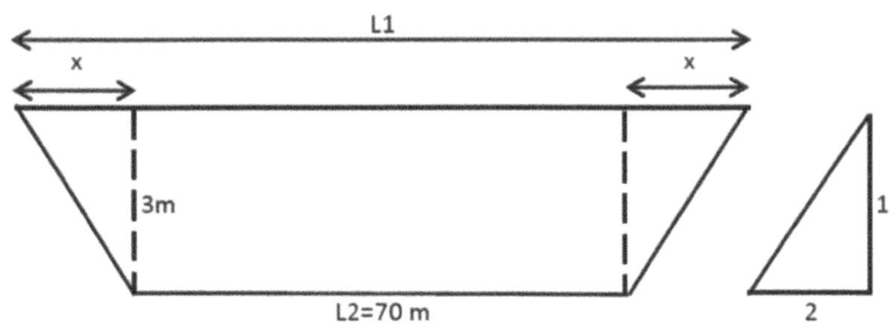

By ratio,

$$\frac{x}{2} = \frac{3}{1}$$

$$x = 6\ \text{m}$$

Hence,

$$L1 = 6\ \text{m} + 70\ \text{m} + 6\ \text{m}$$

$$L1 = 82\ \text{m}$$

Compute the volume using Equation 2.19,

$$W1 = 62 \text{ m}, W2 = 50 \text{ m}, L1 = 82 \text{ m}, L2 = 70 \text{ m}$$

$$V = \frac{1}{6}(3)\left[(W1 \times L1) + (W2 \times L2) + (W1 + W2)(L1 + L2)\right]$$

$$V = \frac{1}{6}(3)\left[(62 \times 82) + (50 \times 70) + (62 + 50)(82 + 70)\right]$$

$$= 12{,}804 \text{ m}^3$$

(I) RETENTION TIME (*t*)

From Equation 2.15,

$$\text{Retention time } (t) = \frac{V}{Q_{\text{ave}}}$$

$$(t) = \frac{12{,}804 \text{ m}^3}{2{,}500 \text{ m}^3/\text{day}}$$

$$= 2.5 \text{ days}$$

(II) VOLUMETRIC ORGANIC LOADING

From Equation 2.17,

$$\text{Aerial Organic Loading (AOL)} = \frac{Q_{\text{ave}}(\text{BOD})}{V} \text{ kg BOD/m}^3\text{.day}$$

$$V = \text{Volume}$$

$$\text{BOD}_5 = 250 \text{ mg/L} = 0.25 \text{ kg/m}^3$$

$$\text{AOL} = \frac{2{,}500(0.25)}{12{,}804} \text{ kg BOD/m}^3\text{.day}$$

$$= 0.0488 \text{ kg BOD/m}^3\text{.day}$$

(III) *F:M*

$$\text{SS} = 500 \text{ mg/L} = 0.5 \text{ kg/m}^3$$

From Equation 2.18,

$$F:M = \frac{Q_{\text{ave}}(\text{BOD})}{V(\text{SS})} \text{ kg BOD/kg SS.day}$$

$$F:M = \frac{2{,}500(0.25)}{12{,}804(0.5)} \text{ kg BOD/kg SS.day}$$

$$= 0.098 \text{ kg BOD/kg SS.day}$$

2.3 FIXED CULTURES/FILM

2.3.1 Rotating Biological Contactor

Rotating biological contactor (RBC) is one of the treatment units that use a rotating media, which allows the attachment of microorganisms on its surface to form slimes that will decompose organic materials on contact. Developed in the late 1960s, the media rotates on a horizontal shaft at a very slow speed of 1–2 RPM (Figure 2.9). Typically, the medium is submerged in wastewater (about 40%–50% in the water), and the remaining part is on air. The microorganisms which are attached to the disc will absorb the organic matter when it touches the wastewater and oxidizes them upon exposure to air as the shaft rotates. When slimes are thick enough, they will fall and be deposited in the secondary sedimentation tank (Figure 2.10).

The type of media used determines the surface area in contact with wastewater. The typical shape made of perforated plastic is shown in Figure 2.11. The calculation of the surface area of media for this kind of material can be assumed to represent a round plate or disc attached to a middle shaft. Hence, the active surface area should consider both sides of the disc. The distance between discs is normally 1 inch.

The RBC can be designed in multiple stages and be arranged either in series or parallel (Figure 2.12).

A typical process flow diagram of the RBC unit in Malaysia is shown in Figure 2.13. The design criteria for the RBC system in accordance with MSIG, Volume 4, 2009 [1], is given in Table 2.6.

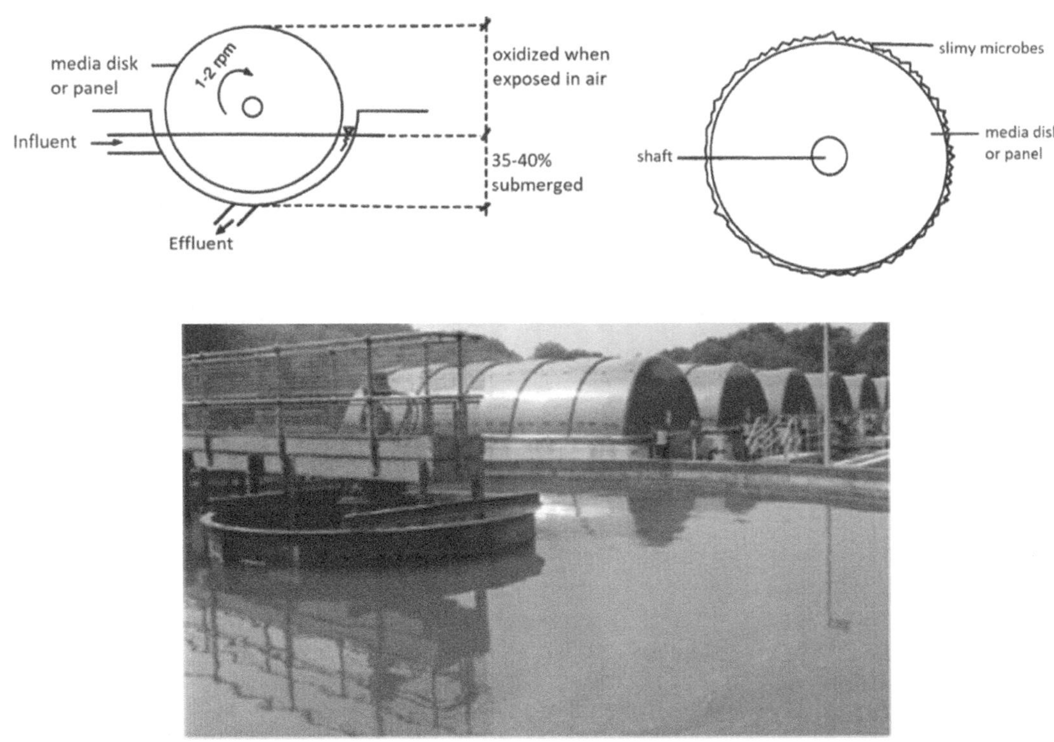

FIGURE 2.9 Typical RBC.

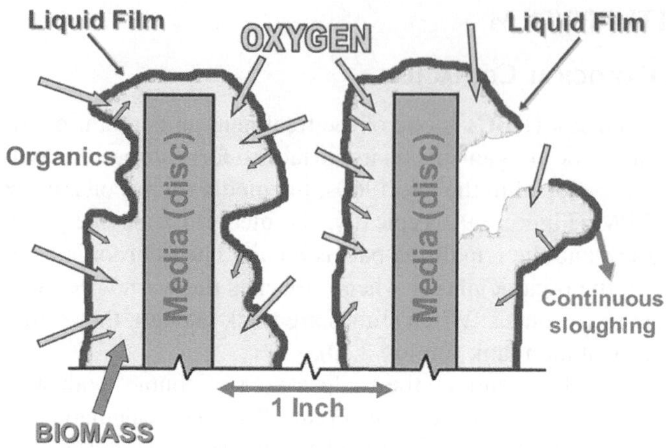

FIGURE 2.10 Removal mechanisms in RBC.

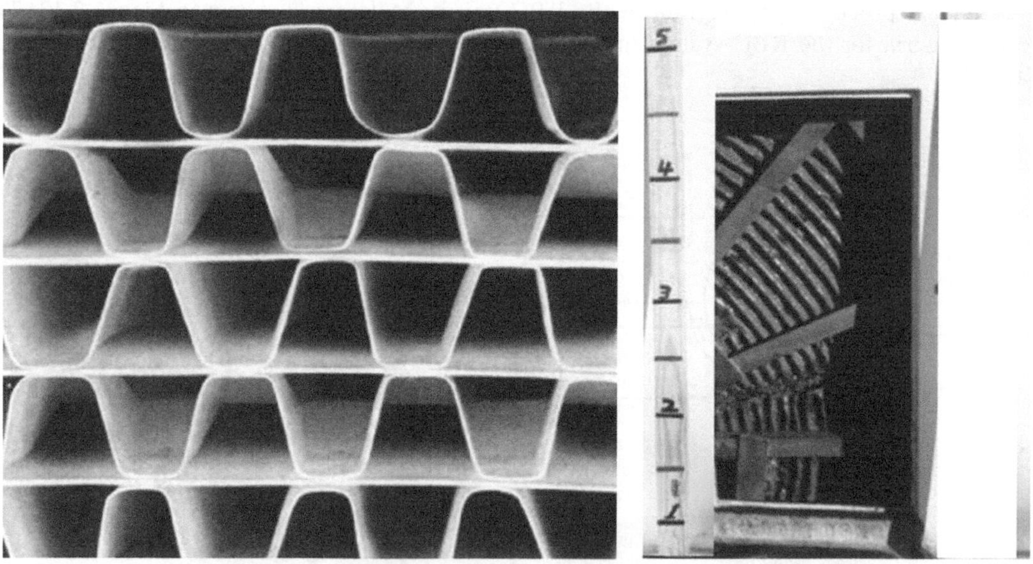

FIGURE 2.11 Typical shape of RBC media attached to a shaft.

$$\text{Organic Loading (O.L.)} = \frac{Q_{\text{ave}} \times \text{BOD}}{\text{Surface area of media}}$$

$$\text{Organic Loading (O.L.)} = \frac{\text{DWF} \times \text{BOD}}{A}\ \text{g/m}^2\text{.day} \quad (2.20)$$

$$\text{Retention time } (t) = \frac{V}{Q_{\text{ave}}} \quad (2.21)$$

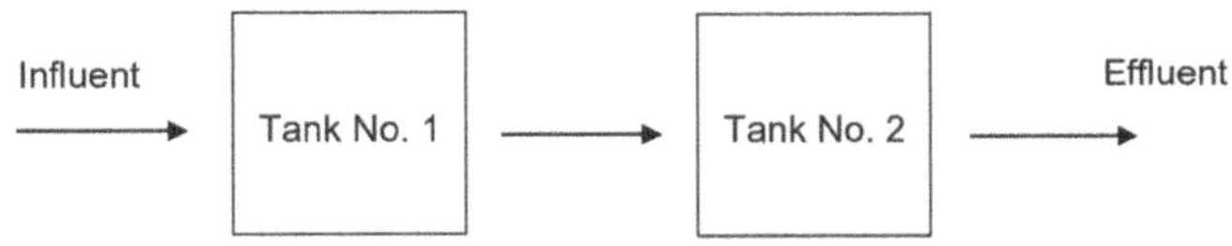

(a) RBC in series

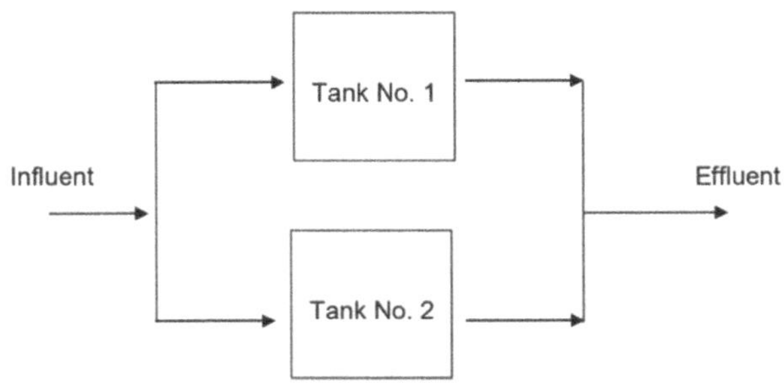

(b) RBC in parallel

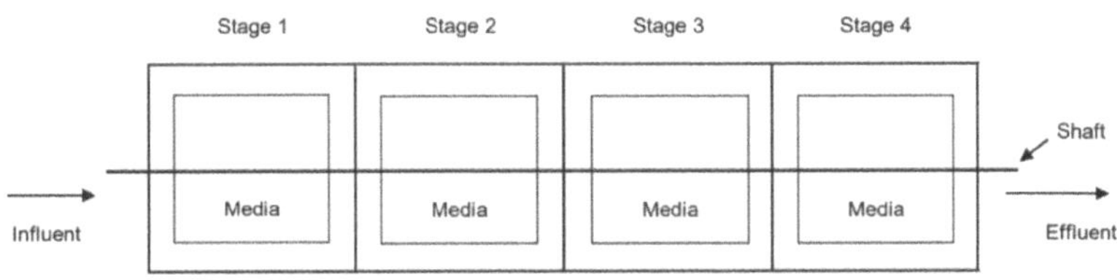

(c) Single shaft, 4-stages RBC

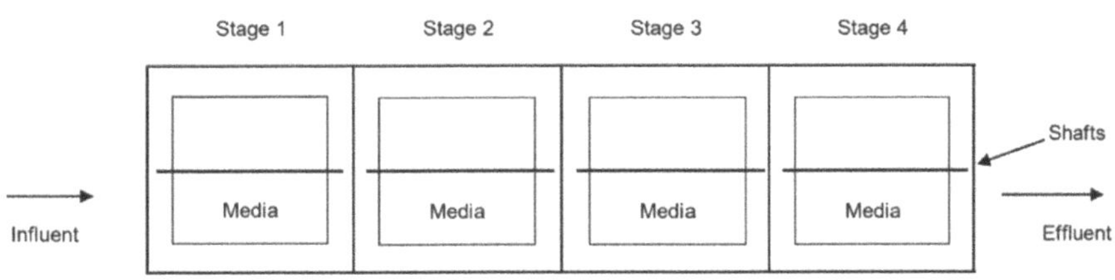

(d) 4-shaft, 4-series RBC

FIGURE 2.12 Stages in RBC design.

Example 2.3

Wastewater with a flow rate of 1,000 m^3/day and the BOD_5 value of 250 mg/L is being treated with a three-stage RBC in series. The diameter of media is 5 m made of a dense plastic medium, 50% submerged, and the shaft is 20 m long for each stage. Each meter of the shaft contains 20 discs which allow microbes to grow on both sides. Calculate the organic load and retention time of this plant.

Solution:

Q = DWF = 1,000 m^3/day

BOD_5 = 250 mg/L = 0.25 kg/m^3

Three stages of RBC

Diameter (D) = 5 m

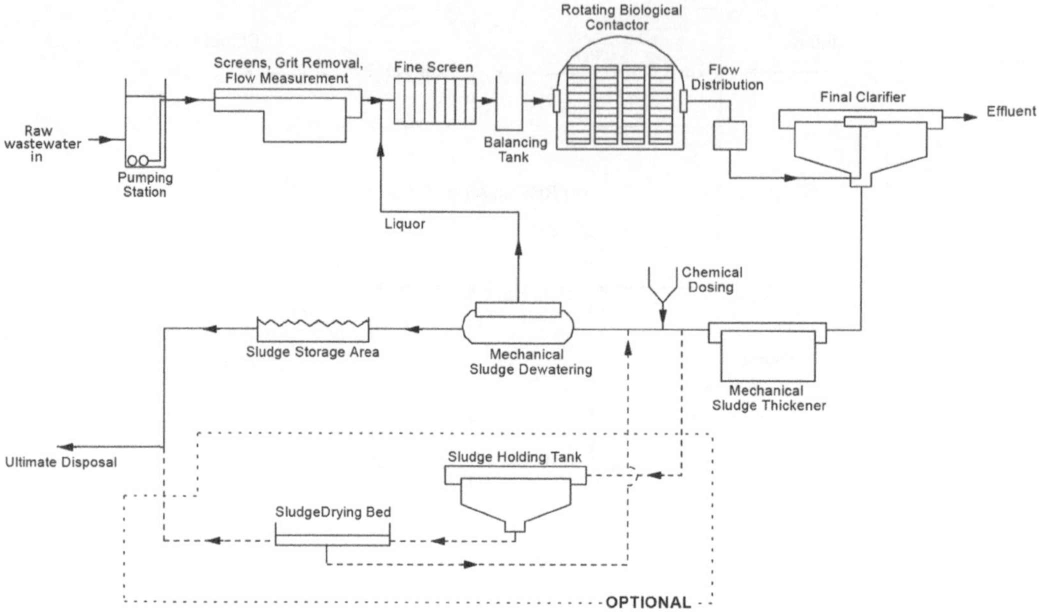

FIGURE 2.13 A typical process flow diagram of the RBC unit in Malaysia (MSIG, Volume 4, 2009) [1].

TABLE 2.6
Design Criteria of RBC

Description	Unit	Design Criteria
Minimum number of stages		3
Organic Loading (O.L.)[a]	g/m^2.day	5–10
Total tank volume[b]		Based on 2 hours at Q_{ave}
Sludge yield	kg excess sludge/kg BOD_5 consumed	0.9
Disc diameter	m	2.5–3.5
Speed of rotation	Rev/min	0.5–1.0
Maximum peripheral velocity	m/s	0.3
The depth of disc submergence	%	40–90

Source: MSIG, Volume 4, 2009 [1].
[a] Refer Equation 2.20.
[b] Refer Equation 2.21.

Length of shaft = 20 m
Number of discs = 20 per m of shaft
Refer Figure 2.12 as guidance

(I) ORGANIC LOADING

$$\text{Surface area of medium} = A = \pi r^2$$

$$r = 2.5 \text{ m}$$

$$\text{Length of shaft} = h = 20 \text{ m}$$

Stage = 3

$$A = \pi r^2 (\text{Number of disc per m})(\text{shaft length})(\text{two sides of disc})(\text{number of stage})$$

$$A = \pi(2.5)^2 \times 20 \times 20 \text{ m} \times 2 \text{ sides} \times 3 \text{ stages}$$

$$A = 47{,}124 \text{ m}^2 \text{ for three stages}$$

From Equation 2.20,

$$\text{Organic Loading (O.L.)} = \frac{\text{DWF} \times \text{BOD}}{A} \text{ g/m}^2\text{.day}$$

$$= \frac{1{,}000 \text{ m}^3/\text{day } (0.25 \text{ kg/m}^3)}{47{,}124 \text{ m}^2}$$

$$= 0.0053 \text{ kg/m}^2\text{.day}$$

$$= 5.3 \text{ g/m}^2\text{.day within range of Figure 2.6}$$

(II) RETENTION TIME (*t*)

$$\text{Retention time } (t) = \frac{\text{Volume of tank}}{\text{DWF}}$$

Since the media is 50% submerged:

$$\text{Depth } (D) = \frac{\text{Diameter}}{2}$$

Width (W) = Diameter of media
Length = length of shaft

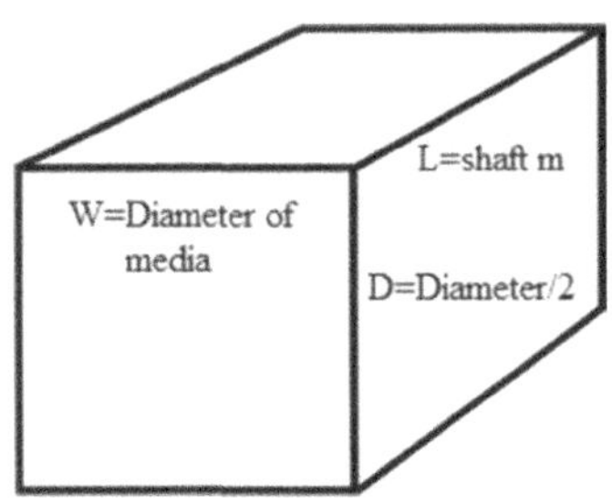

$$\text{Volume of tank, } V = \frac{5 \text{ m}}{2} \times 5 \text{ m} \times 20 \text{ m}$$

$$= 250 \text{ m}^3$$

For three series,

$$t = \frac{3 \times 250 \text{ m}^3}{1{,}000 \text{ m}^3/\text{day}}$$

$$= 0.75 \text{ day}$$

$$= 18 \text{ hours}$$

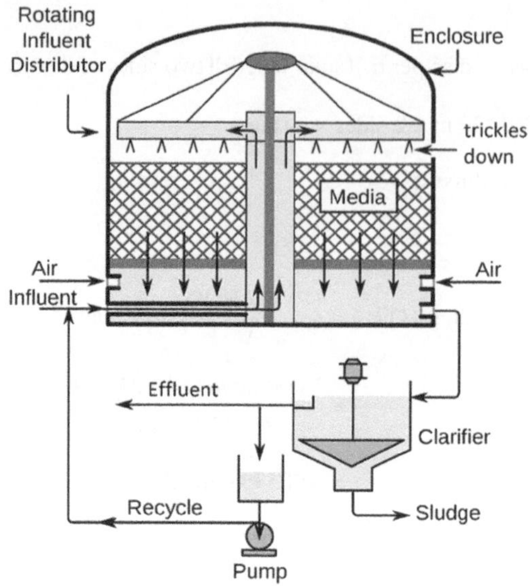

FIGURE 2.14 Trickling filter.

2.3.2 Trickling Filter

Trickling filter systems are typically constructed as beds of media through which pre-settled wastewater trickles with the use of a rotating sprinkler under (mostly) aerobic conditions. It was established in 1900. Natural or forced ventilation normally provides oxygen. Flow distributors or sprayers distribute the wastewater evenly onto the surface of the medium (Figure 2.14). The trickling medium is normally made of a high specific surface area material such as rocks, gravel, shredded PVC bottles, or special pre-formed filter material. A material with a specific surface area between 30 and 900 m^2/m^3 is desirable. The filter is usually 1–3 m deep, but filters packed with lighter plastic filling can be up to 12 m deep. The size of filter media is normally from 25 to 100 mm. Pre-treatment is essential to prevent clogging and to ensure effective treatment. As the wastewater moves by gravity through the medium, soluble and colloidal organic matter is oxidized by the biofilm that forms on the medium. Excess biomass sloughs from the medium and is carried with the treated wastewater to the clarifier, where the solids settle and separate from the treated effluent. At this point, the treated wastewater may be discharged or recycled back into the filter medium for further treatment.

The rate of trickling flow determines its category, either low rate, intermediate rate, or high rate. The trickling filter can remove 65%–90% BOD_5 and SS. Many variations of the trickling filters have been constructed. However, the recommended flow and designs are given in Table 2.7 and Figure 2.15.

The typical process flow diagram of a trickling filter in Malaysia and the design criteria in accordance with MSIG, Volume 4, 2009 [1] is given in Table 2.7.

$$\text{Organic Loading (O.L.)} = \frac{\text{DWF} \times \text{BOD}}{V} \text{ kg/m}^3\text{.day} \tag{2.22}$$

$$\text{Hydraulic loading (H.L.)} = \frac{\text{DWF}}{A} \text{ m}^3/\text{m}^2\text{.day} \tag{2.23}$$

where

V = Void volume (where the wastewater passes)

A = Surface area where the wastewater trickles

TABLE 2.7
Design Parameters for Trickling Filter

Description	Unit	Design Criteria
Design flow	m^3/day	Q_{ave}
Organic loading (OL) (depending on filter type)[a]	kg BOD_5/m^3.day	
Low rate		0.08–0.15
Intermediate rate		0.15–0.5
High rate		0.5–2.0
Recirculation of flow to the head of plant (to maintain wetting rate and improve flow), $Q_{recycle}/Q_{inflow}$		>1.0
Acceptable media		HDPE, PVC, stone, slag, coke, etc. (random or standard arrangement)
Hydraulic loading (HL)[b]	m^3/m^2.day	
Low rate		1–4
Intermediate rate		4–10
High rate		10–40
Sludge yields	kg sludge/kg BOD_5 influent	
Low rate		0.5
Intermediate rate		0.6–0.8
High rate		1.0
Minimum depth of media	m	1.5

Source: MSIG, Volume 4, 2009 [1].
[a] Refer Equation 2.22.
[b] Refer Equation 2.23.

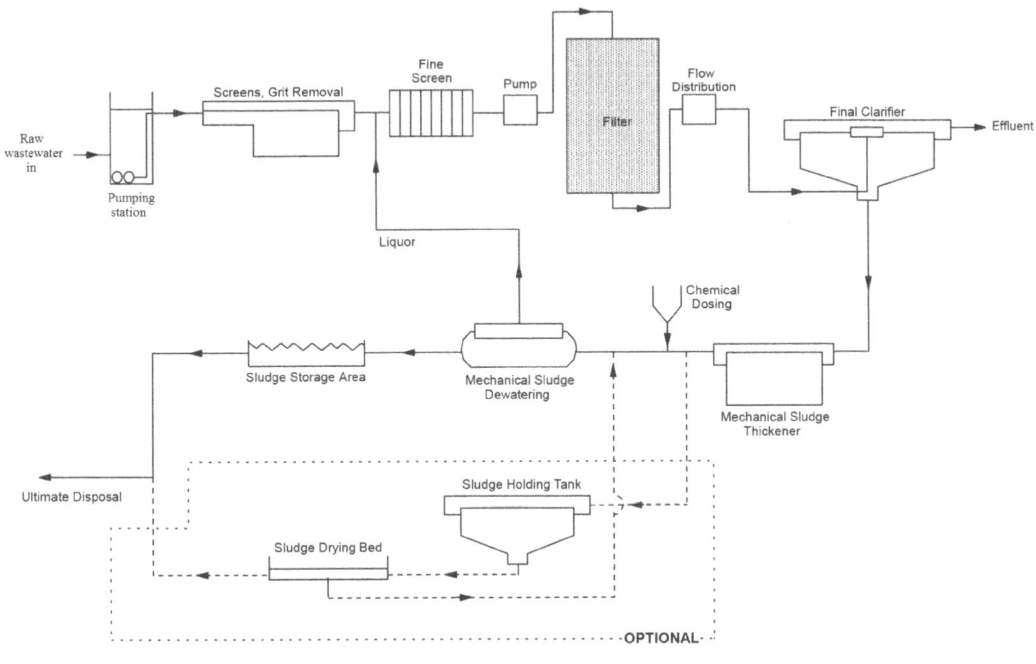

FIGURE 2.15 A typical process flow diagram of a trickling filter in Malaysia (MSIG, Volume 4, 2009) [1].

Example 2.4

Settled wastewater with flow rate 500 m^3/day, BOD_5 200 mg/L is treated with a small cylindrical-shaped slow-rate trickling filter. If the diameter and height of the tower are 5 and 5 m, respectively, calculate the hydraulic loading and organic loading of this filter. Assume the medium is 50% void.

Solution:

Q = DWF = 500 m^3/day
BOD_5 = 200 mg/L = 0.2 kg/m^3
Diameter (D) = 5 m
Height = 5 m

$$\text{Surface area} = A = \pi r^2$$

$$A = \pi(2.5)^2 \text{ m}^2$$

$$= 19.63 \text{ m}^2$$

$$\text{Volume of void} = \text{surface area} \times \text{height of tower} \times \% \text{ void}$$

$$V = \text{surface area} \times \text{height of tower} \times \% \text{ void}$$

$$V = (19.63 \text{ m}^2)(5 \text{ m})(0.5) \text{ m}^3$$

$$= 49.08 \text{ m}^3$$

(I) HYDRAULIC LOADING (HL)

From Equation 2.23,

$$\text{Hydraulic loading (H.L.)} = \frac{\text{DWF}}{A} \text{ m}^3/\text{m}^2.\text{day}$$

$$\text{HL} = \frac{500}{19.63} \text{ m}^3/\text{m}^2.\text{day}$$

$$= 25.47 \text{ m}^3/\text{m}^2.\text{day} < 40 \text{ m}^3/\text{m}^2.\text{day, acceptable}$$

(II) ORGANIC LOADING

From Equation 2.22,

$$\text{Organic Loading (O.L.)} = \frac{\text{DWF} \times \text{BOD}}{V} \text{ kg/m}^3.\text{day}$$

$$\text{OL} = \frac{500 \text{ m}^3/\text{day} \times 0.20 \text{ kg/m}^3}{49.08 \text{ m}^3}$$

$$= 2.0 \text{ kg/m}^3.\text{day, acceptable}$$

2.4 REMOVAL OF NITROGEN (N) AND PHOSPHORUS (P)

Tertiary treatment may be needed should the effluent flow into the sensitive receiving water with stringent discharge standards. In most cases, it involves processes to further remove nutrients, primarily nitrogen and phosphorous, from wastewater. In such cases, tertiary treatment can

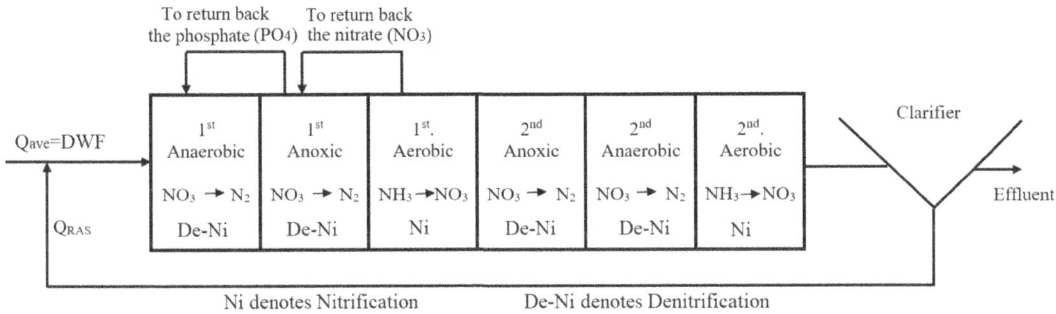

FIGURE 2.16 Nitrification and de-nitrification process to remove N&P (MSIG, Volume 4, 2009) [1].

be employed to further reduce the solids and organic content of the effluent. This treatment can employ conventional processes with an increased detention time to allow for greater removals, or the operations installed for tertiary treatment can involve more exotic and expensive equipment such as electrodialysis units or ion exchange columns. In tertiary treatment, the emphasis is placed on absorptive processes, such as the use of activated carbon; more efficient oxidation, as with ozone; foam separation of impurities; and demineralization using reverse osmosis or distillation. Chemical coagulation and precipitation processes are also involved, especially for metal removals.

Nutrient removal (mainly N and P) is required for effluent discharge to open water bodies such as lakes to prevent eutrophication. Biological, physical, and chemical methods have been widely applied for this purpose. The biological method for removing ammonia is normally done via nitrification and de-nitrification processes in series. This reaction has been explained in Figure 1.3. As the wastewater passes one tank to another, the amount of N will be reduced as it has been converted into N_2. Another common method for nitrogen removal includes air stripping and adsorption process.

The biological phosphorus removal mechanism is based on the role of bacteria, which stores excess phosphorus as polyphosphate and removes simple fermentation substrates produced in the anaerobic zone and assimilates them into storage products within their cells. Another method for phosphorus removal includes chemical precipitation and the adsorption process.

MSIG, Volume 4, 2009 [1] has recommended a typical process flow diagram and the design criteria for removing N&P by the biological process (Figure 2.16 and Table 2.8).

2.5 DISINFECTION

Disinfection refers to the selective destruction of disease-causing organisms in wastewater effluent. Methods of disinfection can be physical, chemical, or radiation. The methods under MSIG, Volume 4, 2009 [1] are given in Table 2.9. Detailed configurations and design may be referred to in the same guidelines.

2.6 SECONDARY SEDIMENTATION TANK OR CLARIFIER

The clarifier is aimed to settle out flocs (MLSS/solids) from a biological process. Clear supernatant is discharged as the effluent or recycled back for further treatment. Some of the settled microorganisms (sludge) are returned to the biological treatment system to maintain the MLSS concentration, and excess microorganisms are removed as sludge, as shown in Figure 2.17.

Under MSIG, Volume 4, 2009 [1], all clarifiers must be equipped with sludge scrapers to skim sludge from the bottom unless they are designed with a 60° hopper bottom. A scum skimmer is also needed to remove scum from the surface (Table 2.10).

TABLE 2.8
Design Requirement for Biological N&P Removal System

Item	Design Parameters	Retention Time (Hours)	MLSS (mg/L)	Recycling
1	First stage anaerobic	1–2	2,000–6,000	RAS from clarifier
2	First stage anoxic	2–4		2 to 1 (MLSS recirculation ratio)
3	First stage aerobic (oxic)	8–12		3 to 2 (MLSS recirculation ratio)
4	Second stage anoxic	2–4		Mixing
5	Second stage anaerobic	1–2		Mixing
6	Second stage aerobic (oxic)	4		—

TABLE 2.9
Requirements for Disinfection Facility

Description	Continuous	Intermittent
Class of STP[a]	Class 4	Class 1 Class 2 Class 3
Type of Disinfection	Chlorination Ultra-violet (UV) Ozone	Chlorination
Facility	(1 duty/1 standby) for equipment	Basic facility structure.

Source: MSIG, Volume 4, 2009 [1].

[a] The Commission may impose separate requirements on a case-by-case basis.

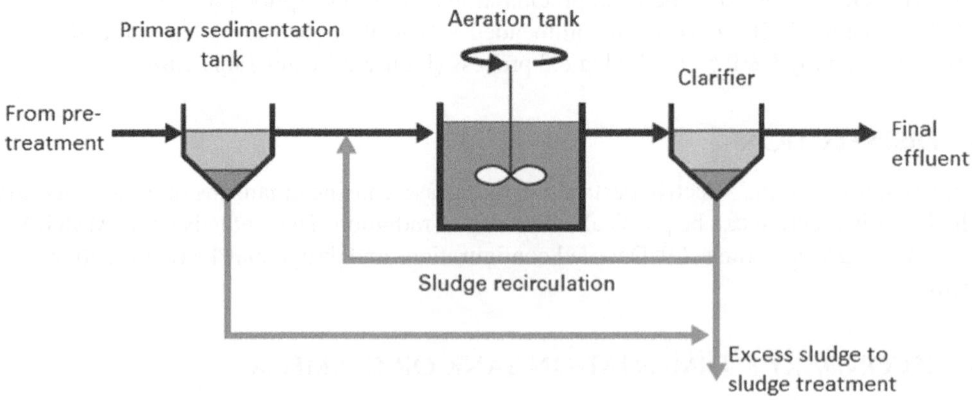

FIGURE 2.17 Secondary sedimentation tank/clarifier.

$$\text{SOR} = \frac{Q_p}{\text{Surface area (top)}}$$

$$\text{SOR} = \frac{Q_p}{A_s}\ \text{m}^3/\text{m}^2.\text{day} \tag{2.24}$$

$$\text{SOR} = \frac{Q_p}{W \times L}\ \text{m}^3/\text{m}^2.\text{day}$$

TABLE 2.10
Design Parameters for Secondary Sedimentation Tank/Clarifier

Description	Unit	Design Criteria ≤5,000 PE	Design Criteria >5,000 PE
Minimum number of tanks		2[a]	2
Tank configuration		Square Circular Rectangular	Circular Rectangular[b]
Minimum side water depth	m	3[c]	3
Minimum hydraulic retention time (HRT) at Q_{peak}	Hours	2	2
[d]Surface overflow rate (SOR) at Q_{peak}	m^3/m^2.day	≤30	≤30
[e]Solids loading rate (SLR) at Q_{peak}	kg/m^2.day	<150	<150
[f]Solids loading rate (SLR) at Q_{ave}	kg/m^2.day	<50	<50
[g]Weir loading rate (WLR) at Q_{peak}	m^3/m.day	<180	<180
Return activated sludge (RAS) pumping rate		Continuous	Continuous
Waste activated sludge (WAS) pumping rate		Continuous or batch	Continuous or batch
Sizing of Rectangular Tanks			
Length (*L*):Width (*W*)		3:1 or greater	
Maximum side water depth (D)	m	3.0	
Width (*W*):Depth (*D*)		1:1 to 2.5:1	
Sizing of Circular Tanks			
Side water depth, minimum	m	3.0[c]	
Floor slope wall		1:12	

Source: MSIG, Volume 4, 2009 [1].

[a] For PE less than or equal to 1,000, a single clarifier is acceptable.
[b] Rectangular tanks are acceptable if equipped with automatic scraping and desludging devices.
[c] For square clarifier with 600 slope minimum 1 m side water depth shall be provided.
[d] Refer Equation 2.24 and Figure 2.18.
[e] Refer Equation 2.25 and Figure 2.18.
[f] Refer Equation 2.26 and Figure 2.18.
[g] Refer Equation 2.27 and Figure 2.18.

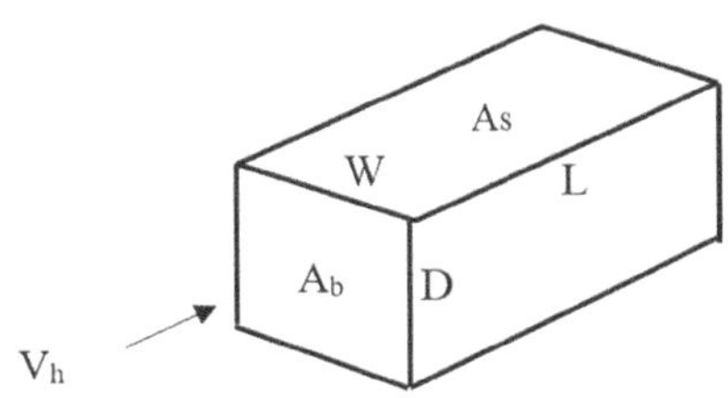

FIGURE 2.18 Formulation of surface overflow rate (SOR), weir loading rate (WLR), horizontal velocity (V_h)

where
W = Width (m)
L = Length (m)
D = Depth (m)
A_b = Approaching surface area (m^2)
A_s = Surface area (top)(m^2)
V_h = Horizontal velocity (mm/s)

$$\text{Solid Loading Rate at } Q_{\text{peak}}\text{, SLR} = \frac{Q_p(\text{MLSS})}{A_s}\ \text{kg/m}^2\text{.day} \tag{2.25}$$

$$\text{Solid Loading Rate at } Q_{\text{ave}}\text{, SLR} = \frac{Q_{\text{ave}}(\text{MLSS})}{A_s}\ \text{kg/m}^2\text{.day} \tag{2.26}$$

$$\text{Weir Loading Rate, WLR} = \frac{Q_P}{L}\ \text{m}^3\text{/m/day} \tag{2.27}$$

where

L = Length of weir (where the wastewater overflows)

$= w$

Example 2.5

Design a rectangular clarifier with the following data:

Design flow: $(Q_{\text{peak}}) = 7{,}000\ \text{m}^3\text{/day}$, $Q_{\text{ave}} = 2{,}000\ \text{m}^3\text{/day}$
Retention time = 2.5 hours
Length (L):Width (W) = 3:1
Depth = 3 m
MLSS = 3,000 mg/L

Solution:

$$\text{Retention time } (t) = V/Q_p$$

$$V = t(Q_p)$$

$$= \left(\frac{2.5}{24}\right)\text{day}\left(7{,}000\ \text{m}^3\text{/day}\right)$$

$$= 729.2\ \text{m}^3$$

Given:

$L : W = 3:1$ and depth $(D) = 3$ m, as follows:

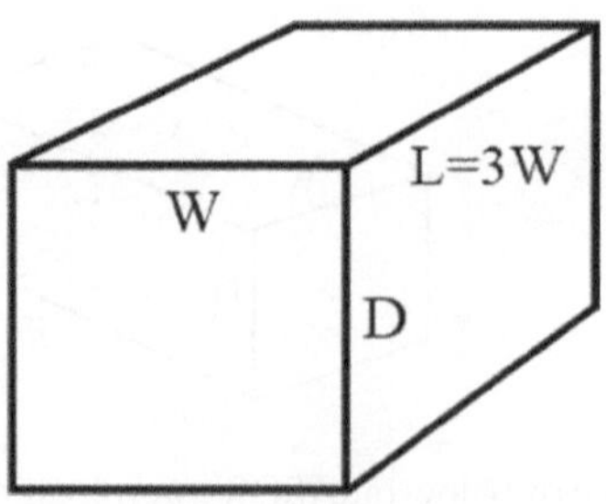

$$V = W \times 3W \times D$$

$$729.2 = 3W^2 \times 3$$

$$W = 9\ \text{m}$$

$$L = 3W = 3(9)$$

$$= 27\ \text{m}$$

The new dimension (excluding freeboard) is as follows:

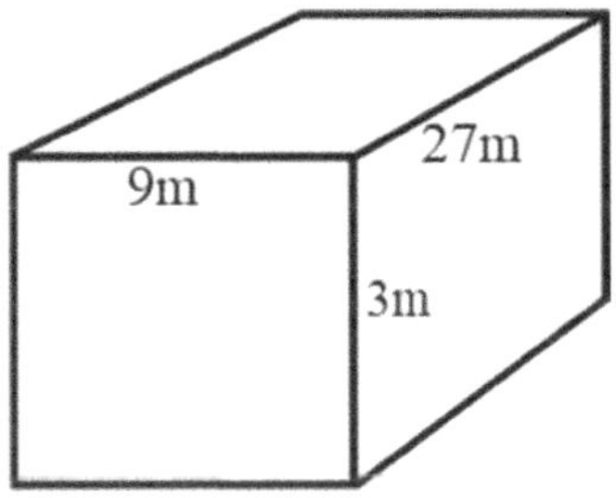

From Equation 2.24–2.27,

a. Surface Overflow Rate, $SOR = \frac{Q_P}{W \times L}\ m^3/m^2.day$

$$SOR = \frac{7{,}000}{9 \times 27}\ m^3/m^2.day$$

$$SOR = 28.8\ m^3/m^2.day \ll 30, \text{acceptable}$$

b. Solid Loading Rate at Q_{peak}, $SLR = \frac{Q_p(MLSS)}{A_s}\ kg/m^2.day$

$$SLR = \frac{7{,}000(3)}{9 \times 27}\ kg/m^2.day$$

$$SLR = 86.42\ kg/m^2.day \ll 150, \text{ acceptable}$$

c. Solid Loading Rate at Q_{ave}, $SLR = \frac{Q_{ave}(MLSS)}{A_s}\ kg/m^2.day$

$$SLR = \frac{2{,}000(3)}{9 \times 27}\ kg/m^2.day$$

$$SLR = 24.7\ kg/m^2.day \ll 50, \text{ acceptable}$$

d. Weir Loading Rate, $WLR = \frac{Q_P}{L}\ m^3/m^2.day$, L is where the wastewater overflows = W

$$WLR = \frac{7{,}000}{9}\ m^3/m^2.day$$

$$= 778\ m^3/m^2.day > 180, \text{ not acceptable}$$

Hence, apply five layers of weirs, each at 9 m.

$$\text{New WLR} = \frac{7{,}000}{9 \times 5}\ m^3/m^2.day$$

$$= 156\ m^3/m^2.day < 180, \text{ acceptable}$$

2.7 FLOW MEASUREMENT

Flow measuring devices are necessary for monitoring plant operation and process control continuously. The requirements under MSIG, Volume 4, 2009 [1] are given in Table 2.11.

TABLE 2.11
Design Parameters for Flow Devices

	Design Criteria	
Description	**≤5,000 PE**	**>5,000 PE**
Location of the flow metre	Inlet or outlet	Inlet and outlet
Type of flow measurement	Closed conduit or open channel	Closed conduit or open channel
Type of flow metre	V-notch (outlet only) rectangular weir (outlet only) flumes Electromagnetic Ultrasonic	V-notch (outlet only) rectangular (outlet only) flumes Electromagnetic Ultrasonic
Method of measurement	Automated or manual measurement of staff gauge to measure the height of crest with calibration curves/tables	Automated devices linked to data logging with an integrator and transmitted to a chart recorder (minimum 7 days chart time)
Measurement times	Continuous or intermittent	Continuous

Source: MSIG, Volume 4, 2009 [1].

FIGURE 2.19 Typical shape of oxidation pond.

2.8 POND PROCESSES

Waste stabilization ponds are simple trapezoidal ponds designed and built for wastewater treatment to reduce the organic content and remove pathogens from wastewater. It is used worldwide for wastewater treatment and is especially suitable for developing countries that have warm climates, which allow the photosynthesis process to occur (Figure 2.19). The influent enters on one side of the waste stabilization pond and exits on the other side as effluent after spending several days in the pond, during which treatment processes take place. The pond is normally constructed after screening and pumping station. Oxidation pond is quite common in Malaysia, especially for small communities because of their low construction and operating costs. Oxidation ponds represented 12% (500 numbers) of all wastewater treatment plants (WWTPs) in Malaysia in 2018. However, its usage is now not recommended as it requires a bigger footprint and may be costly in an urban area.

Waste stabilization ponds involve natural treatment processes that take time because removal rates are slow. Therefore, larger areas are required. Oxidation ponds may comprise one or more shallow ponds in a series. The natural processes of algae and bacteria growth exist in a mutually dependent relationship. Oxygen is supplied from natural surface aeration and by algal photosynthesis. Bacteria present in the wastewater use the oxygen to feed on organic material, breaking it down into nutrients and carbon dioxide. These are, in turn, used by the algae. Other microbes in the pond, such as protozoa, remove additional organics and nutrients to polish the effluent.

There are three types of ponds: (a) anaerobic, (b) facultative, and (c) aerobic (maturation), each with different treatments and design characteristics. There are normally at least two ponds constructed. The first pond reduces the organic material using aerobic digestion, while the second pond polishes the effluent and reduces the pathogens present in wastewater. Wastewater enters a large pond after passing through a settling and screening chamber. After retention for several days, the flow is often passed into a second pond for further treatment before it is discharged into a drain. Bacteria already present in wastewater act to break down organic matter using oxygen from the surface of the pond. Oxidation ponds need to be de-sludged periodically to work effectively. Depending upon the design, oxidation ponds must be de-sludged approximately every 10 years. Figure 2.20 shows different types of oxidation ponds.

The main advantage of the oxidation pond is it is low in cost (if land space is available), does not require mechanical components, except the input pump, and requires minimum maintenance. However, the setback is that if not properly operated and maintained, it may cause an odour problem. It also requires large space and long retention time (minimum 5 days), and some of the algae may escape as effluent.

Various factors affect pond design. These include wastewater characteristics and fluctuations, environmental factors (solar radiation, light, temperature), algal growth patterns and their diurnal and seasonal variation, bacterial growth patterns and decay rates, solids settlement, gasification,

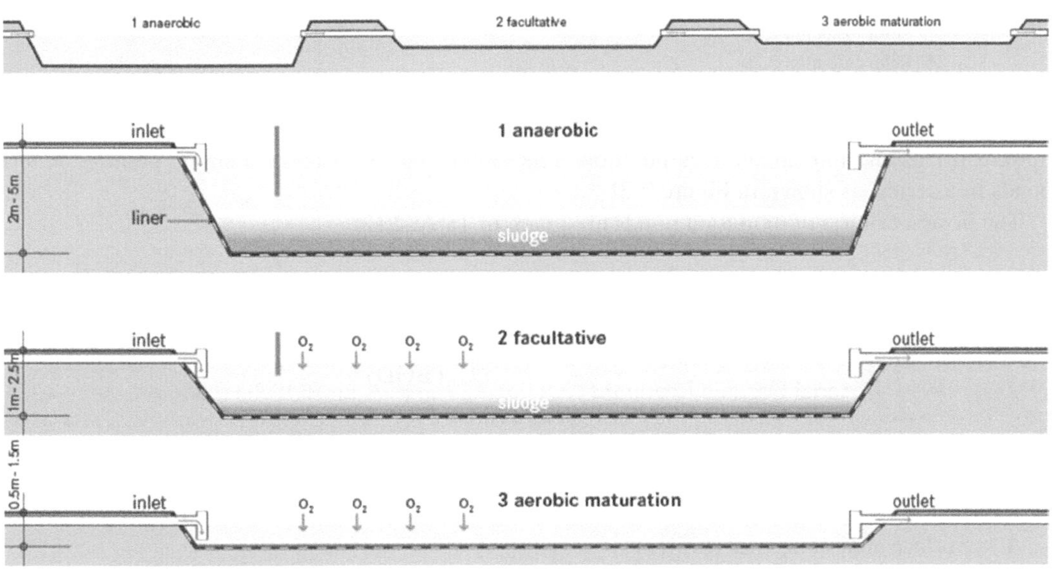

FIGURE 2.20 Types of oxidation ponds [36].

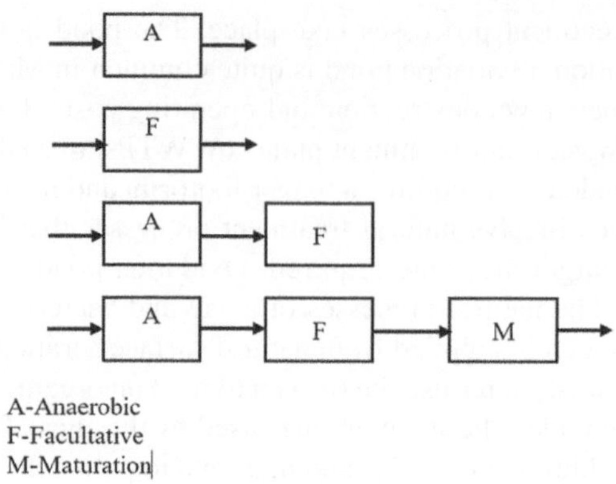

FIGURE 2.21 The arrangement of aeration ponds.

TABLE 2.12
Design Parameters for Oxidation Pond

	Type of Ponds		
Parameters	**Anaerobic**	**Facultative**	**Aerobic Matured**
Q_{design}	1 DWF	1 DWF	1 DWF
Depth (m)	3.0–5.0	1.0–2.0	0.3–1.0
Area (ha)	1	1–20	0.2–0.3
Retention time (day)[a]	8–40	>14 (Standard A) >9–13 (Standard B)	2–6
BOD_5 removed	70%–80%	70%–85%	80%–90%
Sludge generation (m^3/capita.day)	0.04	0.1–0.3	—
Areal organic load (kg BOD_5/ha.day)[b]	<4,000	<330 (std A) 331–505 (std B)	<100
L:W ratio	—	3:1	—

[a] Refer Equation 2.28.
[b] Refer Equation 2.29.

upward diffusion, and sludge accumulation. The system may consist of a single pond or several ponds in a series, as shown in Figure 2.21.

The design criteria of oxidation ponds are given in Table 2.12.

$$\text{Retention time } (t) = \frac{V}{Q_{\text{ave}}} \text{ day} \tag{2.28}$$

$$\text{Aerial Organic Loading (AOL)} = \frac{Q_{\text{ave}}(\text{BOD})}{A_s} \text{ kg BOD/ha.day} \tag{2.29}$$

where

$Q_{\text{ave}} = \text{DWF}\left(\text{m}^3/\text{day}\right)$

$A_s = \text{Surface area of lagoon (top)}\left(\text{m}^2\right)$

$V = \text{Volume of lagoon}\left(\text{m}^3\right)$

2.8.1 Anaerobic Pond

Anaerobic ponds are used as pretreatment of high-strength wastes with a BOD load of up to 4,000 kg BOD_5/ha.day. Such ponds have a smaller surface area compared to facultative ponds and are also deeper (usually 3.0–5.0 m) as light penetration is unimportant. The depth decreases the influence of oxygen production by photosynthesis, leading to anaerobic conditions. Depending on loading and climatic conditions, these ponds can remove between half to two-thirds of the influent BOD. This significantly decreases the load of organic matter that goes to the facultative ponds and thus decreases their required size. Anaerobic ponds have the disadvantage of potentially releasing malodorous gases, which includes hydrogen sulphide with an odour of rotten eggs.

The design criteria of an anaerobic pond are the same as the design of a facultative pond. The difference is the organic loading and the retention time.

Example 2.6

Design an anaerobic pond with the following data:

Q_{ave} = 1,000 m^3/day
Depth = 5 m
Retention time (t) = 8 days
BOD_5 = 250 mg/L. Ignore side slope

Solution:

From Equation 2.28,

$$\text{Retention time } (t) = \frac{V}{Q_{ave}}$$

$$\text{Retention time } (t) = \frac{A_s(D)}{Q_{ave}}$$

$$A_s = \frac{Q_{ave}(t)}{D}$$

$$A_s = \frac{1{,}000 \text{ m}^3/\text{day } (8 \text{ day})}{5 \text{ m}}$$

$$A_s = 1{,}600 \text{ m}^2$$

Aerial Organic Loading (AOL) (Equation 2.29):

$$\text{Aerial Organic Loading (AOL)} = \frac{Q_{ave}(\text{BOD})}{A_s}$$

$$\text{AOL} = \frac{1{,}000(0.25)}{1{,}600} \text{ kg BOD/m}^2\text{.day}$$

$$\text{AOL} = 0.1566 \text{ kg BOD/m}^2\text{.day}$$

$$\text{AOL} = 1{,}560 \text{ kg BOD/ha.day} < 4{,}000, \text{ acceptable (Figure 2.12).}$$

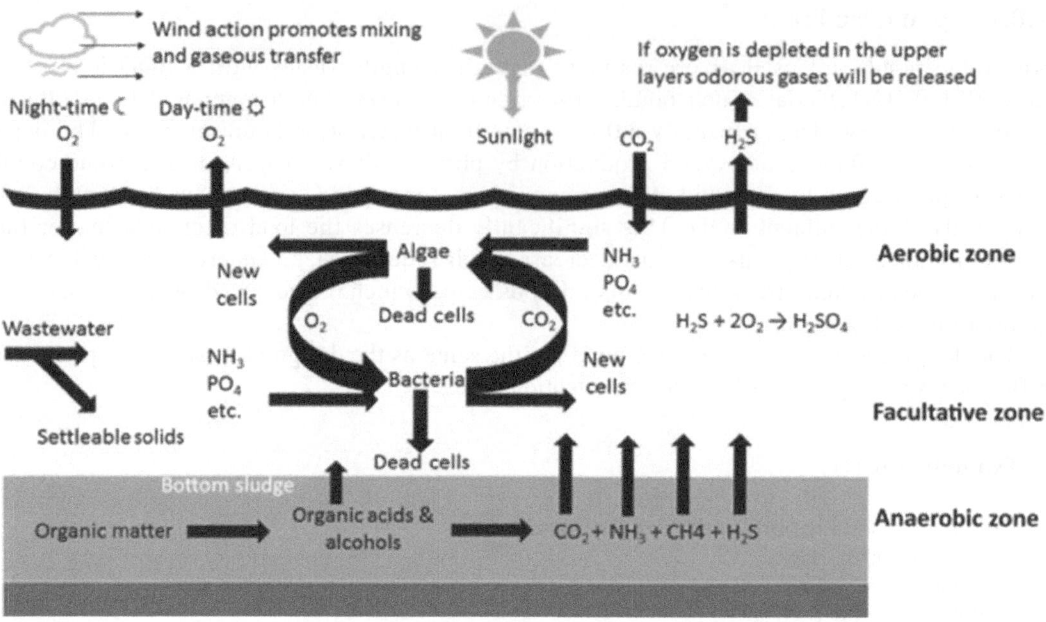

FIGURE 2.22 Typical mechanisms in facultative pond. (Reprinted from Milledge et al. [3]. Novel developments in biological technologies for wastewater processing, Chapter 8, Sustainable Water and Wastewater Processing, 2019, 239–278 with permission from Elsevier.)

2.8.2 Facultative Pond

Facultative pond functions aerobically at the surface while anaerobic conditions prevail at the bottom. They are often about 1–2 m in depth. The shallowness of the pond (1–2 m) caused the occurrence of both reactions. At the top part of the pond (30–50 cm from the surface layer), there will be many algae that will influence the wastewater stabilization process. Algae produce oxygen (photosynthesis) and also consume oxygen (respiration), but they leave an excess of oxygen that can then be used by aerobic bacteria for respiration and the processes of oxidation (or stabilization) of the organic matter in the wastewater. The settled sludge mass originating from raw waste and microbial synthesis in the aerobic layer and dissolved and suspended organics in the lower layers undergo anaerobic stabilization through conversion to methane which escapes the pond in the form of bubbles. The biosolids (sludge after digestion) require de-sludging every 5 years. The mechanisms within the facultative pond are summarized in Figure 2.22.

There are two basic approaches in the design of a facultative pond:

2.8.2.1 Simplistic Approach

This approach has been detailed in Equations 2.28 and 2.29.

Example 2.7

A square facultative pond with a base of 70 m×70 m and a surface 80 m×80 m receives an average wastewater flow of 1,000 m³/day. Calculate the retention time and AOL of this pond if the BOD_5 is 250 mg/L. Take the water depth as 2 m.

Solution:

Determine the volume of the pond based on the trapezoid shape in Figure 2.8 and Equation 2.19:

$$W1 = 80\text{ m}, W2 = 70\text{ m}, L1 = 80\text{ m}, L2 = 70\text{ m}$$

$$V = \frac{1}{6}(2)\left[(W1 \times L1) + (W2 \times L2) + (W1 + W2)(L1 + L2)\right]$$

$$V = \frac{1}{6}(2)\left[(80 \times 80) + (70 \times 70) + (80 + 70)(80 + 70)\right]$$

$$= 11{,}267\text{ m}^3$$

(I) RETENTION TIME

From Equation 2.28,

$$\text{Retention time } (t) = \frac{V}{Q_{\text{ave}}}\ (\text{day})$$

$$t = \frac{11{,}267\text{ m}^3}{1{,}000\text{ m}^3/\text{day}}$$

$$= 11.3\text{ day, complies with Standard B } (9-13\text{ days})$$

(II) AERIAL ORGANIC LOADING

From Equation 2.29,

$$\text{Aerial Organic Loading (AOL)} = \frac{Q_{\text{ave}}(\text{BOD})}{A_s}\ \text{kg BOD/ha.day}$$

$$\text{AOL} = \frac{\left(1{,}000\text{ m}^3/\text{day}\right)\left(0.25\text{ kg BOD/m}^3\right)}{80 \times 80\text{ m}^2}$$

$$= 0.039\text{ kg BOD/m}^2\text{.day}$$

$$= 390\text{ kg BOD/ha.day, complies with Standard B } (331-505\text{ kg BOD/ha.day})$$

2.8.2.2 Design Based on BOD Rate of Reaction

2.8.2.2.1 Design of Facultative Pond

Assume that the reaction in the oxidation pond follows the first-order kinetic:

2.8.2.2.1.1 Retention Time Retention time can be determined using Equation 2.30:

$$\frac{\text{BOD}_e}{\text{BOD}_i} = \frac{1}{1 + kt} \tag{2.30}$$

where

BOD_e = BOD value at effluent (mg/L)

BOD_i = BOD value at influent (mg/L)

k = BOD rate of coefficient (day^{-1})

t = Retention time (day)

2.8.2.2.1.2 Surface Area The total area of the pond can be determined using Equation 2.31:

$$A_s = \frac{Q_{\text{ave}}}{Dk}\left[\frac{\text{BOD}_i}{\text{BOD}_e} - 1\right] \tag{2.31}$$

where

A_s = Surface area of pond (m^2)

D = Depth (m)

Q_{ave} = DWF (m^3/day)

The BOD rate of the coefficient (k) in Equations 2.30 and 2.31 is 0.3 day^{-1} at 20°C. For other temperatures, a correction is necessary, as given in Equation 2.32.

$$k_T = 0.30\ (1.085)^{T-20} \tag{2.32}$$

where

T = Temperature in °C

2.8.2.2.1.3 Aerial Organic Loading The Organic Load can be determined using Equation 2.29:

$$\text{Aerial Organic Loading (AOL)} = \frac{Q_{\text{ave}}(\text{BOD})}{A_s}\ \text{kg BOD/ha.day} \tag{2.29}$$

where

Q_{ave} = DWF (m^3/day)

A_s = Surface area of lagoon (top) (m^2)

Example 2.8

Design suitable facultative pond based on the following data:

Average flow, Q_{ave} = DWF = 1,000 m^3/day
Influent $\text{BOD}_5 = \text{BOD}_i$ = 250 mg/L
Effluent $\text{BOD}_5 = \text{BOD}_e$ = 20 mg/L
Depth = 2 m
BOD rate of reaction, k, at 30°C = 0.3 day^{-1}
Assume no side slope.

Solution:

(I) RETENTION TIME (*t*)

From Equation 2.30,

$$\frac{\text{BOD}_e}{\text{BOD}_i} = \frac{1}{1+kt}$$

Since k is at 30°C, a conversion is needed (Equation 2.32),

$$k_T = 0.30\,(1.085)^{T-20}$$

$$k_{30} = 0.30\,(1.085)^{30-20}$$

$$= 0.678\ \text{day}^{-1}$$

$$\frac{20}{250} = \frac{1}{1+(0.678)t}$$

$$1+0.68t = \frac{1}{0.08}$$

$t = 16.9$ days, compare with value in Table 2.12

(II) SURFACE AREA

From Equation 2.31,

$$A_s = \frac{Q_{\text{ave}}}{Dk}\left[\frac{\text{BOD}_i}{\text{BOD}_e} - 1\right]$$

$$A_s = \frac{1{,}000}{2(0.678)}\left[\frac{250}{20} - 1\right]$$

$$= 8{,}480.8\ \text{m}^2$$

(III) AERIAL ORGANIC LOADING

From Equation 2.29,

$$\text{Aerial Organic Loading (AOL)} = \frac{Q_{\text{ave}}(\text{BOD})}{A_s}$$

$$\text{AOL} = \frac{1{,}000(0.25)}{8{,}480.8}\ \text{kg BOD/m}^2\text{.day}$$

$$\text{AOL} = 0.0295\ \text{kg BOD/m}^2\text{.day}$$

AOL = 295 kg BOD/ha.day, compare with value in Figure 2.12

2.8.3 Maturation Pond

This is an aerobic pond, usually used after facultative ponds. It gives the purpose to increase the effluent quality and to destroy the algae emitted from the facultative pond. The depth is between 1.0 and 1.5 m or less. Shallow ponds benefit from high photosynthetic activity arising from the penetration of solar radiation with a great surface area so that more oxygen can dissolve into the water giving the bacteria enough oxygen to properly function. These ponds can be placed in serial (more than one unit). For instance, two mature ponds with a detention time of 7 days can produce 25 mg/L effluents. Sludge accumulation is very low in maturation ponds.

The design of the maturation pond and the required number of ponds take into consideration the number of faecal coliform (FC) present in the wastewater and at the effluent. The reduced number of FC follows the first-order kinetic and can be determined at 20°C using Equation 2.33:

$$N_e = \frac{N_i}{1+k_b t} \tag{2.33}$$

For n ponds in series at 20°C, the equation can be written as (Equation 2.34):

$$N_e = \frac{N_i}{(1+k_b t_1)(1+k_b t_2)\cdots(1+k_b t_n)} \tag{2.34}$$

where

N_e = Number of FC/100 mL of effluent
N_i = Number of FC/100 mL of influent
k_b = First-order rate constant for FC removal $\left(\text{day}^{-1}\right)$

For design purposes, the following data are set:

$$N_i = 4\times10^7 \text{ FC/100 mL}$$

$$N_e < 5{,}000 \text{ FC/100 mL}$$

The k_b in Equations 2.33 and 2.34 is at 20°C. For other temperatures, a correction is necessary as given in Equation 2.35:

$$k_b = 2.6\,(1.19)^{T-20} \tag{2.35}$$

where

T = Temperature in °C

Example 2.9

For the question in Example 2.8, determine the number of maturation pond after the facultative pond to maintain FC of fewer than 5,000/100 mL at effluent. Take $N_i = 4\times10^7$ FC/100 mL, and BOD is 250 mg/L. Assume the k_b for the facultative pond the same as the k in Example 2.8.

Solution:

For the first facultative pond, calculate the number of FC at effluent (Equation 2.33):

$$N_e = \frac{N_i}{1+k_b t}$$

Since the temperature is 30°C, a correction is necessary (Equation 2.35):

$$k_b = 2.6\,(1.19)^{T-20}$$

$$k_{30} = 2.6\,(1.19)^{30-20}$$

$$= 14.81 \text{ day}^{-1}$$

$$N_e = \frac{N_i}{1+k_b t}$$

$$N_e = \frac{4\times10^7 \text{ FC/100 mL}}{1+14.81 \text{ day}^{-1}\left(16.9 \text{ days}\right)}$$

$$= 159{,}220.2 \text{ FC/100 mL} > 5{,}000 \text{ FC/100 mL, not acceptable}$$

Try one maturation pond after the facultative pond with the same retention time (16.9 days) (Equation 2.28):

$$N_e = \frac{N_i}{(1+k_b t_1)(1+k_b t_2)\cdots(1+k_b t_n)}$$

$$N_e = \frac{N_i}{(\text{Facultative pond})(\text{Maturation pond 1})}$$

$$N_e = \frac{4\times 10^7 \text{ FC/100 mL}}{\left[1+14.81/\text{day}(16.9 \text{ days})\right]\left[1+14.81/\text{day }(16.9 \text{ days})\right]}$$

$$= 633.45 \text{ FC/100 mL, hence} < 5{,}000 \text{ FC/mL, acceptable}$$

Dimension the maturation pond in series:

For the pond, Q_{ave} = 1,000 m³/day and depth (D) 2 m, retention time 16.9 days

From Equation 2.28,

$$\text{Retention time}(t) = \frac{V}{Q_{\text{ave}}}$$

$$\text{Retention time}(t) = \frac{A_s\ (D)}{Q_{\text{ave}}}$$

$$A_s = \frac{Q_{\text{ave}}(t)}{D}$$

$$A_s = \frac{1{,}000 \text{ m}^3/\text{day }(16.9 \text{ day})}{2 \text{ m}}$$

$$A_s = 8{,}450 \text{ m}^2$$

Build a maturation pond with an area of 8,450 m².

AOL per pond (Equation 2.29):

$$\text{Aerial Organic Loading (AOL)} = \frac{Q_{\text{ave}}(\text{BOD})}{A_s}$$

$$\text{AOL} = \frac{1{,}000(0.25)}{8{,}450} \text{ kg BOD/m}^2\text{.day}$$

$$\text{AOL} = 0.0296 \text{ kg BOD/m}^2\text{.day}$$

$$\text{AOL} = 296 \text{ kg BOD/ha.day}$$

2.9 TREATMENT SYSTEMS FOR SMALL COMMUNITIES

2.9.1 Septic Tank

Individual septic tank is one of the simplest forms of wastewater treatment and date to the sewerage system development in France in 1860.

Septic tank may be used to serve small installations where the effluent can be disposed of through subsurface filter systems or by percolation into the existing soil layer on-site. It is a cheap method and is suitable to be used in remote areas, school, or other premises which is far from the public sewer line. These are called communal septic tank (CST). There is also an individual septic tank (IST) which is meant for individual houses or premises. It is quite common to place it about 15 m from the house. A septic tank combines two processes. Sedimentation takes place in the upper portion of the tank, and the accumulated solids are digested by anaerobic decomposition in the lower portion. As wastewater from a building enters a septic tank, its rate of flow is reduced so that the heavier solids sink to the bottom and the lighter solids, including fats and grease, rise to the surface.

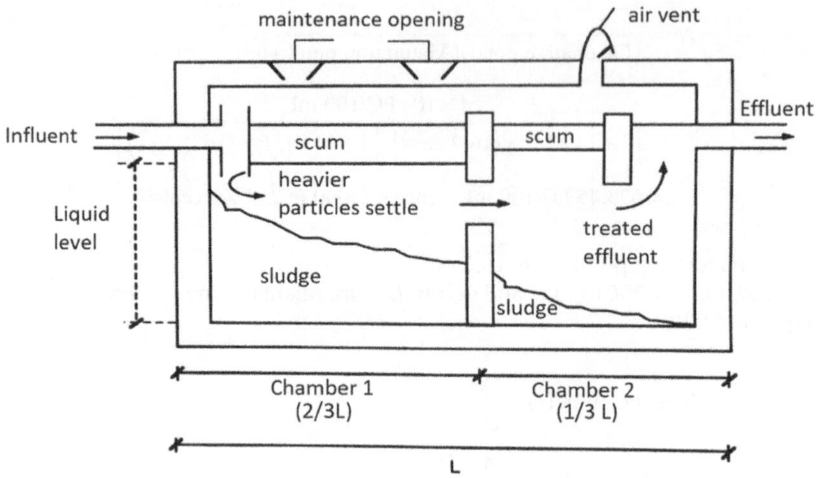

FIGURE 2.23 Typical cast in situ septic tank.

These solids are retained in the tank, and the clarified effluent is discharged. With good care and efficient operation, removal of solids may be as high as 60%. However, septic tanks do not accomplish a high degree of bacterial removal (40%–50% BOD reduction only). Although the wastewater undergoes treatment in passing through the tank, this does not mean that the infectious agents will be removed; hence, septic tank effluent cannot be considered safe. Further treatment of the effluent, including the removal of pathogens, is affected by percolation through the soil. Septic tanks only partially treat wastewater, discharging an effluent still rich in organic material. This has the potential to create a public health and environmental problem, particularly in urban areas.

Septic tanks can be cast in situ or purchased from many package suppliers. Manholes should be provided over the inlet and outlet pipes for observation and maintenance. An in situ cast unit normally comprises two chambers, the volume of the first chamber is doubled for the second one, as shown in Figure 2.23. In the first chamber, solids from the incoming wastewater settle, forming a "sludge", while greases and oils float to the surface forming a "scum" layer. Effluent from between the scum and sludge layers then passes into the second chamber, where further sedimentation occurs. Finally, the effluent leaves the second chamber and is discharged into a drain or allowed to percolate into the soil. A septic tank servicing an average size home of five people will need to be de-sludged every 3–5 years.

The sludge in the tank undergoes anaerobic digestion and is converted into more stable organic compounds and gases such as carbon dioxide (CO_2), methane (CH_4), and hydrogen sulphide (H_2S). Septic tanks are usually designed for 24-hour retention time. Enough storage capacity is provided so that scum and sludge can be deposited in the tank for up to 2 years, after which it must be de-sludged to keep the tank operating satisfactorily.

Under MS1228 [4], the septic tank is suitable for 9–150 PE where there are no central sewerage systems and where effluent discharges will not adversely affect the environment. It is a cheap solution to disposing of wastewater. However, septic tanks only partially treat wastewater, and concentrated groups of tanks can overload the capacity of the receiving environment, creating health and odour problems. There are currently over one million ISTs in Malaysia, making it by far the most common type of wastewater treatment system.

Detailed guidelines on a septic tank are given in the SPAN's Malaysian Sewerage Industry Guidelines, Volume 5, Septic Tanks, 3rd Edition, 2009 [5].

The design criteria of the septic tank are given in Table 2.13 and Equation 2.36.

TABLE 2.13
Design Criteria of Septic Tank

Parameters	Design Criteria
Depth of liquid (m)	1.25–2.0
Width (m)	<0.75
Length (m)	<2 × width
Ratio of L:W (normal)	3:1
Volume of chamber (V) (m^3)	Chamber 1 = 2/3V Chamber 2 = 1/3V
BOD_5 removed	0%–50%
SS removed	0%–70%
Population (PE)	9 < PE < 150

$$C = 225 \text{ PE} \tag{2.36}$$

where
C = Volume, (L)

Example 2.10

Design a septic tank for 100 PE.

Solution:

From Equation 2.36,

$$C = 225 \text{ PE}$$

$$C = 225\,(100)\,\text{L}$$

$$= 22.5 \text{ m}^3$$

Dimension of tank
From Figure 2.13, assume L:W = 3:1 and take D = 1.5 m
Take W = x (as shown in figure below)

$$\text{Volume} = C = 22.5 = 3x(x)(1.5) \text{ m}^3$$

$$x = 2.24 \text{ m}$$

Hence, $D = 2.24$ m, say 2.5 m, and

$$L = 3(2.5) = 7.5 \text{ m}$$

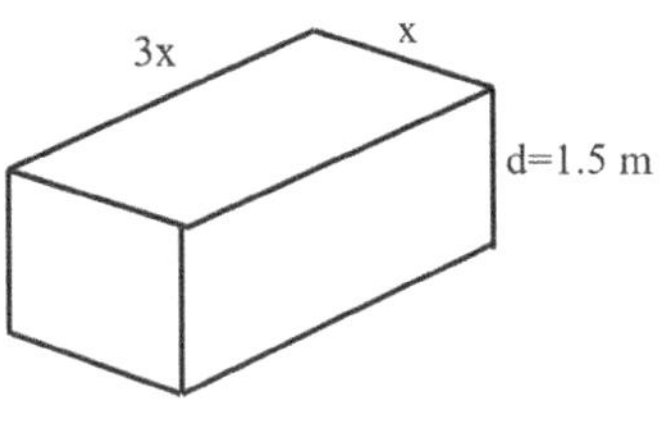

The final dimension is shown as follows:

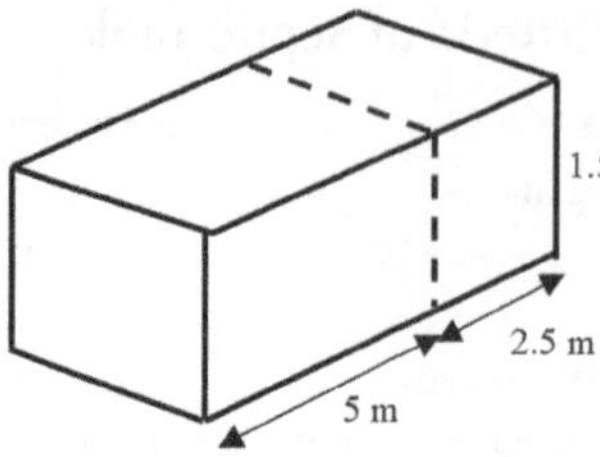

2.9.2 Imhoff Tank

Imhoff tanks are also a simple form of WWTPs with no mechanical equipment involved. Imhoff tank operates in the same manner as a septic tank, except for the configuration of the compartments. Imhoff tank comprises two chambers positioned one above the other. In the upper compartment, sedimentation occurs with solids passing through an opening into the lower chamber. Settled solids as sludge in the lower chamber will undergo anaerobic digestion. Gases from the lower tanks are discharged into the air. Scum is accumulated in the upper tank. Usually, the sedimentation process in the upper chamber is followed by percolating effluent over a coarse stone media before discharge to receiving water. It consists of an upper compartment for settling out solids from slowly flowing wastewater and a lower compartment for anaerobic digestion of the sludge. Accumulated solids are removed periodically through a sludge draw-off pipe.

Imhoff tanks are normally used to serve small communities up to a PE of 1,000. They are relatively cheap to install, operate, and maintain. However, Imhoff tanks, like ISTs, only partially treat wastewater. The effluent from these tanks will not meet the environmental discharge standard. Small package treatment plants have more recently supplanted Imhoff tanks as the popular method of servicing small communities.

Imhoff tanks constituted 24% (800 numbers) of all WWTPs in Malaysia in 2018 and were the second most common form of the treatment plant. They provide limited treatment of wastewater and are not a suitable long-term solution. The effluent from Imhoff tanks can rapidly deteriorate if the tanks are not properly maintained.

A typical Imhoff tank is given in Figure 2.24.

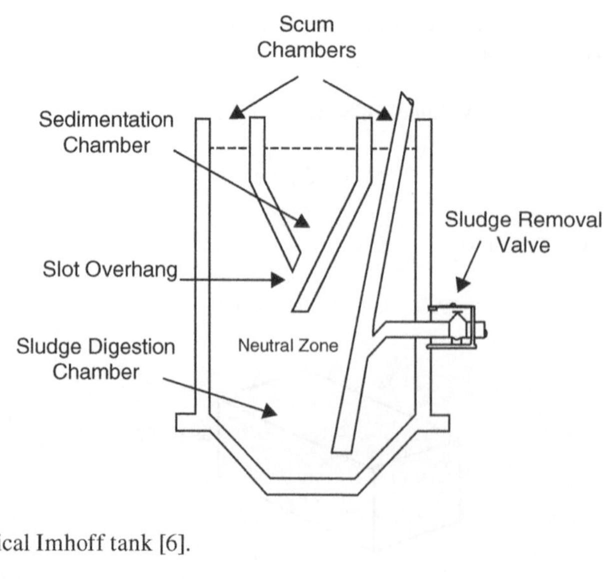

FIGURE 2.24 A typical Imhoff tank [6].

2.10 SLUDGE MANAGEMENT

2.10.1 Introduction

The solids removed from wastewater in both primary and secondary treatment units, together with the water removed with them, constitute wastewater sludge. It requires proper handling and disposal methods. The main source of sludge is from the primary sedimentation tank, secondary sedimentation tank (clarifier), oxidation ponds, septic, and Imhoff tank. The sludge comprises essentially inert and organic matters that are biodegradable and non-biodegradable present in wastewater and bacterial cells generated by the biological treatment processes. The treated sludge, often referred to as biosolids, are ready for safe disposal or reuse.

All sludge needs to be treated for safe disposal back to the environment. The minimum requirement for sludge treatment is to achieve sludge stabilization with a 20% dry solid content (80% water) (MSIG, Volume 4, 2009) [1].

2.10.2 Chemical Compositions of Sludge

Sludge is the material that contains mainly water and a solid fraction (1%–6% solids [7]). Sludge also contains organic, inorganic, and other forms of contaminants. A typical composition of sludge is given in Table 2.14.

2.10.3 Quantity of Sludge

There is a significant amount of sludge produced in Malaysia. In 2020, about 7 million metric tons of sewage sludge was produced in Malaysia and the amount increased over the years [8]. Almost 50% of the sewerage systems in Malaysia uses mechanized plants while others still use septic tanks and oxidation ponds. Some modern mechanized plants have been upgraded with anaerobic digesters whereby the sewage sludge is treated anaerobically to produce CH_4 which is a valuable

TABLE 2.14
Sludge Compositions [7]

	Sedimentation Tank		Treated by an
Parameter	**Primary**	**Secondary**	**Anaerobic Digester**
Specific gravity	1.04	1.02	1.03
pH	5.5–6.5	—	6.8–7.6
Alkalinity	1,200	—	4,400
Volatile acid (mg/L)	3,000	—	<200
Water content (%)	92–96	96–98	94–98
Mineral (%)	20–30	25–35	35–45
Organic concentration (%)	70–80	65–75	55–65
Grease (%)	12–16	—	4–12
Protein (%)	4–14	—	10–20
Carbohydrate (%)	8–10	—	5–8
Nitrogen (%)	2–5	—	2–4
Phosphate (%)	0.5–1.5	—	0.5–1.5
Potash (K_2O) (%)	0.3	—	—
Zn, Pb, & Cu (%)	0.05–0.5	—	0.05–0.5
Pathogenic bacteria (per 100 mL)	103–105	—	10–100
Parasite organism (per 100 mL)	8–12	—	1–3

TABLE 2.15
Sludge Generation Rates

Treatment System	Unit Generation Rates	Comments
	Primary Sludge	
Primary clarifier	0.5 kg sludge/kg solids input	Based on continuous sludge withdrawal
Imhoff tank	0.15 kg sludge/kg SS input	Based on average 6-month desludging period
	Secondary Sludge	
Conventional activated sludge system	0.8–1.0 kg sludge/kg BOD_5 removed	Standard A/B
Extended aeration or oxidation ditch	0.4–0.6 kg sludge/kg BOD_5 removed	Standard A/B
RBC/SBC/High-rate trickling filter system	0.8 kg sludge/kg BOD_5 removed	Standard A/B
Hybrid system	0.4 kg sludge/kg BOD_5 removed	Standard A/B

Source: MSIG, Volume 4, 2009 [1].

energy source to generate electricity (Kumaran et al. [9]). It is estimated that mechanical WWTP in Malaysia produced about 0.01–0.02 kg/person.day of dry sludge in 2018. Most of the sludge is water with 1%–6% solid.

The quantity of dry sludge can be determined as in Equations 2.38 and 2.39:

For primary sedimentation tank:

$$\text{Dry weight} = \text{DWF} \times \text{SS}\,(\text{mg/L}) \times \%\ \text{SS removal} \tag{2.37}$$

For secondary sedimentation tank (clarifier):

$$\text{Dry weight} = \text{DWF} \times \text{BOD}\,(\text{mg/L}) \times \%\ \text{BOD removal} \tag{2.38}$$

MSIG, Volume 4, 2009 [1] has given recommended values of dry sludge produced, based on the treatment systems. This is shown in Table 2.15.

Based on the above assumptions, the quantity of waste sludge requiring treatment and disposal can be estimated. Refer also to design guides related to each of the above individual processes.

Example 2.11

A primary sedimentation tank with DWF 1,000 m³/day, 400 mg/L SS, and 250 mg/L BOD_5 is 60% efficient in removing SS and 40% in removing BOD_5. After an EA system, the flocs are settled in a clarifier which is 70% efficient in removing BOD_5. Calculate in m³/day the dry sludge produced if the Specific Gravity (S.G.) of sludge is 1.04.

Solution:

$$\text{DWF} = 1{,}000\ \text{m}^3/\text{day}$$

$$\text{SS} = 400\ \text{mg/L} = 0.4\ \text{kg/m}^3$$

$$BOD_5 = 250 \text{ mg/L} = 0.25 \text{ kg/m}^3$$

$$\text{S.G.} = 1.04, \text{ hence, density } (\rho) = 1{,}040 \text{ kg/m}^3.$$

Calculate the sludge produced from the primary sedimentation tank:
From Equation 2.38,

$$\begin{aligned}\text{Dry weight} &= 1{,}000 \text{ m}^3\text{/day} \times \text{SS (mg/L)} \times \% \text{ SS removal}\\ &= 1{,}000 \text{ m}^3\text{/day } (0.4 \text{ kg/m}^3)(0.6)\\ &= 240 \text{ kg/day}\end{aligned}$$

Calculate the sludge produced from the clarifier:

$$\text{DWF} = 1{,}000 \text{ m}^3\text{/day}$$

BOD_5 entering clarifier $= 250$ mg/L $\times 0.6$ (as 40% have been removed in primary sedimentation tank)

$$\begin{aligned}&= 150 \text{ mg/L}\\ &= 0.15 \text{ kg/m}^3\end{aligned}$$

From Equation 2.39,

$$\begin{aligned}\text{Dry weight} &= \text{DWF} \times \text{BOD (mg/L)} \times \% \text{ BOD removal}\\ &= 1{,}000 \text{ m}^3\text{/day } (0.15 \text{ kg/m}^3)(0.7)\\ &= 105 \text{ kg/day}\end{aligned}$$

The total amount of sludge produced from primary sedimentation and clarifier:

$$\begin{aligned}&= 240 \text{ kg/day} + 105 \text{ kg/day}\\ &= 345 \text{ kg/day}\end{aligned}$$

Convert kg/day to m^3/day

$$\text{Density } (\rho) = \frac{\text{mass (kg)}}{\text{volume } (V)} = 1{,}040 \text{ kg/m}^3$$

$$1{,}040 \text{ kg/m}^3 = \frac{345 \text{ kg/day}}{V}$$

$$V = 0.332 \text{ m}^3\text{/day}$$

Example 2.12

A clarifier of a conventional activated sludge system produces a dry sludge of 0.9 kg sludge/kg BOD_5 removed. If the DWF is 1,000 m^3/day and BOD_5 250 mg/L, calculate the dry sludge produced per day in kg/day if the tank is 80% efficient.

Solution:

$$\text{DWF} = 1{,}000 \text{ m}^3\text{/day}$$

$$\text{BOD}_5 = 250 \text{ mg/L} = 0.25 \text{ kg/m}^3$$

$$\text{Removal} = 80\%$$

From Equation 2.39:

$$\begin{aligned}\text{Dry weight of sludge} &= \text{DWF} \times \text{BOD}\left(\text{mg/L}\right) \times \%\ \text{BOD removal}\\ &= 1{,}000\ \text{m}^3/\text{day}\left(0.25\ \text{kg/m}^3\right)(0.8)\\ &= 200\ \text{kg/day (BOD removed)}\end{aligned}$$

$$1\ \text{kg BOD}_5\ \text{removed produces}\ \ 0.9\ \text{kg sludge}$$

Hence,

$$\begin{aligned}200\ \text{kg/day BOD}_5\ \text{removed produces}\ 0.9 \times 200\ \text{kg sludge/day}\\ = 180\ \text{kg/day}\end{aligned}$$

2.10.4 Management Options

Typically, sludge treatment and disposal strategy consist of three main stages, as shown in Figure 2.25 (MSIG, Volume 4, 2009) [1].

a. **Stage 1:** Preliminary treatment and digestion
b. **Stage 2**: Conditioning and dewatering
c. **Stage 3**: Utilization and disposal.

2.10.5 Sludge Stabilization and Treatment

It is generally necessary to subject sludge to some treatment to prepare or condition it for ultimate disposal. Such treatment has two objectives – the removal of part or all the water in the sludge to reduce its volume and the decomposition of the putrescible organic solids to mineral solids or relatively stable organic solids. There are three stages in managing the sludge as detailed in MSIG, Volume 4, 2009 [1]:

2.10.5.1 Stage 1: Preliminary Treatment/Thickening and Digestion

Preliminary treatment normally includes a reception or holding facility for screened sludge, primary thickening, and digestion facilities. For sludge brought in from other facilities, the reception facility may comprise an unloading area, screen chamber, reception tank, and transfer pump(s).

The purpose of thickening the sludge is to reduce its volume, which can be several folds. Typical sludge comes at 1% solid fraction (99% water). Thickening it to 2% halves the volume, to 10% reduces the volume by a factor 10 (1% solid fraction ≈ 10,000 mg/L). In Malaysia, thickening equipment, such as centrifuge, drum thickener, or gravity belt thickener, is provided to thicken the raw screened sludge from about 1% dry solids content to about 6% dry solids content (94% water). To assist in the thickening process, an "in-line" polymer dosing system or chemical conditioning shall be provided.

Various thickened percentages have been reported. Gravity thickening and belt thickening with the polymer can achieve up to 10% and 6%, respectively, solid fractions. Dissolved air flotation can achieve up to 6% solid fraction, while centrifuge can achieve up to 8% solid fraction.

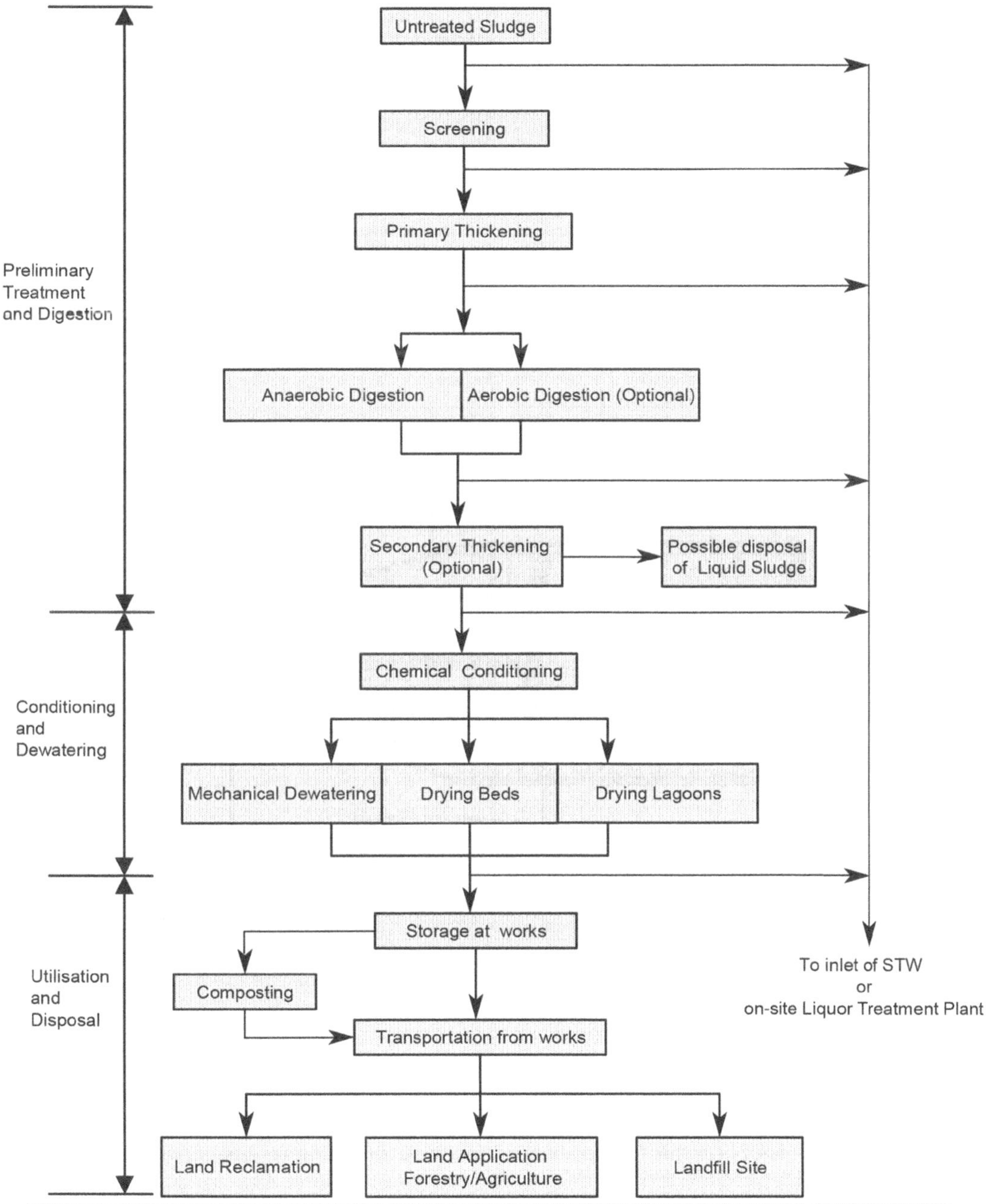

FIGURE 2.25 Sludge management in Malaysia (MSIG, Volume 4, 2009) [1].

Two types of digestion facilities are available for digestion of thickened sludge: aerobic and anaerobic digestion. Secondary thickening is recommended to reduce the volume of digested sludge further.

2.10.5.1.1 Thickening

Thickening is a process used to increase the solids content of sludge by removing water, mainly by physical means, gravity settling, flotation, centrifugation, gravity belt, vacuum filter, and rotary drum. The examples of the vacuum filter and centrifuge are shown in Figures 2.26 and 2.27.

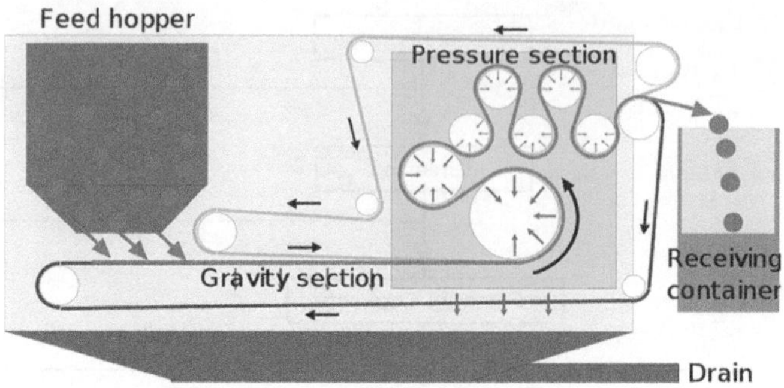

FIGURE 2.26 Vacuum filter, able to get rid of >70%–85% water [10].

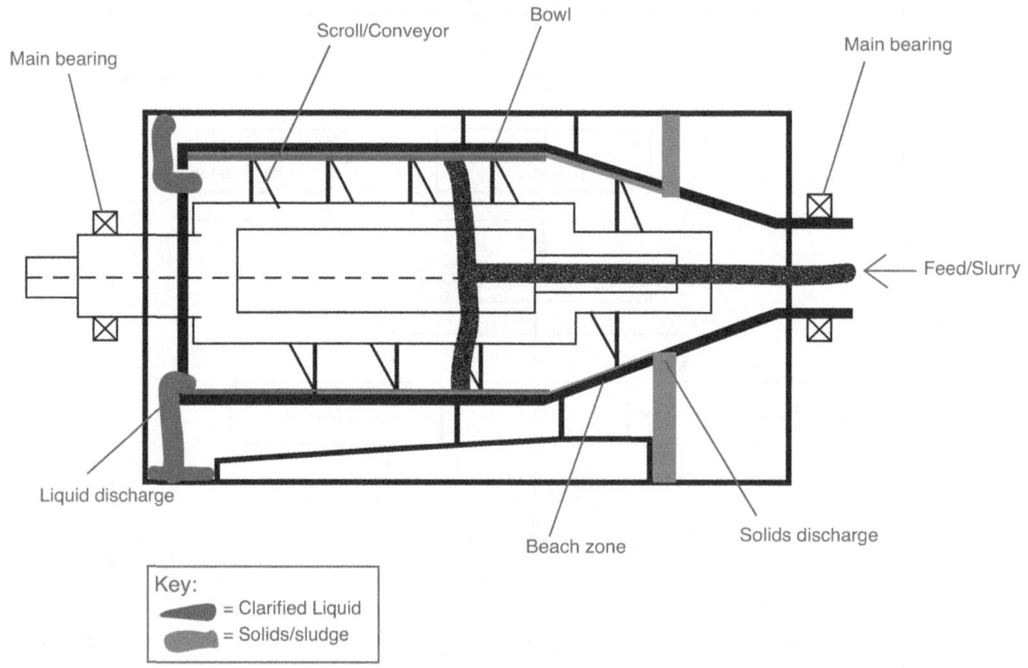

FIGURE 2.27 A typical horizontal centrifuge. Able to produce more than 20% solids (>300 cycles/s) [11].

The design parameters for thickening sludge equipment shall follow Table 2.16.

a. Mechanical thickener shall be designed for 8 hours/day and 5 days/week operation.
b. For the belt, drum, and centrifuge thickener, three polymer injection points shall be provided.
c. Potable water to be provided for a polymer mixing system.

2.10.5.1.2 Anaerobic Digester

Raw sludge from primary and secondary treatments is rich in pathogenic organisms, easily putrescible and rapidly developing unpleasant smells. It must be treated to stabilize either biologically, chemically, or thermally. The most widely used stabilization process is biological stabilization via anaerobic and aerobic digestion. However, the former is the most common. Anaerobic digestion of municipal wastewater sludge has been widely practised since the early 1900s and is the most widely

TABLE 2.16
Design Parameters for Sludge Thickening

Type of Thickening	%Dry Solids	Polymer System	The Speed of Sludge Feed Pump	Backwash Water System
Picket fence gravity thickener	1.5	n/a	n/a	n/a
Dissolved air flotation	2			
Belt thickener	4	Yes with appropriate polymer turndown ratio	<300 rpm	Yes
Drum thickener	4			
Centrifuge	4			

Source: MSIG, Volume 4, 2009 [1].

used biological stabilization of sludge. It is the process for the reduction of sludge and wastewater, which is high in soluble BOD and/or COD without the use of air or elemental oxygen.

In anaerobic digestion process, a large fraction of the organic matter (cells) is broken down into carbon dioxide (CO_2) and methane (CH_4), and this is accomplished in the absence of oxygen. The mechanisms and reactions have been detailed in Chapter 3. About half of the amount is then converted into gases, while the remainder is dried and becomes a residual soil-like material. It is a very low-energy process that produces a fraction of the waste sludge generated through aerobic biological processes.

The treatment of wastewater sludge, from both primary and secondary treatment steps, consists of two main phases.

i. In the first step, all incoming flows of sludge are combined, and the mixture is heated to a mild temperature (about body temperature) to accelerate biological conversion. The residence time here ranges from 10 to 20 days.
ii. In the second tank, the mixture can undergo further digestion. There is no longer active mixing to promote separation, and there is no need for heating as the process generates its own heat.
iii. Next, the settled sludge is dewatered and thickened. The goal is to separate as much water as possible to decrease the volume of material. Finally, a phase known as sludge stabilization reduces the level of pathogens in the residual solids, eliminates offensive odours, and reduces the potential for putrefaction.

Overall, the process converts about 40%–60% of the organic solids to methane (CH_4) and carbon dioxide (CO_2). The chemical composition of the gas is 60%–65% methane, 30%–35% carbon dioxide, plus small quantities of H_2, N_2, H_2S, and H_2O. Of these, methane is the most valuable because it is a hydrocarbon fuel (giving 36.5 MJ/m^3 in combustion). The residual organic matter is chemically stable, nearly odourless, and contains significantly reduced levels of pathogens. The SS are also more easily separated from water relative to the incoming sludge or aerobically treated sludge (such as in an outdoor pond). A two-stage anaerobic digester is shown in Figure 2.28.

In Malaysia, MSIG, Volume 4, 2009 has recommended the design parameter of digesters as given in Table 2.17.

$$\text{Organic Loading Rate (OLR)} = \frac{Q \times \text{VSS}}{V} \text{ kg VSS/m}^3\text{.day} \tag{2.39}$$

FIGURE 2.28 A typical two-stage anaerobic digester [12].

TABLE 2.17
Design Parameters for Aerobic and Anaerobic Digestion

Description	Unit	Design Criteria	
		Aerobic Digestion	Anaerobic Digestion
Number of tanks, minimum	No.		2
Min. solids retention time (*t*)	Days	10	18
Organic loading rate (OLR)[a]	kg VSS/m^3.day	1.6–4.8	0.8–1.6
Typical feed solids concentration	%	2	2–6
Type of mixing		Aerators Diffusers	Gas injection Mechanical stirring Mechanical pumping
Min. water depth, minimum	m	3	7.5
Tank shape		Cylindrical Rectangular	Cylindrical Egg-shaped
Tank dimension, maximum	m	25 diameters 25 length	25 diameters
Dissolved oxygen	mg/L	1–2	—

Source: MSIG, Volume 4, 2009 [1].
[a] Refer Equation 2.40.

Example 2.13

An anaerobic digester receives a sludge sample with DWF 10 m³/day with volatile SS (VSS) of 15,000 mg/L. If the retention time is 18 days, determine the volume and the organic loading rate (OLR) of this digester.

Solution:

$$\text{DWF} = 10\ \text{m}^3/\text{day}$$

$$\text{VSS} = 15{,}000\ \text{mg/L} = 15\ \text{kg/m}^3$$

$$t = 18\ \text{days}$$

(I) DIGESTER VOLUME

$$\text{Retention time } (t) = \frac{V}{\text{DWF}}$$

$$18\ \text{days} = \frac{V}{10\ \text{m}^3/\text{day}}$$

$$V = 180\ \text{m}^3$$

(II) ORGANIC LOADING RATE

From Equation 2.39,

$$\text{Organic Loading Rate (OLR)} = \frac{Q \times \text{VSS}}{V}\ \text{kg VSS/m}^3.\text{day}$$

$$\text{OLR} = \frac{10 \times 15}{180}\ \text{kg VSS/m}^3.\text{day}$$

$$= 0.83\ \text{kg VSS/m}^3.\text{day}$$

2.10.5.2 Stage 2: Conditioning and Dewatering

Dewatering is a process to remove the water, and this can be done by mechanical dewatering and non-mechanical dewatering devices.

a. Mechanical dewatering-belt filter press, centrifuge or filter press. The polymer may be added to optimize the process.
b. Non-mechanical dewatering-drying beds or sludge lagoons.

For a facility serving ≥10,000 PE, the drying beds must be designed to support mechanical/machine lift to clear the dried sludge.

Sludge lagoons of about 2 m depth are also used for sludge stabilization and drying. The sludge lagoons shall be sized to receive sludge for at least 6 months and can undergo stabilization through evaporation and drying for another 6 months period.

Table 2.18 provides Recommended Design Parameters for Sludge Dewatering under MSIG, Volume 4, 2009 [1].

2.10.5.2.1 Drying Bed

Sludge is dewatered using conventional sludge drying beds by evaporating from the surface that is exposed to air and by draining through the sludge pile [13,14]. Although the use of drying beds

TABLE 2.18
Recommended Design Parameters for Sludge Dewatering

		Design Criteria	
Description	**Unit**	**≤2,000 PE**	**>2,000 PE**
Type of device		Belt press centrifuge Filter press Drying bed	Belt press centrifuge Filter press Drying bed[a]
Minimum dry solids (content after dewatering)	%	20	20
The operating period of the mechanical thickening and dewatering facility		5 days/week[b] 8 h/day 250 days/year	5 days/week[b] 8 h/day 250 days/year
The handling capacity of the drying bed		4 weeks cycle on 450 mm thick feed[c]	4 weeks cycle on 450 mm thick feed[c]
Covered storage area		1-month holding	1-month holding

Source: MSIG, Volume 4, 2009 [1].
Note: Access ramp of at least 1.5 m wide shall be provided at all sludge drying beds.
[a] Drying beds must be designed to support mechanical/machine lift for more than 10,000 PE.
[b] Design to be based on one full-time working shift only.
[c] The example for drying bed is given in Figure 2.29.

might be expected in smaller WWTPs and in the warmer sunny regions [15], they are also used in several large facilities.

In computing the area requirements of a sludge drying bed, the designer may assume a maximum 240 mm depth of sludge feed to the bed. The actual quantity of sludge from the upstream unit processes needs to be computed before sizing the bed. The filtration layer is made from gravel and sand at various gradations, as shown in Figure 2.31. Digested sludge is placed on drying beds where the liquid may evaporate or drain into the soil. The dried sludge is a porous humus-like cake that can be used as a fertilizer.

The drying bed may be designed to handle a maximum of 7 days of continuous feed. The next feed to the same bed shall only be after a minimum of 21 days from the last feed. In Malaysia, transparent roofing is necessary to protect the dried sludge cage from the rain.

Example 2.14

A WWTP has produced 1,800 kg/day of dried sludge. Design a suitable drying bed to dewater the sludge if the sludge feed thickness is 240 mm for 7 days continuous feed. Take sludge S.G. as 1.03.

Solution:

$$\text{Thickness } (D) = 0.24 \text{ m}$$

$$\text{Retention time } (t) = 7 \text{ days}$$

$$\text{Dry weight of sludge } (W) = 1{,}800 \text{ kg/day}$$

$$\text{S.G.} = 1.03\text{, hence, density } (\rho) = 1{,}030 \text{ kg/m}^3$$

$$\text{Weight of sludge in } 7 \text{ days} = W \times t$$

$$= 1{,}800 \text{ kg/day}(7 \text{ days})$$

$$= 12{,}600 \text{ kg}$$

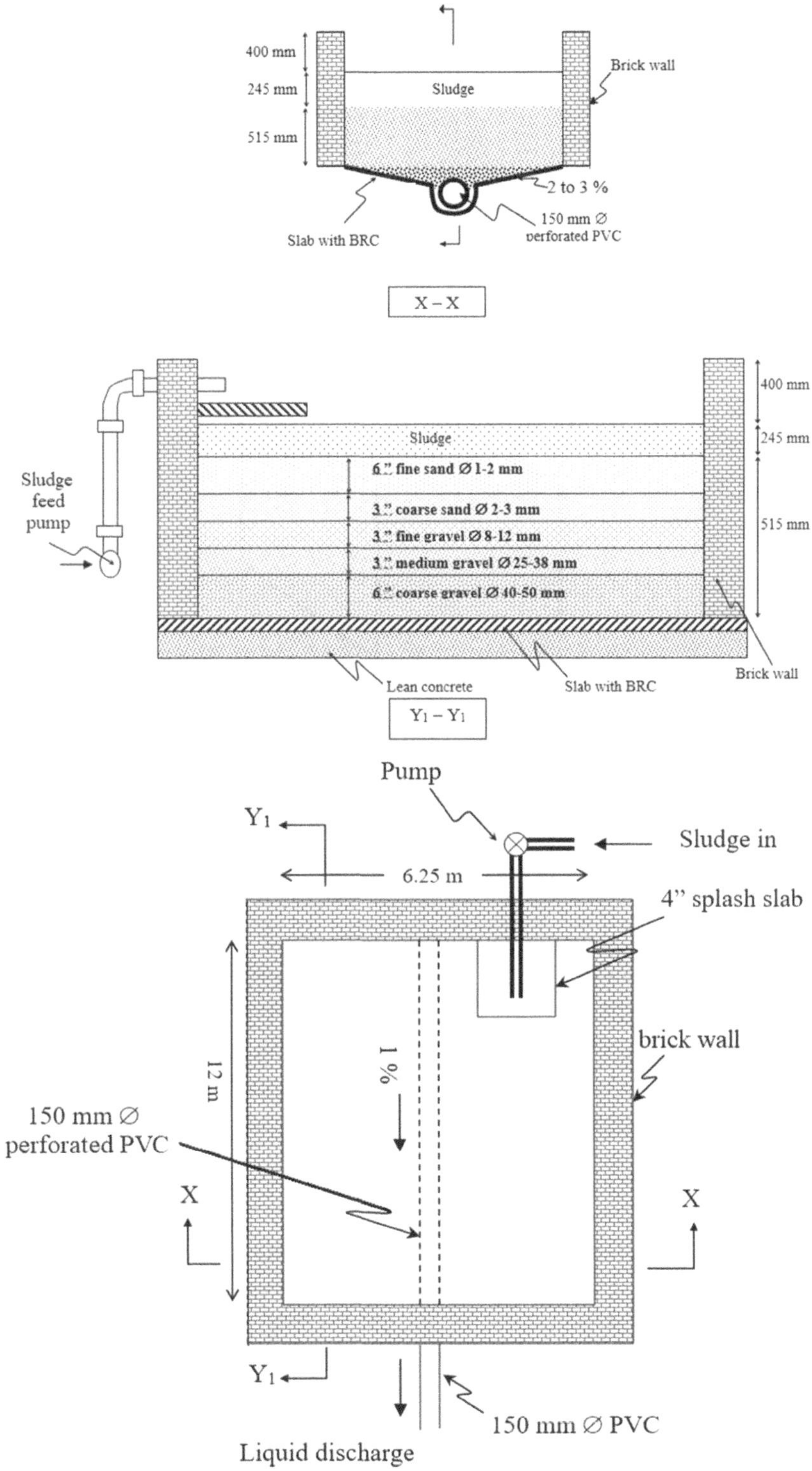

FIGURE 2.29 Drying bed.

Determine the volume of sludge:

$$\text{Density}\ (\rho) = \frac{\text{Weight}}{\text{Volume}}$$

$$1{,}030\ \text{kg/m}^3 = \frac{12{,}600\ \text{kg}}{V}$$

$$V = 12.23\ \text{m}^3$$

Assume, Length (L):Width (W) = 3:1
D = 0.24 m

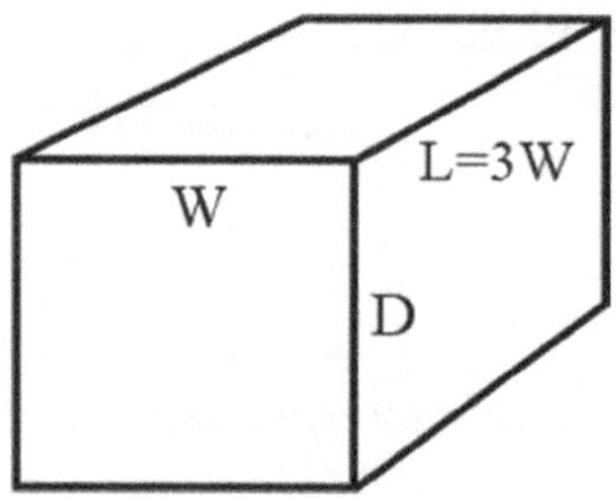

From the above figure,

$$V = 3W^2(D)$$

$$12.23\ \text{m}^3 = 3W^2(0.24)\ \text{m}^2$$

$$W = 4.12\ \text{m}$$

$$L = 4.12 \times 3\ \text{m}$$

$$= 12.36\ \text{m}$$

2.10.5.2.2 Filter Press

This pressing technique filter comprises a set of vertical plates, pressed against each other by hydraulic jacks at one end of the set. The pressure applied to the joint face of each filtering plate must withstand the chamber internal pressure developed by the sludge pumping system. This vertical plate layout forms watertight filtration chambers allowing easy mechanization for the discharge of cakes. Solid sludge gradually accumulates in the filtration chamber until the final compacted cake is formed. The filtrate is collected at the back of the filtration support and carried away by internal ducts (Figure 2.30). A typical filter press is shown in Figure 2.30.

2.10.5.3 Stage 3: Provision of Sludge Holding, Treatment, and Disposal

The final disposal facility of the sludge shall be provided depending on the amount of dry sludge produced. It is quite common to provide a sludge storage tank with a minimum capacity to hold for 30 days with the sludge thickened to 1% solids (MSIG, Volume 4, 2009) [1]. Different types of WWTPs produce different quantities of bio-solid. The principal assumptions adopted on waste generation rates in Malaysia are summarized in Figure 2.15.

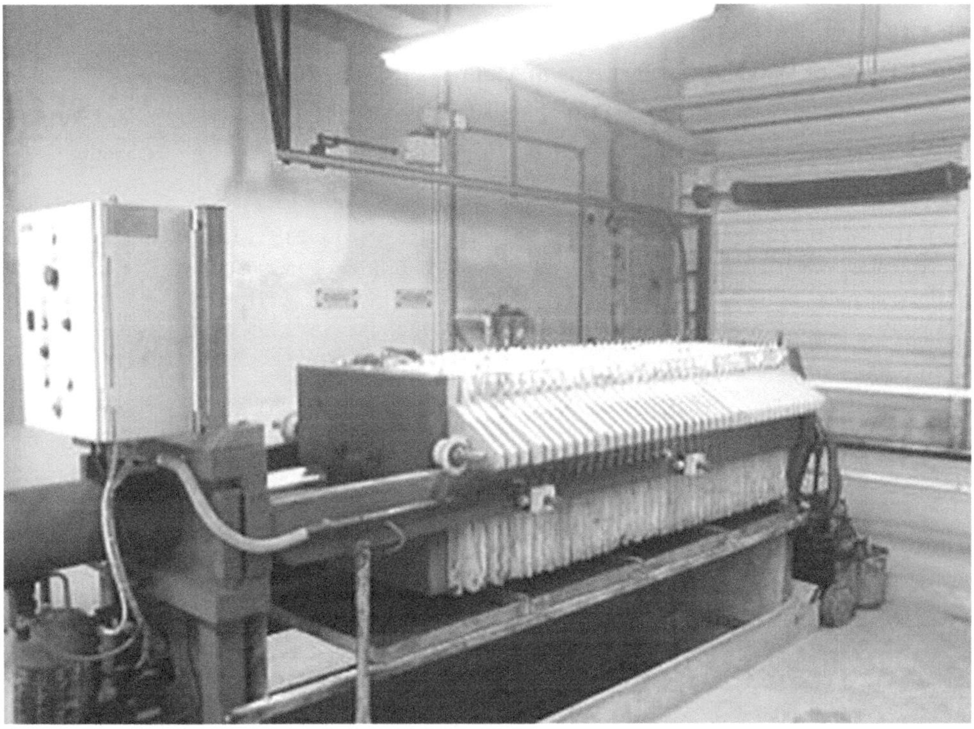

FIGURE 2.30 Filter press [16].

2.11 WASTEWATER TREATMENT PLANT IN MALAYSIA

2.11.1 Common Treatment Systems

In Malaysia, 38% of public WWTPs in the country are mechanical plants, such as EA, OD, rotating biological contactors (RBCs), SBRs and trickling filters.

CSTs and Imhoff tanks, and low-cost secondary systems such as oxidation ponds have been widely used, especially for a small community. However, one of the most critical sources of pollution is septic tanks and pour flush latrines. In 2018, there were over 1.2 million septic tanks in Malaysia, and only 30% were well maintained, i.e. sludge is removed regularly from the tank. Many users of this system neglected their responsibility to ensure that the sludge is removed regularly. Hence, accumulated sludge pollutes the waterways.

According to SPAN's Annual Report 2016 [17], domestic wastewater system facilities in Peninsular Malaysia and FT Labuan consist of public and private WWTPs, CSTs, ISTs, and traditional systems such as pour flush. In urban areas, ISTs, CSTs, and Imhoff tanks will be phased out in stages. The number of facilities and PE for each type of system in 2015–2016 is given in Table 2.19.

2.11.2 Package Units

The term "package units" is used in the field to describe equipment that has been put on the market by several manufacturers that are intended to provide wastewater treatment by the use of prefabricated or modular units. Package units can also refer to complete installation, including both mechanisms and prefabricated containers. This term is also applied to installations where only the mechanisms are purchased and the containers are constructed by the purchaser in accordance with plans and specifications prepared by the manufacturer.

TABLE 2.19
Different Types of Wastewater Systems in Malaysia (SPAN, 2016) [17]

	2014[a]		2015		2016	
Wastewater Systems	**Quantity**	**PE**	**Quantity**	**PE**	**Quantity**	**PE**
		Connected Services				
Public WWTP			6,571	23,517,185	6,752	24,789,450
a. Public centralized WWTP			6,481	16,296,052	6,659	16,964,923
b. Public regional plant			90	7,221,133	93	7,824,527
Private WWTP			3,158	2,795,877	3,338	3,009,095
		Septic Tank & Pour Flush				
Communal septic tank (CST)	3,625	405,432	4,386	532,051	4,386	532,076
Individual septic tank (IST)			1,321,856	6,747,774	1,343,439	6,859,823
Pour flush			894,859	4,474,293	894,859	4,474,293
		Specific Plant				
Imhoff tank	679	507,648				
Oxidation ponds	403	1,681,176				
Mechanical plants	4,902	18,665,408				
		Other				
Network pumping station	982	4,852,844	1,078		1,117	
Length of sewer network (km)			18,689		18,689	

[a] www.iwk.com.my.

Though specific limitations have not been established, individual package units have, in general, been small installations serving a limited population. In Malaysia, commercially available prefabricated treatment plants known as "package plants" are often used to serve small communities up to a PE of 5,000. Care must be taken in using package plants where large variations in flow (hydraulic shock) are experienced; in addition, adequate provision must be made for sludge removal, scum and grease removal, and the proper control of air supply. The most common types of package plants used in Malaysia are EA, contact stabilization, bio-filter, SBRs, and RBC processes.

SPAN has published a technical specification specifically for package units. The Technical Specification on Wastewater Treatment System, SPAN TS 1401:2010 (A1:2013), Part 1: Prefabricated Tanks – Packaged Plants, First Edition, October 2010, Revision March 2013 [18] shall be referred to.

2.11.3 Modern Wastewater Treatment Plants in Malaysia

In this chapter, three modern WWTPs in Malaysia are discussed.

2.11.3.1 Pantai 2 Sewage Treatment Plant

Indah Water Konsortium Sdn Bhd (IWK) has upgraded its Pantai 2 Regional Sewage Treatment Plant (P2STP) to accommodate the needs of 1.423 million people in the surrounding area. P2STP was previously a large aerated lagoon with a capacity that could cover a population of 877,000 [19]. It covers 17 ha of site area and serves as one of the most critical sewage treatment facilities in Kuala Lumpur. It was built under the tenth Malaysia Plan. The RM983 million P2STP was built with a new state-of-the-art mechanized underground regional sewage treatment plant that proved to be a model for innovation in concept, design, engineering, and construction of new standards for sewage treatment in Malaysia and transforming its surrounding environment [20]. The plant is designed for Standard A effluent discharge quality as prescribed in the Malaysia's Environmental Quality (Sewage) Regulations 2009.

The Pantai catchment area covers 6,700 ha in the central and south-western parts of Kuala Lumpur, including Bandar Baru Sentul, Sentul Raya, the Bukit Kiara central business district, Taman Botani, and parts of Petaling Jaya with a 320,000 m^3/day capacity [20] to serve an ultimate PE of 1.8 million population until 2035. With P2STP, about 140 small and inefficient sewage treatment plants that produce substandard effluent can be rationalized, reducing pollution in the Klang River [5,20]. The P2STP is Asia Pacific's biggest STP and Malaysia's first fully underground facility with a 12-ha community centre above the STP containing sports and recreational facilities. Odour scrubber systems are also provided at this plant to treat odorous gases from the sewage and sludge treatment.

The mega sewage treatment plant comprises two main parts: the sludge treatment facility above ground and the fully underground multi-layered sewage treatment facility that goes 17 m deep. At the ground level, there is also a 12-ha public eco park [20], complete with waterways and abundant greenery, as well as recreational facilities, such as a jogging track, playground, courts for futsal, tennis, sepak takraw, volleyball and badminton, community centre, cafeteria, surau, and covered parking.

The design concept adopts a green policy promoting a sustainable living environment via low-energy consumption, low pollution, and low emission. The P2STP was built with sustainability in mind, capitalizing on renewable resources for energy and water to reduce wastage and pollution. It has a Green Building Index (GBI) silver rating. A bio-gas generator converts the methane gases produced during the sludge treatment process to generate up to 700 kW of auxiliary power. Together with solar panels installed atop the parking lots, it supplies 10%–15% of the plant's power needs [20]. Rainwater harvesting produces about 80 m^3 of water a day for toilet flushing and landscaping, while a multi-stage filtration and reverse osmosis membrane system recycles bio-effluent into 2,460 m^3 of treated water daily for plant operations [20]. The aquatic skylight over the underground passageway provides natural lighting for the space below with energy savings of up to 30 kW a day, while the wastewater source heat pump facilitates heat exchange with effluent to generate a cooling load of 1,200 kW [20], which drives the air-conditioning system for the administration building and community centre.

The P2STP unit process for liquid stream consists of pre-treatment, biological treatment (advanced A2O process), clarifier, and wastewater reclamation plant (Figure 2.31). To reduce hydraulic loading during peak flow, excess sewage during peak flow will be stored at the balancing tank and returned back into the system during non-peak flow. It is equipped with an ultrasonic flow sensor for flow monitoring. Rubbish in the incoming raw sewage is removed by using the latest screening technology. Grit, sand, and other solid material are removed in the grit chamber.

This state-of-the-art P2STP uses Advanced Anaerobic-Anoxic-Oxic (A2O) (Figure 2.32) liquid treatment process that uses microorganisms to break down sewage material, which results in better quality effluents, more efficient and effective removal of contaminants such as nitrogen and phosphorus from wastewater, more stable operating system, minimal production of sludge, and a smaller footprint. The A2O process is designed using a smart inflow system and flexible recirculation system to improve the distribution and utilization of carbon sources and removal of carbon, nitrogen, and phosphorus. The aeration process is supported by a total of eight turbo blowers connected to a series of fine bubble diffusers in the aeration tank.

The water phase and sludge are separated in the clarifier. Scum and other floating debris are removed by scrapper. The clarified supernatant overflows through the weir and flows as treated effluent. Sludge that settled at the bottom of the clarifier is pumped back to the A2O tank as return sludge and to WAS buffer tank as waste sludge.

2.11.3.2 Langat Sewage Treatment Plant

Langat Sewerage Project is one of the "Sewerage Non River" projects under Greater Kuala Lumpur (Figure 2.33). The high-impact project involves building a centralized sewage treatment plant with

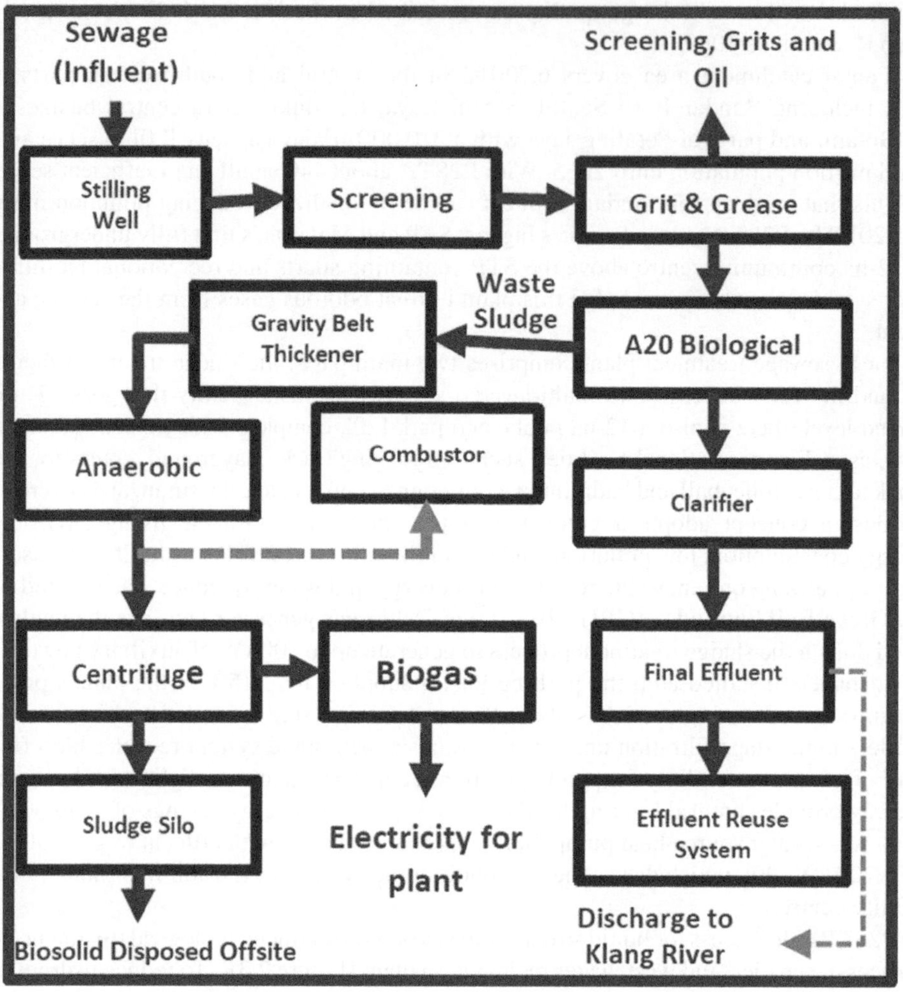

FIGURE 2.31 The treatment process [21].

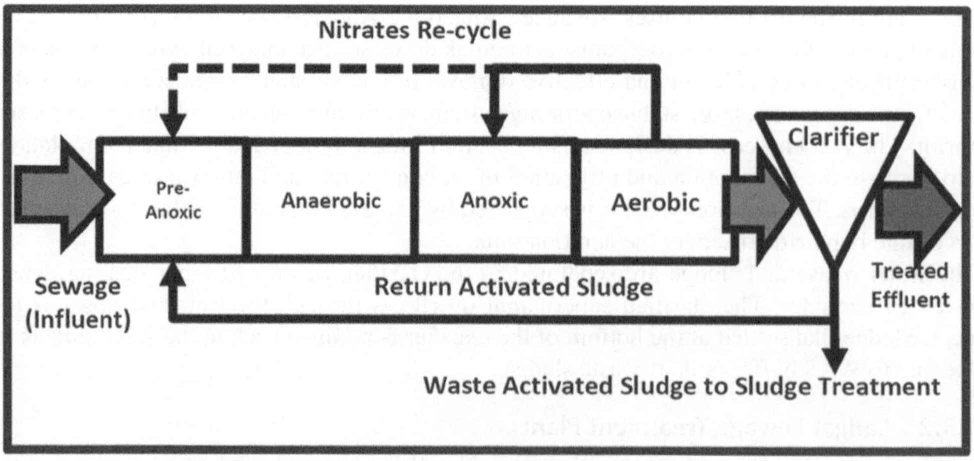

FIGURE 2.32 The A2O biological system [21].

FIGURE 2.33 Langat sewage treatment plant.

920,000 PE, equivalent to 207,000 m^3/day, and laying a sewage piping network of approximately 105 km in total [22,23].

The project covers a catchment area of 60.9 km^2, stretching from Cheras Batu 11 to Kajang (Figure 2.34). The 105 km sewage pipe network connects 1,174 manholes that integrate 164 small, inefficient multipoint sewage treatment plants in various housing areas, which immediately reduces multipoint pollutant discharge into the Langat River [22,23]. It also involves the construction of six new pumping stations and the upgrade of four existing pumping stations that will link directly into one centralized location.

The sewage is treated to Category 1 Standard A, specifically in ammonia removal as stipulated in the Environmental Quality Act (Sewerage) 2009. The plant also functions as an "integrated" space to promote the best use of space and carbon reduction by means of Green Technology-driven initiatives. The wastewater processing is operated underground, utilizing the "deep aeration" method with the Step-Feed Multi-Stage Denitrification Process that will reduce carbon and nitrogen emissions effectively. The Langat Centralized Sewerage Treatment Plant project also provides its communities with a range of public recreation and amenities such as a community hall, landscaped park, and badminton and futsal courts built above the plant.

2.11.3.3 Georgetown Sewage Treatment plant

Georgetown Sewage Treatment plant (previously named The Jelutong sewage treatment plant), is located along Lebuhraya Lim Chong Eu, in Jelutong, Pulau Pinang, Malaysia (Latitude 5.3947126; Longitude 100.3209795) (Figure 2.35). It uses a SBR, the largest type of STP in Malaysia and the second largest type of STP in the world. The plant's ultimate capacity is 1.2 million PE (maximum capacity of 270,000 m^3/day). It mainly caters to wastewater coming from Jelutong, Greater Georgetown, Tanjung Tokong, Air Itam, Gelugor, and Tanjung Bungah.

The process flow diagram of the plant is shown in Figure 2.36.

The inlet works consist of two pump systems (force and gravity mains). The primary screen is equipped for gravity mains, while for force mains, the screening is done at the pump station upstream before the wastewater reaches the treatment plant.

The wastewater then flows to the preliminary treatment, where a secondary screen, grit chamber (Figure 2.37), and oil & grease removal systems (Figure 2.38) are installed in sequence. The wastewater then flows to the SBR system (Figure 2.39) where standard Fill, Aerate, Idle, and Decant sequences are set.

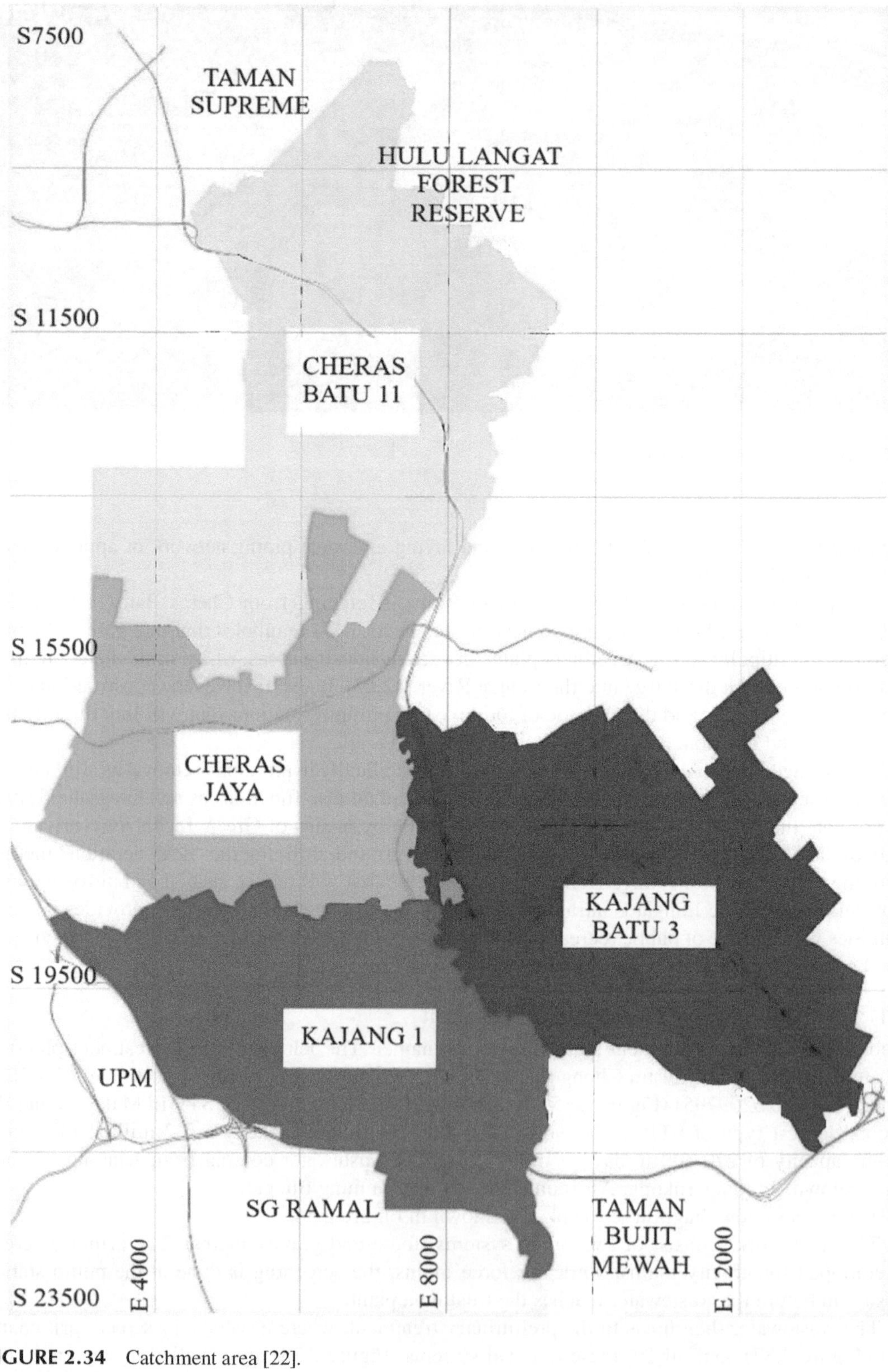

FIGURE 2.34 Catchment area [22].

FIGURE 2.35 Georgetown wastewater treatment plant, Penang, Malaysia [24].

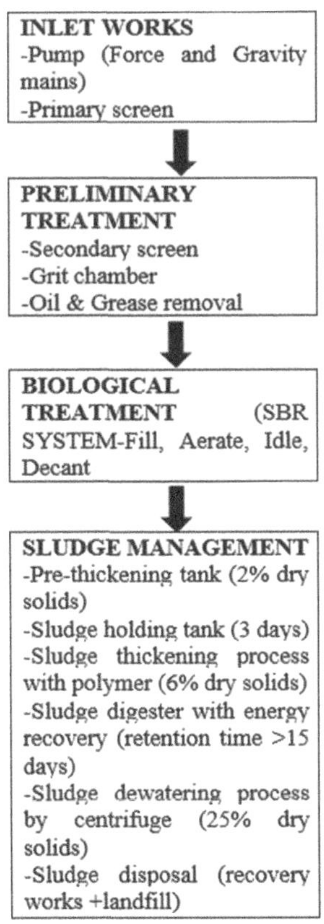

FIGURE 2.36 Process flow diagram of Georgetown sewage treatment plant [24].

FIGURE 2.37 Grit chamber [24].

FIGURE 2.38 Oil and grease removal systems [24].

2.12 WASTEWATER AUTHORITIES

2.12.1 Sewerage Service Department (SSD)

Sewerage Service Department (SSD) was established on March 1, 1994, after the Parliament approved the Sewerage Service Act 1993 (Act 508). The main purpose of SSD establishment at the time was to regulate the services for the existing sewerage systems, which are operated by either the government or private parties throughout the country.

FIGURE 2.39 Aeration in the SBR system [24].

SSD's function as a regulatory agency ended on December 31, 2007, and was taken over by National Water Services Commission (SPAN). Currently, SSD focuses on projects implementation and providing technical advice to the Ministry of Water, Land, and Natural Resources in sewerage-related issues.

2.12.2 National Water Services Commission (SPAN)

National Water Services Commission (Suruhanjaya Perkhidmatan Air Negara, SPAN), which was established in 2008, is a technical and economic regulatory body for the water supply and sewerage services in Peninsular Malaysia and Federal Territories of Putrajaya and Labuan. The Commission regulates all entities in the water supply and sewerage services industry, including public and private water supply and sewerage services operators, water supply and sewerage contractors, permit holders, and suppliers of water and sewerage products. SPAN regulates the water services industry in accordance with the Water Services Industry Act 2006 (Act 655), which was enforced on January 1, 2008. Since inception in 2008, SPAN has been striving to institute improvements in term of standards and performance in the country's water and sewerage services sector.

2.12.3 Indah Water Konsortium (IWK)

Indah Water Konsortium (IWK), a company owned by the Minister of Finance Incorporated, is Malaysia's national sewerage company which has been entrusted with the task of developing and maintaining a modern efficient sewerage system for all Malaysians. A modern and efficient sewerage system is vital for the country to ensure that wastewater is treated before being discharged into our rivers. This will help preserve the country's waste resources, protect public health, and provide a cleaner and safer environment through a proper and well-maintained sewerage system.

In 1994, the Federal Government awarded the company the concession for nationwide sewerage services, which before that was under the responsibility of local authorities. Since then, IWK has taken over the sewerage services from local authorities in all areas except the States of Kelantan, Sabah, Sarawak, Majlis Bandaraya, Johor Bahru, and Majlis Perbandaran Pasir Gudang.

In June 2000, as a testimony of the Government's seriousness in ensuring that a proper and efficient sewerage system will be successfully put in place and maintained, the Government, through the Minister of Finance Incorporated, took over the entire equity in Indah Water from its previous private owners.

2.13 GUIDELINES

2.13.1 Malaysia Standard (MS1228) [4]

In 1991, the Standards & Industrial Research Institute of Malaysia (SIRIM) had produced the MS 1228, Code of Practice for the design and installation of sewerage systems, Section 1 – General. The Code deals with planning, design, construction, installation, and testing of the sewerage system, which includes sewers and sewer appurtenances, wastewater pumping stations, wastewater treatment works, and all the other works necessary to collect, convey, treat, and finally dispose of domestic wastewater and permitted amount of industrial wastewater. The code does not deal with the treatment of industrial effluents and operation and maintenance.

The code is intended to indicate the minimum requirements for the design of sewerage systems and good practices under normal conditions.

The information in MSIG, Volume 4, 2009 [1] shall supersede MS 1228, where similar aspects are covered in these documents or where there is conflicting information between the two documents.

2.13.2 SPAN's Guidelines

SPAN has published five volumes of The Malaysian Sewerage Industry Guidelines up to August 2018.

These are:

Volume 1 Sewerage Policy for New Developments, 2009 [25].
Volume 2 Sewerage Works Procedures, 2nd Edition (revised), July 2013 [26].
Volume 3 Sewer Networks and Pump Stations, 3rd Edition, 2009 [27].
Volume 4 Wastewater Treatment Plants, 3rd Edition, 2009 [1].
Volume 5 Septic Tanks, 3rd Edition, 2009 [5].

In 2016, Upgrading procedures for a public sewerage system for the developer (applicant) had also been published [28].

2.13.3 SPAN's Technical Specifications

Up to August 2018, the Technical Standards and Compliance Division, SPAN, has published the following technical specifications for different applications:

1. Technical Drawing (SPAN DW 0101:2011) – Standard and Typical Drawing for Sewerage System, Part 1: Sewer and Appurtenances – Gravity Sewer, First Edition, February 2012 [29].
2. Technical Specification (SPAN TS 1401:2010 A1:2013) – Wastewater Treatment System, Part 1: Prefabricated Tanks – Packaged Plants, First Edition, October 2010, Revision March 2013 [18].
3. Technical Specification (SPAN TS 1402:2010 A1:2013) – Wastewater Treatment System, Part 2: Construction and Installation, First Edition, October 2010, Revision March 2013 [30].

2.14 SEWERAGE CHARGES

Sewerage rates have been updated by IWK for various locations, as shown in Tables 2.20–2.24, and took effect from January 1, 2023. The tariff charges are in accordance with the Water Services Industry Act 2006: Water Services Industry (Sewerage Services Charges) Regulations 2022 [31].

TABLE 2.20
Domestic Customers (Connected) [32]

Category	Connected Charge Per Month (RM)
Low-cost houses and government quarters in categories F, G, H, and I	4.00
Houses in Kampung, New Villages, and Estates	5.00
Premises and government quarters in categories A, B, C, D, and E	12.00

TABLE 2.21
Industrial Premises[a] [33]

Category	Rate Based on Number of Employees (RM)
Premises receiving individual septic tank service	2.00 per head per month
Premises with connected sewerage services	2.50 per head per month

[a] "Industrial Premises" means any premises in which the principal activity carried out involves the making, altering, blending, ornamenting, finishing, or otherwise treating or adapting any article or substance with a view to its use, sale, transport, delivery, or disposal and includes the assembly of parts and ship repairing but does not include any activity normally associated with retail or wholesale trade.

TABLE 2.22
Government Premises (Monthly Basic Charge) [34]

Sewerage Services	Basic Charge (RM)
Connected	40.00
Septic tank	25.00

TABLE 2.23
Government Premises (Monthly Excess Charge) [34]

Water Usage	Excess Charge (RM)
Up to $100\,m^3$	No charge
More than $100\,m^3$	$0.45/100\,m^3$
More than $200\,m^3$	$0.95/100\,m^3$

TABLE 2.24
Commercial Premises[a] [35]

Band Group	Annual Value (RM)	Basic Charges (RM) Connected	Basic Charges (RM) Septic Tank
1	0–2,000	8.00	7.00
2	2,001–5,000	14.00	8.00
3	5,001–10,000	20.00	14.00
4	10,001–20,000	26.00	19.00
5	20,001–30,000	29.00	21.00
6	30,001–40,000	32.00	23.00
7	40,001–50,000	35.00	25.00
8	50,001–60,000	38.00	27.00
9	60,001–70,000	41.00	29.00
10	70,001–80,000	44.00	31.00
11	80,001–90,000	47.00	33.00
12	90,001–100,000	50.00	35.00
13	100,001–200,000	180.00	120.00
14	200,001–400,000	495.00	330.00
15	400,001–600,000	522.00	348.00
16	600,001–800,000	1,980.00	1,320.00
17	800,001–1,000,000	2,160.00	1,440.00
18	1,000,001–3,000,000	4,320.00	2,880.00
19	3,000,001–5,000,000	8,800.00	5,400.00
20	5,000,001–7,000,000	9,200.00	6,000.00
21	More than 7,000,001	9,600.00	6,600.00
The rate of an excess volume of water usage per month		Water usage up to 100 m^3 – no charge Water usage more than 100 m^3 but less than 200 m^3 – RM 0.30 per m^3 Water usage more than 200 m^3 – RM 0.45 per m^3	

[a] "Commercial premises" means any premises used wholly or partly for trade, business, provision of services, or any other activity, whether for profit or otherwise.

Monthly sewerage service charges for Government premises are a sum of the fixed basic charge, based on the type of sewerage services and excess charge based on average water consumption in excess of 100 m^3.

Monthly sewerage service charges for commercial premises are a sum of the basic charge based on the premise's annual value and the excess charge based on average water consumption in excess of 100 m^3.

NOTE

1 In accordance with Malaysian Sewerage Industry Guidelines, National Water Services Commission (SPAN), and with design examples.

REFERENCES

1. Suruhanjaya Perkhidmatan Air Negara, *Malaysian Sewerage Industry Guidelines (MSIG), Volume 4 Sewage Treatment Plants*, 3rd Edition, Kuala Lumpur, 2009.
2. Suruhanjaya Perkhidmatan Air Negara (SPAN) (2016). *Technical Specification for Standard Sewage Treatment Plant: Part 1-Extended Aeration (150–5000 PE), (SPAN TS 1402:2016)*, Kuala Lumpur.
3. Milledge, J.J., Thompson, E.P., Sauvêtre, A., Schroeder, P., and Harvey, P.J. (2019). Novel developments in biological technologies for wastewater processing. In: *Sustainable Water and Wastewater Processing*, C.M. Galanakis and E. Agrafioti (eds.), pp. 239–278, Elsevier, Amsterdam.
4. Standards & Industrial Research Institute of Malaysia (SIRIM), Malaysian Standard. (1991). *MS 1228, Code of Practice for Design and Installation of Sewerage Systems*, Kuala Lumpur.
5. Suruhanjaya Perkhidmatan Air Negara (SPAN) (2009). *Malaysian Sewerage Industry Guidelines (MSIG), Volume 5 Septic Tanks*, 3rd Edition, Kuala Lumpur.
6. Mikelonis, A., Herrera, A., Adams, E.E., and Hodge, M. (2010). Honduran Imhoff tanks: Potentials and pitfalls. *J. Water Manag. Model.* 18:R236–22, CHI JWMM 2010;R236–22.
7. Nicoll, E.H. (1988). *Small Water Pollution Control Works: Design and Practice*, Ellis Horwood Ltd.
8. Hanum, F., Yuan, L.C., Kamahara, H., Aziz, H.A., Atsuta, Y., Yamada, T., and Daimon, H. (2019). Treatment of sewage sludge using anaerobic digestion in Malaysia: Current state and challenges. *Front. Energy Res.* 7:19.
9. Kumaran, P., Hephzibah, D., Sivasankari, R., Saifuddin, N., and Shamsuddin, A.H. (2016). A review on industrial scale anaerobic digestion system deployment in Malaysia: opportunities and challenges. *Renew. Sustain. Energy Rev.* 56:929–940.
10. Wikipedia. (2024). Belt Filter. Retrieved from https://en.wikipedia.org/wiki/Filter_press (accessed date 14 March 2024).
11. Wikipedia. (2024). Decanter centrifuge. Retrieved from https://en.wikipedia.org/wiki/Decanter_centrifuge (accessed date 14 March 2024).
12. Wikipedia. (2024). Sewage sludge treatment. Retrieved from https://en.wikipedia.org/wiki/Sewage_sludge_treatment (accessed date 14 March 2024).
13. Wang, L.K. and Wang, M.H.S. (2022). Drying beds design. In: *Evolutionary Progress in* Science, *Technology, Engineering, Arts and Mathematics (STEAM),* L.K. Wang and H.P. Tsao (eds.), 4(9C), STEAM-VOL4-NUM9C-SEPT2022, 45 pages, Lenox Institute Press, New York, NY, USA. https://doi.org/10.17613/ykv5-z712
14. Wang, L.K. and Wang, M-H.S. (2024). Sludge drying beds design. In: *Science, Technology, Engineering, Arts and Mathematics*, L.K. Wang and H.-P. Tsao (eds.), Vol. 6, No. 3A, 43 pages, Lenox Institute Press, New York, NY, USA.
15. Wang, L.K. and Wang, M.-H.S. (2024). Sludge drying beds design using international metric units. In: *Science, Technology, Engineering, Arts and Mathematics*, L.K. Wang and H.-P. Tsao (eds.), Vol. 6, No. 3B, 48 pages, Lenox Institute Press, New York, NY, USA. https://doi.org/10.17613/7ydw-1j78
16. Wikipedia. (2024). Filter Press. Retrieved from https://en.wikipedia.org/wiki/Filter_press (accessed date 14 March 2024).
17. Suruhanjaya Perkhidmatan Air Negara (SPAN) (2016). Annual Report, Kuala Lumpur.
18. Suruhanjaya Perkhidmatan Air Negara (SPAN) (2010). *Technical Specification (SPAN TS 1401:2010 A1:2013)-Sewage Treatment System, Part 1: Prefabricated Tanks-Packaged Plants*, 1st Edition, Revised March 2013, Kuala Lumpur.
19. New Straits Times. (2017). Pantai 2 Regional Sewage Treatment Plant upgraded to accommodate 1.423 million residents in KL and PJ. Retrieved from https://www.nst.com.my/news/nation/2017/10/287759/pantai-2-regional-sewage-treatment-plant-upgraded-accommodate-1423 (accessed date 13 March 2024).
20. Construction. (2024). Pantai 2 Sewage Treatment Plant. Retrieved from https://www.constructionplusasia.com/my/pantai-2-sewage-treatment-plant/ (accessed date 13 March 2024).
21. Indah Water Konsortium (IWK). (2024). Pantai 2 Regional Sewage Treatment Plant (RSTP) Kuala Lumpur. Retrieved from https://www.iwk.com.my/cms/upload_files/files/English%20Brochure-Pantai%202.pdf (accessed date 13 March 2024).
22. MMC Pembetungan Langat Sdn. Bhd. (2024). Langat Sewerage Project. Retrieved from https://www.langatsp.com/ (accessed date 13 March 2024).
23. MMC Pembetungan Langat Sdn Bhd. (2024). MMC Pembetungan Langat Sdn Bhd. Retrieved from https://www.mmc.com.my/page53.html (accessed date 13 March 2024).
24. Indah water konsortium promotion video on Jelutong Sewage Treatment Plant, 2023. Unpublished.

25. Suruhanjaya Perkhidmatan Air Negara (SPAN) (2009). *Malaysian Sewerage Industry Guidelines (MSIG), Volume 1 Sewerage Policy for New Developments*, Suruhanjaya Perkhidmatan Air Negara (SPAN), Kuala Lumpur, 2009.
26. Suruhanjaya Perkhidmatan Air Negara (SPAN) (2013). *Malaysian Sewerage Industry Guidelines (MSIG), Volume 2 Sewerage Works Procedures*, 2nd Edition (revised), SPAN, Kuala Lumpur.
27. Tebbutt, T.H.Y. (1997). *Principles of Water Quality Control*, 5th Edition, Butterworth-Heinemann London.
28. Suruhanjaya Perkhidmatan Air Negara (SPAN) (2016). *Upgrading Procedures for a Public Sewerage System for a Developer (Applicant)*, Kuala Lumpur.
29. Suruhanjaya Perkhidmatan Air Negara (SPAN). *Technical Drawing (SPAN DW 0101:2011)-Standard and Typical Drawing for Sewerage System, Part 1: Sewer and Appurtenances-Gravity Sewer,* 1st Edition, Kuala Lumpur.
30. Suruhanjaya Perkhidmatan Air Negara (SPAN) (2010). *Technical Specification (SPAN TS 1402:2010 A1:2013) Sewage Treatment System, Part 2: Construction and Installation*, 1st Edition, Revision March 2013, Kuala Lumpur.
31. Government of Malaysia (2023). *Water Services Industry Act 2006 (Act 655): Water Services Industry (Sewerage Services Charges) Regulations 2022*, MDC Publisher, Kuala Lumpur.
32. Indah Water Konsortium (IWK). (2024). Charges-Connected (Domestic). Retrieved from https://www.iwk.com.my/charges-domestic (accessed date 13 March 2024).
33. Indah Water Konsortium (IWK). (2024). Charges-Connected (Industrial). Retrieved from https://www.iwk.com.my/charges-industrial (accessed date 13 March 2024).
34. Indah Water Konsortium (IWK). (2024). Charges-Connected (Government). Retrieved from https://www.iwk.com.my/charges-government (accessed date 13 March 2024).
35. Indah Water Konsortium (IWK). (2024). Charges-Connected (Commercial). Retrieved from https://www.iwk.com.my/charges-commercial (accessed date 13 March 2024).
36. Wikipedia. (2024). Waste stabilization pond. Retrieved from https://en.wikipedia.org/wiki/Waste_stabilization_pond (accessed date 14 March 2024).

3 Wastewater Engineering Design

Physical-Chemical Treatment

Hamidi Abdul Aziz

LIST OF ABBREVIATIONS

ANN	Ammoniacal nitrogen
ASP	Activated sludge process
BOD	Biochemical oxygen demand
COD	Chemical oxygen demand
Commission	See SPAN
CST	Communal septic tank
DN	Nominal internal diameter
DWF	Dry weather flow
EA	Extended aeration
HRT	Hydraulic retention time
IST	Individual septic tank
IWK	Indah Water Konsortium
NH_3-N	Ammoniacal nitrogen
NO_3-N	Nitrate nitrogen
Q_{ave}	Average flow or DWF
Q_{max}	Maximum flow
Q_{min}	Minimum flow
Q_{peak}	Peak flow
RAS	Return activated sludge
RBC	Rotating biological contactor
SBR	Sequential batch reactor
SLR	Solids loading rate
SOR	Surface overflow rate
SPAN	Suruhanjaya Perkhidmatan Air Negara
SS	Suspended solids, also referred as TSS
SVI	Sludge volume index
TOC	Total organic carbon
TSS	Total suspended solids
WAS	Waste activated sludge
WLR	Weir loading rate
WWTP	Wastewater treatment plant

DOI: 10.1201/9781003532354-3

3.1 INTRODUCTION

This chapter details physical treatment methods as the first step in the multiple processes in wastewater treatment plant (WWTP) design. The design will be based on the Malaysian Sewerage Industry Guidelines, Suruhanjaya Perkhidmatan Air Negara (SPAN), Volume 4: Wastewater Treatment Plant, 3rd Edition, 2009, referred hereafter as MSIG, Volume 4 [1]. However, MSIG does not cover all aspects of the design and construction of WWTP. Where information is not covered in the volume, the designer shall follow the requirements set in MS 1228 [2]. However, the information in MSIG, Volume 4 shall supersede MS 1228, where similar aspects are covered in these documents or where there is conflicting information between the two documents.

The primary purpose of the treatment of wastewater is to prevent the pollution of the receiving waters, and the overall objective is to get the quality of final effluent to be at a good standard (Standard A or B of the Environmental Quality Act, 1974) [3]. To reach this standard, the design of the treatment unit plant must comply with the incoming load. Hence, it is important for the engineers to select a suitable method that can fulfil this requirement. As shown in Figure 3.1, suppose that the incoming wastewater contains biochemical oxygen demand (BOD) and suspended solids (SS) values of 280 and 360 mg/L, respectively. The engineer in charge should design a system that could treat the pollutant to the standard discharge limit of 20 mg/L for BOD and 50 mg/L for SS.

Preliminary and/or primary treatment refers to physical unit operations and is the first stage of the treatment system. Secondary treatment refers to biological and chemical unit processes, while tertiary refers to combinations of various treatment systems to mainly remove nutrients.

Physical methods of wastewater treatment accomplish the removal of mainly settleable or floatable particles by use of naturally occurring forces, such as gravity and floatation, as well as by use of physical barriers. In general, the mechanisms involved in physical treatment do not result in changes in the chemical structure of the target substances. Many techniques have been devised to accomplish this aim for both small and large quantities of wastewater. Physical wastewater treatment processes include the use of screens, comminutors, oil and grease device, grit removal, and clarifiers (sedimentation).

3.2 EXPECTED PERFORMANCES

The performances of physical and biological treatment units differ, depending on the types of wastewater. The latter will be discussed in Chapter 5. Table 3.1 highlights the common characteristics of domestic wastewater, which is tested for the organics content and particle. It has been highlighted in Section 1.5.2 that there are various forms of solids, some are settleable and others are dissolved. The solids are also organics in nature, except the inert such as grit. Some of the BOD is also contributed by the organics in solids. The particles are mainly removed in the physical unit, while the organics are mainly eradicated at the biological stage.

Some notes from Table 3.1:

i. 54/90% = 60% of settleable SS can be removed by settlement in physical treatment units. The remaining 40% of solids (mainly in non-settleable solids or in dissolved solids containing organics) needs to be further treated in the biological process.

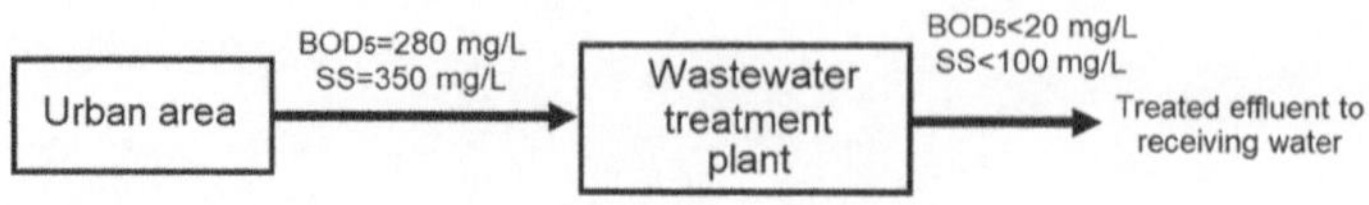

FIGURE 3.1 Expected removal performance of a treatment plant.

TABLE 3.1
Removal Performances of Organics and Solids at the Treatment Plant

	Organics and Inorganics (mg/L)			
Parameter	**Organics in Solids (1)**	**Inorganics (2)**	**Total Organics (3)**	**BOD_5**
Suspended solids, SS (settleable)	39	15	54	19
Suspended solids, SS (non-settleable)	26	10	36	23
Total suspended solids (total as settleable and non-settleable)	Total		90	42
Solids (dissolved)	80	80	160	12

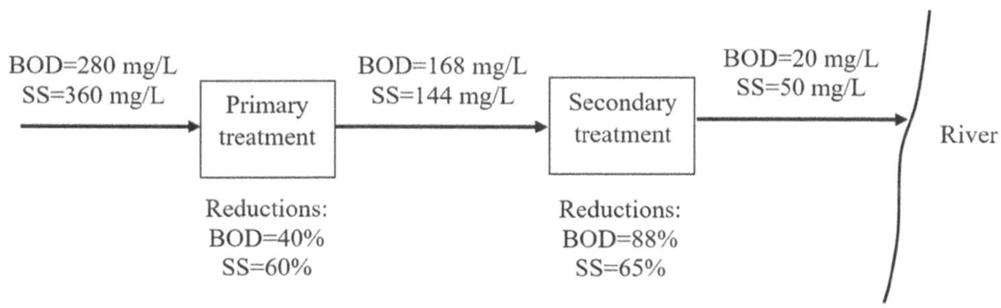

FIGURE 3.2 Expected efficiencies of physical and biological treatment units.

ii. 19/42% = 45% of BOD_5 can be removed through settlement in physical treatment units. The remaining 55% BOD_5 (mainly in dissolved organics form) needs to be further treated in the biological process. In other words, physical systems will not be adequate to remove organics from the wastewater even though the supernatant may look clear. The biological treatment processes will be discussed in Chapter 5.

Consider a case in Figure 3.2 where the influent's BOD and SS are 280 and 360 mg/L, respectively, in comparison to the standard discharge limit of 20 and 50 mg/L, respectively. If we assume the performance of the physical treatment unit is 40% for BOD reduction (normal range 25%–40%) and 60% for SS (normal range 50%–70%), then the biological plant should perform as follows:

For BOD:

$$\left(\frac{168-20}{168}\right)\times 100\% = 88.1\%$$

For SS:

$$\left(\frac{144-50}{144}\right)\times 100\% = 65.3\%$$

So, appropriate treatment units that are capable of meeting the above efficiencies need to be provided.

3.3 PHYSICAL DESIGN REQUIREMENTS

3.3.1 Buffer Zone

Suitable buffer distances should separate a WWTP from its surrounding areas. Recent buffer zone requirements recommended by the Department of Environment (DoE) and by other relevant planning authorities also should be referred. The MSIG, Volume 4, 2009 [1] has recommended the plant siting to comply with an appropriate buffer zone, as shown in Figure 3.3.

a. A minimum distance of 30 m from the fence of the treatment plant to the nearest habitable building property line within the residential and commercial development.
b. A minimum distance of 20 m from the fence of the treatment plant to the nearest property line within the industrial development.
c. A minimum distance of 10 m from the fence of the treatment plant to the nearest habitable building property line if the proposed treatment plant is fully enclosed.
d. A minimum distance of 10 m from the fence of the treatment plant to the nearest habitable building property line if the proposed treatment plant is covered or buried. However, this reduction in buffer requirement does not apply if the nearby habitable buildings are of high-rise type.
e. Plants with population equivalent (PE) less than 150 but are provided with proper odour and noise mitigation measures may have a 10 m reduced buffer at the discretion of the Commission.

The buffer zone can be used for any purpose except permanent habitable buildings. For example, the buffer zone may be used as a drainage reserve, road or highway reserve, transmission reserve, utility reserve, or public park.

In the case where buffer area is to be regularly used by the residents, such as car parks and playgrounds, proper precautions during the design stage must be taken to minimize nuisances such as odour, noise, and an unpleasant sight to the surrounding environment. Adequate and proper screening, odour containment, and treatment facilities must be provided at the WWTP to address these issues.

3.3.2 Siting Criteria

Some of the important siting criteria as detailed in MSIG, Volume 4, 2009 [1] shall be observed. The plant:

a. Should not cause a nuisance to the surrounding, hence, shall be located as far as possible from the habitable building.
b. Due to gravitational force, it shall be located at the lowest point of a sewerage catchment basin.
c. Shall be located near to a suitable receiving water that can receive and assimilate treated effluent from the plant without reducing beneficial uses of the watercourse downstream.
d. Shall be located on an area that is relatively flat or with a relatively mild slope for efficient hydraulics.
e. Shall not be in an area that will result in long-term operational problems or rapid deterioration of the assets.
f. Shall have proper access road leading to it.
g. Shall be located such that sewers are easily connected/conveyed to the proposed site.
h. For safety reasons, it shall be located away from children's playgrounds.
i. Shall be sited away from the following:

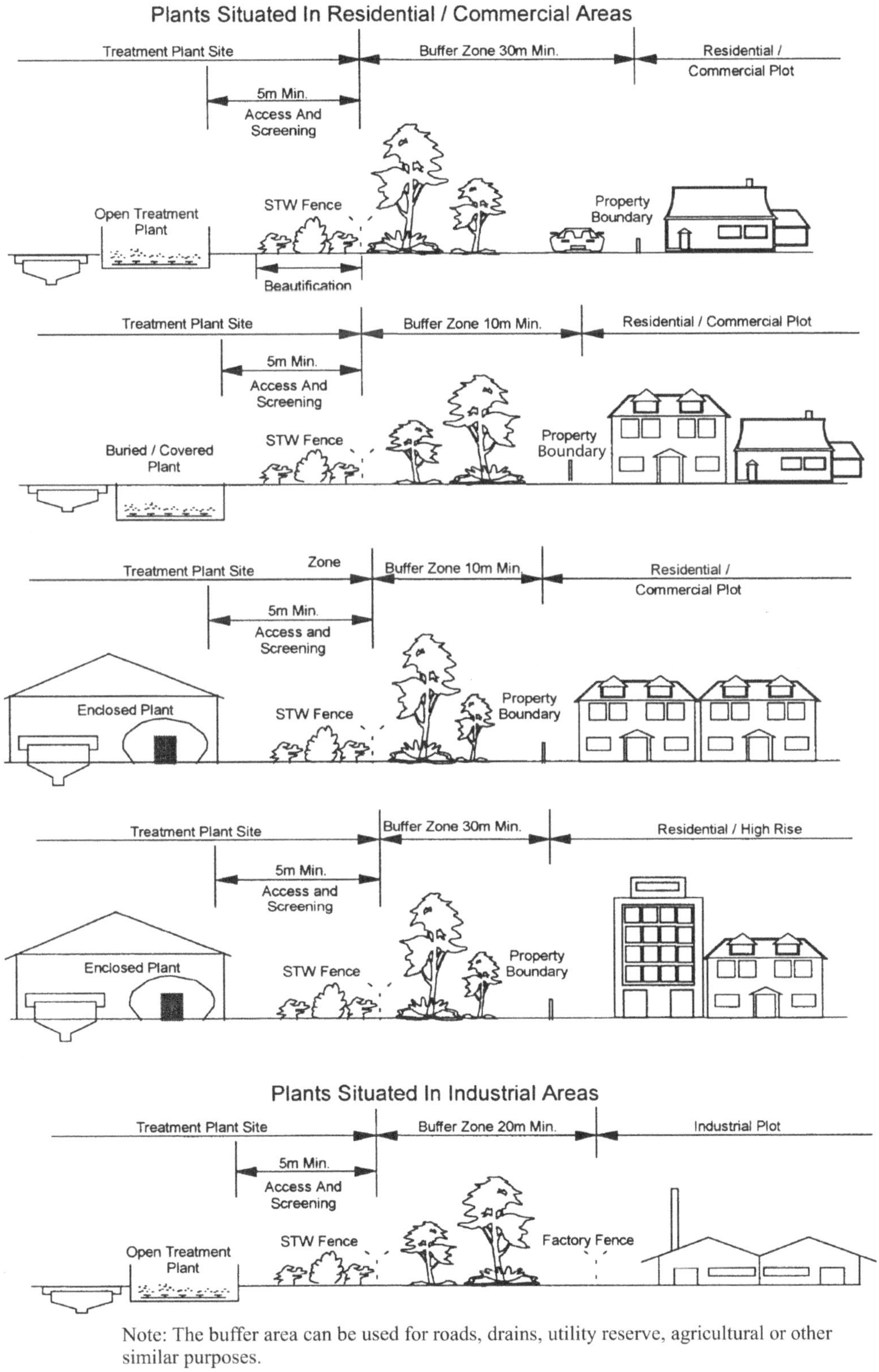

FIGURE 3.3 Buffer zone requirement (MSIG, Volume 4, 2009) [1].

i. Existing cemeteries and gazetted reserves for the cemetery.
ii. Religious centres.
iii. Eating places.

Emergency bypass shall be provided either at the last manhole or wet-well. The bypass shall discharge to the nearest drain, which shall have enough capacity to cater for the discharge during rainfall.

3.3.3 Environmental Impact Assessment (EIA) Requirement

An environmental appraisal or environmental impact assessment (EIA) study shall follow the Environmental Quality (Prescribed Activities) (Environmental Impact Assessment) order, 1987 under Section 34A of the Environmental Quality Act, 1974 [3] (the EIA order, 1987 and the EQA, 1974, respectively). The type and scope of the study will depend on the classification of the plant and the environmental sensitivity of the locality being considered. EIA shall be conducted in compliance with Volume 2 of the MSIG, 2009. EIA is required for PE>20,000.

3.3.4 Basic Design Parameters and Effluent Standard

The MSIG, Volume 4, 2009 [1], has set the design influent and effluent values in compliance with values in Tables 2.3 and 3.2, respectively.

Numbers in Bold refer to the Environmental Quality Act 1974 [3], under the Environmental Quality (Wastewater and Industrial Effluents) Regulations, 1979, regulations 8 (1), 8 (2) and 8 (3).

In cases where treatment plant discharge capacity is higher than the receiving river flow rates, the final effluent quality must be designed for minimal environmental impact.

TABLE 3.2
Absolute and Design Effluent Values in Accordance with Environmental Quality Act 1974 [3] under Environmental Quality (Wastewater) Regulations 2009, Second Schedule, Regulation 7 (for New WWTP)

	Effluent Discharge to Rivers/Stream				Effluent Discharge to Stagnant Water Bodies Such as Lakes, Ponds, and Slow-Moving Watercourses Where Dead Zone Occurs			
	Standard A (mg/L)		Standard B (mg/L)		Standard A (mg/L)		Standard B (mg/L)	
Parameter	Absolute	Design	Absolute	Design	Absolute	Design	Absolute	Design
BOD_5	20	10	50	20	**20**	**10**	**50**	**20**
SS	50	20	100	40	**50**	**20**	**100**	**40**
COD	120	60	200	100	**120**	**60**	**200**	**100**
NH_3-N	10	5	20	10	5	2	5	2
NO_3-N	20	10	50	20	10	5	10	5
Total phosphorus	5		10		**5**	**2**	**10**	**5**
Oil and grease	5	2	10	5	**5**	**2**	**10**	**5**

TABLE 3.3
Classification by Treatment Plant Capacity (MSIG, Volume 4, 2009) [1]

Classification	Population Equivalent (PE)
Class 1	Up to 1,000
Class 2	1,001–5,000
Class 3	5,001–20,000
Class 4	Above 20,000

TABLE 3.4
Modularization Requirements (MSIG, Volume 4, 2009) [1]

STP Classifications	No. of Modules	No. of Trains[a]
Class 1 (<1,000 PE)	1	n/a
Class 2 (1,001–5,000 PE)	1	Max 2
Class 3 (5,001–20,000 PE)	Min. 2, Max. 3	Max 2 for each module
Class 4 (>20,000 PE)	Min. 4, Max. 10	Max 2 for each module

[a] Trains refer to the number of lines or the number of series or backup unit.

3.4 SIZING OF TREATMENT UNIT

3.4.1 Classification by Treatment Plant Capacity

MSIG, Volume 4, 2009 [1] stipulates that WWTP is classified by the design capacity in terms of PE, as shown in Table 3.3.

3.4.2 Modular, Trains, and Backup System

MSIG, Volume 4, 2009 [1] also requires the need to follow modular and standby/backup requirements. Stage development of a WWTP is governed by the timeframe of the overall development plan of the catchments and the size, shape, and soil condition of the land reserved for the WWTP. Modular units (a complete system according to phases) will be constructed to cater to the stage development until the whole development project is completed. In determining the appropriate number of modules and corresponding timing for a staged development, it is crucial for the designer to estimate the flow capacity build-up (ultimate PE) over the entire development phase. The modules must have enough capacity to treat the wastewater. Table 3.4 indicates the modularization requirements in accordance with WWTP classes to attain an efficient modularization of WWTP development.

3.4.3 Backup Units

To avoid significant downtime in WWTP treatment and overloading of the process units, standby units shall be provided for the critical processes. The backup capacity provided shall be such that when the one-unit process is taken out of operation, the remaining units shall not be overloaded beyond 50% of their rated capacities.

The backup facilities shall be provided:

a. Inlet Works/Pumps
b. Screen Facilities
c. Grit Chambers
d. Biological Treatment
e. Secondary Clarifiers
f. Sludge Facilities
g. Pumping Facilities and Blowers

3.4.4 Design Flow

It is recommended that all treatment units from sewer networks to an equalization tank and the secondary sedimentation tank/clarifier shall be designed based on peak flows. Other units that mainly involve biological systems are designed at average flow or dry weather flow (DWF). The calculations for an average and peak flow have been detailed in Section 2.11.4.

3.5 LAND AREA REQUIREMENT

3.5.1 Standard Land Requirements

The recommended land area requirements for various WWTP capacities are derived from relevant treatment process concepts and taken into consideration in other design criteria. It varies from one system to another. The MSIG, Volume 4, 2009 [1] has detailed the requirements for each type of treatment system.

The example of land area requirement for Class 4 (20,001–450,000 PE) is given in Table 3.5. This is followed by Table 3.6, which provides the land area requirements for stabilization pond and aerated lagoon treatment systems (2,000–200,000 PE). Please refer to MSIG, Volume 4, 2009 [1], of land requirements for other treatment units.

3.5.2 Land Area Approval and Structure Plan

The flow diagram for the approval of planning layout and requirements for structural plans as detailed in MSIG, Volume 4, 2009 [1], are given in Figures 3.4 and 3.5, respectively.

3.6 DESIGN STAGES

MSIG, Volume 4, 2009 [1], lists some of the important steps in the design process of WWTP. The design comprises the following stages, namely process design, functional design, and detailed design.

3.6.1 Process Design

Process design involves the selection of a suitable sequence of processes to meet stipulated final effluent requirements for the plant concerned.

3.6.2 Functional Design

In the functional design, calculation of capacities required is conducted for all major units, channels, pumps and pipework, and the definition of control requirements. These include designs for hydraulic, organic, and solid loadings.

TABLE 3.5
Land Area Requirements for Mechanized Class 4 Plants (MSIG, Volume 4, 2009) [1]

	Land Area Requirement (Excluding Buffer Zone)	
Population Equivalent	**(ha)**	**(Acre)**
20,001	1.19	2.95
25,000	1.37	3.38
30,000	1.53	3.79
35,000	1.81	4.48
40,000	1.97	4.88
45,000	2.12	5.25
50,000	2.23	5.52
55,000	2.37	5.84
60,000	2.52	6.22
65,000	2.67	6.61
70,000	2.93	7.23
75,000	3.27	8.07
80,000	3.49	8.61
85,000	3.69	9.12
90,000	3.89	9.61
95,000	4.07	10.06
100,000	4.25	10.49
110,000	4.57	11.29
120,000	4.87	12.02
130,000	5.14	12.70
140,000	5.39	13.32
150,000	5.63	13.90
160,000	5.84	14.44
170,000	6.05	14.95
180,000	6.25	15.43
190,000	6.43	15.89
200,000	6.60	16.32
250,000	7.36	18.20
300,000	7.98	19.73
450,000	9.36	23.14

3.6.3 Detailed Design

In the detailed design stage, structural design of units and channels; detailing of pipelines, fittings, and control valves; and selection of mechanical, electrical, and control equipment are conducted. Structural requirements and constructional characteristics will also be prepared at this stage.

The following data are also considered in the detailed design:

3.6.3.1 Biochemical and Physical Characteristics

Biochemical and physical characteristics are chemical characteristics of wastewater, availability of suitable nutrients, the effect of filamentous growth and sludge bulking, the effect of dissolved oxygen, minimum and maximum residence times, sludge production and frequency of desludging, an effective material balance analysis, floc formation and settlement, length to width ratio, weir loadings, and overflow rate.

TABLE 3.6
Land Area Requirements for Stabilization Pond and Aerated Lagoons (MSIG, Volume 4, 2009) [1]

Population Equivalent	Standard A		Standard B	
	(ha)	(Acre)	(ha)	(Acre)
2,000	0.48	1.18	0.45	1.10
3,000	0.69	1.69	0.59	1.45
4,000	0.89	2.20	0.71	1.75
5,000	1.09	2.68	0.82	2.04
10,000	2.03	5.01	1.31	3.24
15,000	2.92	7.2	1.72	4.25
20,000	3.78	9.3	2.09	5.16
25,000	4.62	11.4	2.42	5.99
30,000	5.45	13.5	2.74	6.77
35,000	6.26	15.5	3.04	7.50
40,000	7.05	17.4	3.32	8.2
45,000	7.85	19.4	3.59	8.9
50,000	8.63	21.3	3.86	9.5
55,000	9.40	23.2	4.11	10.2
60,000	10.16	25.1	4.36	10.8
65,000	10.92	27.0	4.60	11.4
70,000	11.68	28.9	4.83	11.9
75,000	12.42	30.7	5.06	12.5
80,000	13.17	32.5	5.28	13.1
85,000	13.91	34.4	5.50	13.6
90,000	14.64	36.2	5.72	14.1
95,000	15.37	30.0	5.93	14.6
100,000	16.10	39.8	6.13	15.2
110,000	17.54	43.3	6.54	16.2
120,000	18.97	46.9	6.93	17.1
130,000	20.38	50.4	7.31	18.1
140,000	21.79	53.8	7.69	19
150,000	23.18	57.3	8.05	19.9
160,000	24.57	60.7	8.4	20.8
170,000	25.95	64.1	8.75	21.6
180,000	27.32	67.5	9.09	22.5
190,000	28.68	70.9	9.43	23.3
200,000	30.04	74.2	9.76	24.1

3.6.3.2 Construction Characteristics

These involve the examination of ground conditions and soil characteristics, land availability, distance to nearest habitation, delivery and construction time, recommended maintenance requirements, testing and commissioning procedures.

3.6.3.3 Structural Requirements

The requirements include but are not limited to:

a. The wall, slab, beams, columns, and structure for WWTP shall be in reinforced concrete.
b. The wall shall have a minimum thickness of 225 mm.

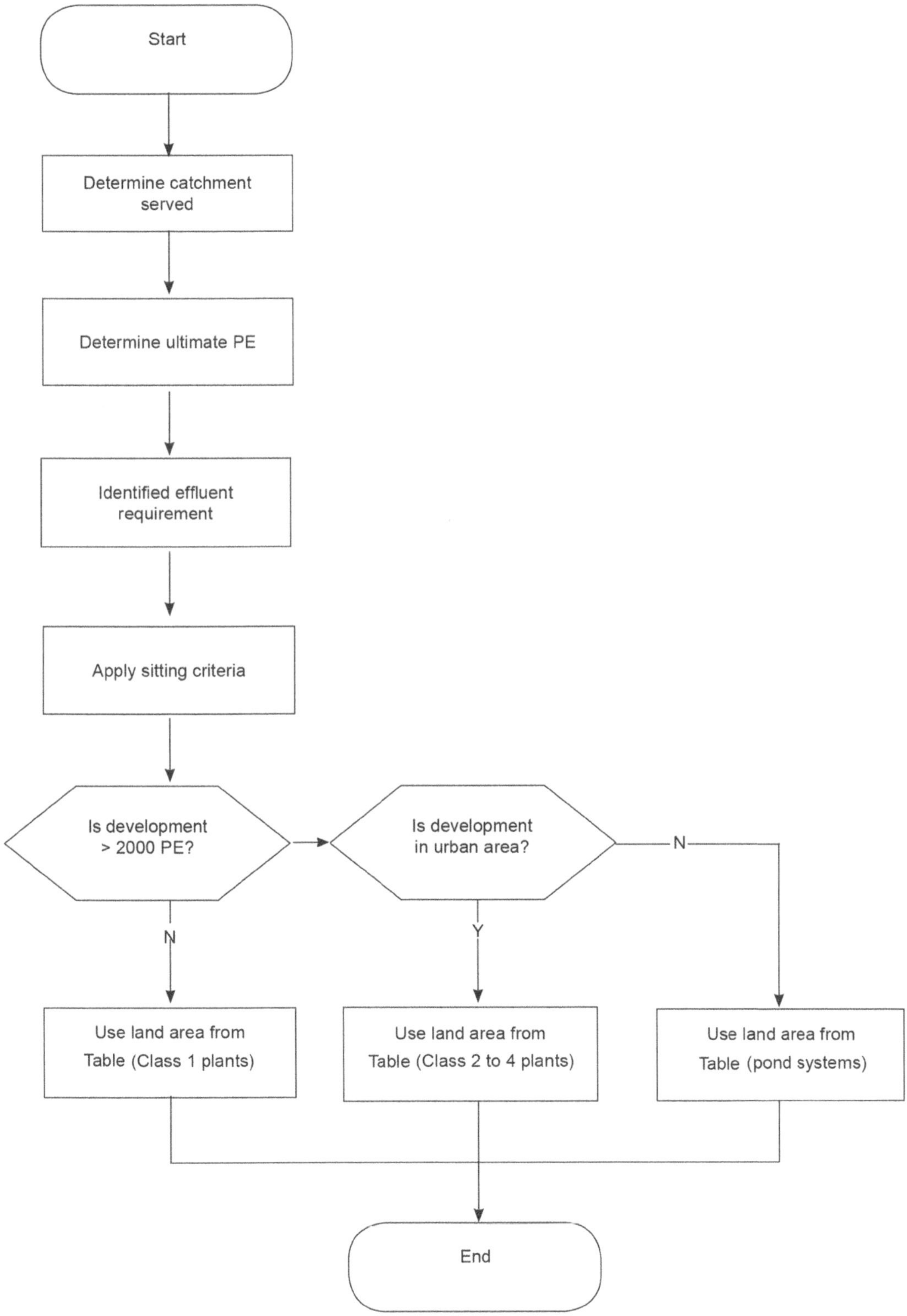

FIGURE 3.4 WWTP land area requirements for planning layout approval for new development (MSIG, Volume 4, 2009) [1].

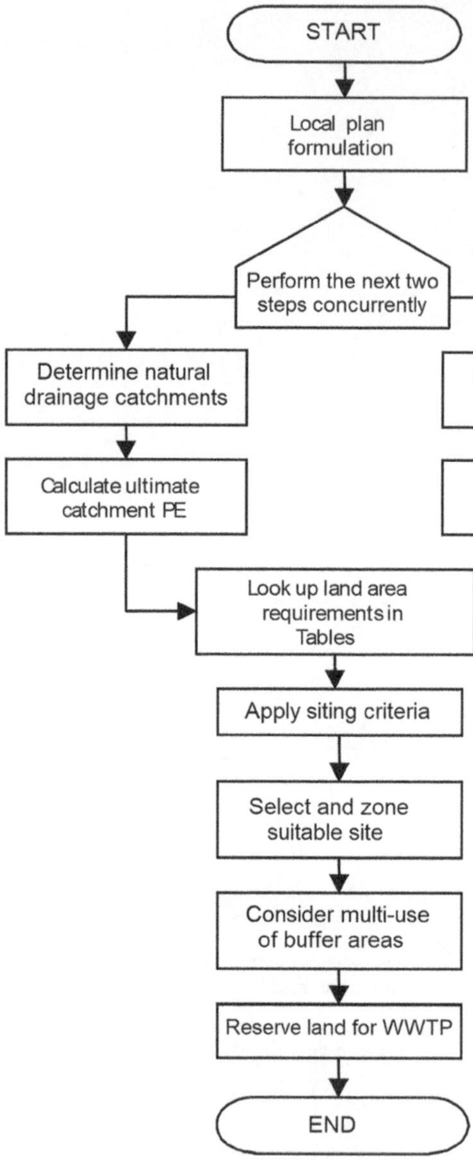

FIGURE 3.5 WWTP land area requirements for structure plans (MSIG, Volume 4, 2009) [1].

c. A special foundation shall be provided where necessary.
d. Proper jointing to prevent breakage and leakage.
e. Water retaining and slope protection where applicable.

3.6.3.4 Mechanical and Electrical Requirements

The requirements include but are not limited to:

a. The design shall simplify the equipment required, control system, maintenance and operational procedures while fulfilling the intended performance and standard of service.
b. Equipment selected shall be from manufacturers (and models) approved by SPAN.

c. Equipment, cable, and cabling design and installation shall follow TNB requirements.
d. Equipment shall be equipped with safety protection (i.e. emergency stop button, warning signage, etc.). See Section 3.5.
e. The appropriate type of control system is provided for the treatment plant.
f. Construction materials are to be protected against corrosion due to high humidity.
g. Earthing and protection against lightning.
h. System manuals, plant function diagrams, electrical system, electrical circuit, and instrument loop diagrams shall be provided before the plant is pre-commissioned.
i. Detailed and shop drawing for equipment, instrumentation, cable, and cabling shall be provided.

3.7 TYPES OF PHYSICAL TREATMENT UNITS

Various methods of wastewater treatment systems have been developed in recent years to meet the need to protect public health and the environment. Sophisticated treatment systems have evolved, which produce a high-quality effluent.

In general, wastewater treatment processes are divided into three stages: preliminary (physical), primary (physical) treatment, and secondary (biological) treatment. More advanced processes (advanced or tertiary treatment) may be required for special wastes, especially when the effluent from secondary treatment is unacceptable.

The primary treatment plays an important role to remove most of the solids and some organics before biological processes. In this treatment, most of the settleable solids are separated or removed. The biological activity of the wastewater in primary treatment is of negligible importance. The purpose of primary treatment is to reduce the velocity of the wastewater sufficiently to permit solids to settle and the material floatable to the surface.

Physical treatment units consist of preliminary and primary treatments. Preliminary treatment is defined as the removal or cutting up of heavy inorganic solids and wastewater constituents that may cause maintenance or operational problems with the treatment operations, such as larger suspended and floating solids in the form of debris, rags, and grit. It also removes excessive amounts of oils or greases. At most plants, preliminary treatment or pre-treatment is used to protect pumping equipment and facilitate subsequent treatment processes. The following devices are commonly used as primary and secondary screens, comminutor, grit chamber, oil and grease removal, and equalization tank. Primary sedimentation tank and floatation constitute the main primary treatment system.

MSIG, Volume 4, 2009 [1] details a few factors that need to be considered in selecting appropriate treatment systems. Among the important criterion include but are not limited to the process, flow pattern and variations, characteristics on incoming wastewater, past performances and reliability of a long-term record, chemical and energy requirements, sludge handling and treatment, regulatory requirement, odour and noise control, land space availability, aesthetics, safety factor, ease of operation and maintenance, the ability of process units to be expanded in tandem with flow increases (modulation), and the overall cost. Some of the common physical, chemical, and biological treatment units are shown in Table 3.7.

A typical master flow diagram of WWTP according to the category of treatment and functions of each unit is shown in Figure 3.6, with connection details in Figure 3.7.

3.7.1 Inlet Chamber

Provision for inlet chamber before the primary screen channel is necessary for proper operation and maintenance of WWTP. The requirements for the inlet chamber under the MSIG, Volume 4, 2009 [1] are summarized in Table 3.8.

Typical details of double penstock are shown in Figure 3.8.

TABLE 3.7
Different Types of Wastewater Treatment Units

Treatment Systems	Pre-treatment	Primary	Secondary Treatment	Tertiary
Purpose	Removal of rags, rubbish, grit, oil, grease	Removal of settleable and floatable materials	Biological treatment to remove organic and suspended solids	Biological and chemical treatment to remove nutrients and pathogens
Examples	Screening	Sedimentation	Activated sludge	Filtration
	Grit removal	Floatation	Rotating biological contactor (RBC)	Disinfection
	Grease tank		Aerated lagoon	Nitrification-denitrification
	Pre-aeration		Trickling filter	Chemical treatments
	Flow measurement		Biofiltration	Adsorption
	Flow equalization		Secondary sedimentation/ clarifier	

3.7.2 Primary Screen

3.7.2.1 Principles

The screen is designed to remove rags, rubbish, grit, oil, grease, and large debris that may have entered the treatment plant and should be removed to avoid problems for the subsequent treatment units. It consists of a grating of steel bars spaced about 2–4 cm on its centre and placed at an angle to the flow of wastewater through an open channel. There are two types of screen, manually raked (Figure 3.9) and mechanically raked (Figure 3.10). For automatic type, skimmers rotate around the surface of the basin to remove oils that may have been flushed into the system. The screen removes coarse and floating solids from the wastewater. The screen must be cleaned regularly, and the removed solids must be disposed of safely. In the design, the velocity through the screen is controlled so that it is not too fast to flush the screenings or too slow, which allows the debris to settle at the screen's face and block the flow.

Provisions and design parameters of the primary screen according to MSIG Volume 4, 2009 [1] are given in Tables 3.9 and 3.10, respectively.

3.7.2.2 Screen Chamber

According to MSIG, Volume 4, 2009 [1], screen chambers must be of open channel construction with proper ventilation to consider the necessary health and safety aspects. The chamber must also be an adequate gradient to prevent the settlement of solids in the chamber. Figure 3.11 shows the typical arrangement of screen chambers at depth over 5 m.

Provision of Primary Screens and the design requirement by MSIG, Volume 4, 2009 [1] are given in Tables 3.9 and 3.10.

3.7.2.3 Design of Primary Screen

There are many approaches to the design of the screen. The simplest is given in Equation 3.1, with a layout shown in Figure 3.13.

$$W = \frac{(B+S)}{S} \times \frac{F}{V(D)} \tag{3.1}$$

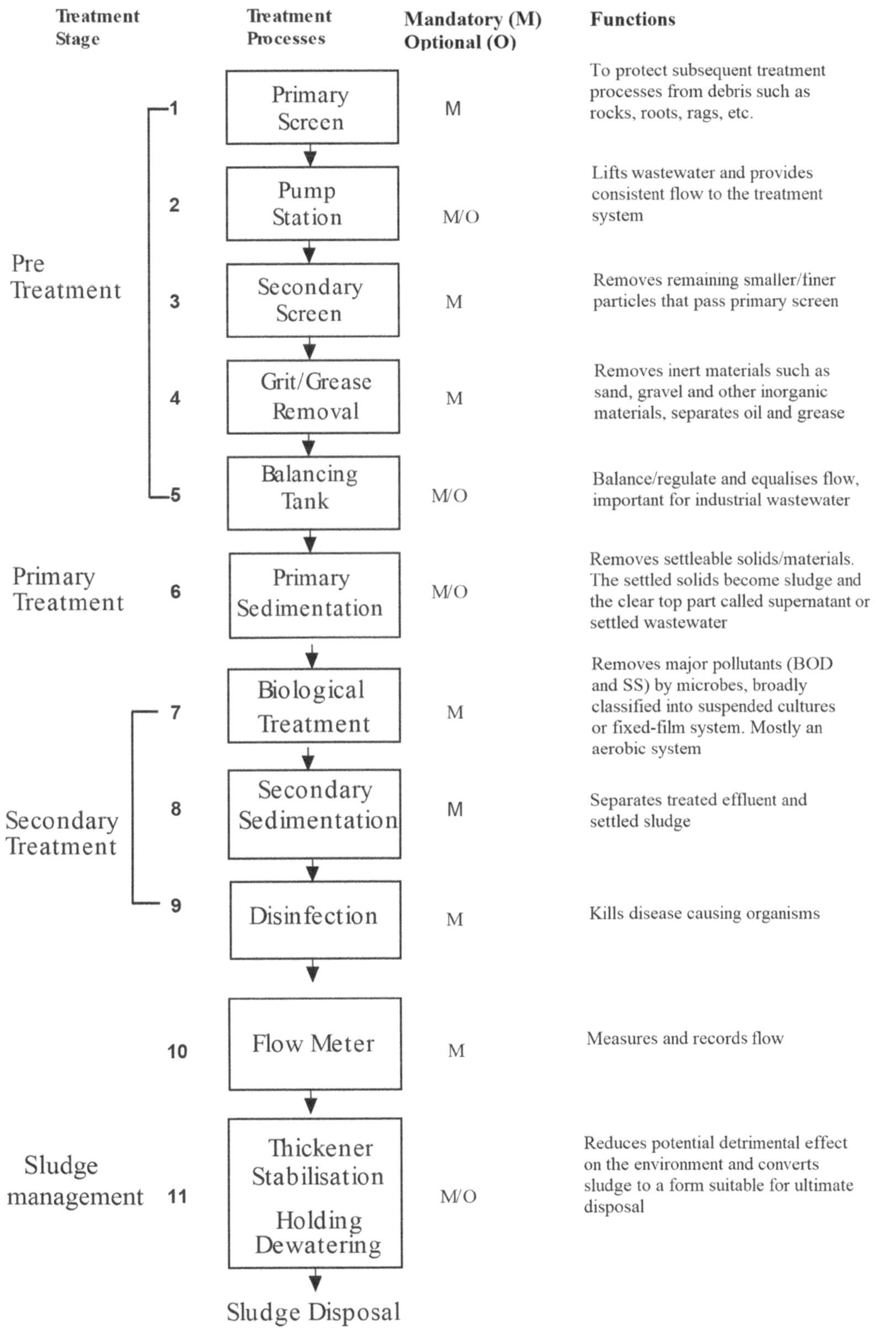

FIGURE 3.6 A typical master flow diagram of WWTP according to the category of treatment and functions (MSIG, Volume 4, 2009) [1].

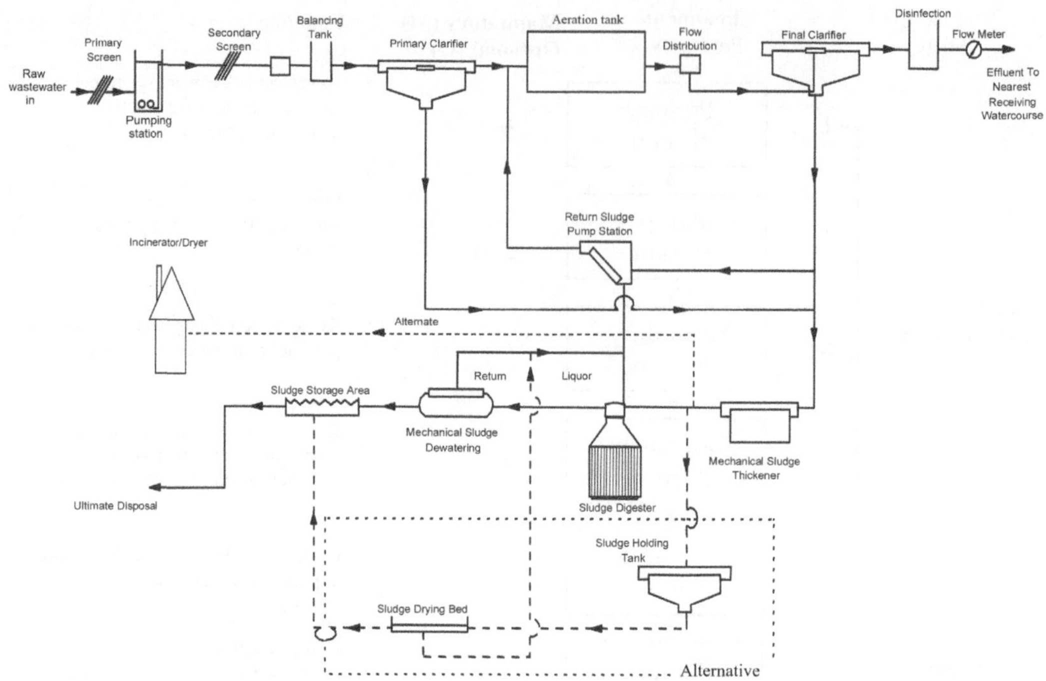

FIGURE 3.7 A master flow diagram for a typical WWTP (MSIG, Volume 4, 2009) [1].

TABLE 3.8
Requirements for Inlet Chamber (MSIG, Volume 4, 2009) [1]

Unit Process	Type	Requirements (Mandatory)			Notes
		≤20,000 PE	>20,000 PE	>50,000 PE	Single and dual penstocks refer to members of penstocks required. For more than 50,000 PE, the main penstock must be motorized.
Inlet chamber	Single	Yes	n/a	n/a	
	Dual	n/a	Yes	Yes	
	Motorized	n/a	No	Yes	

where

W = width of chamber (m)
B = width of blade (mm)
S = opening size (mm)
$F = Q$ peak (m^3/s)
V = maximum flow through velocity at the screen face (m/s)
D = flowing depth at a peak flow rate (m)

$$\text{Number of blades} = \frac{\text{Chamber width}}{\text{Blade width} + \text{spacing}} \quad (3.2)$$

Example 3.1

A manually raked primary screen at a WWTP is designed to treat approximately 50,000 residents. Based on the given data, determine the suitable width of the chamber and the appropriate blades needed. Calculate also the quantity of screenings if the storage period is 7 days.

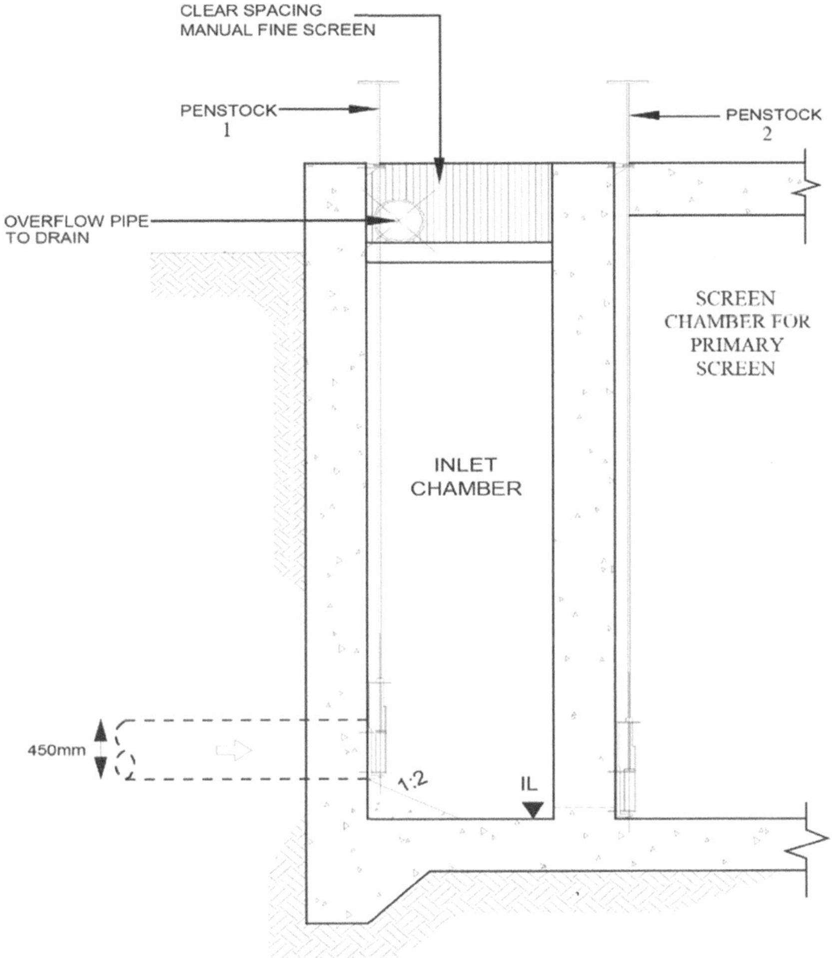

FIGURE 3.8 Double penstock (MSIG, Volume 4, 2009) [1].

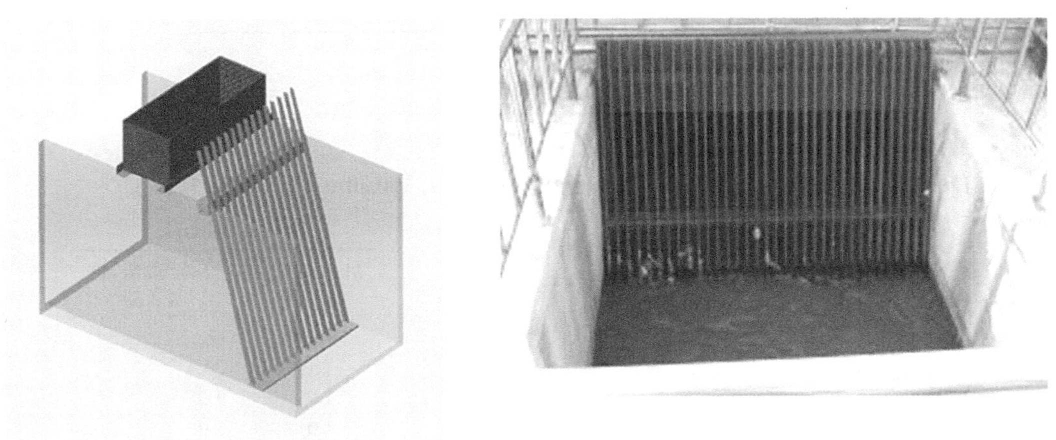

FIGURE 3.9 Typical example of a manual screen.

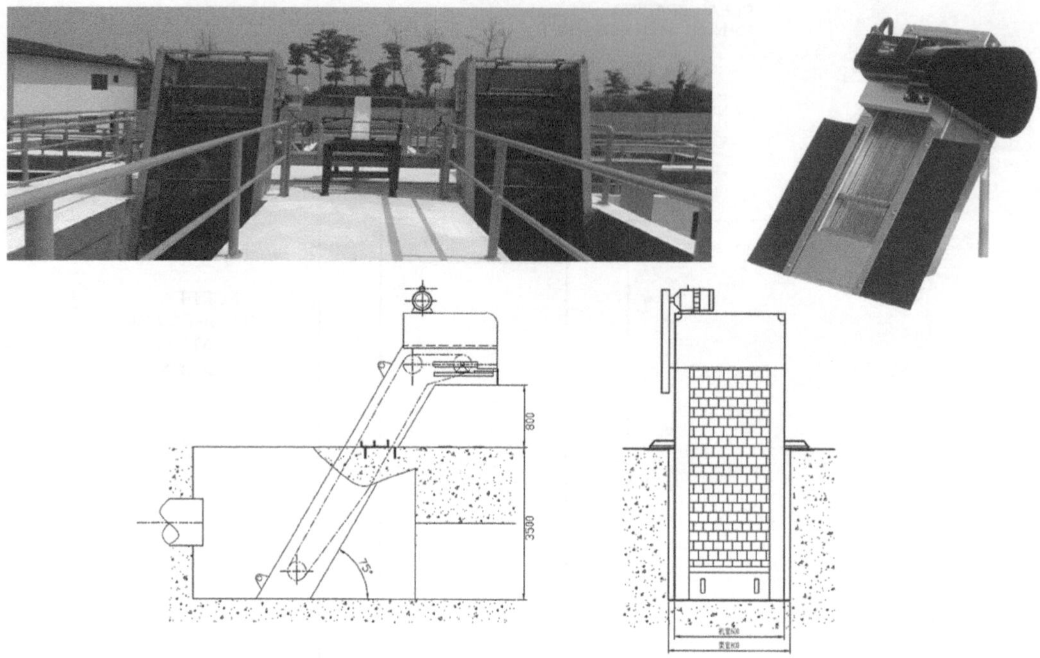

FIGURE 3.10 Typical example of an automatic screen.

TABLE 3.9
Provision of Primary Screens (MSIG, Volume 4, 2009) [1]

		Numbers of Primary Screen	
Requirements		**≤5,000 PE**	**>5,000 PE**
Duty	Manual	1 unit	—
	Mechanical	—	1 unit
Standby	Manual	—	
	Mechanical	—	1 unit
By-pass		1 unit	1 unit

TABLE 3.10
Design Parameters for Primary Screens (MSIG, Volume 4, 2009) [1]

		Design Criteria	
Description	**Unit**	**Manually Raked**	**Mechanically Raked[a]**
Design flow	m^3/s	Q_{peak}	Q_{peak}
Maximum clear spacing	mm	25	25
Maximum approach velocity at the feed channel	m/s	1.0	1.0
Maximum flow-through velocity at the screen face	m/s	1.0	1.0
Minimum freeboard	mm	150[b]	150

(*Continued*)

TABLE 3.10 (*Continued*)
Design Parameters for Primary Screens (MSIG, Volume 4, 2009) [1]

Description	Unit	Design Criteria: Manually Raked	Design Criteria: Mechanically Raked[a]
Estimated volume of screenings per volume of wastewater	$m^3/10^6 m^3$	30	See Figure 3.12
Screenings storage capacity	Day	7	7
Minimum channel width	mm	500	500
Minimum channel depth	mm	500	500
RC Staircase with riser detail	1 unit	Anti-skid and non-corrosive	Anti-skid and non-corrosive

[a] Washing and dewatering of screenings shall be provided.

[b] Designer shall ensure that with 50% of blockage at the face of the screen, enough freeboard is provided to prevent the approach channel from overflowing.

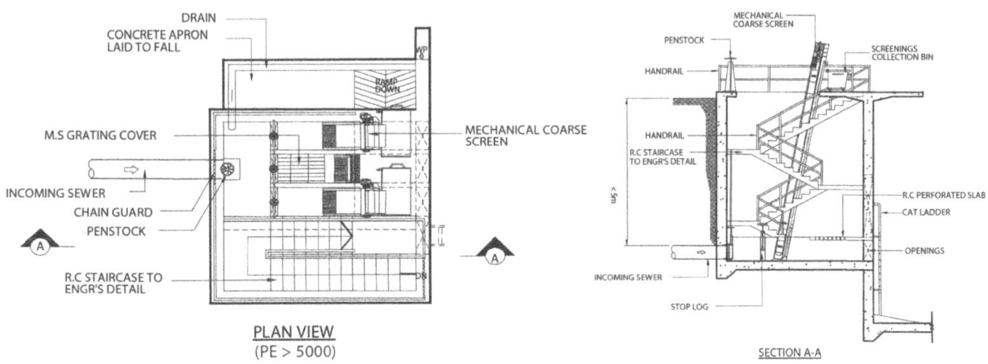

FIGURE 3.11 Typical drawing of screen chamber based on depth (>5 m for different PE) (MSIG, Volume 4, 2009) [1].

PE = 50,000 people
Size of opening = 25 mm
Blade thickness = 10 mm
Water consumption rate, $q = 225$ L/capita.day
Q peak = $4.7p^{-0.11}$ DWF
Maximum flow-through velocity at the screen face = 0.9 m/s
Depth of wastewater in chamber = 0.85 m
Storage period of screenings = 7 days

Solution:

Given data:

$B = 10$ mm
$V = 0.9$ m/s
$F = 4.7p^{-0.11}$ DWF
$q = 225$ L/capita.day
$S = 25$ mm
$D = 0.85$ m
PE = 50,000 people
Storage period of screenings = 7 days

(I) DETERMINE THE WIDTH OF THE SCREEN CHAMBER

$$\text{DWF} = q(\text{PE})$$

$$= \frac{225 \times 10^{-3}\ \text{m}^3/\text{s}\ (50{,}000)}{24 \times 3{,}600}$$

$$= 0.130\ \text{m}^3/\text{s}$$

$Q_{\text{peak}} = 4.7 p^{-0.11}\text{DWF}$; since PE=50,000, p=50

$$Q_{\text{peak}} = 4.7(50^{-0.11})\left(0.130\ \text{m}^3/\text{s}\right)$$

$$= 0.397\ \text{m}^3/\text{s}$$

Insert all the data into Equation 3.1:

$$W = \frac{(B+S)}{S} \times \frac{F}{V(D)}$$

$$W = \frac{(10+25)}{25} \times \frac{0.397}{0.9(0.85)}$$

$$= 0.727\ \text{m}$$

From Equation 3.2,

$$\text{Number of blades} = \frac{\text{Chamber width}}{\text{Blade width} + \text{spacing}}$$

$$\text{Number of blades} = \frac{727\ \text{mm}}{(10+25)\ \text{mm}}$$

$$= 21$$

Arrange accordingly, therefore, the new blade width,

$$= (21)\,(10+25)$$

$$= 0.735\ \text{m (refer figure)}$$

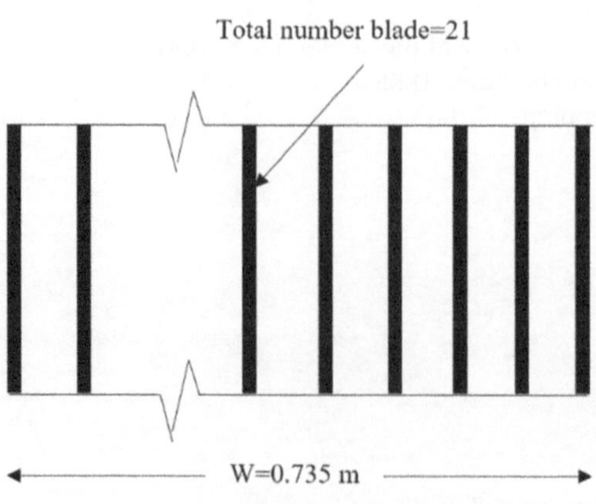

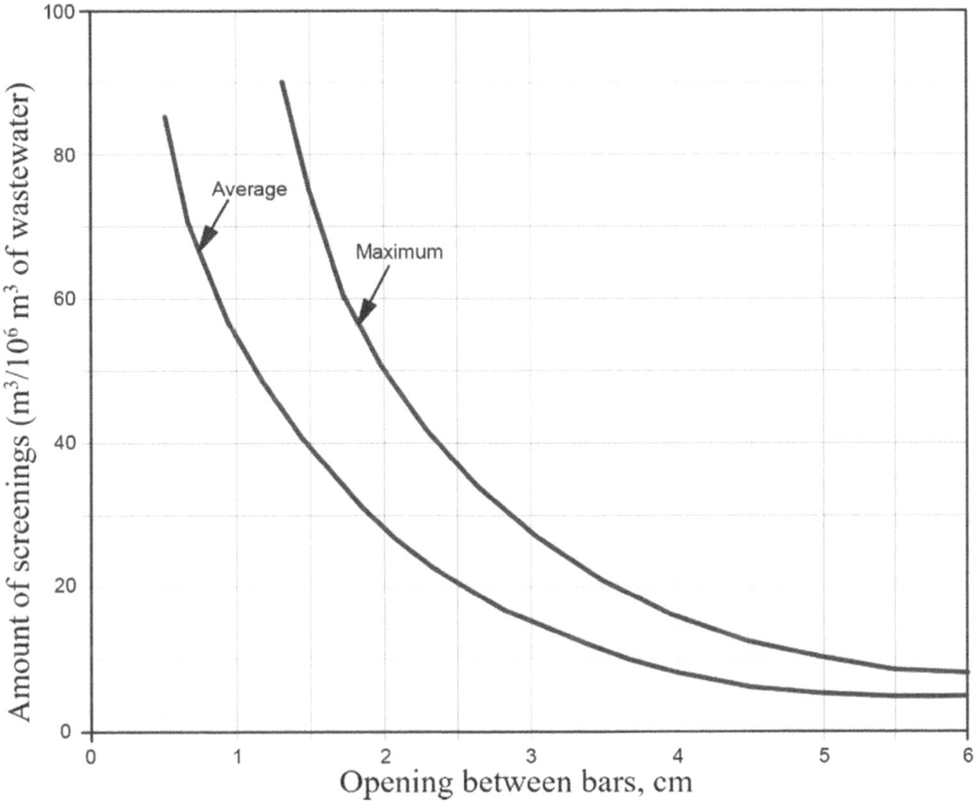

FIGURE 3.12 Screenings chart (MSIG, Volume 4, 2009) [1].

(II) CALCULATE THE QUANTITY OF SCREENINGS

From Table 3.10, the quantity of screening accumulated = 30 $m^3/10^6 m^3$ wastewater (manually raked). If automatically raked, please use the value in Figure 3.12.

$$Q_p = 0.397 \text{ m}^3/\text{s}$$

For 7 days, the total amount of wastewater,

$$= Q_p (7 \text{ days})$$

$$= 0.397 \text{ m}^3/\text{s}(7 \times 24 \times 60 \times 60)$$

$$= 240{,}105.6 \text{ m}^3$$

$10^6 m^3$ wastewater produces 30 m^3 of screenings; hence, 240,105.6 m^3 of wastewater produces,

$$= \frac{240{,}105.6 \text{ m}^3}{10^6 \text{ m}^3} \times 30 \text{ m}^3$$

$$= 7.203 \text{ m}^3 \text{ of screenings}$$

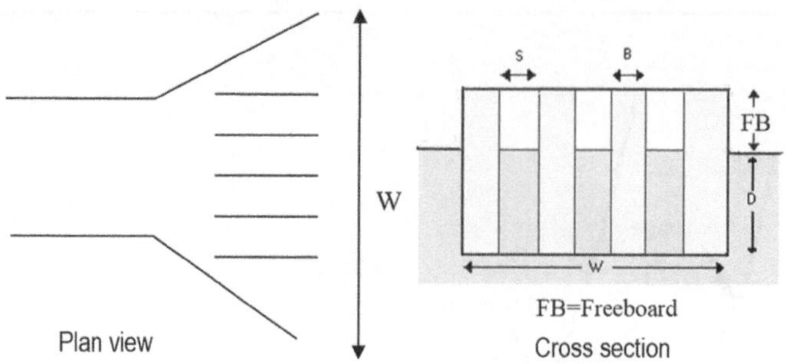

FIGURE 3.13 A typical plan view and a cross section of screen.

3.7.3 Wastewater Pumping

3.7.3.1 Design of the Pump Sump

Inlet pump stations must be installed after primary screens to protect the pumps from being damaged or clogged. The pump station is necessary to lift wastewater to a higher point for treatment, to provide consistent inlet flows to the treatment system, and to prevent overflow of raw wastewater. For safety, good ventilation and lighting are necessary. Either floats or ultrasonic level controller may be used for the start-stop levels of pumps; the ultrasonic level control is recommended due to it being clog-free.

There are two types of pump stations, i.e. dry-well and wet-well; a typical example of each is given in Figures 3.14 and 3.15, respectively. Recommended design parameters for inlet pump stations in accordance with MSIG, Volume 4, 2009 [1], is given in Table 3.11.

Example 3.2

A rectangular wet-well pump sump receives wastewater from 3,000 PE and water consumption rate (q) 225 L/capita.day. Dimension the sump. Take L:W 1:1 and depth (D) 2 m.

Solution:

$$\begin{aligned}\text{DWF} &= q(\text{PE})\\ &= 225\times10^{-3}\times(3{,}000)\ \text{m}^3/\text{day}\\ &= 675\ \text{m}^3/\text{day}\\ &= Q_{\text{ave}}\end{aligned}$$

$Q_{\text{peak}} = 4.7p^{-0.11}\text{DWF}$; since PE=3,000, p=3

$$\begin{aligned}Q_{\text{peak}} &= 4.7(3^{-0.11})\left(675\ \text{m}^3/\text{day}\right)\\ &= 2{,}811\ \text{m}^3/\text{day}\end{aligned}$$

From Table 3.11, retention time (t) is 30 minutes at Q_{peak}

$$t = \frac{V}{Q_p},\quad \text{where}\quad V = \text{volume of tank}$$

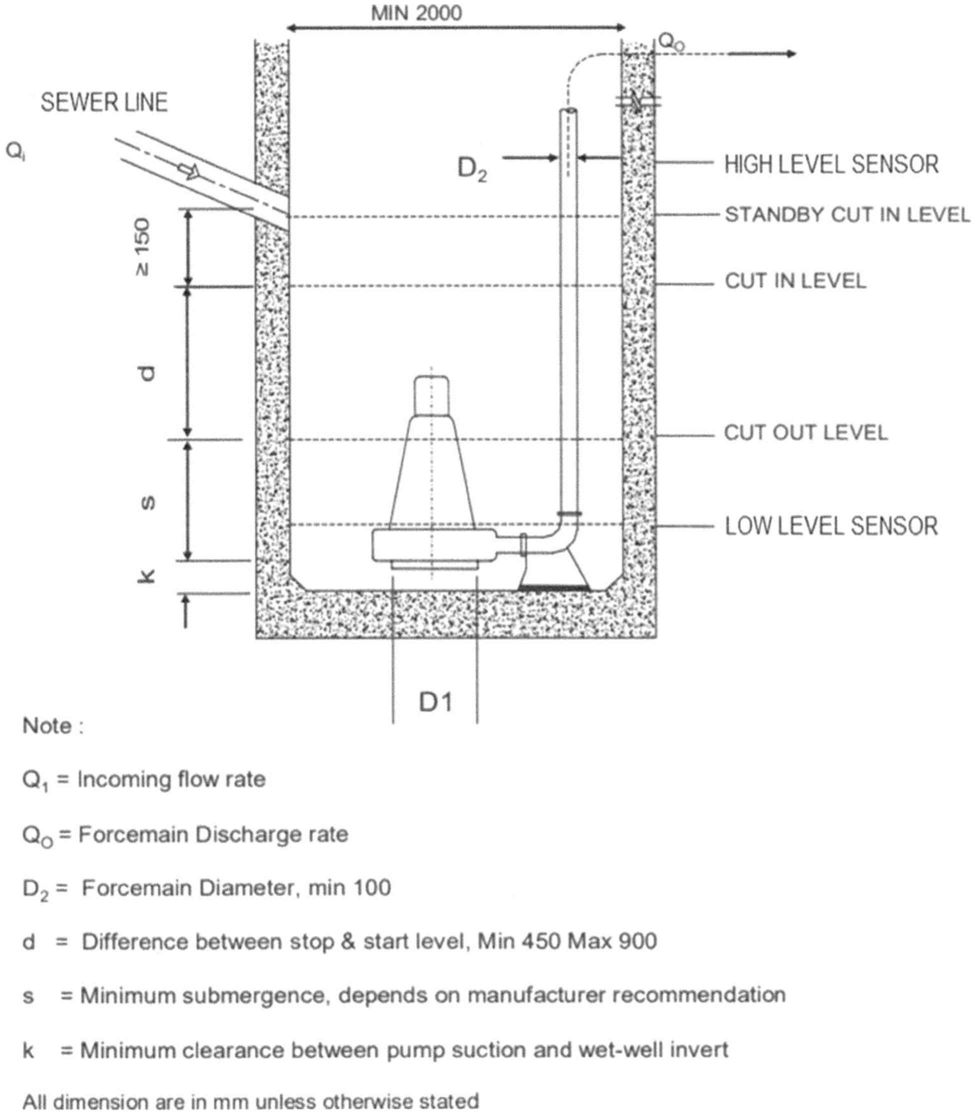

FIGURE 3.14 Typical dimensions of wet-well submersible pump station (MSIG, Volume 4, 2009 [1].

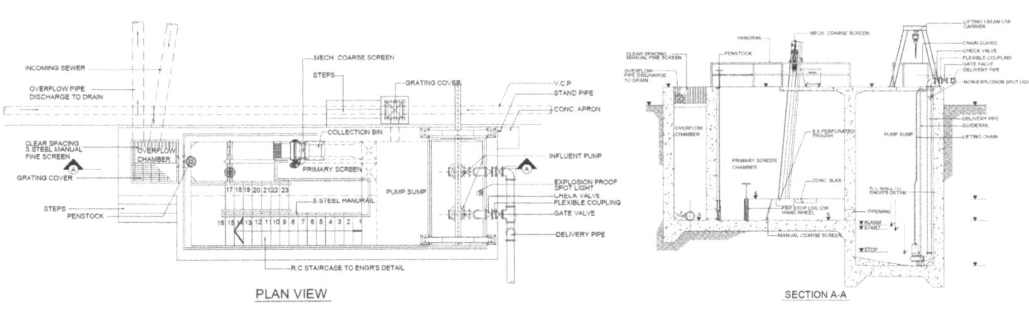

FIGURE 3.15 Typical details of wet-well pump station, MSIG, Volume 4, 2009 [1].

TABLE 3.11
Recommended Design Parameters for Inlet Pump Stations (MSIG, Volume 4, 2009) [1]

Description	Unit	PE≤1,000	1,000 < P ≤ 5,000	5,000 < P ≤ 20,000	PE > 20,000
Type of station		Wet-well	Wet-well	Wet-well or dry-well up to 10,000 PE 10,000 PE above-wet-well and dry-well	Wet-well and dry-well
Number of pumps (all identical and work sequentially)		2 1 duty, 1 standby (100% standby)	2 1 duty, 1 standby (100% standby)	4 (2 sets) 1 duty, 1 assist, per set (100% standby)	6 (3 sets) 1 duty, 1 assist, per set (50% standby)
Pumps design flow (for pump sizing)		Each at Q_{peak}	Each at Q_{peak}	Each at 0.5 Q_{peak}	Each at 0.25 Q_{peak}
Retention time at Q_{peak}	min	30	30	30	30
Min pass through openings	mm	75	75	75	75
Minimum suction and discharge openings	mm	100	100	100	100
Pumping cycle (average flow conditions)	Start/hour	6 minutes 15 max	6 minutes 15 max	6 minutes 15 max	6–15
Lifting device[a]		Lifting davit	Lifting beam and block	Mechanical and block	Mechanical

[a] Motorized hoists shall be provided when the lifting weight exceeds 100 kg.

$$V = t(Q_p)$$

$$= \left(\frac{30}{60 \times 24}\right) \text{day} \left(2{,}811 \text{ m}^3/\text{day}\right)$$

$$= 58.6 \text{ m}^3$$

L:W = 1:1 and depth (D) =2 m, as follows:

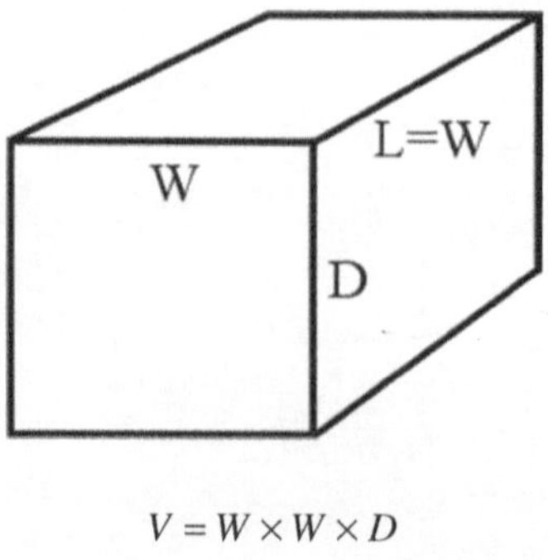

$$V = W \times W \times D$$

$$58.6 = W^2 \times 2$$

$$W = 5.41 \text{ m}$$

The new dimension (excluding freeboard) is as follows:

(I) TYPES OF PUMPS

There are many types of pumps used in the WWTP. Their main location of use is shown in Figure 3.16. Some of the common pumps in use are shown in Figures 3.17–3.21. Details of their functions and workability are available in the hydraulics textbook.

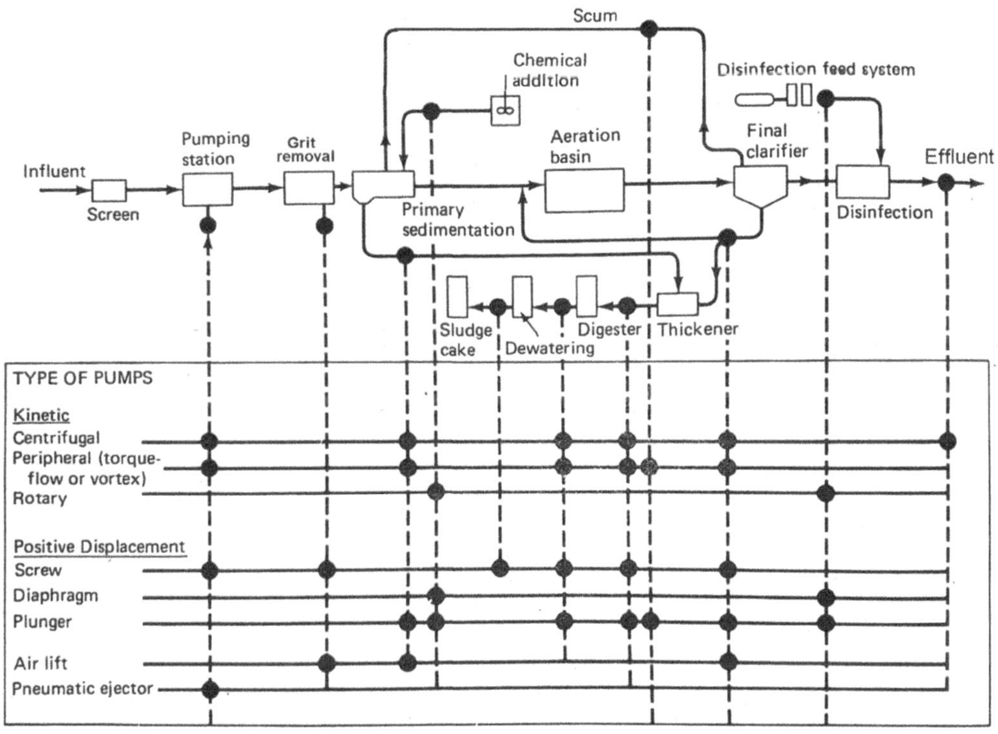

FIGURE 3.16 Typical locations of pumps at WWTP, Qasim, 1998 [4].

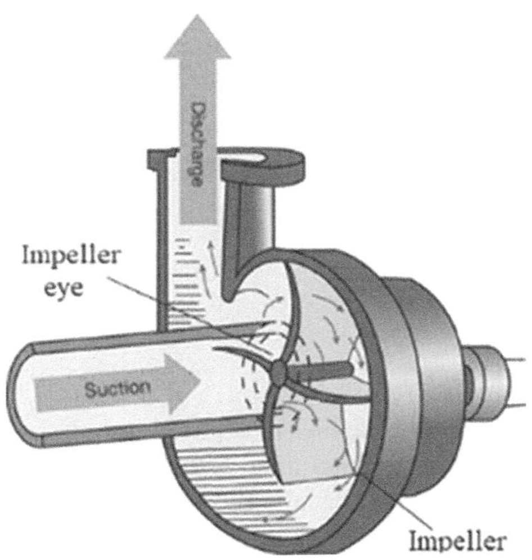

FIGURE 3.17 Centrifugal force pump.

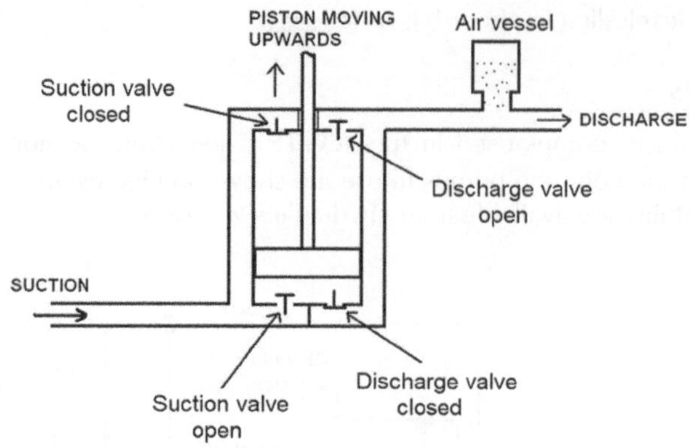

FIGURE 3.18 Positive displacement plunger pump.

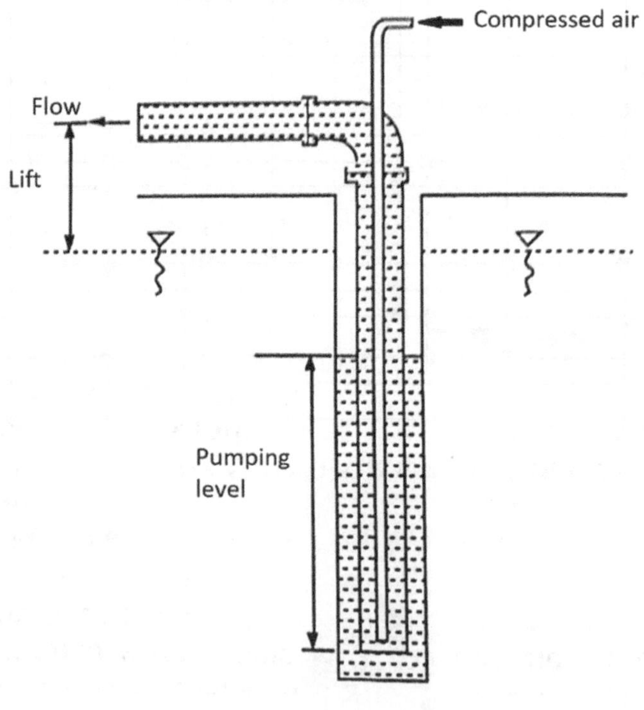

FIGURE 3.19 Air lift pump.

3.7.4 Design of Secondary Screen

After the inlet pump station, further screening is required to reduce the remaining floating matter (plastic, paper, cloth, and other particles) and finer particles in the wastewater that will disrupt the treatment process downstream. It is also installed to minimize blockages in sludge handling and treatment facilities.

Provisions and design parameters of the primary screen according to MSIG, Volume 4, 2009 [1] are given in Tables 3.12 and 3.13, respectively.

FIGURE 3.20 Submersible pump.

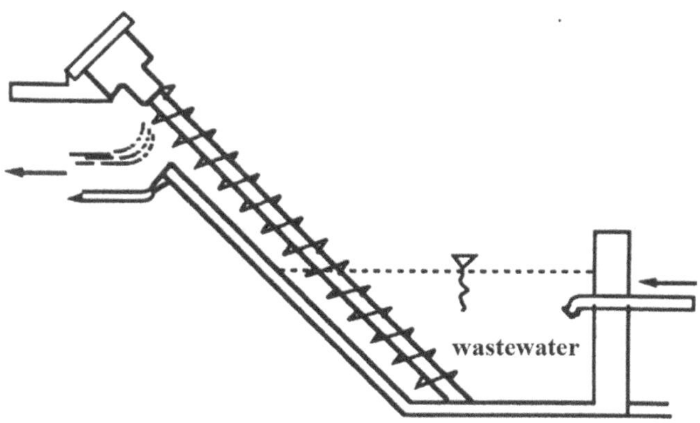

FIGURE 3.21 Screw pump.

TABLE 3.12
Provision Requirement of Secondary Screens (MSIG, Volume 4, 2009) [1]

		Numbers of Secondary Screen	
Requirements		**≤5,000 PE**	**>5,000 PE**
Duty	Manual	1 unit	—
	Mechanical	—	1 unit
Standby	Manual	—	—
	Mechanical	—	1 unit
Bypass	Screen	1 unit	1 unit

TABLE 3.13
Design Parameters for Secondary Screens (MSIG, Volume 4, 2009) [1]

Description	Unit	Manually Raked	Mechanically Raked[a]
Design flow	m^3/s	Q_{peak} or pump flow, whichever is higher	Q_{peak} or pump flow, whichever is higher
Maximum clear spacing	mm	12	12
Slope to the vertical		30°–45°	15°–45°
Maximum approach velocity at the feed channel	m/s	1.0	1.0
Maximum flow-through velocity at the screen face	m/s	1.0	1.0
Minimum freeboard	mm	150[b]	150
Estimated volume of screenings per volume of wastewater (at $Q_{average}$)	$m^3/10^6 m^3$	See Figure 3.12	
Screenings skip storage capacity	Day	7	7
Minimum channel width	mm	500	500
Minimum channel depth	mm	500	500
RC Staircase with riser detail	1 unit	Anti-skid and non-corrosive	Anti-skid and non-corrosive

[a] Washing and dewatering of screenings shall be provided.

[b] Designer shall ensure that with 50% of blockage at the face of the screen, sufficient freeboard is provided to prevent the approach channel from overflowing.

3.8 COMMINUTOR

Comminutors, also known as grinders, are shredding and screening devices widely used in WWTPs and pumping stations. As opposed to blocking the debris in the screen, comminutors break down wastewater solids in plant headworks and pump stations. It works just like a home blender. The equipment gets heavy use and is a critical part of the wastewater treatment process. The advantage is that it does not require raking systems to clean the screen face; however, the main setback is it just passes all the solids into the subsequent treatment processes. The maintenance is mainly required for wear and tear. Its usage is not recommended under MSIG, Volume 4, 2009 [1], comminutors as replacements for primary screens are generally not recommended. It may be considered if the consultant can provide good engineering reasons for its application. A typical example of comminutor is shown in Figure 3.22.

3.9 GRIT REMOVAL SYSTEM

This unit process is important to minimize problems associated with grit, especially to pumps and also sludge digestion and dewatering facilities. The particles must be removed at an early stage of the process because grit may interfere with biological treatment due to its inert property. Grits are abrasive and wear down equipment.

Grit is normally removed either by being removed by gravitating settlement or centrifugal separation. For example, at a chamber, the velocity of waste flow is reduced to a point where the grit (denser sand, pebble, rock, eggshells, and other grit) will settle out, but the organic solids will remain in suspension and flow to the next treatment system. A grit chamber is designed to remove inorganic matter, particularly sand, with a diameter ≥0.2 mm. Generally, grit with 0.2 mm and specific gravity of 2.65 will have a velocity of the settlement of 1.2 m/min, which is higher than the velocity of settlement for organic load in wastewater. If the speed is too low, organic matter will

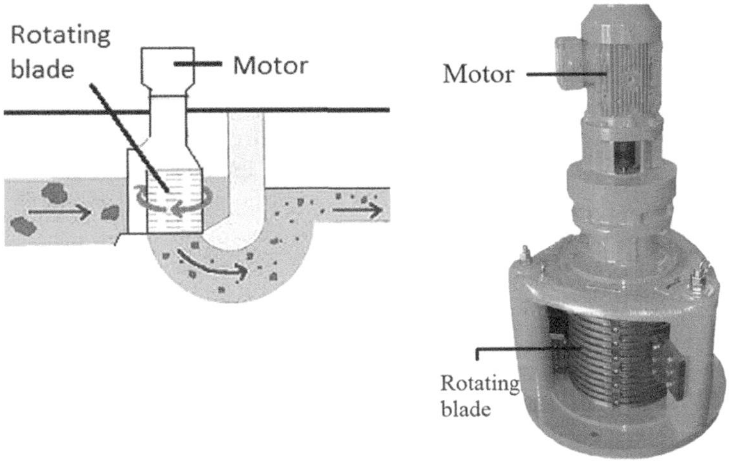

FIGURE 3.22 Comminutor.

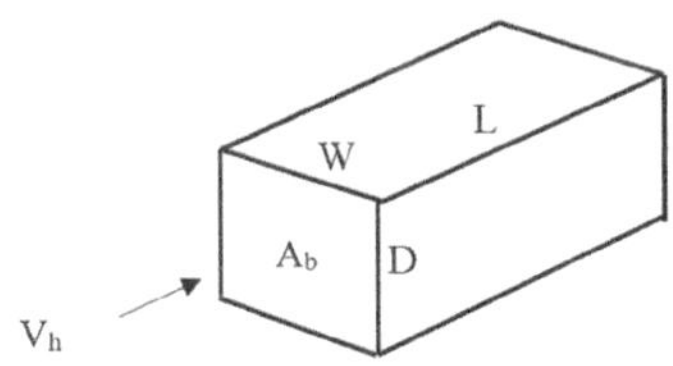

FIGURE 3.23 Maximum gravity flow through velocity/horizontal velocity at Q_{peak}.

be settled, while if it is too fast, grit will flow through the tank. There are a few types of the grit removal system. The most common are grit chamber, rectangular (Figures 3.23 and 3.24), constant velocity grit channel, detroiter, and vortex grit chamber.

Provisions and design parameters of the grit removal system according to MSIG, Volume 4, 2009 [1] are given in Table 3.14.

$$\text{Maximum gravity flow through velocity at } Q_{peak}\ (V_h) = \frac{Q_p}{A_b}\ (\text{m/s}) \tag{3.3}$$

Example 3.3

A square, aerated horizontal flow grit chamber needs to be designed to treat wastewater from 5,000 PE, with a water consumption rate (q) of 225 L/capita.day. Determine the volume of this chamber and the quantity of grit accumulated every day if the retention time at Q_{peak} is 3 minutes.

Solution:

Given data:

PE = 5,000 people

q = 225 L/capita.day

Retention time = 3 minutes

Estimated grit quantity for 5,000 PE = 0.03 $m^3/10^3\,m^3$ of wastewater

TABLE 3.14
Provision Requirement and Design Parameter of Grit Removal System (MSIG, Volume 4, 2009) [1]

		Population		
Operation		**≤5,000 PE**	**>5,000 PE**	
Duty	Manual	1 unit @ design flow	—	
	Mechanical	—	–1 unit @ design flow (up to 10,000 PE) –2 units @ 50% design flow each (>10,000 PE)	
Standby	Manual	1 unit @ design flow	—	
	Mechanical	—	—	
Bypass		—	Yes	
Design criteria	Unit		Population	
		≤5,000 PE	>5,000 PE	>5,000 PE
Design flow	m^3/day	Q_{peak}	Q_{peak}	Q_{peak}
Grit removal	—	Manual (tanker)	Mechanical	Mechanical
Chamber type	—	Horizontal flow (2 units required for duty and standby during cleaning) vortex also acceptable	Square, rotary or vortex type simple mechanized grit chamber	Aerated
Minimum detention time at Q_{peak}	minute	3	3	3
Maximum gravity flow through velocity at Q_{peak} ($V_h = Q_p$ approaching surface area)[a]	m/s	0.20	0.20	0.20
Maximum centrifugal flow through velocity	m/s	n/a	<1.0	<1.0
Head loss (at parshall flume)	—	35% of depth	—	
Aeration requirement	L/s/meter length of tank	—	—	10.0
Chamber dimension:	—			
Depth: Width		1:2	Manufacturer's specification	Manufacturer's specification
Length: Width		2:1		
Estimated grit quantity	$m^3/10^3 m^3$ of wastewater	0.03	0.03	0.03
Washing and dewatering of grit	—	No	Yes	Yes

[a] Refer Equation 3.3 and Figure 3.23.

(I) DETERMINE THE VOLUME OF THE TANK

$$\begin{aligned} \text{DWF} &= q(\text{PE}) \\ &= 225 \times 10^{-3} \text{ m}^3/\text{day}(5{,}000) \\ &= 1{,}125 \text{ m}^3/\text{day} \end{aligned}$$

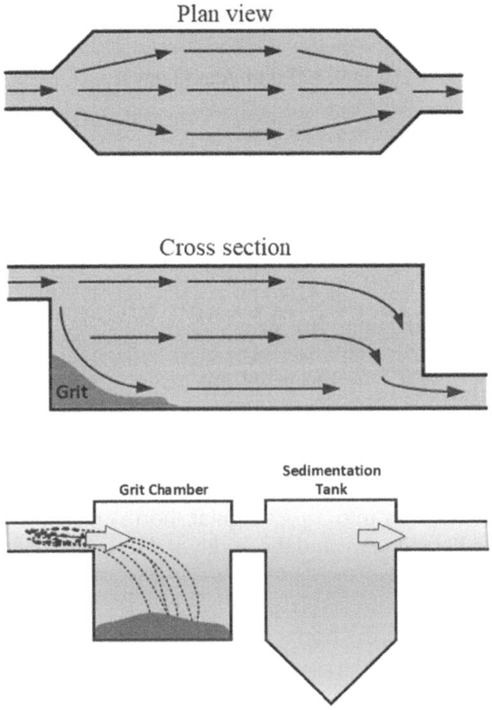

FIGURE 3.24 An example of a rectangular grit chamber.

$$Q_{\text{peak}} = 4.7p^{-0.11}\text{DWF}; \text{ since PE}=5{,}000,\ p=5$$

$$Q_{\text{peak}} = 4.7(5^{-0.11})\left(1{,}125\ \text{m}^3/\text{day}\right)$$
$$= 4{,}429.6\ \text{m}^3/\text{day}$$

Retention time (t) = 3 minutes

$$t = \frac{V}{Q_p},\quad \text{where}\quad V = \text{volume of tank}$$
$$V = t\left(Q_p\right)$$
$$= \left(\frac{3}{60 \times 24}\right)\text{day}\ \left(4{,}429.6\ \text{m}^3/\text{day}\right)$$
$$= 9.23\ \text{m}^3$$

Then, the tank can be dimensioned accordingly by assuming suitable depth (2–3 m).

(II) ESTIMATE QUANTITY OF GRIT

Estimated grit quantity for 5,000 PE = 0.03 $m^3/10^3\ m^3$ of wastewater

For 1 day, the total amount of wastewater,

$$= Q_p(1 \text{ day})$$
$$= 4{,}429.6 \text{ m}^3/\text{day}(1 \text{ day})$$
$$= 4{,}429.6 \text{ m}^3$$

10^3 m^3 wastewater produces 0.03 m^3 of grit; hence, 4,429.6 m^3 of wastewater

$$= \frac{4{,}429.6 \text{ m}^3}{10^3 \text{ m}^3} \times 0.03 \text{ m}^3$$
$$= 0.13 \text{ m}^3 \text{ of grit}$$

Example 3.4

A peak flow from 25,000 PE flows into a rectangular, horizontal flow grit chamber. If the water consumption rate (q) is 225 L/capita.day and the width of the tank is 1.5 m, using data in Table 3.14, determine:

i. The depth of this chamber, if the horizontal velocity, V_h=0.2 m/s.
ii. The amount of oxygen required to aerate the chamber.
iii. The amount of grit to be disposed of in a week.

Solution:

Consider the shape as shown in the following figure.

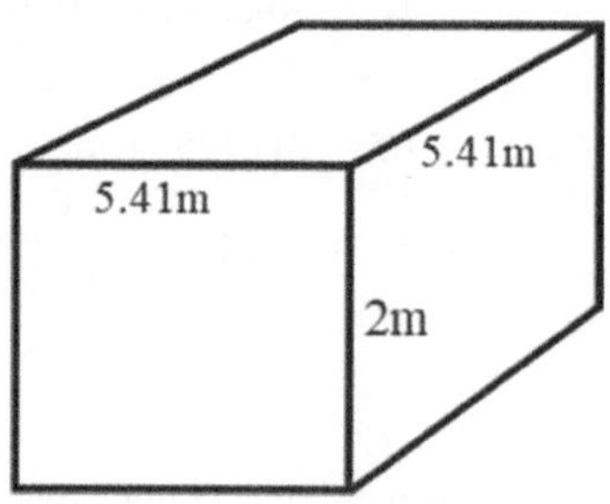

(I) DEPTH OF CHAMBER

From Table 3.14, the maximum gravity flow through velocity is also called horizontal velocity, V_h

$$V_h = \frac{Q_P}{A_b} \text{ where } A_b = \text{approaching surface area} = \text{Width} \times \text{Depth}$$

$$\text{DWF} = q(\text{PE})$$

$$= 225 \times 10^{-3} \text{ m}^3/\text{day}(25{,}000)$$
$$= 5{,}625 \text{ m}^3/\text{day}$$

$Q_{\text{peak}} = 4.7p^{-0.11}\text{DWF}$; since PE=25,000, p=25

$$Q_{\text{peak}} = 4.7(25^{-0.11})\left(5{,}625\ \text{m}^3/\text{day}\right)$$

$$= 18{,}554\ \text{m}^3/\text{day}$$

$$\text{Width } (W) = 1.5\ \text{m}$$

$$V_h = \frac{Q_P}{A_b} \quad \text{where} \quad A_b = \text{approaching surface area} = \text{Width} \times \text{Depth}$$

$$V_h = \frac{Q_P}{(W \times D)}$$

$$0.2\ \text{m/s} = \frac{18{,}554\ \text{m}^3/\text{day}}{(1.5\ \text{m} \times D)}$$

$$0.2\ \text{m/s} = \frac{18{,}554\ \text{m}^3/\text{day}}{(1.5\ \ \text{m} \times D)(24 \times 60 \times 60)}$$

$$D = 0.72\ \text{m}$$

(II) OXYGEN REQUIRED

From Table 3.14, the aeration required is 10 L/s/m length of the tank. From the same table, the retention time (t) is 3 minutes.

Determine the length of the tank:

$$t = \frac{V}{Q_P}, \text{ where } V = \text{volume of tank} = W \times D \times \text{Length of tank } (L)$$

$$W \times D \times L = t(Q_P)$$

$$1.5 \times 0.72 \times L = 3\ \text{minutes}\left(18{,}554\ \text{m}^3/\text{day}\right)$$

$$1.5 \times 0.72 \times L = \frac{3\ \ \text{day}\left(18{,}554\ \text{m}^3/\text{day}\right)}{24 \times 60}$$

$$L = 35.8\ \text{m}$$

Aeration required=10 L/s/m length of tank

For 35.8 m, aeration required

$$= 10\ \text{L/s}\ (35.8\ \text{m})$$

$$= 358\ \text{L/s}$$

The dimension of the tank (excluding freeboard) is shown in the following figure.

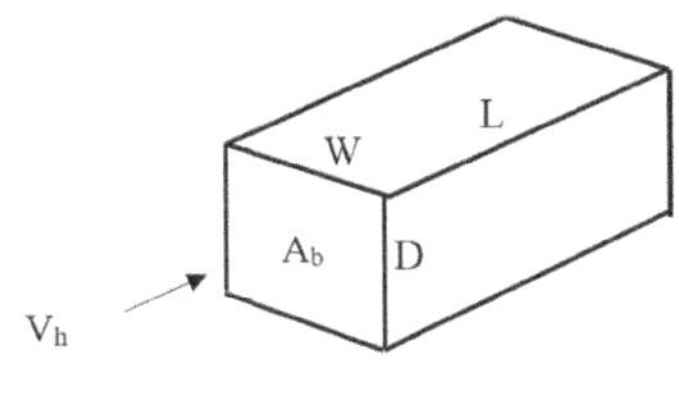

(III) THE ESTIMATED QUANTITY OF GRIT IN 7 DAYS

From Table 3.14, estimated grit quantity for 25,000 PE = 0.03 $m^3/10^3 m^3$ of wastewater

For 7 days, the total amount of wastewater,

$$= Q_P(7 \text{ days})$$

$$= 18{,}554 \text{ m}^3/\text{day}(7 \text{ day})$$

$$= 129{,}878 \text{ m}^3$$

$10^3 m^3$ wastewater produces 0.03 m^3 of grit; hence, 129,878 m^3 of wastewater produces,

$$= \frac{129{,}878 \text{ m}^3}{10^3 \text{ m}^3} \times 0.03 \text{ m}^3$$

$$= 3.9 \text{ m}^3 \text{ of grit}$$

3.10 GREASE REMOVAL

Removing oil and grease from wastewater is essential to the proper functioning of wastewater systems as they create problems at the clarifier and is carried over in the final effluent. Approximately over 80% of oil and grease will be removed from greasy wastes by properly maintained grease interceptors and traps. As oil and grease reach WWTP, they are mainly separated via density differences through gravitation or floatation. The treatment principle normally allows the oily matter to be floated for a certain detention time. Free oil and grease globules lighter than water rose through the liquid and skimmed from the top surface where they are normally removed to sandpit before ultimate disposal. The drainage from the sandpit shall be returned to the inlet channel for further treatment. There are a few available technologies such as floating tanks, tilted plate oil separators, grease traps, etc. A typical example of a floatation tank is shown in Figure 3.25.

Provisions and design parameters of the grease removal system according to MSIG, Volume 4, 2009 are given in Table 3.15.

Example 3.5

Using data in Table 3.15, determine the suitable volume of an aerated floatation tank for oil and grease removal for a housing scheme with 2,500 houses. Take water consumption rate (q) as 225 L/capita.day.

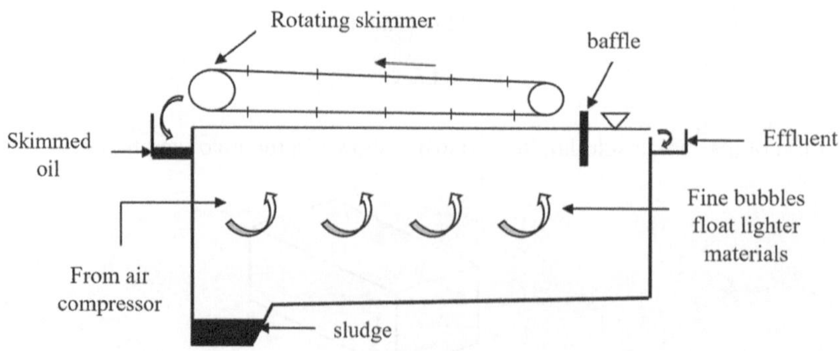

FIGURE 3.25 Floatation tank for oil and grease removal.

TABLE 3.15
Provision Requirement and Design Parameter of Grease Removal System (MSIG, Volume 4, 2009) [1]

Operation	Type	≤5,000 PE	>5,000 PE	
Duty	Manual	1 unit	—	
	Mechanical	—	• 1 unit @ design flow (up to 10,000 PE) • 2 units @ 50% design flow each (>10,000 PE)	
Standby	Manual	1 unit @ design flow	—	
	Mechanical	—	—	
Bypass		—	Yes	
Design criteria	Unit	≤5,000 PEa	>5,000 PE	
Design flow	m^3/day	Q_{peak}	Q_{peak}	
Grease removal	—	Simple manual	Manual interceptor	Mechanical
Chamber type	—	Rectangular	Baffled tank	Aerated type
Minimum detention time (Q_{peak})	min	3	3	3
Grit and grease storage period before off-site disposal	Day	30	7	7

[a] Combined grit and grease chamber is allowed. If combined, then total detention time shall comply to 6 minutes at Q_{peak}.

Solution:

In Malaysia, per house = 5 PE

$$\text{Hence, total PE} = 2{,}500 \times 5$$

$$\text{PE} = 12{,}500$$

$$\text{DWF} = q(\text{PE})$$

$$= 225 \times 10^{-3}\ \text{m}^3/\text{day}(12{,}500)$$

$$= 2{,}812.5\ \text{m}^3/\text{day}$$

$Q_{peak} = 4.7p^{-0.11}\text{DWF}$; since PE=12,500, p=12.5

$$Q_{peak} = 4.7(12.5^{-0.11})\left(2{,}812.5\ \text{m}^3/\text{day}\right)$$

$$= 10{,}012\ \text{m}^3/\text{day}$$

For PE>10,000, apply two units of tank at half of the Q_{peak} for each tank
From Table 3.15, retention time (t)=3 minutes

$$t = \frac{V}{Q_p}, \text{ where } V = \text{volume of tank}$$

For each tank, $V = t\left(Q_p\right)$

$$= \left(\frac{3}{60 \times 24}\right)\text{day}\ \left(10{,}012/2\ \text{m}^3/\text{day}\right)$$

$$= 10.43\ \text{m}^3$$

Then, the tank can be dimensioned accordingly by assuming a suitable depth (2–3 m).

3.11 EQUALIZATION/BALANCING TANK

Providing consistent flow and loading to WWTP, especially a biological process, is important to maintain optimal treatment. Equalization (EQ) tanks are aimed to provide consistent influent flow to downstream processes by retaining high flow fluctuations, in other words, to prevent flow variations from entering secondary treatment processes. By regulating the flow, it reduces hydraulic loading into secondary treatment processes and potential overflows that may cause a health hazard and pollution. Due to the additional retention time, aeration and mixing are mandatory in equalization basins to prevent the raw wastewater from becoming anaerobic and to maintain solids in suspension. A typical equalization tank is shown in Figure 3.26.

In MSIG, Volume 4, 2009, equalization/balancing tanks are mandatory for all treatment processes that are not designed at peak flow. However, for extended aeration plants that are designed with a retention time of more than 18 hours and clarifiers designed at peak flow, the use of balancing tanks is not required. The tank is designed at peak flow and regulates the flow by pumping at average flow (DWF). This allows a balance water level of 0.6–1.0 m (dead water) as recommended by MSIG, Volume 4, 2009 [1].

Provisions and design parameters of the equalization system according to MSIG, Volume 4, 2009 [1] are given in Table 3.16.

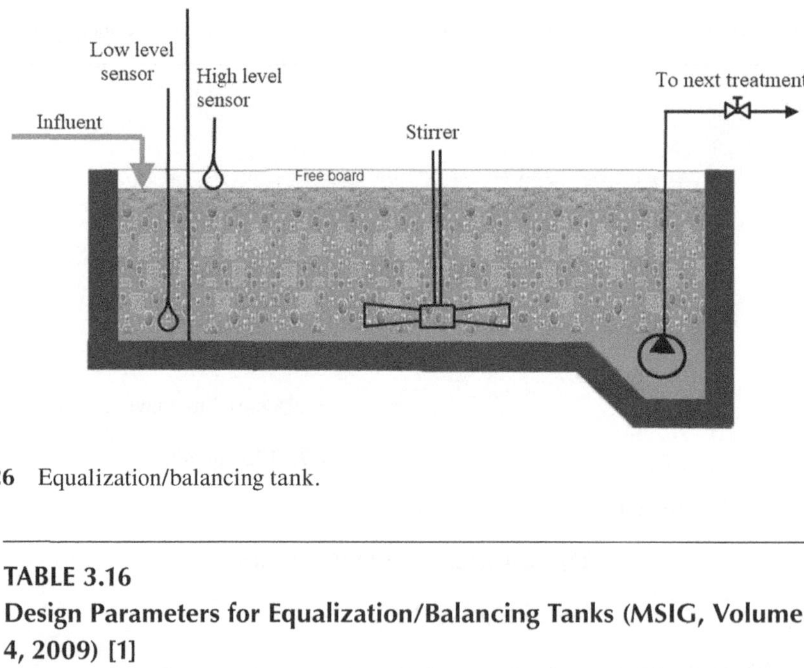

FIGURE 3.26 Equalization/balancing tank.

TABLE 3.16
Design Parameters for Equalization/Balancing Tanks (MSIG, Volume 4, 2009) [1]

Description	Unit	Design Criteria
Design flow	m^3/day	Q_{peak}
Volume of tanks	m^3	1.5 hours detention at Q_{peak}
Mixing power requirements	W/m^3 of wastewater	5 at top water level (TWL)
Aeration	m^3air/hour/m^3 wastewater	1 m^3 of air supply for every m^3 of wastewater stored per hour at TWL
Overflow bypass to downstream unit requirement	—	Yes

Example 3.6

An equalization tank is required to regulate flow from a housing development project with 1,000 houses, a small factory with 100 workers, and 20 shop lots with a floor area of 200 m^2 each. Using data in Table 3.16, design a suitable tank for this wastewater. Take water consumption rate (q) as 225 L/capita.day.

Solution:

(I) CALCULATE THE FLOW RATE FROM PE

From Table 2.5, the recommended PE is 5 per house, 0.3 per factory worker, 3 per 100 m^2 gross area of shops (commercial premises).

$$\text{Total PE} = (1{,}000 \times 5) + (100 \times 0.3) + \left(\frac{20 \times 200}{100} \times 3\right)$$

PE = 5,150

$$\begin{aligned}\text{DWF} &= q(\text{PE})\\ &= 225 \times 10^{-3}\ \text{m}^3/\text{day}(5{,}150)\\ &= 1{,}158\ \text{m}^3/\text{day}\end{aligned}$$

$Q_{\text{peak}} = 4.7p^{-0.11}\text{DWF}$; since PE = 5,150, p = 5.15

$$\begin{aligned}Q_{\text{peak}} &= 4.7\left(5.15^{-0.11}\right)\left(1{,}158\ \text{m}^3/\text{day}\right)\\ &= 4{,}545\ \text{m}^3/\text{day}\end{aligned}$$

(II) DETERMINE THE VOLUME AND DIMENSION OF THE TANK

From Table 3.16, retention time (t) is 1.5 hours at Q_{peak}.

$$\begin{aligned}t &= \frac{V}{Q_p},\ \text{where } V = \text{volume of tank}\\ V &= t(Q_p)\\ &= \left(\frac{1.5}{24}\right)\text{day}\ \left(4{,}545\ \text{m}^3/\text{day}\right)\\ &= 284\ \text{m}^3\end{aligned}$$

Then, the tank can be dimensioned accordingly:

Assume as follows:
Top Water Level (TWL) = 3 m = D
Length to width ratio (3:1)

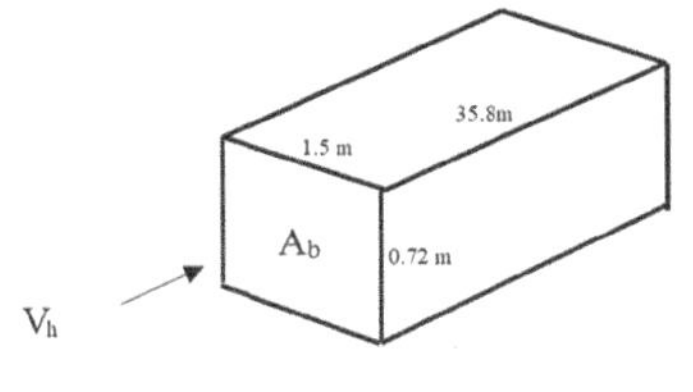

$$\text{Volume}, V = W \times D \times L$$

$$284 \text{ m}^3 = W \times 3 \text{ m} \times 3W$$

$$9W^2 = 284$$

$$W = 5.62 \text{ m, say } 6 \text{ m}$$

$$\text{Hence}, L = 3W$$

$$L = 18 \text{ m}$$

$$\text{New volume}(V) = 6 \times 3 \times 18 \text{ m}^3$$

$$V = 324 \text{ m}^3$$

The new dimension (excluding freeboard) is as follows:

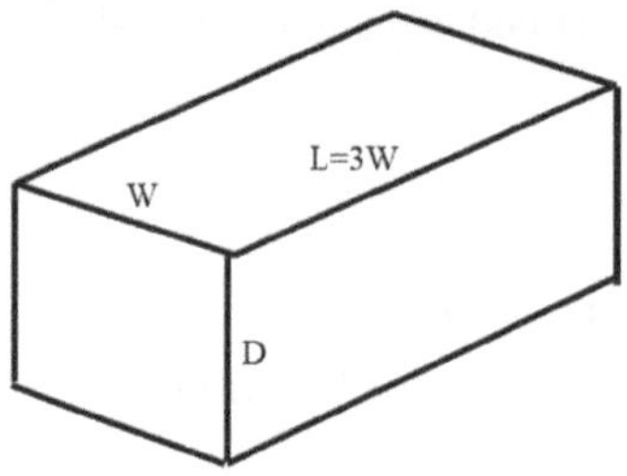

(III) DETERMINE MIXING REQUIREMENT

From Table 3.16, the mixing requirement is 1 m³ of air supply per hour for every m³ of wastewater stored at TWL.

Calculate the quantity of O_2 required:
The volume of wastewater at $Q_{peak} = 324 \text{ m}^3$
For every m³ of wastewater, 1 m³/h of O_2 is required
Hence, the O_2 required for 345 m³ of wastewater $= 324 \times 1 \text{ m}^3$

$$= 324 \text{ m}^3/\text{h}$$

(IV) DETERMINE THE POWER REQUIREMENT

From Table 3.16, the mixing requirement is 5 W at TWL

Since the new volume is 324 m³, the power needed (*P*),

$$P = 324 \times 5 \text{ W}$$

$$P = 1{,}620 \text{ W}$$

$$= 1.62 \text{ KW}$$

3.12 PRIMARY SEDIMENTATION TANK

Sedimentation is a very important solid-liquid separation process in wastewater treatment. The main function of the primary sedimentation tank is to reduce the organic loadings on the secondary treatment units. These are usually done in large tanks in which the velocity of wastewater is reduced to allow solids to settle out of the water by gravity, where the settleable solids are pumped away (as primary sludge). Usually, this consists of simply holding wastewater for a short period in a tank under quiescent conditions, allowing the heavier solids to settle, and removing the "clarified" effluent as settled wastewater, as shown in Figure 1.3. The rectangular (Figure 3.27) and circular types of the tank (Figure 3.28) are the most common.

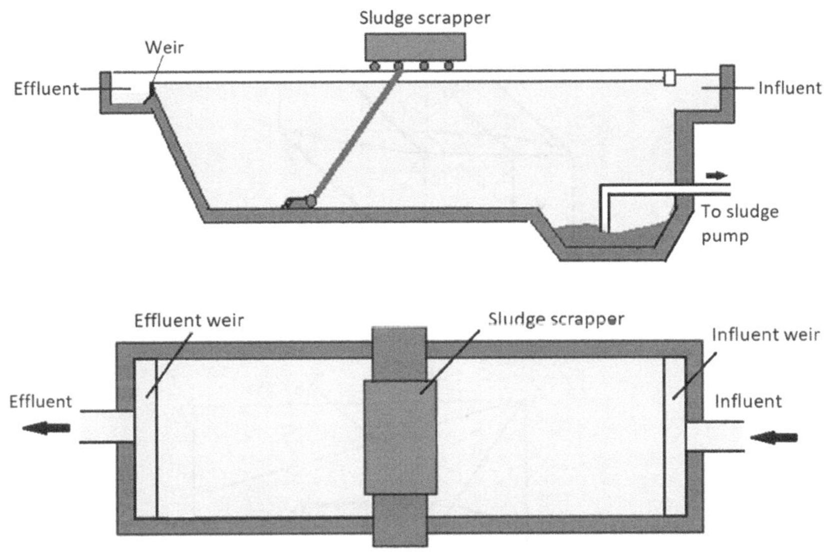

FIGURE 3.27 Rectangular primary sedimentation tank with horizontal flow.

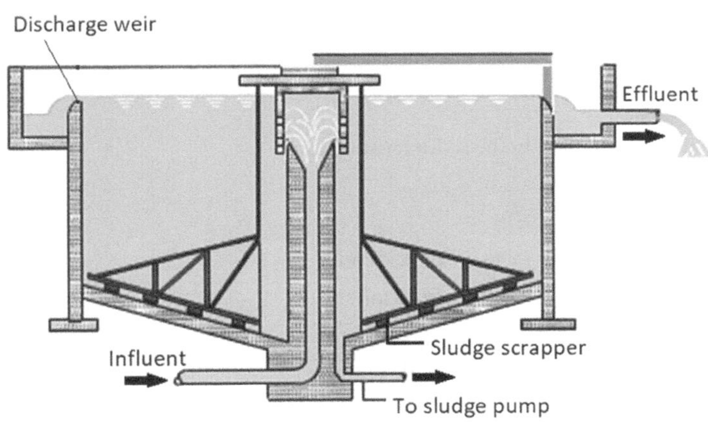

FIGURE 3.28 Circular primary sedimentation tank with radial flow.

In the design of the primary sedimentation tank, the horizontal velocity (V_h) and settling velocity (V_o) are controlled so that the particles will settle ideally.

Consider a rectangular sedimentation tank as shown in Figure 3.29.

The design of the sedimentation tank is based on the moving of a referred particle from the top of the tank (point A) and flows to the bottom of the tank (point C). For ideal settling, both V_o and V_h should be similar in this settling phenomenon [5]. There are three possible conditions in Figure 3.29:

1. If V_o is higher than V_h, the particle will settle in front of the tank. This is not favourable.
2. If V_h is greater than V_o, the particle will be flushed out of the tank without proper settling. This is not favourable.
3. If V_h is approximately like V_o, the particle settles ideally.

However, the settling is not influenced by the depth of the tank. Consider a rectangular shape, as shown in Figure 3.29. Assume the settling/terminal velocity is the depth over time taken for a particle to settle from point A to B.

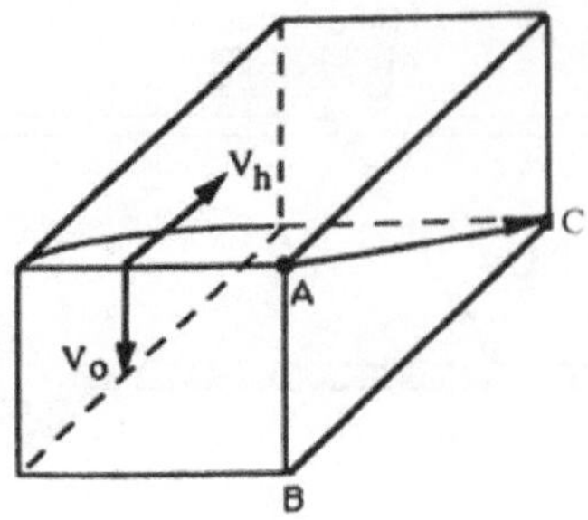

(a) The 3-D view

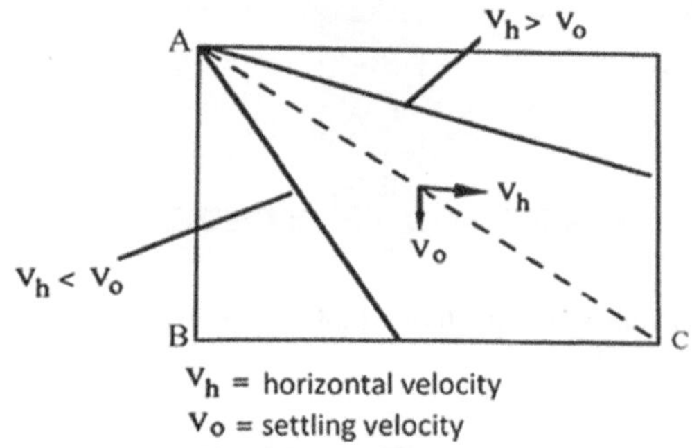

(b): Cross-section

FIGURE 3.29 Ideal sedimentation in the sedimentation tank.

$$V_o = \frac{\text{tank depth}}{\text{time taken}}$$

$$= \frac{\text{tank depth}}{\dfrac{\text{tank volume}}{\text{flow rate}}}$$

$$= \frac{\text{depth}}{\dfrac{\text{depth} \times \text{surface area}}{\text{flow rate}}}$$

$$= \frac{\text{flow rate}}{\text{area}}$$

$$= \frac{Q}{A_s}$$

Hence, settling is not influenced by the depth of the tank.

3.12.1 Sedimentation Theory

Settling basins and clarifiers are designed to retain water so that SS can settle by gravity. By sedimentation principles, suitable treatment technologies should be chosen depending on the specific gravity, shape, size, and shear resistance of particles. There are four types of sedimentation processes depending on the size and density of particles and physical properties of the solids:

Type 1: Dilutes, non-flocculent, free-settling, and unhindered (every particle settles independently).
Type 2: Dilute, flocculent (particles can flocculate as they settle).
Type 3: Concentrated suspensions, zone settling, hindered settling.
Type 4: Concentrated suspensions, compression (sludge).

These are summarized in Figures 3.30–3.32 and Table 3.17.

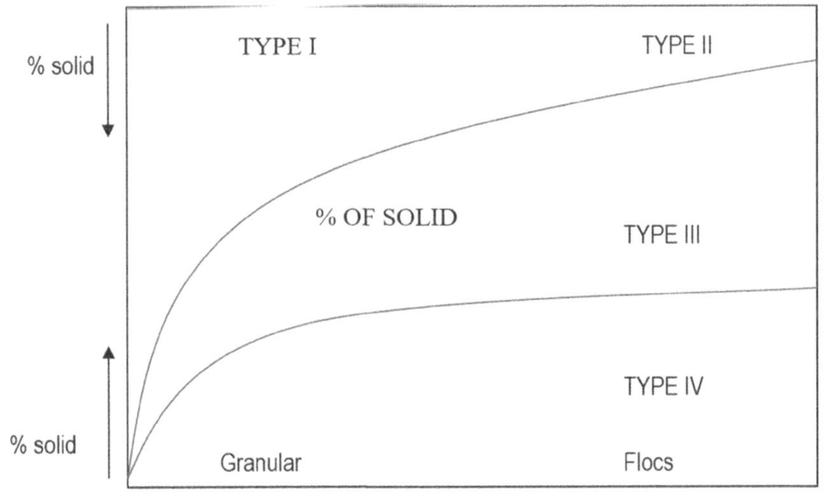

FIGURE 3.30 Classes of settlements.

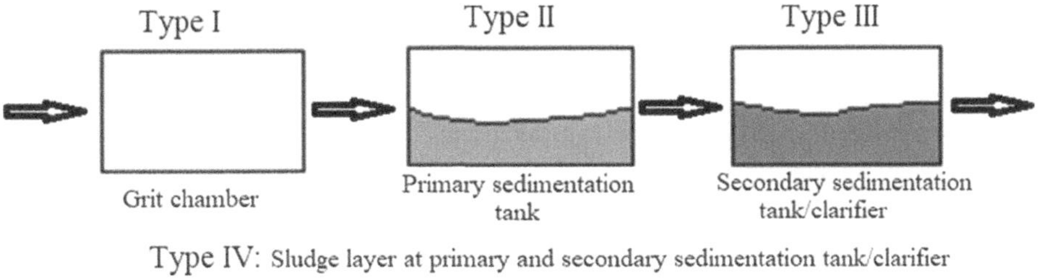

FIGURE 3.31 Primary location of settlements.

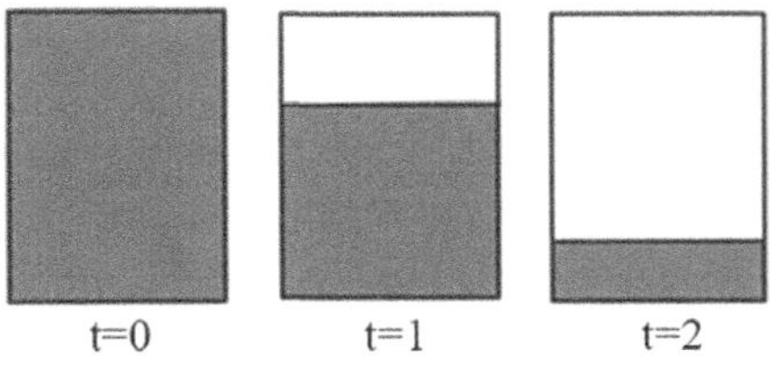

FIGURE 3.32 Zone settling in clarifier.

TABLE 3.17
Classes of Sedimentation

Classes of Settlement	Common Name	Characteristics	Example	Location of Occurrence (Figure 3.30)
Type I	Unhindered settling	Low solids – no flocs. Granular, dilute, non-flocculent, free-settling, and unhindered (every particle settles independently with no interactions between particles).	Settling of grit such as sand	Grit chamber
Type II	Flocs settling	Less solid – some flocs. Dilute and occurs at a low concentration of solid; settlement of flocs happens when particles collide and some of them are attached to each other (particles can flocculate as they settle).	Flocs	Primary sedimentation tank
Type III	Zone settling or hindered settling	Medium solids – high flocs. Particles are too close to each other; they no longer settle independently (settled as zone or blanket, as shown in Figure 3.32)	Flocs	Clarifier/secondary sedimentation tank
Type IV	Compression settling	High solids – sludge. Compression zone, the sludge is compressed by gravity (the weight of solids), as the settled solids are compressed under the weight of overlying solids.	Sludge settlement	Sludge zone in primary and secondary sedimentation tank/clarifier

3.12.2 Ideal Settling of a Discrete Particle (Type I)

When a particle enters the treatment plant, it has its own kinetic energy, whether it can be settled down (gravity force) or pushed upwards by wastewater flow. However, for a grit particle, its settling behaves as a discrete particle (Type I). There is a balance between the gravitational force and the buoyancy force (uplifting force). There is a kinetic model to relate this settling behaviour by assuming the particle is round and the Reynolds number ($R_e < 1$) (Stoke's Law). The discrete settling or terminal velocity of the perfect sphere can be expressed as in Equations 3.4–3.6:

$$v = \frac{gd^2(S_s - 1)}{18v} \tag{3.4}$$

$$v = \frac{gd^2(\rho_s - \rho_w)}{18\mu} \tag{3.5}$$

$$\text{For perfect sphere, } R_e = \frac{\rho_w \times v \times d}{\mu} \tag{3.6}$$

where

v = setting velocity (m/s)
g = gravitational acceleration (m/s^2)
d = diameter of particle (mm)
S_s = specific gravity of particles ($\frac{\rho_s}{\rho}$)
υ = kinematic viscosity of fluid (m^2/s)

$$\upsilon = \frac{\mu\ (\text{kg/ms})}{\rho\ (\text{kg/m}^3)}$$

μ = dynamic viscosity of fluid (kg/m.s)
ρ = density of fluid (kg/m^3)
ρ_s = density of particle (kg/m^3)
ρ_w = density of water (kg/m^3)
R_e = Reynolds number.

Example 3.7

Calculate the settling velocity of grit with sphere diameter 0.06 mm and specific gravity of 2.5 in wastewater. Take a dynamic viscosity (μ) at 20°C as 1.011×10^{-3} (Ns/m^2 or Pa.s or kg/ms).

Solution:

Given:

$$d = 0.06 \text{ mm}$$

$$g = 9.81 \text{ m/s}^2$$

$S_s = 2.5$, hence $\rho_s = 2{,}500 \text{ kg/m}^3$

$$\mu = 1.011 \times 10^{-3} \text{ Ns/m}^2$$

From Equation 3.5,

$$v = \frac{gd^2(\rho_s - \rho_w)}{18\mu}$$

$$= \frac{\left[(9.81 \text{ m/s}^2)(0.06 \times 10^{-3})^2 \text{ m}^2\right](2{,}500 - 1{,}000) \text{ kg/m}^3}{18 \times 1.011 \times 10^{-6} \text{ Ns/m}^2}$$

$$= 2.91 \times 10^{-3} \text{ m/s}$$

$$= 2.91 \text{ mm/s}$$

3.12.3 Design Approach

Primary sedimentation may remove SS in the range of 45%–65% and BOD in the range of 25%–40%.

Under MSIG, Volume 4, 2009 [1] has put the design criteria as given in Table 3.18. Also, provisions should be made on the following:

i. Removal of sludge daily.
ii. Holding tanks for the wasted sludge must be provided.
iii. V-notch weirs with baffle shall be provided at the outlet.
iv. Scum skimming shall be provided to remove both floating materials and scum. These materials can be either discharged to a biosolid holding tank or sand drying bed. They shall not be returned to the preliminary treatment units.
v. Flow distribution channel/chamber shall be provided for flow isolation or equal flow distribution.

where

W = Width(m)
L = Length(m)

TABLE 3.18
Design Parameters for Primary Sedimentation (MSIG, Volume 4, 2009) [1]

Description	Unit	Design Criteria
Sedimentation Followed by Secondary Treatment		
Detention time at Q_{peak}, $t=(V/Q_{peak})$	Hour	>2
Surface overflow rate (SOR) at Q_{peak}	m^3/m^2.day	<45
Circular or rectangular (maximum) = (Q_{peak}/top surface area) (Refer Figure 3.33 and Equation 3.7)		
Weir loading rate (WLR)[a] at Q_{peak} (Q_{peak}/length of the weir) (Refer Figure 3.34 and Equation 3.8)	m^3/m.day	<150
Horizontal or upward flow rate at Q_{peak} (Q_{peak}/approaching surface area) (Refer Figure 3.33 and Equation 3.9)	mm/s	<5[b]
Sizing of Rectangular Tanks		
Length : Width		> 3:1
Min water depth	m	2.5
Width : Depth		1:1 to 2.5:1
Sludge hopper		≥60° from horizontal with sludge pump
Sizing of Circular Tanks		
Water depth	m	>3.0
Floor slope wall		1:12 (8.33)
Diameter	m	≤50 Perimeter walk away if ≥50

[a] Refer Figure 3.33.
[b] Guidelines for Developers [6].

$$D = \text{Depth}(\text{m})$$
$$A_b = \text{Approaching surface area}\left(\text{m}^2\right)$$
$$A_s = \text{Surface area (top)}\left(\text{m}^2\right)$$
$$V_h = \text{horizontal velocity}(\text{mm/s}).$$

3.12.3.1 Surface Overflow Rate (SOR)

$$\text{SOR} = \frac{Q_p}{\text{Surface area (top)}}$$

$$\text{SOR} = \frac{Q_p}{A_s}\ \text{m}^3/\text{m}^2.\text{day}$$

$$\text{SOR} = \frac{Q_p}{W \times L}\ \text{m}^3/\text{m}^2/\text{day} \tag{3.7}$$

3.12.3.2 Weir Loading Rate (WLR)

$$\text{WLR} = \frac{Q_p}{L}\ \text{m}^3/\text{m.day} \tag{3.8}$$

where

L = length of weir (where the wastewater overflows)

$= W$

3.12.3.3 The Horizontal or Upward Flow Velocity

Horizontal velocity (V_h) is for the rectangular tank; the up-flow rate is for the circular tank.

$$V_h = \frac{Q_p}{A_b} \text{ mm/s} \tag{3.9}$$

Example 3.8

A primary sedimentation tank receives peak wastewater of 5,000 m³/day. If the SOR is 45m³/m².day, design:

(i) A rectangular tank
(ii) A circular tank with bottom 1:12.

Solution:

$$Q_{\text{peak}} = 5{,}000 \text{ m}^3/\text{day}$$

(I) RECTANGULAR TANK

(A) THE DIMENSION OF THE TANK

Assume the following data from Table 3.18:

$$\text{Length}(L) : \text{Width}(W) = 4:1$$

$$\text{Width}(W) : \text{Depth}(D) = 2.5:1$$

From Equation 3.7 and Figure 3.33,

$$\text{SOR} = \frac{Q_p}{A_s} \text{ m}^3/\text{m}^2/\text{day}$$

$$45 \text{ m}^3/\text{m}^2/\text{day} = \frac{5{,}000 \text{ m}^3/\text{day}}{A_s}$$

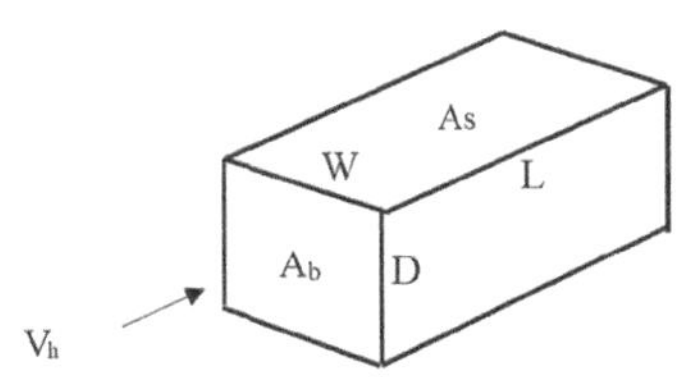

FIGURE 3.33 Formulation of surface overflow rate (SOR), weir loading (WL), horizontal velocity (V_h), and settling velocity (V_o).

$$A_s = 111.11\ \text{m}^2$$

$$A_s = L \times W$$

From the assumption,

$$L : W = 4 : 1, \text{ hence } L = 4W$$

So, $A_s = 4W \times W$

$$A_s = 111.11\ \text{m}^2 = 4W^2$$

$W = 5.27$ m, say 6 m

$$\text{Hence, } L = 4 \times 6$$

$$= 24\ \text{m}$$

From the next assumption,

$$W : D = 2.5 : 1, W \text{ from above} = 6\ \text{m}$$

$$D = \frac{6}{2.5}\ \text{m}$$

$$= 2.4\ \text{m, say } 3\ \text{m}$$

The new dimension is as follows:

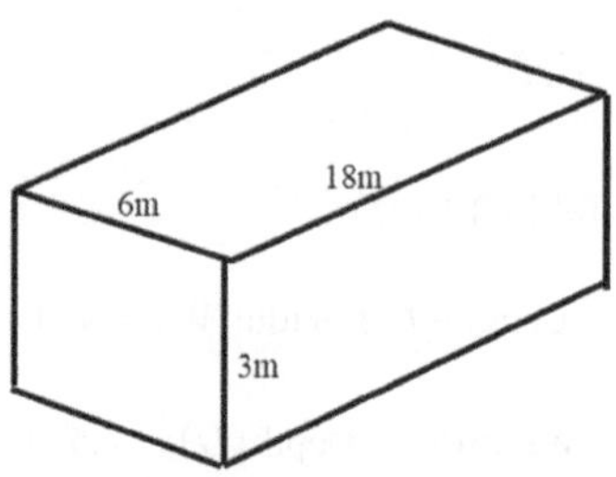

(b) Detention Time (t)

$$t = \frac{V}{Q_p}$$

$$\text{Volume} = V = 6 \times 3 \times 24\ \text{m}^3$$

$$V = 432\ \text{m}^3$$

$$t = \frac{432\ \text{m}^3}{5{,}000\ \text{m}^3/\text{day}}$$

$$= 0.0864\ \text{day}$$

$$= 2.07 \text{ hours} > 2 \text{ hours, acceptable (Table 4.18)}$$

(c) Weir Loading Rate

From Equation 3.8

$$\text{WLR} = \frac{Q_P}{L}\ \text{m}^3/\text{m/day}$$

FIGURE 3.34 Weir to control settled particles from escaping the tank.

$$L = \text{width of tank where the wastewater overflows}$$
$$= 6 \text{ m}$$
$$\text{WLR} = \frac{5{,}000}{6} \text{ m}^3/\text{m/day}$$
$$= 833 \text{ m}^3/\text{m/day} > 150, \text{ not acceptable (Table 4.18)}$$

Apply multiple weirs in series (six lines as an example in Figure 3.34), then

$$\text{WLR} = \frac{5{,}000}{6 \times 6} \text{ m}^3/\text{m/day}$$
$$= 139 \text{ m}^3/\text{m/day} < 150, \text{ acceptable (Table 4.18)}$$

(D) Horizontal Velocity (V_H)

From Equation 3.9,

$$V_h = \frac{Q_P}{A_b} \text{ mm/s}$$
$$V_h = \frac{Q_P}{W \times D} \text{ mm/s}$$
$$V_h = \frac{5{,}000 \text{ m}^3/\text{day}}{6 \text{ m} \times 3 \text{ m}} \text{ m/day}$$
$$V_h = 277.78 \text{ m/day}$$
$$= 3.22 \text{ mm/s}, < 5 \text{ mm/s, acceptable (Table 3.18)}$$

(II) A CIRCULAR TANK WITH A BOTTOM SLOPE OF 12:1

(A) Dimension of Tank

Consider the shape of the tank as follows:

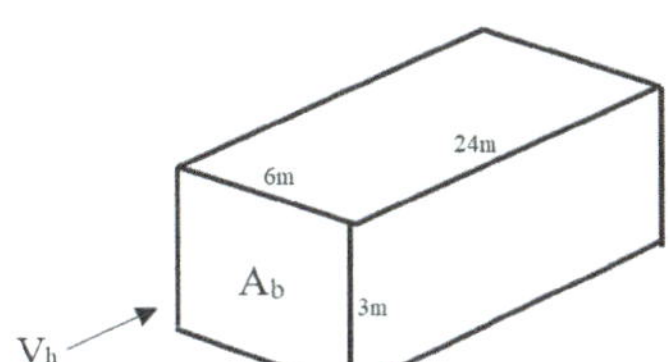

We know from previous calculation, surface area of tank, $A_s = 111.11\ m^2$

$$A_s = \pi r^2$$

where r is the radius

$$111.11 = \pi r^2$$

$r = 5.95$ m, say 6 m

Assume depth of the tank, $D = 4$ m (from Table 3.18).

Calculate the volume of the tank: Consider the shape as follows:

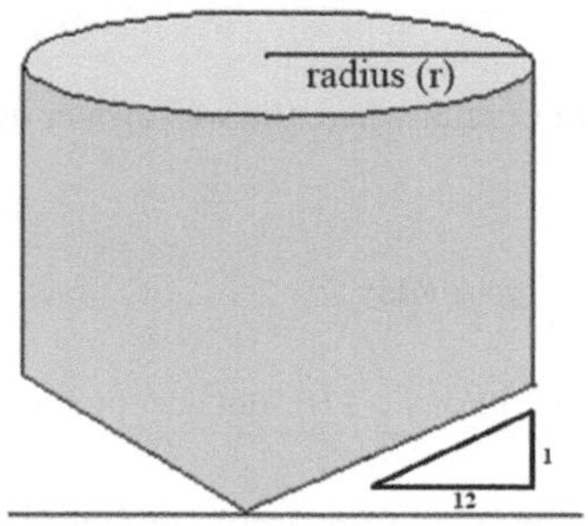

Total volume = V = Volume of cylinder (ABCDE) + Volume of cone (CDEFG)

(i) The volume of cylinder ABCDE (V_1)

$$V_1 = (\text{cylinder surface area})(\text{height})$$
$$= \pi r^2 (\text{AE})$$
$$= \pi (6)^2 (4)\ \text{m}^3$$
$$= (113.1\ \text{m}^2)(4)\ \text{m}^3$$
$$= 452.4\ \text{m}^3$$

Cylinder surface area = 113.1m²

(ii) The volume of cylinder CDEF (V_2)

Since the base to height ratio is 12:1, for 6 m radius (CD), the height is 0.5 m (DF) as follows:

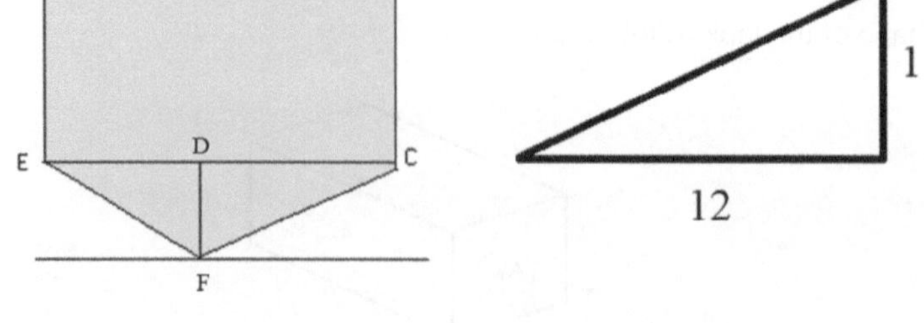

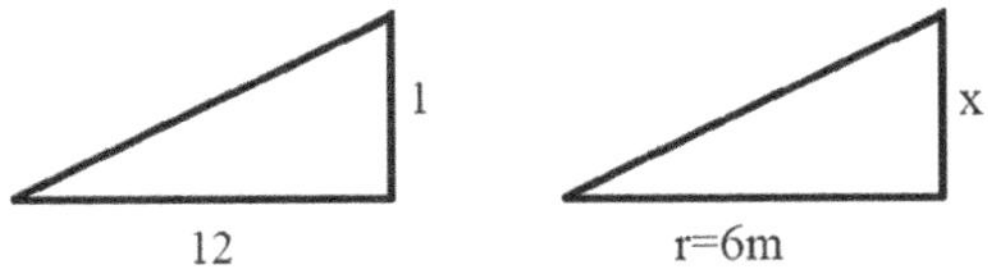

$$\frac{x}{1} = \frac{6}{12}$$

$$x = 0.5 \text{ m} = \text{DF}(\text{height of cone})$$

$$\text{Volume of cone } (V_2) = \frac{1}{3}(\text{base area})(\text{height})$$

$$V_2 = (\pi r^2)(0.5 \text{ m}^3)$$

$$V_2 = (\pi \times 6^2)(0.5 \text{ m}^3)(1/3)$$

$$= 18.85 \text{ m}^3$$

$$V_1 + V_2 = 452.4 + 18.85 \text{ m}^3$$

$$V = 471.25 \text{ m}^3$$

(b) Detention Time (t)

$$t = \frac{V}{Q_p}$$

$$t = \frac{471.25 \text{ m}^3}{5{,}000 \text{ m}^3/\text{day}}$$

$$= 0.09425 \text{ days}$$

$$= 2.262 \text{ hours} > 2 \text{ hours, acceptable (Table 3.18)}$$

(c) Weir Loading Rate

From Equation 3.8

$$\text{WLR} = \frac{Q_p}{L} \text{ m}^3/\text{m/day}$$

L = perimeter of cylinder where the wastewater overflows

$$= 2\pi r \text{ m}$$

$$= 2\pi\,(6) \text{ m}$$

$$= 37.7 \text{ m}$$

$$\text{WLR} = \frac{5{,}000}{37.7} \text{ m}^3/\text{m/day}$$

$$= 132.6 \text{ m}^3/\text{m/day} < 150, \text{ acceptable(Table 3.18)}$$

(d) Upward Flow Velocity (V_{up})

From Equation 3.9,

$$V_{up} = \frac{Q_p}{\text{Surface area of cylinder}} \text{ mm/s}$$

$$V_{up} = \frac{Q_p}{\pi r^2}\ \text{mm/s}$$

$$V_{up} = \frac{5{,}000\ \text{m}^3/\text{day}}{\pi\,(6)^2}\ \text{m/day}$$

$$V_{up} = 44.21\ \text{m/day}$$

$$= 0.511\ \text{mm/s},\ < 5\ \text{mm/s},\ \text{acceptable (Table 3.18)}$$

REFERENCES

1. *Suruhanjaya Perkhidmatan Air Negara (SPAN),* Malaysian Sewerage Industry Guidelines (MSIG), *Volume* 4 *Sewage Treatment Plants,* SPAN, 3rd Edition, Kuala Lumpur, 2009.
2. Standards & Industrial Research Institute of Malaysia (SIRIM), MS 1228, *Code of Practice for Design and Installation of Sewerage Systems*, Kuala Lumpur, 1991.
3. Government of Malaysia, Environmental Quality Act 1974 (Act 127), *Regulations, Rules & Orders, International Law Book Services*, Kuala Lumpur, 2022.
4. Qasim, S.R. *Wastewater Treatment Plants: Planning, Design, and Operation*, 2nd Edition, CRC Press, 1998.
5. Tebbutt, T.H.Y. *Principles of Water Quality Control*, 5th Edition, Butterworth-Heinemann, 1997.
6. Guidelines for Developers, *Volume* IV, *Sewage Treatment Plants*, 2nd Edition, Ministry of Housing and Local Government, Sewerage Services Department, 1998.

4 Fundamentals and Innovations of Iron and Manganese Control

Mu-Hao Sung Wang and Lawrence K. Wang

ACRONYM

ABF	Automatic backwash filtration
ADT	Air dissolving tube
AOP	Advanced oxidation process
AOP-SBR	Advanced oxidation process-sequencing batch reactor
APHA	American Public Health Association
AWWA	American Water Works Association
CES	Cooperative Extension Service
DAF	Dissolved air flotation
DAFF	Dissolved air flotation and filtration
DAOFF	Dissolved air-ozone flotation and filtration
DBP	Disinfection byproducts
DGF	Dissolved gas flotation
DO	Dissolved oxygen
DOM	Dissolved organic matter
EF	Electroflotation
Fe	Iron
Fe-Mn	Iron and manganese
GAC	Granular activated carbon
GPG	Grain per gallon
HAL	Health Advisory Level
IAF	Induced air flotation
IGF	Induced gas flotation
IX	Ion exchange
LIWT	Lenox Institute of Water Technology
MF	Microfiltration
Mn	Manganese
NDSU	North Dakota State University
NF	Nanofiltration
O&M	Operation and maintenance
PAC	Powdered activated carbon
PC	Physicochemical
PC-SBR	Physicochemical sequencing batch reactor
POE	Point-of-entry
POU	Point-of-use
R&D	Research and development
RO	Reverse osmosis

DOI: 10.1201/9781003532354-4

SBR	Sequencing batch reactor
SMCL	Secondary Maximum Contaminant Levels
STEAM	Science, Technology, Engineering, Arts, and Mathematics
STEM	Science, Technology, Engineering, and Mathematics
TOC	Total organic carbon
TTHM	Total trihalomethane
UCMR	Unregulated Contaminant Monitoring Rule
UF	Ultrafiltration
UIUC	University of Illinois at Urbana-Champaign
UNIDO	United Nations Industrial Development Organization
USEPA	US Environmental Protection Agency
WEF	Water Environment Federation
WQA	Water Quality Association
WTP	Water treatment plant

4.1 INTRODUCTION AND FUNDAMENTALS OF IRON AND MANGANESE CONTROL

4.1.1 Iron and Manganese in Groundwater and Surface Water

Natural sources of iron and manganese are more common in deeper wells where the natural water has been in contact with rock for a longer time. In coal mining regions, the two metals may also occur from both deep and surface mining activities. Iron and manganese often occur together in groundwater. Occasionally, iron pipes also may be a source of iron in drinking water. Groundwater with high iron and/or manganese may appear colorless initially but orange-brown (iron) or black (manganese) stains or particles quickly appear as the water is exposed to oxygen. Iron and manganese also cause operational problems in industrial processes. The following is a summary: (a) groundwater source: Iron and manganese are often found in groundwater in their dissolved forms, Fe^{2+} and Mn^{2+}, due to the reducing conditions in aquifers; and (b) surface water source: These elements can also be present in surface water, typically in particulate forms due to oxidation.

4.1.2 Health and Aesthetic Concerns

While not typically harmful at low concentrations, high levels of iron and manganese can cause health issues and are regulated in drinking water standards [1,2].

Iron and manganese can cause staining of laundry and plumbing fixtures, and impart a metallic taste to water.

4.1.3 Conventional Treatment Methods

There are four basic conventional water treatment methods for iron and manganese control: (a) oxidation and filtration; (b) sequestration by phosphate addition; (c) ion exchange (IX) process; and (d) biological filtration. This publication introduces both conventional and innovative processes but places emphasis on the innovative processes.

4.1.3.1 Oxidation and Filtration

Oxidation converts dissolved iron and manganese into their particulate forms. Common oxidants include chlorine, potassium permanganate, and ozone.

Filtration removes the oxidized particles from the water. Filtration media can include sand, anthracite, or specialized media like greensand.

4.1.3.2 Sequestration by Phosphate Addition

Phosphates can be added to water to keep iron and manganese in solution, preventing them from precipitating and causing stains. There are some limitations, however. Sequestration is generally used for low concentrations of iron and manganese and is not a removal method.

4.1.3.3 Ion Exchange

IX resins can be used to remove iron and manganese, particularly in water softening processes. Without IX resin regeneration, the process will be expensive. Therefore, the resins need to be periodically regenerated with a salt solution to maintain their effectiveness.

4.1.3.4 Biological Filtration

Biological filtration process is effective for iron and manganese removal due to microbial action. Certain bacteria can oxidize iron and manganese, forming particles that can then be filtered out by conventional filtration process. This biological method can be more sustainable and cost-effective in the long term.

4.1.4 Industrial Applications for Iron and Manganese Removal

There are two major industrial applications: (a) Boiler Feed Water Systems: High-purity water is essential to prevent scaling and corrosion in boilers. Iron and manganese removal is critical in this context; and (b) Cooling Systems: Iron and manganese should be controlled for preventing fouling and maintaining efficiency in industrial cooling systems.

4.1.5 US Cooperative Extension Service and UN International Service

Cooperative Extension Service (CES) is an educational organization usually affiliated with a State-funded university. It is directed toward the rural population in the State. CES members are usually the State University professors who offer their advice and information voluntarily to help the rural population solve their problems. CES also aims to increase the efficiency of the family farm, increase production and generally increase the standard of living of the farm family.

CES empowers farmers, ranchers and rural communities of all sizes to meet the challenges they face, adapt to changing technology, improve nutrition and food safety, prepare for and respond to emergencies, improve water quality and protect our environment.

The purpose of a "cooperative extension service" is to realize the economic, cultural and social needs of the rural population and its surrounding community, and to provide assistance in fulfilling their needs. Cooperatives often have a strong commitment to their community and a focus on strengthening the community they exist in or serve.

Writing this publication is for updating the authors' previous CES service at the University of Illinois at Urbana-Champaign (UIUC), and for disseminating technical information to agricultural countries through the United Nations Industrial Development Organization (UNIDO).

This publication introduces the state-of-the-art control technologies and terminologies for removing iron and manganese from groundwater. It is an updated academic spin-off from the authors' previous technical manuals of the Lenox Institute of Water Technology (LIWT) [1–3], the UNIDO [4] and the UIUC, CES [5]. Both author Mu-Hao Sung Wang and author Lawrence K. Wang were LIWT professors. Author LK Wang was also a UNIDO Senior Adviser and a Professor and Extension Specialist of UIUC.

Specifically the old technical information [1–5] was updated with the available data sources from the US Environmental Protection Agency (USEPA) [6], the Great Lakes–Upper Mississippi River Board of State and Provincial Public Health and Environmental Managers [7], the University of Nebraska-Lincoln CES [8,9], New Hampshire State Department of Environmental Service

[10], the North Carolina State University CES [11], North Dakota State University (NDSU), North Dakota Extension Service, South Dakota Extension Service, Minnesota Extension Service [12], Wilkes University [13], Pennsylvania State University Extension Service [14], American Water Works Association [15,16], process equipment manufacturers [17–21] and the authors' own LIWT [22–46]. Since both authors were professors and public servants, it is the authors' sincere hope that this publication may be useful to all State CES in this nation, and all other agricultural nations.

4.1.6 Summary

This publication covers the subjects of: (a) the sources, environmental engineering significance, environmental impact, types, concentrations and problems of soluble iron Fe(II) and manganese Mn(II) in groundwater, (b) the CES, and the Secondary Maximum Contaminant Levels (SMCL) recommended by the USEPA for iron and manganese in the drinking water; (c) the state-of-the-art technologies for control or removal of iron and manganese; (d) the recommended dosages of various oxidants for conversion of soluble Fe(II) and Mn(II) to their insoluble forms of Fe(III) and Mn(IV), respectively; (e) the US 10 States Standards for design of iron and manganese removal facilities; (f) the advantages and disadvantages of various iron and manganese control or removal processes; (g) the rules of thumb for process system selection; (h) the latest research and development (R&D) and call for further research; and (i) all related technical terminologies, such as sequestering chemicals, polyphosphate, sodium silicate, potassium permanganate, oxygen, ozone, chlorine, chlorine dioxide, catalytic oxidation, filtration, oxidizing filtration, manganese greensand, birm, manganese dioxide filtration media, Pyrolox, Pro-OX, IX, softener, dissolved air flotation (DAF), induced air flotation (IAF), electroflotation (EF), physicochemical sequencing batch reactor (PC-SBR), Mangazur, Ferazur, Advanced Oxidation Process (AOP), etc.

For removal of soluble iron and manganese from groundwater, a CES Extension Specialist may recommend to the farmers one of the following process systems for their consideration depending on the raw groundwater quality, drinking water standards and the farmers' needs:

a. **Sequestration by Polyphosphate or Sodium Silicates:** This treatment process is only for treating low levels of dissolved iron and manganese (Fe-Mn) at combined concentrations up to 3 mg/L; USEPA is discouraging this practice;
b. **IX Water Softener:** This treatment process is normally for treating low to moderate levels of dissolved Fe-Mn, concentrations at less than 5 mg/L. The NDSU has recommended the upper limit Fe-Mn concentration = 10 mg/L [12]; IX will not work where the Fe-Mn have turned to a rusty color; the use of IX to treat only Fe-Mn is not recommended due to many disadvantages. Where Fe-Mn concentrations are high (almost 5 mg/L) and hardness is also high (over 150 mg/L as $CaCO_3$), then IX is an appropriate treatment process;
c. **Oxidizing Filtration Alone (such as Natural Manganese Greensand or Manufactured Zeolite Coated with Manganese Oxide, or Manganese Dioxide Media)**: This treatment system is an option normally for treating moderate levels of dissolved iron and manganese at combined concentrations up to 15 mg/L although NDSU has recommended an upper limit Fe-Mn concentration = 10 mg/L; raw water NOM/TOC concentrations should be less than 5 mg/L; Water Quality Association (WQA) has set the voluntary performance standards for oxidizing filtration: "An oxidizing filter shall reduce 10 mg/L plus or minus 1.0 mg/L soluble iron to not more than 0.2 mg/L, and 2.0 mg/L plus or minus 0.2 mg/L soluble manganese to not more than 0.05 mg/L";
d. **Aeration Oxidation (Including Venturi Aeration, Induced Air Flotation Aeration, etc.) Followed by Filtration:** High levels of dissolved iron and manganese at combined concentrations up to 25 mg/L can be oxidized to a solid form by aeration (mixing with air) if the concentrations of natural organic matters (NOM), total organic carbon (TOC) and ammonia are lower than 5 mg/L;

e. **Chemical Oxidation Followed by Filtration:** This treatment system can treat groundwater containing high levels of dissolved or oxidized iron and manganese greater than 10 mg/L, and if iron is combined with NOM/TOC forming organometallic complex, or when Fe-Mn bacteria is present, potassium permanganate, hydrogen peroxide, chlorine, chlorine dioxide or ozone is effective chemical oxidant for chemical oxidation prior to filtration. Although chemical oxidation and filtration will work at all levels of soluble iron, it is recommended only for levels above 10 mg/L of soluble iron;
f. **Biological Oxidation-Filtration:** Commercial Ferazur and Mangazur (which are both pressure biological oxidation filters seeded with Fe bacteria and Mn bacteria, respectively) have successfully treated raw water containing soluble iron up to 1.71 mg/L and soluble manganese up to 1.00 mg/L, respectively, and their effluents have met the Canadian Drinking Water Quality Guides of 0.3 mg/L Fe and 0.05 mg/L Mn [21];
g. **Lime-Soda Ash Softening Process System Including Oxidation and Filtration:** It is recommended by the 10 States Standards for simultaneous removal of hardness, iron and manganese, coliforms, total trihalomethane (TTHM) at all their concentration levels [7];
h. **Lime-Sodium Aluminate Softening Process System Including Oxidation and Filtration:** It is developed by the LIWT for simultaneous removal of hardness, iron and manganese, coliforms, TTHM at all their concentration levels [26];
i. **Advanced Oxidation Process (AOP) Followed by Filtration:** This treatment system is well established. It can treat the contaminated groundwater under the direct influence of surface water. The raw water may contain high concentrations of Cryptosporidium and coliforms, some NOM/TOC, and high levels of dissolved or oxidized iron and manganese greater than 10 mg/L, and if iron is combined with some NOM/TOC forming organometallic complex, or when Fe-Mn bacteria is present [3–5];
j. **Dissolved Air Flotation Clarification (DAF) Followed by Oxidation and Filtration:** This treatment system is well established. It can treat the contaminated groundwater under the direct influence of surface water. The raw water may contain high concentrations of Cryptosporidium, coliforms, Fe-Mn bacteria, volatile organic compounds (VOC), NOM/TOC, organometallic complex, THM and high levels of dissolved or oxidized iron and manganese greater than 10 mg/L;
k. **Physicochemical (PC) Sequencing Batch Reactor (SBR) Followed by Filtration:** This system was developed by the LIWT, now well established although its applications are new. There are many SBR process equipment manufacturers. The system uses one or more oxidants, such as air, ozone, oxygen, hydrogen peroxide, chlorine, potassium permanganate, etc. The filtration can be sand filtration, multi-media filtration, granular activated carbon (GAC) filtration, oxidizing filtration, cloth filtration, bag filtration, diatomaceous earth precoat filtration, membrane filtration, etc.; this treatment system can treat contaminated groundwater under the direct influence of surface water. The raw water may contain high concentrations of Cryptosporidium, coliforms, Fe-Mn bacteria, NOM/TOC, organometallic complex, THM and high levels of dissolved or oxidized iron and manganese greater than 10 mg/L [3–5];
l. **Biological Oxidation Contactor Followed by Oxidizing Filtration:** (a) It is a case history of Gilbert (population = 1,100), IA, USA; (b) Raw water quality: temperature = 13.3°C; pH = 7.63; alkalinity = 410 mg $CaCO_3$/L; total hardness = 280 mg $CaCO_3$/L; As = 23 pg/L; Fe = 2.91 mg/L; Mn = 0.08 mg/L; NH_4 = 2.91 mg N/L; NO_2 = 0.01 mg N/L; NO_3 = 0.02 mg N/L; TOC = 2.68 mg/L; DO = 1.1 mg/L; (c) Water quality analysis: Raw water had high hardness concentration; toxic arsenic, high iron concentration, moderate manganese concentration, high ammonia concentration, and high TOC concentration; (d) biological oxidation contactor design: gravel = 0.5 inch; bed depth = 55 inches; air flow = 2.5 L/min; loading rate = 2.2 gpm/ft^2; backwash schedule = monthly; (e) oxidizing filtration design: anthracite bed depth = 10 inches on the top; Mn oxide-coated sand = 30 inches at bottom;

(f) Final treatment results: (f1) Biological oxidation contactor and oxidizing filter are both biologically active; (f2) Contactor is continuously aerated; (f3) Filter removes particles; (f4) Nitrification requirement = 4.5 mg O_2 per mg NH_3-N; (f5) Fe oxidized by air and bacteria; (f6) bacteria oxidize As(III) As(V), which then sorbs with Fe oxides; (f7) bacteria oxidize Mn [6].

4.2 IRON AND MANGANESE CONTROL TECHNOLOGIES AND TERMINOLOGIES

Advanced Oxidation Process (AOP): It is an advanced physicochemical (PC) process system involving the use of both ultraviolet (UV) light and a strong chemical oxidant (such as ozone, potassium permanganate, hydrogen peroxide, chlorine, chlorine dioxide, etc.) for oxidation of iron, manganese and other organic substances or microorganisms. AOP is a highly effective oxidation process and disinfection process, but it is very expensive. AOP can be either a continuous process or a sequencing batch reactor AOP process (AOP-SBR). The AOP-SBR is developed by the LIWT for small water treatment plants (WTPs). The AOP-SBR is used only when the groundwater has high concentrations of iron and manganese and in the presence of high concentrations of ammonia, or NOM, such as humic acid, lignin, tannin, etc. AOP-SBR is a PC-SBR process system.

Advanced Oxidation Process (AOP) and Filtration: It is a complete process system for initial oxidation of iron and manganese (Fe-Mn) by AOP and subsequent removal of Fe-Mn by filtration when Fe-Mn concentrations and NOM concentrations in groundwater are high. Filtration unit can be sand filtration, multi-media filtration, GAC filtration or oxidizing filtration. However, the use of membrane filtration (MF, UF, NF, RO) for final polishing is not recommended.

Aeration: Aeration process units may work by cascading, bubbling, air-stripping, surface aeration, submerged aeration, DAF, or IAF to add air to the water in order to accomplish oxidation by oxygen in the air. Aeration may be advantageous because it does not add chemicals to the water. Maintenance costs are low for aeration units but the initial purchase costs are often higher than other treatment options. Aeration units also require a filter for the removal of the oxidized iron and manganese which must be backwashed. The water should also be disinfected to keep bacteria from colonizing the aerator.

Aeration Followed by Filtration for Iron and Manganese Removal: (a) High levels of dissolved iron and manganese at combined concentrations up to approximately 5,000–10,000 μg/L (5–10 mg/L) can be removed with aeration (mixing with air) followed by filtration. In this process, air is mixed with water to oxidize the iron and manganese into particles. The air-saturated water with saturated dissolved oxygen (DO) then enters the vessel, usually a pressure-type aerator for domestic use, where air is separated from the water. The water then flows through a filter to remove the iron and manganese particles. More time and DO are required for removing manganese compared to iron with this type of system. Periodically backwashing of the filter is the most important maintenance step involved in operation to ensure proper performance. (b) Aeration is not recommended for water containing organic complexes of iron/manganese or iron/manganese bacteria that will clog the aspirator and filter. (c) Venturi aeration is commonly used by WTPs. Where DO in the water is low, air is added to the water by a special Venturi nozzle. The mixture of air and water then passes into a detention tank where the oxygen dissolves into the water for oxidation reactions. The aerated water then enters the filtration unit for final polishing. (d) Air dissolving tube (ADT) is an innovative aeration process equipment. The mixture of air and water passes into an ADT under pressure where the air (containing oxygen and nitrogen) dissolves into the water under supersaturated condition. The ADT effluent enters a water tank where oxidation occurs. The tank effluents then enter a filtration unit for final polishing.

Air Flotation + Physical Separation for Iron and Manganese Removal: (a) Either DAF or IAF plus filtration may be an effective physical-chemical process for iron removal. In this system, the DAF or IAF added air is an oxidant for converting soluble Fe(II) to insoluble Fe(III) which can then be removed by sand filtration or multi-media filtration, or GAC filtration; (b) No additional oxidant will be required if the oxygen in air is sufficient; (c) No additional sedimentation clarifier is needed if DAF or IAF is used; (d) For combined Fe-Mn removal, the DAF/IAF required chorine disinfectant will be sufficient for manganese removal if ammonia/TOC/DOM concentrations in water are not elevated; (e) For combined Fe-Mn removal, low dosage of permanganate will be needed for manganese removal if ammonia/TOC/DOM concentrations in water are elevated. The process system can be a continuous system, or a sequencing batch system. Do not use any type of membrane filtration (such as MF, UF, NF or RO) for physical separation because a membrane filter will become clogged by rust particles.

Air Oxidation (Aeration) for Iron and Manganese Removal: (a) Aeration process is simple physical-chemical process using air as an oxidant; (b) It converts soluble iron ions (II) to iron particles (III) that are easier to remove via filtration; but it is not effective for Mn ions removal, unless an additional oxidant, such as ozone, hydrogen peroxide, chlorine, potassium permanganate, etc. is used.

Anion Exchange: An IX process involves the use of anion exchanger (anionic IX resin) mainly for removal of negatively charged ions, such as sulfate, phosphate, chloride, and negatively charged organic ions.

Bacterial Iron: Slime depositing in toilet tanks, bath tubs or fouling water filters and softeners is an indication of bacterial iron's presence. It is also called "iron biofouling." Bacterial iron attacks wells and water systems in all sorts of aquifer environments, both contaminated and pristine. In some places, it causes great damage; in others, it is considered a minor nuisance.

Bag Filtration: It is a simple low-cost filtration process device using bags as the filter media. Bag filtration is often used by small WTPs to remove the fully oxidized iron and manganese as the water comes from the well. The particles are removed from the water by passage through bag filters. The bags must be manually cleaned which creates higher operational costs.

Biological Oxidation for Iron and Manganese Removal: (a) It is a biological process using bacteria naturally present in the groundwater to oxidize soluble Fe and Mn ions; (b) It has longer filter run times, lower chemical cost and can simultaneously remove Fe, Mn and NH_3; (c) Mn-oxidizing bacteria have long acclimation time (3–6 months); (d) Ferazur-Mangazur Filtration System is a commercially available pressure biological filtration system supplied by ONDEO Degrémont Ltd. of Dorval, QC, Canada, specifically for iron and manganese removal; (e) Conventional slow sand filtration (by gravity) after seeding with iron and manganese bacteria can also remove iron and manganese; (f) The LIWT has developed a biological pressure GAC filtration system for removing iron and manganese when it is properly seeded with iron and manganese bacteria.

Biological Oxidation Contactor Followed by Oxidizing Filtration: (a) It is a case history of Gilbert (population = 1,100), IA, USA; (b) Raw water quality: temperature = 13.3°C; pH = 7.63; alkalinity = 410 mg $CaCO_3$/L; total hardness = 280 mg $CaCO_3$/L; As = 23 pg/L; Fe = 2.91 mg/L; Mn = 0.08 mg/L; NH_4 = 2.91 mg N/L; NO_2 = 0.01 mg N/L; NO_3 = 0.02 mg N/L; TOC = 2.68 mg/L; DO = 1.1 mg/L; (c) Water quality analysis: Raw water had high hardness concentration; toxic arsenic, high iron concentration, moderate manganese concentration, high ammonia concentration, and high TOC concentration; (d) biological oxidation contactor design: gravel = 0.5 inch; bed depth = 55 inches; air flow = 2.5 L/min; loading rate = 2.2 gpm/ft^2; backwash schedule = monthly; (e) oxidizing filtration design: anthracite bed depth = 10 inches on the top; Mn oxide-coated sand = 30 inches at bottom; (f) Final treatment results: (f1) Biological oxidation contactor and oxidizing filter are both biologically active; (f2) Contactor is continuously aerated; (f3) Filter removes particles; (f4) Nitrification requirement = 4.5 mg O_2 per mg NH_3-N; (f5) Fe oxidized by air and bacteria; (f6) bacteria oxidize As(III) As(V), which then sorbs with Fe oxides; (f7) bacteria oxidize Mn.

Birm Catalyst Oxidizing Filtration Advantages and Disadvantages: (a) Advantages are: (a1) No "foreign" chemicals are added to the water; (a2) Its labor cost is low; (a3) It can handle a wide range of Fe-Mn levels; and (a4) It can often reduce some objectionable odor. (b) There are few disadvantages.

Birm Catalyst Oxidizing Filtration Operation with Pre-Venturi Aeration: (a) Where DO in the water is low, ordinary air is added to the water by a special Venturi nozzle. The mixture of air and water then passes into a detention tank where the oxygen dissolves into the water. An air release valve allows the unused nitrogen and excess air to be bled off. The water then enters the birm filtration media filter tank. The coating on the birm media acts as a catalyst to force the completion of the chemical reaction between the DO and soluble Fe(II)-Mn(II) so as to form insoluble Fe(III)-Mn(IV) rust particles. (b) The insoluble Fe-Mn precipitate is then physically caught by the birm filtration media; (c) The birm filter media is backwashed periodically to remove the precipitates, but no chemical rejuvenation of the birm is needed. This method is effective for very high levels of Fe-Mn; (d) There are some constraints when using birm media. Where the DO level of the raw water is above 15%, there is no pressing need to install the Venturi nozzle; however, such installation is often still done in view of the low cost of the device and in consideration of possible source water variation. The pH should be over 7.0+ for iron and 8.0+ for manganese removal. If both Fe-Mn are substantially present, the pH should be under 8.5 so as not to produce colloidal iron. There are minimum flow requirements for backwash. Other manufacturer comments recommend that tannins and hydrogen sulfide should be low. These other contaminants will produce a precipitate that fouls the media. Organic material should be less than 5 mg/L. Effectiveness of the birm media will be reduced by long-term exposure to chlorine; however, short-term chlorination to kill Fe-Mn bacteria is acceptable. No appreciable phosphate should be present. The precipitate of Fe-Mn tends to lower the pH and thus the alkalinity or pH adjustment may be needed. An "enriched" media is often necessary to remove manganese.

Birm Filter: (a) Birm filter is one of three oxidizing filters (manganese greensand filter, birm filter and manganese oxide filter). Birm is a trade name of the Clack Corp. The media is manufactured from natural pumice mineral which is then covered with a very thin layer of manganese dioxide. The filter medium acts as a catalyst to promote the reaction between the oxygen and iron dissolved in the water. A brim filter does not require regeneration because it utilizes DO present in the raw water to oxidize the iron and manganese. As a result, the raw water must contain a certain amount of DO and the pH should be at least 6.8 for iron removal and 7.5 for manganese removal. The oxygen acts as a catalyst to change the dissolved iron to a dense rust particle that the birm can then remove. The iron rust contaminants become trapped in the birm filter media and are then regularly removed by an automated backwashing of the iron filter. (b) It absolutely cannot be used in the presence of hydrogen sulfide gas or if chlorine is in the water. (c) Because most groundwaters are generally low in DO, some type of aeration is needed to boost the DO in the water. While lowest in cost, birm may need to be changed every few years.

Catalytic Granular Activated Carbon (GAC) Adsorption and Filtration for Iron and Manganese Removal: GAC adsorbs and then oxidizes and filters dissolved iron in 1 unit. It is effective for concentrations of dissolved iron or dissolved manganese less than 1.0 mg/L. Maintenance requirements are less than oxidizing filters because no chemicals are added, but backwashing is still necessary. Catalytic GAC requires a minimum of 4.0 mg/L of DO in the source water. Some raw water supplies may need pre-aeration, Venturi aeration, DAF or IAF to increase the DO concentration.

Catalytic Oxidation: Catalytic oxidation of iron and manganese is achieved by the use of a suitable catalytic filter medium (oxidizing filtration medium), such as manganese greensand, birm, manganese oxide (Pyrolox, etc.). An oxidizing agent is added to the water to be treated. This may be atmospheric oxygen, chlorine or potassium permanganate depending upon the catalytic medium used. It is then passed through the filtration device containing the appropriate medium. Given the sufficient degree of oxidation of the water, proper pH and sufficient detention time with the catalytic medium, iron and manganese are removed.

Cation Exchange: An IX process involving the use of cation exchanger (cationic IX resin) is mainly used for removal of calcium, magnesium, iron and other positively charged metal ions, or positively charged organic ions.

Chemical Coagulation: This is a chemical process in which one coagulating chemical coagulates (reacts) with another colloidal or suspended solid forming insoluble fine flocs due to charge neutralization and agglomeration.

Chemical Oxidation: Chemical oxidation processes include (a) air oxidation, (b) free chlorination using chlorine (Cl_2) or sodium hypochlorite, (c) ozonation using ozone (O_3), (d) permanganation using potassium or potassium permanganate (MnO_4^-), (e) hydrogen peroxide oxidation, (f) pure oxygen oxidation and (g) AOP, such as UV+one of the above oxidants.

Chemical Oxidation Followed by Filtration: (a) In this method, the water is treated with an oxidizing agent such as ozone, chlorine or potassium permanganate to convert any dissolved iron and manganese into solid, oxidized forms that can then be filtered through a sand trap filter. Significant retention or contact time is required to allow oxidation to take place. For this reason, a storage tank may be used. (b) This option is commonly applied to public water supplies and is less frequently applied to private water supplies due to their complexity and cost. (c) High levels of dissolved or oxidized iron and manganese greater than 10,000 µg/L can be treated by chemical oxidation followed by a sand trap filter to remove the precipitated particles. This treatment is particularly useful when iron is combined with organic matter or when iron bacteria are present.

Chemical Precipitation: A chemical process in which one soluble chemical agent precipitates another soluble chemical agent.

Chlorination and Filtration: (a) It is one of the chemical oxidation and filtration options for iron and manganese removal. (b) The pH of the water supply plays an important role when choosing an oxidizing agent. Chlorine bleach is most effective for oxidizing iron at pH level 6.5–7.5. If the pH of water is less than 6.5, a neutralization of water is needed before chemical oxidation. Furthermore, chlorine is not recommended as an oxidant for very high manganese levels, as a very high pH (pH of 9.5 or greater) is necessary to completely oxidize manganese. Potassium permanganate is more effective in oxidizing manganese at pH values above 7.5.

Chlorination for Iron and Manganese Removal: (a) It is an effective chemical process using chlorine as an oxidant for removing Fe and Mn ions if filter media is coated with MnO_x. (b) It is not effective if elevated NH_3 is present. (c) It could form disinfection byproducts (DBPs) if TOC or dissolved organic matter (DOM) is elevated. (d) A chlorine feeder is used to feed the chlorine (sodium hypochlorite solution) into the water upstream. (e) The mixing tank or pipe coil is necessary to provide contact time for the insoluble iron and manganese precipitates to form. Chlorine is not recommended as an oxidant for very high manganese levels because a very high pH is necessary to completely oxidize the manganese.

Chlorine Dioxide Dosage for Iron and Manganese Oxidation: 1.21 mg chlorine dioxide per mg Fe(II); 2.45 mg chlorine dioxide per mg Mn(II). Almost all the common oxidants except chloramines will convert Fe(II) ferrous iron ions to insoluble ferric hydroxide Fe(III) state and will convert Mn(II) manganese ions to insoluble manganese dioxide Mn(IV) state.

Chlorine Dosage for Iron and Manganese Oxidation: 0.62 mg chlorine per mg Fe(II); 0.77 mg chlorine per mg Mn(II). Almost all the common oxidants except chloramines will convert Fe(II) ferrous iron ions to insoluble ferric hydroxide Fe(III) state and will convert Mn(II) manganese ions to insoluble manganese dioxide Mn(IV) state.

Cooperative Extension Service (CES): (a) CES is an educational organization usually affiliated with a State-funded university. It is directed toward the rural population in the State. CES members are usually State university professors who offer their advice and information voluntarily to help the rural population solve their problems. CES also aims to increase the efficiency of the family farm, increase production and generally increase the standard of living of the farm family; (b) CES empowers farmers, ranchers and rural communities of all sizes to meet the challenges they face, adapt to changing technology, improve nutrition and food safety, prepare for and respond to emergencies, improve water quality and protect our environment; (c) The purpose of a "cooperative extension service" is to realize the economic, cultural and social needs of the rural population and its surrounding community, and to provide assistance in fulfilling their needs. Cooperatives often have a strong commitment to their community and a focus on strengthening the community they exist in or serve.

Dissolved Air Flotation (DAF): DAF is a process involving pressurization of air at 25–95 psig for dissolving air into water, and subsequent release of pressure (to 1 atm) under laminar flow conditions for generating extremely fine air bubbles (normally 20–80 μm) which become attached to the impurities to be removed. The air flow rate is about 1% of influent liquid flow rate. The attachment of air bubbles to the impurities can be a result of physical entrapment, electrochemical attraction, surface adsorption and/or gas stripping. The specific gravity of bubble-impurity agglomerate is less than 1, resulting in buoyancy or non-selective flotation (i.e. a save-all process). In the past 40 years, DAF has been used mainly for sludge thickening and fiber recovery. Now DAF becomes a high technology for potable water purification and wastewater treatment.

Dissolved Air Flotation (DAF) Followed by Oxidation and Filtration: It is a process system developed by the LIWT for removing iron and manganese from ground water or cooling tower water. DAF is one of the dissolved gas flotation (DGF) processes if air is used for flotation. DAF will give high DO concentration in the water, so no extra oxidant will be needed for dissolved iron Fe(II) oxidation and removal. For removal of dissolved manganese Mn(II), extra oxidant, such as potassium permanganate, is needed for manganese oxidation, if the subsequent filtration is a conventional sand filtration unit. The post-filtration can be sand filtration, multi-media filtration or oxidizing catalytic filtration. This PC process system can be operated as a continuous system or a SBR system.

Dissolved Air-Ozone Flotation and Filtration (DAOFF) Process System: DAOFF is an innovative process system developed by the LIWT for purification of ground water and cooling tower water. The system is similar to the combined dissolved air flotation and filtration (DAFF), except that both air and ozone are injected into the flotation unit for oxidation, disinfection and clarification, so turbidity, color, heavy metals, hardness, iron and manganese, coliform bacteria, etc. can all be simultaneously removed. Both oxygen (in air) and ozone are oxidants for iron and manganese removal. The filtration unit can be sand filtration, multi-media filtration or oxidizing catalytic filtration.

Drinking Water Standards of Iron and Manganese: The USEPA has recommended that drinking water have no more than 0.3 mg/L (or 0.3 ppm) of iron and less than 0.05 mg/L of manganese. USEPA has also set a Health Advisory for manganese of 0.3 mg/L. Private water systems serving individual homes are not subject to State or federal drinking water standards. Levels of iron can be much higher and are not associated with health concerns but manganese concentrations above 300 μg/L may pose a health issue for rice or soy formula-fed infants. Infants are at a greater risk than older children and adults due to higher manganese adsorption rates and their brains and bodies are rapidly developing. Higher levels, over 1,000 μg/L, may pose a risk for all consumers.

Electroflotation: It is a process involving the generation of hydrogen and oxygen bubbles in a dilute electrolytic aqueous solution by passing a direct current between two electrodes: (a) anode and (b) cathode. Anode reaction generates oxygen bubbles and hydrogen ions; while cathode reaction generates hydrogen bubbles and hydroxide ions. Either aluminum or steel sacrificial electrodes can be employed for generating the gas bubbles as well as coagulants at the same time. Non-sacrificial electrodes are employed for generating the gas bubbles only, and can be made of titanium (as the carrier material) and lead dioxide (as the coating material). Electrical power is supplied to the electrodes at a low voltage potential of 5–20 V DC by means of a transformer rectifier. Small bubbles in the range of 20–50 μm are produced under laminar hydraulic flow conditions feasible for flotation separation of fragile flocs from water in a small system. The floats on the water surface are the impurities/pollutants removed from water. The clarified water is discharged from the flotation clarifier's bottom. There can be unexpected advantages and disadvantages when EF is employed. For instance, chlorine bubbles may be generated as a water disinfectant if the water contains significant amounts of chloride ions. Certain unexpected gas bubbles may be generated and may be undesirable.

Electroflotation (EF) Followed by Oxidation and Filtration: It is a process system developed by LIWT for removing iron and manganese from ground water or cooling tower water. EF generates hydrogen bubbles and oxygen bubbles for flotation and oxidation. So EF will give high DO concentration in the water, and no extra oxidant will be needed for dissolved iron Fe(II) oxidation and removal. For removal of dissolved manganese Mn(II), extra oxidant, such as potassium permanganate, is needed for manganese oxidation, if the subsequent filtration is a conventional sand filtration unit. The post-filtration can be sand filtration, multi-media filtration or oxidizing catalytic filtration. This PC process system can be operated as a continuous system or a SBR system.

Electrolytic Flotation: Same as EF.

Environmental Engineering Significance of Iron and Manganese: (a) Iron and manganese (Fe-Mn) occur naturally in earth's geology both in the bedrock and unconsolidated sand/gravel deposits. Fe-Mn dissolve into water as acidic rainfall percolates through the soil and bedrock. Iron Fe(II) and manganese Mn(II) are two similar naturally occurring water contaminants that are not health concerns in drinking water. Instead, they both have secondary or recommended drinking water standards because they cause aesthetic problems that make the water undesirable to use in the home and a bitter metallic taste that can make the water unpleasant to drink for both humans and farm animals. Iron can also cause an orange or brown stain in sinks and in the laundry. Manganese often results in a dense black stain or solid. In higher concentrations Fe-Mn cause the following problems of staining, metallic or vinyl taste, oily appearing crusty sheen to the water surface, rotten egg odor, growing Fe-Mn bacteria and clogging plumbing system. The two metals are associated with the above problems when iron levels exceed 300 μg/L and manganese levels exceed 50 μg/L. Iron levels can be much higher and are still not associated with health concerns but manganese levels above 300 μg/L may pose a health issue for rice or soy formula-fed infants. Higher manganese levels, over 1,000 μg/L, may pose a risk for all consumers. Neither iron nor manganese are regulated under the Safe Drinking Water Act but the USEPA does have a lifetime health advisory for manganese of 300 μg/L. (b) Iron and manganese deposits can build up in pipes, pressure tanks, water heaters and water softeners. This reduces the available quantity and pressure of the water supply. Iron and manganese accumulation becomes an economic problem when water supply or water softening equipment must be replaced. There also are increases in energy costs associated with pumping water through constricted pipes or heating water with electric heating rods coated with iron or manganese deposits. (c) A problem that frequently results when there is high iron or manganese levels in water is the presence of iron or manganese bacteria. An indicator of iron and manganese bacteria is a foul odor, similar to a rotten egg smell. The odor is a byproduct of the bacteria feeding on the iron and/or manganese. These bacteria occur in soil, shallow aquifers and some surface water. They may be introduced into a well or water system when it is constructed or repaired. The bacteria are not a health concern but may form red-brown (iron) or black-brown (manganese) slime layers, often detected in toilet tanks, and may clog water systems.

Fe-Mn Removal by Oxidation and Filtration (10 States Standards): (a) Oxidation may be by aeration and/or by chemical oxidation with chlorine, potassium permanganate, sodium permanganate, ozone or chlorine dioxide. (b) Chemical feed points shall be provided prior to filtration. (c) A minimum detention time of 30 minutes shall be provided following aeration to ensure that the oxidation reactions are as complete as possible. This minimum detention may be omitted only where a pilot plant study indicates a reduced need for detention. The reaction tank/detention basin should be designed to prevent short circuiting. The reaction tank/detention basin shall be provided with an overflow, vent and access hatch. (d) Sedimentation basins may be required when treating water with high iron and/or manganese content, or where chemical coagulation is used to reduce the load on the filters. Provisions for sludge removal shall be made. (e) Filters shall be provided and shall conform to the 10 States Standards.

Fe-Mn Removal by IX (10 States Standards): This process of iron and manganese removal should not be used for water containing more than 0.3 mg/L of iron, manganese or combination thereof. This process is not acceptable where either the raw water or wash water contains DO or other oxidants.

Fe-Mn Removal by Manganese-Coated Media Filtration (10 States Standards): This process consists of a continuous or batch feed of oxidant to the influent of a manganese-coated media filter; (a) Provisions should be made to apply the oxidant as far ahead of the filter as practical and to a point immediately before the filter; (b) Aeration may be used prior to the addition of the oxidant to reduce the amount of the chemical needed; (c) An anthracite media cap of at least 6 inches or more as required by the reviewing authority should be provided over manganese-coated media; (d) Typical filtration rate is 4 gallons/min/ft^2 (9.5 m/h); (e) Typical wash rate is 8–10 gallons/min/ft^2 (20–24 m/h) with manganese greensand and 15–20 gallons/min/ft^2 (37–49 m/h) with manganese-coated media; (f) Air washing should be provided.

Fe-Mn Removal Process Selection Considerations: (a) Level of process complexity and chemical safety (strong chemical oxidants, backwash, etc.); (b) residual (e.g. backwash waste stream) disposal (acceptable route such as sanitary sewer or receiving water discharge); (c) water use restrictions (relatively frequent backwash necessary); (d) water chemistry and water quality (concentrations of soluble iron and manganese, pH, acidity, alkalinity, hardness, DO, NOM (such as humic acid, tannin, lignin, etc.), TOC, ammonia, and the presence of iron and manganese bacteria); (e) well-understood process; (f) operation and maintenance (O&M) costs; (g) presence of co-occurring contaminants; (h) potential unintended consequences; (i) available facility space; and (j) compatibility of the existing water treatment system.

Fe-Mn Water Treatment Sampling Taps (10 States Standards): Smooth-nosed sampling taps shall be provided for control purposes. Taps shall be located on each raw water source, each treatment unit influent and each treatment unit effluent, and should be provided at points between the anthracite media and the manganese-coated media.

Ferazur-Mangazur Filtration System: It is a biological oxidation-filtration system supplied by ONDEO Degrémont Ltd. of Dorval, QC, Canada. The presence of both iron and manganese in the water of Waterloo, QC, Canada, requires the use of two biological filtration steps: (a) Ferazur for the removal of iron and (b) Mangazur for the removal of manganese. The Ferazur-Mangazur system consists of two separate single media pressure biological filters, and includes an intermediate aeration column to adjust the pH and provide oxygen to the process. After bio-mass seeding, the concentrations of iron and manganese at the outlet of the full-scale plant are both significantly lower than the Canadian Drinking Water Quality Guidelines of 0.3 mg/L for iron and 0.05 mg/L for manganese. During normal operation the turbidity is below 0.5 NTU at the filter outlet. The biological system has great potential for small- and medium-sized communities because of the low capital and operating costs, simple operation without chemical addition, and the stable, reliable quality of treated water.

Ferric Iron (Red Water Iron): (a) The ferric iron Fe(III) is basically ferrous iron which has been exposed to oxidizing agent (such as oxygen, ozone, chlorine, etc.), but usually is oxidized by the oxygen from the air. As carbon dioxide leaves the water, oxygen combines with the iron to form ferric ions (Fe^{3+}). These oxidized particles are visible insoluble suspended solids in water; (b) Ferric iron generally stick together to form large flakes and can easily be removed by filtration; (c) Some ferric iron may be present in colloidal form. Their large surface area and charge relative to their mass causes the individual colloidal particles to repel one another. As a result, they will not stabilize and cannot be removed by filtration easily.

Ferrous Iron (Clearwater Iron): (a) The ferrous iron Fe(II) is soluble and not visible in poured water. It is found in water which contains no oxygen, such as water from deep wells or groundwater. Carbon dioxide reacts with iron in the ground to form water-soluble ferrous bicarbonate, which, in the water, produces ferrous ions (Fe^{2+}); (b) There are a variety of ways for removing ferrous iron, each with its own strengths and limitations. These methods fall into the following categories: IX, Oxidation-filtration, and DAF.

Filox: It is one type of manganese oxide filter, or oxidizing filter which are made from greater than 80% pure manganese dioxide ore for iron removal without the use of any oxidizing chemicals.

Filter Run Length (Run Time): The time in filtration service between filter cleanings is termed the run length or run time.

Filtration: Dual media filtration by gravity is one of the most economical forms of granular media filtration. Dual media filtration involves the use of both sand and anthracite as filter media, with anthracite being placed on top of the sand. Single medium, such as sand only, is called sand filtration. If only GAC is used, it is called GAC filtration. Granular media filtration involves the passage of water through a bed of filter media with resulting deposition of solids. Eventually, the pressure drop across the bed becomes excessive or the ability of the bed to remove suspended solids is impaired. Cleaning is then necessary to restore operation head and the filter effluent quality. The time in service between cleanings is termed the run length or run time. The head loss at which filtration is interrupted for cleaning is called the terminal head loss, and this head loss is maximized by the judicious choice of media sizes.

Grain per Gallon (GPG): One grain per gallon equals approximately 17.2 mg/L.

Greensand Catalytic Oxidizing Filtration Advantages and Disadvantages: Advantages are: (a1) Filter can be backwashed and recharged for re-use; (a2) It can consistently remove high concentrations of Fe/Mn to extremely low levels; (a3) It adds no sodium to the water; (a4) It can significantly reduce "rotten egg" odors; and (a5) There are minimum flow rates required for backwash. (b) Disadvantages and limitations of this oxidizing filter are: (b1) The pH of the water is critical to the filter's effective operation. The raw water pH should be 7.0 or higher for iron (pH=8 or higher for manganese) or pH adjustment will be necessary. If you have a shallow (dug) well, an inexpensive way to raise the pH is to add marble chips (also called calcite chips) to the well. Don't forget to disinfect the well after this operation. Calcite will somewhat increase the water's hardness. (b2) The rate of required backwash water is high. Keep this in mind in selecting the size of the oxidizing filter suitable to your source water output. Where well output is low and the required filter size is large, two smaller filters might be substituted so that each can be backwashed separately. This duplication would create additional expense.

Greensand Filtration (or Manganese Greensand Filtration) Operation: (a) $KMnO_4$ greensand filtration using greensand media is effective for catalytic oxidation of very high levels of Fe-Mn. In greensand filtration operation, the pH needs, at a minimum, to be over 7 and preferably over 7.5 to ensure full precipitation of Fe-Mn. Where manganese concentration is high, a pH greater than 8 is recommended. If the pH of raw water is low, a pH adjustment is needed to raise the pH; (b) The entire greensand filtration bed shall be made up of manganese greensand with, perhaps, a layer of anthracite on top; (c) The greensand bed residence time of 15 minutes or longer is recommended. Actual bed residence time may be determined by a pilot plant or full-scale plant testing; (d) The $KMnO_4$ can be added either continuously (at a very dilute concentration) or in a more concentrated form at the time of backwash (batch mode). In the latter option, the $KMnO_4$ enhances a special oxygen-rich coating on the greensand media that causes precipitation of the Fe-Mn within the media during normal filter operation; (e) A contact/detention tank is sometimes installed to allow more time for the oxygen and the Fe-Mn contaminants to produce a sizable rust particle; (f) Where the manganese level is high, feeding a dilute chlorine solution may improve manganese removal. $KMnO_4$, chlorine or aeration can also remove hydrogen sulfide odor; (g) Periodically the greensand filter is backwashed. The rusty water is typically discharged into an approved dry well, septic tank or sewer. This discharge of Fe-Mn to a septic field does not have the groundwater pollution consequences of water softener brine.

Health Advisory: In the USA, a Health Advisory is a non-enforceable drinking water standard which is meant to alert consumers to possible health effects from a drinking water constituent. For instance, the 0.3 mg/L Health Advisory for manganese was created due to concerns about various neurological health effects from regularly consuming water above 0.3 mg/L.

Induced Gas Flotation (IGF): It is a process involving introduction of gas directly into the water through a revolving impeller, a diffuser system or an ejector, or a combination of them, at low pressure (slightly higher than 1 atm) for generating big gas bubbles (80 μm to over 1 mm) in large volume under turbulent hydraulic flow conditions. The gas flow rate is about 400% of the influent water flow rate. Physical entrapment and electrochemical attraction play minor roles in an IGF system. The attachment of gas bubbles to the impurities is mainly a result of surface adsorption, gas stripping and oxidation. Surface active substances (inks, detergents, ores, soaps, etc.) together with impurities are selectively separated in a foam phase at the water surface. The foam containing the surfactant and the impurities are removed by suction device, volatile substances are removed by gas stripping action. The clarified waster is discharged from the flotation clarifier's bottom. Reducing agents, such as ferrous ions, can be oxidized to ferric ions for subsequent separation in ferric hydroxide form if air is used as a gas.

Induced Air Flotation (IAF) Followed by Oxidation and Filtration: It is a process system developed by the LIWT for removing iron and manganese from ground water. IAF is one of IGF processes if air is used for flotation. IAF will give high DO concentration in the water, so no extra oxidant will be needed for dissolved iron Fe(II) oxidation and removal. For removal of dissolved manganese Mn(II), extra oxidant, such as potassium permanganate, is needed for manganese oxidation, if the subsequent filtration is a conventional sand filtration unit. The post-filtration can be sand filtration, multi-media filtration or oxidizing catalytic filtration. This PC process system can be operated as a continuous system or a SBR system.

Ion Exchange (IX) (Water Softener): (a) IX is a PC process involving the use of synthetic resins where a presaturated ion in the solid phase (the "adsorbent," usually sodium) is exchanged for the unwanted ions in the water. The IX process is reversible for water or wastewater treatment in which ions from an ion exchanger (i.e. an insoluble permanent solid medium or a resin) are exchanged for ions in a solution or fluid mixture surrounding the ion exchanger. The IX process is reversed during resin regeneration; (b) The IX process should be considered only for the removal of small quantities of hardness, iron and manganese (such as the upper layer of iron is about 5–7 ppm); (c) A major difficulty in using IX for controlling iron and manganese is that if any oxidation occurs during the process, the resulting precipitate can coat and foul the IX resins; (d) Do not use IX where the concentration of iron is above 5 or 6 ppm, or when there is both dissolved and precipitated iron or manganese in the water; (d) There are two types of IX process: (d1) Cation exchange and (d2) Anion exchange; (e) Only cation exchange is used for removing hardness, iron and manganese. Cation IX process removes Ca^{2+}, Mg^{2+}, Fe^{2+} and Mn^{2+}, maintains original low DO, but generates waste brine. Cation IX process can also remove other cationic metals (such as zinc, lead, silver, etc.) at the same time.

Ion Exchange Water Softener Operation and Regeneration for Iron and Manganese Removal: (a) Conventional water softeners can be effective for removing dissolved iron concentrations below 5,000 μg/L and dissolved manganese concentrations below 500 μg/L. Iron and manganese removal is accomplished in the same way as calcium and magnesium (hardness) removal in water. Iron and manganese are exchanged with sodium on the IX resin. The resin can then be backwashed to remove the trapped iron and manganese and then can be regenerated for reuse. Softening with potassium chloride regeneration, however, may be more effective than sodium chloride in removing manganese; (b) Some vendors recommend using special softener salt that contains additives (e.g. food-grade acids) that remove accumulated iron from the resin during regeneration. Not all water softeners are able to remove iron and manganese from water. Be sure to check the manufacturer's specifications; (c) The raw water should not come in contact with any oxidizing agent like air or chlorine before entering the softener. If oxidized iron and manganese are present in raw water, filtration should be used for the removal [47].

Iron (Fe) and Manganese (Mn) Treatment Processes and Considerations: (a) *Sequestration by polyphosphate or sodium silicates:* This treatment process is only for treating low levels of dissolved iron and manganese at combined concentrations up to 3 mg/L; USEPA is **discouraging this practice**; (b) *IX water softener:* This treatment process is only for treating low to moderate levels of dissolved Fe-Mn concentrations at less than 5 mg/L (NDSU recommended upper limit Fe-Mn concentration = 10 mg/L); IX will not work where the Fe-Mn have turned to a rusty color.

The use of IX to treat only Fe-Mn is not recommended due to many IX disadvantages. Where Fe-Mn concentrations are high (almost 5 mg/L), and hardness is also high (over 150 mg/L as $CaCO_3$), then IX is an appropriate treatment process; (c) *Oxidizing filtration alone (such as natural manganese greensand or manufactured zeolite coated with manganese oxide, or manganese dioxide media):* This treatment system is an option for treating moderate levels of dissolved iron and manganese at combined concentrations up to 15 mg/L (NDSU recommended upper limit of Fe-Mn concentration = 10 mg/L); raw water NOM/TOC concentrations should be less than 5 mg/L; WQA has set voluntary performance standards for oxidizing filtration, "An oxidizing filter shall reduce 10 mg/L plus or minus 1.0 mg/L soluble iron to not more than 0.2 mg/L, and 2.0 mg/L plus or minus 0.2 mg/L soluble manganese to not more than 0.05 mg/L"; (d) *Aeration oxidation (including Venturi aeration, and IAF aeration, etc.) followed by filtration:* high levels of dissolved iron and manganese at combined concentrations up to 25 mg/L can be oxidized to a solid form by aeration (mixing with air) if the concentrations of NOM and ammonia are lower than 5 mg/L; (e) *Chemical oxidation followed by filtration:* this treatment system can treat groundwater containing high levels of dissolved or oxidized iron and manganese greater than 10 mg/L, and if iron is combined with NOM/TOC forming organometallic complex, or when Fe-Mn bacteria is present, potassium permanganate, hydrogen peroxide, chlorine, chlorine dioxide, or ozone is an effective chemical oxidant for chemical oxidation prior to filtration. Although chemical oxidation and filtration will work at all levels of soluble iron, it is recommended only for levels above 10 mg/L of soluble iron; (f) *Biological oxidation-filtration:* commercial Ferazur and Mangazur have successfully treated raw water containing soluble iron up to 1.71 mg/L and soluble manganese up to 1.00 mg/L, respectively, and their effluents have met the Canadian Drinking Water Quality Guides of 0.3 mg/L Fe and 0.05 mg/L Mn; (g) *Lime-soda ash softening process system including oxidation and filtration:* It is recommended by the 10 States Standards for simultaneous removal of hardness, iron and manganese, coliforms, TTHM at all their concentration levels; (h) *Lime-sodium aluminate softening process system including oxidation and filtration:* It is developed by the LIWT for simultaneous removal of hardness, iron and manganese, coliforms, TTHM at all their concentration levels; (i) *AOP followed by filtration:* This treatment system is well established. It can treat the contaminated groundwater under direct influence of surface water. The raw water may contain high concentrations of Cryptosporidium and coliforms, some NOM/TOC, and high levels of dissolved or oxidized iron and manganese greater than 10 mg/L, and if iron is combined with some NOM/TOC forming organometallic complex, or when Fe-Mn bacteria is present (j) *DAF clarification followed by oxidation and filtration:* This treatment system is well established. It can treat the contaminated groundwater under direct influence of surface water. The raw water may contain high concentrations of Cryptosporidium, coliforms, Fe-Mn bacteria, VOC, NOM/TOC, organometallic complex, THM and high levels of dissolved or oxidized iron and manganese greater than 10 mg/L. (k) *PC-SBR followed by filtration:* This system is developed by the LIWT, now well established although its applications are new. There are many SBR process equipment manufacturers. The system uses one or more oxidants, such as air, ozone, oxygen, hydrogen peroxide, chlorine, potassium permanganate, etc. The filtration can be sand filtration, multi-media filtration, GAC filtration, oxidizing filtration, cloth filtration, bag filtration, diatomaceous earth precoat filtration, membrane filtration, etc.; this treatment system can treat contaminated groundwater under the direct influence of surface water. The raw water may contain high concentrations of Cryptosporidium, coliforms, Fe-Mn bacteria, NOM/TOC, organometallic complex, THM and high levels of dissolved or oxidized iron and manganese greater than 10 mg/L.

Iron (Fe) Water Quality Criteria: (a) 0.3 mg/L of iron in water for domestic water supplies (welfare) and (b) 1.0 mg/L of iron in water for freshwater aquatic life.

Iron and Manganese (Fe-Mn) Appearance in Water: Fe-Mn will often give an oily appearing, "crusty" sheen to the water surface. (Actual oil does not appear "crusty" when disturbed, but "feathers out" like a rainbow.)

Iron and Manganese (Fe-Mn) Clogging Plumbing System: Fe-Mn support the growth of Fe-Mn bacteria. These non-health-related bacteria can clog strainers, pumps and valves. Periodic or continuous chlorination is the best means to control Fe-Mn bacteria. Once present, Fe-Mn bacteria are difficult to eradicate from a well.

Iron and Manganese (Fe-Mn) Health Effects and Concern: Both Fe-Mn are essential nutrients for all life, and a typical daily dietary intake for humans is approximately 5 mg for each mineral. There is a health concern with high levels of manganese in drinking water. Manganese may affect neurological and muscle function in humans. While USEPA is developing their health-based drinking water standard, the Bureau of Health Risk Assessment of the New Hampshire Department of Health and Human Services has adopted an interim health-based standard for manganese of 0.84 mg/L.

Iron and Manganese (Fe-Mn) Staining: Fe-Mn stain laundry and water use fixtures. Where the concentration of iron is high, the color of the staining tends toward orange-brown. Where manganese predominates, the discoloration is more black or gray.

Iron and Manganese (Fe-Mn) Sulfur Odor: Hydrogen sulfide, which causes a characteristic "rotten egg" odor, can also be liberated from the soil by the same conditions (i.e. low DO and low pH) that cause Fe-Mn to dissolve in water. Hydrogen sulfide is frequently encountered in water with excessive Fe-Mn. Some of the treatment methods used to remove iron and manganese will also "remove" hydrogen sulfide gas.

Iron and Manganese (Fe-Mn) Taste: Fe/Mn cause a metallic or vinyl type taste in water.

Iron and Manganese Bacteria Control: (a) Iron and manganese bacteria are not a health issue but a nuisance issue. The most common approach to control iron and manganese bacteria is with shock chlorination. This involves introducing a concentrated chlorine solution into the entire water distribution system, letting the system remain idle for at least two to three hours, and then flushing the system to remove the chlorine. It is not possible to kill all iron and manganese bacteria in a system but shock chlorination is a method for minimizing their occurrence. Therefore, shock chlorination will most likely need to be repeated from time to time. (b) If bacteria re-growth is rapid, repeated shock chlorination becomes time consuming. Continuous application of low levels of chlorine may be more effective. An automatic liquid chlorine injector pump or a dispenser that drops chlorine pellets into the well is a common choice. (c) Chlorine will change dissolved iron into oxidized iron that will precipitate. A filter may be needed to remove oxidized iron if continuous chlorination is used to control iron bacteria.

Iron and Manganese Chemistry: (a) Mostly present in groundwater as Fe(II) (Fe^{2+}) and Mn(II) (Mn^{2+}) which are soluble; (b) Oxidized forms Fe(III) (Fe^{3+}) and Mn(IV) (Mn^{4+}) are relatively insoluble (a solid form); (c) Some possible co-contaminants, such as ammonia, arsenic and TOC, and DOM, in the water will form organometallic complex with soluble iron and manganese, and will increase oxidant dosage, processing detention time and the overall treatment difficulty.

Iron and Manganese Control (10 States Standards): Iron and manganese control, as used herein, refers solely to treatment processes designed specifically for this purpose. The treatment process used will depend upon the character of the raw water. The selection of one or more treatment processes must meet specific local conditions as determined by engineering investigations, including chemical analyses of representative samples of water to be treated, and receive the approval of the reviewing authority. It may be necessary to operate a pilot plant in order to gather all information pertinent to the design. Consideration should be given to adjusting pH of the raw water to optimize the chemical reaction.

Iron and Manganese Health Benefits: (a) Iron is an essential mineral that is naturally found in many foods such as lean meat, poultry, seafood, nuts, beans, vegetables and iron fortified grain products such as: breads, cereals and pastas. Iron's primary purpose is to carry oxygen throughout the body within the hemoglobin of the red blood cells. It is also necessary for cellular function, neurological development and physical growth; (b) Manganese is an essential nutrient found in many foods such as tea, leafy green vegetables, grains, fruit, nuts, legumes, fish, some meats and some infant formulas, especially rice and soy. It is essential for bone development and many processes in the body; (c) Like all minerals, it can have adverse health effects when humans and/or animals ingest too little or too much. Food and drinking water are the primary source of manganese for adults and children. Breast milk, soy and rice-based formula and cereals are the primary sources of manganese for infants.

Iron Bacteria: (a) They are bacteria which rely on iron to survive and grow; (b) Iron bacteria can be controlled by periodic well chlorination or it can be treated in the building. The treatment involves the following: chlorination, retention, filtration; (c) GAC is usually used as the filter media if the excess chlorine needs to be removed at the same time.

Iron Environmental Impact: (a) Iron is the fourth most abundant, by weight, of the elements that make up the earth's crust. Common in many rocks, it is an important component of many soils, especially the clay soils where usually it is a major constituent. Iron in water may be present in varying quantities dependent upon the geology of the area and other chemical components of the waterway. (b) Iron is an essential trace element required by both plants and animals. In some waters it may be a limiting factor for the growth of algae and other plants; this is true especially in some marl lakes where it is precipitated by the highly alkaline conditions. It is a vital oxygen transport mechanism in the blood of all vertebrate and some invertebrate animals. (c) The ferrous, or bivalent (Fe^{2+}), and the ferric, or trivalent, irons are the primary forms of concern in the aquatic environment, although other forms may be in organic and inorganic wastewater streams. The ferrous (Fe^{2+}) form can persist in waters void of DO and originates usually from groundwaters or mines when these are pumped or drained. For practical purposes the ferric (Fe^{3+}) form is insoluble. Iron can exist in natural organometallic or humic compounds and colloidal forms. Black or brown swamp waters may contain iron concentrations of several mg/L in the presence or absence of DO, but this iron form has little effect on aquatics. The excessive amount of iron in the water may cause staining, red water, taste, accumulation in distribution system.

Iron Types in Water: There are three main forms of iron in water: (a) ferrous iron, or clearwater iron; (b) ferric iron, or red water iron; and (c) bacterial iron.

Lenox Institute of Water Technology (LIWT): It is a non-profit college in Massachusetts, USA, with expertise in environmental STEAM (science, technology, engineering, arts and mathematics) education, R&D, invention, process development, monitoring system/methods development, patent application, licensing, fund raising, engineering design and project management. LIWT teams up with Krofta Engineering Corporation (KEC) for technology transfer, equipment design and voluntary humanitarian global service through free education, training and academic publications.

Lime-Soda Softening Process (10 States Standards): This is a process system approved by the 10 States Standards for simultaneous removal of hardness, iron and manganese.

Lime-Sodium Aluminate Softening Process (LIWT): This is a process system developed by the LIWT for simultaneous removal of hardness, iron and manganese.

Limestone Contactor: It is process unit full of limestones. For acidic water with low redox potential, the limestone contactor increases the pH of aggressive water by binding carbon dioxide in the limestone.

Magnetic/Electronic Iron and Manganese Removal Devices: The WQA, USA, has indicated that there is no proof that magnetic/electronic Fe/Mn removal devices are effective.

Management of Fe-Mn Present in Water and Process Selection: (a) For clearwater iron as high as 0.6 mg/L, or clearwater manganese as high as 0.10 mg/L: The undesirable staining caused by Fe/Mn can sometimes be minimized by adding a food-grade polyphosphate chemical to the water to prevent the formation of the rusty colored forms of Fe-Mn. These chemicals "coat" and "tie up" the dissolved Fe/Mn ions preventing a reaction with oxygen and subsequent precipitation. (An ion is the dissolved form of an atom or compound.) Sequestering prevents the staining effect but does not remove Fe-Mn. USEPA allows the use of sequestering chemicals, but is not recommending it; (b) For water containing clearwater iron and/or clearwater manganese generally less than 2–5 mg/L of dissolved Fe-Mn: IX (cation exchange) is a recommended process for small point-of-entry (POE), or point-of-use (POU), or very small water treatment plants (WTPs); oxidation followed by filtration, or oxidizing filtration (catalytic oxidation-filtration) is recommended for small or large WTP; (c) For clearwater containing higher than 5 mg/L of dissolved Fe-Mn or for rusty colored water from the tap: only oxidation followed by filtration, or oxidizing filtration (catalytic oxidation-filtration) is recommended for any water treatment systems, small or large; (d) For clearwater containing hardness, higher than 5 mg/L of dissolved Fe-Mn, rusty colored water, significant amount of ammonia, and DOM or TOC, the recommended process system is the combination of coagulation/flocculation, clarification (sedimentation or DAF), oxidation, recarbonation and filtration (sand filtration, multi-media filtration, GAC filtration or oxidizing filtration), in which if DAF is used instead of sedimentation for clarification, DAF has a built-in oxidation. The chosen process system can be either a continuous system or a PC-SBR system.

Manganese (Mn) Water Quality Criteria: (a) SMCL=0.05 mg/L of manganese in water; (b) Many systems target 0.01–0.02 mg/L of manganese in water.

Manganese Environmental Impact: (a) Manganese does not occur naturally as a metal but is found in various salts and minerals, frequently in association with iron compounds. The principal manganese-containing substances are manganese dioxide (MnO_2), pyrolusite, manganese carbonate (rhodochrosite) and manganese silicate (rhodonite). The oxides are the only important minerals mined. Manganese is not mined in the United States except when manganese is contained in iron ores that are deliberately used to form ferro-manganese alloys. (b) The primary uses of manganese are in metal alloys, dry cell batteries, micro-nutrient fertilizer additives, organic compounds used in paint driers and as chemical reagents. Permanganates are very strong oxidizing agents of organic materials. (c) Manganese is a vital micro-nutrient for both plants and animals. When manganese is not present in sufficient quantities, plants exhibit chlorosis (a yellowing of the leaves) or failure of the leaves to develop properly. Inadequate quantities of manganese in domestic animal food result in reduced reproductive capabilities and deformed or poorly maturing young. Livestock feeds usually have sufficient manganese, but beef cattle on a high corn diet may require a supplement. (d) Excessive amount of manganese in water may cause staining, accumulation in distribution system. (e) There are health concerns of neurotoxin. (f) The Lifetime Health Advisory Level (HAL)=0.3 mg/L of manganese. (g) Acute Lifetime Health Advisory Level (HAL)=1 mg/L. (h) Health Canada maximum allowable concentration=0.12 mg/L which is included in Unregulated Contaminant Monitoring Rule (UCMR) 4.

Manganese Greensand Filter: (a) It is one of three oxidizing filters (manganese greensand filter, birm filter and manganese oxide filter). The greensand filter medium is a processed material consisting of nodular grains of the zeolite mineral glauconite coated with manganese oxide. The IX properties of the glauconite facilitate the bonding of the coating. The filter media is periodically treated with potassium permanganate to form a coating that oxidizes the dissolved iron and manganese so that it is pulled out of solution in the form of a red/brown (iron) or black (manganese) flakes or particles. These larger particles of iron and manganese can then be filtered out by the greensand. (b) The greensand media is treated with potassium permanganate to form a regenerated coating that oxidizes the dissolved iron and manganese and then filters them out of the water. Because these units combine oxidation and filtration, they can be used to treat raw water with dissolved and/or oxidized iron and manganese. (c) The coating responsible for iron and manganese removal is maintained through regeneration with potassium permanganate at about 1.5–2 oz./ft^3 of greensand; (d) When properly maintained manganese greensand filters are extremely efficient for moderate levels of both dissolved and oxidized iron and manganese. They are generally recommended when the combined iron and manganese concentration is in the range of 3–10 mg/L. (e) Significant maintenance is required, such as: (e1) frequent regeneration with a potassium permanganate solution as it is consumed during oxidation of the dissolved metals; and (e2) regular backwashing to remove the oxidized iron and manganese particles. (f) The greensand filter works well right down to a pH of 6.5 but works better and lasts longer if the pH is in the range of 6.8–7.5. (g) Greensand's advantage is that it is lightweight and requires less backwash water, and is lower in cost than solid manganese oxide media filters (see below). It does need to be changed every few years though and does not last as long as manganese oxide media. (h) An alternative choice to potassium permanganate can be to inject some chlorine bleach upstream of the greensand filter which then allows the filter to remove iron bacteria, odors and enable the greensand filter to remove virtually all iron as well as manganese.

Manganese Dioxide Filters: (a) Manganese oxide filter is one of three oxidizing filters (manganese greensand filter, birm filter and manganese oxide filter). Pro-OX, Pyrolox, Filox, Catalox and MangOX are all manganese oxide filters. Pro-OX, Filox, MangOX and Pyrolox are all brand names for solid manganese dioxide filter media. These types of media are made from greater than 80% pure manganese ore and are collected and processed specifically for water treatment due to its purity. It is a naturally mined and processed ore and is independently tested and certified by NSF for drinking water. (b) Natural manganese dioxide ore oxidizes and then filters the resulting insoluble iron. It does not need other chemicals. However, it needs ideally to backwash at 25–30 gallons/ft^2, because the media is extremely heavy. (c) These types of well water filters work best when the water has been injected with chlorine bleach upstream of the manganese dioxide filtration. (d) Manganese oxide media can be utilized down to a pH of 6.2 theoretically, but work best if the pH is between 6.8 and 7.5. (e) The filter media can last 10–15 years. (f) The filter media will remove other metals such as arsenic under the right conditions. (g) For manganese oxide to work correctly, the Oxidation-Reduction Potential (ORP) level (which can be easily measured) must be over 200. Depending on the application manganese oxide filter media works great if air or potassium permanganate is injected/added, or the water is chlorinated or ozonated prior to the iron filter, which ensures the ORP level is high enough for the oxidation process to reliably occur.

Manganese Water Quality Criteria: (a) 50 μg/L of manganese in water for domestic water supplies (welfare); (b) 100 μg/L of manganese in water for protection of consumers of marine molluscs.

Mangazur® Filtration System: Mangazur is a biological filtration system supplied by ONDEO Degrémont Ltd. (of Dorval, QC, Canada) for manganese removal from a groundwater source. One of the new systems (in Woodstock, New Brunswick, Canada), it consists of two, 10-ft diameter pressure filters filled with the specially designed filter media "Biolite." It is equipped with a raw water backwash system and control panel. This biological process allows the backwashing of the filters with raw water in certain cases. The system can be designed with automatic valves and sensors to limit operator intervention during operation and backwashing. In one case study, after a short 5-week seeding period, manganese removal efficiency had reached 90%, and the concentration of manganese in the filtered water was, and continues to be, 0.02–0.03 mg/L which is lower than the Canadian Drinking Water Guidelines Aesthetic Objective of 0.05 mg/L.

MangOX: It is one type of manganese oxide filters, or oxidizing filters which are made from greater than 80% pure manganese ore for iron removal without the use of any chemicals.

MnO_x-Coated Media Filtration for Iron and Manganese Removal: Both manganese greensand and birm filter media are the MnO_x-coated filter media. Either greensand filtration or birm filtration can improve Mn ions removal, but generate residuals in backwash wastewater.

Multistage Treatment: If water has high levels of iron and manganese in both dissolved and solid forms, a multistage treatment operation is necessary. For example, the water could be chlorinated to oxidize dissolved iron and kill iron bacteria, and filtered through to remove particles. This can be followed by activated carbon filtration to remove excess chlorine and then water softening for hardness control and to remove any residual, dissolved iron or manganese.

Oxidant Dosage, Chlorine Dioxide, for Iron and Manganese Removal: (a) 1.21 mg chlorine dioxide per mg Fe(II); and (b) 2.45 mg chlorine dioxide per mg Mn(II).

Oxidant Dosage, Chlorine, for Iron and Manganese Removal: (a) 0.62 mg chlorine per mg Fe(II); and (b) 0.77 mg chlorine per mg Mn(II).

Oxidant Dosage, Oxygen, for Iron and Manganese Removal: (a) 0.14 mg oxygen per mg Fe(II); and (b) 0.29 mg oxygen per mg Mn(II).

Oxidant Dosage, Ozone, for Iron and Manganese Removal: (a) 0.43 mg ozone per mg Fe(II); and (b) 0.88 mg ozone per mg Mn(II) at pH 8–8.5.

Oxidant Dosage, Potassium Permanganate, for Iron and Manganese Removal: (a) 0.94 mg potassium permanganate per mg Fe(II); and (b) 1.92 mg potassium permanganate per mg Mn(II).

Oxidation: Before iron and manganese can be filtered, they need to be oxidized to a state in which they can form insoluble complexes. Oxidation involves the transfer of electrons from the iron, manganese or other chemicals being treated to the oxidizing agent. Oxidation methods fall into two groups: those using additives like chlorine, ozone or air; and those using an oxidizing filter media.

Oxidation Followed by Filtration Process System: Oxidation followed by filtration is a relatively simple process. The oxidant chemically oxidizes the iron or manganese (forming a particle), and kills iron bacteria and any other disease-causing bacteria that may be present. The filter then removes the iron or manganese particles. When combined levels of iron and manganese exceed 10 mg/L, the most effective treatment involves oxidation followed by filtration. In this process, an oxidant is added to convert any dissolved iron and manganese into the solid, oxidized forms that can then be easily filtered from the water. Chlorine gas, sodium hypochlorite, potassium permanganate, hydrogen peroxide, air, oxygen, ozone can all be the oxidant. Filtration can be sand filtration, multi-media filtration, GAC filtration, bag filtration or oxidizing filtration. The process system can be either a continuous system or a PC-SBR. It should be noted that membrane filtration (such as reverse osmosis, MF, UF, NF, etc.) is not recommended for iron and manganese removal because it can easily be clogged.

Oxidizing Filter Operation: In oxidizing filters, all soluble Fe-Mn is first purposely converted to the insoluble rusty-colored form. These larger particles are then strained out on the filter media. Periodically the filter media is backwashed to remove the collected rust particles. The Fe-Mn-laden backwash is discharged into a dry well, leach field or sewer.

Oxidizing Filters: (a) Oxidizing filters both oxidize and filter iron and manganese in one unit. There are three types of oxidizing filters: (a1) manganese-treated greensand filtration; (a2) birm filtration and (a3) manganese oxide filtration (Pro-OX, Pyrolox, Filox, Catalox, Katalox and MangOX are all brand names for solid manganese dioxide filter media). (b) An oxidizing filter typically contains a filter media of natural manganese greensand or manufactured zeolite coated with manganese oxide. The filter media is periodically treated with potassium permanganate to form a coating that oxidizes the dissolved iron and manganese so that it is pulled out of solution in the form of a red/brown (iron) or black (manganese) flakes or particles. These larger particles of iron and manganese can then be filtered out by the greensand. Caution must be exercised when using potassium permanganate as it is a strong skin, eye and lung irritant. (c) Oxidizing filters, such as manganese greensand filters, require significant maintenance including frequent regeneration with a potassium permanganate solution. In addition, these units require regular backwashing to remove the oxidized iron and manganese particles. The potassium permanganate solution used for regeneration must be handled and stored with care. Compared to manganese greensand, synthetic zeolite requires less backwash water and softens the water as it removes iron and manganese. (d) Oxidizing filters, which oxidize and filter iron and manganese, are the most widely used option for managing moderate levels of dissolved iron and manganese. Because these units combine oxidation and filtration, they can be used to treat raw water with dissolved and/or oxidized iron and manganese. (e) An oxidizing filter (natural manganese greensand, manufactured silica gel zeolite coated with manganese dioxide, plastic resin beads or pumicite) is recommended by the North Carolina CES when the total combined iron and manganese concentration is less than 15 mg/L.

Oxidizing Filtration Media (Catalytic Oxidizing Media): They include: (a) manganese greensand media is zeolite coated with manganese dioxide. The whole of the greensand bed is made up of manganese greensand with perhaps a layer of anthracite on top; (b) birm filtration media is man-made media coated with manganese dioxide; and (c) manganese dioxide filtration media (such as Pro-OX, Pyrolox, Filox, Catalox and MangOX) is of un-coated natural manganese dioxide ores specially selected for their catalytic activity and precisely screened to a size range of 20/40 mesh.

This size range is chosen so that, having regard to the different densities of manganese dioxide and silica sand, the material will distribute itself evenly throughout a bed of 18/35 mesh sand when added to it to produce a catalytic filter (a minimum of 20% manganese dioxide by volume for municipal application and up to 100% manganese dioxide for smaller filtration system application). For example, when manganese dioxide is used for a municipal WTP, the normal practice is to employ a mixture of 20% by volume manganese dioxide and 80% by volume 18/35 mesh sand (0.5–1.0 mm) with a layer of anthracite, if required.

Oxygen Dosage for Iron and Manganese Oxidation: (a) 0.14 mg oxygen per mg Fe(II); 0.29 mg oxygen per mg Mn(II). Almost all the common oxidants except chloramines will convert Fe(II) ferrous iron ions to insoluble ferric hydroxide Fe(III) state and will convert Mn(II) manganese ions to insoluble manganese dioxide Mn(IV) state.

Oxygenation for Iron and Manganese Removal: It is a process similar to aeration, except that pure oxygen is used instead of air. (a) Usually a nitrogen-oxygen separator is needed. The separator is fed with air, but produces nitrogen gas and oxygen gas. The oxygen gas from the separator is then bubbled through the water stream for soluble iron and manganese oxidation. The oxidized metals are subsequently removed by filtration. (b) Alternatively, electrocoagulation and/or EF can be used to generate oxygen and hydrogen bubbles, and the oxygen bubbles can be used for oxygenation.

Ozonation: It is a PC oxidation process using ozone as an oxidant. An ozone generator is used to produce ozone that is then fed by a gas injector into the water stream to convert ferrous iron into ferric iron. Ozone has the greatest oxidizing potential of the common oxidizers. This is followed by a contact time tank and then by a catalytic medium or an inert multilayered filter for removing the suspended particles.

Ozonation for Iron and Manganese Removal: (a) It is a rapid but expensive PC process using ozone as an oxidant for removing iron and manganese ions; (b) Ozonation forms colloidal iron and manganese that are occasionally difficult to remove via filtration; (c) Ozone is a strong oxidant but it is a much more unstable gas that must be generated on-site using an ozone generator. Once the ozone is produced, it is injected into the water where its oxides dissolved iron and manganese which must then be filtered; (d) Ozonation may be useful where multiple water quality problems must be treated (i.e. bacteria and metals).

Ozone Dosage for Iron and Manganese Oxidation: (a) 0.43 mg ozone per mg Fe(II); 0.88 mg ozone per mg Mn(II) at pH 8–8.5. Almost all the common oxidants except chloramines will convert Fe(II) ferrous iron ions to insoluble ferric hydroxide Fe(III) state and will convert Mn(II) manganese ions to insoluble manganese dioxide Mn(IV) state.

Permanganate Oxidation (Permanganation) for Iron and Manganese Removal: (a) It is chemical process using permanganate (sodium or potassium) for effective soluble Fe and Mn removal even if ammonia and TOC are elevated; (b) Overdosing will be problematic, and will cause pink water and may even generate increased Mn residuals.

Phosphate Addition Treatment (Sequestration): See Polyphosphate addition treatment.

Physicochemical (PC) Sequencing Batch Reactor (SBR): PC-SBR process is operated in a sequencing batch mode, and uses one or more oxidants (such as air, ozone, oxygen, hydrogen peroxide, chlorine, potassium permanganate, etc.) + filtration (such as sand filtration, multi-media filtration, GAC filtration, oxidizing filtration, cloth filtration, bag filtration, diatomaceous earth precoat filtration, etc.) for removal of hardness, iron and manganese. Do not use any membrane filtration (such as MF, UF, NF or RO) to avoid clogging problems.

Point-of-Entry (POE) Treatment Methods: (a) A POE device treats water and removes water contaminants from all water entering the home or apartment complex. So POE is a whole home or whole business water treatment system. Since excess iron and manganese are aesthetic problems that affect all potential uses of the water (e.g. stains on fixtures or laundry), in many cases these are removed from all water entering the home using POE treatment devices. (b) The four most commonly applied POE methods for treating water containing dissolved iron and manganese are: IX water softeners; oxidizing filters; aeration (pressure type) followed by filtration; and chemical oxidation followed by filtration. An additional method that may apply to some situations is phosphate treatment. (c) Treatment techniques are most effective for iron and manganese removal in water that has an almost neutral pH (approximately 7.0). The phosphate compound treatment is an exception and is most effective in the pH range of 5.0–8.0. An individual should work with a reputable water treatment professional to select the best treatment method.

Point-of-Use (POU) Treatment Methods: A POU device treats water and removes contaminants at the location of usage, such as a faucet, or shower. POU devices such as reverse osmosis, distillation and carbon filtration can remove dissolved manganese (and often iron), and are good options for providing safe drinking water. Note that some filters, including pour-through pitchers and in-line facet units, include carbon and other adsorbents engineered to remove manganese. When manganese concentrations are higher than 200 μg/L, the effective life of the filter may be significantly shortened, and replacement, more frequent than that indicated by the manufacturer, may be necessary.

Polyphosphate Addition Treatment (Sequestration): (a) Water containing dissolved iron less than 1,000 μg/L may be simply treated using a food-grade polyphosphate compound. The addition of phosphate, however, generally does not work well for manganese removal. This process involves the addition of phosphate to the water using a chemical feed pump, the phosphate chemical sequesters ("coats" and "ties up") dissolved iron but does not actually remove iron. Water treated with phosphate may retain a metallic taste but helps to prevent staining of fixtures, dishes and laundry. Since manganese is not removed, this process is not recommended for high levels of manganese. In addition, a high phosphate concentration may cause diarrhea. In some cases, it can also increase the corrosion rate of copper plumbing. (b) Phosphate compounds are not stable at high temperatures, and if the water is heated or boiled, the phosphates will break down and release iron, which can react with oxygen and precipitate out. (c) The addition of phosphate should not be used where the use of phosphate in cleaning products is banned. Phosphate, from any source, that gets into surface water, can, like nitrates, contribute to nutrient pollution of the water body.

Potassium Permanganate ($KMnO_4$): It is a purple crystal or liquid oxidant which can be used to oxidize and precipitate (meaning to convert from soluble form to a solid form) soluble Fe-Mn. It is also a good chemical for disinfection, and taste and odor control of drinking water.

Potassium Permanganate Demand due to Iron Fe(II) and Manganese Mn(II) Presence: (a) 0.94 (approximate 1.00) mg potassium permanganate per mg Fe(II); 1.92 (approximate 2.00) mg potassium permanganate per mg Mn(II); (b) Example: For raw water having 0.7 mg/L Fe(II), and 0.4 mg/L Mn(II), potassium permanganate demand $= 1 \times 0.7$ mg/L $+ 2 \times 0.4$ mg/L $= 1.5$ mg/L.

Potassium Permanganate Dosage for Iron and Manganese Oxidation: (a) 0.94 (approximate 1.00) mg potassium permanganate per mg Fe(II); 1.92 (approximate 2.00) mg potassium permanganate per mg Mn(II). Almost all the common oxidants except chloramines will convert Fe(II) ferrous iron ions to insoluble ferric hydroxide Fe(III) state and will convert Mn(II) manganese ions to insoluble manganese dioxide Mn(IV) state.

Pro-OX: It is one type of manganese oxide filters, or oxidizing filters which are made from greater than 80% pure manganese ore for iron removal without the use of any chemicals.

Pyrolox: It is one type of manganese oxide filters, or oxidizing filters which are made from greater than 80% pure manganese ore for iron removal without the use of any chemicals.

Secondary Maximum Contaminant Level (SMCL): Iron and manganese are currently both classified under the SMCL standards, indicating that removing these contaminants to the SMCL is recommended, but not enforced by either federal or state drinking water regulations. The USEPA SMCLs are 300 μg/L (0.3 mg/L) iron and 50 μg/L (0.05 mg/L) manganese.

Sequencing Batch Reactor (SBR), Physicochemical (PC): PC-SBR is a new process developed by Lawrence K. Wang and Mu-Hao Sung Wang of the LIWT for removal of calcium, magnesium, iron and/or manganese. The batch process sequence includes chemical addition, clarification (sedimentation or air flotation), oxidation and filtration. When air flotation is used for clarification, no extra oxidation is needed for removing iron alone because air flotation itself is an aeration oxidation process. For removing both iron and manganese, potassium permanganate is an effective oxidant for extra oxidation although other chemical oxidants can also be used. Air flotation can be either DAF or IAF. Filtration can be sand filtration, multi-media filtration or oxidizing catalytic filtration.

Sequestering Chemicals: (a) The undesirable staining caused by Fe-Mn can sometimes be minimized by adding a food-grade polyphosphate chemical to the water to prevent the formation of the rusty colored forms of Fe-Mn, if clearwater iron concentration is as high as 0.6 mg/L and clearwater manganese concentration is as high as 0.10 mg/L, and if there is no rusty colored water observed. These chemicals "coat" and "tie up" the dissolved Fe-Mn ions preventing a reaction with oxygen and subsequent precipitation. Sequestering prevents the staining effect but does not remove Fe-Mn; (b) Advantages of sequestering are: (b1) Low-cost equipment and chemicals; (b2) Small space requirements; and (b3) Other water quality factors are not too important; (c) Disadvantages of sequestering are: (c1) The effectiveness of the treatment is reduced at higher water temperatures or when the water is aerated or when bleach is added; (c2) Sequestering chemicals loses effectiveness with time. They revert from the "poly" to the "ortho" form of phosphate; (c3) Phosphate compounds could foster bacterial growth within your home plumbing system and residual phosphate in septage causes weed and algae growth in lakes; (c4) Not effective for that portion of the Fe-Mn that has already turned rusty; and (c5) Not commonly used in a private family home well.

Sequestration (Polyphosphate Addition): (a) It is a chemical process of adding polyphosphate to disperse Fe^{3+} and Mn^{4+} colloidal particles; (b) Sequestration process does not remove the soluble Fe or Mn ions, so soluble Mn still poses a health concern; (c) Polyphosphates revert or break down over time, still get accumulated in the distribution system and still get customer complaints, so polyphosphate is not recommended for control of iron, manganese or corrosion.

Sequestration by Polyphosphates (10 States Standards): This process is not recommended when iron, manganese or combination thereof exceeds 0.5 mg/L and shall not be used when it exceeds 1.0 mg/L. The total phosphate applied shall not exceed 10 mg/L as PO_4. Where phosphate treatment is used, satisfactory chlorine residuals shall be maintained in the distribution system. Possible adverse effects on corrosion must be addressed when phosphate addition is proposed for iron sequestering. Polyphosphate treatment may be less effective for sequestering manganese than for iron; (a) Feed equipment shall conform to the requirements of the 10 States Standards; (b) Stock phosphate solution must be kept covered and disinfected by carrying approximately 10 mg/L free chlorine residual unless the phosphate is not able to support bacterial growth and the phosphate is being fed from the covered shipping container. Phosphate solutions having a pH of 2.0 or less may also be exempted from this requirement by the reviewing authority; (c) Polyphosphates shall not be applied ahead of iron and manganese removal treatment. The point of application shall be prior to any aeration, oxidation or disinfection if no iron or manganese removal treatment is provided; (d) The polyphosphate feed point shall be located as far ahead of the oxidant feed point as possible.

Sequestration by Sodium Silicates (10 States Standards): Sodium silicate sequestration of iron and manganese is appropriate only for groundwater supplies prior to air contact. On-site pilot tests are required to determine the suitability of sodium silicate for the particular water and the minimum feed needed. Rapid oxidation of the metal ions such as by chlorine or chlorine dioxide must accompany or closely precede the sodium silicate addition. Injection of sodium silicate more than 15 seconds after oxidation may cause detectable loss of chemical efficiency. Dilution of feed solutions much below 5% silica as SiO_2 should also be avoided for the same reason. Sodium silicate treatment may be less effective for sequestering manganese than for iron; (a) Sodium silicate addition is applicable to waters containing up to 2 mg/L of iron, manganese or combination thereof; (b) Chlorine residuals shall be maintained throughout the distribution system to prevent biological breakdown of the sequestered iron; (c) The amount of silicate added shall be limited to 20 mg/L as SiO_2, but the combined amount of added and naturally occurring silicate shall not exceed 60 mg/L as SiO_2; (d) Sodium silicate shall not be applied ahead of iron or manganese removal treatment.

Sources of Iron and Manganese in Drinking Water: Natural sources of iron and manganese are more common in deeper wells where the natural water has been in contact with rock for a longer time. In coal mining regions, the two metals may also occur from both deep and surface mining activities. Iron and manganese often occur together in groundwater. Occasionally, iron pipes also may be a source of iron in drinking water. Groundwater with high iron and/or manganese may appear colorless initially but orange-brown (iron) or black (manganese) stains or particles quickly appear as the water is exposed to oxygen.

Types of Fe-Mn Present in Water: (a) Clearwater iron or clearwater manganese: The water is totally clear when drawn from the tap. Fe-Mn is present in dissolved form. The terms "clearwater iron" and "clearwater manganese" are often used to describe this form. The scientific name for clearwater iron is called "ferrous" and for manganese, "manganous." (b) Water is rusty colored when drawn from the tap: After exposing to oxygen or other oxygen-like chemicals, clearwater Fe-Mn will precipitate to form fine brownish (ferric) or blackish (manganic) "rust" particles. The tendency to precipitate these minerals is also influenced by changes of water.

US Drinking Water Standards: Public drinking water standards established by USEPA fall into two categories—Secondary Standards and Primary Standards. Primary Standards are based on health considerations and are designed to protect human health. Primary Standards are enforceable. Secondary Standards are based on aesthetic factors such as taste, odor, color, corrosivity, foaming and staining properties of water that may affect the suitability of a water supply for drinking and other domestic uses. Secondary Standards are recommended, but are not enforceable.

Venturi Aeration: It is commonly used by WTPs. Where DO in the water is low, air is added to the water by a special Venturi nozzle. The mixture of air and water then passes into a detention tank where the oxygen dissolves into the water for oxidation reactions. The aerated water then enters the filtration unit for final polishing;

Water Sampling for Iron and Manganese Testing: (a) A water analysis should be done to determine the source of the iron and manganese. Iron and manganese may be present in the water supply or be caused by corroding pipes (iron or steel). Iron from pipe corrosion indicates low pH that may need to be corrected; water samples should be taken from the faucet closest to the pump. Allow the water to run for 5–10 minutes before sampling to obtain fresh water that has not been exposed to air. Do not sample water that has gone through a water heater or a water treatment unit such as a softener. If the water is clear when first drawn, but red or black particles appear after the water settles, dissolved iron and manganese are present. If the water has a red tint but no particles settle out after a time, colloidal iron is the cause. Reddish brown or black-brown slimy masses inside the toilet tank indicate iron or manganese bacteria. Laboratory tests are recommended in all cases to determine iron and manganese concentrations.

Water Softener (IX): (a) Conventional water softeners are sometimes effective for removing hardness, iron and small amounts of manganese. Water softeners are typically used to remove calcium and magnesium hardness in water by an exchange process. The calcium and magnesium are removed from the water and sodium is added in their place. Iron and manganese removal is accomplished in the same way by exchanging the iron and manganese for sodium. The iron and manganese are then removed from the softener resin bed through backwashing and regeneration. (b) Removal efficiencies by softeners will vary depending on the iron concentration, water hardness and pH. Softeners are generally only recommended when the water pH is greater than 6.7, the water hardness is between 3 and 20 grains per gallon (50–350 mg/L) and the dissolved iron concentration is less than 5 mg/L. (c) Oxidized forms of iron and manganese will foul the softener resin. Thus, it is critical that the raw water not come in contact with any oxidizing agents like air or chlorine before entering the softener.

Water Softener and IX Advantages and Disadvantages: (a) Advantages of IX water softeners are: (a1) IX water softener resin can be regenerated for reuse; (a2) IX can consistently remove Ca-Mg and Fe-Mn ions; (a3) IX water softeners have lower backwash water requirements than oxidizing filters; and (a4) The Fe-Mn removal is not appreciably affected by the water pH or other factors; (b) Disadvantages of IX water softeners are: (b1) Softening will not operate satisfactorily if iron bacteria or rusty colored water exists, even if occasionally. If particles are present, a sediment prefilter is often placed before the resin tank; (b2) A water softener will not remove hydrogen sulfide odor; and (b3) Water softeners produce waste brine that must be disposed of. If a sewer is not available, and disposal of the waste brine will likely be into the ground, it creates the potential of polluting the groundwater and subsequently the user's own well or the wells of the neighbors downhill.

Water Softener and IX Applications: (a) Water softener is an IX process (cation exchange) for removing mainly hardness (calcium ions and magnesium ions) but also clearwater iron and clearwater manganese from water by the POE, the POU, and small WTP users. (b) IX softening is described as effective for water containing less than 2–5 mg/L of dissolved Fe-Mn. IX will not work at all where the Fe-Mn have turned to a rusty color. Other aspects of water quality such as pH or alkalinity are not important in the efficiency of IX. (c) In general, the use of water softeners to treat only Fe-Mn is not recommended. Where both Fe/Mn and hardness are high, the IX water softener can be considered.

Water Softener and IX Demand Regeneration: (a) Normally clock timers backwash a water softener whether IX resin needs regeneration or not, such as during vacation periods. This excessive backwashing needlessly increases salt use and the generation of waste brine. The demand regeneration approach allows the frequency of IX softener resin regeneration cycles to be reduced. The controls on these devices include those that measure the conductivity or those that measure the volume of water treated. In each case, resin regeneration is triggered based on actual need rather than the passage of time. (b) In addition, users can also set the IX water softener to use a high or low level of salt during each regeneration cycle. Use of a low level of salt (6–7 lbs) per cubic foot of resin gives sufficient efficiency of contaminant removal, per pound of salt used.

Water Softener and IX Operations: IX treatment process consists of passing the water through a cationic IX resin media bed. The Fe-Mn ions and also calcium and magnesium ions in the water are "exchanged" for sodium (Na^+) ions which have been temporarily stored in the resin material. As the hardness and Fe-Mn are removed from the water, sodium is added proportionally. For every 10 mg/L of hardness and Fe-Mn removed, approximately 5 mg/L of sodium will be added to the treated water. For those concerned with elevated sodium levels in their drinking water, potassium chloride (KCl) can be used in place of sodium chloride (NaCl). Eventually the removal capacity of the IX resin material will become exhausted and the media will need to be regenerated. The regeneration process begins with a physical backwash of the media to remove fine particles. The resin is then immersed in a strong salt brine solution. The sodium (or potassium) from the salt enters the resin and displaces the previously removed Fe-Mn and hardness (Ca-Mg). After a period of time (approximately 20 minutes), the remaining brine, along with the displaced Fe-Mn and Ca-Mg, are flushed out of the device and disposed of into a dry well, septic tank or sewer. Studies by the WQA indicate that waste brine does not injure leach fields or septic tanks.

REFERENCES

1. Wang, LK and Wang, MHS (1988). *Standards and Guides of Water Treatment and Water Distribution Systems, Volume I.* US Department of Commerce, National Technical Information Service, Springfield, VA, USA. NTIS Publication No. PB88–177902, 345 pages.
2. Wang, LK and Wang, MHS (1989). *Standards and Guides of Water Treatment and Water Distribution Systems, Volume II.* US Department of Commerce, National Technical Information Service, Springfield, VA, USA. NTIS Publication No. PB89-102289/AS, 256 pages.
3. Krofta, M and Wang, LK (2000). *Flotation Engineering*, 1st edition. Lenox Institute of Water Technology, Lenox, MA, USA. Technical Manual No. Lenox/1–06–2000/368, January 2000.
4. Wang, LK (1995). The State of the Art Technologies for Water Treatment and Management. United Nations Industrial Development Organization (UNIDO), Technical Report No. 8-8-95, 1995, Vienna, Austria.

5. Wang, LK (1994). Technologies and Terminologies of Iron and Manganese Control. Lecture material of Illinois Extension Specialist LK Wang. The University of Illinois at Urbana-Champaign, Cooperative Extension Service Urbana, Illinois 61801, USA (Teaching lecture material).
6. Keithley, A (2020). *Iron and Manganese Control in Groundwater Systems.* US Environmental Protection Agency, Office of R&D, Washington, DC, USA.
7. Editor (2018). *Recommended Standards for Water Works (10 States Standards)*, 18th edition. Great Lakes - Upper Mississippi River Board of State and Provincial Public Health and Environmental Managers, St. Paul, MN, USA.
8. Dvorak, B and Schuerman, B (2021). *Sources of Iron and Manganese.* The University of Nebraska-Lincoln, Cooperative Extension Service, Lincoln, NE, USA. https://extensionpublications.unl.edu/assets/html/g1714/build/g1714.htm
9. Dvorak, B and Schuerman, B (2021). *Drinking Water: Iron and Manganese.* The University of Nebraska–Lincoln Extension, Lincoln, NE, USA.
10. NHDES (2000). *Removal of Iron and Manganese from Drinking Water.* Publication WD-WSEB-3–7, New Hampshire Department of Environmental Services, Concord, NH, USA.
11. Herman, GM (1996). *Iron and Manganese in Household Water.* North Carolina State University, Cooperative Extension Service, Raleigh, NC, USA.
12. Seelig, B, Derickson, R, and Bergsrud, F. (2009). *Iron and Manganese Removal, AE-1030*, North Dakota State University, North Dakota Extension Service, Fargo, ND, USA.
13. Manager (2009). *Iron and Manganese in Drinking Water.* Wilkes University, Center for Environmental Quality, Wilkes-Barre, PA, USA.
14. Swistock, B and Sharpe, W. (2022). *Iron and Manganese in Private Water Systems.* Pennsylvania State University, University Park, PA, USA. https://extension.psu.edu/iron-and-manganese-in-private-water-systems
15. APHA/AWWA/WEF (2017). *Standard Methods for the Examination of Water and Wastewater.* American Public Health Association, Washington, DC, USA.
16. AWWA (1999). *Water Quality and Treatment.* American Water Works Association, Denver, CO, USA.
17. Clear Water Store Editor (2022). Birm, Greensand, and Pro-OX Iron Filters: A Primer, Clear Water Store, USA. https://www.cleanwaterstore.com/blog/birm-greensand-and-mangox-iron-filters-a-primer/
18. Zabel (2000). *Iron and Manganese Removal from Potable Waters.* Zabel and Associates, Inc., Destin, FL, USA.
19. Wong, JM, Hess, RJ, and Wang, LK (2021). Operation and performance of the AquaDAF process system for water purification. In: *Environmental Flotation Engineering*, Wang, LK, Wang, MHS, Shammas, NK, and Aulenbach, DB (editors), Springer Nature, Cham, Switzerland, pp. 301–342.
20. Wong, JM, Farmerie, JE, and Wang, LK (2021). Operation and performance of the Clari-DAF process system for water purification. In: *Environmental Flotation Engineering*, Wang, LK, Wang, MHS, Shammas, NK, and Aulenbach, DB (editors), Springer Nature, Cham, Switzerland, pp. 343–370.
21. Williams, P (2002). Biological Iron and Manganese Removal as a Viable Alternative for Groundwater Treatment. ONDEO Degremont, Canada. *Environmental Science and Engineering Magazine.*
22. Krofta, M and Wang, LK (1985). Application of dissolved air flotation to the Lenox, Massachusetts water supply: Water purification by flotation. *Journal New England Water Works Association*, 99(3), pp. 245–264.
23. Krofta, M and Wang, LK (1985). Application of dissolved air flotation to the Lenox, Massachusetts water supply: Sludge thickening by flotation or lagoon. *Journal New England Water Works Association*, 99(3), pp. 265–284.
24. Wang, LK, and Wang, MHS (2021). First wave of flotation technology evolution: Once the world's largest DAF-filtration plant and its hydroelectric facility. In: *Integrated Natural Resources Research*, Wang, LK, Wang, MHS, and Hung, YT (editors), Springer Nature, Cham, Switzerland, pp. 449–480.
25. Wang, LK, Wang, MHS, and Fahey, EM (2021). Innovative dissolved air flotation potable water filtration plant in Lee, Massachusetts, USA. In: *Environmental Flotation Engineering*, Wang, LK, Wang, MHS, Shammas, NK, and Aulenbach, DB (editors), Springer Nature, Cham, Switzerland, pp. 73–94.
26. Wang, LK and Wang, MHS (2025). An innovative process system consisting of dissolved air flotation, recarbonation and filtration, using lime and sodium aluminate for removal of hardness, iron and manganese from groundwater. In: *Control of Heavy Metals in the Environment, Volume 1*, Wang, LK, Wang, MHS, Hung, YT, and Chen, JP (editors), CRC Press, Boca Raton, FL, USA, pp. 345–357.

27. Wang, LK and Wang, MHS (2025). Removal of iron, copper, coliform bacteria and hardness from cooling tower water by dissolved air-ozone flotation. In: *Control of Heavy Metals in the Environment, Volume 2*, Wang, LK, Wang, MHS, Hung, YT, and Chen, JP (editors), CRC Press, Boca Raton, FL, USA, pp. 287–302.
28. Wang, LK (2021). Humanitarian engineering education of the Lenox Institute of Water Technology and its new potable water flotation processes. In: *Environmental Flotation Engineering*, Wang, LK, Wang, MHS, Shammas, NK, and Aulenbach, DB (editors), Springer Nature, Cham, Switzerland, pp. 1–72.
29. Wang, LK and Wang, MHS (2021). A new wave of flotation technology advancement for wastewater treatment. In: *Environmental Flotation Engineering*, Wang, LK, Wang, MHS, Shammas, NK, and Aulenbach, DB (editors), Springer Nature, Cham, Switzerland, pp. 143–166.
30. Wang, LK, Wang, MHS, and Wong, JM (2021). Innovative circular gravity flotation and fiber detection for fiber separation. In: *Environmental Flotation Engineering*, Wang, LK, Wang, MHS, Shammas, NK, and Aulenbach, DB (editors), Springer Nature, Cham, Switzerland, pp. 167–188.
31. Wang, LK and Wang, MHS (2026). Cocoa processing waste pretreatment and mobile flotation plants. In*: Industrial Water and Waste Treatment: Technology Developments*, Wang, LK, Wang, MHS, and Hung, YT (editors), CRC Press, Boca Raton, FL, USA.
32. Wang, LK and Wang, MHS (2025). Treatment of food ingredients manufacturing effluent by physico-chemical-biological two-stage wastewater treatment system. In: *Industrial Water and Waste Treatment: Technology Developments*, Wang, LK, Wang, MHS, and Hung, YT (editors), CRC Press, Boca Raton, FL, USA.
33. Wang, LK and Wang, MHS (2026). Hazardous wastewater treatment in petroleum refining industry using best available flotation and filtration technologies. In: *Control of Heavy Metals in the Environment, Volume 1*, Wang, LK, Wang, MHS, Hung, YT, and Chen, JP (editors), CRC Press, Boca Raton, FL, USA, pp. 249–274.
34. Wang, LK and Wang, MHS (2026). Wastewater treatment and resources recovery in paper and pulp industry using high rate dissolved air flotation. In: *Industrial Water and Waste Treatment: Technology Developments*, Wang, LK, Wang, MHS, and Hung, YT (editors), CRC Press, Boca Raton, FL, USA.
35. Wang, LK, Kurylko, L, and Wang, MHS (1992). Gas dissolving and releasing liquid treatment system. US Patent No. 5,167, 806. US Patent and Trademarks Office, Washington, DC, USA.
36. Wang, LK, Wang, MHS, Wang, P, and Clesceri, NL (2021). Biological and physicochemical sequencing batch reactors using sedimentation or flotation. In: *Environmental Flotation Engineering*, Wang, LK, Wang, MHS, Shammas, NK, and Aulenbach, DB (editors), Springer Nature, Cham, Switzerland, pp. 397–417.
37. Wang, LK, Kurylko, L, and Wang, MHS (1994). Sequencing batch liquid treatment. US Patent No. 5,354,458. US Patent and Trademarks Office, Washington, DC, USA.
38. Wang, LK, Vaccari, DA, Li, Y, and Shammas, NK (2004). Chemical precipitation. In: *Physicochemical Treatment Processes*, Wang, LK, Hung, YT, and Shammas, NK (editors), Humana Press, Totowa, NJ, USA, pp. 141–198.
39. Wang, MHS and Wang, LK (2021). Glossary of water quality, treatment and recovery. In: *Integrated Natural Resources Research*, Wang, LK, Wang, MHS, and Hung, YT (editors), Springer Nature, Cham, Switzerland, pp. 569–630.
40. Shammas, NK and Wang, LK (2016). *Water Engineering: Hydraulics, Distribution and Treatment.* John Wiley & Sons, Hoboken, NJ, USA, 806 pages.
41. Krofta, M and Wang, LK (1982). Potable water treatment by dissolved air flotation filtration. *Journal American Water Works Association*, 74(6), pp. 304–310.
42. Wang, LK, Wu, JS, Shammas, NK, and Vaccari, DA (2004). Recarbonation and softening. In: *Physicochemical Treatment Processes*, Wang, LK, Hung, YT, and Shammas, NK (editors), Humana Press, Totowa, NJ, USA, pp. 199–228.
43. Shammas, NK, Hahn, HH, Wang, MHS, and Wang, LK (2021). Fundamentals of chemical coagulation and precipitation. In: *Environmental Flotation Engineering*, Wang, LK, Wang, MHS, Shammas, NK, and Aulenbach, DB (editors), Springer Nature, Cham, Switzerland, pp. 95–142.
44. Wang, LK and Wang, MHS (2025). Reduction of high color, humic substances, turbidity, coliform bacteria, THMMFP, UV absorbance, cysts, chlorine demand, and metals from unpredictable water source by dissolved air flotation and filtration. In: *Control of Heavy Metals in the Environment, Volume 2*, Wang, LK, Wang, MHS, Hung, YT, and Chen, JP (editors), CRC Press, Boca Raton, FL, USA, pp. 379–395.

45. Wu, BC (1982). Separation of Organics and Inorganics from Water by Bubble Separation Process. Lenox Institute of Water Technology (formerly Lenox Institute for Research), Master Thesis, 108 pages. M. Krofta, LK Wang and MHS Wang (advisors), Lenox, MA, USA.
46. Wang, LK and Wang, MHS (2025). Reduction of color, turbidity, odor, humic acid, metals, EDB, coliform and TTHM by adsorption, flotation and filtration. In: Control of *Heavy Metals in the Environment, Volume 2*, Wang, LK, Wang, MHS, Hung, YT, and Chen, JP (editors), CRC Press, Boca Raton, FL, USA, pp. 379–395.
47. Wang, MHS and Wang, LK (2026). Ion exchange engineering fundamentals. In: Industrial *Water and Waste Treatment: Fundamentals*, Wang, LK, Wang, MHS, and Hung, YT (editors), CRC Press, Boca Raton, FL, USA. pp. 427–453.

5 Fundamentals and Innovations of Calcium and Magnesium Control

Mu-Hao Sung Wang and Lawrence K. Wang

ACRONYM

ABF	Automatic backwash filtration
APHA	American Public Health Association
AWWA	American Water Works Association
Ca-NCH	Calcium non-carbonate hardness
DAF	Dissolved air flotation
DAOFF	Dissolved air-ozone flotation and filtration
DOM	Dissolved organic matter
Fe-Mn	Iron and manganese
IX	Ion exchange
LIWT	Lenox Institute of Water Technology
NCH	Non-carbonate hardness
O&M	Operation and maintenance
PAC	Powdered activated carbon
PC-SBR	Physicochemical sequencing batch reactor
SBR	Sequencing batch reactor
STEAM	Science, Technology, Engineering, Arts, and Mathematics
STEM	Science, Technology, Engineering, and Mathematics
TDS	Total dissolved solids
TTHM	Total trihalomethane
WEF	Water Environment Federation
WTP	Water treatment plant

5.1 INTRODUCTION

5.1.1 Hardness Sources

Natural sources of hardness principally are limestones which are dissolved by percolating rainwater made acid by dissolved carbon dioxide. Industrial and industrially related sources include the inorganic chemical industry and discharges from operating and abandoned mines. Hardness in fresh water frequently is distinguished in carbonate and non-carbonate fractions. The carbonate fraction is chemically equivalent to the bicarbonates present in water. Since bicarbonates generally are measured as alkalinity, the carbonate hardness usually is considered equal to the alkalinity.

5.1.2 Types of Hardness

Technically speaking there are calcium hardness, magnesium hardness, carbonate hardness, and non-carbonate hardness (NCH).

DOI: 10.1201/9781003532354-5

Calcium hardness is caused by the calcium ions dissolved in water, expressed as mg/L $CaCO_3$, while magnesium hardness is caused by the magnesium ions dissolved in water, expressed as mg/L $CaCO_3$.

Carbonate hardness is caused by cations from the dissolution of calcium or magnesium carbonate and bicarbonate in water. Since most of the alkalinity in natural waters is caused by the bicarbonate and carbonate ions, chemically carbonate hardness is equivalent to the alkalinity.

NCH is caused by cations mainly from calcium and magnesium compounds of chloride, sulfate, or silicate that are dissolved in water. NCH equals the total hardness minus the carbonate hardness.

5.1.3 Degrees and Maximum Levels of Hardness

Soft water is the water containing 0–75 mg/L (as $CaCO_3$) of total hardness.

Moderately hard water is the water containing 76–150 mg/L (as $CaCO_3$) of total hardness.

Very hard water is the water containing over 300 mg/L (as $CaCO_3$) of total hardness.

The maximum levels accepted by industry as a raw water source are: (a) Electric Utilities Industry=5,000 mg/L as $CaCO_3$; (b) Textile Industry=120 mg/L as $CaCO_3$; (c) Pulp and Paper Industry=475 mg/L as $CaCO_3$; (d) Chemical Industry=1,000 mg/L as $CaCO_3$; (e) Petroleum Industry=900 mg/L as $CaCO_3$; (f) Primary Metals Industry=1,000 mg/L as $CaCO_3$; (g) Agricultural Industry=Hardness is not a determination of concern for irrigation use of water. The concentrations of the calcium and magnesium cations, which comprise hardness, are important in determining the exchangeable sodium in a given water.

5.1.4 Softening Process

Softening process is any physicochemical process that reduces the total hardness of "hard water" or "very hard water" to the "moderately hard" level of 76–150 mg/L (as $CaCO_3$) for domestic consumption, or to the "soft" level of 0–75 mg/L (as $CaCO_3$) for specific industrial applications, or to any other levels requested by the industrial, agricultural, or commercial clients.

5.2 ENVIRONMENTAL ENGINEERING SIGNIFICANCE OF HARDNESS REMOVAL

Hardness is mainly caused by the polyvalent metallic ions dissolved in water. In fresh water, these are primarily calcium and magnesium although other metals such as iron, strontium, and manganese contribute to the extent that appreciable concentrations are present. Hardness commonly is reported as an equivalent concentration of calcium carbonate ($CaCO_3$). The concept of hardness comes from water supply practice. It is measured by soap requirements for adequate lather formation and as an indicator of the rate of scale formation in hot water heaters and low-pressure boilers.

Because hardness concentrations in water have not been proven to be health related, the final level achieved principally is a function of economics. Since hardness in water can be removed with treatment by such processes as lime-soda ash softening and zeolite or ion exchange (IX) systems, a criterion for raw waters used for public water supply is not practical.

5.3 LITERATURE REVIEW

Wang and Wang [1] have introduced some innovations in the areas of hardness control and Fe-Mn removal: (a) a new softening and Fe-Mn removal process system consisting of dissolved air flotation (DAF), recarbonation, and filtration has been demonstrated by operating a continuous water treatment plant (WTP); (b) a new chemical combination including lime, sodium aluminate, and carbon dioxide can be used for simultaneous removal of Ca-Mg hardness, iron, and manganese from groundwater; (c) DAF (3–15 minutes detention time) replaces conventional sedimentation (4–6 hours detention) which is much more cost-effective; (d) DAF does not need an additional aeration process step for Fe-Mn removal; (e) sodium aluminate replaces conventional soda ash for better removal of permanent hardness, or NCH and better removal of insoluble manganese particles (note: produced after oxidation by DAF) due to its coagulant nature.

In this research, groundwater having 15 NTU of turbidity, 216 mg/L as $CaCO_3$ of total hardness, 3.56 mg/L of iron, and 1.57 mg/L of manganese at pH 7.8 was successfully treated by a continuous WTP consisting of a rectangular static hydraulic flocculator (detention time=5.6 minutes), a DAF clarifier (detention time=3 minutes), a recarbonation facility, and three sand filters (sand bed=11 inches of silica sand; filtration rate=2.5 gpm/ft^2=101.75 Lpm/m^2). With lime for initial chemical precipitation and pH adjustment to 11.2, and 20 mg/L (as Al_2O_3) of sodium aluminate for enhanced precipitation and coagulation, excellent product water meeting the U.S. Drinking Water Standards was produced. The quality of the final filter effluent was: turbidity, 0.72 NTU; pH, 6.5; total hardness, 60 mg/L as $CaCO_3$; iron, 0.04 mg/L; and manganese, 0.02 mg/L. It appears that the new water treatment system including flocculator, DAF clarifier, recarbonation facility, and filter is technically feasible for groundwater treatment. Lime, sodium aluminate, and carbon dioxide are excellent chemical aids for simultaneous removal of turbidity, hardness, iron, and manganese.

An ozone-assisted physical-chemical treatment process involving the use of chemical coagulation-precipitation, clarification (either sedimentation or flotation), and filtration for cooling tower raw make-up water treatment, side-stream recirculating water treatment, or blowdown water treatment was investigated by the Lenox Institute of Water Technology (LIWT) [2] using bench-scale laboratory tests, continuous pilot plant demonstration, and full-scale cooling tower operation. Only the preliminary laboratory data are introduced at this stage. The results show that 95% of iron ions, 67.8% of copper ions, 100% of total coliform bacteria, and over 98% of total hardness can be removed by the ozone-assisted dissolved air-ozone flotation and filtration (DAOFF) process system. The DAOFF system may be adopted for (a) production of cooling tower make-up water using raw reservoir/stream water or raw groundwater water as the source water; (b) treatment of cooling tower side-stream recirculation water for scale removal, new scale prevention, corrosion, bacteria disinfection, algae control, and water purification; or (c) treatment of cooling tower blowdown water for final discharge to a deep well (without groundwater contamination), an ocean, or a receiving water (without thermal pollution or excessive TDS). Ozone kills all bacterial slimes which commonly bind the scale on the metal surface, and keeps the hardness in suspension, so the suspended hardness will not form scale and may be effectively removed by filtration. Wang and Wang [2] are calling for further research to determine the chemistry of hardness removal by ozonation and filtration.

5.4 INNOVATIVE DAF SOFTENING PROCESSES AND NEW CHEMICAL COMBINATIONS

Conventional softening processes for hardness are well established. They include: (a) single-stage excess lime softening process; (b) single-stage excess lime-soda ash softening process; (c) single-stage lime softening process; (d) single-stage lime-soda ash softening process; (e) softening process for simultaneous hardness and dissolved organic matter (DOM) removal; (f) softening process for simultaneous hardness, iron, and manganese removal; (g) two-stage excess lime treatment process; (h) two-stage lime/soda ash softening process; and (i) IX softening process. The brief process descriptions of conventional softening processes can be found in Section 5.5.

In recent years, full-scale DAF clarification has successfully replaced conventional sedimentation clarification for both water treatment [3–10] and wastewater treatment [11–16] due to the shorter detention time, smaller footprint, and lower construction and operation and maintenance (O&M) costs of DAF.

DAF is a process involving pressurization of air at 25–95 psig for dissolving air into water, and subsequent release of pressure (to 1 atm) under laminar flow conditions for generating extremely fine air bubbles (normally 20–80 μm) which become attached to the impurities to be removed. The air flow rate is about 1% of influent liquid flow rate. The attachment of air bubbles to the impurities can be a result of physical entrapment, electrochemical attraction, surface adsorption, and/or gas stripping. The specific gravity of bubble-impurity agglomerate is less than one, resulting in buoyancy or non-selective flotation (i.e. a save-all process). In the past 40 years, DAF has been used mainly for sludge thickening and fiber recovery. Now DAF has become a high technology for potable water purification and wastewater treatment [3–16].

DAF water and wastewater treatment process systems can be either continuous systems or sequencing batch reactors (SBRs) [17–19]. The authors also have developed various physicochemical SBR (PC-SBR) including both DAF-SBR and physicochemical sedimentation-SBR (PC sedimentation-SBR) for water purification [17–19].

The innovative DAF clarification softening process systems are introduced below:

1. **Innovative Single-Stage Excess Lime-Sodium Aluminate Softening Process for Simultaneous Removal of Total Hardness, Iron, Manganese, and DOM:** It includes the process sequence of rapid mixing (excess lime+sodium aluminate addition), flocculation, DAF clarification, recarbonation (pH=8.6), and filtration. This single-stage lime-sodium aluminate softening process is suitable for treating a source water having high calcium, high magnesium carbonate hardness (higher than 40 mg/L as $CaCO_3$), some calcium non-carbonate hardness (Ca-NCH), and some DOM. The innovative DAF softening process system can be either a continuous system or a DAF-SBR system.
2. **Innovative Single-Stage Lime-Soda Ash Softening Process:** It includes the process sequence of rapid mixing (lime+soda ash addition), flocculation, DAF clarification, recarbonation (pH=8.6), and filtration. This single-stage lime-soda ash softening process is suitable for treating a source water having high calcium, low magnesium carbonate hardness (less than 40 mg/L as $CaCO_3$), and some Ca-NCH. The innovative DAF softening process system can be either a continuous system or a DAF-SBR system.
3. **Innovative Single-Stage Lime-Sodium Aluminate Softening Process for Simultaneous Removal of Total Hardness, Iron, Manganese, and DOM:** It includes the process sequence of rapid mixing (lime+sodium aluminate addition), flocculation, DAF clarification, recarbonation (pH=8.6), and filtration. This single-stage lime-sodium aluminate softening process is suitable for treating a source water having high calcium, low magnesium carbonate hardness (less than 40 mg/L as $CaCO_3$), some Ca-NCH, and some DOM. The innovative DAF softening process system can be either a continuous system or a DAF-SBR system.
4. **Innovative Single-Stage Excess Lime Softening Process:** It includes the process sequence of rapid mixing (excess lime addition), flocculation, DAF clarification, recarbonation (pH=8.6), and filtration. This single-stage excess lime process is suitable for a raw source water having no NCH, high calcium, and high magnesium carbonate hardness. The innovative DAF softening process system can be either a continuous system or a DAF-SBR system.
5. **Innovative Single-Stage Excess Lime-Soda Ash Softening Process:** It includes the process sequence of rapid mixing (excess lime+soda ash addition), flocculation, DAF clarification, recarbonation (pH=8.6), and filtration. This single-stage lime-soda ash softening process is suitable for treating a source water having high calcium, high magnesium carbonate hardness (higher than 40 mg/L as $CaCO_3$), and some Ca-NCH exceeding water quality standards. The innovative DAF softening process system can be either a continuous system or a DAF-SBR system.
6. **Innovative Single-Stage Lime Softening Process:** It includes the process sequence of rapid mixing (lime addition), flocculation, DAF clarification, recarbonation (pH=8.6), and filtration. This single-stage lime process is suitable for a raw source water having no NCH, high

calcium, and low magnesium carbonate hardness (less than 40 mg/L as $CaCO_3$). The innovative DAF softening process system can be either a continuous system or a DAF-SBR system.

7. **Innovative Softening Process for Simultaneous Hardness and DOM Removal:** It can be any innovative softening process with lime and an extra sodium aluminate addition to the rapid mixing and flocculation units. DAF is used for clarification. The innovative DAF softening process system can be either a continuous system or a DAF-SBR system.
8. **Innovative Softening Process for Simultaneous Hardness, Iron, and Manganese Removal:** It can be any innovative softening process using DAF for clarification because no additional aeration process step is needed ahead of rapid mixing and flocculation units.
9. **Innovative Two-Stage Excess Lime Treatment Process:** It includes the process sequence of rapid mixing (excess lime addition), flocculation, primary DAF clarification, recarbonation, secondary DAF clarification, recarbonation, and filtration. Disinfection and corrosion control are also required for potable water application. The process is suitable for treating a source water with high calcium, high magnesium carbonate hardness, but no NCH. The innovative DAF softening process system can be either a continuous system or a DAF-SBR system.
10. **Innovative Two-Stage Lime/Soda Ash Softening Process:** It includes the process sequence of rapid mixing (lime addition; $pH > 10.8$), flocculation, first-stage DAF clarification, first-stage recarbonation, rapid mixing (soda ash addition), flocculation, second-stage DAF, second-stage recarbonation ($pH = 8.6$), and filtration. Disinfection and corrosion control are also required for potable water application. The process is suitable for treating a source water with high calcium, low magnesium carbonate hardness (less than 40 mg/L as $CaCO_3$), and some NCH. The innovative DAF softening process system can be either a continuous system or a DAF-SBR system.
11. **Adoption of Innovative PC-SBR for All Conventional Sedimentation Softening Processes and IX Softening Process Listed Here:** (a) single-stage excess lime softening process; (b) single-stage excess lime-soda ash softening process; (c) single-stage lime softening process; (d) single-stage lime-soda ash softening process; (e) softening process for simultaneous hardness and DOM removal; (f) softening process for simultaneous hardness, iron, and manganese removal; (g) two-stage excess lime treatment process; (h) two-stage lime/soda ash softening process; and (i) IX softening process.

5.5 HARDNESS REMOVAL TECHNOLOGIES AND TERMINOLOGIES [1,17–29]

Anion exchange: An IX process involving the use of anion exchanger (anionic IX resin) mainly for the removal of negatively charged ions, such as sulfate, phosphate, chloride, and negatively charged organic ions.

Calcium hardness: It is caused by the calcium ion dissolution in water, expressed as mg/L $CaCO_3$.

Carbonate hardness: (a) It is caused by cations from the dissolution of calcium or magnesium carbonate and bicarbonate in water; (b) Since most of the alkalinity in natural waters is caused by the bicarbonate and carbonate ions, chemically carbonate hardness is equivalent to the alkalinity.

Carbonation: (a) Carbonation is the chemical reaction of carbon dioxide to give carbonates, bicarbonates, and carbonic acid. In chemistry, the term is sometimes used in place of carboxylation, which refers to the formation of carboxylic acids; (b) Carbonation is the saturation of a liquid with carbon dioxide gas. In other words, it is a term used to describe the dissolution of carbon dioxide gas in water utilizing pressure and temperature.

Cation exchange: An IX process involving the use of cation exchanger (cationic IX resin) mainly for the removal of calcium, magnesium, iron, and other positively charged metal ions, or positively charged organic ions.

Chemical coagulation: A chemical process in which one coagulating chemical coagulates (reacts) with another colloidal or suspended solid forming insoluble fine flocs due to charge neutralization and agglomeration.

Chemical precipitation: A chemical process in which one soluble chemical agent precipitates another soluble chemical agent.

Conventional single-stage excess lime softening process: It includes the process sequence of rapid mixing (excess lime addition), flocculation, sedimentation clarification, recarbonation (pH=8.6), and filtration. This single-stage excess lime process is suitable for a raw source water having no NCH, high calcium, and high magnesium carbonate hardness (more than 40 mg/L as $CaCO_3$).

Conventional single-stage excess lime-soda ash softening process: It includes the process sequence of rapid mixing (excess lime+soda ash addition), flocculation, sedimentation clarification, recarbonation (pH=8.6), and filtration. This single-stage lime-soda ash softening process is suitable for treating a source water having high calcium, high magnesium carbonate hardness (more than 40 mg/L as $CaCO_3$), and some Ca-NCH exceeding water quality standards.

Conventional single-stage lime softening process: It includes the process sequence of rapid mixing (lime addition), flocculation, sedimentation clarification, recarbonation (pH=8.6), and filtration. This single-stage lime process is suitable for a raw source water having no NCH, high calcium, and low magnesium carbonate hardness (less than 40 mg/L as $CaCO_3$).

Conventional single-stage lime-soda ash softening process: It includes the process sequence of rapid mixing (lime+soda ash addition), flocculation, sedimentation clarification, recarbonation (pH=8.6), and filtration. This single-stage lime-soda ash softening process is suitable for treating a source water having high calcium, low magnesium carbonate hardness (less than 40 mg/L as $CaCO_3$), and some Ca-NCH.

Conventional softening process for simultaneous hardness and DOM removal: It can be any conventional softening process with lime and an extra alum addition to the rapid mixing and flocculation units.

Conventional softening process for simultaneous hardness, iron, and manganese removal: It can be any conventional softening process with an additional aeration process step usually ahead of rapid mixing and flocculation units. Sedimentation is used for clarification.

Conventional two-stage excess lime treatment process: It includes the process sequence of rapid mixing (excess lime addition), flocculation, primary sedimentation clarification, recarbonation, secondary sedimentation clarification, recarbonation, and filtration. Disinfection and corrosion control are also required for potable water application. The process is suitable for treating a source water with high calcium, high magnesium carbonate hardness, but no NCH.

Conventional two-stage lime/soda ash softening process: It includes the process sequence of rapid mixing (lime addition; pH>10.8), flocculation, first-stage sedimentation clarification, first-stage recarbonation, rapid mixing (soda ash addition), flocculation, second-stage sedimentation, second-stage recarbonation (pH=8.6), and filtration. Disinfection and corrosion control are also required for potable water application. The process is suitable for treating a source water with high calcium, low magnesium carbonate hardness (less than 40 mg/L as $CaCO_3$), and some NCH.

Dissolved air flotation (DAF): DAF is a process involving pressurization of air at 25–95 psig for dissolving air into water, and subsequent release of pressure (to 1 atm) under laminar flow conditions for generating extremely fine air bubbles (normally 20–80 μm) which become attached to the impurities to be removed. The air flow rate is about 1% of influent liquid flow rate. The attachment of air bubbles to the impurities can be a result of physical entrapment, electrochemical attraction, surface adsorption, and/or gas stripping. The specific gravity of bubble-impurity agglomerate is less than 1, resulting in buoyancy or non-selective flotation (i.e. a save-all process). In the past 40 years, DAF has been used mainly for sludge thickening and fiber recovery. Now DAF has become a high technology for potable water purification and wastewater treatment.

Excess lime softening process: It is any softening process that requires to add excess lime for treating a source raw water with high calcium and high magnesium carbonate hardness (higher than 40 mg/L as $CaCO_3$). It can be either a one- or two-stage process.

Filter run length (run time): The time in filtration service between filter cleanings is termed the run length or run time.

Filtration: Dual media filtration by gravity is one of the most economical forms of granular media filtration. Dual media filtration involves the use of both sand and anthracite as filter media, with anthracite being placed on top of the sand. Single medium, such as sand only, is called sand filtration. If only granular activated carbon (GAC) is used, it is called GAC filtration. Granular media filtration involves the passage of water through a bed of filter media with resulting deposition of solids. Eventually, the pressure drop across the bed becomes excessive or the ability of the bed to remove suspended solids is impaired. Cleaning is then necessary to restore the operation head and the filter effluent quality. The time in service between cleanings is termed the run length or run time. The head loss at which filtration is interrupted for cleaning is called the terminal head loss, and this head loss is maximized by the judicious choice of media sizes.

Hard water: It is the water containing 151–300 mg/L (as $CaCO_3$) of total hardness.

Hardness classification: (a) soft water = 0–75 mg/L $CaCO_3$ of total hardness; (b) moderately hard = 75–150 mg/L $CaCO_3$ of total hardness; (c) hard = 150–300 mg/L $CaCO_3$ of total hardness; (d) very hard = 301 mg/L $CaCO_3$ of total hardness and up.

Hardness environmental engineering significance: It is caused by the polyvalent metallic ions dissolved in water. In fresh water, these are primarily calcium and magnesium although other metals such as iron, strontium, and manganese contribute to the extent that appreciable concentrations are present. Hardness commonly is reported as an equivalent concentration of calcium carbonate ($CaCO_3$). The concept of hardness comes from water supply practice. It is measured by soap requirements for adequate lather formation and as an indicator of the rate of scale formation in hot water heaters and low-pressure boilers.

Hardness maximum levels: The maximum levels accepted by industry as a raw water source are: (a) Electric Utilities Industry = 5,000 mg/L as $CaCO_3$; (b) Textile Industry = 120 mg/L as $CaCO_3$; (c) Pulp and Paper Industry = 475 mg/L as $CaCO_3$; (d) Chemical Industry = 1,000 mg/L as $CaCO_3$; (e) Petroleum Industry = 900 mg/L as $CaCO_3$; (f) Primary Metals Industry = 1,000 mg/L as $CaCO_3$; (g) Agricultural Industry = Hardness is not a determination of concern for irrigation use of water. The concentrations of the calcium and magnesium cations, which comprise hardness, are important in determining the exchangeable sodium in a given water.

Hardness removal rationale: Because hardness concentrations in water have not been proven to be health related, the final level achieved principally is a function of economics. Since hardness in water can be removed with treatment by such processes as lime-soda ash softening and zeolite or IX systems, a criterion for raw waters used for public water supply is not practical.

Hardness sources: Natural sources of hardness principally are limestones which are dissolved by percolating rainwater made acid by dissolved carbon dioxide. Industrial and industrially related sources include the inorganic chemical industry and discharges from operating and abandoned mines. Hardness in fresh water frequently is distinguished in carbonate and non-carbonate fractions. The carbonate fraction is chemically equivalent to the bicarbonates present in water. Since bicarbonates generally are measured as alkalinity, the carbonate hardness usually is considered equal to the alkalinity.

Innovative single-stage excess lime-sodium aluminate softening process for simultaneous removal of total hardness, iron, manganese, and DOM: It includes the process sequence of rapid mixing (excess lime + sodium aluminate addition), flocculation, DAF clarification, recarbonation (pH = 8.6), and filtration. This single-stage lime-sodium aluminate softening process is suitable for treating a source water having high calcium, high magnesium carbonate hardness (higher than 40 mg/L as $CaCO_3$), some Ca-NCH, and some DOM. The innovative DAF softening process system can be either a continuous system or a DAF-SBR system.

Innovative single-stage lime-soda ash softening process: It includes the process sequence of rapid mixing (lime + soda ash addition), flocculation, DAF clarification, recarbonation (pH = 8.6), and filtration. This single-stage lime-soda ash softening process is suitable for treating a source water having high calcium, low magnesium carbonate hardness (less than 40 mg/L as $CaCO_3$), and some Ca-NCH. The innovative DAF softening process system can be either a continuous system or a DAF-SBR system.

Innovative single-stage lime-sodium aluminate softening process for simultaneous removal of total hardness, iron, manganese, and DOM: It includes the process sequence of rapid mixing (lime + sodium aluminate addition), flocculation, DAF clarification, recarbonation (pH = 8.6), and filtration. This single-stage lime-sodium aluminate softening process is suitable for treating source water having high calcium, low magnesium carbonate hardness (less than 40 mg/L as $CaCO_3$), some Ca-NCH, and some DOM. The innovative DAF softening process system can be either a continuous system or a DAF-SBR system.

Innovative single-stage excess lime softening process: It includes the process sequence of rapid mixing (excess lime addition), flocculation, DAF clarification, recarbonation (pH = 8.6), and filtration. This single-stage excess lime process is suitable for a raw source water having no NCH, high calcium, and high magnesium carbonate hardness. The innovative DAF softening process system can be either a continuous system or a DAF-SBR system.

Innovative single-stage excess lime-soda ash softening process: It includes the process sequence of rapid mixing (excess lime + soda ash addition), flocculation, DAF clarification, recarbonation (pH = 8.6), and filtration. This single-stage lime-soda ash softening process is suitable for treating a source water having high calcium, high magnesium carbonate hardness (higher than 40 mg/L as $CaCO_3$), and some Ca-NCH exceeding water quality standards. The innovative DAF softening process system can be either a continuous system or a DAF-SBR system.

Innovative single-stage lime softening process: It includes the process sequence of rapid mixing (lime addition), flocculation, DAF clarification, recarbonation (pH = 8.6), and filtration. This single-stage lime process is suitable for a raw source water having no NCH, high calcium, and low magnesium carbonate hardness (less than 40 mg/L as $CaCO_3$). The innovative DAF softening process system can be either a continuous system or a DAF-SBR system.

Innovative softening process for simultaneous hardness and DOM removal: It can be any innovative softening process with lime and an extra sodium aluminate addition to the rapid mixing and flocculation units. DAF is used for clarification.

Innovative softening process for simultaneous hardness, iron, and manganese removal: It can be any innovative softening process using DAF for clarification because no additional aeration process step is needed ahead of rapid mixing and flocculation units.

Innovative two-stage excess lime treatment process: It includes the process sequence of rapid mixing (excess lime addition), flocculation, primary DAF clarification, recarbonation, secondary DAF clarification, recarbonation, and filtration. Disinfection and corrosion control are also required for potable water application. The process is suitable for treating a source water with high calcium, high magnesium carbonate hardness, but no NCH. The innovative DAF softening process system can be either a continuous system or a DAF-SBR system.

Innovative two-stage lime/soda ash softening process: It includes the process sequence of rapid mixing (lime addition; pH > 10.8), flocculation, first-stage DAF clarification, first-stage recarbonation, rapid mixing (soda ash addition), flocculation, second-stage DAF, second-stage recarbonation (pH = 8.6), and filtration. Disinfection and corrosion control are also required for potable water application. The process is suitable for treating a source water with high calcium, low magnesium carbonate hardness (less than 40 mg/L as $CaCO_3$), and some NCH. The innovative DAF softening process system can be either a continuous system or a DAF-SBR system.

IX softening process: (a) It is a reversible physicochemical process for water or wastewater treatment in which ions from an ion exchanger (i.e. an insoluble permanent solid medium or a resin) are exchanged for ions in a solution or fluid mixture surrounding the ion exchanger; (b) IX process removes Ca^{2+}, Mg^{2+}, Fe^{2+}, and Mn^{2+}; maintains original low dissolved oxygen (DO), but generates waste brine [29].

Magnesium hardness: It is caused by the magnesium ions' dissolution in water, expressed as mg/L $CaCO_3$.

Moderately hard water: It is the water containing 76–150 mg/L (as $CaCO_3$) of total hardness.

Non-carbonate hardness (NCH): (a) NCH is caused by cations mainly from calcium and magnesium compounds of chloride, sulfate, or silicate that are dissolved in water; (b) NCH equals the total hardness minus the carbonate hardness.

Recarbonation: (a) It is a physicochemical process of adding carbon dioxide gas to lime-softened, clarified water for the purpose of reducing the pH and promoting chemical stability. Carbon dioxide reduces pH through the conversion of hydroxide ions to carbonate and bicarbonate alkalinity; (b) A process in which carbon dioxide gas is bubbled into the water being treated to lower the pH, or into the wastewater to precipitate organic matter (such as dissolved protein), or inorganic matter.

Soda ash softening process: It is any softening process that requires to add soda ash for treating a source raw water with some noncarbonate hardness that exceeds the water quality standards.

Soft water: It is the water containing 0–75 mg/L (as $CaCO_3$) of total hardness.

Softening process: It is any physicochemical process that reduces the total hardness of "hard water" or "very hard water" to the "moderately hard" level of 76–150 mg/L (as $CaCO_3$) for domestic consumption, or to the "soft" level of 0–75 mg/L (as $CaCO_3$) for specific industrial applications.

Split softening process: Any softening process that treats only a portion of the raw water with chemical(s) throughout the entire unit process chain, and lets the remaining portion of raw water (without chemical addition) bypass the rapid mixing-flocculation, and go to a clarifier directly.

Total hardness: (a) Total hardness is caused by the sum of multivalent metallic cations (such as Ca^{2+}, Mg^{2+}, Sr^{2+}, Fe^{2+}, Mn^{2+}, etc.) from the dissolution of these cations in water (expressed as mg/L $CaCO_3$) although normally only calcium ions and magnesium ions are considered; (b) Normally only the total hardness and calcium hardness in water are analyzed, magnesium hardness is then calculated as the difference between total hardness and calcium hardness; (c) Total hardness is the sum of calcium hardness and magnesium hardness in mg/L as $CaCO_3$; (d) Total hardness minus the carbonate hardness is equal to noncarbonate hardness.

Very hard water: It is the water containing over 300 mg/L (as $CaCO_3$) of total hardness.

REFERENCES

1. Wang, LK and Wang, MHS (2025). An innovative process system consisting of dissolved air flotation, recarbonation and filtration, using lime and sodium aluminate for removal of hardness, iron and manganese from groundwater. In: *Control of Heavy Metals in the Environment, Volume 1*, Wang, LK, Wang, MHS, Hung, YT, and Chen, JP (editors), CRC Press, Boca Raton, FL, USA, pp. 233–248.
2. Wang, LK and Wang, MHS (2025). Removal of iron, copper, coliform bacteria and hardness from cooling tower water by dissolved air-ozone flotation. In: *Control of Heavy Metals in the Environment, Volume 2*, Wang, LK, Wang, MHS, Hung, YT, and Chen, JP (editors), CRC Press, Boca Raton, FL, USA, pp. 287–302.
3. Krofta, M and Wang, LK (1985). Application of dissolved air flotation to the Lenox, Massachusetts water supply: Water purification by flotation. *Journal New England Water Works Association*, 99(3), pp. 245–264.
4. Krofta, M and Wang, LK (1985). Application of dissolved air flotation to the Lenox, Massachusetts water supply: Sludge thickening by flotation or lagoon. *Journal New England Water Works Association*, 99(3), pp. 265–284.
5. Wang, LK and Wang, MHS (2021). First wave of flotation technology evolution: Once the world's largest DAF-filtration plant and its hydroelectric facility. In: *Integrated Natural Resources Research*, Wang, LK, Wang, MHS, and Hung, YT (editors), Springer Nature, Cham, Switzerland, pp. 449–480.
6. Wang, LK and Wang, MHS (2026). America's first potable water flotation-filtration plant: The award winning Lenox Water Treatment Plant. In: *Industrial Water and Waste Treatment: Resources Conservation*, Wang, LK, Wang, MHS, and Hung, YT (editors), CRC Press, Boca Raton, FL, USA.
7. Wang, LK, Wang, MHS, and Fahey, EM (2021). Innovative dissolved air flotation potable water filtration plant in Lee, Massachusetts, USA. In: *Environmental Flotation Engineering,* Wang, LK, Wang, MHS, Shammas, NK, and Aulenbach, DB (editors), Springer Nature, Cham, Switzerland, pp. 73–94.
8. Wong, JM, Hess, RJ, and Wang, LK (2021). Operation and performance of the AquaDAF process system for water purification. In: *Environmental Flotation Engineering*, Wang, LK, Wang, MHS, Shammas, NK, and Aulenbach, DB (editors), Springer Nature, Cham, Switzerland, pp. 301–342.
9. Wong, JM, Farmerie, JE, and Wang, LK (2021). Operation and performance of the Clari-DAF process system for water purification. In: *Environmental Flotation Engineering*, Wang, LK, Wang, MHS, Shammas, NK, and Aulenbach, DB (editors), Springer Nature, Switzerland, pp. 343–370.
10. Wang, LK (2021). Humanitarian engineering education of the Lenox Institute of Water Technology and its new potable water flotation processes. In: *Environmental Flotation Engineering*, Wang, LK, Wang, MHS, Shammas, NK, and Aulenbach, DB (editors), Springer Nature, Cham, Switzerland, pp. 1–72.
11. Wang, LK and Wang, MHS (2021). A new wave of flotation technology advancement for wastewater treatment. In: *Environmental Flotation Engineering*, Wang, LK, Wang, MHS, Shammas, NK, and Aulenbach, DB (editors), Springer Nature, Cham, Switzerland, pp. 143–166.
12. Wang, LK, Wang, MHS, and Wong, JM (2021). Innovative circular gravity flotation and fiber detection for fiber separation. In: *Environmental Flotation Engineering*, Wang, LK, Wang, MHS, Shammas, NK, and Aulenbach, DB (editors), Springer Nature, Cham, Switzerland, pp. 167–188.
13. Wang, LK and Wang, MHS (2026). Cocoa processing waste pretreatment and mobile flotation plants. In: *Industrial Water and Waste Treatment: Technology Developments*, Wang, LK, Wang, MHS, and Hung, YT (editors), CRC Press, Boca Raton, FL, USA.
14. Wang, LK and Wang, MHS (2026). Treatment of food ingredients manufacturing effluent by physicochemical-biological two-stage wastewater treatment system. In: *Industrial Water and Waste Treatment: Technology Developments*, Wang, LK, Wang, MHS, and Hung, YT (editors), CRC Press, Boca Raton, FL, USA.
15. Wang, LK and Wang, MHS (2025). Hazardous wastewater treatment in petroleum refining industry using best available flotation and filtration technologies. In: *Control of Heavy Metals in the Environment, Volume 1*, Wang, LK, Wang, MHS, Hung, YT, and Chen, JP (editors), CRC Press, Boca Raton, FL, USA, pp. 249–274.
16. Wang, LK and Wang, MHS (2026). Wastewater treatment and resources recovery in paper and pulp industry using high rate dissolved air flotation. In: *Industrial Water and Waste Treatment: Technology Developments*, Wang, LK, Wang, MHS, and Hung, YT (editors), CRC Press, Boca Raton, FL, USA.
17. Wang, LK, Kurylko, L, and Wang, MHS (1992). Gas dissolving and releasing liquid treatment system. US Patent No. 5,167, 806. US Patent and Trademarks Office, Washington, DC, USA.

18. Wang, LK, Wang, MHS, Wang, P, and Clesceri, NL (2021). Biological and physicochemical sequencing batch reactors using sedimentation or flotation. In: *Environmental Flotation Engineering*, Wang, LK, Wang, MHS, Shammas, NK, and Aulenbach, DB (editors), Springer Nature, Cham, Switzerland, pp. 397–417.
19. Wang, LK, Kurylko, L, and Wang, MHS (1994). Sequencing batch liquid treatment. US Patent No. 5,354,458. US Patent and Trademarks Office, Washington, DC, USA.
20. Wang, LK, Vaccari, DA, Li, Y, and Shammas, NK (2004). Chemical precipitation. In: *Physicochemical Treatment Processes*, Wang, LK, Hung, YT, and Shammas, NK (editors), Humana Press, Totowa, NJ, USA, pp. 141–198.
21. Wang, MHS and Wang, LK (2021). Glossary of water quality, treatment and recovery. In: *Integrated Natural Resources Research*, Wang, LK, Wang, MHS, and Hung, YT (editors), Springer Nature, Cham, Switzerland, pp. 569–630.
22. Wang, LK (2021). Humanitarian engineering education of the Lenox Institute of Water Technology and its new potable water flotation processes. In: *Environmental Flotation Engineering*, Wang, LK, Wang, MHS, Shammas, NK, and Aulenbach, DB (editors), Springer Nature, Cham, Switzerland, pp. 1–72.
23. Shammas, NK and Wang, LK (2016). *Water Engineering: Hydraulics, Distribution and Treatment.* John Wiley & Sons, Hoboken, NJ, USA, 806 pages.
24. AWWA (1999). *Water Quality and Treatment.* American Water Works Association, Denver, CO, USA.
25. Krofta, M and Wang, LK (1982). Potable water treatment by dissolved air flotation filtration. *Journal American Water Works Association*, 74(6), pp. 304–310.
26. APHA/AWWA/WEF (1992). *Standard Methods for the Examination of Water and Wastewater*, 18th Edition. American Public Health Association, Washington, DC, USA.
27. Wang, LK, Wu, JS, Shammas, NK, and Vaccari, DA (2004). Recarbonation and softening. In: *Physicochemical Treatment Processes*, Wang, LK, Hung, YT, and Shammas, NK (editors), Humana Press, Totowa, NJ, USA, pp. 199–228.
28. Shammas, NK, Hahn, HH, Wang, MHS, and Wang, LK (2021). Fundamentals of chemical coagulation and precipitation. In: *Environmental Flotation Engineering*, Wang, LK, Wang, MHS, Shammas, NK, and Aulenbach, DB (editors), Springer Nature, Cham, Switzerland, pp. 95–142.
29. Wang, MHS and Wang, LK (2025). Ion exchange engineering fundamentals. In: *Industrial Water and Waste Treatment: Fundamentals*, Wang, LK, Wang, MHS, and Hung, YT (editors), CRC Press, Boca Raton, FL, USA. pp. 427–453.

6 Biodegradation of Tannery Solid Wastes

Victória Vieira Kopp, Mariliz Gutterres, João Henrique Zimnoch dos Santos, Yung-Tse Hung, and Lawrence K. Wang

ACRONYM

Al	Aluminum
CH_4	Methane
CO_2	Carbon dioxide
$K2HPO_4$	Dipotassium hydrogen phosphate
KH_2PO_4	Potassium dihydrogen phosphate
MgAl	Magnesium and aluminum
NH3	Ammonia
NPK	Nitrogen, phosphorus, and potassium
PLGA	Poly(lactide co-glycolide)
Ti	Titanium
TiO_2	Titanium dioxide
Zr	Zirconium

6.1 INTRODUCTION

Leather is a material used in the production of various useful and lifestyle accessories, such as shoes, bags, garments, upholstery, personal protective equipment (gloves), and other products. It is a bio-based and biodegradable material [1]. Its raw material is skin/hide, one of the by-products of the meat industry. Brazil has the biggest cattle herd in the world [2]. Leather processing supports the environment by turning a waste product into a fashionable item. This unique material shows excellent vapor permeability, strength, thermal, and hydrothermal stability. Currently, the world trade in leather exceeds US$ 80 billion annually and is expected to grow as the human population increases. The supply of hides for the leather industry will be constant due to the significant increase in the demand for meat [3,4]. The leather supply chain is globalized, consisting of breeders, slaughterhouses, tanneries, and manufacturers; it could include retailers, chemical suppliers and technology/machinery providers, and service providers, such as logistics companies, marketing, information technology, and final consumers [5]. Considerable solid waste is generated during the production of leather, where up to 70% of the wet weight of the original hides can be wasted [4]. In this chapter, approaches to achieve a more biodegradable leather and the methods to recover biodegradable materials or energy from tannery solid waste by a biological technique are discussed.

6.2 LEATHER PROCESSING

In contrast to hides, leather is resistant to microbial attack and thermally stable. Leather manufacturing transforms hides into leather during three main stages of processing: beamhouse, tanning, and finishing. First, the beamhouse stage is for cleaning and preparation of the hide. Tanning is the

DOI: 10.1201/9781003532354-6

most important step, where the tanning agents react with the collagen matrix, stabilizing the protein and transforming the hide into leather. Leather demonstrates resistance toward chemical, thermal, and microbiological degradation. Different tanning agents produce leathers with different properties and uses. Tanning processes are classified according to the type of tanning reagent used to crosslink the collagen fibers [6]. The more important examples of these agents are chromium salts, vegetable tannins, and synthetic tannins. Chrome tanning with chromium salts is the most employed tanning agent due to the distinctive characteristics that it provides to the leather, such as high hydrothermal stability and good elasticity; it also results in a blue/green coloration [7]. Vegetable tanning agents are extracted from various plant materials and are classified as hydrolyzable (chestnut and myrobalan) or condensed polyphenolics (hemlock and wattle) [8]. The leather from vegetable tanning is rigid, thicker, able to accept a stamp, and is brownish in color. It is typically used as shoe soles and belts or combined with other tanning agents to obtain the required characteristics. Finishing steps at the end provide resistance and other characteristics to the leather. In the current linear leather manufacturing process, tanneries generate solid waste materials before and after the tanning stage.

6.2.1 Tannery Solid Wastes

Every year, 4 million tons of solid wastes are generated by the leather industry worldwide [9]. It is estimated that about 1,000 kg of rawhide produces 250 kg of finished leather, and about 450–730 kg of solid waste is generated during the conversion [10]. The amount of solid waste produced during each main step is: 80% during beamhouse, 19% during tanning, and 1% during finishing [11]. The solid wastes come from the different steps of the process (Figure 6.1).

The waste from untanned skin/hides is easier to treat than tanned waste. The tanning process makes the leather more rigid and resistant toward chemical, thermal, and microbial degradation [12].

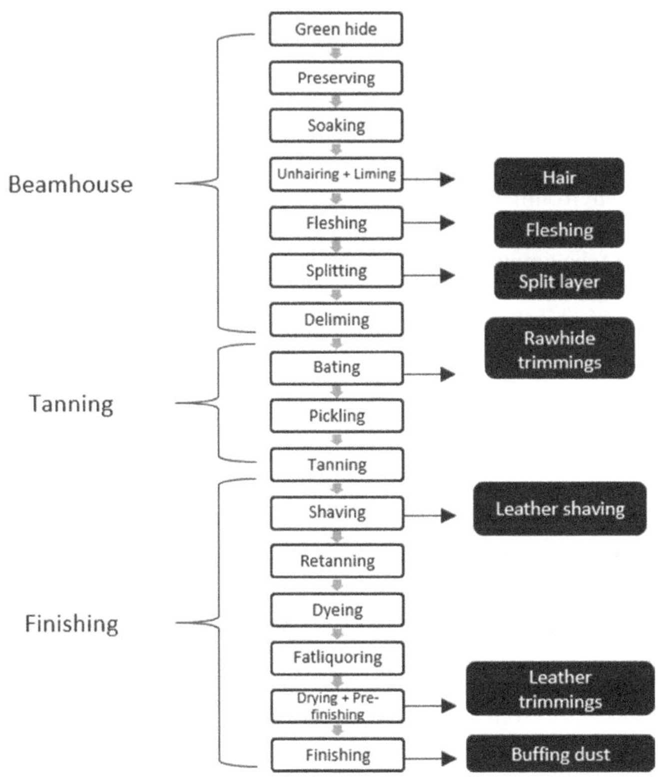

FIGURE 6.1 Solid waste generated during leather processing.

Waste containing chrome as the tanning agent is the most problematic due to the possible oxidation of trivalent chromium salt into the hexavalent form [13]. In Brazil, this waste containing chromium is classified as a dangerous solid residue (Class I) [14]. Every year, the Brazilian tanning industry spends around US$ 3.2 million for the disposal of chromium leather shavings [15].

The safe disposal of solid wastes is needed. Despite recovery options for generated waste, most leather waste is disposed of in landfill sites, without material or energy recovery; thus, the carbon cycle is not completed. The decomposition rates of biodegradable materials in landfills are very slow due to the absence of exposure to sunlight, humidity, and air, and there is the risk of producing leachates over the years. Most European countries have introduced landfill restrictions and taken measures to reduce biodegradable waste going to landfills [16]. However, when solid waste is managed correctly, it can be a valuable resource. After different methods of biological treatment, it is possible to obtain valuable products. The biodegradability of leather is a complex process and is influenced by the nature of the tanning agents and post-treatments applied [17].

6.3 SOLUTION FOR THE LEATHER INDUSTRY: BIODEGRADATION

6.3.1 Biodegradable Leather

Before hides/skins are tanned, they are naturally biodegradable. The tanning process changes the chemistry inside the leather fibers and makes it more difficult for the enzymes from bacteria and fungi to break down the collagen molecules [18]. Leather is made to be durable material; however, it is necessary to think about its end-of-life cycle. Vegetable tanned leather is the most common type of heritage leather found in museums and collections [19].

Leather is made from a natural and biodegradable material. Thus, it is possible to achieve a more biodegradable leather using alternative tanning agents. Tanned leather undergoes biodegradation at different rates. When considering combinations of tannin substances, leathers tanned with oxazolidine and vegetable tannin are more biodegradable than leathers tanned with resorcinol-oxazolidine and vegetable tannins. When considering mineral chemical agents, leather tanned with Ti-Al showed a higher rate of biodegradation than leather tanned with Ti-Zr or chrome [20]. Leather tanned with dialdehyde polysaccharides showed better biodegradation than chrome and vegetable tanned leathers [21]. An organic tanning agent was developed based on amino acids, which biodegrades easily and is an environmentally friendly leather chemical material [22]. Leather tanned with titanium salts biodegrades by composting. The leather degrades and reduces the carbon content by approximately 10%, and the amount of the waste was reduced by 41% in 45 days. Meanwhile, in this comparison study, chrome and vegetable tanned leather were inalterable [17].

For a cleaner process, biodegradable materials have been used as supplements for the leather industry. A biodegradable nanocomposite was developed with hydroxyapatite nanoparticles that had a surface that was tailored with a poly(lactide co-glycolide) (PLGA) co-polymer and utilized as a retanning agent for leather [23]. A fatliquor agent based on castor oil was prepared using nano-TiO_2 as an emulsifying agent. It is more biodegradable than conventional modified castor oil [24]. A MgAl layered double hydroxide nanomaterial was synthesized by using a bio-based caprolactam-modified casein, which is a completely biodegradable protein, as the template. Next, it was blended with a modified casein emulsion and applied as a flame retardant during leather finishing [25].

6.3.2 Biological Treatment of Tannery Solid Wastes

6.3.2.1 Composting

One environmentally friendly solid waste treatment is composting [17]. Composting is an aerobic biological treatment that produces a nutrient-rich biomass that can be used for soil enhancement [18]. Composting of biodegradable materials generates carbon dioxide, water, and some form of compost, which can be used as a fertilizer. However, certain biodegradable materials may also produce hazardous by-products that are soil contaminants. If biodegradable products are to be composted, they must meet stringent quality criteria [16]. The solid digestate can be used on agricultural land as a fertilizer. Vermicomposting is a bio-oxidative process that involves the action of epigeic earthworm species and different micro-organisms to accelerate the biodegradation and stabilization of organic materials. This type of composting improves the nutrient fertilizer value of the waste materials [26].

6.3.2.2 Anaerobic Digestion

The collagenous materials present in leather wastes are digested by biodegradation through anaerobic process to recover chemical energy, fuel energy [27], and biogas. Biogas, a gaseous fuel, is considered renewable and produces carbon-free emissions since the basic raw material is classified as biomass [28]. Biogas is composed of 55%–70% methane (CH_4), 30%–45% carbon dioxide (CO_2), and traces of other gases [6]. The obtained biogas has a high calorific value in quantities that can supplement the energy demand of tanneries [28].

6.3.3 Tannery Solid Wastes: Biodegradable Products and Biological Treatment

The disposal of tannery solid waste has become more difficult and will no longer be sustainable. Also, the poor utilization of solid wastes threatens the environment. Thus, to prevent the disposal of tons of untreated solid waste from tanneries in landfills, the waste can be considered as a source of raw materials for biodegradable products or go through a biological treatment, as shown in Figure 6.2. As solid waste is produced during different leather processing steps, it is a source of biodegradable materials and has different biodegradation solutions.

6.3.3.1 Fleshing

Fleshing is the removal of the tissue that adheres to the hide and is mainly composed of protein and fat [29]. Fleshing waste was biodegraded by fermentation to produce alkaline protease. Bacterial isolates, such as *Bacillus licheniformis, Bacillus coagulans, Bacillus polymyxa*, and *Bacillus amyloliquefaciens*, have been used [30]. Fleshing waste was vermicomposted using the epigeic earthworm *Eisenia foetida*, which was able to convert this waste into nutrient-rich products [31]. Vermicomposting generated value-added products, such as indole 3-acetic acid, gibberellic acid, and kinetin [32]. The compost and vermicompost produced from tannery waste after animal fleshing improved the growth and yield of a commercial tomato (*Lycopersicon esculentum L.*) plant [33]. Fleshing is a source of biogas [29]. Co-digestion of limed animal fleshing waste with a biodegradable fraction of municipal solid waste was studied to minimize ammonia and to maximize biogas yield. The highest observed specific biogas yield was at a nearly neutral pH of 6.5, and this pH value is achievable naturally by maintaining a high organic loading rate. Blending of co-digested fleshing waste and municipal solid waste minimizes NH_3 toxicity and thus maximizes the renewable energy recovery through the anaerobic co-digestion process [34].

6.3.3.2 Rawhide Trimmings

The amount of trimmings is much more than the demand for them [35]. The trimmings contain proteins, lipids, and carbohydrates [11]. Low-cost gelatin can be obtained from these abundant waste materials. Gelatin, obtained from raw hides, was used to fabricate eco-friendly and cost-effective biodegradable films. The films had good mechanical performance when compared to that of traditional low-density polyethylene films; they also had good water resistance, excellent light transmittance, and good biodegradability in soil. One potential application is an eco-friendly mulch film for agricultural applications [36]. Highly flexible and transparent bioplastic films were developed from the hydrothermal hydrolysis of raw trimmings using citric acid as a non-toxic crosslinker. These bioplastics represent a friendly alternative to fossil-derived plastics and can be used in packaging, wound healing, and other biocompatible applications [37]. Trimmings can also produce biodegradable retanning agents [38].

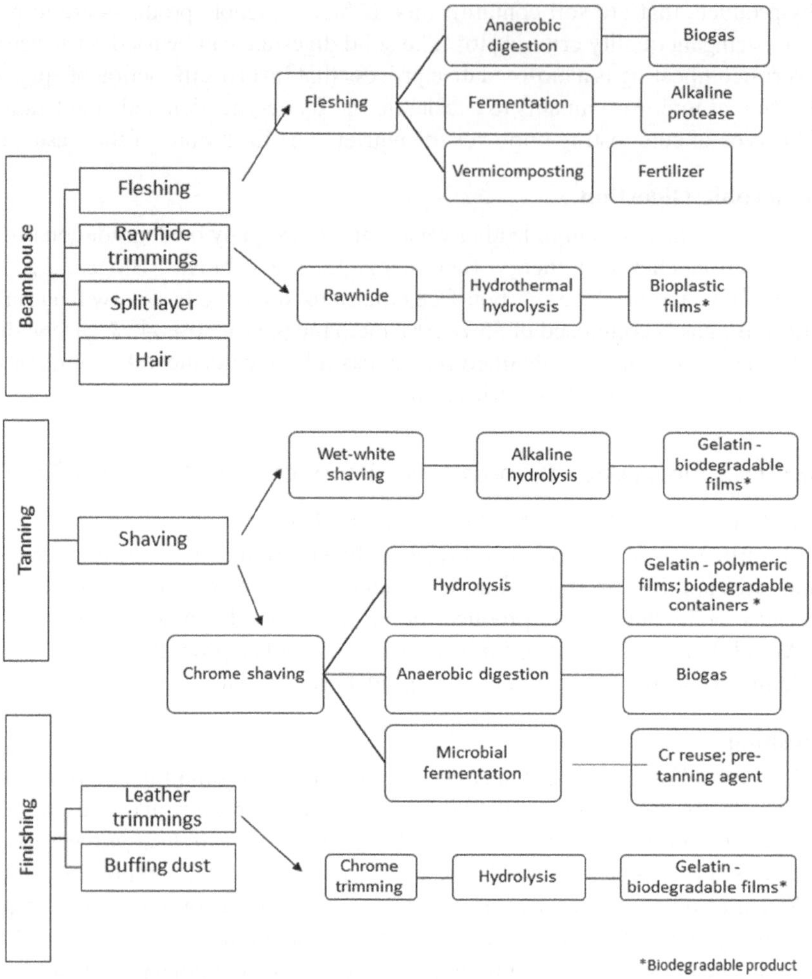

FIGURE 6.2 Solid waste from leather processing, biodegradable products, and biological treatments.

6.3.3.3 Leather Shavings

Leather shavings come from the thickness adjustment of leather and are mainly composed of protein and tanning agents. Gelatin was extracted from wet-white leather shavings by alkaline hydrolysis and was blended with poly(L-lactic) acid to obtain biodegradable films. The biodegradable films had good thermal processability, transparency, flexibility, and mechanical properties; thus, they could be used to produce low-cost biodegradable films for packaging applications [39]. A new bio-based material was produced by the alkaline hydrolysis of leather chrome shavings consisting of protein hydrolysates that can be used as biodegradable containers and pot for plant nursery [40]. Chrome shavings were used as protein source for biodegradation using the bacteria *Bacillus subtilis* P13. The liquefaction of the proteinaceous waste, recovery of a dehairing protease as a concentrated product, and chromium recovery for reuse in tanning occurred [41]. The fungi *Aspergillus carbonarius* was isolated and selected from an established microbial culture collection. About 97% liquefaction of the tannery waste was achieved, and the obtained liquid was used to recover chromium [42]. The bacteria *Bacillus subtilis* was used to biodegrade solid wastes containing chromium [43]. Microbial growth was achieved, and a valuable protease by-product, which has applications in the pre-tanning process, was produced [4]. Also, *Aspergillus carbonarius* was used to recycle chromium [42].

As leather waste containing chromium is hard to degrade, treatments are necessary. A pre-treatment that consisted of the removal of leather tanning agents improved the biodegradability [27]. The influence of the tanning agent and thermal pretreatment was studied [44]. The origin of the shavings from vegetable or chrome tanning was not significant to any parameter due to the small proportion of this residue in the assays. The thermal pre-treatment was beneficial only for assays with vegetable tannins, since there was evidence that the heating mechanism degraded part of the phenolic organic matter and consequently reduced its toxicity. The macronutrients ratio present in tannery wastes is normally not ideal for anaerobic digestion, mainly due to the excessive amount of nitrogen. Hence, nutrients added to the co-anaerobic digestion of leather shavings and sludge from a tannery were investigated. When the nutrients consisting of yeast extract, K_2HPO_4, and KH_2PO_4 were used, it was possible to reduce 71% of the electric or 35% of the thermal energy consumption [45].

After anaerobic digestion, biosolid residues are rich in nutrients and can be used as an organic fertilizer [46]. Fertilizer is obtained as a by-product of composting and anaerobic digestion. Bio-based fertilizers can be a choice for better crop production. Wet-blue leather and potato peels were transformed into an eco-friendly bio-organic NPK fertilizer. Chromium was extracted from the leather using acids and recovered [47]. Purified protein hydrolysate with zero chromium was extracted from chrome-tanned leather and used as a nitrogen source for fertilizer formulation. The fertilizer was employed as a nitrogen source for soybean production. A plant growth study demonstrated that the formulated fertilizer produced similar effects to that of plants treated with a commercial fertilizer [48]. China added waste containing chromium to its hazardous waste list in 2016, prohibiting it from being disposed of in landfills and through incineration. Thus, a solution is to liquefy leather waste and turn it into organic fertilizer or construction materials, such as adhesives, after researchers remove the heavy metal chromium from the leather sludge [49].

6.3.3.4 Leather Trimmings

The geometrization of finished leather generates trimmings that mainly contain protein, tanning agents, and dye [50]. Gelatin extracted from chrome leather scraps (trimmings) by hydrolysis was the base of biodegradable films. The chrome content in the gelatin was 30 mg/kg, less than the required (200 mg/kg). The extracted gelatin was modified with β-cyclodextrin to prepare biodegradable films. The viscosity, biodegradability, thermal stability, and physical properties of the blend were significantly increased when compared to those for a gelatin film without β-cyclodextrin. This biodegradable film can be used in commercial fields [51]. Gelatin extracted from chromed leather waste by hydrolysis was used with cornstarch and glycerol to produce polymeric films. The combination of starch and protein (gelatin) resulted in a material classified as a "biodegradable polymer obtained from biomass". These films were compared with commercial gelatin films. The recovered gelatin, without any post-extraction treatment, made the starch films more hygroscopic, more soluble in water, less resistant, and more flexible; films with a higher water vapor permeability were also produced [52].

6.4 ALTERNATIVES TO LEATHER

Currently, there are many leather substitutes. Synthetic materials that imitate the optical appearance of leather are made of synthetic polymers, such as polyester textiles coated by polyvinyl chloride or polyurethane films, which makes them an entirely fossil fuel-based material [1]. In Brazil, Law 4888 prohibits the use of expressions such as "synthetic leather" or "ecological leather". Only products from animal hide can be called leather.

Due to the low price, these synthetic materials compete with leather. The use of these materials is due to their ability to be customized and high performance; they can be used in gears, conveyor belts, vessels, upholstery, shoes, and clothing. However, they cannot compete with the unique properties of leather, such as its strength, elasticity, water vapor permeability, abrasion resistance, durability, and longevity [1]. Unlike leather, most of them are nonbiodegradable [53]. Customers have concerns about sustainability, and many want to use natural, fully biodegradable, and animal-free materials. Trendy alternative materials are made from mushrooms, apples, pineapple leaves, and cactus leaves, for example.

The tensile strength and tear strength are the most important mechanical properties for materials used in shoes, gloves, and apparel goods. When leather and nine alternative materials were tested, leather showed the highest value for the tensile strength and tear strength. Restricted substances were identified in some of the samples. The leather alternatives did not achieve the properties of leather, which are due to the rich structure of the leather [1].

6.5 CHALLENGES AND OPPORTUNITIES

The raw material for leather is the hide, a biodegradable material. Leather processing makes the hide resistant to microbial attack and thermally stable. Therefore, the biodegradability of leather is a complex process due to tanning agents and post-treatments. A leather that is more biodegradable could be one solution for the reduction of solid waste from tanneries. The challenge is developing a metal-free tanning agent that provides strength and thermal and hydrothermal stability to leather and can also biodegrade. Also, certain chemicals added to leather are an obstacle to obtaining a more biodegradable leather. Any additives should be biodegradable, too, as shown in the literature from the steps of retanning, fatliquoring, and finishing. Thus, these solid wastes will not adversely impact the environment, and the conversion to by-products will be easy.

Currently, tannery solid wastes are available in great quantities and at a low cost. Thus, payment for disposal of these wastes in landfills and conversion of them into high value-added biodegradable products needs to be encouraged. Fleshing, trimmings, and shavings are considered as significant resource for the production of biodegradable by-products, such as biofilms, enzymes, fertilizers, and biogas. The large-scale utilization of tannery solid wastes into various useful biodegradable products and for green energy sources should be the focus of the leather industry.

GLOSSARY

Beamhouse: Is the process of preparing the hides for tanning and occurs between preservation and tanning.

Leather Shavings: Come from the thickness adjustment of leather and are mainly composed of protein and tanning agents.

Leather Trimmings: Generated from the geometrization of finished leather and mainly contain protein, tanning agents, and dye.

Rawhide Trimmings: They are the waste of hide before tanning step.

Wet-blue Leather: Leather tanned using chromium salts.

Wet-white Leather: White in color, tanned with aluminum, zirconium, formaldehyde, or glutaric aldehyde that has not undergone further processing and remains wet.

REFERENCES

1. M. Meyer, S. Dietrich, H. Schulz, A. Mondschein, Comparison of the technical performance of leather, artificial leather, and trendy alternatives, *Coatings*. 11, 1–15 (2021).
2. E. Hansen, P. Monteiro, D. Aquim, A. Witt, J. Kern, A. Luiza, M. Gutterres, Impact of post-tanning chemicals on the pollution load of tannery wastewater, *J. Environ. Manage.* 269, 110787 (2020).
3. V. Sivakumar, R. Mohan, C. Muralidharan, Alternative methods for salt free/less salt short term preservation of hides and skins in leather making for sustainable development – A review, *Text. Leather Rev.* 2, 46–52 (2019).
4. T. Pringle, M. Barwood, S. Rahimifard, The challenges in achieving a circular economy within leather recycling, *Procedia CIRP*. 48, 544–549 (2016).

5. O. Omoloso, K. Mortimer, W.R. Wise, L. Jraisat, Sustainability research in the leather industry: A critical review of progress and opportunities for future research, *J. Clean. Prod.* 285, 125441 (2021).
6. C. Agustini, F. Spier, M. da Costa, M. Gutterres, Biogas production for anaerobic co-digestion of tannery solid wastes under presence and absence of the tanning agent, *Resour. Conserv. Recycl.* 130, 51–59 (2018).
7. S. Lv, Q. Zhou, Y. Li, Y. He, H. Zhao, Tanning performance and environmental effects of nanosized graphene oxide tanning agent, *Clean Technol. Environ. Policy.* 18, 1997–2006 (2016).
8. P. Auad, F. Spier, M. Gutterres, Vegetable tannin composition and its association with the leather tanning effect, *Chem. Eng. Commun.* 0, 1–11 (2019).
9. K. Koppiahraj, S. Bathrinath, S. Saravanasankar, Leather waste management scenario in developed and developing nations, *Int. J. Eng. Adv. Technol.* 9, 852–857 (2019).
10. C.B. Agustini, M. Gutterres, Biogas production from solid tannery wastes, in: A. Vico, N. Artemio (Eds.), *Biogas: Production, Applications and Global Developments*, Nova Science Publishers, New York, pp. 79–102 (2017).
11. J. Kanagaraj, K.C. Velappan, N.K. Chandra Babu, S. Sadulla, Solid wastes generation in the leather industry and its utilization for cleaner environment – A review, *J. Sci. Ind. Res. (India).* 65, 541–548 (2006).
12. M.J.H. Dowlath, S.K. Karuppannan, P. Rajan, S.R. S.B. Mohamed Khalith, K.D. Arunachalam, Application of advanced technologies in managing wastes produced by leather industries— An approach toward zero waste technology, in: C.M. Hussain (Ed.), *Concepts of Advanced Zero Waste Tools*, Elsevier, Amsterdam, pp. 143–179 (2021).
13. M. Gutterres, B. Mella, Chromium in tannery wastewater, in: S.K. Sharma (Ed.), *Heavy Metals in Water: Presence, Removal and Safety*, Royal Society of Chemistry, Cambridge, pp. 315–344 (2015).
14. ABNT, NBR 10.004, Brazil, 2004. https://analiticaqmcresiduos.paginas.ufsc.br/files/2014/07/Nbr-10004-2004-Classificacao-De-Residuos-Solidos.pdf.
15. C. Manera, A.P. Tonello, D. Perondi, Adsorption of leather dyes on activated carbon from leather shaving wastes: Kinetics, equilibrium and thermodynamics studies, *Environ. Technol.* 0, 1–39 (2018).
16. T. Staikos, R. Heath, B. Haworth, S. Rahimifard, End-of-life management of shoes and the role of biodegradable materials, in: *Proceedings of the 13th CIRP International Conference on Life Cycle Engineering, LCE 2006*, Leuven, Belgium, pp. 497–502 (2006).
17. E. Zuriaga-Agustí, M.V. Galiana-Aleixandre, A. Bes-Piá, J.A. Mendoza-Roca, V. Risueño-Puchades, V. Segarra, Pollution reduction in an eco-friendly chrome-free tanning and evaluation of the biodegradation by composting of the tanned leather wastes, *J. Clean. Prod.* 87, 874–881 (2015).
18. Nera, Biodegradability and disintegration of leather (2021). https://www.neratanning.com/knowledge/whitepaper-biodegradability-and-disintegration-of-leather/
19. Z. Sebestyén, E. Jakab, E. Badea, E. Barta-rajnai, C. Şendrea, Z. Czégény, Thermal degradation study of vegetable tannins and vegetable tanned leathers, *J. Anal. Appl. Pyrolysis.* 138, 178–187 (2019).
20. R.R. Constantinescu, V. Deselnicu, M. Crudu, G. Macovescu, L. Albu, Comparative study regarding leather biodegradability, *Leather Footwear J.* 15, 73–84 (2015).
21. N. Ariram, B. Madhan, Development of bio-acceptable leather using bagasse, *J. Clean. Prod.* 250, 119441 (2020).
22. X. Wu, X. Qiang, D. Liu, L. Yu, X. Wang, An eco-friendly tanning process to wet-white leather based on amino acids, *J. Clean. Prod.* 270, 122399(2020).
23. S. Selvaraju, S. Ramalingam, J.R. Rao, Preparation and application of biodegradable nanocomposite for cleaner leather processing, *J. Clean. Prod.* 158, 225–232 (2017).
24. B. Lyu, H. Di Wang, J.Z. Ma, D.G. Gao, P. Jin, Preparation and application of castor oil/nano-TiO_2 composite fatliquoring agent via a pickering emulsion method, *J. Clean. Prod.* 126, 711–716 (2016).
25. W. An, J. Ma, Q. Xu, Bio-template synthesis of MgAl layered double hydroxide with enhanced flame retardant property for leather finishes, *Appl. Surf. Sci.* 551, 149409 (2021).
26. H.A. Mupambwa, P.N.S. Mnkeni, Optimizing the vermicomposting of organic wastes amended with inorganic materials for production of nutrient-rich organic fertilizers: A review, *Environ. Sci. Pollut. Res.* 25, 10577–10595 (2018).
27. K. Dhayalan, N.N. Fathima, A. Gnanamani, J.R. Rao, B.U. Nair, T. Ramasami, Biodegradability of leathers through anaerobic pathway, *Waste Manag.* 27, 760–767 (2007).
28. G.P.S. Priebe, M. Gutterres, Special review: Anaerobic digestion of leather industry wastes - An alternate source vof energy, *J. Am. Leather Chem. Assoc.* 112, 59–71 (2017).
29. R. Kandasamy, S.K. Venkatesan, M.I. Uddin, S. Ganesan, Anaerobic biovalorization of leather industry solid waste and production of high value-added biomolecules and biofuels, In *Biovalorisation of wastes to renewable chemicals and biofuels* (pp. 3–25). Elsevier Inc., 2020.
30. J. Ahmad, T.A. Ansari, Alkaline protease production using proteinaceous tannery solid waste, *J. Pet. Environ. Biotechnol.* 04, 1–4 (2013).

31. B. Ravindran, S.L. Dinesh, L.J. Kennedy, G. Sekaran, Vermicomposting of solid waste generated from leather industries using epigeic earthworm Eisenia foetida, *Appl. Biochem. Biotechnol.* 151, 480–488 (2008).
32. B. Ravindran, S.M. Contreras-Ramos, G. Sekaran, Changes in earthworm gut associated enzymes and microbial diversity on the treatment of fermented tannery waste using epigeic earthworm Eudrilus eugeniae, *Ecol. Eng.* 74, 394–401 (2015).
33. B. Ravindran, S.R. Lee, S.W. Chang, D.D. Nguyen, W.J. Chung, B. Balasubramanian, H.A. Mupambwa, M.V. Arasu, N.A. Al-Dhabi, G. Sekaran, Positive effects of compost and vermicompost produced from tannery waste-animal fleshing on the growth and yield of commercial crop-tomato (Lycopersicon esculentum L.) plant, *J. Environ. Manage.* 234, 154–158 (2019).
34. P. Shanmugam, N.J. Horan, Optimising the biogas production from leather fleshing waste by co-digestion with MSW, *Bioresour. Technol.* 100, 4117–4120 (2009). https://doi.org/10.1016/j.biortech.2009.03.052.
35. Y. Tang, J. Zhao, Y. Zhang, J. Zhou, B. Shi, Conversion of tannery solid waste to an adsorbent for high-efficiency dye removal from tannery wastewater: A road to circular utilization, *Chemosphere.* 263, 127987 (2021).
36. L. Chen, T. Qiang, X. Chen, W. Ren, H.J. Zhang, Fabrication and evaluation of biodegradable multi-cross-linked mulch film based on waste gelatin, *Chem. Eng. J.* 419, 129639 (2021).
37. V. Muralidharan, M.S. Arokianathan, M. Balaraman, S. Palanivel, Tannery trimming waste based biodegradable bioplastic: Facile synthesis and characterization of properties, *Polym. Test.* 81, 106250 (2020).
38. M. Sathish, B. Madhan, J. Raghava Rao, Leather solid waste: An eco-benign raw material for leather chemical preparation – A circular economy example, *Waste Manag.* 87, 357–367 (2019).
39. H. Elsayed, S. Ibrahim, Biodegradable package films of poly(L-Lactic) acid/extracted gelatin blend from white leather fibers, *Egypt. J. Chem.* 63, 3059–3074 (2020).
40. L. Sartore, E. Schettini, F. Bignotti, S. Pandini, G. Vox, Biodegradable plant nursery containers from leather industry wastes, *Polym. Compos.* 39, 2743–2750 (2018).
41. P. Pillai, G. Archana, A novel process for biodegradation and effective utilization of chrome shavings, a solid waste generated in tanneries, using chromium resistant Bacillus subtilis P13, *Process Biochem.* 47, 2116–2122 (2012).
42. E.A. Katsifas, E. Giannoutsou, M. Lambraki, M. Barla, A.D. Karagouni, Chromium recycling of tannery waste through microbial fermentation, *J. Ind. Microbiol. Biotechnol.* 31, 57–62 (2004).
43. M.N. Aftab, A. Hameed, Ikram-Ul-Haq, C. Run, Biodegradation of leather waste by enzymatic treatment, *Chinese J. Process Eng.* 6, 462–465 (2006).
44. C. Agustini, M. Meyer, M. Da, M. Gutterres, Biogas from anaerobic co-digestion of chrome and vegetable tannery solid waste mixture: Influence of the tanning agent and thermal pretreatment, *Process Saf. Environ. Prot.* 118, 24–31 (2018).
45. T. Simioni, C.B. Agustini, A. Dettmer, M. Gutterres, Nutrient balance for anaerobic co-digestion of tannery wastes: Energy efficiency, waste treatment and cost-saving, *Bioresour. Technol.* 308, 123255 (2020).
46. N. Vasudevan, A.D. Ravindran, Biotechnological process for the treatment of fleshing from tannery industries for methane generation, *Curr. Sci.* 93, 1492–1494 (2007).
47. S. Majee, G. Halder, D.D. Mandal, O.N. Tiwari, T. Mandal, Transforming wet blue leather and potato peel into an eco-friendly bio-organic NPK fertilizer for intensifying crop productivity and retrieving value-added recyclable chromium salts, *J. Hazard. Mater.* 411, 125046 (2021).
48. A. Pati, R. Chaudhary, Soybean plant growth study conducted using purified protein hydrolysate-based fertilizer made from chrome-tanned leather waste, *Environ. Sci. Pollut. Res.* 22, 20316–20321 (2015).
49. C. Daily, New tech turns leather waste into fertilizer (2017). **https://www.chinadaily.com.cn/china/2017-09/26/content_32511527.htm**
50. S. Tahiri, M. de la Guardia, Treatment and valorization of leather industry solid wastes: A review, *JALCA.* 104, 52–67 (2009).
51. X. Dang, Z. Shan, H. Chen, Biodegradable films based on gelatin extracted from chrome leather scrap, *Int. J. Biol. Macromol.* 107, 1023–1029 (2018).
52. B.S. Scopel, M.E. Ribeiro, A. Dettmer, C. Baldasso, Cornstarch-gelatin films: Commercial gelatin versus chromed leather waste gelatin and evaluation of drying conditions, *J. Polym. Environ.* 26, 1998–2006 (2018).
53. J. Bustillos, A. Loganathan, R. Agrawal, B.A. Gonzalez, M.G. Perez, S. Ramaswamy, B. Boesl, A. Agarwal, Uncovering the mechanical, thermal, and chemical characteristics of biodegradable mushroom leather with intrinsic antifungal and antibacterial properties, *ACS Appl. Bio Mater.* 3, 3145–3156 (2020).

7 Waste Stabilization Pond Fundamentals and Innovations

Mu-Hao Sung Wang and Lawrence K. Wang

ACRONYM

ASL	Anaerobic sludge lagoons
ASP	Activated sludge process
BAT	Best available technology
BOD	Biochemical oxygen demand
BOD_5	Five-day biochemical oxygen demand
COD	Chemical oxygen demand
DAF	Dissolved air flotation
DO	Dissolved oxygen
DT	Detention time
FSL	Facultative sludge lagoon
gpd	Gallon per day
HRSP	High rate stabilization ponds
LIWT	Lenox Institute of Water Technology
MG	Million gallons
MGD	Million gallons per day
MLVSS	Mixed liquor volatile suspended solids
MN	Minnesota
MPCA	Minnesota Pollution Control Agency
NPDES	National Pollutant Discharge Elimination System
O&M	Operation and maintenance
PETO	Pond enhanced treatment and operation
SBR	Sequencing batch reactor
SDL	Sludge drying lagoons
TKN	Total Kjeldahl nitrogen
TOC	Total organic carbon
TSS	Total suspended solids
UASB	Up-flow anaerobic sludge blanket
UNIDO	United Nations Industrial Development Organization
URI	University of Rhode Island
USDA	US Department of Agriculture
USEPA	US Environmental Protection Agency
VS	Volatile solids
WSP	Waste stabilization pond
WWTP	Wastewater treatment plants

DOI: 10.1201/9781003532354-7

7.1 INTRODUCTION

7.1.1 Summary

This publication is an updated version of a United Nations Industrial Development Organization (UNIDO) technology transfer lecture material for transferring US environmental technologies to developing countries. It is also one of the authors' professional memoirs dedicated to the late Dr. Calvin P.C. Poon. Possible WSP pond systems for both warm and cold weather operations are introduced in detail. Many design examples and case histories including a milk waste treatment case history in Albany, MN, USA, and a poultry processing waste treatment facility design are presented.

This technology transfer publication covers the following subjects: (a) waste stabilization pond (WSP) ecology, biology, and process reactions; (b) types of WSPs; (c) other types of waste treatment ponds or lagoons; (d) historical and recent developments in pond applications; (e) algae, phosphorus, BOD and TSS removals by air flotation; (f) non-aerated facultative pond analysis; (g) upgrading a non-aerated facultative pond by installation of mechanical aerators; (h) converting a non-aerated facultative pond to a partially aerated high-rate aerobic pond system; (i) alternative pond effluent treatment technologies; (j) determination of WSP minimum depth; (k) food processing waste treatment by WSP; (l) winter operation of WSPs; (m) analysis of an existing WSP; (n) upgrading a WSP system using activated sludge process and dissolved air flotation for a poultry processing plant; and (o) design of an anaerobic lagoon for treating a poultry processing wastewater containing high strength of soluble organic matters.

The updated Cost Index allows the old cost data to be updated to the current years. Future research direction involving the use of dissolved air flotation (DAF) for algae harvesting from the high-rate aerobic pond is recommended. A complete glossary section defines the WSP biotechnologies for the purpose of academic education.

7.1.2 Concepts and Physical Behavior

One of the simplest forms of biological treatment processes is the stabilization pond or stabilization lagoon. It is also the most common industrial wastewater treatment facility. This versatile installation serves many basic purposes, including: (a) storage or impoundment of wastewater; (b) settling and removal of suspended solids; (c) storage or impoundment of settled solids; (d) equalization; (e) aeration; (f) biological treatment; and (g) evaporation. The relative simplicity and low operating costs of a stabilization pond make it the preferred technology for handling, treatment and disposal of agricultural wastewater as well as municipal wastewater for small communities.

The main advantages of WSP are: (a) WSP's operational and maintenance (O&M) costs are low; (b) WSP provide effective treatment with minimal threat to the environment; and (c) WSP work well in clay soils where conventional subsurface on-site absorption fields will not work [1].

There are some noticeable disadvantages, however: (a) WSP must be constructed in clay soil or be lined to prevent leakage; (b) WSP may overflow occasionally during extended periods of heavy rainfall; (c) If there are extended periods of overcast windless days, offensive odors may occur for a brief time, although WSP usually recover rapidly if this occurs; (d) WSP cannot be installed on a small lot because it takes up a relatively large space; (f) WSP are not aesthetically acceptable to some people because some people consider lagoons unsightly and unsafe; and (g) As with any other open body of water, there is some potential danger even with fence because this does not always prevent access by people or pets [1]

Based on a survey by the US Federal government [2], treatment systems in the general category of "stabilization ponds" usually serve small communities of whom 90% have populations of 10,000 persons or fewer.

7.2 WASTE STABILIZATION POND ECOLOGY, BIOLOGY AND PROCESS REACTIONS

The simplicity of a stabilization pond belies the fact that complicated reactions – both chemically and biologically – often occur in a pond. A highly complex ecological system exists that is subject to diurnal and seasonal variations. In the authors' previous publication [1], the authors have depicted the major reactions taking place in a WSP: Eight major processes can be identified in active stabilization ponds. They are: (a) sedimentation; (b) aerobic decomposition; (c) anaerobic fermentation; (d) bacterial and algal symbiosis; (e) oxygen transfer across the water surface; (f) sulfur bacteria actions; (g) evaporation; and (h) seepage. The readers are referred to the literature [1–78] for the detailed technical information on the theories and principles of WSP.

The biological population depends on the type of stabilization pond, the influent wastewater characteristics, the time of the day, and the season of the year. Because of limited control in operation, variations of the biological population in stabilization ponds are much more apparent than in any other biological process.

Unlike activated sludge processes, stabilization ponds are operated at a very low bacterial population. One seldom finds a bacterial concentration higher than 100 mg/L in stabilization ponds. Total plate counts of bacteria generally fall in the range of 106–107/mL.

Coli-aerogenes group is predominant in stabilization ponds. If carbohydrates are limited in the wastewater because of pretreatment or decomposition in the sewer line before reaching the pond, proteolytic and lipolytic bacteria will predominate. Achromobacter, Pseudomonas, and Flavobacterium are dominant in an aerobic pond, accounting for 90% of the total bacterial count. The abundant number of Achromobacter and Bacillus (lipolytic) and other proteolytic groups in anaerobic lagoons. Gram-positive bacteria, including Streptococcus faecalis and Bacillus, commonly occur in aerobic stabilization ponds, although they are never present in significant numbers [1].

Bacteria die-off in ponds. Although individual algal species exert little influence on the dieoff of enteric bacteria, mixtures of axenic algal cultures increase the die-off rates. Escherichia, Pseudomonas, and Serratia exhibit after growth potential whereas Proteus, Alcaligenes, Enterobacter, Salmonella, Shigella, and Vibrio do not. The die-off rates of these bacteria are increased by anaerobic pretreatment of the wastewater.

Desulfovibrio sp. are abundant in stabilization ponds when substantial amounts of sulfate are present in wastewater. The green Chlorobacteriacea and the purple Thirohodaceae then oxidize H_2S to sulfur elements and sulfates in the presence of light. Both can use short infrared radiation according to Hutchinson [12] and Rabinowitch [13]. High H_2S concentration can promote an overgrowth of the purple sulfur bacteria to the extent that the entire pond is covered by a pinkish-violet or rose red scum. As much as 83 mg/L of pigment consisting mainly of bacterial chlorophylls and xanthophylls has been found in such pond water [13].

Algal population varies from one pond to another. Algal counts can be as high as 15 million/mL. Green algae usually dominate because they can adapt better to environmental changes such as extreme temperature and dissolved oxygen (DO).

In most ponds, coccoid green algae and green flagellates dominate the plankton throughout the year, while pennate diatoms and filamentous blue-green algae dominate the benthic flora. Green flagellates dominate in the winter and early spring, while coccoid greens, especially Ankistrodesmus convolutus, Chlorella ellipsoidea, and Chlorella vulgaris dominate in the fall, late spring, and summer. The population of some species such as Chlorella, Chalamydomonas, and Euglena may be reduced suddenly. Some evidence of zooplankton grazing and accumulation of inhibitory substance such as chlorellin provide partial explanation of the sudden (overnight) decrease of algal population in stabilization ponds. This algal autoinhibitor is possibly the same toxic substance that increases the rate of bacterial die-off in lagoons [1].

Protozoa and rotifers can be found abundantly following the significant increase of bacterial and algal population. Their function in stabilization ponds is believed to be the same as in activated sludge and trickling filter processes in that they help to stabilize the prey population and therefore the treatment performance.

There are over 50 species of aquatic insects in stabilization lagoons in California and Central Missouri lagoons [1]. The species compositions are similar in these two reports. Midge larvae actively pump water in ponds; in doing so they remove settled algae and help in the circulation of large quantities of oxygenated water at the mud-water interface. Consequently, they facilitate BOD removal in the pond. One or more of the three species of midges, Glyptotendipes barbipes, Chironomus plumosus, and Tanypus punctipennis comprise more than 94% of the total number of insects collected in the Midwestern lagoons [1].

Mosquito breeding takes place in any stabilization pond that provides a protected area for oviposition. Emergent vegetation, overflow structures, and swamp areas created by lagoon effluent are principal factors conducive to mosquito breeding. Predominant species found in lagoons by researchers are Culex pipiens and Cules tarsalis [14].

Some benthic species of insects, such as Planthemis lydia and Laccophilus spp. serve as predators of midge larvae and pupae in stabilization ponds. With vegetation in poorly managed lagoons, predaceous insects may include Dytiscid larvae and adults, Anax junius, Tropisternus lateralis nimbatus larvae, Belostoma sp., Odonata naiads and Hemiptera. Relative densities and successive seasonal changes in aquatic insect populations are believed to be the result of climatic conditions as well as intraspecific and interspecific competition.

7.3 TYPES OF WASTE STABILIZATION PONDS AND LAGOONS

WSPs are constructed with earthen dikes for holding and treating wastewater before final effluent discharge to receiving water. According to the type of biological transformation, oxygen supply methods, and the dominant biological and physicochemical reactions that occur in the WSPs, WSP can be divided into four principal types: (a) facultative (aerobic-anaerobic) ponds; (b) aerated ponds: (b1) aerated aerobic ponds; and (b2) aerated facultative ponds; (c) aerobic ponds; and (d) anaerobic ponds. The US Environmental Protection Agency (USEPA) introduces and recommends the basic design criteria (Table 7.1) for all WSP except anaerobic ponds. However, anaerobic ponds are still

TABLE 7.1
Design Criteria for Stabilization Ponds

Criteria	Anaerobic Pond	Facultative Pond	Aerated Ponds		High-Rate Aerobic
			Aerobic-Anaerobic	Aerobic	
Organic loading, lb BOD_5 per ac-d	200–1,000	15–75[a]	30–100	20–400	60–200
Detention time, *d*	20–180	20–180[b]	7–20	3–20	3–6
Depth, ft	8–20	3–8[b]	3–8	6–20	2–4
BOD_5 conversion, %	50–85	80–95	80–95	80–95	80–95
Treatment application	Untreated wastewater and sludge	Untreated to partially treated wastewater	Untreated to partially treated wastewater	Primary effluent; tertiary pond	Treatment of soluble organic; nutrient removal; tertiary pond
Effluent Characteristic					
BOD concentration, mg/L	100–300[a]	20–50	20–70	20–70	20–30
Suspended solids, mg/L	80–160	160–400	100–350	250–300	150–350
Algae concentration, mg/L	—	40–160	10–40	—	80–200

[a] Usually followed by a facultative or aerobic pond for further treatment.

[b] In arctic and sub-arctic regions, 20 lb/acres/day organic loading, 6–8 months detention time, and 6–8 ft. deep.

introduced in this publication due to its importance. Anaerobic ponds are occasionally used for wastewater treatment, but frequently used for sludge treatment.

"Controlled discharge pond" is a special form of WSP. Controlled discharge ponds have long hydraulic detention times (DTs), and the effluent is discharged once or twice per year when the effluent quality is satisfactory.

7.3.1 Facultative (Aerobic-Anaerobic) Ponds

Facultative lagoons are customarily contained within earthen dikes. Depending on soil characteristics, lining with various impervious materials such as rubber, plastic or clay may be necessary. They are intermediate depth (3–8 ft.) ponds in which the wastewater is stratified into three zones. These zones consist of an anaerobic bottom layer, an aerobic surface layer, and an intermediate zone. Stratification is a result of solids settling and temperature-water density variations.

Sludge at the bottom of facultative lagoons will undergo anaerobic digestion producing carbon dioxide, methane and cells. Facultative lagoons are often and for optimum performance should be operated in series. When three or more cells are linked, the effluent from either the second or third cell may be recirculated to the first. Recirculation rates of 0.5–2.0 times the plant flow have been used to improve overall performance.

They are fully demonstrated and in use especially for the treatment of relatively weak municipal wastewater in areas where real estate costs are not a restricting factor and when land costs are low and operation and maintenance costs are to be minimized. In very cold climates, facultative lagoons may experience reduced biological activity and treatment efficiency. Since ice formation can also hamper operations, extra long holding time is needed in order to discharge the lagoon effluent in warm weather.

There are two types of facultative lagoons or ponds: (a) non-aerated facultative lagoons (20–180 days hydraulic DT; 3–8 ft. depth; 40–160 mg/L algae as oxygen source; 20–50 mg/L effluent BOD_5; 160–400 mg/L effluent TSS); and (b) aerated facultative lagoons (10–20 days hydraulic DT; 3–8 ft. depth; surface aerators as oxygen source; 20–70 mg/L effluent BOD_5; sedimentation clarifier or dissolved air flotation clarifier is needed for TSS removal) (see Table 7.1).

Aerobic-anaerobic ponds are same as facultative ponds, or facultative lagoons, which can be either aerated or non-aerated.

7.3.2 Aerated Pond or Lagoon

There are two types of aerated lagoons: (a) aerated aerobic lagoon (completely mixed; 3–20 days hydraulic DT; 6–20 ft. depth); and (b) aerated facultative lagoon (surface layer mixed with aerators; 7–20 days hydraulic DT; 3–8 ft. depth). They are all medium-depth basins designed for the biological treatment of wastewater on a continuous basis. In contrast to non-aerated stabilization ponds, which obtain oxygen from photosynthesis and surface reaeration, the aerated lagoons employ aeration devices which supply supplemental oxygen to the system. The aeration devices may be mechanical (i.e., surface aerator), or diffused air systems. Surface aerators are divided into two types: cage aerators and the more common turbine and vertical shaft aerators. Many diffused air systems utilized in lagoons consist of plastic pipes supported near the bottom of the cells with regularly spaced sparger holes drilled in the tops of the pipes. They are used for domestic and industrial wastewater of low and medium strength. Commonly used where land is inexpensive and costs and operational control are to be minimized. It is relatively simple to upgrade existing oxidation ponds, lagoons, and natural bodies of water with this type of treatment (Table 7.1).

7.3.2.1 Aerated Aerobic Lagoon or Pond

An aerated aerobic lagoons lagoon (completely mixed; 3–20 days hydraulic DT; 6–20 ft. depth; 20–400 lb/acre/day BOD_5 loading; 800 mg/L MLVSS) is normally designed to achieve complex mixing as an activated sludge reactor. The lagoons may be lined with conditions and environmental regulations. It uses various types of aeration, devices to achieve completely mixed (all aerobic) conditions, and requires a final settling tank. In very cold climates aerated lagoons may experience reduced biological activity and treatment efficiency, and the formation of ice (Table 7.1).

7.3.2.2 Aerated Facultative Lagoon or Pond

Aerated facultative lagoons (surface layer mixed with aerators; 7–20 days hydraulic DT; 3–8 ft. depth; surface aerators as oxygen source; 20–70 mg/L effluent BOD_5; sedimentation clarifier or dissolved air flotation clarifier is needed for TSS removal) are normally designed to achieve partial mixing only, aerobic-anaerobic stratification will occur, and a large fraction of the incoming solids and a large fraction of the biological solids produced from waste conversion settle to the bottom of the lagoon cells. As the solids begin to build up, a portion will undergo anaerobic decomposition. Volatile toxics can potentially be removed by the aeration process, and incidental removal of other toxics can be expected to be similar to an activated sludge system. Several smaller aerated lagoon cells in series are more effective than one large cell. Tapering aeration intensity downward in the direction of flow promotes settling out of solids in the last cell.

A non-aerated polishing cell following the last aerated cell is an optional, but recommended, design technique to enhance suspended solids removal prior to discharge. Oxygen in the surface stabilization zone of an aerated facultative lagoon is provided by mechanical surface aeration. This is in contrast to a non-aerated facultative lagoon in which algae is the source of oxygenation for photosynthesis in order to create aerobic surface conditions. In general, the aerobic surface layer serves to reduce odors while providing treatment of soluble organic by-products of the anaerobic processes operating at the bottom. The use of supplemental top layer surface aeration can improve overall treatment capacity, particularly in northern climates where icing over facultative lagoons is common in the winter.

7.3.3 Aerobic Ponds, Non-aerated

They are also called "high-rate aerobic ponds" which maintain DO throughout their entire depth.

They are one of "aerobic ponds". Specifically, it should be technically named "partially-aerated high-rate aerobic ponds", which maintains DO throughout their entire shallow depth, usually between 2 and 4 ft. deep, allowing light to penetrate the full depth. The design criteria are: (a) 5-day BOD organic loading L=60–200 mg/L; (b) hydraulic DT t=3–6 days; (c) pond depth d=2–4 ft.; (d) DO concentration=aerobic throughout entire shallow pond depth; (e) minimum algae concentration in pond water=80–200 mg/L; (f) mixing condition=allowing to expose all algae (80–200 mg/L algae) to sunlight and to prevent deposition and subsequent anaerobic conditions; and (g) high-rate aerobic ponds are limited to warm, sunny climates. Oxygen is provided by both algal photosynthesis and surface reaeration with very limited mechanical aeration, and aerobic bacteria stabilize the waste. It is suitable for the removal of soluble organic and nutrients.

The pond's effluent 5-day BOD is expected to be excellent (20–30 mg/L). However, operation is somewhat more complex than for a facultative pond and, unless an algae removal step is provided, the effluent will contain high TSS (150–350 mg/L). Because of their shallow depths, paving or covering the bottom is required to prevent weed growth.

7.3.4 Anaerobic Pond or Lagoon

Anaerobic ponds or lagoons are relatively deep (up to 20 ft.) ponds with steep sidewalls in which anaerobic conditions are maintained by keeping loading so high that complete deoxygenation is prevalent. Although some oxygenation is possible in a shallow surface zone, once greases form an impervious surface layer, complete anaerobic conditions develop. Treatment or stabilization results from thermophilic anaerobic digestion of organic wastes.

The treatment process is analogous to that occurring in single-stage untreated anaerobic digestion of sludge in which acid-forming bacteria break down organics. The resultant acids are then converted to carbon dioxide, methane, cells, and other end products. In the typical anaerobic lagoon, raw wastewater enters near the bottom of the pond (often at the center) and mixes with the active microbial mass in the sludge blanket, which is usually about 6 ft deep. The discharge is located near one of the sides of the pond, submerged below the liquid surface. Excess undigested grease floats to the top, forming a heat retaining and relatively air-tight cover.

Wastewater flow equalization and heating are generally not practiced. Excess sludge is washed out with the effluent. Recirculation of waste sludge is not required. Anaerobic lagoons are capable of providing treatment of high-strength wastewaters and sludge and are resistant to shock loads. They are customarily contained within earthen dikes. Depending on soil characteristics, lining with various impervious materials such as rubber, plastic or clay may be necessary. Pond geometry may vary, but surface area to volume ratios are minimized to enhance heat retention. Although anaerobic processes are common for sludge digestion, anaerobic lagoons for wastewater treatment have found some agricultural applications. The process is well demonstrated for the stabilization of highly concentrated organic wastes such as manure.

Typically anaerobic lagoons are used in series followed by aerobic or facultative lagoons. They are effective as roughing units prior to aerobic treatment of high-strength wastes. For efficient operation, water temperatures above 75°F should be maintained. BOD removals of 50%–70% are achievable depending on loading and temperature conditions. TSS concentrations may increase, especially if the influent BOD is primarily dissolved. Generally does not produce an effluent suitable for direct discharge to receiving waters. In anaerobic lagoons excess sludge is usually washed out in the effluent. Anaerobic lagoons are often used for preliminary treatment..

7.4 OTHER TYPES OF WASTE TREATMENT PONDS OR LAGOONS

7.4.1 Other Waste Stabilization Ponds

There are also WSPs called tertiary ponds, multiple-stage ponds, and integrated ponds. Considering the biological actions and methods of oxygen supply, these ponds or their components can fit into either one or different combinations of the four classifications previously described.

7.4.2 Evaporation Lagoon or Pond

Evaporation lagoon (which is not a WSP) is one type of total containment pond that is applicable in climates where the evaporative losses exceed the rainfall. The evaporation lagoon may be described as an open holding facility that depends solely on climatic conditions such as evaporation, precipitation, temperature, humidity, and wind velocity to affect dissipation (evaporation) of on-site wastewater. Individual lagoons may be considered as an alternate means of wastewater disposal on individual pieces of property. The basic impetus to consider this system is to allow building and other land uses on properties that have soil conditions not conducive to the workability and acceptability of the conventional on-site drain field or leaching bed disposal systems. Generally, if the annual evaporation rate exceeds the annual precipitation, this method of disposal may at least be considered. The deciding factor then becomes the required land area and holding volume. It should be noted that for unlined on-site installations such as homes and small industrial applications, there may also be a certain amount of infiltration or percolation in the initial period of operation. However, after a time, it may be expected that solids deposition will eventually clog the surface to the point where infiltration is eliminated. The potential impact of wastewater infiltration to the groundwater, and particularly on-site water supplies, should be evaluated in any event and, if necessary, lagoon lining may be utilized to alleviate the problem. Evaporation lagoons are often preceded by septic tanks or aerobic units in order to provide a more acceptable influent to and minimize sludge removal from the lagoon. The on-site utilization of evaporation lagoons for the disposal of domestic wastewater, from homes, farms, and smaller agricultural, industrial or commercial facilities may be applicable where access to a municipal sanitary sewer is not available; where subsurface methods are not feasible and where effluent polishing for surface discharge is not practical.

7.4.3 Waste Storage Lagoons

Agricultural waste storage lagoons store high-strength wastes from food processing operations such as vegetable canneries and beet sugar plants. The waste may have associated nuisance odors. Complete biological treatment usually is not accomplished in a lagoon, but may later be realized by an additional process such as land application or mechanical treatment. Generally, the main purpose of a storage lagoon is for the removal of settleable and floating solids.

7.4.4 Sludge Lagoons

WSPs are constructed with earthen dikes for holding and treating wastewater before final effluent discharge to receiving water. Sludge lagoons are also constructed with earthen dikes, but are for holding and dewatering digested sludge before final sludge disposal.

A sludge lagoon operates as a sequencing batch reactor (SBR) in which (a) sludge is periodically drawn from an anaerobic digester, placed in a lagoon for drying by evaporation and decanting supernatant liquor, and removed after a period of drying; and (b) the SBR cycle is then repeated. Lagoons, however, are not typically provided with an underdrain system because most of the drying is accomplished by water evaporation and decanting supernatant liquor. Lagoons may be of any shape, but a rectangular shape facilitates rapid sludge removal, Lagoon dikes should have a slope of 1:3, vertical to horizontal, and should be of a shape and size to facilitate maintenance, mowing, and passing of maintenance vehicles atop the dike, and access for entry of trucks and front-end loaders into the lagoon. Surrounding areas should be graded to prevent surface water from entering the lagoon. Public access to the sludge lagoons should be limited.

There are three types of sludge lagoons: (a) sludge drying lagoons; (b) facultative sludge lagoons; and (c) anaerobic sludge lagoons.

7.4.4.1 Sludge Drying Lagoons

Sludge drying lagoons (SDL) may be used as a substitute for drying beds for the dewatering of digested sludge. Lagoons are not suitable for dewatering untreated sludge, limed sludge, or sludge with a high-strength supernatant because of their odor and nuisance potential. The performance of the lagoons, like that of drying beds, is affected by climate; precipitation and low temperatures inhibit dewatering. Lagoons are most applicable in areas with high evaporation rates.

Solids loading rates suggested for SDL are 2.2–2.4 lb/yr/cu ft of lagoon capacity. Other designers have made recommendations ranging from 1 ft.^{2}/capita for primary digested sludge in an arid climate to as high as 3–4 ft.^{2}/capita for activated sludge wastewater treatment plants (WWTP) where the annual rainfall is 36 inches. A dike height of about 2 ft. with the depth of applied sludge between 24 and 48 inches and the depth of sludge after decanting of 15 inches has been used. Sludge depths of 2.5–4.0 ft. may be used in warmer climates where longer drying periods are possible. Sludge may be, dewatered from 5% solids to 40%–45% solids in 2–3 years, using sludge depths of 2–4 ft. Sludge drying lagoons may also be used as temporary sludge storage facilities when spreading on agricultural land cannot be carried out due to such factors as wet ground, frozen ground or snow cover.

Sludge lagoons as a means of dewatering digested sludge will be permitted only upon proof that the character of the digested sludge and the design mode of operation are such that offensive odors will not result. Where sludge lagoons are permitted, adequate provisions shall be made for other sludge dewatering facilities or sludge disposal in the event of upset or failure of the sludge digestion process.

The soil must be reasonably porous and the bottom of the lagoons must be at least 1.2 m above the maximum groundwater table. Surrounding areas shall be graded to prevent surface water entering the lagoon. In some critical instances, the government may require a lagoon to be lined with plastic or rubber material; Lagoons should be at least 1 m in depth while maintaining a minimum of 0.6 m of freeboard.

Adequate provisions shall be made to seal the sludge lagoon bottom and embankments to prevent leaching into adjacent soils or groundwater. The area required will depend on local climatic conditions. Not less than two lagoons should be provided. Consideration shall be given to prevent pollution of ground and surface water.

The cycle time for lagoons varies from several months to several years. Typically, sludge is pumped to the lagoon for 18 months and then the lagoon is rested for 6 months. Sludge is removed mechanically, usually at a moisture content of about 70%.

Provisions shall be made to maintain sufficient continuity of service so that sludge may be dewatered without accumulation beyond storage capacity. Unless standby facilities are available, adequate storage facilities shall be provided. The storage capacity should be sufficient to handle at least 4 days of sludge production volume.

7.4.4.2 Facultative Sludge Lagoons

Facultative sludge lagoons (FSLs) are designed to maintain an aerobic surface layer free of scum or membrane-type film build-up. The aerobic layer is maintained by keeping the annual organic loading to the lagoon at or below a critical area loading rate of 20 lb of volatile solids (VS) per 1,000 ft.^{2}/day and by using floating surface aerators/mixers to provide agitation and mixing of the aerobic surface layer. A FSL with surface areas of from 4 to 7 acres requires the operation of two surface mixers from 6 to 12 h/day to maintain scum-free surface conditions, and to provide DO. Aerobic surface layer of the FSL is usually from 1 to 3 ft. in depth and supports a population of between 50×10^3 and 6×10^6 organisms/mL of algae (Chorella). DO is mainly supplied to this layer by algal photosynthesis, by direct surface transfer from the atmosphere, and by the surface aerators/mixers. The DO is used by the bacteria in the aerobic degradation of colloidal and soluble organic matter in the digested sludge liquor, while the digested sludge solids settle to the lagoon bottom and continue their anaerobic decomposition. Supernatant is periodically returned to the plant's headwork for reprocessing. Most of the water in the lagoon is evaporated, so the influent sludge is dewatered.

7.4.4.3 Anaerobic Sludge Lagoons

Anaerobic sludge lagoons (ASL) are designed for sludge decomposition by anaerobic microorganisms under anaerobic conditions in the absence of oxygen. Depths of an ASL can vary between 15 and 35 ft. Anaerobic decomposition takes place in varying sludge layers utilizing at least a 5-ft. water cap as the top liquid layer to contain sludge odors. No surface agitation is used in an ASL operation. In a typical ASL operation, a feed sludge of 57% volatile solids (VS) is loaded to different ASL at rates of 36–50 lbs VS per 1,000 ft.^{2}/day. These typical ASL systems accomplish an overall VS reduction of over 17%. The ASL supernatant is disposed of on 1,320 acres of alfalfa-brome hay fields by land application. Average annual quantity to dispose equals 700,000 wet tons with an average ammonia content of 110 mg/L and an average TKN content of 156 mg/L.

7.5 HISTORICAL AND RECENT DEVELOPMENTS IN PONDS APPLICATIONS

There have been very few new developments in stabilization pond processes. Much of the effort in recent years has been devoted toward understanding the mechanism of substrate removal in the complex biological community and to improving methods of algal removal for better control of effluent quality. In this section the areas of recent development in the use of stabilization ponds are discussed.

7.5.1 Nutrient Removal and Controlled Eutrophication

It is well-recognized by environmental engineers that nitrogen and phosphorus uptake by algae in stabilization ponds contributes significantly to nutrient removal from wastewater. Nitrogen and phosphorus are removed by the amount of removal by each of these removal mechanisms, which is dictated by the dominant pond activity, which in turn is determined by the environmental conditions of the pond, e.g., degree of mixing, radiation availability, pH, temperature, organic loading, and so on. With a better understanding of pond activities today, attempts have been made to optimize the processes of nutrient removal in stabilization ponds [30,31].

Significant nutrient removal from ponds, ranges from 30% to 95%. The optimum removal of N or P depends on the BOD/N/P loading ratio of the pond, with N expressed in total Kjeldahl-N and P in total phosphorus. The BOD/N/P ratio for optimum removal of N ranges from 78/13/1 to 16/8/1, while the same ratio for optimum removal of P ranges from 52/10/1 to 21/5/1. Compromising between the optimum ratio for N and P removal, the loadings are approximately 52 lb BOD/ac./d, 12 lb N/ac./d, and 1.6 lb P/ac./d.

A high NH_4^+ nitrogen concentration in excess of 45 mg/L or NO_3^- nitrogen concentration above 35 mg/L will inhibit algal growth. The maximum algal concentration is obtained when NH_4^+ nitrogen concentration is 12–20 mg/L [1].

To take advantage of the ability of algae to remove nutrients, a controlled eutrophication technique can be applied to stabilization ponds. In essence, the technique is to regulate the assimilation of nutrients by algae in mass culture. Furthermore, the algal mass produced can be biologically removed by linking the system to a commercially valuable culture of plankton-feeding invertebrates such as oysters, clams, mussels, brine shrimp, or fish. Although the feasibility of such a controlled eutrophication system has been successfully demonstrated, the scale-up presents engineering problems. In these nutrient-algae-aquaculture systems, one should not lose sight of the primary objective of nutrient removal while mass culture of algae and balancing of the ecosystem with invertebrates are merely viable means to achieve the goal of nutrient removal. This is a very important consideration because it has been found that various requirements needed to meet the separate objectives of nutrient removal and producing algae for aquaculture are not entirely compatible. Increasingly higher removals of nitrogen through algal assimilation will occur as the dilution rate or the fraction of wastewater in the influent is decreased. On the other hand, the maximum yield of algae will occur at a relatively high dilution rate, leaving a relatively high nitrogen concentration [1].

To optimize nutrient removal in the nutrient-algae-aquaculture system, it is important to maintain NH_4^+ nitrogen as the dominant N form in the wastewater. In a massive algal growth system, the pH of the water will reach as high as 10.5 at times during the day. Stripping of NH_3 to the atmosphere occurs at the high values of pH typically attained in these ponds. High nitrogen removal (70%–75%) occurs because of the combination of algal assimilation and NH_3 stripping. Therefore, for maximum nitrogen removal, nitrification of the wastewater at the treatment plant preceding the algal system must be avoided. Removal of phosphorus is also best achieved at high pH in fresh as well as in saline water environment because of precipitation. Only 25%–50% of the phosphorus in a typical secondary effluent could be expected to be removed through incorporation into algal cells. The precipitated phosphorus, if not physically removed, could revert into solution when it comes in contact with receiving water at a lower pH.

On average, the maximum attainable yield of algae is 6 g of particulate carbon/m^2/day or approximately 12 g dry weight of algal cells/m^2/day. Local temperature and sunlight effects will control the practical limit of algal production and in turn the yield of aquaculture products.

7.5.2 Integrated Anaerobic-Facultative-Aerobic Pond Systems

An integrated pond system consists of an anaerobic pond, several facultative ponds, and an aerobic pond in series. The anaerobic pond is isolated at the center of the series and is surrounded by the facultative ponds. All these ponds in turn are surrounded by a buffer pond of indefinite size and shape designed mainly for water dissipation and esthetic appeal. Effluent is drawn from near the bottom of the anaerobic pond to retain grease and heat. Moving through the facultative ponds in series, the effluent is retained in the buffer pond. A shallow well can be installed in the berm that separates the last facultative pond and the buffer pond. Water from the well could be used for park irrigation or to maintain the water level in the ponds.

The integrated pond system may replace septic tank systems for subdivisions. A hydraulic balance had to be maintained so that the inflow was equal to the net evaporation and net percolation. There was no overflow from the system. Hydrological data on precipitation and evaporation as well as percolation rates through soil at the site were analyzed and carefully evaluated.

The required area for the system was less than that of septic tank systems. It is an attractive solution for subdivision wastewater disposal in areas where sewerage is not yet available and will not be available within several years. In many areas, septic tanks are not feasible because of experienced failures or because percolation tests show that required leaching field areas are greater than the size of existing lots. However, the system should be installed on a fixed-population basis and in areas where local surface water or groundwater supply is not used for drinking water.

The buffer pond, through skillful design by landscape architects, can contribute immensely to community acceptance of the system. Operating experience at the City of Esparto and Santee, both in California, USA, or similar ponds in series show that no sight or odor nuisance, nor hazard to health would result from using the integrated pond-park concept.

7.5.3 Activated Sludge Process Integration

The activated sludge process (ASP) can be successfully integrated into the ponding system in the framework of the pond enhanced treatment and operation (PETAO) concept [32]. The PETAO system, an ASP-variant, was designed to combine oxidation ponding as a low-tech primary stage and a polishing facility as a secondary stage [33]. The full scale system has two variants in which the secondary facility can be either a trickling filter or an activated sludge process. A series of oxidation ponds treat the bulk of organic load (up to 70%), which substantially decreases the size of the relatively high-tech secondary facility.

It was demonstrated that microalgae, acting as heterotrophs in the dark, contribute to the mucilage production in the trickling filter. Biofilm slime, predominantly exopolysaccharide, was shown to be produced by the microalgae experiencing stress transfer from the mixotrophic (growth on organics in the light) conditions in the ponds to the heterotrophic conditions in the trickling filter (metabolism in the dark). Microalgae are thought to use low molecular weight organics such as amino acids, monosaccharides, VFA, etc. Microalgae appear to play a prominent part in the system and more importantly, help produce the superior quality of its final effluent.

The relative importance of the different mechanisms involved in the removal of microalgae in the trickling filter and activated sludge process varies. Removal by rotifers and protozoa plays a greater role in the trickling filter. The removal in the trickling filter is characterized by both flocculation (owing to algal and bacterial exopolysaccharide production) and degradation (through bacterial activity) of algae. In the activated sludge process, algae removal is achieved primarily through stress-induced exopolysaccharide production (both algal and bacterial) and subsequent flocculation (embedding of the algal biomass in the sludge flocs).

It was suggested that using the PETAO ASP as a secondary treatment produce better sludge flocculation. Reciprocally, better flocculation would lead to higher retention of slow-growing nitrifiers enhancing the nitrification potential of the ASP. The higher nitrification rate owing to lower heterotrophic conditions is particularly important at lower temperatures (below 15°C). At the lower substrate concentrations, a nitrifying population is built up at shorter sludge ages in PETAO ASP than in a conventional ASP.

7.5.4 Integrated Duckweed and Stabilization Pond

Posttreatment of effluent from an up-flow anaerobic sludge blanket (UASB) reactor, which was fed with domestic wastewater, was conducted in an integrated pond system [34]. The system consisted of a series of shallow duckweed and stabilization ponds. The main objective of post-treatment is the removal of bacterial pathogens and further polishing of effluent quality. Rapid and efficient pathogen removal can be achieved in shallow stabilization ponds but their effluent BOD and TSS are relatively high, owing to the presence of algae. Passing stabilization pond effluent through duckweed ponds was expected to remove algae owing to reduced light penetration. Duckweed ponds have revenue-generating potential because the produced biomass can be used as animal fodder. However, when applied separately, their pathogen removal is poor. At a pilot plant system with an overall retention time of 4.2 days consisting of ten ponds in series, two duckweed ponds in the first stage, three stabilization ponds in the second stage and five duckweed ponds in the third stage, rapid removal took place in the stabilization ponds. Increasing the retention time of the stabilization ponds to 3–4 days is suggested for consistently satisfying the WHO criterion for unlimited irrigation. A first-order fecal coliform decay constant K_d was calculated for each of the three stages with the values of 0.7–3.2, 4.0–5.9, and about 1.4/day, respectively. The shading by the duckweed cover in the last stage proved to be able to remove practically all algae. Excellent effluent quality with respect to TSS (11 mg/L) and wastewater treatment for reuse in irrigation could be achieved in one simple system.

Ammonium removal was due to uptake by the duckweed plants, nitrification, sedimen¬tation, combined volatilization of NH_3 and denitrification [35]. The duckweed plants were shown to release oxygen to the pond water remained completely aerobic and could remove 95%–99% of the influent BOD [36]. The duckweed ponds can be situated in the treatment scheme before and after the algae pond treatment; in the former case, to benefit from high nutrient concentrations, and in the latter, to remove the algae from the algae pond effluent and for additional nutrient conversion. Further information on nitrogen removal, attached growth WSPs, and algal ponds can be found in refs. [37,38].

7.5.5 Deep Self-Regeneration and Anoxic Waste Stabilization Ponds

WSPs are becoming the most popular forms of secondary treatment owing to their low cost of construction, operation, and maintenance. However, as stated above, the presence of algae in the effluent makes its quality unacceptable in terms of total suspended solids and biochemical oxygen demand. Stringent environmental regulations concerning effluent discharges make it necessary to remove algae from pond effluent.

Anoxic WSPs operate in the so-called "gray area" of organic loading, which lies between fully anaerobic and facultative conditions. In anoxic WSPs, a removal efficiency of suspended solids of 60%–80% and biochemical oxygen demand and chemical oxygen demand (COD) of 70%–80%, algae of 76%–98% can be achieved. Regarding algal species in anoxic pond effluent, the motile flagellate algae (Euglena and Chlamydomonas) were the only species found to exist.

In pond systems, it is known that algae stimulate bacteria, and bacteria stimulate algae. In this way, the chemical units that comprised the organic waste eventually become incorporated into the algae as stable organic components of living cells. The death and decomposition of large numbers of algae lead to undesirable conditions similar to those caused by the original wastewater in the pond. Consequently, it needs to be avoided to keep the trophic structure well balanced for optimum performance. Deep treatment ponds offer theoretical advantages over conventional stabilization ponds, which have greater surfaces and are shallower. They occupy less surface area and, in arid and semi-arid areas, summer evaporation losses are reduced. Deep ponds could become an inexpensive, simple way to improve the sanitary and chemical quality of water and, at the same time, be used as reservoirs for agricultural irrigation. Arauzo et al. [39] carried out the investigation of an experimental deep self-regeneration pond involving the in-depth examination of the role of algae in a deep urban wastewater self-regeneration pond. Evaluation was made of the phytoplankton community, the dynamics of its populations and its relationship with the performance and the trophic structure of the pond. The aim of their design was to improve the quality of wastewater effluent for its reuse.

Despite their fertilizing value and protein content, algae are required to be removed owing to their deleterious effect on the receiving environment. Accordingly, if pond treatment is to remove BOD from the influent, then algae should be removed from the effluent [40]. The USEPA considered the removal of algae from ponds effluent as a condition for successful operation and production of good quality effluent [41]. Different treatment processes were applied to upgrade effluents from WSPs.

Filtration is commonly employed to upgrade WSP effluent [42,43]. Disadvantages of slow sand filtration include the deep penetration of small algal species. If these fine particles are present in large amounts, cleaning of the filter cannot be achieved by normal scraping methods. Furthermore, anaerobic conditions may develop in these filters, especially during darkness or at low temperatures, owing to the decomposition of algae which will result in the production of soluble BOD, deficiency of oxygen, production of odor, and release of nutrients. Owing to the increasingly restrictive legal requirements, the rising cost of liquid waste disposal, and the growing need for innovative sources of water supply, liquid waste reduction, and recovery of valuable products, cross-flow membrane separation processes are gaining considerable prominence in many sectors of the industry [44].

Cross-flow filtration is a process in which the formation of a filter cake is either limited or, under certain conditions, almost completely suppressed by a flow of the suspension parallel to the filtration surface and, because this system is pressurized, water is forced through the filter. In contrast, in the case of conventional cake filtration, the suspension flows at right angles to the filter medium under the applied pressure. The particles are retained by the filter medium while the liquid flows through the filter cake and through the medium. It is clear that, in the case of conventional filtration, clogging will occur in a short time owing to the build-up of filtered cake, while in cross-flow filtration particles deposited on the filter medium are swept away by the cross-flow velocity actions.

7.5.6 Combination with Constructed Wetlands

Recently, the use of constructed wetlands [46] has been receiving more attention for various types of wastewater treatment alternatives [47]. The inclusion of constructed wetland technology as a component of an overall wastewater management system has the potential to help address the issues of agricultural wastewater treatment and the protection of water quality. An integrated wastewater treatment facility, consisting of upper (solids separators, anaerobic lagoons, and aerobic ponds) and lower (wetland cells) subsystems, has been built to replace the lagoon [48]. The demonstration of the system for dairy treatment is that the collection sump of the new waste treatment facility collects all dairy wastewater outflows. Wastewater is then pumped to solids separators, and flows by gravity to anaerobic ponds and aerobic ponds. The upper subsystem is expected to treat the water sufficiently so that the wetland cells may achieve further pollutant reductions. The lower subsystem, comprised of eight surface wetland cells with an approximate surface of $5{,}000\,m^2$, receives outflow from the ponds. The cells were planted with cattail, soft-stem bulrush, and reed. After treatment is completed through the lagoons and ponds followed by the wetland cells, the wastewater can be reused to flush barns or to irrigate crops.

An integrated aquaculture-wetland ecosystem using tertiary-treated municipal wastewater in Los Angeles County, California was proposed for food production and nitrogen removal [49,50]. The ecosystem connected polyculture aquaculture ponds with in-pond aquatic plant systems, a solar energy aeration system, and an artificial wetland. Ponds were stocked with hybrid tilapia, common carp, mosquito fish, and red swamp crayfish, and were flushed weekly with new wastewater. The concept of using tertiary-treated wastewater for aquatic food production may be attractive in the urban areas of many meagcities, both for fish markets and to stem the growing discharges of wastewater that are causing coastal pollution.

An index of biotic integrity based on crayfish, fish, and amphibian assemblages was developed to assess vernal ponds and palustrine wetland habitats along the southern shore of Lake Michigan [51]. The other experimental study of constructed wetlands included multi-stage systems [52]. Pretreated water from an anaerobic stabilization pond and treated water from the last pond of a lagoon system were used to test the system's suitability as a complementary system for removing nitrogen and phosphorus [53].

Seepage losses from constructed wetlands, wildlife refuges, wastewater lagoons, runoff collection ponds, and other engineered surface impoundments of water or aquatic ecosystems in areas with deep groundwater levels can be minimized with artificial liners (plastic, compacted earth, etc.) or with natural processes like sediment accumulation and microbiological and chemical processes [54]. Sediment accumulation can be increased by deliberately adding soil slurries or muddy (turbid) water. When these are applied to the water surface, settling of the particles to the bottom then creates a graded sediment layer with the coarsest particles on the bottom and the finest particles on top. Column studies in the laboratory showed that, for a given amount of soil added, such a graded layer gives more seepage control than a compacted earth liner. Also, split slurry applications to create a layered lining gave more seepage control than when the same total amount of soil was given in one single slurry application. Subsequent addition of sodium carbonate to the water reduced seepage even more, giving a total seepage reduction in the columns from 1,000 to 0.2 cm/day. The study also indicated the importance of avoiding turbid inflows into infiltration basins for groundwater recharge, as well as soil erosion within the basins themselves, if infiltration rates need to be maximized.

7.5.7 Synopsis of Major Developments

WSPs design practice today is much different from what used to be 20 years ago. There is now a new generation of WSPs that are smaller, more efficient, more reliable, with less or no odor problems, and which release effluents of better quality than the previous generation. In summary, numerous different types of ponds have been developed during the years: anaerobic ponds (open or covered); aerobic ponds (maturation or polishing); facultative ponds; ponds with or without recirculation; high rate stabilization ponds (HRSP); partially mixed and bi-dimensional plug-flow ponds; ponds for both municipal and industrial effluents, etc. [55]. To locate names and websites of manufacturers and products for stabilization ponds and lagoons consult ref. [56].

The main new concepts and tools available for the designers of modern WSPs are in the fields of hydraulics, kinetics, out-of-the-pond elements, and design of the wastewater treatment series [55].

a. Hydraulics:
1. Actual hydraulic pattern (tracer studies) and development of tools to improve it;
2. Pond shape;
3. Wind effect on wastewater circulation within the pond;
4. Stratification control;
5. Nonsteady-state approach to reactor analysis [57];
6. Use of both horizontal and vertical baffles to force plug-flow conditions [58];
7. Sequential batch operation.

b. Kinetics:
1. Pond acceleration by low-energy mixing, aeration, or re-circulation;
2. Pond acceleration by addition of fixed biomass;
3. Combination of ponds with more intensive treatment units;

c. Out-of-the-Pond Elements:
1. Different types of pond covers and devices to avoid smell release;
2. Intensive anaerobic pretreatment to reduce pond size;
3. Posttreatment for algae removal (rock filters, constructed wetlands) and other requirements.

d. Design of the Wastewater Treatment Series:

1. Ponds of different sizes instead of modularization.
2. Use of wastewater storage reservoirs for irrigation, controlled discharge, and other purposes.

7.6 ALGAE, PHOSPHORUS, BOD AND TSS REMOVALS BY AIR FLOTATION

Algae have many of the properties of fine near-colloidal particles, ranging in size from the submicron to about 2 micrometers in diameter or length. They can be flocculated and floated using an induced air flotation system. Tests have been done on water from a wastewater maturation pond [45]. The flocculant of iron chloride was added at an average dose of 10 mg/L. It seems to be very effective in the removal not only of the algae but also of the phosphorus compounds. At a pH of 9 in water, the iron salt hydrolyses to form mainly ferric hydroxide, which has excellent flocculating properties. The soluble phosphates in the water reacted with the iron salts to form an insoluble precipitate, which was flocculated with ferric hydroxide produced by a stoichiometric excess of iron chloride [42,43]. The removal efficiencies of algae cell count and total phosphorus were 98.8% and 94.2%, respectively. The excess of iron chloride addition could be simultaneously removed with flotation.

Dissolved air flotation (DAF) is one of the best processes for the removal of algae, phosphate, biochemical oxygen demand (BOD), chemical oxygen demand (COD), total organic carbon (TOC), total suspended solids (TSS), and coliform bacteria from wastewater [59–61]. Example 7.9 illustrates the case histories on how a WSP system can be upgraded by DAF.

Example 7.1: Non-aerated Facultative Pond Analysis

In this section, an example of process analysis and design is presented. There is an existing non-aerated facultative pond system with the following given information of the pond facility, flow, and influent 5-day BOD as follows: (a) pond area $A = 872{,}356\,\text{ft.}^2$ or 20 acres; (b) depth $d = 3$ ft.; (c) wastewater flow $Q = 1.1$ MGD or 1.1 million gallons/day; and (d) influent 5-day BOD concentration = 305 mg/L. Determine the DT (t) of the wastewater within the facultative pond and the influent organic loading for the pond (5-day BOD lb/acre/day).

Solution:

1. Volume of the facultative pond $V = A \times d = 20\ \text{acre} \times 3\ \text{ft.} = 60$ acre-ft.
   Detention time t
   $= V/Q$
   $= [(60\ \text{acre-ft})\ (7.48\ \text{gal/ft.}^3)\ (43{,}560\ \text{ft.}^2/\text{acre})]/[1.1 \times 10^6\ \text{gpd}]$
   = 17.77 days
   For the existing non-aerated facultative pond, Table 7.1 (USEPA design criteria) shows that a 17.77-day DT for the biological process by the pond is not within the acceptable 20–180-day range.
2. 1 mg/L = 8.34 lb/MG;
   Therefore, Influent BOD $C_i = 305\ \text{mg/L} = 305 \times 8.34$ lb/MG
   Organic loading L
   = (Flow Q) × (Influent BOD C_i)/(Area A)
   = (1.1 MGD) × (305 × 8.34 lb/MG)/(20 acre)
   = 140 lb/acre/day of 5-day BOD

By checking Table 7.1, the organic loading of the existing non-aerated facultative pond is higher than normally expected (15–75 BOD_5 lb/acre/day). If the pond is not performing as desired, Table 7.1 or additional engineering data in the literature should be consulted for possible solutions to remedy this environmental engineering problem.

Example 7.2: Upgrading a Non-aerated Facultative Pond by Installation of Mechanical Aerators

Assuming you are the environmental engineer in charge, try to solve the engineering problem of an existing non-aerated facultative pond described in the "Section 7.6 Example 7.1" based on the US Environmental Protection Agency (USEPA) Design Criteria in Table 7.1.

Solution:

The existing non-aerated facultative pond's engineering conditions from Example 7.1" are: (a) pond area A=872,356 ft^2 or 20 acres; (b) depth d=3 ft; (c) wastewater flow Q=1.1 MGD or 1 million gallons per day; (d) influent 5-day BOD concentration=305 mg/L; (e) calculated hydraulic DT t=17.77 days; (f) calculated organic loading L=140 lb/acre/day of 5-day BOD.

According to the updated USEPA design criteria shown in Table 7.1, it appears that the conversion of the existing "non-aerated facultative pond" to an upgraded "aerated aerobic pond" will solve the organic overloading problem. The new design hydraulic DT t=3–20 days (O.K. for 17.77 days); and the new design 5-day BOD organic loading=20–400 lb/acre/day (O.K. for 140 lb/acre/day). However, the effluent 5-day BOD would be worse (20–70 mg/L BOD_5); and the TSS a new completely mixed bioreactor effluent would be too high.

It would be necessary to install mechanical aerators for (a) supplying oxygen under completely mixed hydraulic condition; (b) operating the system similar to an activated sludge complex bioreactor. Besides, secondary clarifier such as sedimentation and dissolved air flotation will be needed to reduce the effluent TSS concentration.

It should also be noted that the DT requirement is much longer in extremely cold regions, such as MN, USA [62,63].

Example 7.3: Converting a Non-aerated Facultative Pond to a Partially Aerated High-Rate Aerobic Pond System

Assuming you are the environmental engineer in charge, try again to solve the engineering problem of an existing non-aerated facultative pond described in Example 7.1" based on the technical data in the literature.

Solution:

The existing non-aerated facultative pond's engineering conditions from Example 7.1" are: (a) pond area A=872,356 ft^2 or 20 acres; (b) depth d=3 ft; (c) wastewater flow Q=1.1 MGD or 1.1 million gallons/day; (d) influent 5-day BOD concentration=305 mg/L; (e) calculated hydraulic DT t=17.77 days; (f) calculated organic loading L=140 lb/acre/day of 5-day BOD.

From this book, it appears that a "partially-aerated high-rate aerobic pond" described in Section 7.16 might be another engineering solution.

High-rate aerobic ponds are one of "aerobic ponds". Specifically, it should be technically named "partially-aerated high-rate aerobic ponds", which maintains DO throughout their entire shallow depth, usually between 2 and 4 ft. deep, allowing light to penetrate the full depth. The design criteria are: (a) 5-day BOD organic loading L=60–200 lb/acre/day; (b) hydraulic DT t=3–6 days; (c) pond depth=2–4 ft.; (d) DO concentration=aerobic throughout entire shallow pond depth; (e) minimum algae concentration in pond water=80–200 mg/L; (f) mixing condition=allowing to expose all algae (80–200 mg/L algae) to sunlight and to prevent deposition and subsequent anaerobic conditions; and (g) high-rate aerobic ponds are limited to warm, sunny climates. Oxygen is provided by both algal photosynthesis and surface reaeration with very limited mechanical aeration, and aerobic bacteria stabilize the waste. It is suitable for the removal of soluble organic and nutrients. The pond's effluent 5-day BOD is expected to be excellent (20–30 mg/L). However, operation is somewhat more complex than for a facultative pond and, unless an algae removal step is provided, the effluent will contain high TSS (150–350 mg/L). Because of their shallow depths, paving or covering the bottom is required to prevent weed growth.

It appears that the existing non-aerated facultative pond can be converted to the above partially-aerated high-rate aerobic pond system because the existing pond has a shallow depth $d=3$ ft. (meeting the shallow depth requirement of 2–4 ft.), an acceptable influent 5-day BOD organic loading of 140 lb/acre/day (in the range of 60–200 lb/acre/day of organic loading requirement), and an acceptable hydraulic DT of $t=17.7$ days (higher than the required 3–6 days DT).

Conversion of the existing non-aerated facultative pond system to a partially-aerated high-rate aerobic pond system would be technically feasible if the remaining process and environmental conditions can be met: (a) DO concentration = aerobic throughout entire shallow pond depth; (b) minimum algae concentration in pond water = 80–200 mg/L; (c) mixing condition = allowing to expose all algae (80–200 mg/L algae) to sunlight and to prevent deposition and subsequent anaerobic conditions; and (d) high-rate aerobic ponds are limited to warm, sunny climates. In addition, the pond effluent will have a high TSS concentration of 150–350 mg/L, so the pond effluent should be further treated for TSS removal.

Dissolved air flotation (DAF) is a highly efficient clarification process for the separation and recovery of algae [9,23,24,27,28]. A combination of DAF and the high-rate aerobic pond could be a win-win solution for pollution control and energy conversion.

Example 7.4: Alternative Pond Effluent Treatment Technologies

Discuss alternative pond effluent treatment technologies available for total suspended solids (TSS) removal.

Solution:

Conventional sedimentation process, innovative dissolved air flotation (DAF), tertiary multi-media filtration, and various natural biological processes [64,65] are all feasible for TSS removal from the pond effluent. The readers are referred to the literature for the process descriptions of sedimentation clarification, DAF, tertiary filtration, and natural processes [40–43, 46,47, 59, 62, 64–71].

Minnesota Pollution Control Agency (MPCA), St. Paul, Minnesota, USA publishes an excellent manual for preliminary planning, design, and O&M of various WSPs [62]. The MPCA recommends some solutions: (a) polishing pond; (b) infiltration pond; (c) waste storage pond; and (d) spray irrigation pond for storage or polishing the WSP effluent [62]. Two alternative filtrations: (a) intermittent sand filtration; and (b) rock filtration are also introduced as possible solutions.

POLISHING POND

A polishing pond system follows a secondary mechanical treatment system and is considered a form of tertiary treatment. Its DTs vary between three and 210 days, depending on their purpose. A polishing pond typically is a non-aerated WSP system. Its purpose is to: (a) Further reduce the biochemical oxygen demand (BOD) and total suspended solids (TSS) before discharge; (b) provide storage for controlled discharges or land application; and (c) possibly remove phosphorus.

INFILTRATION POND OR LAGOON

Infiltration ponds or lagoons are a series of shallow ponds or lagoons about 0.5–1.5 ft. deep used to dispose of pre-treated wastewater. Operations normally consist of periodic controlled amounts to alternate shallow ponds with subsequent drying periods. Any remaining pollutant removal occurs by filtration, biological activity, and absorption as the effluent percolates through the soil. Depending on the location of the groundwater, infiltration ponds may be designed with an underlying network of drains to collect and return the water back to the head works of the treatment system.

WASTE STORAGE LAGOONS

Agricultural waste storage lagoons store high-strength wastes from food processing operations such as vegetable canneries and beet sugar plants. The waste may have associated nuisance odors. Complete biological treatment usually is not accomplished in a lagoon, but may later be realized by an additional process such as land application or mechanical treatment. Generally, the main purpose of a storage lagoon is for the removal of settleable and floating solids.

SPRAY IRRIGATION POND

A spray irrigation pond is identical to a stabilization pond, except for two differences: (a) Treated wastewater is discharged to agricultural land when a cover crop that is not used for human consumption is actively growing; (b) The minimum DT for a spray irrigation pond increases to at least 210 days to allow for winter storage.

INTERMITTENT SAND FILTRATION

The intermittent sand filter is an outdoor, gravity, filtration system that capitalizes on the availability of land area. It is a biological and physical wastewater treatment mechanism consisting of an underdrained bed of granular material, usually sand. The filter surface is flooded intermittently with lagoon effluent at intervals which permit the surface to drain between applications. It is recommended that the flow be directed to one filter for 24 hours. That filter is then allowed to drain and dry for 1–2 days, and the flow goes to an adjacent filter. It is preferable to have three filter beds where good operation and treatment may be accomplished over a 3-day cycle. The filter contains drainage pipes that are laid with open joints at depths of 3–4 ft. and surrounded with layers of coarse stone and gravel graded from coarse to fine to keep sand out. When the filter is located in natural sand deposits, the percolating waters may reach the groundwater table, and no effluent may come to view. In areas where percolation to the groundwater is not permitted, the filter must be provided with an impermeable base or lining. Influent wastewater is piped to the beds for discharge into a protective stone or concrete apron or into a concrete flume distributor. Surface accumulations of solids are periodically removed and disposed of as fill or by some other means. Filters must be re-sanded when they become too shallow as a result of the periodic surface cleaning. In cold weather, the beds are plowed into ridges and furrows (1–1.5 ft. deep) to keep them from freezing and opening up cracks through which the applied wastewater can escape with little treatment. To form protective sheets of ice spanning the furrows and keep the beds warm, furrowed beds are dosed deeply on cold nights. Lagoon effluent may be applied to filter beds of progressively smaller size of filter media; e.g., #1 sand e.s. = 0.72 mm, #2 sand e.s. = 0.40 mm, #3 sand e.s. = 0.17 mm, and gravel support. The technology has been practiced with proven performance for many years where land is available. The filter is used successfully for (a) polishing of domestic wastewater facultative stabilization pond effluent; (b) final treatment of bio-degradable industrial wastewater stabilization pond effluent where minimum maintenance is required. It is unsuitable for anaerobic lagoon effluents, and but is suitable for aerated lagoons if aerated lagoons are followed by facultative lagoons. The filter effluent quality (treatment of domestic wastewater stabilization pond) is expected to be excellent: less than 10 mg/L filter effluent BOD_5 and less than 10 mg/L filter effluent TSS.

ROCK FILTRATION

The rock filtration is an alternative for the removal of algae from non-aerated WSP effluents. The non-aerated WSPs, either aerobic or facultative, rely on algae to provide oxygen source for photosynthesis biological reactions. The filter consists of a submerged bed of rocks (5–20 cm diameter) through which the lagoon effluent is passed vertically or horizontally, allowing the algae to become attached to the rock surface and thereby be removed. The basic simplicity of operation and maintenance are the key advantages of this process. The effluent quality achievable and the dependability of long-term operation, however, have not yet been proven. The lagoon effluent enters the bottom of the filter through an influent channel. The lagoon effluent then rises from the influent channel and moves horizontally toward the discharge weirs where it flows into a covered effluent channel. Finally, the flow from each side of the rock filter is combined, chlorinated, and discharged. The rock surface is approximately 0.30 m (1.0 ft.) above the water elevation to prevent the growth of algae on the rock filter. The rock filter can be used for polishing domestic wastewater stabilization pond effluent and also for final treatment of biodegradable industrial wastewater stabilization pond effluent where low maintenance is required. There is a demonstrated case history of a full-scale rock filter which was designed and constructed as part of a lagoon expansion and upgrading project at Veneta, Oregon, USA, in operation since 1975. The rock filter system treats wastewater from a population of approximately 2,200 with no industrial wastewater contribution.

Example 7.5: Determination of WSP Minimum Depth

The design of a WSP system (such as an aerobic pond, or an anaerobic pond) should be decided by many factors. Previous Examples illustrate how to design a pond based on the flow and the amount of 5 days added per day. Table 7.1 shows the acceptable hydraulic loading rates and BOD loading rates of various WSPs. Accordingly, the pond surface area at the average operating depth can be either sized for a new pond, or analyzed for an existing pond, so that both the acceptable hydraulic loading rate and the acceptable BOD organic loading rate are not exceeded. Discuss how the minimum depth of a WSP should be determined.

Solution:

Even though a WSP is designed on the basis of surface area, it must have enough capacity to accommodate the waste volume and sludge volume. In addition, depth must be provided to accommodate (a) the normal precipitation less evaporation on the pond surface, (b) the 25-year, 24-hour storm precipitation on the pond surface, and (c) the freeboard. Should the government regulations not permit an emergency outflow or for some other reason, one is not used, the minimum freeboard is 1 ft. above the top of the required volume.

AEROBIC POND AS AN EXAMPLE

Figure 7.1 illustrates these volume depth requirements for an aerobic pond. An aerobic pond should never be overloaded with oxygen-demanding material, such as organic pollutants causing high BOD, COD, TOC, etc. The pond should be filled to the minimum operating level, generally 2 ft., before being loaded with waste, as shown in Figure 7.1 and Appendix 7.1. The maximum wastewater level should not exceed 5 ft. The wastewater level must be maintained within the designed operating range. Sludge should be removed when it exceeds the designed sludge storage capacity.

Both the normal precipitation less evaporation on the pond surface, and the 25-year, 24-hour storm precipitation on the pond surface may be determined based on the procedures outlined by Wang and Wang [66].

ANAEROBIC POND AND STORAGE POND AS ADDITIONAL EXAMPLES

The US Department of Agriculture (USDA, 2009) [4] has recommended the volume depth requirements for an anaerobic pond (Figure 7.2, Appendix 7.2) and a storage pond (Figures 7.3 and 7.4). The readers are referred to Wang et al. [67] for the detailed explanations and solutions.

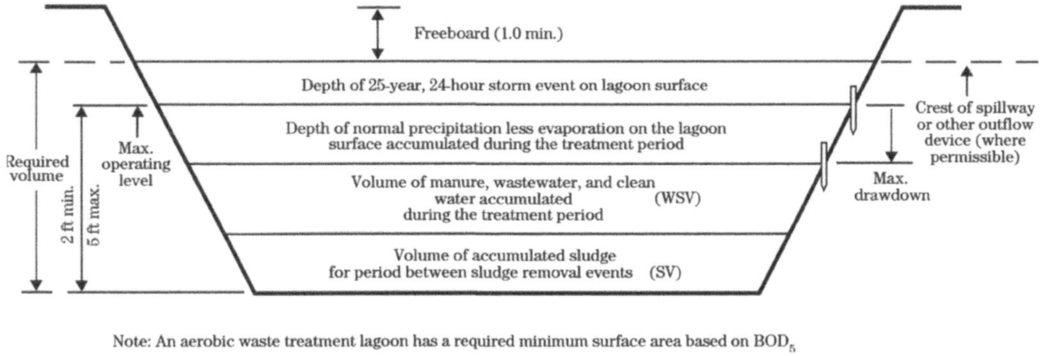

FIGURE 7.1 Aerobic pond cross section. (Source: USDA, 2009.) [4]

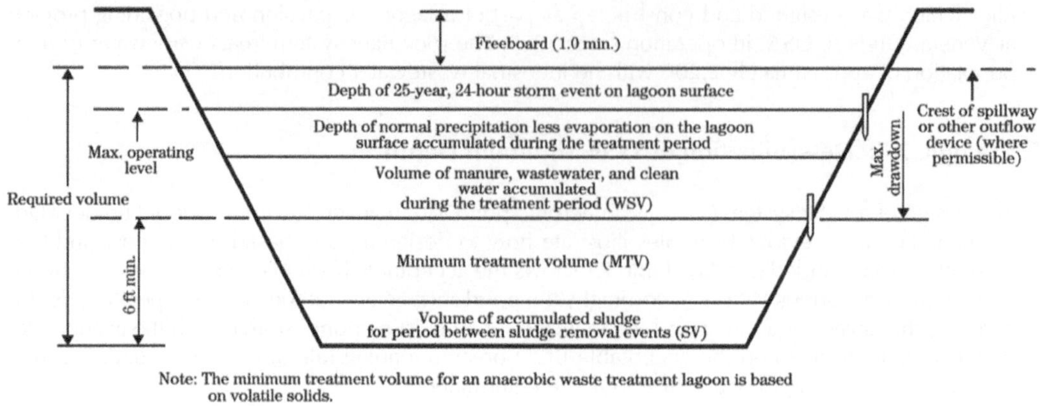

FIGURE 7.2 Anaerobic pond cross section. (Source: USDA, 2009.) [4]

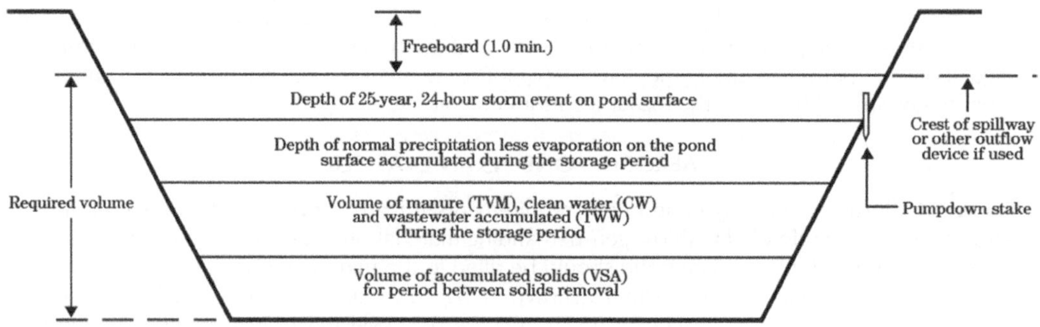

FIGURE 7.3 Storage pond cross section without a watershed. (Source: USDA, 2009.) [4]

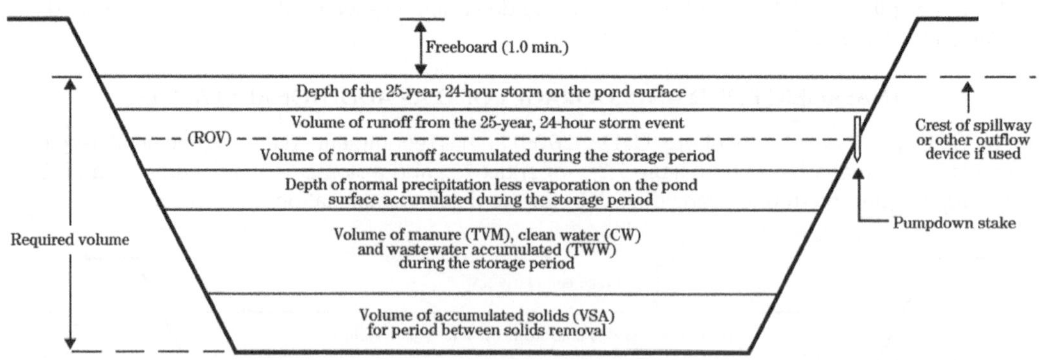

FIGURE 7.4 Storage pond cross section with a watershed. (Source: USDA, 2009.) [4]

Example 7.6: Food Processing Waste Treatment by WSP

Table 7.1 shows the minimum design criteria recommended by the US Environmental Protection Agency (USEPA) for warm weather regions. WSPs still work well in cold weather regions if much more stringent design criteria required by the local government agencies are followed. Search for the Internet information to find a typical case history of food processing waste treatment using WSP biotechnology in cold weather region.

TABLE 7.2
Typical Effluent Discharge Limits for Waste Stabilization Ponds in MN, USA

BOD	25 mg/L
TSS	45 mg/L
Fecal coliforms[a]	200 colony forming units (CFU)/100 mL
pH range	6.0–9.0
Phosphorus[b]	1.0 mg/L

[a] If the system cannot produce effluent quality meeting fecal limits, disinfection may be necessary (stabilization ponds typically do not need disinfection).

[b] In addition, some pond systems also may have a maximum mass kg/year limit [107].

Solution:

A WSP system at Albany, MN, USA was constructed in 1955 and was the first such system in the State of Minnesota. The system was upgraded in 1984 with the construction of the two pond cells in the background [62]. The pre-aeration pond was added in 2001. Their Albany-MN WSP system has been operated and performed successfully since 1955. The Albany-MN plant is of special interest because milk waste constituted about 75% of the total strength of the wastewater it received at the time of its installation. The Minnesota Department of Health, USA has been continuously monitoring the plant. Results of samples collected from the ponds, together with physical observations, have shown that operation is satisfactory based on conformance to design criteria, so that WSPs no longer are considered experimental, but are approved on the same basis as those for conventional mechanical treatment plants. Today there are over 600 municipal wastewater treatment systems in Minnesota, USA. Of those, a little less than half are WSPs. These pond installations range considerably in size from a total of about 335 acres to less than 1 acre with an average size of around 15 acres [62]. Minnesota pond systems have shown that when properly designed, constructed, operated, and maintained, a WSP system is capable of producing an effluent compliant with National Pollutant Elimination System (NPDES) permit limits. The only exception to this is meeting limits for total phosphorus and ammonia nitrogen. Biological conditions, such as cold temperatures that inhibit nitrification, will prevent a pond system from meeting ammonia nitrogen permit limits. Chemicals are needed to reduce total phosphorus to below NPDES permit limits.

Table 7.2 shows the NPDES permit limits details for WSPs. The National Pollutant Discharge Elimination System (NPDES) permit establishes effluent quality limits for all wastewaters discharged to surface waters of the state. The limits were calculated to prevent violation of water-quality standards for the receiving water. Ponds are allowed a controlled discharge to receiving waters only during the spring and fall periods when stream flow is usually higher. Some pond systems may have additional NPDES limits.

Example 7.7: Winter Operation of WSPs

In cold climates, aerobic oxygenation and/or anaerobic fermentation biological reaction rates will be significantly reduced during the winter and early spring and the pond effluent quality will be reduced to the equivalent of primary effluent when an ice cover persists on the pond water surface. As a result, many states in the northern United States, Canada, northern China, etc. prohibit discharge from WSPs during the winter season. Introduce the current design criteria of WSPs required by a local government in the USA where the winter weather is extremely cold to a typical Example. Introduce the concept of "controlled discharge" and "hydrograph controlled release (HCR)".

Solution:

The winter weather in Minnesota (MN), USA is extremely cold, so the website of https://www.pca.state.mn.us/sites/default/files/wq-wwtp5-53.pdf has been searched [62,63]. Only a small portion of the design criteria of the State of Minnesota is presented below for the purpose of illustration of the differences between the USEPA general design criteria (Table 7.1) and the State of Minnesota's specific design criteria for WSPs. The readers are referred to the original sources of information for more details.

ORGANIC LOADING

Non-aerated system pond design should insure a maximum 5-day Biochemical Oxygen Demand (BOD_5) loading of 0.5 lbs/day/1,000 ft.^{2} at the mean operating depth in the primary cells. This is equivalent to 22 lbs BOD_5/acre/day.

Aerated pond system loading rates are to be determined on a peak monthly basis which may be characterized by industrial users or other sources.

DETENTION TIME

The non-aerated system should have at least 180 days of DT between the 2 and a maximum 6 ft. depth ("operating range") in the entire pond system should be provided. For those facilities north of an east/west line approximately at Brainerd and spray irrigation facilities an average of 210 days of storage should be provided. If total maximum daily load studies require reduced discharge periods in the spring and fall, further evaluation of DT requirements must be performed. The Minnesota Pollution Control Agency (MPCA) states that the storage capacity of all ponds should be calculated as the volume between 2 ft. from the pond bottom and the 6 ft. maximum elevation ("operating range"). Design flow must be determined by the following MPCA Design Flow Determination Guidelines http://www.pca.state.mn.us/publications/wq-wwtp5-20.pdf.

The aerated pond system design for minimum DT for BOD removal may be estimated using a design formula recommended by MPCA. In no case shall the total DT of the aerated cells be less than 25 days for systems required to meet 25 mg/L BOD or 36 days for systems required to meet 15 mg/L BOD. A critical temperature of 1°F (0.5°C) shall be used to determine the DT required in the system unless otherwise justified and 68°F (20°C) is to be used to determine oxygen requirements of the aerated pond system.

INDUSTRIAL WASTE TREATMENT CONSIDERATION

Industrial waste treatment consideration should be given to the type and effects of industrial wastes on the treatment facility and process, and in some cases, it may be necessary to pre-treat industrial or other discharges. Industrial wastes should not be disposed of in WSPs without assessment of the effect such substances may have upon the treatment facility and process or other discharge requirements in accordance with applicable regulations. These wastes include, but are not limited to, the following.

1. Any toxic chemicals which may inhibit biological or bacteriological process;
2. Any strong oxidizing agents or disinfectants in quantities sufficient to inhibit the growth of microorganisms;
3. Metal plating wastes or other wastes containing heavy metals and/or toxic inorganic chemicals, such as cyanide, reduced sulfur compounds, arsenic, selenium and free halogens Recommended Design Criteria December 2009 Minnesota Pollution Control Agency 5;
4. Detergent wastes or other waste containing excessive phosphorus or surfactants;
5. Plastics, pharmaceutical wastes and/or other synthetic organic chemicals not amenable to biological treatment;
6. Any wastes containing excessive amounts of non-biodegradable oil, grease, or tar;
7. Any wastes containing in excess of one mg/one phenol;

8. Any acidic or alkaline wastes which because of quantity, strength or un-equalized flow may upset the biological process or damage the physical facility;
9. Any wastes containing radioactive chemicals;
10. Nutrient-deficient wastes (such as paper mill wastes);
11. Any other waste which may be defined as an incompatible pollutant;
12. If the pH is outside the range of 6.0–9.0 evaluate the possible impact on the treatment

AERATED POND SYSTEM SOLIDS REMOVAL

Continuous discharging Aerated pond systems will generate algae during the treatment process which will likely exceed the Total Suspended Solids (TSS) limit in the permit. In order to avoid violating the solids effluent limitation due to these algae, either the algae must be removed by some process (such as filtration, chemical treatment, etc.) or a 180 or 210 day controlled discharge must be provided, which does not entirely ensure compliance.

AERATED SYSTEM OXYGEN AND MIXING CAPABILITIES

Oxygen requirements generally depend on the BOD loading, the degree of treatment required, the concentration of suspended solids to be maintained, and the temperature of wastewater being treated. Aeration equipment shall be capable of supplying a minimum of 2 lb oxygen/lb BOD applied. A minimum of 2–3 mg/L of oxygen should be maintained in all ponds. Tapered aeration, in conjunction with influent configurations discussed (see MPCA Technical Criteria "Aerated Pond Review Checklist") shall be considered in order to reduce the critical oxygen demand in localized areas. Data on pumping, circulation, and mixing capabilities shall be supplied. This data should include both direct and induced flows created by equipment.

DISINFECTION

Disinfection may be required for continuous discharging systems. The design checklist for disinfection needs to be submitted with plans and specifications.

POND LAYOUT OF NON-AERATED SYSTEM

At a minimum, a pond system should consist of three cells, (two primary cells and one secondary), designed to facilitate both series and parallel operation. When the horizontal surface area of the dike(s), toe-to-toe, of a three-cell system would be greater than or equal to the mean pond surface area, a two-cell is permissible. However, two cell systems shall not exceed 10 acres total pond(s) surface area at mean operating depth. Three cell systems provide more operational flexibility than two cell systems. The design of all systems shall provide for a secondary cell(s) with an absolute minimum hydraulic capacity of one-third the total hydraulic capacity of the total system. This is a very important factor in designing pond systems to allow adequate time to discharge enough water for the following storage period (Figure 7.5) [62,63].

CONTROLLED DISCHARGE AND HYDROGRAPH CONTROLLED RELEASE

Controlled discharge is one type of operational modes of WSPs. When any types of WSP are operated under the "controlled discharge" mode, the pond discharge is prohibited during the winter months in cold climates and/or during peak algal growth periods in the summer. In this approach, each WSP cell in the system is isolated, and then discharged sequentially when the weather condition is appropriate.

Controlled discharge pond is a form of WSP. Controlled discharge ponds have long hydraulic DTs and the effluent is discharged once or twice per year when the effluent quality is satisfactory.

Hydrograph controlled release (HCR) is a special operational mode of WSPs. The liquid in the WSP is retained until flow volume and conditions in the receiving water are adequate for the WSP effluent discharge.

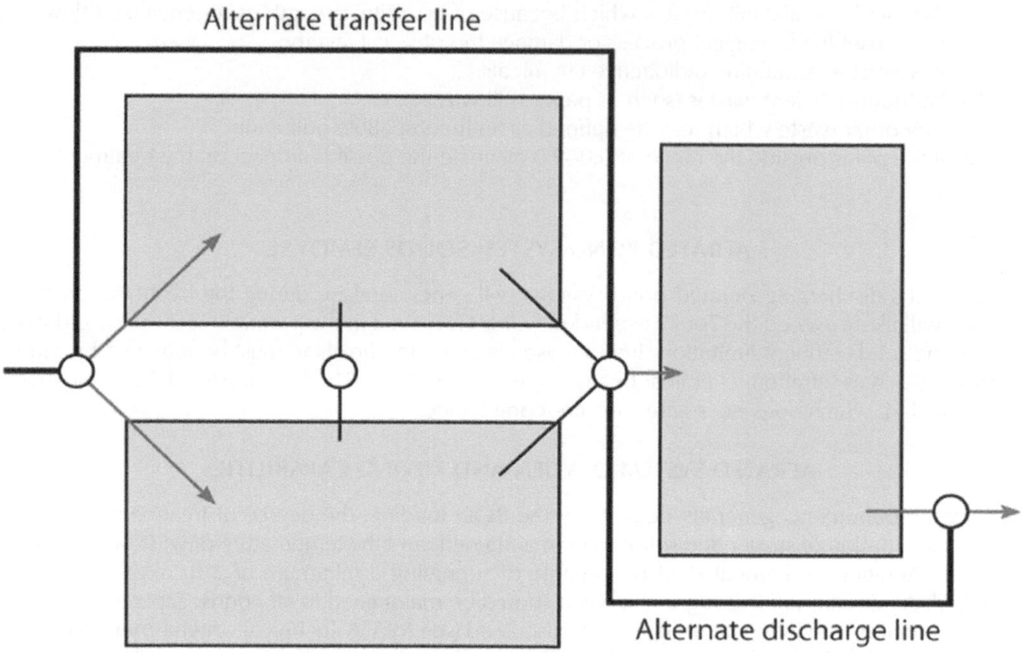

FIGURE 7.5 Flow diagram of non-aerated stabilization pond system [62–63]. (Source: MCPA.)

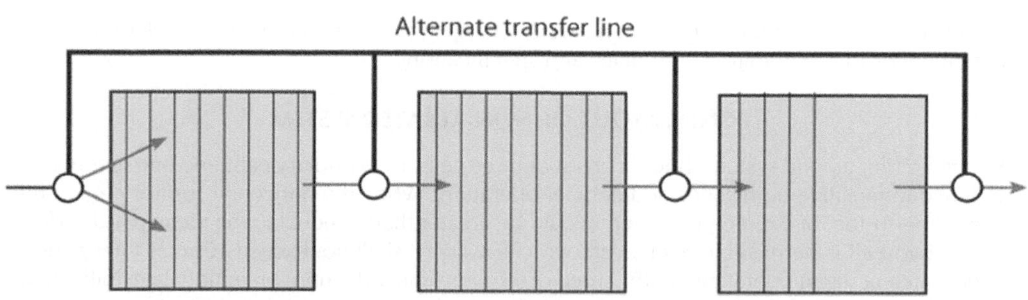

FIGURE 7.6 Flow diagram of aerated pond system. (Source: MCPA.) [59]

POND LAYOUT OF AERATED SYSTEM

At a minimum, an aerated pond system shall consist of three aerated cells along with a non-aerated settling area. The non-aerated secondary cell or the non-aerated portion of the aerated cell used for settling should not be included in the volume calculation of the detention requirement. If a non-aerated settling area is used it must have a volume of at least one-third of the total aerated pond system. For systems serving a population equivalent of less than 750 two aerated cells with a non-aerated settling area may be permitted. Flexibility shall be provided for both series and parallel operations. In the alternate/emergency mode of operation (usually parallel), the aeration capabilities should be adequate to prevent overloading. Design for cell isolation and/or de-watering should reflect considerations such as frost heaving of the seal, groundwater degradation of the seal while dewatered, aeration equipment preservation and general seal protection for whatever mode of isolation is employed (Figure 7.6) [62].

POND BOTTOM ELEVATIONS DIFFERENCES FOR NON-AERATED SYSTEM

The design must have adequate elevation difference between primary and secondary ponds to permit gravity filling of the secondary ponds from the primary. Secondary pond bottom elevations must be 4 ft. lower than primary pond bottom elevations. Where this is not feasible, a portable pump and appropriate hoses must be provided. This pump shall have the capacity to refill the secondary cell at 6 inches/day. The capacity of the pump shall be such that the discharge is spread evenly over the 24-hour period and existing structures must be used to pump into and out of.

POND DEPTH OF NON-AERATED POND SYSTEM

Liquid depths for pond systems shall be designed for a maximum of 6 ft. Six feet is the maximum depth for sunlight to penetrate to adequately maintain algae growth and thus aerobic conditions. Except for maintenance in no instance is it recommended to drain any pond below the minimum 2 ft. level. Water levels less than 2 ft. during the summer can promote aquatic weed growth which could impact the seal. Keeping at least 2 ft. of water in the pond will also discourage vegetative growth, prevent scouring of the bottom deposits and avoid freezing problems with piping, etc. The designed high water depth shall be at least 4.5 ft. in order to discourage. To avoid freezing of piping after the fall discharge and prior to ice formation, the level in the secondary pond(s) should be increased to at least 3 ft. by the retention of winter flows. In the spring, pond levels can be lowered again at a time when surface runoff dilution waters are generally at a maximum and increased once again to maximum depth with summer flows.

POND DEPTH OF AERATED POND SYSTEM

Aeration will be required when the water depth is above 6 ft. in order to maintain 2–3 mg/L DO. Normal design water depth is 10–15 ft. This depth limitation may be altered depending on the aeration equipment, waste strength and climatic conditions.

POND CONSTRUCTION REGARDING EMBANKMENT AND DIKE MATERIAL COMPACTION

Embankment and dike cores shall be compacted to at least 95% Standard Proctor Density to form a stable structure. Vegetation and other unsuitable materials should be removed from the area upon which the embankment is to be placed. Trees, boulders, rocks, etc. shall not be disposed of by burial on the pond site or in the pond dikes.

POND CONSTRUCTION REGARDING DIKE TOP WIDTH AND SURFACE MATERIAL

The minimum dike width is 8 ft. to permit access of maintenance vehicles. The surface must be compact and able to support vehicle traffic around the perimeter of the pond(s). Acceptable material includes Class IV or a combination of crushed concrete and Class I. Material used for top-of-dike construction must be comparable to Class IV and must be approved by the MPCA engineer.

POND CONSTRUCTION REGARDING MAXIMUM AND MINIMUM DIKE SLOPE

Inner slopes shall be no steeper than a 3:1 slope (run to rise). Mowing and other maintenance may not be possible on slopes greater than this and may be difficult on 3:1 slopes. Conversely, inner slopes shall be no flatter than a 4:1 (run to rise) slope. Slopes flatter than 4:1 are not allowed as the greater shallow water area is conductive to emergent vegetation. Outer dikes and maintenance roads need to be designed and constructed to preclude surface runoff from entering the ponds.

POND CONSTRUCTION REGARDING FREEBOARD

Minimum freeboard between the maximum water level and the top of the dike is 3 ft. measured vertically (as opposed to slope length).

HYDRAULIC CAPACITY OF NON-AERATED POND SYSTEM

The hydraulic capacity for controlled discharge systems shall permit transfer or discharge of water at a minimum rate of 6 inches of pond water depth per day at the minimum head. For calculation purposes, minimum head over the 2 ft. riser may be assumed to be 1 ft.

HYDRAULIC CAPACITY OF AERATED POND SYSTEM

The Hydraulic capacity for constant discharge structures and piping shall allow for a minimum of the peak instantaneous wet weather flow of the system.

Example 7.8: Analysis of an Existing WSP

There is an existing WSP system with the following given information of its facility, flow and influent 5-day BOD as follows: (a) primary pond area A=750 ft. long×407 ft. wide=305,250 ft.^{2} (b) secondary pond area A=750 ft. long×203 ft. wide=152,250 ft.^{2}; (c) maximum depth d=6 ft; (d) minimum depth d=2 ft; (e) wastewater flow Q=76,000 gpd; and (f) influent 5-day BOD concentration $C_{i=}$220 mg/L.

Determine (a) the area, in acres, of the primary and secondary ponds.; (b) the storage volume, in gallons and acre-feet of the primary and secondary ponds; (c) the total DT, in days, between the maximum and minimum depths; (d) the BOD loading, in pounds per day on the primary pond; and (e) the total primary pond organic loading, in pounds of BOD per acre per day.

Solution:

1. Primary pond area=305,250 ft.^{2}=(305,250 ft.^{2})×(1 acre/43,560 ft.^{2})=7 acres.
2. Secondary pond area=152,250 ft.^{2}=(152,250 ft.^{2})×(1 acre/43.560 ft.^{2})=3.5 acres
3. Effective depth=6 ft.−2 ft.=4 ft.
4. Primary pond volume
   = (7 acres)×(4 ft.)
   = (28 acre-ft.) (326,000 gal/1 acre-ft.)
   = 9,128,000 gal.
5. Secondary pond volume
   = (3.5 acres)×(4 ft.)
   = (14 acre-ft.) (326,000 gal/1 acre-ft.)
   = 4,564,000 gal.
6. Total detention time t
   = (total volume V)/(flow rate Q)
   = (9,128,000 gal+4,564,000 gal)/(76,000 gpd)
   = 180 days
7. Primary pond organic loading L
   = (220 mg/L)×[(8.34 lb/MG)/(1 mg/L)]×(0.076 MGD)
   = 139.44 lb/day of 5-day BOD

Example 7.9: Upgrading a WSP System Using Activated Sludge Process and Dissolved Air Flotation for a Poultry Processing Plant

WSPs are the most common process of treating poultry wastes at private waste-treatment installations in the United States as well as many developing countries due to the availability of low-cost land when most of the poultry-processing facilities were constructed a long time ago in rural areas. Table 7.3 shows the average characteristics of poultry processing wastewater. Successful operation of these WSPs depends to a high degree on favorable climatic conditions: warm, clear, and sunny. Large storage ponds keep the wastewater in winter for warm weather operation in warm seasons. Table 7.4 indicates that a properly designed aerobic lagoon can be expected to achieve about a 90% reduction in BOD and TSS. When anaerobic lagoons are used, expected BOD and

TABLE 7.3
Typical Poultry Wastewater Characteristics

Analysis	Unit	Range	Average
pH	unit	6.3–7.4	6.9
DO	ppm	0–2.0	0.5
BOD	ppm	370–620	473
TSS	ppm	120–296	196
Total solids	ppm	—	650
Volatile solids	ppm	—	486
Fixed solids	ppm	—	164
Settleable solids	mL/L	15–20	17.5
Oil and grease	ppm	170–230	201

Source: USEPA.

TABLE 7.4
Poultry Processing Wastewater Treatment Costs

Type of Treatment	Capital Costs ($1,000)	Level of Treatment, % BOD Reduction)	Operation and Maintenance Cost (Cents/1,000 Gal.)	Land Requirements (Acres)	Total Treatment Costs (Cents/1000 Gal.)	Total Treatment Cost (Cents/Bird)
Conventional activated sludge	825	90+	10.63	3.0	15	0.15
High-rate activated sludge	660	80+	6.81	2.0	11	0.11
Extended aeration	1,569	95+	13.61	3.0	25	0.25
Contact stabilization	473	90+	7.44	2.5	9.4	0.094
Standard rate trickling filter	750	90+	6.00	8.5	12	0.12
High rate trickling filter	555	90+	12.27	5.5	13	0.13
Aerobic lagoons	740	90+	NEG.	74	8	0.08
Anaerobic lagoons	300	70–80+	NEG.	15	3.3	0.033
Anaerobic/ Aerobic lagoons	450	90+	NEG.	30	4.9	0.049

Note: Based on a flow of 1.0 MGD having a BOD load of 450 ppm. Values given are calculated by the present worth method based on an interest rate of 5% and an expected facility life of 25 years. Total treatment costs do not reflect land costs.

TSS removals are only about 70%–80%. An existing poultry processing plant in Florida, USA is using a WAP system consisting of (a) preliminary treatment of poultry processing waste by mechanical screens to remove offal and feathers; (b) primary treatment of preliminary effluent by anaerobic lagoon (i.e. equivalent to an anaerobic septic tank); and (c) secondary treatment of anaerobic lagoon effluent by an aerobic lagoon. Odor problems are frequently associated with anaerobic lagoons, although chemical additives can usually control the problem. Flies, insects, and excessive growth of bordering vegetation have been a problem. Figure 7.7 shows one of the possible solutions recommended by the US Environmental Protection Agency (USEPA) [72] for this Florida poultry processing plant. Please offer your explanations using Table 7.4, Figure 7.7, and any other available information in the literature.

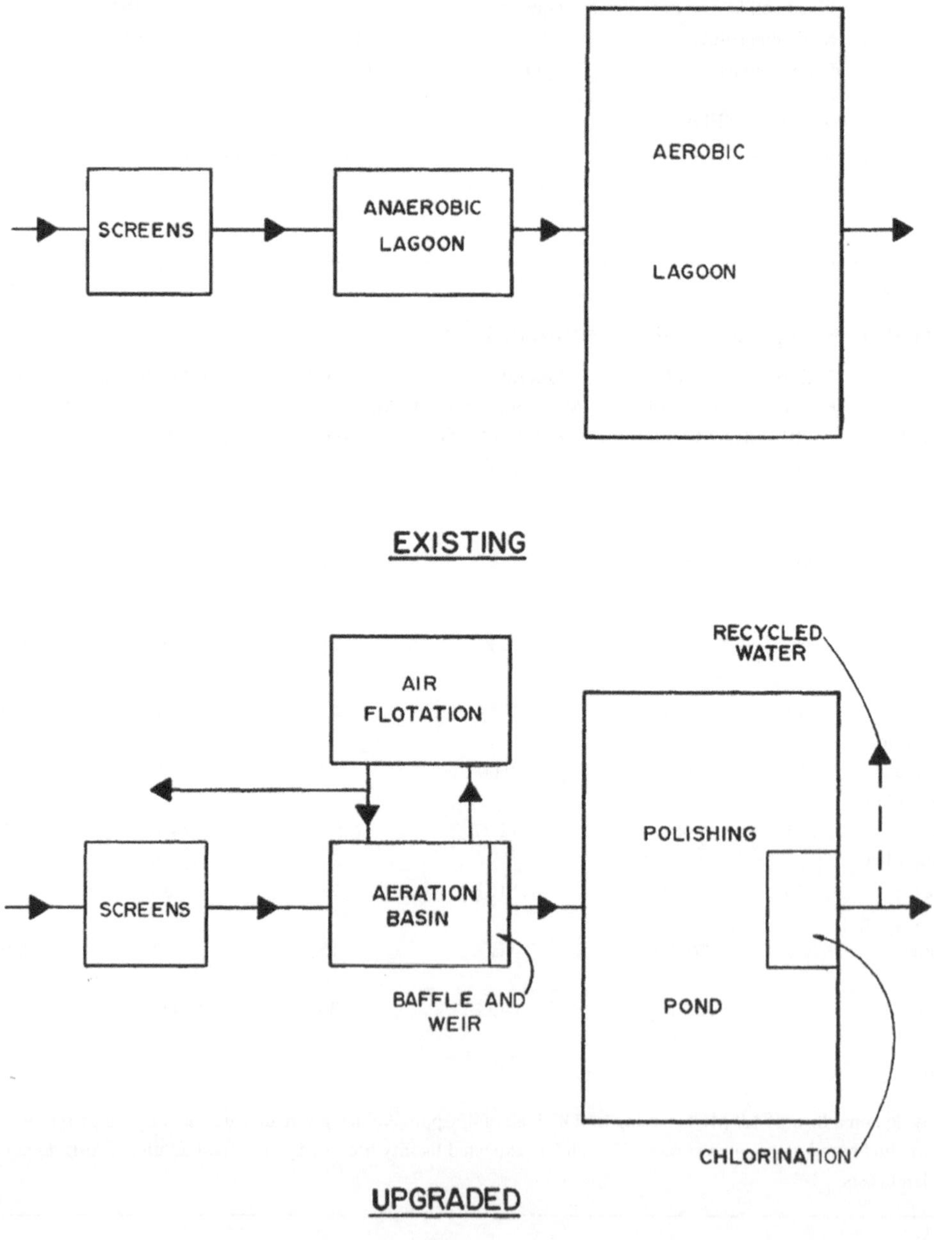

FIGURE 7.7 A lagoon system upgrade recommended by the US Environmental Protection Agency. (Source: USEPA.)

Solution:

The costs for waste treatment are difficult to project due, in large part, to their dependency upon local conditions. Some of the local conditions include design codes and climatic and geological considerations. Table 7.3 provides a rough approximation of the costs of waste treatment as applied to the poultry-processing industry. The USEPA's Table 7.3 [72] was constructed by adjusting the average costs of similar municipal waste-treatment facilities to fit poultry-processing requirements. The treatment costs presented do not reflect land-acquisition costs. The values tabulated are 1967 prices and are based on the present-value method of calculating costs. The selected interest rate is 5% and the expected life of the structure is 25 years. Considering that the actual useful life of the facility probably will be 40–50 years and that the costs are based on equivalent municipal facilities, this approach can be considered conservative Appendix 7.3 is a Cost Index table by which the readers may convert the 1967 costs to the 2022 costs for the purpose of comparison. It is important to note that the cost data in Table 7.3 do not reflect land costs. The total treatment costs per bird of the existing WAP system (anaerobic lagoon + aerobic lagoon, shown at the top of Figure 7.7) is almost the cheapest in comparison with any conventional biological processes (such as conventional activated sludge, high-rate activated sludge, extended aeration, contact stabilization, and trickling filter), although the land requirement of WSP biotechnology is extremely high.

The USEPA recommended solution is shown in Figure 7.7's bottom which fully utilizes the existing wastewater treatment facilities: (a) the existing screens of preliminary treatment are kept; (b) the existing anaerobic lagoon is converted to an aerated lagoon (equivalent to an aeration basin of the extended aeration process) plus dissolved air flotation (DAF) as the secondary treatment; and (c) the existing aerobic lagoon is converted to a tertiary polishing pond with chlorination disinfection. The effluent of an upgraded wastewater treatment system (shown at the bottom of Figure 7.7) will solve the odor problem and meet the NPDES secondary discharge limitations on BOD and TSS. Although the overall wastewater treatment costs after upgrading are higher, there are many advantages: (a) the old aerobic lagoon for secondary treatment could not consistently meet the NPDES effluent limitations on BOD and TSS, but now the converted tertiary polishing pond can; (b) the added chlorination disinfection is required by the NPDES effluent discharge standards; (c) the old anaerobic lagoon's odor problem is solved; (d) the high-quality polishing pond effluent can be recycled for poultry cleaning operations representing some cost saving; (e) the recommended DAF secondary flotation clarification is much more cost-effective than conventional sedimentation clarification, again, representing some cost saving; (f) DAF is the best available technology (BAT) for algae removal [60,73], so the algae problem is solved.

7.7 DESIGN OF AN ANAEROBIC LAGOON FOR TREATING A POULTRY PROCESSING WASTEWATER CONTAINING HIGH STRENGTH OF SOLUBLE ORGANIC MATTERS

There are two types of anaerobic lagoons or ponds: (a) anaerobic wastewater lagoons (AWL) for wastewater treatment; and (b) anaerobic sludge lagoon (ASL) for sludge dewatering and stabilization. Normally an anaerobic wastewater lagoon is simply called "anaerobic lagoon".

An AWL treats high-strength wastewater by anaerobic microorganisms under anaerobic conditions in the absence of oxygen. AWL must be followed by an aerated, aerobic, or a facultative pond; the reason is that the effluent of AWL alone will not meet any effluent discharge limitations on BOD, TSS, etc. AWL, however, is a good pre-treatment of non-toxic industrial wastewater (such as food processing wastewater) with high organic concentrations.

Anaerobic lagoon (i.e. AWL) function without DO throughout any of its depth. Normally it has a depth of 8–20 ft., and a typical detention time (DT) of 20–50 days. A deep sludge blanket covers the bottom of these lagoons, while a scum layer covers the surface. The scum layer is important to the lagoon operation because: (a) it helps minimize offensive odors; (b) it blocks transfer of DO from air to the lagoon through surface contact; and (c) it helps insulate the pond or lagoon to ensure ideal conditions for waste solids fermentation.

Different states in the USA [61,62,74], or different countries [59,75,76] will have different design criteria for designing an anaerobic lagoon, or any lagoon. Assume you are designing an anaerobic lagoon for treatment of a high-strength poultry processing wastewater in the State of Pennsylvania (PA), introduce the PA design criteria for anaerobic lagoon and design one, if the poultry processing plant's wastewater flow is 0.076 MGD.

Solution:

1. The design criteria of anaerobic lagoons (anaerobic wastewater lagoons) are found in the literature [74]: (a) lagoon DT=20–50 days; (b) lagoon size=0.5–2 acres; (c) lagoon depth=8–16 ft; (d) lagoon liquid pH=6.5–7.2 units; (e) lagoon liquid temperature=6–50°C; (f) optimum temperature range=30°C; (g) 5-day BOD loading=200–500 lb/acre/day.
2. A typical poultry processing wastewater characteristics can be found from the literature [72]: (a) total solids (TS)=650–910 mg/L; (b) total suspended solids (TSS)=120–300 mg/L (average 200 mg/L); (c) 5-day biochemical oxygen demand (BOD_5)=370–620 mg/L (average 473 mg/L); (d) pH=6.3–7.4 units (average 6.9 units); (e) DO=0–2 mg/L (average 0.5 mg/L); (f) volatile solids (VS)=average 486 mg/L; (g) fixed solids (FS)=average 164 mg/L; (h) settleable solids=15–20 mL/L (average 17.5 mL/L); (i) oil and grease (O&G)=170–230 mg/L (average 201 mg/L).
3. It is given that the plant flow Q=0.076 MGD. The highest 5-day BOD concentration is chosen for design=620 mg/L. Percent BOD reduction by anaerobic lagoon=70%–80% according to the USEPA [72]. The lowest percent BOD reduction chosen for design=70%. The expected anaerobic lagoon effluent BOD=620 mg/L×(100−70)/100=186 mg/L. Select 5-day BOD loading=280 lb/acre/day for design from the PA requirement.
4. The organic loading to the anaerobic lagoon system
   = (620 mg/L)×[(8.34 lb/MG)/(1 mg/L)]×(0.076 MGD)
   = 393 lb/day of 5-day BOD
5. Since the selected 5-day BOD loading=280 lb/acre/day for design (from the PA requirements), the required anaerobic lagoon area=(393 lb/day)/(280 lb/acre/day)=1.4 acre, which is in the range of PA recommended lagoon size of 0.5–2 acres.
6. The final design of an anaerobic lagoon system is summarized as follows:
   - 6a. Anaerobic lagoon design flow=0.076 MGD
   - 6b. The highest 5-day BOD concentration is chosen for design=620 mg/L.
   - 6c. Percent BOD reduction by anaerobic lagoon=70%–80% according to the USEPA [72]. The lowest percent BOD reduction chosen for design=70%.
   - 6d. The expected anaerobic lagoon effluent BOD=620 mg/L×(100−70)/100=186 mg/L.
   - 6e. Anaerobic lagoon design BOD loading rate=250 lb/acre/day for design (PA requirements=200–500 lb/acre/day).
   - 6f. Anaerobic lagoon area=1.4 acre (PA recommended lagoon size=0.5–2 acres), Okay.
   - 6g. Anaerobic lagoon depth selected=8 ft. PA recommended lagoon depth=8–16 ft.) Okay
   - 6h. Anaerobic lagoon volume=(1.4 acre)×(8 ft.)=11.2 acre-ft., since 1 acre-ft.=325,850 gallon=0.325850 MG, lagoon volume=(11.2 acre-ft.)(0.325850 MG/1 acre-ft.)=3.65 MG
   - 6i. DT=(3.65 MG)/(0.076 MGD)=48 days (PA recommended lagoon DT=20–50 days), so okay. Slightly over 50 days is okay, but below 20 days is not okay.
   - 6j. Since anaerobic lagoon is only a pre-treatment for treating the target poultry processing wastewater, it can not be used alone and must be followed by a facultative pond, an aerated facultative pond (aerated partial-mix pond), an aerated complete-mix pong, or a non-aerated aerobic pond.

The top portion of Figure 7.7 shows the minimum WSP system required for treatment of the poultry processing wastewater: (a) screens for preliminary treatment; (b) anaerobic lagoon for primary treatment; and (c) aerobic lagoon for secondary treatment.

6k. Since it is assumed that the percent BOD reduction by anaerobic lagoon = 70%, the expected anaerobic lagoon effluent BOD = 620 mg/L × (100 − 70)/100 = 186 mg/L.

Example 7.8" shows that a primary lagoon (A = 7 acres; effective depth = 4 ft.; V = 9,128,000 gallons) and a secondary lagoon (A = 3.5 acres; effective depth − 4 ft.; V = 4,564,000 gallons) with a combined total detention of 180 days will be able to treat 0.076 MGD of an influent with 220 mg/L. (Note: 220 mg/L is higher than 186 mg/L; both wastewater flow rates are 0.076 MGD).

Therefore a better WSP system recommended for treating the target poultry wastewater will be: (a) screens for preliminary treatment; (b) anaerobic lagoon for primary treatment; (c) a first aerobic or facultative lagoon for secondary treatment; and (d) a second aerobic or facultative lagoon for final polishing or tertiary treatment. The final lagoon is also called the tertiary lagoon or polishing lagoon.

7.8 GLOSSARY OF WSPS AND LAGOONS [46,47,59,78]

Aerated Aerobic Lagoon or Pond: An aerated aerobic lagoons lagoon (completely mixed; 3–20 days hydraulic DT; 6–20 ft. depth; 20–400 lb/acre/day BOD_5 loading; 800 mg/L MLVSS) is normally designed to achieve complex mixing as an activated sludge reactor. The lagoons may be lined with conditions and environmental regulations. It uses various types of aeration, devices to achieve completely mixed (all aerobic) conditions, and requires a final settling tank. In very cold climates aerated lagoons may experience reduced biological activity and treatment efficiency, and the formation of ice.

Aerated Complete-Mix Lagoon (ACML): It is also called "aerated aerobic lagoon or pond". An ACML is similar to the complete-mix activated sludge process except that ACML does not recycle cellular material, resulting in lower mixed liquor suspended solids (MLSS). Accordingly, ACML requires a longer hydraulic DT than complete-mix activated sludge process.

Aerated Facultative Lagoon or Pond: Aerated facultative lagoons (surface layer mixed with aerators; 7–20 days hydraulic DT; 3–8 ft. depth; surface aerators as oxygen source; 20–70 mg/L effluent BOD_5; sedimentation clarifier or dissolved air flotation clarifier is needed for TSS removal) are normally designed to achieve partial mixing only, aerobic-anaerobic stratification will occur, and a large fraction of the incoming solids and a large fraction of the biological solids produced from waste conversion settle to the bottom of the lagoon cells. As the solids begin to build up, a portion will undergo anaerobic decomposition. Volatile toxics can potentially be removed by the aeration process, and incidental removal of other toxics can be expected to be similar to an activated sludge system. Several smaller aerated lagoon cells in series are more effective than one large cell. Tapering aeration intensity downward in the direction of flow promotes settling out of solids in the last cell. A non-aerated polishing cell following the last aerated cell is an optional, but recommended, design technique to enhance suspended solids removal prior to discharge. Oxygen in the surface stabilization zone of an aerated facultative lagoon is provided by mechanical surface aeration. This is in contrast to a non-aerated facultative lagoon in which algae is the source of oxygenation for photosynthesis in order to create aerobic surface conditions. In general, the aerobic surface layer serves to reduce odors while providing treatment of soluble organic by-products of the anaerobic processes operating at the bottom. The use of supplemental top layer surface aeration can improve overall treatment capacity, particularly in northern climates where icing over of facultative lagoons is common in the winter. Aerated facultative lagoon or pond is also called "aerated partial-mix lagoon".

Aerated Partial-Mix Lagoon: It is also called "aerated facultative lagoon".

Aerated Lagoons or Ponds: There are two types of aerated lagoons: (a) aerated aerobic lagoon (completely mixed; 3–20 days hydraulic DT; 6–20 ft depth) which is also called "aerated complete-mix lagoon"; and (b) aerated facultative lagoon (surface layer mixed with aerators; 7–20 days hydraulic DT; 3–8 ft. depth), which is also called "aerated partial-mix lagoon". They are all medium-depth basins designed for the biological treatment of wastewater on a continuous basis. In contrast to non-aerated stabilization ponds, which obtain oxygen from photosynthesis and surface reaeration, the aerated lagoons employ aeration devices that supply supplemental oxygen to the system. The aeration devices may be mechanical (i.e., surface aerator), or diffused air systems. Surface aerators are divided into two types: cage aerators and the more common turbine and vertical shaft aerators. Many diffused air systems utilized in lagoons consist of plastic pipes supported near the bottom of the cells with regularly spaced sparger holes drilled in the tops of the pipes. They are used for domestic and industrial wastewater of low and medium strength. Commonly used where land is inexpensive and costs and operational control are to be minimized. It is relatively simple to upgrade existing oxidation ponds, lagoons, and natural bodies of water to this type of treatment.

Aerated Pond System (MN, USA): It is similar to a stabilization pond system, except the DO is supplied by mechanical aeration. DTs will vary with the design of the overall system. It is constructed to provide either a continuous discharge or controlled discharge.

Aerobic-Anaerobic Ponds: It is the same as a facultative pond, or facultative lagoon, which can be either aerated or non-aerated.

Aerobic Ponds, Non-aerated: They are also called "high rate aerobic ponds" which maintain DO throughout their entire depth. They are usually installed with design criteria of: 10 acre size in multiples series or parallel; 3–4 ft. depth; 10–40 days hydraulic DT; 3–5 in/day hydraulic loading; 60–120 lb/acre/day 5-day BOD loading; and 80–200 mg/L algae, allowing light to penetrate the full depth for photosynthesis. Mixing is often provided to expose all algae to sunlight and to prevent deposition and subsequent anaerobic conditions. Oxygen is provided by photosynthesis and surface reaeration, and aerobic bacteria stabilize the waste. DT is short, 3–5 days being usual.

Algae: Algae are microscopic plants or phytoplankton that live either floating or suspending in water or being attached to a submerged surface. Algae contain chlorophyll which has the capability of using sunlight energy to degrade and convert carbon-based organic matter into simpler compounds for growth by the so-called photosynthesis process. DO is produced as a by-product of photosynthesis during active sunlight hours. A portion of DO is used by algae for respiration, however, during night hours. There are two types of algae: (a) good green algae which is indicative of a healthy pond; and (b) bad blue-green algae which is indicative of poor pond condition.

Anaerobic Lagoons or Ponds: There are two types of anaerobic lagoons or ponds: (a) anaerobic wastewater lagoons (AWL) for wastewater treatment; and (b) anaerobic sludge lagoon (ASL) for sludge dewatering and stabilization. They are relatively deep (up to 20 ft.) ponds with steep sidewalls in which anaerobic conditions are maintained by keeping loading so high that complete deoxygenation is prevalent. Although some oxygenation is possible in a shallow surface zone, once greases form an impervious surface layer, complete anaerobic conditions develop. Treatment or stabilization results from thermophilic anaerobic digestion of organic wastes. The treatment process is analogous to that occurring in single-stage untreated anaerobic digestion of sludge in which acid-forming bacteria break down organics. The resultant acids are then converted to carbon dioxide, methane, cells, and other end products. In the typical anaerobic lagoon, raw wastewater enters near the bottom of the pond (often at the center) and mixes with the active microbial mass in the sludge blanket, which is usually about 6 ft. deep. The discharge is located near one of the sides of the pond, submerged below the liquid surface. Excess undigested grease floats to the top, forming a heat retaining and relatively air-tight cover. Wastewater flow equalization and heating are generally not practiced. Excess sludge is washed out with the effluent. Recirculation of waste sludge is not required. Anaerobic lagoons are capable of providing treatment of high-strength wastewaters and sludge and are resistant to shock loads. They are customarily contained within earthen dikes. Depending on soil characteristics, lining with various impervious materials such as rubber, plastic or clay may be necessary. Pond geometry may vary, but surface area to volume ratios are minimized to enhance heat retention. Although anaerobic processes are common for sludge digestion, anaerobic lagoons for wastewater treatment have found some agricultural applications. The process is well demonstrated for the stabilization of highly concentrated organic wastes such as manure. Typically anaerobic lagoons are used in series with aerobic or facultative lagoons. They are effective as roughing units prior to aerobic treatment of high-strength wastes. For efficient operation, water temperatures above 75°F should be maintained. BOD removals of 50%–70% are achievable depending on loading and temperature conditions. TSS concentrations may increase, especially if the influent BOD is primarily dissolved. Generally does not produce an effluent suitable for direct discharge to receiving waters. In anaerobic lagoons excess sludge is usually washed out in the effluent. Since anaerobic lagoons are often used for preliminary treatment recirculation, removal of sludge is not generally required.

Anaerobic Wastewater Lagoon (AWL) or Pond: An AWL treats high-strength wastewater by anaerobic microorganisms under anaerobic conditions in the absence of oxygen. AWL must be followed by an aerated, aerobic, or a facultative pond; the reason is that the effluent of AWL alone will not meet any effluent discharge limitations on BOD or TSS. AWL, however, is a good pre-treatment of industrial wastewater with high organic concentrations.

Blue-Green Algae: The presence of blue-green filamentous algae in pond water is an indication of poor pond conditions. They are associated with low pH and low nutrient levels in water. They compose most algal blooms and cause high total suspended solids in the pond effluent.

Controlled Discharge: When any type of WSP is operated under the "controlled discharge" mode, the pond discharge is prohibited during the winter months in cold climates and/or during peak algal growth periods in the summer. In this approach, each WSP cell in the system is isolated and then discharged sequentially when the weather condition is appropriate.

Controlled Discharge Pond: It is a form of WSP. Controlled discharge ponds have long hydraulic DTs and the effluent is discharged once or twice per year when the effluent quality is satisfactory.

Evaporation Lagoon or Pond: It is one type of total containment ponds which are applicable in climates where the evaporative losses exceed the rainfall. The evaporation lagoon may be described as an open holding facility which depends solely on climatic conditions such as evaporation, precipitation, temperature, humidity, and wind velocity to affect dissipation (evaporation) of on-site wastewater. Individual lagoons may be considered as an alternate means of wastewater disposal on individual pieces of property. The basic impetus to consider this system is to allow building and other land uses on properties which have soil conditions not conducive to the workability and acceptability of the conventional on-site drain field or leaching bed disposal systems. Generally, if the annual evaporation rate exceeds the annual precipitation, this method of disposal may at least be considered. The deciding factor then becomes the required land area and holding volume. It should be noted that for unlined on-site installations such as homes and small industrial applications, there may also be a certain amount of infiltration or percolation in the initial period of operation. However, after a time, it may be expected that solids deposition will eventually clog the surface to the point where infiltration is eliminated. The potential impact of wastewater infiltration to the groundwater, and particularly on-site water supplies, should be evaluated in any event and, if necessary, lagoon lining may be utilized to alleviate the problem. Evaporation lagoons are often preceded by septic tanks or aerobic units in order to provide a more acceptable influent to and minimize sludge removal from the lagoon. The on-site utilization of evaporation lagoons for the disposal of domestic wastewater, from homes, farms, and smaller agricultural, industrial or commercial facilities may be applicable where access to a municipal sanitary sewer is not available; where subsurface methods are not feasible and where effluent polishing for surface discharge is not practical.

Facultative Lagoons or Ponds: There are two types of facultative lagoons or ponds: (a) non-aerated facultative lagoons (7–30 days hydraulic DT; 3–6 ft. depth; 40–160 mg/L algae as oxygen source; 15–40 mg/L effluent BOD_5; 25–50 mg/L effluent TSS); and (b) aerated facultative lagoons (7–20 days hydraulic DT; 3–8 ft. depth; surface aerators as oxygen source; 20–70 mg/L effluent BOD_5; sedimentation clarifier or dissolved air flotation clarifier is needed for TSS removal). Facultative lagoons are customarily contained within earthen dikes. Depending on soil characteristics, lining with various impervious materials such as rubber, plastic or clay may be necessary. They are intermediate depth (3–8 ft.) ponds in which the wastewater is stratified into three zones. These zones consist of an anaerobic bottom layer, an aerobic surface layer, and an intermediate zone. Stratification is a result of solids settling and temperature-water density variations. Sludge at the bottom of facultative lagoons will undergo anaerobic digestion producing carbon dioxide, methane, and cells. Facultative lagoons are often and for optimum performance should be operated in series. When three or more cells are linked, the effluent from either the second or third cell may be recirculated to the first. Recirculation rates of 0.5–2.0 times the plant flow have been used to improve overall performance. They are fully demonstrated and in use especially for the treatment of relatively weak municipal wastewater in areas where real estate costs are not a restricting factor and when land costs are low and operation and maintenance costs are to be minimized. In very cold climates, facultative lagoons may experience reduced biological activity and treatment efficiency. Since ice formation can also hamper operations, extra long holding time is needed in order to discharge the lagoon effluent in warm weather.

Green Algae: Green algae, such as Chlorella, are mobile near the water surface. The presence of green algae is an indication of a healthy WSP. Green algae are associated with high pH and nutritional value.

High-Rate Aerobic Ponds: They are one of "aerobic ponds". Specifically, it should be technically named "partially-aerated high-rate aerobic ponds", which maintains DO throughout their entire shallow depth, usually between 3 and 4 ft. deep, allowing light to penetrate the full depth. The design criteria are: (a) 5-day BOD organic loading $L = 60–200$ mg/L; (b) hydraulic DT $t = 3–6$ days; (c) pond depth $d = 2–4$ ft.; (d) DO concentration = aerobic throughout entire shallow pond depth; (e) minimum algae concentration in pond water = 80–200 mg/L; (f) mixing condition = allowing to expose all algae (80–200 mg/L algae) to sunlight and to prevent deposition and subsequent anaerobic conditions; and (g) high-rate aerobic ponds are limited to warm, sunny climates. Oxygen is provided by both algal photosynthesis and surface reaeration with very limited mechanical aeration, and aerobic bacteria stabilize the waste. It is suitable for removal of soluble organic and nutrients. The pond's effluent 5-day BOD is expected to be excellent (20–30 mg/L). However, operation is somewhat more complex than for a facultative pond and, unless an algae removal step is provided, the effluent will contain high TSS (150–350 mg/L). Because of their shallow depths, paving or covering the bottom is required to prevent weed growth.

Hydrograph Controlled Release (HCR): HCR is a special operational mode of WSPs. The liquid in the WSP is retained until flow volume and conditions in the receiving water are adequate for the WSP effluent discharge.

Infiltration Pond or Lagoon: Infiltration ponds or lagoons are a series of shallow ponds or lagoons about 0.5–1.5 ft. deep used to dispose of pre-treated wastewater. Operations normally consist of periodic controlled amounts to alternate shallow ponds with subsequent drying periods. Any remaining pollutant removal occurs by filtration, biological activity, and absorption as the effluent percolates through the soil. Depending on the location of the groundwater, infiltration ponds may be designed with an underlying network of drains to collect and return the water back to the head works of the treatment system.

Intermittent Sand Filtration: The intermittent sand filter is an outdoor, gravity, filtration system that capitalizes on the availability of land area. It is a biological and physical wastewater treatment mechanism consisting of an underdrained bed of granular material, usually sand. The filter surface is flooded intermittently with lagoon effluent at intervals which permit the surface to drain between applications. It is recommended that the flow be directed to one filter for 24 hours. That filter is then allowed to drain and dry for 1–2 days, and the flow goes to an adjacent filter. It is preferable to have three filter beds where good operation and treatment may be accomplished over a 3-day cycle. The filter contains drainage pipes that are laid with open joints at depths of 3–4 ft. and surrounded with layers of coarse stone and gravel graded from coarse to fine to keep sand out. When the filter is located in natural sand deposits, the percolating waters may reach the groundwater table, and no effluent may come to view. In areas where percolation to the groundwater is not permitted, the filter must be provided with an impermeable base or lining. Influent wastewater is piped to the beds for discharge into a protective stone or concrete apron or into a concrete flume distributor. Surface accumulations of solids are periodically removed and disposed of as fill or by some other means. Filters must be re-sanded when they become too shallow as a result of the periodic surface cleaning. In cold weather, the beds are plowed into ridges and furrows (1–1.5 ft. deep) to keep them from freezing and opening up cracks through which the applied wastewater can escape with little treatment. To form protective sheets of ice spanning the furrows and keep the beds warm, furrowed beds are dosed deeply on cold nights. Lagoon effluent may be applied to filter beds of progressively smaller size of filter media; e.g., #1 sand e.s. = 0.72 mm, #2 sand e.s. = 0.40 mm, #3 sand e.s. = 0.17 mm, and gravel support. The technology has been practiced with proven performance for many years where land is available. The filter is used successfully for (a) polishing of domestic wastewater facultative stabilization pond effluent; (b) final treatment of bio-degradable industrial wastewater stabilization pond effluent where minimum maintenance is required. It is unsuitable for anaerobic lagoon effluents, and but is suitable for aerated lagoons if aerated lagoons are followed by facultative lagoons. The filter effluent quality (treatment of domestic wastewater stabilization pond) is expected to be excellent: less than 10 mg/L filter effluent BOD_5 and less than 10 mg/L filter effluent TSS.

Non-aerated Facultative Lagoons: Oxygen in the surface stabilization zone of a non-aerated facultative lagoon (7–30 days hydraulic DT; 3–6 ft. depth; 40–160 mg/L algae as oxygen source; 15–40 mg/L effluent BOD_5; 25–50 mg/L effluent TSS) is provided by reaeration and photosynthesis. The photosynthetic activity at the lagoon surface produces oxygen diurnally, increasing the DO during daylight hours, while surface oxygen is depleted at night. This is in contrast to aerated facultative lagoons in which mechanical surface aeration is used to create aerobic surface conditions.

Oxidation Pond: It is the same as a facultative pond, or facultative lagoon.

Photosynthetic Pond: It is same as a facultative pond, or facultative lagoon.

Polishing Pond System (MN, USA): A polishing pond system is an aerated or non-aerated pond system that follows a secondary treatment system and is considered a form of tertiary treatment. Its DTs vary between three and 210 days, depending on their purpose. A polishing pond typically provides storage prior to a discharge or land application. Its secondary purposes are to: (a) further reduce the Biochemical Oxygen Demand (BOD) and Total

Suspended Solids (TSS) before discharge; (b) possibly remove phosphorus and some coliform bacteria. It is also called tertiary pond.

Pre-aeration Pond System (MN, USA): It is an aerated pond system designed to reduce the influent Biochemical Oxygen Demand (BOD) loading prior to a discharge to a secondary treatment system.

Rock Filtration: The rock filter is an alternative for the removal of algae from non-aerated WSP effluents. The non-aerated WSPs, either aerobic or facultative, rely on algae to provide oxygen source for photosynthesis bioreaction. The filter consists of a submerged bed of rocks (5–20 cm diameter) through which the lagoon effluent is passed vertically or horizontally, allowing the algae to become attached to the rock surface and thereby be removed. The basic simplicity of operation and maintenance are the key advantages of this process. The effluent quality achievable and the dependability of long-term operation, however, have not yet been proven. The lagoon effluent enters the bottom of the filter through an influent channel. The lagoon effluent then rises from the influent channel and moves horizontally toward the discharge weirs where it flows into a covered effluent channel. Finally, the flow from each side of the rock filter is combined, chlorinated, and discharged. The rock surface is approximately 0.30 m (1.0 ft.) above the water elevation to prevent the growth of algae on the rock filter. The rock filter can be used for polishing domestic wastewater stabilization pond effluent and also for the final treatment of biodegradable industrial wastewater stabilization pond effluent where low maintenance is required. There is a demonstrated case history of a full-scale rock filter which was designed and constructed as part of a lagoon expansion and upgrading project at Veneta, Oregon, USA, in operation since 1975. The rock filter system treats wastewater from a population of approximately 2,200 with no industrial wastewater contribution.

Sludge Drying Lagoon: Digested sludge has often been applied to sludge lagoons adjacent to or in the proximity of treatment facilities. These sludge lagoons are primarily designed to accomplish long-term drying of the digested sludge through the physical processes of percolation and evaporation, primarily the latter. This method of sludge processing has been extremely popular in the U.S. due to its relatively low cost (when inexpensive land is plentiful) and minimal O&M requirements, especially at smaller wastewater treatment facilities. The process is relatively simple, requiring periodic decanting of supernatant back to the head of the plant and occasional mechanical excavation of dewatered or dried sludge for transportation to its ultimate disposal location. Lagoons can be a very useful process step. Supernatant is far better (low TSS) than supernatant from a secondary digester or even a thickener, ultimate disposal of the product solids often is as a soil conditioner or for landfilling. Sludge lagoons may also be used as contingency units at treatment plants to store and/or process sludge when normal processing units are either overloaded or out of service. The drying time for 30% solids is generally quite lengthy and may require years. Climatic conditions and pre-lagoon sludge processing greatly influence lagoon performance. In warmer, drier climates well-digested sludge is economically and satisfactorily treated by sludge-drying lagoons because of their inherent simplicity of operation and flexibility. Complete freezing causes sludge to agglomerate so when it thaws supernatant decants or drains away easily. Well-digested sludge minimizes potential odor problems which are inherent in this type of system. Multiple cells are required for efficient operation.

Spray Irrigation Pond: A spray irrigation pond is identical to a stabilization pond, except for two differences: (a) Treated wastewater is discharged to agricultural land when a cover crop that is not used for human consumption is actively growing; (b) The minimum DT for a spray irrigation pond increases to at least 210 days to allow for winter storage.

Spray Irrigation Pond Treatment System (MN, USA): It is the same as a stabilization pond system however discharge is to the land with a cover crop not used for human consumption.

Stabilization Pond Systems (MN, USA): It is a controlled type of a facultative/aerobic pond system. The DO is generally supplied by algae and/or wind action. DT is from 180 to 210 days with a controlled discharge to avoid a release of water during the summer and winter months.

Tertiary Pond: It is also a polishing pond. It can be an aerated or non-aerated pond system that follows a secondary treatment system and is considered a form of tertiary treatment. Its DTs vary between three and 210 days, depending on their purpose. A polishing pond typically provides storage prior to a discharge or land application. Its secondary purposes are to: (a) further reduce the Biochemical Oxygen Demand (BOD) and Total Suspended Solids (TSS) before discharge; (b) possibly remove phosphorus and some coliform bacteria.

Total Containment Pond: The total containment pond can be a form of facultative pond or simply an evaporation lagoon. It is applicable in climates where the evaporative losses exceed the rainfall.

Waste Stabilization Ponds: The most basic classification involves the description of the dominant biological reactions or reactions that occur in the pond. Four principal types are: (a) facultative (aerobic-anaerobic) ponds; (b) aerated ponds: (b1) aerated aerobic ponds; (b2) aerated facultative ponds; (c) aerobic ponds; and (d) anaerobic ponds.

Waste Storage Lagoons: Agricultural waste storage lagoons store high-strength wastes from food processing operations such as vegetable canneries and beet sugar plants. The waste may have associated nuisance odors. Complete biological treatment usually is not accomplished in a lagoon, but may later be realized by an additional process such as land application or mechanical treatment. Generally, the main purpose of a storage lagoon is for the removal of settleable and floating solids.

ACKNOWLEDGEMENT AND DEDICATION

This chapter is an updated version of its first edition [1], which was originally written by Dr. Calvin P. C. Poon and the authors (Mu Hao Sung Wang, and Lawrence K. Wang), and an updated clear production technology transfer module of the United Nations Industrial Development Organization (UNIDO) publication [79].

This book chapter is dedicated to late Dr. Calvin P. C. Poon, who was the authors' Academic Advisor at the Department of Civil and Environmental Engineering, University of Rhode Island (URI), Kingston, RI, USA, and was the authors' lifetime mentor and dear friend..

NOMENCLATURE

- A surface area of pond, cm^2 or ft^2
- C concentration at any time
- Ci influent pollutant concentration, mg/L
- d depth of the pond, cm
- L organic load, lb/acres/day
- Q pond flow rate, MGD
- t detention time, d
- t time
- V volume, acre-ft

APPENDIX 7.1: ALLOWABLE BOD LOADING RATE FOR DESIGN OF AEROBIC LAGOON IN THE USA (SOURCE: USDA)

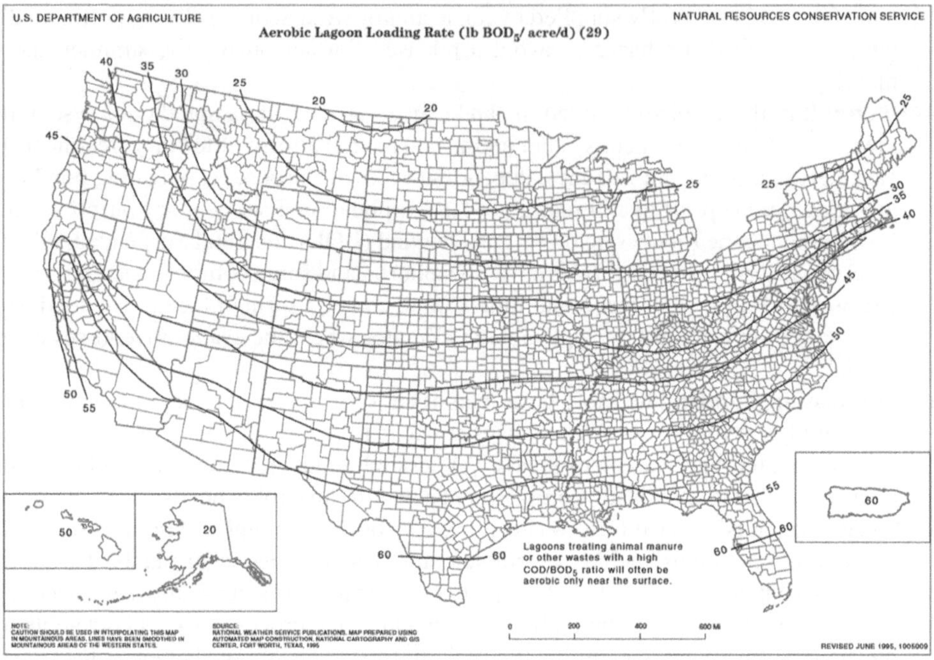

APPENDIX 7.2: ALLOWABLE VOLATILE SOLIDS LOADING RATE FOR DESIGN OF ANAEROBIC LAGOON IN THE USA (SOURCE: USDA)

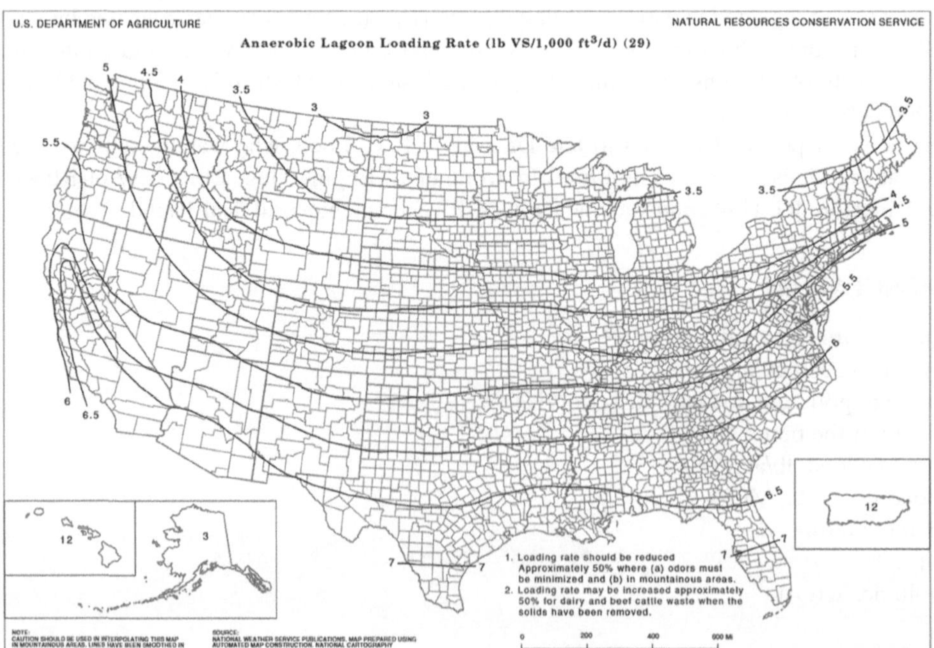

APPENDIX 7.3: UNITED STATES YEARLY AVERAGE COST INDEX FOR UTILITIES [29,64,65]

Year	Index	Year	Index
1967	100	1995	439.72
1968	104.83	1996	445.58
1969	112.17	1997	454.99
1970	119.75	1998	459.40
1971	131.73	1999	460.16
1972	141.94	2000	468.05
1973	149.36	2001	472.18
1974	170.45	2002	486.16
1975	190.49	2003	497.40
1976	202.61	2004	563.78
1977	215.84	2005	605.47
1978	235.78	2006	645.52
1979	257.20	2007	681.88
1980	277.60	2008	741.36
1981	302.25	2009	699.70
1982	320.13	2010	720.80
1983	330.82	2011	758.79
1984	341.06	2012	769.30
1985	346.12	2013	776.44
1986	347.33	2014	791.59
1987	353.35	2015	786.32
1988	369.45	2016	782.46
1989	383.14	2017	803.93
1990	386.75	2018	841.84
1991	392.35	2019	866.18
1992	399.07	2020	867.71
1993	410.63	2021	893.02[a]
1994	424.91	2022	918.91[a]

[a] Projected future cost index values.

REFERENCES

1. P. C. C. Poon, L. K. Wang and M. H. S. Wang, Waste stabilization ponds and lagoons, in *The Handbook of Environmental Engineering*, Volume 3. (Wang, L. K. and Pereria, N. C. editors). Humana Press, Clifton, NJ, pp. 305–360 (1986).
2. Federal Water Quality Administration (FWQA), *Federal Water Quality Administration Municipal Waste Facilities in the United States*, No. CWT-6. FWQA, Washington, DC (1970).
3. L. K. Wang and M.H.S. Wang, Standards and Guides of Wastewater Treatment and Sewer Systems, Technical Manual No. LIR/01-87/227, 314 PAGES, Lenox Institute, Massachusetts, USA (1987).
4. US Department of AgrIculture (USDA). Agrucultural Waste Management Field Handbook, USDA, Washington DC, USA (2009).
5. J. C. Goldman, *The Effect of Cadimum on Algal Growth: Its Relationship to Eutrophication*. Occasional Paper 6. Utah Water Research Lab., Logan, UT (1971).
6. P. C. Kerr, The Interrelation of Carbon and Phosphorus in Regulating Heterotrophic and Autotrophic Populations in Aquatic Ecosystems. FWQA Research Series 16050 FGS 07/70 (1970).
7. R. Oakley. Design and Performance of Earth Lined Containment Systems, ASCE Conference Geotechnology Practice Waste Disposal, pp. 117-136, American Society of Civil Engineers, New York, NY, USA (1987).

8. Oklahoma Department of Environmental Quality, Lagoons Oxidation Ponds, Oklahoma City OK, USA. https://www.deq.state.ok.us/factsheets/local/Lagoon.pdf (2005).
9. N. K. Shammas and L. K. Wang, *Water Engineering: Hydraulics, Distribution and Treatment.* John Wiley & Sons, Hoboken, NJ, 806 pages (2016).
10. C. Rohwer, *Evaporation from Free Water Surfaces.* Technical Bulletin 271, US Department of Agriculture, Washington DC (1931).
11 S. Davis, Dairy Waste Ponds Effectively Self-Scaling. Paper No. 72-222, *Annual Meeting of the American Society of Agricultural Engineers* (1972).
12. G. E. Hutchinson, *A Treatise on Limnology Geography,* Physics and Chemistry. John Wiley & Sons, New York, p. 1015 (1957). https://www.scirp.org/reference/referencespapers?referenceid=1671366
13. E. I. Rabinowitch, The photochemical storage of energy. *Transactions of the Conference on the Use of Solar Energy*, Tucson, Arizona, USA, Vol. 19, pp. 182–187 (1955).
14 L. D. Beadel, Sewage lagoons and mosquito problems. *Proceedings Symposium on Waste Stabilization Lagoons*, Kansas City, MO, US Dept. HEW, p. 41 (1960).
15. MAF General Description of Oxidation Pond Functions, Processes, and Performance, Ministry of Agriculture and Forestry, Wellington, New Zealand (2005). https://www.maf.govt.nz/mafnet/rural-nz/sustainable-resource-use/resource-management/dairy-shed-wastewater/dairyef4.htm
16. H. R. Fleckseder, Design Guides for Biological Wastewater Treatment Processes, Performance of the Aerated Lagoon Process. Tech Report GHE-70-22, CRWR-71, University of Texas, Austin, TX (1970).
17. N. K. Shammas and L. K. Wang, Principles and kinetics of biological processes, in *Advanced Biological Treatment Processes*, Wang, L. K., Shammas, N. K. and Hung, Y. T. (editors). The Humana Press, Totowa, NJ, pp. 1–58 (2009).
18. Metcalf and Eddy, *Wastewater Engineering Treatment and Reuse*, 4th Edition. McGraw Hill, New York (2003).
19. J. L. Mancini, *Advances in Water Quality Improvement.* The University of Texas, Austin, TX (1966).
20. E. L. Barnhart, The treatment of chemical wastes in aerated lagoons. *Chemical Engineering Progress Symposium Series No. 90*, New York, NY, Vol. 64 (1968).
21. L. K. Wang and M. H. S. Wang, Dissolved oxygen model, vapor pressure, evaporation lagoon, and flotation post aeration, in *Industrial Water and Waste Treatment: Technology Developments*, Wang, L. K., Wang, M. H. S and Hung, Y. T. (editors), CRC Press, Boca Rotan, FL (2025).
22. G. V. Levin, Froth flotation for harvesting algae and its possible application to sewage treatment. *9th Industrial Waste Conference, Purdue University, Part 1*, pp. 421–434 (1964). Lafayette, IN, USA.
23. M. G. McGaray, Water reclamation and algae harvesting. *J. Water Poll. Control Fed.* 43, 824 (1971).
24. W, J. O'Brien, *Two Methods for Algae Removal from Oxidation Pond Effluents.* US EPA Technology Transfer Seminar Publication (1973).
25. R. E. Mickinney, Waste treatment lagoons-state of the art. *Missouri Basin Engineering Health Council, USEPAWPCRS*, 17090EHX (1971).
26. W. J. Oswald, Designing waste ponds to meet water quality criteria. *2nd International Symposium for Waste Treatment Lagoons, Missouri Basin Engineering Health Council and FWQA*, Washington, DC, USA, pp. 186–194 (1970).
27. L. K. Wang, M. H. S. Wang, N. K. Shammas and D. B. Aulenbach, *Environmental Flotation Engineering.* Springer Nature, Cham, 433 pages (2021).
28. L. K. Wang, N. K. Shammas, W. A. Selke and D.B. Aulenbach, *Flotation Technology.* Humana Press, Totowa, NJ, 680 pages (2010).
29. US ACE, Civil Works Construction Cost Index System Manual, 110-2-1304, U.S. Army Corps of Engineers, 2020-Tables, Revised 31 March, Washington, DC, p. 44 (2020).
30. R. J. Craggs, J. P. Sukias, C. T. Tanner and R. J. Daviescolley, Advanced pond system for dairy-farm effluent treatment. *New Zealand J. Agric. Res.* 47, 449–460 (2004).
31. J. Garaa, M. Hernandez-Marine and R. Mujeriego, Analysis of key variables controlling phosphorus removal in high rate oxidation ponds provided with clarifiers. *African J. OnLine (AJOL)* 28, 1 (2002). https://www.ajol.info
32. O. V. Shipin, P. G. J. Meiring, R. Phaswana and H. Kluever, Integrating ponds and activated sludge process in the PETRO concept. *Wat. Res.* 33, 8, 1767–1774 (1999).
33. O. V. Shipin, P. G. J. Meiring and P. D. Rose, PETRO system: A low tech approach to the removal of wastewater organics (incorporating an effective removal of microalgae by the trickling filter). *Water SA.* 24, 2, 46–52 (1998).
34. P. Van der Steen, A. Brenner, J. Van Buuren and G. Oron, Post-treatment of UASB reactor effluent in an integrated duckweed and stabilization pond system. *Wat. Res.* 33, 3, 615–620 (1999).

35. P. Van der Steen, A. Brenner and G. Oron, An integrated duckweed and algae pond system for nitrogen removal and renovation. *Wat. Sci. Tech.* 38, 1, 335–343 (1998).
36. G. J. Alaerts, M. D. Rahaman Mahbubar and P. Kelderman, Performance analysis of a full-scale duckweed-covered sewage lagoon. *Wat. Res.* 30, 4, 843–852 (1996).
37. A. Rakkoed, S. Danteravanich and U. Puetpaiboon, Nitrogen removal in attached growth waste stabilization ponds of wastewater froma rubber factory. *Wat. Sci. Tech.* 40, 1, 45–52 (1999).
38. N. N. Bich, M. I. Yaziz and N. A. K. Bakti, Combination of Chlorella vulgaris and Eichhornia Crassipes for wastewater nitrogen removal. *Wat. Res.* 33, 10, 2357–2362 (1999).
39. M. Arauzo, M. F. Colmenarejo, E. Martinez and M. G. Garcia, The role of algae in a deep wastewater self-regeneration pond. *Wat. Res.* 34, 14, 3666–3674 (2000).
40. L. K. Wang, *Removal of Algae from Lagoon Effluent.* Technical Report #LIR/01-88-167, Lenox Institute of Water Technology, Lenox, MA (1988).
41. USEPA, *Process Design Manual for Upgrading Existing Wastewater Treatment Plants.* U.S. Environmental Protection Agency, US EPA-625/1-71-004a, Office of Technology Transfer, Cincinnati, OH (1974).
42. L. K. Wang, Y. T. Hung and N. K. Shammas (editors), *Physicochemical Treatment Processes*, The Humana Press, Totowa, NJ (2005).
43. L. K. Wang, Y. T. Hung and N. K. Shammas (editors), *Advanced Physicochemical Treatment Processes*, The Humana Press, Totowa, NJ (2006).
44. M. H. Al-Malack, G. K. Anderson and A. Almasi, Treatment of an oxicpond effluent using cross-flow microfiltration. *Wat. Res.* 32, 12, 3738–3746 (1998).
45. G. J. Jameson, Hydrophobicity and floc density in induced-air flotation for water treatment. *Colloids Surf. A Physicochem. Eng. Asp.* 151, 269–281 (1999).
46. N. K. Shammas and Y. T. Hung (editors), *Advanced Biological Treatment Processes.* The Humana Press, Totowa, NJ (2009).
47. L. K. Wang, M. H. S. Wang and Y. T. Hung, *Waste Treatment in the Biotechnology, Agricultural and Food Industries, Volume 1*, Springer Nature, Cham, 485 pages (2022).
48. M. M. Karpiscak, R. J. Freitas, C. P. Gerba, L. R. Sanchez and E. Shamir, Management of dairy waste in the Sonoran desert using constructed wetland technology. *Wat. Sci. Tech.* 40, 3, 57–65 (1999).
49. D. T. Hill, V. W. E. Payne, J. W. Rogers and S. R. Kown, Ammonia effects on the biomass production of five constructed wetland plant species. *Bioresour. Technlol.* 62, 109-113 (1997).
50. B. A. Costa-Pierce, Preliminary investigation of an integrated aquaculture-wetland ecosystem using tertiary-treated municipal wastewater in Los Angeles County, California. *Ecol. Eng.* 10, 341–354 (1998).
51. T. P. Simon, R. Jankowski and C. Morris, Modification of an index of biotic integrity for assessing vernal ponds and small palustrine wetlands using fish, crayfish, and amphibian assemblages along southern Lake Michigan. *Aquat. Ecosyst. Health Manag.* 3, 407–418 (2000).
52. T. P. Simon and P. M. Stewart, Application of an index of biotic integrity for dunal, palustrine wetlands-emphasis on assessment of nonpoint source landfill effects on the Grand Calumet Lagoons. *Aquat. Ecosyst. Health Manag.* 1, 63–74 (1998).
53. R. G. Cerezo, M. L. Suarez and M. R. Vidal-Abarca, The performance of a multi-stage system of constructed wetlands for urban wastewater treatment in a semiarid region of SE Spain. *Ecol. Eng.* 16, 501–517 (2001).
54. H. Bouwer, J. Ludke and R. C. Rice, Sealing pond bottoms with muddy water. *Ecol. Eng.* 18, 233–238 (2001).
55. Juanico-Environmental Consultants Ltd, Advanced Waste Stabilization Pond-New Concepts for the Design of Advanced Waste Stabilization Ponds (Lagoons). https://www.juanico.co.il (2005).
56. Google, Lagoons and Oxidation Ponds, List of Products and Manufacturers. https://directory.google.com/Top/Science/Environment/Water_Resources/Wastewater/Products_ and_Services/Lagoons_and_Oxidation_Ponds/ (2004).
57. M. Juanico and E. Friedler, Hydraulic age distribution in perfectly mixed non steady-state reactors. *ASCE J. Environ. Eng.* 120, 6, 1427–1445 (1994).
58. M. Juanico, Should waste stabilization ponds be designed for perfect-mixing or plug-flow? *Wat. Sci. Tech.* 23, 1495–1502 (1991).
59. USEPA, *Municipal Wastewater Stabilization Ponds.* US Environmental Protection Agency, Cincinnati, OH, 342 pages (1983).
60. L. K. Wang, N. K. Shammas, W. A. Selke and D. B. Aulenbach, *Flotation Technology*, Humana Press, NJ 680 pages (2010).
61. R. M. Sharman, *Sludge Lagoons*, Linn-Benton Community College, Albany, OR (1982).

62. MPCA, *Stabilization Pond Systems: Operations, Maintenance, Management.* Minnesota Pollution Control Agency, St. Paul, MN 196 pages (2013).
63. P. Meyer, G. Erickson, D. Sahli, N. Groh, B. Smith, V. Sathyaseelan and B. Olafson, *Recommended Pond Design Criteria.* Minnesota Pollution Control Agency, St. Paul, MN 27 pages (2009).
64. L. K. Wang, M. H. S. Wang and N. K. Shammas, Innovative pact activated sludge, captor activated sludge, activated bio-filter, vertical loop reactor, and phostrip processes, in *Waste Treatment in the Biotechnology, Agricultural and Food Industries, Volume 1,* Wang, L. K., Wang, M. H. S. and Hung, Y. T. (editors), Springer Nature, Cham, pp. 241–275 (2022).
65. L. K. Wang, M. H. S. Wang and N. K. Shammas, Agricultural waste treatment by water hyacinth aquaculture, wetland aquaculture, evapotranspiration, rapid rate land treatment, slow rate land treatment, and subsurface infiltration, in *Waste Treatment in the Biotechnology, Agricultural and Food Industries, Volume 1,* Wang, L. K., Wang, M. H. S., and Hung, Y. T. (editors). Springer Nature, Cham, pp. 277–316 (2022).
66. L. K. Wang and M. H. S. Wang, Understanding evaporation, transpiration, evapotranspiration, precipitation and runoff volume for agricultural waste management, in *Evolutionary Progress in Science, Technology, Engineering, Arts, and Mathematics (STEAM),* Wang, L. K., and Tsao, H. P. (editors), Lenox Institute Press, Auburndale, Massachusetts, pp. 1–81 (2022). DOI: 10.17613/m8tf-zd10
67. L. K. Wang, Y. T. Hung and J. P. Chen, Animal wastes treatment using anaerobic lagoons, in *Biosolids Treatment Processes,* Wang, L. K., Shammas, N. K. and Hung, Y. T. (editors). Humana Press, Totowa, NJ, pp. 431–450 (2007).
68. L. K. Wang, N. K. Shammas, G. K. Evanylo and M. H. S. Wang, Engineering management of agricultural land application for watershed protection, in *Modern Water Resources Engineering,* Wang, L. K. and Yang, C. T. (editors). Humana Press, Totowa, NJ, pp. 571–642 (2014).
69. M. H. S. Wang and L. K. Wang, Glossary of natural resources and environmental pollution control, in *Environmental and Natural Resources Engineering,* Wang, L. K., Wang, M. H. S., Hung, Y. T. and Shammas, N. K. edit(ors). Springer Nature, Cham, pp. 421–494 (2021).
70. M. H. S. Wang and L. K. Wang, Glossary of water quality, treatment, and recovery, in *Integrated Natural Resources Research,* Wang, L. K., Wang, M. H. S. and Hung, Y. T. (editors). Springer Nature, Cham, pp. 569–629 (2021).
71. M. H. S Wang and L. K. Wang, Environmental water engineering glossary, in *Advances in Water Resources Engineering,* Yang, C. T. and Wang, L. K. (editors). Springer Nature, Cham, pp. 471–556 (2015).
72. USEPA, *Waste Treatment: Upgrading Existing Poultry-Processing Facilities to Reduce Pollution.* US Environmental Protection Agency, Cincinnati, OH 48 pages (1973).
73 L. K. Wang, M. H. S. Wang, N. K. Shammas and D. B. Aulenbach, *Environmental Flotation Engineering.* Springer Nature, Cham, 433 pages (2021).
74. PDEP, *Treatment Ponds and Lagoons,* Pennsylvania Department of Environmental Protection, PA (2024).
75. H. A. Aziz, Wastewater engineering, in Environmental Science, Technology, *Engineering, and Mathematics (STEM),* Wang, L. K., Wang, M. H. S. and Pankivskyi, Y. I. (series editors). Lenox Institute Press, MA, 209 pages (2023). https://doi.org/10.17613/9g1r-6886
76. Y. I. Pankivskyi and O. Y. Oshurkevych-Pankivska, Water supply and sewerage systems, in *Environmental Science, Technology, Engineering, and Mathematics (STEM),* Wang, L. K., Wang, M. H. S. and Pankivskyi, Y. I. (series editors). Lenox Institute Press, Auburndale, MA, 261 pages (2023). https://doi.org/10.17613/x5qn-d460
77. D. H. Vanderholm, D. L. Day, A. J. Muehling, L. K. Wang, Y. T. Hung, E. Butler, M. H. S. Wang and H. Yehya, Treatment and management of livestock wastes, in *Waste Treatment in the Biotechnology, Agricultural and Food Industries, Volume 1,* Wang, L. K., Wang, M. H. S. and Hung, Y. T. (editors). Springer Nature, Cham, pp. 1–78 (2022).
78. L. K. Wang, J. V. Krouzek, and U. Koungtson. *Case Studies of Cleaner Production and Site Remediation,* United Nations Industrial Development Organization (UNIDO), Vienna, TT-5-4-95, 136 pages (1995).

8 Fundamentals and Applications of Nanomaterials for Industrial Wastewater Treatment

Md. Abu Bakar Siddiquc, Shaifa Abrarin, Abu Reza Md. Towfiqul Islam, Rahat Khan, and Mohammed Mahbubul Matin

NOMENCLATURE

AOPs	Advanced oxidation processes
AOX	Adsorbable organic halides
BG	Brilliant green
BOD	Biochemical oxygen demand
$CaCO_3$	Calcium carbonate
CNS	Central nervous system
CNTs	Carbon nanotubes
COD	Chemical oxygen demand
CS	Chitosan
CSg-PA	Chitosan-graft-polyacrylamide
CVD	Chemical vapor deposition
DBP	Dibutyl phthalate
DMP	Dimethyl phthalate
DTPA	Diethylenetriaminepentaacetic acid
ECs	Emerging contaminants
EDTA	Ethylenediaminetetraacetic acid
EPA	Environmental Protection Agency
ETP	Effluent treatment plant
Fe_3O_4	Magnetite
FeO	Wüstite
G	Graphene
GC-MS	Gas chromatography mass spectrometry
GFH	Granular ferric hydroxide
GO	Graphene oxide
GO-PANI	Graphene oxide polyaniline nanocomposite
H_2SO_4	Sulfuric acid
HCl	Hydrochloric acid
HGMS	High-gradient magnetic separation
HNO_3	Nitric acid
M-RGO	Magnetite-reduced graphene oxide
MWCNT	Multi-walled carbon nanotubes
NaOH	Sodium hydroxide
NCC	Nanocellulose

DOI: 10.1201/9781003532354-8

NMOs	Nanosized metal oxides
NMs	Nanomaterials
NPs	Nanoparticles
NZVI	Nanoscale zero-valent iron
PAEs	Phthalic acid esters
PAHs	Polycyclic aromatic hydrocarbons
PCBs	Polychlorinated biphenyl
PCDDs	Polychlorinated dibenzo-*p*-dioxins
PCDFs	Polychlorinated dibenzofurans
PCP	Pentachlorophenol
PFOA	Perfluorooctanoic acid
PhACs	Pharmaceutical active compounds
QDs	Quantum dots
rGO	Reduced GO
SWCNT	Single-walled carbon nanotubes
TBT	Tri-*n*-butyltin
TCMP	Trichloromethylpropanol
TDS	Total dissolved solid
TSS	Total suspended solid
USEPA	United States Environmental Protection Agency
VOCs	Volatile organic compounds
α-Fe_2O_3	Hematite
α-FeOOH	Goethite
β-Fe_2O_3	Maghemite
β-FeOOH	Akaganeite
γ-FeOOH	Lepidocrocite

8.1 INTRODUCTION

8.1.1 Nanotechnology Fundamentals and Environmental Sustainability

8.1.1.1 Nanotechnology

Nanotechnology involves manipulating matter at the atomic and molecular scale, typically below 100 nanometers. At this scale, materials exhibit unique physical, chemical, and biological properties that differ significantly from their bulk counterparts. These properties include:

1. Increased Surface Area: Nanoparticles have a high surface area-to-volume ratio, enhancing their reactivity and interaction with other substances.
2. Quantum Effects: At the nanoscale, quantum mechanical effects become significant, affecting optical, electrical, and magnetic properties.
3. Enhanced Mechanical Properties: Nanomaterials can exhibit superior strength, flexibility, and durability.

8.1.1.2 Environmental Sustainability and Research Objectives

Since the beginning of the modern scientific era, globalization has spread rapidly throughout the world for the betterment of human beings. However, with rapid industrialization, the world is getting polluted, which affects the health of all organs of the environment such as air, soil, sediment, and water, as well as both ecosystems and biodiversity due to the uncontrolled discharge of industrial pollutants [1]. For this reason, pollution is a global concern for all living beings [2]. Several industrial units such as tanning, textiles, pharmaceuticals, paints and varnish, pulp and paper, sugar, chemicals, fertilizers, pesticides, garments, oils, power stations, ship repairing docks, and cement are located in and around the globe, which has been growing rapidly. Most industries have an established effluent treatment plant (ETP). However, in many cases, industries are unable to utilize their ETP properly for many reasons, such as a lack of effective techniques and management, suitable chemicals or materials, good infrastructure, willingness for appropriate use, and so on. This is alarming to the environment, which includes air, water bodies, and soil. Because industries routinely discharge massive amounts of untreated or improperly treated effluent directly into the environment, the environment severely deteriorates [2].

Industrial effluents discharged by industries are major contaminants, causing environmental pollution and causing serious disruption to people's livelihoods by polluting adjacent water, soil, and plants, which can have an adverse impact on human health and the environment [3,4]. Therefore, effective treatment procedures with appropriate materials and techniques are necessary to save the environment from extensive contamination and to achieve the ultimate goal of environmental sustainability. Some key fundamentals of environmental sustainability are: (a) Ecosystem Health: Ensuring the stability and resilience of ecosystems by protecting biodiversity and natural habitats; (b) Resource Management: Efficiently using natural resources like water, minerals, and forests to prevent depletion and ensure long-term availability; (c) Pollution Prevention: Reducing emissions and waste to minimize environmental impact, including air, water, and soil pollution; (d) Energy Efficiency: Promoting the use of renewable energy sources (solar, wind, and hydro) and improving energy efficiency to reduce reliance on fossil fuels; (e) Sustainable Agriculture: Implementing farming practices that protect the environment, conserve resources, and support biodiversity; (f) Waste Management: Developing strategies for reducing, reusing, and recycling waste to minimize landfill use and environmental contamination; (g) Climate Change Mitigation: Taking actions to reduce greenhouse gas emissions and enhance carbon sinks to combat global warming; (i) Environmental Policy and Ethics: Creating and enforcing regulations that protect the environment and promote sustainable practices while considering ethical implications. These fundamentals are essential for addressing global challenges such as climate change, resource depletion, and pollution.

In recent years, nanomaterials (NMs) have played an important role in this regard. Various kinds of NMs, which have a large surface area and functional active sites on the surface, can be used effectively for the treatment of industrial pollutants from contaminated industrial wastewater. The main objectives of this chapter are to (a) provide some general information on NMs, (b) introduce some promising NMs for the industrial wastewater treatment, (c) identify the types of various pollutants (organic and inorganic) discharged by the selected industries, (d) present the current NMs used for the treatment of the diversified industrial pollutants, and (e) propose some future research dimension for the industrial wastewater treatment. This chapter will be beneficiary for researchers, environmentalists, policymakers, and academics for taking proper remediation steps regarding the industrial pollutions and conducting more advanced research in the relevant field.

8.1.2 General Information and Classifications of Nanomaterials

8.1.2.1 General Information of Nanomaterials

The term "nano" is derived from the Greek word nannos, meaning "very short man." "Nano" should be considered a different state of aggregation of matter in addition to the solid, liquid, gas, and plasma states. The reason for this is that nanoparticles (NPs) of a given material have quite a different physical (optical, magnetic), chemical, and mechanical properties than the same material in bulk form [5,6].

A NM is defined as a material with one or more external dimensions on the nanoscale, or an internal structure or surface structure on a scale ranging from 1 to 100 nm [7,8]. According to this definition, most of the materials around us would qualify as NMs, as their internal structure is modulated at the nanoscale [9]. The European Commission suggests that a NM should consist of

50% or more particles having a size of between 1 and 100 nm [10]. The agencies of the European Union think that the term "nanomaterial" should be defined as follows [10]:

"Nanomaterial" means a natural, incidental, or manufactured material containing particles, in an unbound state or as an aggregate or as an agglomerate, and where, for 50% or more of the particles in the size distribution, one or more external dimensions is in the size range of 1–100 nm [11,12]. In specific cases and where warranted by concerns for the environment, health, safety, or competitiveness, the number size distribution threshold of 50% may be replaced by a threshold between 1% and 50%. Based on the above statement, fullerenes, graphene flakes, and single-wall carbon nanotubes (CNTs) with one or more external dimensions below 1 nm should be considered NMs.

The International Organization for Standardization (ISO) defines NM as a material with any external dimension in the nanoscale or with an internal or surface structure in the nanoscale [13], and NP as a nano-object with all three external dimensions in the nanoscale (1–100 nm) [7].

A NP is defined as a particle with at least one external dimension on the nanoscale (1–100 nm). NPs usually consist of an agglomeration of particles with sizes in a particular distribution, some larger and some smaller than the 100 nm limit. A NP can be thought of as an aggregate composed of a relatively small number of atoms. The smaller NPs, sometimes called clusters, consist of only a few atoms. The larger NPs can have more than 105 atoms [14]. NPs are defined as objects of an intermediate size between the domain of atoms, which must be described using quantum mechanics, and the bulk materials that can be described by classical electrodynamics and solid-state physics [14]. Therefore, NPs are objects at a mesoscopic scale, i.e., pertaining to a scale between the microscopic and macroscopic worlds or an intermediate scale. Because of this, some NPs exhibit a number of new phenomena.

The unique properties and phenomena of the materials at the nanoscale (1–100 nm) are due to the fact that, at this scale, materials behave very differently from when they are in bulk form [15–22]. The extraordinary properties of NMs are based on the size effect. Reduction of size induces various extraordinary properties and phenomena that cannot be seen at the bulk level, i.e., at the micron-scale and up. The origin of the size effect is quantum confinement, which changes the electronic properties of materials such as density of states, discrete energy bands, changes in the edges of conduction, and valence bands. The change in the electronic properties of the materials leads to the drastic changes in the properties of materials to what they show at the bulk scale. For example, opaque substances become transparent (copper); inert materials become catalysts (platinum); stable materials turn combustible (aluminum); solids turn into liquids at room temperature (gold); and insulators become conductors (silicon). Since the last decade, nano-sized adsorbents such as CNTs, nanofibers, polymer-based NMs, metal oxide NMs, etc. have been used in water purification and desalination [20,21,23,24].

"Nanotechnology can be broadly defined as the creation, processing, characterization, and utilization of materials, devices, and systems with dimensions on the order of 0.1–100 nm, exhibiting novel and significantly enhanced physical, chemical, and biological properties, functions, phenomena, and processes due to their nanoscale size" [25]. Nanobiotechnology, nanosystems, nanoelectronics, and nanostructured materials, of which nanocomposites are a big part, are all areas of interest in nanotechnology right now.

8.1.2.2 General Classification of Nanomaterials

Generalized NMs are generally classified based on their morphology, dimensionality [26], uniformity, agglomeration, and chemical composition [27]. Depending on the mode of synthesis, the morphology or shape of NMs is different. The main characteristics of morphology include flatness, sphericity, and aspect ratio. Small-aspect-ratio morphologies are generally in the shape of spherical, oval, cubic, prism, helical, or pillar. High-aspect-ratio morphologies attain the shapes of zigzags, helices, and belts [28]. NPs with different morphologies include nanorods, nanozigzags, nanohooks, nanostars, nanocubes, nanohelices, and nanoplates. Based on dimension, NMs are classified into four classes as 0-dimensional (0D), 1-dimensional (1D), 2-dimensional (2D), and 3-dimensional

(3D). Zero-dimensional nanoparticles (0D-NPs) include atomic clusters, filaments, cluster assemblies, spherical NMs, cubes, nanorods, polygons, hollow spheres, metals, and core-shell NMs as well as quantum dots (QDs) [29]. 1D NMs are typically metallic, polymeric, ceramic, nanorods, nanowires [30], nanotubes [31], and nanofibers. 2D NMs include nanosheets, nanofilms [32], and nanoribbons [33] (ultrafine-grained overlayers or buried layers), single-layered and multi-layered, crystalline or amorphous, thin films, nanoplates, and nanocoating. 3D NMs include nanophase materials consisting of equiaxed nanometer-sized grains. 3D nanostructures are not confined to the nanoscale in all dimensions, and they are generally known as bulk NMs. 3D NMs combine multiple nanocrystals in different directions. Examples of the same are foams, fibers, carbon nanobuds, nanotubes, fullerenes, pillars, polycrystals, honeycombs, and layer skeletons [34,35]. As per uniformity, NPs can be classified as isometric and inhomogeneous or dispersed and agglomerate. This agglomeration relies on NPs' electromagnetic properties, magnetism, and surface charge. Further, NPs' agglomeration in a liquid relies on morphology and functionalization, resulting in hydrophobicity or hydrophilicity. NMs are further classified according to size: in terms of diameter, NMs are sized between 1 and 100 nm [9]. Based on composition, NMs are organized into four main types: (a) carbon-based NMs, (b) organic-based NMs, (c) inorganic-based NMs, and (d) composite-based NMs.

8.1.2.2.1 Carbon-Based NMs

These NMs are mainly composed of carbon and usually take the form of hollow spheres, ellipsoids, or tubes. Carbon is the 15th most abundant element in the earth's crust and the fourth most abundant element in the universe. The allotropes of carbon include amorphous carbon, graphite, diamond, graphene, CNT, and fullerene (Figure 8.1). Carbon is considered as unique, diverse, and completely different from a single element, in terms of structures and properties, such as 3D diamond crystal and graphite, 2D graphite sheets, 1D CNTs, and 0D fullerene molecules (Figure 8.2). Spherical and ellipsoidal carbon NMs are referred to as fullerenes, while cylindrical ones are called as CNTs [36]. Carbon-based NMs include fullerene, CNTs, nanodiamond, and graphene [37]. In the last few decades, another carbon-based NM, nanodiamond (size range: 4–10 nm) of sphere-like shape has attracted much attention among scientists. Nanodiamond (diamond NPs) is a special form of carbon NPs distinguished by special characteristics, such as its lattice structure, its large surface, as well as unique optical and magnetic properties.

8.1.2.2.2 Organic-Based NMs

Organic-based NMs are mainly composed of organic compounds, such as lipids, carbohydrates, or polymeric substances, ranging from 10 nm to 1 m [38]. Some polymeric substances are dendrimers, liposomes, and micelles [39]. These NMs are polymer NPs built from branched units. The surface of a dendrimer has numerous chain ends, which can be tailored to perform specific chemical functions, and this property makes them useful for catalysis. The 3D dendrimers contain interior cavities into which other molecules could be placed, and are successfully used for drug delivery. For the synthesis of organic-based NMs, various techniques were used, including mechanical grinding, laser ablation, solvent replacement, reduction in solution, and ion association [40,41].

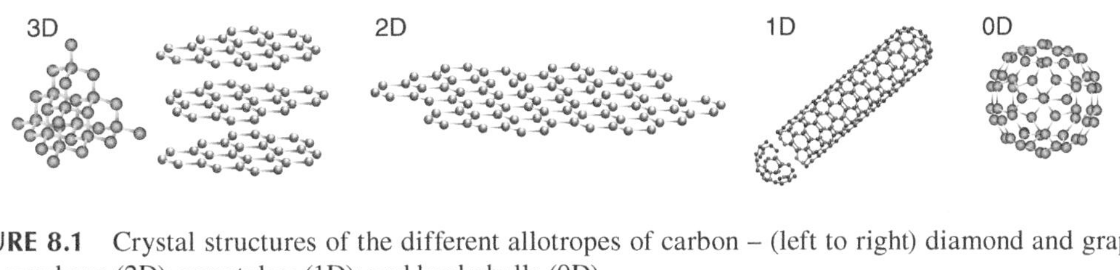

FIGURE 8.1 Crystal structures of the different allotropes of carbon – (left to right) diamond and graphite (3D); graphene (2D); nanotubes (1D); and buckyballs (0D).

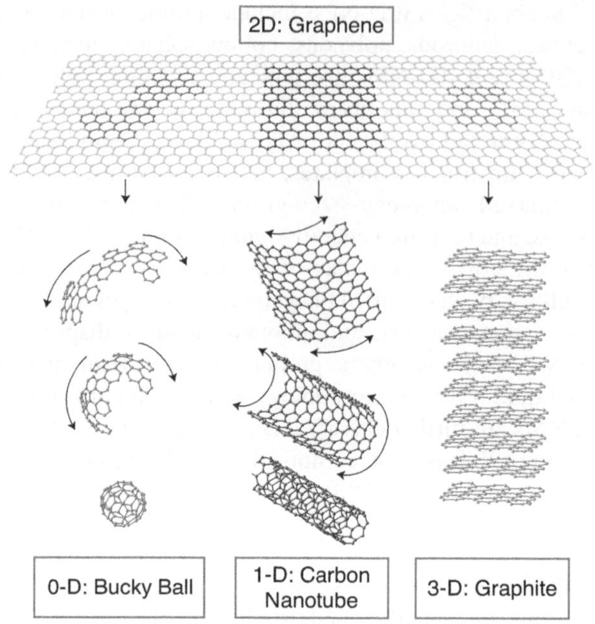

FIGURE 8.2 Structures of different forms of carbon: bucky balls (Fullerene-0D); carbon nanotube (1D); graphite (3D).

8.1.2.2.3 Inorganic-Based NMs

Metals, metal oxides, and ceramic-based NMs are some examples of inorganic-based NMs that exhibit interesting optical and electronic properties as their size approaches the nanoscale [42]. The clays occurring through the crystal growth in the diverse chemical conditions of the Earth's crust and the pigments are regarded as natural inorganic NMs. Bailey et al. [43] report that inorganic QDs are nanocrystals of semiconducting materials that are between 2 and 10 nm in size.

8.1.2.2.4 Composite-Based NMs

A composite is a combination of two or more different materials that are mixed in an effort to blend the best properties of both. A nanocomposite is a multiphase solid material where one of the phases has at least one dimension on the nanoscale (100 nm), or the structure has a nanoscale repeat distance between the different phases that make up the material. For example, the combination of NPs with multiple materials to form a unique NM is called composite-based NMs (Figure 8.3). Nanocomposites are found in nature, for example, in the structure of the bone, eggshell, and abalone shell. The idea behind nanocomposite is to use NPs or nanostructures as building blocks within a hosting matrix to create new materials with unprecedented flexibility and improvement in their mechanical, physical, chemical, or biological properties. There is a wide range of nanocomposite materials, for example, polymer-based nanocomposites; non-polymer-based nanocomposites; metal/metal nanocomposites; ceramic/ceramic nanocomposites; polymer/ceramic nanocomposites; carbon/metal nanocomposites; carbon/polymer nanocomposites; carbon/ceramic nanocomposites; polymer/layered silicate nanocomposites. The benefits of this combination often include greatly improved mechanical [44], thermal barrier, decreased gas permeability [45], and flame-retardant properties [46].

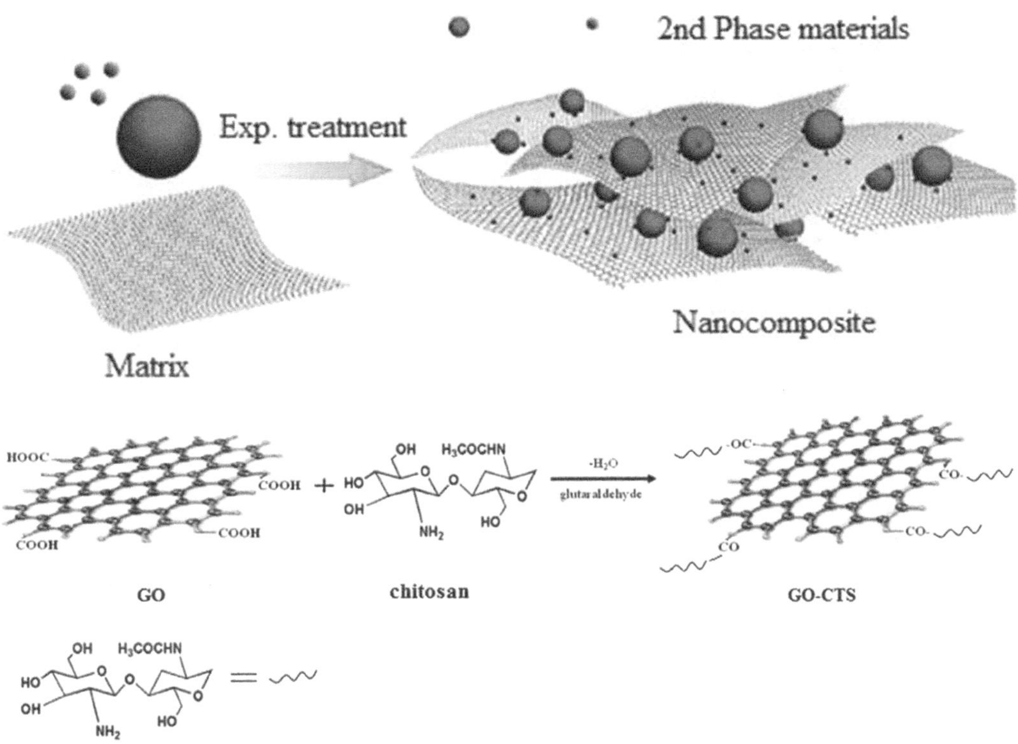

FIGURE 8.3 Illustrative formation of composite.

8.1.3 Promising Carbon-Based NM

Carbon-based NMs are presently considered the most versatile materials that can be used for the improvement of wastewater treatment processes. Extensive research carried out worldwide has resulted in the discovery of new carbon-related materials that have been successfully implemented in wastewater treatment and environmental protection technologies [47,48]. The unique physical and chemical properties of carbon-based NMs determine a wide range of practical applications [49]. The next-generation wastewater treatment and water supply systems relying on carbon NMs and nanotechnologies should be environmentally friendly and economically feasible [50]. The high surface area to volume ratio of NMs enhances reactivity with environmental pollutants [51]. The most commonly investigated carbon-based NMs are: graphene (G), graphene oxide (GO), single-walled carbon nanotubes (SWCNTs), multi-walled carbon nanotubes (MWCNTs), chitosan (CS), and nano-cellulose.

8.1.3.1 Graphene and GO

Graphene is a two-dimensional carbon NM with a single layer of sp^2-hybridized carbon atoms arranged in six-membered rings (Figure 8.4). An important branch of carbon-based materials is graphene derivatives, namely GO. GO is functionalized graphene with varying oxygen-containing groups [52]. The particle consists of an oxygen-containing functional group attached to the graphene layer (Figure 8.5). Such a structure, especially a high density of oxygen functional groups (e.g., carboxyl, hydroxyl, carbonyl, and epoxy) in the carbon lattice, enables GO to form stable suspensions in an aqueous solution [53]. This property enables practical applications that are specific to

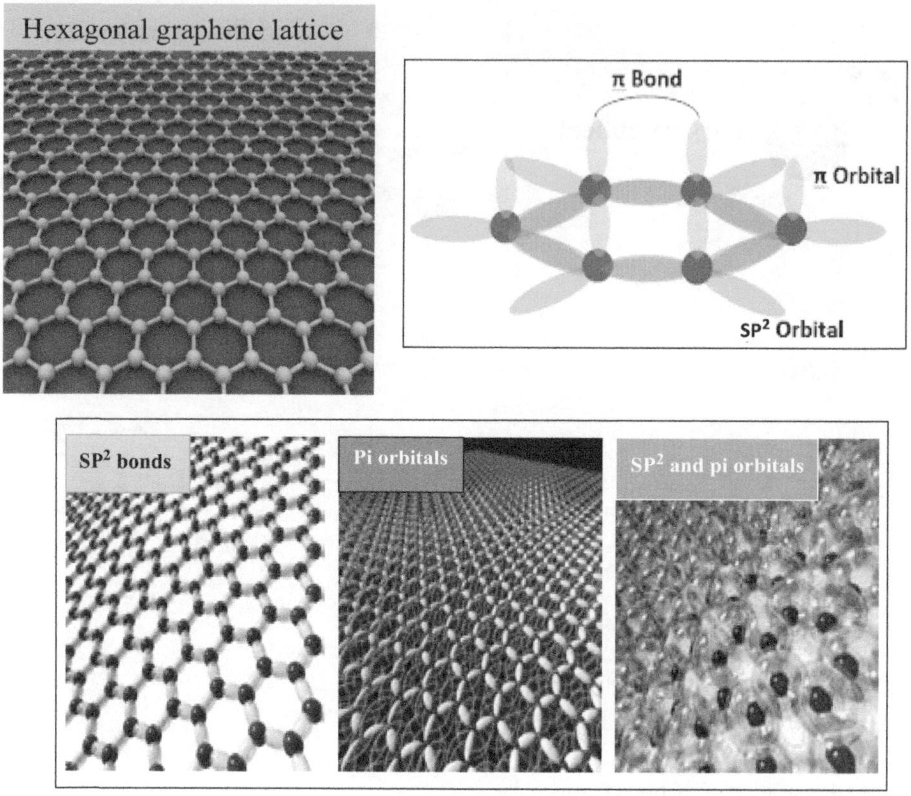

FIGURE 8.4 Hexagonal lattice structure of graphene, and the types of orbitals (sp^2 *and* π), chemical bonds (σ, π, and sp^2 bonds), and hybridization (sp^2) therein.

processes occurring in the water environment. The presence of a wide range of consecutive oxygen functional groups (such as epoxy (C-O-C), hydroxyl (-OH), carboxyl (-COOH), and carbonyl (C=O) groups) on its basal planes and at the edges and a large surface area are characteristics that make GO an excellent adsorbent for the removal of heavy metals (Figure 8.6) and several other pollutants [54]. Taking the oxygen-rich functional groups into account, GO can be well dispersed in both water and organic solvents but also provides more possibilities for the fabrication of graphene-based multifunctional materials [55–58]. Its mechanical strength and high adsorption capacity allow GO to form hybrid nanocomposites with various materials. Importantly, this NM is easy to make and inexpensive to produce [59,60]. The large-scale production of functionalized graphene at a low cost should result in good adsorbents for water purification. This is because these materials have a two-dimensional layer structure, a large surface area, a high pore volume, and surface functional groups. The inorganic NPs also keep the adsorbent from sticking together (as illustrated in the upper portion of Figure 8.3).

8.1.3.1.1 Synthesis of GO and Graphene

The simple steps for the synthesis of GO and graphene are illustrated in Figure 8.7.

GO can be synthesized following the Hummers method [61], a well-known and commonly adopted technique in this regard [62]. As reported, sulfuric acid and potassium permanganate combine together to form a dark red oil of dimanganese heptoxide, which plays a key role in the oxygenation of graphite, and sodium nitrate serves as a catalyst. Further oxidation occurs by permanganate

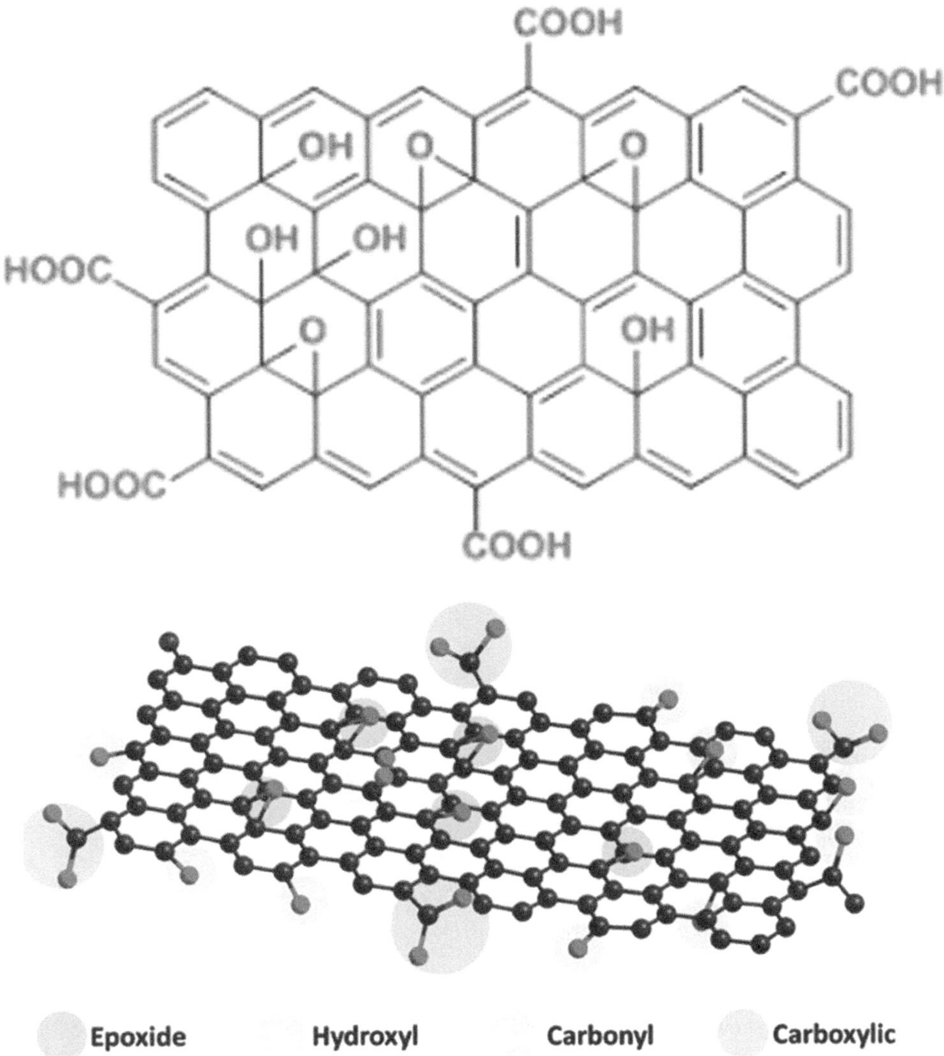

FIGURE 8.5 Graphene oxide (GO).

ion after the addition of water and before the addition of hydrogen peroxide, which is used to reduce residual permanganate and manganese dioxide from earlier steps. The washing with HCl and water results in a GO solution free of impurities [63].

$$KMnO_4(aq) + 3H_2SO_4(aq) \rightarrow K^+(aq) + MnO^{3+}(aq) + H_3O^+(aq) + 3HSO_4^-(aq)$$

$$KMnO_4(s) \rightarrow K^+(aq) + MnO_4^-(aq)$$

$$MnO^{3+}(aq) + MnO_4^-(aq) \rightarrow Mn_2O_7(l)$$

$$\text{Graphite} + Mn_2O_7(l) \rightarrow GO$$

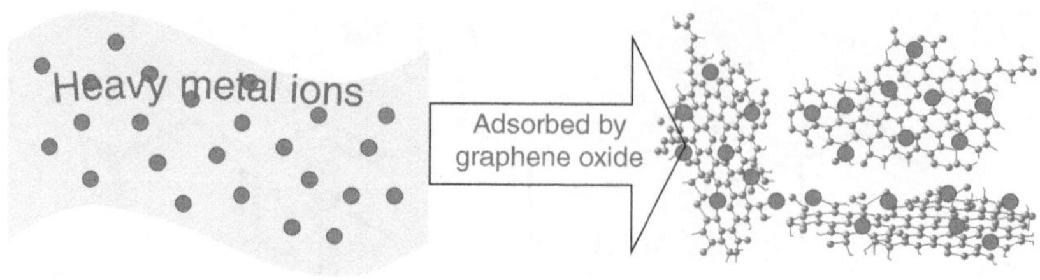

FIGURE 8.6 Adsorption of heavy metal ions by graphene oxide (GO).

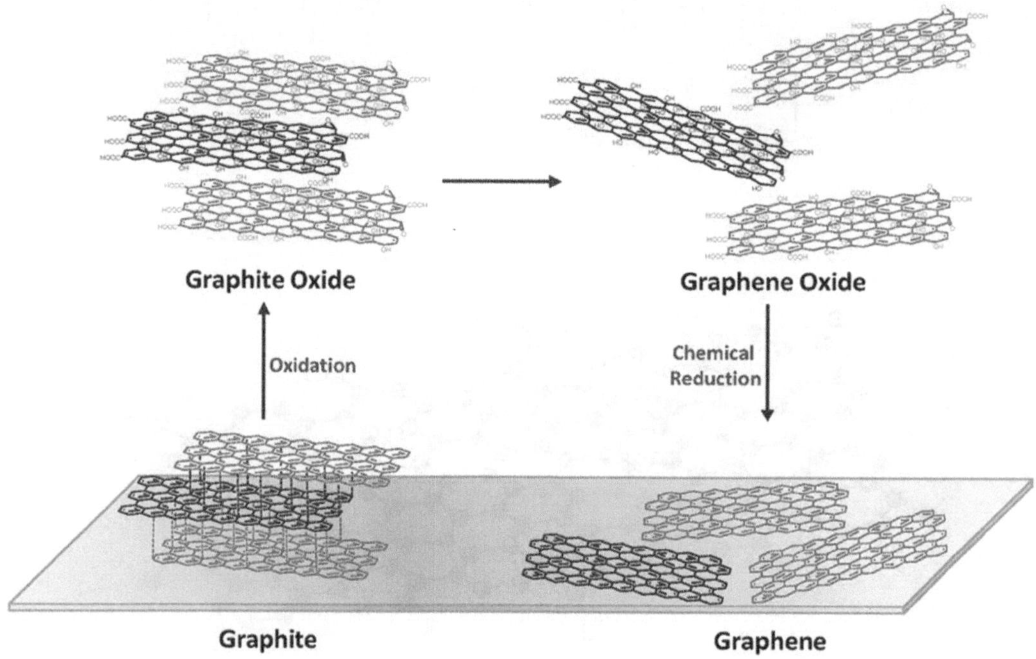

FIGURE 8.7 Schematic showing the synthesis of graphene from graphite.

Due to various advantages, GO is ultimately reduced to graphene, which is generally termed as "reduced GO" (rGO), as the complete removal of the oxygen-containing groups has not yet been achieved. There are a number of strategies for the reduction of GO. Graphene sheets can be obtained by chemical reduction of GO sheets. GO can also be reduced chemically to produce monolayer graphene sheets using hydrazine hydrate as a reducing agent. Several reducing agents have also been used for the reduction of GOs, such as hydrazine hydrate, sodium borohydride, and hydroiodic acid. Chemical reduction is usually carried out at ambient temperature or by using moderate heating, which makes it a low-cost and easy way to produce large-scale graphene compared with thermal annealing reduction. The entire process for the preparation of rGO is shown in the flow chart in Figure 8.8.

8.1.3.2 Carbon Nanotube

A CNT is a tube-shaped material made of carbon, having a diameter measured on the nanometer scale. The graphite layer looks a bit like rolled-up chicken wire, with a hexagonal mesh that doesn't break graphene molecules at the points of the hexagons. CNT is a versatile allotrope of carbon. It has a cylindrical and long tubular structure that consists of rolled-up sheets of graphene. It comprises

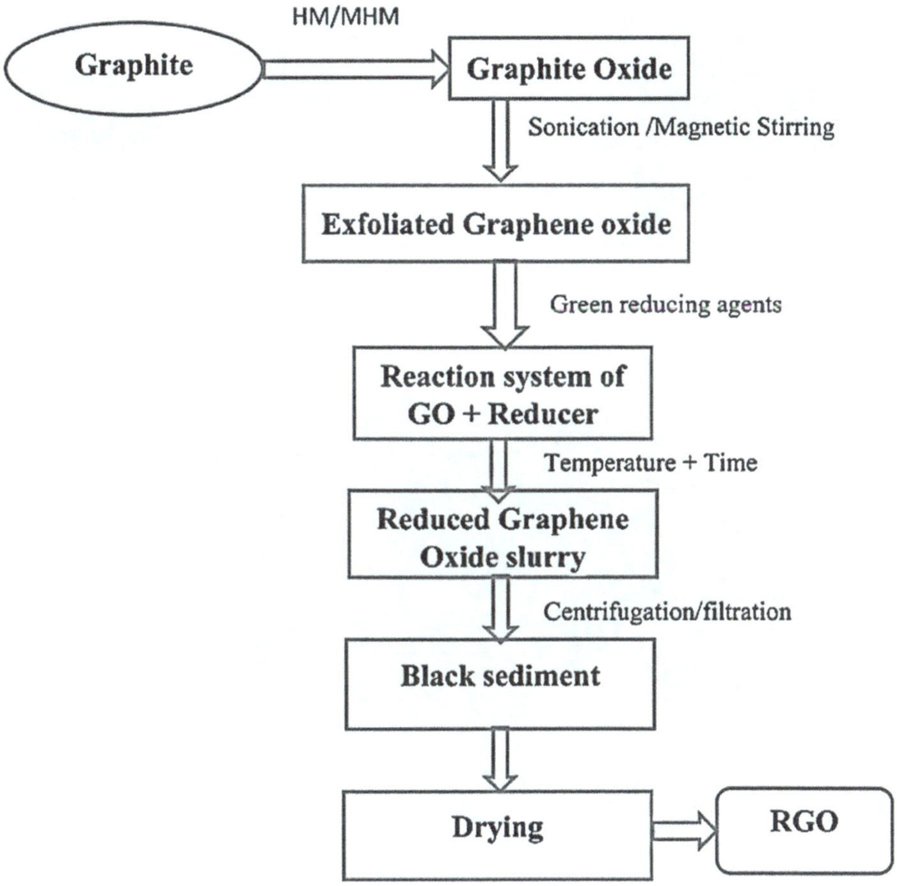

FIGURE 8.8 The flow chart for the preparation of rGO.

SWCNTs), which are made of a single layer of carbon atoms, and MWCNTs, formed by several nanotubes interlinked concentrically. While the diameter of SWCNTs reaches up to 3 nm, for MWCNTs it can even exceed 100 nm. Due to the presence of multiple layers of carbon atoms, MWCNTs have a higher mechanical strength than SWCNTs. Figure 8.9 shows different forms of CNT.

CNTs can be prepared by plasma-based methods such as arc discharge and laser ablation, thermal preparation methods like chemical vapor deposition (CVD), and hydrothermal methods. Out of all the conventional methods used for the preparation of CNTs, CVD is the most preferred as it produces CNTs in different forms like powder, straight, coiled, etc. The growth of the CNT can be easily monitored with changing temperatures [64].

CNTs are one of the most commonly used adsorbents for the treatment of metal ions in water, and they perform with remarkable stability and removal efficiency under extreme conditions [65]. MWCNTs' high surface area, interactions, and superior physicomechanical and chemical properties made them ideal for heavy metal ion adsorption from wastewater, but their low dispersion into the polymer solution limited their use in wastewater treatment. These drawbacks can be diminished by the oxidization of MWCNTs with mineral acids such as HNO_3, H_2SO_4, and/or the mixture of these two oxides [66]. CNTs remove heavy metals from water due to the interaction between heavy metal ions and oxygen-containing functional groups on the surface of CNTs [67]. MWCNTs were modified with CS and used for different applications, such as the adsorption of metal ions [68].

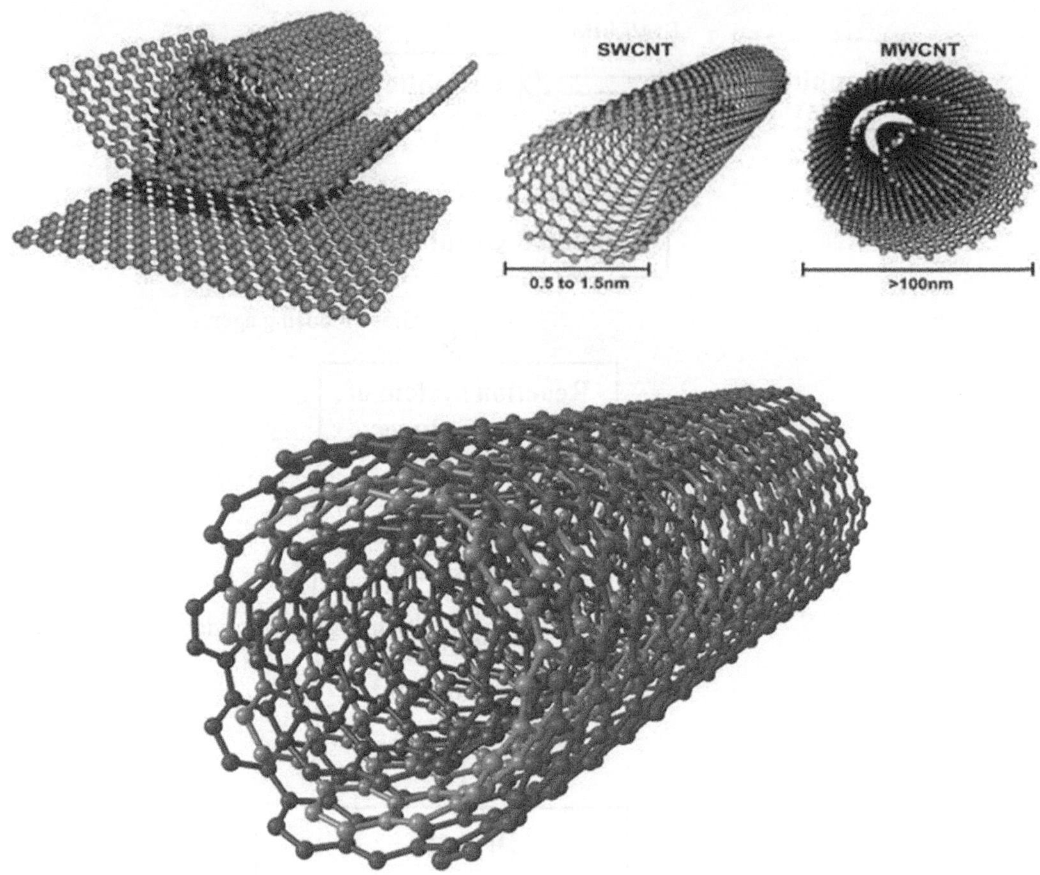

FIGURE 8.9 Pictorial presentations of carbon nanotubes.

8.1.3.3 Chitosan

CS is a natural product derived from chitin, a polysaccharide found in the exoskeletons of shellfish like shrimp and crabs [69]. It is a naturally occurring biopolymer that is extracted from crustacean shells or from fungal biomass [70]. CS is the most abundant component of the shell of crawfish, shrimp, and crab. Due to its non-toxic, non-allergic, and biocompatible nature, CS and its derivatives are used extensively in water treatment, waste treatment, agriculture, food, medicine, and the cosmetics field [71]. The high porosity of this natural polymer results in novel binding properties for metal ions such as cadmium, copper, lead, uranyl, mercury, and chromium [72]. CS has good sorption capacity for several metal ions because it possesses a number of different functional groups such as hydroxyls and amines to which metal ions can bind either by chemical or physical adsorption [73–75]. CS has been used for three decades in the water purification process. When CS is placed over oil spills, it holds the oil mass together, making it easier to clean up the spill. Water purification plants throughout the world use CS to remove oils, grease, heavy metals, and fine particulate matter that cause turbidity in wastewater streams [76]. CS has the potential to reduce and solve some environmental pollution problems, creating a greener environment, and CS is a renewable polymer. Some of the properties which are commercially attractive are polymeric, including natural decomposition, non-toxic to the environment and human beings, with no side effects or allergenic effects if implanted into the body. CS occurs naturally in the environment in large quantities and is second in abundance to cellulose. CS is a water-insoluble cationic biopolymer obtained from chitin via a deacetylation process whereby the acetamide groups are hydrolyzed to produce acetate ions

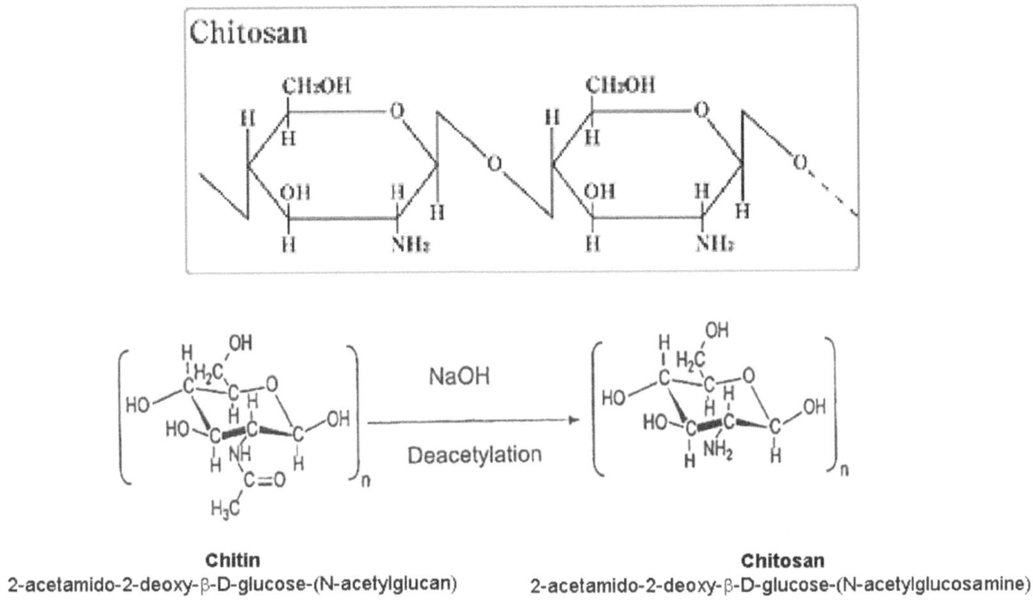

FIGURE 8.10 Chitosan and its preparation from the conversion of chitin by deacetylation process.

and amino (-NH_2) groups (Figure 8.10). In the acidic condition, the amino groups of CS are easily pronated and can bind anionic functional groups or anions. The presence of large numbers of amino groups on the CS backbone plays a key role in the hybridization process.

8.1.3.3.1 Synthesis of CS

CS can be synthesized from the shells of crawfish, shrimp, and crab, which is a three-step process. First of all, the shells are washed thoroughly with tap water as well as deionized water to remove all impurities and then dried in an electric oven at 80°C. The shells were then ground with a grinder and the following steps are performed.

8.1.3.3.2 Demineralization

In the demineralization step, the dried shells are treated with 5% HCl at a solid solvent ratio of 1:15 (W/V) and kept undisturbed overnight to remove $CaCO_3$ from the shells through CO_2 evolution, followed by stirring and ultrasonication for 1 hour and 30 minutes, respectively. Then, it was filtered, washed several times with deionized water, and dried in an electric oven overnight.

$$CaCO_3 \rightarrow CO_2 + H_2O$$

8.1.3.3.3 Deproteinization

The demineralized shell is treated with 5% NaOH with a solid to solvent ratio of 1:10 with continuous stirring at 80°C for 6 hours under reflux. The mixture is then kept undisturbed overnight, filtered, washed several times with deionized water, and dried in an electric oven overnight at 60°C.

8.1.3.3.4 Deacetylation

At this stage, the deproteinized shell is treated with 70% NaOH at a ratio of 1:10 with vigorous magnetic stirring at 80°C under reflux for 6 hours in order to produce CS from chitin through the deacetylation process (Figure 8.10). The product is then washed with deionized water at least 3 times and dried at 60°C for 24 hours.

8.1.3.4 Nanocellulose

Cellulose is a kind of abundant renewable resource in the natural world. It is formed by the repeated bonding of two 1–4-glycosidic bonds through hydrogen bonds (Figure 8.11). It is a biological macromolecule with a polymorphic form. The source of cellulose covers almost all biomass materials. The raw material is subjected to a series of pretreatments to remove non-cellulosic materials such as lignin, hemicellulose, lipid, wax, pectin, alginate, etc., and then nanocellulose with a diameter of 1–100 nm can be produced by chemical, physical, biological, and other separation processes [77]. Cellulose molecules with at least one dimension on the nanoscale (1–100 nm) are referred to as nanocellulose.

8.1.3.4.1 Synthesis of Nanocellulose

The hydrolyzed method is the most widely used method for the preparation of nanocellulose. The cellulose is composed of a crystallization zone and amorphous zone (Figure 8.12), and the amorphous region of the cellulose can be removed by acid hydrolysis to prepare nanocellulose of high crystallinity. The hydrolyzed acid is sulfuric acid, hydrochloric acid, phosphoric acid, hydrobromic acid, formic acid, and the like. In general, the acid dissolves the amorphous regions of cellulose, and the crystalline regions are left alone [78]. The process for preparing nanocellulose by acid hydrolysis is well-rounded, and the obtained nanocellulose has a uniform particle size. Nanocelluloses of different

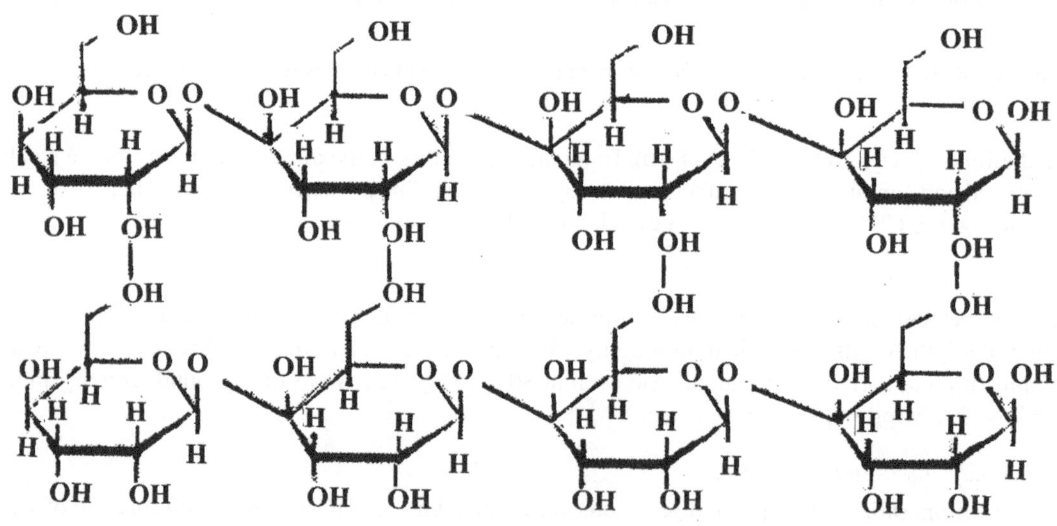

FIGURE 8.11 Structure of cellulose.

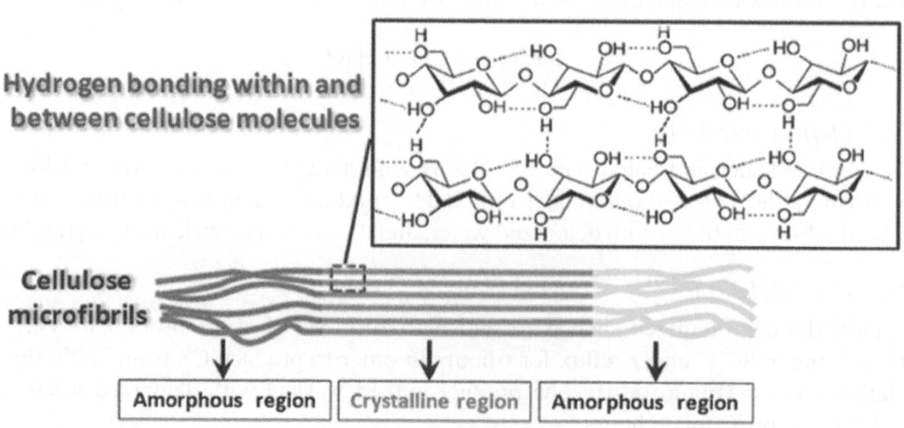

FIGURE 8.12 Illustration of crystallization and amorphous zones in cellulose.

sizes and crystallinities can be prepared by controlling conditions such as acid concentration, the temperature and time of the hydrolysis reaction, and so on. The length and diameter of CNCs commonly vary from a length of 200–500 nm to a diameter of 3–35 nm. Although acid hydrolysis is the easiest and oldest method of CNC preparation, recently, some other methods such as enzymatic hydrolysis treatment have been studied that degrade the amorphous domain from cellulosic fibers [79]. The enzymatic hydrolysis method, similar to the acid hydrolysis method, mainly utilizes the selective action of cellulose on the glycosidic bond in the amorphous region to hydrolyze the nanocellulose considered a more environment-friendly, greener, and efficient method than the acid hydrolysis method.

8.1.4 Metal Oxides NM

The nanosized metal oxides (NMOs), including nanosized iron oxides, manganese oxides, copper oxides, zinc oxides, aluminum oxides, titanium oxides, magnesium oxides, silicon oxides, and cerium oxides, are classified as the most promising ones for heavy metal removal from aqueous systems [80–82]. This is partly because of their large surface areas and high activity caused by the size quantization effect [83,84]. Recent studies suggest that many NMOs exhibit very favorable sorption to heavy metals in terms of high capacity and selectivity, which would result in the deep removal of toxic metals to meet increasingly strict regulations [85]. For example, iron compounds such as hematite, goethite, iron oxide-coated materials, and granular ferric hydroxide (GFH) are the preferred group of substances for arsenic adsorption because they lead to low leaching of adsorbed arsenic from exhausted adsorbents [86,87]. Several types of iron oxides, including hematite (α-Fe_2O_3), maghemite (β-Fe_2O_3), magnetite (Fe_3O_4), and wüstite (FeO); iron oxides/hydroxides, including goethite (α-FeOOH), akaganeite (β-FeOOH), lepidocrocite (γ-FeOOH), feroxyhyte, have been reported. These iron NPs have magnetic properties which facilitate their easy separation from water using the high-gradient magnetic separation (HGMS) technique after adsorption. It is worth mentioning here that Fe_3O_4 magnetic NPs are a widely used, economically viable, and reusable nanoadsorbent for the effective removal of toxic metal ions from water as compared to the nanoparticulates of non-magnetic oxides [88].

Generally, the synthesis methods can be classified into two categories: (a) physical approaches, including inert gas condensation, severe plastic deformation, high-energy ball milling, ultrasound shot peening; and (b) chemical approaches, including reverse micelle (or microemulsion), controlled chemical co-precipitation, chemical vapor condensation, pulse electrode position, liquid flame spray, liquid-phase reduction, gas-phase reduction, etc. [89]. Among these synthesis protocols, co-precipitation [90,91], thermal decomposition and/or reduction [92], and hydrothermal synthesis [93] techniques are widely used and are easily scalable with high yields [94].

8.1.5 Metal NMs

Metal NPs, such as gold, iron, silver, and other metallic NPs, have unusual chemical, optical, and electrical properties as compared to bulk materials [34]. By decreasing the size of metal particles, the surface area of the materials increases. The activated surface area of metal NPs is a crucial property for several applications, such as catalysis and adsorption processes. For instance, zero-valent iron NPs have been reported for the removal of metallic pollutants such as Cd, Zn, Se, As, Cr, Ag, and Pb from drinking water and industrial effluents to prove their potential for environmental applications [95].

8.1.6 Nanocomposite Materials

Nanocomposite materials are multiphase solid materials with at least one dimension in the nanoscale (less than 100 nm) [96–98]. The phases in the composites can be amorphous, semicrystalline, crystalline, or combinations thereof. They can be inorganic or organic, or both, and essentially of any composition. NMs also included in nanocomposites are NPs, nanofibers, and nanoclays. The combination of several materials forming a composite is capable of producing different properties, which makes the materials useful for several applications, such as water treatment.

CS–tripolyphosphate/TiO_2 nanocomposites can be used for reactive orange dye adsorption [99]. Alumina NPs in CS-graft-polyacrylamide (CSg-PA) are used as adsorbents for arsenic removal [100]. There were several reasons for grafting new functional groups: (a) to increase the density of the sorption sites; (b) to change the pH range for metal sorption; and (c) to change the sorption sites and/or the uptake mechanism in order to increase sorption selectivity for the target metal [101].

NMOs have wide applications for water treatment. But when the size of metal oxides gets down to the nanometer range, the increased surface energy makes them less stable. Consequently, NMOs are prone to agglomeration due to Van der Waals forces or other interactions [102], and the high capacity and selectivity of NMOs would be greatly decreased or even lost. Moreover, NMOs are unusable in fixed beds or any other flow-through systems because of the excessive pressure drops (or the difficult separation from aqueous systems) and poor mechanical strength. To improve the applicability of NMOs in real wastewater treatment, they were then impregnated into porous supports of a large size to obtain composite adsorbents [103]. The widely used porous supports include activated carbon, GO, graphene, natural materials such as CS and nanocellulose, synthetic polymeric hosts, etc. Besides traditional NMOs, magnetic NMOs attract increasing attention because they can be easily separated from water under a magnetic field [104]. Also, magnetic NMO-based composite adsorbents allowed easy isolation from aqueous solutions for recycling or regeneration [105]. Such facile separation is essential to improve the operational efficiency and reduce the cost of water/wastewater treatment.

Further, the use of metal composites containing two or more metals inherits the advantages of parent oxides and shows the synergistic effect of higher adsorption capacity than individual metal oxides. Various metal oxide composites such as Fe/Ce [106], Fe/Mn [107], Fe/Ti [108], Ce/Ti [109], Fe/Zr [110], [111], Fe/Cu [112], Mn/Co [113], Fe/Cr [114], and magnetite-GO [4] have been used for arsenic removal [115]. They selected Fe_3O_4, Mn_3O_4, and $MnFe_2O_4$ nanophases for their studies and reported that bimetal oxide showed comparatively higher adsorption capacity. Zhang et al. [113] used $MnFe_2O_4$ and $CoFe_2O_4$ magnetic NPs that were prepared by the chemical co-precipitation method. The paramagnetic nature of these magnetic NMs makes their separation from solution easy with a magnetic field after adsorption or desorption. Ratheesh Chandra et al. [116] used magnetite reduced graphene oxide (M-rGO) composites as adsorbents and reported that arsenic (V) ion removal on the M-rGO surface was due to the electrostatic attraction while As(III) ion removal was due to surface complexation. The incorporation of GO into the usual adsorbents like CS and nanocellulose also increased their compressive strength and produced an efficient and biodegradable monolith. GO nanocomposite with CS and/or nanocellulose will maintain the basic integrity and open-OH groups, improving the capacity for adsorption. In this study, we embedded the polyaniline nanocomposite (PANI) chains at the edge of GO sheets by in situ chemical oxidation polymerization to obtain a GO-PANI nanocomposite and tested it for the adsorption of cationic brilliant green (BG) dye from water. The results showed that GO-PANI is a good way to get rid of cationic BG dyes from water solutions [117]. CS/MWCNT/Fe_3O_4 composite nanofibers demonstrated significantly improved efficiency for Cr(VI) extraction from water and tannery waste [118]. Carboxylated MWCNTs treated with CS are used for the development of nanocomposite for the effective adsorption of Cr (VI) from an acidic aqueous solution, with a maximum adsorption of 140 mg/g or more [67].

Taking material dimensions down to the nanometer level creates interaction phase interfaces, which are very important for the enhancement of material properties. The ratio between surface area and volume of reinforced material used during nanocomposites preparation is directly involved in the understanding of the structure-property relationship [119]. The combination of materials to produce nanocomposites revealed that the surface area of the materials increased greatly. The adsorption capacity of composite materials can be improved through a variety of interactions, including electrostatic interactions, n-interactions between lone pair electrons delocalized into the π-orbital, dipole-dipole hydrogen bonding, and Yoshida H-bonding [120]. Through the nanoscale phase process, the goal of synergy between different constituents is achieved. Material properties change when the material constituent size is below a particular level, known as the "critical size" [121].

8.1.7 Nanotechnology Environmental Applications

Nanotechnology offers innovative solutions for environmental pollution control across various domains:

8.1.7.1 Water Treatment

Nanotechnology is widely applied to modern water treatment. The following are three typical examples.

1. **Nanofiltration Membranes:** These membranes can remove contaminants such as heavy metals, pathogens, and organic pollutants from water with high efficiency.
2. **Nanosorbents:** Nanomaterials like carbon nanotubes and graphene oxide are used to adsorb pollutants due to their large surface area and high reactivity.
3. **Photocatalysis:** Nanoparticles like titanium dioxide (TiO_2) can degrade organic pollutants in water when exposed to light, breaking them down into harmless substances.

8.1.7.2 Air Pollution Control

The following are two common air pollution control applications of nanotechnology:

1. **Nano-catalysts:** These are used in catalytic converters to reduce vehicle emissions by converting harmful gases like NO_x and CO into less harmful substances.
2. **Nanofibers and Nanotubes:** These materials can filter particulate matter from the air, improving air quality.

8.1.7.3 Soil Remediation

Applications of nanotechnology to soil remediation are new and are still in the research stage. Typical examples are:

1. **Nanoscale Zero-Valent Iron (NZVI):** NZVI particles are used to remediate contaminated soil and groundwater by reducing and immobilizing heavy metals and organic pollutants.
2. **Nanoclays:** These are used to stabilize and contain contaminants, preventing their spread in the environment.

8.1.7.4 Sensing and Monitoring

Nanobiosensors sensors can detect and monitor environmental pollutants at very low concentrations, providing real-time data for effective pollution management. Nanotechnology's potential in environmental applications is vast, offering more efficient, cost-effective, and sustainable solutions for pollution control compared to traditional methods.

8.2 IDENTIFICATION OF POLLUTANTS IN THE INDUSTRIAL WASTEWATER

At the end of the 20th century, the global environment became polluted with a number of industrial pollutants. The pollution of the global environment with a complex mixture of organic pollutants has resulted from industrial discharges and anthropogenic activities, as well as the inadvertent formation of by-products of various industrial processes. Inorganic pollutants are usually chemicals of mineral origin, such as metals, salts, and minerals originating from industrial, agricultural, and domestic wastes. They are responsible for water pollution and the serious diseases arising from it. Inorganic pollutants mostly result from various human activities, such as handling, storing, manufacturing, and disposing of chemicals. The most important category of inorganic pollutants is considered to be heavy metals. Heavy metals are natural components of the Earth's crust and soil. Conventionally, they are defined as the elements having an atomic number of 20 and an atomic density of 4 g/cm^3. However, ecologically, heavy metals belong to that group of conservative pollutants that are not subjected to any kind of breakdown or bacterial attack and are therefore perpetual additions to the environment. They are the major pollutants in the marine, sea, ground, industrial, and even treated wastewater environments. Heavy metals are distributed into two groups: (a) Metals that in trace amounts serve as essential nutrients for many organisms but are toxic in higher amounts (e.g., As, Cr, Co, Cu, Ni, Se, Va, and Zn). (b) Metals that have no nutritional value and are highly poisonous (e.g., Pb, Hg, Cd, Ur, Ag, and Be). The increase in urbanization and industrialization has made heavy metals reach alarming levels in the environment, thereby creating a need for their remediation.

In the following sections, some commonly found pollutants discharged from the specific industries to the environment are discussed.

8.2.1 Tanning Industry

Tannery wastewater contains high organic loadings (biochemical oxygen demand [BOD], chemical oxygen demand [COD], and total suspended solid [TSS]), salts (sodium, chloride, and sulfide), phenolic compounds, and endocrine-disrupting chemicals such as nonylphenols and phthalates, and other toxic metals, especially chromium (a highly toxic proven carcinogen) [122]. Tanneries generally employ chrome tanning to process raw skins and generate a huge quantity of tannery effluents which eventually contaminate the surrounding environment. These effluents contain large amounts of carcinogenic hexavalent chromium along with other potentially toxic heavy metals [123]. In addition, the discharged effluents contain significant quantities of toxic organic and inorganic chemicals, including arsenic sulfide, calcium hydrosulfate, chromium sulfate, sodium sulfate, sodium carbonate, dimethylamine, surfactants, sodium bicarbonate, sodium sulfhydrate, sodium hydrosulfite, sodium hydroxide, sulfides, lime, chlorides, ammonia, fats, liquors, sulfonated oils, dyes, biocides, and other organic matters [124–126]. All the pollutants found in the tannery wastewater are either organic or inorganic.

8.2.1.1 Organic Pollutants

Tannery wastewater has a complex mixture of various chemical pollutants and can be identified using gas chromatography mass spectrometry (GC-MS) employing suitable solvents for extraction. However, they are not all possible to extract using a single solvent extraction method. Therefore, combinations of different solvent systems are suggested to extract the majority of organic pollutants for GC-MS analysis. Tannery waste characteristically contains a complex mixture of both organic and inorganic pollutants. Various organic pollutants identified in tannery wastewater are mentioned in Table 8.1 [127,128].

Most organic pollutants identified using the solvents (1–5) in Table 8.1 have three common characteristics: (a) one or more cyclical ring structures of either aromatic or aliphatic nature, (b) a lack of polar functional groups, and (c) a variable number of halogen atoms, usually chlorine. The detection of chemical compounds (Table 8.1) in the treated tannery wastewaters clearly indicates the

TABLE 8.1
Organic Pollutants Identified in Tannery Wastewater Using Different Solvent Systems through GC-MS Analysis

Solvents Used	Identified Organic Pollutants
1. Acetonitrile+acetone	Benzene 2,2,3-Trimethyl oxepane
2. Methanol	1,3-Nitrophthalic acid
	1,2-Benzenedicarboxylic acid, di-isooctyl ester
	1,3-Hexadien-5-yne

(Continued)

TABLE 8.1 (*Continued*)
Organic Pollutants Identified in Tannery Wastewater Using Different Solvent Systems through GC-MS Analysis

Solvents Used	Identified Organic Pollutants
3. Chloroform	2-phenylethanol
	Nonadec-1-ene
	(3*R*,6*S*)-3-Methyl-1-6(-prop-1-en-2-yl)-cyclohex-1-enol-bis(2-methoxyethyl)phthalate
	Hexatriacontane
	Heneicosane
	2,4-Di-tert-butylphenol
	Tricosane
	2,3-Epoxypinane
4. Dichloromethane (DCM)	Trimethyl-1(2,6-ditert-butylphenoxy)silane, Dibutyl phthalate, Octanedioic acid
	Azelaic acid
	Caprolactam
	Oleic acid/Triacontane
	Heptadecane
	Hexadecanoic acid
	Eicosane
	9-methylnonadecane
	Heptadecane
	4-Trimethylsiloxyphenylphenoxysulfone
	Dotriacontane
5. Ethyl acetate	L-(+)-Lactic acid
	Acetic acid
	Benzene
	2-Hydroxy-3-methyl-butanoic acid
6. DCM-diethyl ether (1:1)	Derivatives of fatty acids and organic acids: Hexadecanoic acid, docosane, benzoic acid, 3-(4,-(T-butyl) phenyl) furan-2-5-dione, benzene acetamide, resorcinol, dibutyl phthalate, benzene 1,2,4 triol, 1-pentene 1,3-diphenyl, 2-bromotetradecanoic acid, phosphoric acid, 9-octadecanoic acid, octadecanoic acid, monopalmitin
7. DCM-*n*-hexane (1:1)	Fatty acids and carboxylic acids: benzoic acid, 3-(4-(T-butyl) phenyl) furan-2-5-dione, 2-pentenoic acid, benzene acetamide, dodecanoic acid, resorcinol, dibutyl phthalate, 9-octadecanoic acid, 10-undecynoic acid, docosanoic acid 1,2,3-propanetriyl, acetic acid
8. Others	Chlorinated phenols: chlorophenols, Pentachlorophenols (PCPs, a carcinogen), polychlorinated biphenyls (PCBs)
	Azodyes
	Formaldehyde resins
	Dibutyl phthalate (DBP) (an endocrine-disrupting chemical)
	Benzyl butyl phthalate (BBP) (an endocrine-disrupting chemical)
	Bis(2-ethylhexyl) phthalate (DEHP) (an endocrine-disrupting chemical)
	Short-chain, chlorinated paraffins
	Anthracene (a carcinogen)
	Nonylphenol (an endocrine-disrupting chemical and xenoestrogen)
	N-methylpyrrolidone
	Methylisothiazolinone (a carcinogen)
	Organotin compounds (dibutyltin) (a carcinogen)
	Azo dyes (Orange II) (a carcinogen)
	Chloromethyl benzene or benzyl chloride
	4-Chloro-3-methyl-phenol or *p*-chloro-*m*-cresol (PCMC)
	Hexachlorobenzene (a carcinogen)

less efficacy of the treatment plants. Fatty acids such as hexadecanoic acid, dodecanoic acid, and octadecanoic acid might have originated as a result of the processing of raw hides and skins [129]. Phthalates (such as dibutyl phthalate: DBP), benzoic acid, and resorcinol are used as plasticizers to increase the flexibility and pliability of leather products, as a biocide in raw hide/skin preservation, and as surfactants, respectively, in the leather industry [130–132]. Phthalates such as DBP, diethyl phthalate, butylbenzyl phthalate, and diethylhexyl phthalate and phthalic acid have been listed as priority pollutants by the United States Environmental Protection Agency (USEPA) [133]. Discharge of these phthalates causes water pollution and serious toxicological effects on aquatic organisms.

If some key properties of organic pollutants are known, then environmental chemists can make predictions about their fate and behavior in the natural environment. Yadav et al. [134] list some of these properties, such as solubility in water, vapor pressure, and partition coefficients between solid, liquid, and gaseous phases. Organic pollutants can be biodegraded when the dissolved oxygen is sufficient, but with insufficient oxygen, fish and other aquatic species may die, and without oxygen, decomposition becomes anaerobic, with the formation of methane [135]. The loss of dissolved oxygen is detrimental to aquatic organisms.

In the last few years, leather tanning has adopted eco-friendly, non-toxic, and biodegradable chemicals as per strict regulations to limit the pollution level. But it is clear from many studies [136] that the quality of treated wastewater has not yet gotten better.

8.2.1.2 Inorganic Pollutants

First, the important inorganic pollutant is sodium chloride from the preservation of rawhides. Generally, 1–103 kg of fresh hides needs 300–400 kg of salt for their preservation. Therefore, 70% of salt pollution in the tannery comes from raw hides. The second wastes are sulfide and lime that were not absorbed by the pelts during the liming process. In addition, the most toxic chemical is chromium (VI) oxidized from chromium (III) in tanning. Chromium (VI) has been shown to be carcinogenic, mutagenic, and allergenic. Chromium basic sulfate is the most commonly used tanning substance [137]. The leather industry alone accounts for 40% of global Cr consumption. During the tanning process, the leather takes up about 60%–80% of the applied chromium, and the rest of the metallic salts are usually discharged into effluent waters with serious environmental impact [138,139]. The tannery waste can have between 40,000 and 50,000 mg/L of total chromium [140]. Other metals that might be discharged from the tannery industry in wastewater include Al, Co, Ni, Cu, Sb, Ba, Pb, Se, Hg, Zn, As (a carcinogen), and Zr, Cd, and Pb, sometimes used in yellow pigments (cadmium sulfate and lead chromate pigment), are considered highly toxic. Sodium dichromate and cobalt dichloride are also used as inorganic compounds.

In tanning wastewater, the most common chemicals are ammonia released from deliming process, And sodium sulfide and sodium hydrosulfide, which are used during the unhairing process as well as from the breakdown of hairs. Under alkaline conditions, the sulfides remain largely in solution. If the pH of effluent drops below 9.5, hydrogen sulfide evolves from the wastewater, and the lower the pH, the higher the rate of evolution, which is characterized by the smell of rotten eggs. Hydrogen sulfide is a soluble gas. When absorbed, weak acids can form and cause corrosion. Sulfides can be oxidized into non-toxic compounds by certain bacteria in rivers, but if they generate anaerobic conditions, they cause harm to aquatic life. Sulfides in wastewater can be determined in several ways. One of the most accurate methods relies on the acidification of suitable quantities of wastewater to generate hydrogen sulfide. This is flushed through the apparatus using nitrogen gas, where it is trapped and converted into zinc sulfide, which is then determined by the titration process.

Chrome-tanning powder contains high levels of sodium sulfate. Additionally, while removing the sulfide components from wastewater by aeration, the oxidation process creates a whole range of substances, including sodium sulfate. Sulfate can be removed from the wastewater by adding barium chloride solution to a sample of filtered wastewater. The sulfates get precipitated as barium sulfate and the filtrate is dried.

Tannery wastewater is generally treated by various physicochemical and biological methods or by a combination of both [141]. Physical and chemical processes are frequently employed to treat contaminated sites but often do not destroy contaminants [142].

8.2.2 Textile Industry

Textile wastewater is alkaline in nature and highly colored and often contains harmful residual dyes such as acidic, basic, reactive, dispersed, azo, diazo, anthraquinone-based, and metal complex dyes (some of which are carcinogenic in nature). Many persistent toxic pollutants in wastewater may interfere with the treatment plant's performance; they may pass through unchanged, transformed, generated, or accumulated in the sludge. Both organic and inorganic compounds are included in this category. The Environmental Protection Agency (EPA) prioritizes pollutants list including 129 priority pollutants (organic and inorganic) divided into 65 classes that will be regulated by categorical discharge standards. They are chosen because they are known or thought to cause cancer, genetic changes, birth defects, or high acute toxicity [143].

8.2.2.1 Organic Pollutants

Textile industry wastewater is considered one of the worst polluters of our precious water and soil ecosystems. It causes carcinogenic, mutagenic, genotoxic, cytotoxic, and allergenic threats to living organisms. Textile wastewater contains several types of organic pollutants, such as dyes, formaldehyde, phthalates, phenols, surfactants, perfluorooctanoic acid (PFOA), pentachlorophenol, various azo dyes, dioxins, pesticides, dispersants, and leveling agents. The quality of wastewater varies depending on the process that makes it.

Several studies indicate that a small fraction of new textiles have a very high concentration of polychlorinated dibenzo-*p*-dioxins (PCDDs) and polychlorinated dibenzofurans (PCDFs). Besides dye molecules, textile industry wastewater also contains a high load of salts, alkalis, binders, dispersants, volatile organic compounds (VOCs), aromatic compounds, color content, surfactants, chlorobenzenes, reducing agents, dioxin, phthalates, phenols, pentachlorophenol, and detergents [144–146].

Some other organic compounds identified in the untreated textile wastewater are phenol, hexamine, ethyl-tetramethyl-heptane, 2,6-di-*tert*-butyl-quinone, tetramethyl-dodecane, ionol-(2,6-di-*tert*-buty l-4-methyl-phenol), diethylphthalate, nonylphenol isomer, 3,5-di-*tert*-butyl-4-hydroxybenzoic acid, 2,6-di-*tert*-butyl-4-ethyl-phenol, 2,6-di-*tert*-butyl-4-methoxymethyl-phenol, tetramethyl-butyl-phenol isomer, nonylphenol isomer, tetramethyl-butyl-phenol isomer, octadecane, butyl-2-ethylhexyl-, benzoic acid phenyl ester, 2,2-dimethyl-2-phenyl-acetophenone, butyl-octyl-phthalate, 4,40-methylenebis-phenol, benzyl-quinoline, docosane, tributyl-*O*-acetylcitrate, tricosane, 7-(diethylamino)-4-methyl-2*H*-1-benzopyran-2-one, benzyl-butyl-phthalate, triphenyl-phosphonic acid, diisooctylphthalate, hexacosane, triacontane, and dotriacontane [147].

The main chemicals that are known to be harmful in textiles are the dyes, which are made up of many different chemicals (Table 8.2). Dyes are synthetic organic compounds that are used to add color to substrates like cloth, paper, leather, and other objects. Textile dyes are aromatic, heterocyclic, and persistent in nature and possess different chromophore groups such as azo (NN), nitro (NO), carbonyl (CO), quinoid groups, and auxochrome groups like hydroxyl (OH), carboxyl (COOH), amine (NH_2), sulfonate (SO_3H), etc. [148,149]. Dyes such as methyl orange, Congo red, methylene blue, azure B, reactive dye, direct red, remazol red, scarlet, malachite green, acid orange, and remazol brilliant blue R are extensively used in TIs [150,151]. These dyes are highly resistant to light, water, chemicals, detergents, and microbial activities [148,152].

TABLE 8.2
Application and Toxicity of Different Chemicals Used in Textile Industries [153–155]

Chemicals	Applications
Chlorinated solvents, e.g., Trichloroethane (TCE)	Applied in scouring of fabric (removal of impurities from fibers)
Carbon disulfide (CS_2)	Applied in manufacturing of viscose rayon fibers
Alkylphenols, nonylphenol ethoxylates (NPEs) and nonylphenols (NPs)	Applied as scouring agents (wool), laundry detergents, emulsifiers and dye-dispersing agents for dyes, and also to protect or stabilize polymers
Chlorobenzenes (e.g. hexachlorobenzene (HCB))	Applied as dyeing carrier, dyestuff, and in preparation of azo pigments
Phthalates (DINP: di-isononyl phthalate, and BBP: butyl benzyl phthalate)	Applied in printing and dyeing operations and coating of textiles to make the cloths soften
Heavy metals (Sb: antimony, Cd: cadmium, Pb: lead, Hg: mercury, As: arsenic, Ni: nickel and Cr: chromium = priority pollutants)	Applied in preparation of pigments and dyes in coloring of textiles
Azo dyes (e.g. methyl nitro, orange, remazol, reactive dyes, etc.)	Applied in dyeing of textile fibers, mainly for cotton but also viscose, silk, wool and synthetic fibers
Organotin compounds (DBT: dibutyltin, TBT: tributyltin, and TPhT: triphenyltin)	Applied as biocide to prevent the heavy-duty textiles and reduce body odor posed by the microbial breakdown of sweat
Perfluorinated chemicals (PFCs): perfluorooctane sulfonate (PFOS), a banned persistent organic pollutant (POP) in Stockholm Convention	Applied to make the textiles both water and stain-proof
Formaldehyde	Applied in textile finishing (to make cloths softer, water-resistant, and more crease-resistant and prevent shrinking)
Volatile organic compounds (VOCs: toluene, methyl isobutyl ketone, xylene, methyl ethyl ketone, dichloromethane, 1,1,1-trichloroethane)	Applied in preparation of solvent-based inks or pastes used during printing and drying of textiles
Short-chain chlorinated paraffins (SCCPs): classified as POPs under Stockholm Convention	Applied as a flame retardant, waterproof and rot-proof to heavy textiles, like military tents
Chlorophenols (e.g. pentachlorophenol (PCP)): cauterized as group B2 probable human carcinogen as per USEPA	Applied as a preservative (fungicide) on heavy-duty textiles and fibers
Brominated Flame Retardants (BFRs), polybrominated biphenyls (PBPs), polybrominated diphenyl ethers (PBDEs), and hexabromocyclododecane (HBCD), e.g. hexabromobiphenyls	Applied to make textiles less flammable to prevent burning
Dioxins	Applied as a preservative agent for textile fibers like cotton and other fibers during sea transit, and in cotton bleaching and also used in preparation of dyestuffs
Organophosphorus compounds, e.g. tributylphosphate (TBP)	Applied as a strong wetting agent and in preparation of pigment pastes used in printing
Silver and nano silver compounds	Application of silver nanoparticles as antimicrobial or antibacterial agents in textiles

8.2.2.1.1 Some More Organic Pollutants

8.2.2.1.1.1 Hydrocarbon Solvents: Aliphatic Hydrocarbons The aliphatic hydrocarbons present in the effluent of the textile industry [156] are cyclohexene (C_6H_{10}), cyclohexane (C_6H_{12}), *n*-hexane (C_6H_{14}), *n*-heptane (C_7H_{16}), pentane (C_5H_{12}), and petroleum ether. Most aliphatic compounds are flammable, allowing the use of hydrocarbons as fuel. Aliphatic hydrocarbons are asphyxiants and central nervous system (CNS) depressants. Serious toxic effects of aliphatic hydrocarbons include asphyxia and chemical pneumonitis for many kinds of paraffin, axonal neuropathy for *n*-hexane, and cancer for 1,3-butadiene [157].

8.2.2.1.1.2 Hydrocarbon Solvents: Aromatic Hydrocarbons Effluents of textile-dyeing plants are extremely difficult to treat due to their high content of pollutants such as polycyclic aromatic hydrocarbons (PAHs) [158]. Some of the aromatic hydrocarbons found in the textile effluent are benzene (C_6H_6), naphthalene ($C_{10}H_8$), toluene (C_7H_8), and xylenes or dimethylbenzene (C_8H_{10}). Most PAHs do not dissolve easily in water. They stick to solid particles and settle on the bottoms of lakes or rivers. In soils, PAHs are most likely to stick tightly to particles; certain PAHs move through the soil to contaminate underground water. Some people who have breathed or touched mixtures of PAHs and other chemicals for long periods of time have developed cancer [159].

8.2.2.1.1.3 Oxygenated Solvents (Alcohols/Glycols/Ethers/Esters/Ketones/Aldehydes) An oxygenated solvent contains oxygen molecules with high solubility and low toxicity. These volatile solvents are widely used in textile processing and are found in the effluent of the textile industry [160]. These are methanol (CH_3OH), ethanol (C_2H_5OH), propanol (C_3H_7OH), butanol (C_4H_9OH), ethylene glycol ($C_2H_6O_2$), diethyl ether [$(C_2H_5)_2O$], ethyl acetate (C_4H_8O), methyl ethyl ketone or butanone (C_4H_8O), 1,4-dioxane, ethylene glycol, methyl isobutyl ketone, and so on.

8.2.2.1.1.4 Dyestuffs and Pigments Containing Dangerous Substances The presence of color in wastewater is one of the main problems in the textile industry. Colors are easily visible to the human eye, even at very low concentrations. Hence, color from textile waste carries significant esthetic importance. Most of the dyes are stable and have no effect on light or oxidizing agents. They are also not easily degradable by conventional treatment methods. In most textile industries, dye removal from effluent is a major issue [161,162].

8.2.2.2 Inorganic Pollutants

A number of toxic trace elements can be found in textile materials because they are frequently present in various textile processes. In addition, raw textile materials can also contain several trace elements. Textile wastewater contains a variety of heavy metals like Pb, Cd, As, Cr, Zn, Ni, Sb, and Hg. The pigments containing Pb, Cd, Hg, V, Cr(VI), Co, Ni, As, Sb, or Se involve toxic and hazardous risks [153].

Metal complex dyes (Co, Cu, Cr, Pb), pigments, mordant (Cr), catalyst in synthetic fabric manufacture (antimony oxide), synergists of flame retardants (Sb_2O_3), antimicrobials (NPs of silver, titanium oxide, and zinc oxide), as well as water repellents and odor-preventive agents have been used in textile products and clothing for many years [163]. The relationship between trace elements and textiles may pose an important environmental problem for the textile industry, while the presence of certain toxic trace elements in clothing may also pose health hazards for consumers [164,165].

8.2.3 Pharmaceutical Industry

Pharmaceutical wastewater is acidic in nature, has high COD and total dissolved solid (TDS), and contains many organic solvents, formulations, disinfectants, and many generic drugs such as antibiotics, analgesics, etc. [122].

8.2.3.1 Organic Pollutants

Pharmaceutical active compounds (PhACs) are one of the emerging pollutants of major concern. The main source of the PhACs is the effluents dispensed from wastewater treatment plants due to the continuous entrance of the emerging contaminants into the aquatic systems. The PhACs include antibiotics, tranquilizers, diuretics, and psychiatric drugs [166]. Apart from organic contaminants, pharmaceutical industry wastewater also releases antibiotic, antihistamine, antituberculosis, and analgesic groups of pharmaceuticals depending on their production [167,168]. For example, the quinolone antibiotic nalidixic acid was found at a level of 45 mg/L in wastewater from the pharmaceutical industry [167].

Numerous investigations into the treatment of waste streams coming from the chemical synthesis industry confirm the presence of toxic solvents and pharmaceuticals in water. For instance, Deshpande et al. [169] detected nitromethane, methyl isocyanate, k-salt, furfuryl alcohol, methanol, dimethyl sulfamic oxide, dimethylamine, *p*-formaldehyde, cystamine, and furamine in wastewater coming from the ranitidine-producing industry. Similarly, another study confirmed the presence of dimethyl formamide, *n*-butyl acetate, ethyl acetate, methylene chloride, and ethyl acetate in the waste stream of the bacampicillin and sultamicillin tosylate producing industries [170]. However, the concentration of these pollutants was presented in the form of COD and other elements such as sulfate, sulfite, chloride, and calcium. The use of solvents in the pharmaceutical manufacturing industry results in the discharge of recalcitrant pollutants such as phenol, toluene, nitrophenols, nitroaniline, and trichloromethylpropanol (TCMP) into wastewater. These pollutants are not removed by biological methods effectively and often seem to have an adverse effect on the treatment plant. Because these pollutants cannot be easily removed by biological treatment, biologically treated effluent contains a significant amount of BOD and COD [168].

A variety of pollutants are also released during the manufacture of pharmaceutical products and in the reaction and purification of solvents [171,172]. These include benzene, phenol, toluene, halogenated solvents, and cyanide. Although the EPA has banned or restricted the use of some 23 solvents, including some VOCs and chlorinated solvents, some are still used by the pharmaceutical industry since the relevant drugs cannot be manufactured by using other solvents; for instance, methylene chloride. The major nonconventional solvents used in industry are methanol, ethanol, isopropanol, acetone, and ethyl acetate. Also on this list are heteroaromatics like pyridine and piperidine, which don't change during the reaction process [168].

8.2.3.2 Inorganic Pollutants

Several studies found an elevated concentration of toxic elements of Hg, Cd, and Ni in effluent samples collected from pharmaceutical industries [173].

8.2.4 Pulp and Paper Industry

One of the largest polluters of the environment is pulp and paper industry effluent and wastewater after secondary treatment [174]. The pulp and paper industry is the world's sixth-ranked polluting industry, producing harmful wastewater on a massive scale following the making of paper [175]. Pulp and paper mill wastewater is a highly intense dark brown color and contains recalcitrant dioxins, furans, lignins, adsorbable organic halides (AOX, a group of halogenated organic compounds), phenolic and chlorophenolic compounds, especially pentachlorophenol (highly toxic to living beings and hazardous to the environment) [122]. The released wastewater contains 40%–45% raw materials, including chlorophenols, insecticides, chlorinated phenols, chlorolignins, chlorinated resins, fatty acids, lignocellulose, and biocides, all of which are classified [176,177].

The potential effluents from paper mills are classified into the following four main categories based on their generation phase/state, i.e., solid waste, wastewater, gaseous waste, and particle waste [178]. The wastewater mainly contains suspended solids (bark particles, dirt, pigments, and fibers), dissolved inorganics (such as sodium hydrogen and sodium silicate), toxic chemicals (such as resin acids, tannins, and chlorinated organic compounds), and bleached agents [176]. Effluents also comprise color, dyes, and toxic chemicals [179] Along with them, other pollutants discharged from the paper industry include sizing agents, alcohols, thermal loads, turpentine, and microbial biomass such as coliform groups [180].

Overall, the wastewater discharged from pulp and paper industries is characterized by complex organic as well as inorganic pollutants. These pollutants are discharged mainly from pulping and bleaching processes during paper manufacturing. The main gaseous pollutants are hydrogen sulfides, sodium sulfide, dimethyl sulfide, methyl mercaptan, sulfur, chlorine dioxide, and particulates of sulfur and nitrogen oxides which have been linked to chronic respiratory disorders, skin and eye irritation, and cardiac problems, as well as nausea and headache.

The inorganic pollutants in the discharged pulp and paper industries wastewater include various inorganic salts, and inorganic metallic and non-metallic compounds in the complex form. The major metals discharged from the pulp and paper industries are Cd, Cr, Cu, Fe, Ni, Mn, Pb, Zn, Cu, Hg, As, K, Na, Ca, Mg, and non-metallic elements are primarily chloride, sulfate, and phosphate. These are generated from the pulping and bleaching of pulp and the washing process. The toxicity of different heavy metals varies due to the presence of other inorganic non-metallic components and the hardness of pulp and paper industries effluent along with other metals [181,182]. The discharge of wastewater from pulp and paper industries, along with heavy metals, raises various parameters such as TS, TSS, and TDS, which affect aquatic organism life [183].

The detected major organic and inorganic pollutants are hexadecanoic acids, octacosane, sitosterol trimethylsilyl ether, 1-tetradecane, 2-methoxy phenol, trichlorocatechol, tetrachlorocatechol, chlorophenols, chloroguaiacols, chlorosyringols, chlorocatechols, terpenes, methanol, phenol, alkylated phenols, decalone, benzoic acid, abietic acid, and dehydroabietic acid [182].

The pulp and paper mill effluents often contain dioxins, which are dominated by octachlorinated dioxin. Dioxins have never been intentionally produced but are released to the environment as unwanted by-products from chemical processes such as chlorine bleaching of pulp and paper, the manufacture of pesticides, herbicides, and fungicides. Dioxins can be formed via precursor compounds, such as pentachlorophenol chlorinated diphenylethers, polychlorinated biphenyls, or chlorobenzene, which are used as fungicidal and bacterial agents in wood treatment [184]. Even the trace quantities of PCP in wood or pulp may be a source of chlorinated dioxins in effluents and final products. Dioxins can be removed by, e.g., coagulation, sedimentation, and filtration [185]. Most dioxin congeners can be removed from raw water during the water purification process (87% removal efficiency). However, due to chlorination, their amounts may increase during treatment processes. Therefore, effective methods for their removal from water are still needed. Advanced oxidation processes (AOPs) are methods based on the generation of intermediate radicals (e.g., hydroxyl radicals) that are applied to degrade organic compounds. For purposes of wastewater treatment, methods such as photolysis, photocatalysis, and Fenton oxidation have been studied. In particular, photolysis based on sunlight radiation degrades dioxins and furans effectively.

In the papermaking process, biocides are the main practice in eliminating unwanted microbial growth and are widely used to combat microbial, fungal, and algal growth. The biocide is selected based on knowledge of the process conditions and the types of microorganisms that form the biofilm. For example, carbamate-based biocides (e.g., sodium dimethyl dithiocarbamate and disodium ethylene bisthiocarbamate) and glutaraldehyde are usual biocides to control biofilm formation in pulp and paper mills. Other biocides used in papermaking processes include, e.g., halogenated hydantoin, 2-bromo-2-nitropropane-1,3-diol, and *n*-octylisothiazolin-3-one. Biocides, often termed preservatives, include, e.g., thione, glutaraldehyde, and 1,2-benzo-isothiazolin-3-one. Previously, the most important biocides in papermaking were chlorine and copper sulfate. Later in the 1950s and 1960s, organic mercury compounds, such as phenylmercuric acetate and ethylmercury salts, were used, often in combination with chlorophenates. However, biocides were prohibited because of their adverse environmental effects.

A fraction of biocides will degrade biologically or chemically, but some will remain in the aqueous phase, and the rest will be present in the effluent or solid material. In addition, some biocides can be fiber retentive and accumulate in the final paper product due to their physicochemical properties. Biological methods are preferred for the treatment of biocide-containing effluents. However, toxic compounds inactivate degrading microorganisms. Therefore, a useful approach is to partially pretreat the toxic waste by oxidation. Oxidation treatments include photochemical processes (UV/O_3, UV/H_2O_2), photocatalysis (TiO_2/UV, Fenton, photo-Fenton), and chemical oxidation (O_3, O_3/H_2O_2, H_2O_2/Fe^{2+}).

Some other organic pollutants of pulp and paper industry effluent [182]:

- Trichlorotrihydroxybenzenes,bromomethylpropanylbenzene,dichloromethylpropanylbenzene.
- Octylphenol, nonylphenol, bisphenol A, estrone, estradiol, triclosan, PCBs, PAHs, PCDD/PCDF, trichlorohydroxyfuranone.
- Trichlorocatechol, tetrachlorocatechol, dichlorohydroxyquinone.
- Octadecenoic acid amide, linoleic acid, linoleic acid isomer Juvabione, dehydrojuvabione.
- Methyl dehydroabietatethyl dehydroabietate, ethyl abietate, dehydroabietic acid, abietic acid, chlorinated pterostilbene.
- Dimethyldiphenylmethane, dimethylphenylethane, di-*iso*-propyl naphthalene.
- Chlorophenols, chloroguaiacols, chlorosyringols, chlorocatechols.
- Chlorovanillins, chlorinated dehydro abietic acids.
- Nonylphenolethoxycarboxylates, sitosterol, wood sterols, O-cresol.
- Terpenes, methyl trisulfide, hexachloroethane, formyl thiophene, acetylthiophene, propionyl thiophene.
- Trichlorophenol, dehydroabietic acid, chlorodehydroabietic acid, dichlorodehydroabietic acid.
- Nonylphenolethoxycarboxylates.
- Nitrilotriacetic acid, EDTA, DTPA.
- Terpenes, methanol, phenol, alkylated phenols, decalone, benzoic acid, abietic acid, dehydroabietic acid.

Treatment of all the organic pollutants in the pulp and paper industries effluent is difficult. However, the activated sludge was used for the treatment of organic material [186]. However, the treated effluent still contains biorecalcitrant organics, such as aromatic and lignin-related substances. These compounds mineralize to a lesser extent during the biological treatment. The molecular weight and polarity of substances are important factors for the evaluation of their environmental behavior. Biological treatment is largely used to treat pulp and paper mill wastewaters and remove organic material content [187]. However, it can only partially remove the color of water. The wastewater also contains other organic components, such as lignin and its derivatives, which biodegrade reluctantly [186]. In addition, the transformation of these compounds into other toxic compounds during the biological treatment may occur. The fatty acids in the pulp and paper mill effluents can be degraded anaerobically if their concentrations are kept low enough to avoid them causing significant inhibition to anaerobic bacteria.

8.2.5 Paint and Varnish Industry

8.2.5.1 Dimethyl Phthalate

Dimethyl phthalate (DMP) is a colorless organic compound having a variety of applications. It belongs to the class of refractory organic plasticizer compounds known as phthalic acid esters (PAEs), which are widely used as additives in a variety of polymeric compounds, such as plastic toys, paints, cosmetics, and automobiles. PAEs have been listed as a priority pollutant in the United States, Europe, and China [188,189]. DMP is highly soluble in water and is moderately toxic to humans compared to its metabolic intermediate monomethylphthalate (MMP) [166].

8.2.5.2 Tributyltin

Tri-*n*-butyltin (TBT) belongs to the trialkyl organotin family of chemicals. TBT is used extensively in several products used in daily life. It is used as an antifouling agent in paints.

8.2.6 Sugar Industry

The sugar industry generates about 1,000 L of wastewater for every ton of sugarcane crushed [190]. Furthermore, sugar factories annexed to ethanol distilleries produce, on average, 155 L of spent wash (wastewater), 250 kg of bagasse, 95 kg of sugar, and 12 L of ethanol from 1 ton of sugarcane [191]. Wei & Xu, Saranraj & Stella, Sahu [192–194] found large amounts of pollutants in the wastewater from the sugar industry, such as 5-day biochemical oxygen demand (BOD_5) of 1,090 mg/L, COD of 3,682 mg/L, total solids (1,287 mg/L), chloride (377 mg/L), sulfate (430 mg/L), nitrogen (1,300 mg/L), calcium (393 mg/L), and magnesium (299 mg/L). Usually, sugar industry wastewater contains large amounts of organic and particulate matter, which are above environmental discharge standards.

Currently, industrial wastewater containing various metals, including Cr, Cd, Cu, Ni, Fe, Hg, Mn, Pb, and Zn, is being discharged into the environment in developing countries [195,196]. Many research reports show the presence of high metal concentrations in sugar industry wastewater, particularly in ethanol distillery spent wash [197–200]. When these heavy metals are released into bodies of water, they will be hydrated and become highly toxic, disrupting the normal enzymatic functions of biota (flora and fauna) and enhancing the absorption process of biota [201,202]. Industrial metal pollution is a serious threat to the survival of terrestrial and aquatic life too because toxic metals act slowly and persist for long periods in biotic systems due to their nonbiodegradability [195].

8.3 SOME WORKS ON THE TREATMENT OF INDUSTRIAL POLLUTANTS USING NANOMATERIALS

Table 8.3 shows some recent works on the treatment of industrial pollutants in various industries using NMs. For tanning industries, it is found that most of the researchers used several metals or metal oxide NPs for the treatment of mostly inorganic pollutants such as metals or metalloids. Few researchers worked on the treatment of phenol. They found good removal efficiency for their treatment experiments. For the textile industry, some researchers utilized Fe_2O_3 and MgO NPs for the treatment of dyes and several organic pollutants. In the pharmaceutical industry wastewater, several medicinal residues including antibiotics are reported. Researchers used CNTs and their composites with inorganic NMs such as Al_2O_3, $CoFe_2O_4$, and various composites of inorganic NMs with biomolecules of plant materials for the treatment of various organic pollutants in pharmaceutical industries. Nogueira et al. [203] used nano-TiO_2 and Fe_2O_3 for the treatment of Zn, Al, and Cd while Singh et al. [204] applied iron NPs for the treatment of organic pollutants in the pulp and paper industries' wastewater. However, the pulp and paper industries contain many organic pollutants in the discharged wastewater to be treated before being released into the environment. It is noted that many industries have some common pollutants of both inorganic and organic types, and these can be treated by adopting some common strategies.

8.4 LIMITATIONS AND FUTURE RESEARCH SCOPES ON INDUSTRIAL WASTEWATER TREATMENT

For maximum economic growth of a country and for the betterment of human beings, industrial development is crucial. But with rapid industrialization, the world is getting polluted which affects the health of all organs of the environment such as air, soil, sediment, and water as well as both ecosystem and biodiversity due to the uncontrolled discharge of industrial pollutants [1]. To protect the environment, effective ETP should be ensured in the industries. Most importantly, the development of effective materials can play an important role in industrial wastewater treatment. NM and nanotechnology could be a good choice in this regard due to the exceptional properties of the materials.

TABLE 8.3
Recent Works on the Treatment of Industrial Pollutants Using Nanaomaterials

Industries	Pollutants: Inorganic	Pollutants: Organic	Nanomaterials for Treatment	Removal Efficiency (%)	References
Tanning industry	Co, Cr, Pb, Ni, Cd	—	γ-Fe_2O_3 nanoparticles	66.2, 83.0, 87.9, 74.0, 98.7	[205]
	Co, Pb, Cd, Ni, Cr	—	MgO nanoparticles	63.4, 72.7, 74.1, 70.8, 94.2	[206]
	Cd, Pb, Cu, Ni, As, Cr, Hg, Sn, B, H_2S	Oil, phenols, CN	Fe/Cu nanoparticles	—	[207]
	Cr^{6+}, Fe^{3+}	—	Kaolin/ZnO nanoparticles	100, 99.8	[208]
	—	Phenol	ZnO/montmorillonite	88	[209]
	Co, Cr, Pb, Ni, Cd	—	α-Fe_2O_3 nanoparticles	35.1, 75.4, 13.4, 13.5, 7.3	[210]
	Co, Cr, Pb, Ni, Cd	—	MgO nanoparticles	35.5, 87.06, 15.4, 28.2, 0.6	[210]
Textile industry	—	Dyes	Fe_2O_3 nanoparticles	—	[205]
	—	Dyes	MgO nanoparticles	—	[206]
	—	4,5-Dihydro-3-methyl-2-butenal, 1,2,3,4,5-pentamethylcyclopentene, 3,3,4-trimethyldecane, butyl-cyclohexylmethyl ester, 2,3,4-trimethylpentane, *N*-desmethyltapentadol, 2-ethoxyamphetamine	α-Fe_2O_3 nanoparticles	—	[210]
	—	4,5-Dihydro-3-methyl-2-butenal, 1,2,3,4,5-pentamethylcyclopentene, 3,3,4-trimethyldecane, butyl-cyclohexylmethyl ester, 2,3,4-trimethylpentane, N-desmethyltapentadol, 2-ethoxyamphetamine	MgO nanoparticles	—	[210]

(Continued)

TABLE 8.3 (*Continued*)
Recent Works on the Treatment of Industrial Pollutants Using Nanomaterials

Industries	Pollutants: Inorganic	Pollutants: Organic	Nanomaterials for Treatment	Removal Efficiency (%)	References
Pharmaceutical industry	—	Oxytetracycline	Single-walled carbon nanotubes	—	[211]
	—	—	Double-walled carbon nanotubes	—	[211]
	—	Ciprofloxacin	Multi-walled carbon nanotubes	—	[211]
	—	Tetracycline	Multi-walled carbon nanotubes	—	[212]
	—	Ciprofloxacin	Multi-walled carbon nanotubes	—	[65]
	—	Sulfamethoxazole and Sulfapyridine	Multi-walled carbon nanotubes	—	[213]
	—	Carbamazepine Diclofenac sodium (DS)	Multi-walled carbon nanotubes/Al_2O_3	—	[214]
	—	Pefloxacin	O-multi-walled carbon nanotubes	—	[215]
	—	Sulfamethoxazole	Multi-walled carbon nanotubes/$CoFe_2O_4$	—	[216]
	—	AMOX	Sn/Zn/TiO_2	—	[217]
	—	Acetaminophen	Hydroxylated multi-walled carbon nanotubes	74%	[218]
	—	Acetaminophen	Ozone-treated multi-walled carbon nanotubes	95%	[219]
	—	Acetaminophen, naproxen, ibuprofen, ketoprofen, nadolol, bisoprolol, aniline	Multi-walled carbon nanotubes	100, 70,70,40,70,40, 98.9	[220]
	—	Tetracycline	Multi-walled carbon nanotubes	53.1–99.9 (pH variation)	[221]
	—	Tetracycline	ZrO_2 nanoparticles with *Pseudomonas aeruginosa* bacteria	98.73	[222]
	—	Caffeine	$AgFeO_2$ nanoparticles with *Amaranthus blitum* leaf	99.9	[223]
	—	Crystal violet	Keratine nanoparticles from human hair	75.97	[224]
	—	Ibuprofen, 17b-estradiol, sulfamethoxazole and ciprofloxacin	Pt and Pd nanoparticles with *Desulfovibrio vulgaris*	94, 85, 70	[225]
	—	Rhodamine B	ZnO colloidal nanoparticles of *Cynometra ramiflora*	98	[226]
	—	Methyl violet, 2,4-dichlorophenoxyacetic acid and paracetamol	Fluorescent ZnS nanoparticles of *Aspergillus flavus*	87, 33, 51	[227]
	—	Naproxen	SnO_2/Activated Carbon nano composite from *Corchorus olitorius*	94	[228]
	—	Diclofenac sodium	Se nanoparticles of *S. griseobrunneus*	97.43	[229]
	—	4-Nitrophenol	Ag nanoparticles of Palm date fruit extract	99	[230]
	—	Imatinib and imipenem	Fe_3O_4/Au nanocomposite with *Carum carvi* L. seeds	92 and 97	[231]
Pulp and paper industry	Zn, Al, and Cd	—	Nano-TiO_2 and nano-Fe_2O_3	—	[203]
	—	Organic pollutants	Iron nanoparticle	—	[204]
	—	AOX	Nano zero valent iron	—	[232]

Removal efficiencies (%) are shown according to the order of the respective pollutants.

Further, a large number of pollutants (organic and inorganic) are found in the industrial discharge but only a limited number of materials and techniques are available for their treatment. This opens a room to search for more advanced materials for the treatment of industrial wastewater. Conventionally, from the known types of pollutants of the selected industries, it will be helpful for the scientist to develop some promising materials for the possible effective treatment of the identified pollutants, although regular and continuous monitoring work for the pollutant level in the wastewater samples should be carried out.

Most of the previous works (Table 8.3) so far studied for the book chapter treated the industrial pollutants with only a few NMs, which is insufficient considering the huge number of industrial pollutants. Furthermore, these studies are mostly discrete and incomprehensive. This implies that there is still much opportunity for researchers to work with this issue of wastewater treatment for extensive industrial pollutants by developing suitable and effective treatment materials and methods. In addition, naturally, environmental fading or biodegradation and natural bioaccumulation of the industrial pollutants in wastewater may be another important research topic in this regard.

8.5 CONCLUSIONS

This chapter identified a large number of industrial pollutants in some selected industries and some recent NMs for their possible treatment before discharging to the environment, in addition to some biological and chemical treatments of the pollutants. It is found that many researchers utilized some NMs of both inorganic and organic origin for the treatment of industrial pollutants. However, considering the large number of pollutants discharged by several industries, it is very difficult to treat all the pollutants with the employment of a limited number of NMs. This invokes work on this issue to develop more effective materials for the treatment of industrial wastewater, ensuring an effective ETP. Furthermore, the collective information of this chapter could also be useful for scientists and environmentalists to know about the diverse pollutants discharged with the industrial wastewater and to develop more interesting NMs for the treatment of several industrial pollutants.

GLOSSARY

Aliphatic hydrocarbons: Hydrocarbons based on chains of carbon atoms.

Aromatic hydrocarbons: Aromatic organic molecules made up entirely of carbon and hydrogen with sigma bonds and delocalized pi electrons between carbon atoms forming rings.

Bioaccumulation: The net accumulation of a contaminant in or on an organism from all sources including water, air, and diet.

Biodegradation: Degradation of the materials into environmentally acceptable products such as water, carbon dioxide, and biomass by the action of naturally available microorganisms under normal environmental conditions.

Deacetylation: A process of hydrolysis of acetamide groups (removal of an acetyl group) in chitin using strong NaOH solution at high temperatures (100°C or more) to produce the amino group of the new compounds (CS).

Demineralization: The process of removing mineral matters or salts.

Deproteinization: An act or process of deproteinizing, or removing protein.

Environmental sustainability: Environmental sustainability is a broad and interdisciplinary field that focuses on maintaining the health of our planet for future generations. Some key fundamentals of environmental sustainability are: (a) Ecosystem Health: Ensuring the stability and resilience of ecosystems by protecting biodiversity and natural habitats; (b) Resource Management: Efficiently using natural resources like water, minerals, and forests to prevent depletion and ensure long-term availability; (c) Pollution Prevention: Reducing emissions and waste to minimize environmental impact, including air, water, and soil pollution; (d) Energy Efficiency: Promoting the use of renewable energy sources (solar, wind, and hydro) and improving energy efficiency to reduce reliance on fossil fuels; (e) Sustainable Agriculture: Implementing farming practices that protect the environment, conserve resources, and support biodiversity; (f) Waste Management: Developing strategies for reducing, reusing, and recycling waste to minimize landfill use and environmental contamination; (g) Climate Change Mitigation: Taking actions to reduce greenhouse gas emissions and enhance carbon sinks to combat global warming; (h) Environmental Policy and Ethics: Creating and enforcing regulations that protect the environment and promote sustainable practices while considering ethical implications. These fundamentals are essential for addressing global challenges such as climate change, resource depletion, and pollution.

Hydrocarbon: Any of a class of organic chemicals made up of only the elements carbon and hydrogen.

Industrial wastewater: The wastewater generated by manufacturing or industrial processes.

Nanocomposites: A multiphase solid material where one of the phases has at least one dimension on the nanoscale (100 nm).

Nanomaterials: Material with one or more external dimensions on the nanoscale, or an internal structure or surface structure on a scale ranging from 1 to 100 nm.

Nanoparticle: Particle with at least one external dimension on the nanoscale (1–100 nm).

Nanotechnology: Nanotechnology involves manipulating matter at the atomic and molecular scale, typically below 100 nm. At this scale, materials exhibit unique physical, chemical, and biological properties that differ significantly from their bulk counterparts. These properties include: (a) Increased Surface Area: Nanoparticles have a high surface area-to-volume ratio, enhancing their reactivity and interaction with other substances; (b) Quantum Effects: At the nanoscale, quantum mechanical effects become significant, affecting optical, electrical, and magnetic properties; and (c) Enhanced Mechanical Properties: Nanomaterials can exhibit superior strength, flexibility, and durability.

Oxygenated solvents: An oxygenated solvent contains oxygen molecules with high solubility and low toxicity.

Pollutants: The substances which are emitted directly from the source and persist in the form in which they were added to the environment.

REFERENCES

1. Bhuiyan, M. A. H., Suruvi, N. I., Dampare, S. B., Islam, M. A., Quraishi, S. B., Ganyaglo, S., & Suzuki, S. (2011). Investigation of the possible sources of heavy metal contamination in lagoon and canal water in the tannery industrial area in Dhaka, Bangladesh. *Environmental Monitoring and Assessment*, *175*(1), 633–649. https://doi.org/10.1007/s10661-010-1557-6
2. Ahsan, M., Satter, F., Siddique, M., Bakar, A., Akbor, M., Ahmed, S., & Khan, R. (2019). Chemical and physicochemical characterization of effluents from the tanning and textile industries in Bangladesh with multivariate statistical approach. *Environmental Monitoring and Assessment*, *191*(9), 1–24.
3. Rakib, M. A., Huda, M. E., Hossain, S. M., Naher, K., Khan, R., Sultana, M. S., & Patwary, M. A. (2013). Arsenic content in inactive tissue: human hair and nail. *Journal of Scientific Research and Reports*, *2*(2), 522.
4. Chandra, V., Park, J., Chun, Y., Lee, J. W., Hwang, I. C., & Kim, K. S. (2010). Water-dispersible magnetite-reduced graphene oxide composites for arsenic removal. *ACS Nano*, *4*(7), 3979–3986.
5. Lue, J. T. (2001). A review of characterization and physical property studies of metallic nanoparticles. *Journal of Physics and Chemistry of Solids*, *62*(9–10), 1599–1612.
6. Krahne, R., Morello, G., Figuerola, A., George, C., Deka, S., & Manna, L. (2011). Physical properties of elongated inorganic nanoparticles. *Physics Reports*, *501*(3–5), 75–221.
7. ISO/TS 27867 (2008). Nanotechnologies—terminology and definitions for nano-objects—nanoparticle, nanofibre and nanoplate. Available from: https://www.iso.org/standard/44278.html
8. UNION, P. (2009). Regulation (EC) No 1223/2009 of the European parliament and of the council. *Official Journal of the European Union L*, *342*, 59.

9. Buzea, C., Pacheco, I. I., & Robbie, K. (2007). Nanomaterials and nanoparticles: sources and toxicity. *Biointerphases*, *2*(4), MR17–MR71.
10. Potocnik, J. (2011). Commission recommendation of 18 October 2011 on the definition of nanomaterial (Text with EEA relevance) (2011/696/EU). *Official Journal of the European Union*, European Commission, L 275, 38–40.
11. European Union (2011). Regulation (EU) No 1169/2011 of the European Parliament and of the Council of 25 October 2011 on the provision of food information to consumers. *Official Journal of the European Union*, *304*(19), 46–54.
12. Rauscher, H., Sokull-Klüttgen, B., & Stamm, H. (2012). The European Commission's recommendation on the definition of nanomaterial makes an impact. *Nanotoxicology*, *7*(7), 1195–1197.
13. Iso.org, 80004-1:2010, I. (2017). ISO/TS 80004-1:2010—Nanotechnologies—vocabulary—part 1 Core terms. Iso.org. Available from: https://www.iso.org/standard/51240.html
14. Khlebtsov, N. G., & Dykman, L. A. (2010). Optical properties and biomedical applications of plasmonic nanoparticles. *Journal of Quantitative Spectroscopy and Radiative Transfer*, *111*(1), 1–35.
15. Wilson, M., Kannangara, K., Smith, G., Simmons, M., & Raguse, B. (2002). *Nanotechnology: Basic Science and Emerging Technologies*. Strawberry Hills: ReadHowYouWant.
16. Khlobystov, A. N., Porfyrakis, K., Kanai, M., Britz, D. A., Ardavan, A., Shinohara, H., & Briggs, G. A. D. (2004). Molecular motion of endohedral fullerenes in single-walled carbon nanotubes. *Angewandte Chemie*, *116*(11), 1410–1413.
17. Sun, X. H., Wong, N. B., Li, C. P., Lee, S. T., & Sham, T. K. (2004). Chainlike silicon nanowires: morphology, electronic structure and luminescence studies. *Journal of Applied Physics*, *96*(6), 3447–3451.
18. Hochbaum, A. I., Fan, R., He, R., & Yang, P. (2005). Controlled growth of Si nanowire arrays for device integration. *Nano Letters*, *5*(3), 457–460.
19. Zhong, Z., Chen, F., Subramanian, A. S., Lin, J., Highfield, J., & Gedanken, A. (2006). Assembly of Au colloids into linear and spherical aggregates and effect of ultrasound irradiation on structure. *Journal of Materials Chemistry*, *16*(5), 489–495.
20. Geckeler, K. E., & Rosenberg, E. (Eds.). (2006). *Functional Nanomaterials* (pp. 328–329). Valencia: American Scientific Publishers.
21. Bhushan, B. (2007). Springer *Handbook of Nanotechnology* (Vol. 2). Berlin: Springer.
22. Teo, B. K., & Sun, X. H. (2007). Silicon-based low-dimensional nanomaterials and nanodevices. *Chemical Reviews*, *107*(5), 1454–1532.
23. Edelstein, A. S., & Cammaratra, R. C. (1998). *Nanomaterials: Synthesis, Properties and Applications*. Boca Raton, FL: CRC Press.
24. Rahman, M. R., Bakri, M. K. B., Matin, M. M., & Khui, P. L. N. (2022). Micro and nano effects of recycled plastic waste to reinforce and enhance in biocomposites, in: M. R. Rahman & M. K. B. Bakri (Eds.), *Recycled Plastic Biocomposites* (pp. 195–211). Cambridge, MA: Woodhead Publishing. https://doi.org/10.1016/B978-0-323-88653-6.00001-8
25. Thostenson, E. T., Li, C., & Chou, T. W. (2005). Nanocomposites in context. *Composites Science and Technology*, *65*(3–4), 491–516.
26. Seigel, R.W. (1994). Nanophase materials, in: G. L. Trigg (Ed.), *Encyclopedia of Applied Physics* (pp. 1–27). Weinheim: VCH.
27. Pankaj, K. T., Shruti, T., Charul, V., & Aanchal, R. (2013). Estimation of toxic effects of chemically and biologically synthesized silver nanoparticles on human gut microflora containing Bacillus subtilis. *Journal of Toxicology and Environmental Health Sciences*, *5*(9), 172–177.
28. Sekunowo, O. I., Durowaye, S. I., & Lawal, G. I. (2015). An overview of nano-particles effect on mechanical properties of composites. *International Journal of Aerospace and Mechanical Engineering*, *9*(1), 1–7.
29. Capsal, J. F., Dantras, E., Dandurand, J., & Lacabanne, C. (2011). Molecular mobility in piezoelectric hybrid nanocomposites with 0–3 connectivity: volume fraction influence. *Journal of Non-crystalline Solids*, *357*(19–20), 3410–3415.
30. Wang, Z. L., & Song, J. (2006). Piezoelectric nanogenerators based on zinc oxide nanowire arrays. *Science*, *312*(5771), 242–246.
31. She, G., Zhang, X., Shi, W., Fan, X., & Chang, J. C. (2007). Electrochemical/chemical synthesis of highly-oriented single-crystal ZnO nanotube arrays on transparent conductive substrates. *Electrochemistry Communications*, *9*(12), 2784–2788.
32. Yamano, A., Takata, K., & Kozuka, H. (2012). Ferroelectric domain structures of 0.4-µm-thick Pb(Zr,Ti)O_3 films prepared by polyvinylpyrrolidone-assisted Sol-Gel method. *Journal of Applied Physics*, *111*(5), 054109.

33. Qi, Y., Jafferis, N. T., Lyons Jr, K., Lee, C. M., Ahmad, H., & McAlpine, M. C. (2010). Piezoelectric ribbons printed onto rubber for flexible energy conversion. *Nano Letters*, *10*(2), 524–528.
34. Asghari, F., Jahanshiri, Z., Imani, M., Shams-Ghahfarokhi, M., & Razzaghi-Abyaneh, M. (2016). Antifungal nanomaterials: synthesis, properties, and applications, in: A. M. Grumezescu (Ed.), *Nanobiomaterials in Antimicrobial Therapy* (pp. 343–383). Oxford: William Andrew Publishing.
35. Shiau, B. W., Lin, C. H., Liao, Y. Y., Lee, Y. R., Liu, S. H., Ding, W. C., & Lee, J. R. (2018). The characteristics and mechanisms of Au nanoparticles processed by functional centrifugal procedures. *Journal of Physics and Chemistry of Solids*, *116*, 161–167.
36. Dai, H. (2002). Carbon nanotubes: synthesis, integration, and properties. *Accounts of Chemical Research*, *35*(12), 1035–1044.
37. Yan, Q. L., Gozin, M., Zhao, F. Q., Cohen, A., & Pang, S. P. (2016). Highly energetic compositions based on functionalized carbon nanomaterials. *Nanoscale*, *8*(9), 4799–4851.
38. Kreuter, J. (1994). *Colloidal Drug Delivery Systems* (Vol. 66). Boca Raton, FL: CRC Press.
39. Kumar, R., & Lal, S. (2014). Synthesis of organic nanoparticles and their applications in drug delivery and food nanotechnology: a review. *Nanomaterials & Molecular Nanotechnology*, *3*, 4. *of*, *11*, 2.
40. Utekhina, A. Y., & Sergeev, G. B. (2011). Organic nanoparticles. *Russian Chemical Reviews*, *80*(3), 219.
41. Naahidi, S., Jafari, M., Edalat, F., Raymond, K., Khademhosseini, A., & Chen, P. (2013). Biocompatibility of engineered nanoparticles for drug delivery. *Journal of Controlled Release*, *166*(2), 182–194.
42. Asta, M., Kauzlarich, S. M., Liu, K., Navrotsky, A., & Osterloh, F. E. (2007). Inorganic nanoparticles. Unique properties and novel applications. *Material Matters* (Milwaukee, WI, USA), *2*(1), 3–6.
43. Bailey, R. E., Smith, A. M., & Nie, S. (2004). Quantum dots in biology and medicine. *Physica E: Low-dimensional Systems and Nanostructures*, *25*(1), 1–12.
44. Okamoto, M., & Ray, S. S. (2004). Polymer/clay nanocomposites. *Encyclopedia of Nanoscience and Nanotechnology*, *8*, 791–843.
45. Yano, K., Usuki, A., Okada, A., Kurauchi, T., & Kamigaito, O. (1993). Synthesis and properties of polyimide–clay hybrid. *Journal of Polymer Science Part A: Polymer Chemistry*, *31*(10), 2493–2498.
46. Gilman, J. W. (1999). Flammability and thermal stability studies of polymer layered-silicate (clay) nanocomposites. *Applied Clay Science*, *15*(1–2), 31–49.
47. Chenab, K. K., Sohrabi, B., Jafari, A., & Ramakrishna, S. (2020). Water treatment: functional nanomaterials and applications from adsorption to photodegradation. *Materials Today Chemistry*, *16*, 100262.
48. Madima, N., Mishra, S. B., Inamuddin, I., & Mishra, A. K. (2020). Carbon-based nanomaterials for remediation of organic and inorganic pollutants from wastewater: a review. *Environmental Chemistry Letters*, *18*(4), 1169–1191.
49. Zaytseva, O., & Neumann, G. (2016). Carbon nanomaterials: production, impact on plant development, agricultural and environmental applications. *Chemical and Biological Technologies in Agriculture*, *3*(1), 1–26.
50. Zhu, J., Hou, J., Zhang, Y., Tian, M., He, T., Liu, J., & Chen, V. (2018). Polymeric antimicrobial membranes enabled by nanomaterials for water treatment. *Journal of Membrane Science*, *550*, 173–197.
51. Amin, M. T., Alazba, A. A., & Manzoor, U. (2014). A review of removal of pollutants from water/wastewater using different types of nanomaterials. *Advances in Materials Science and Engineering*, 2014. https://doi.org/10.1155/2014/825910
52. Kong, Q., Zhang, H., Lan, Y., Shi, X., Fang, Z., Chang, Q., Liu, J., & Wei, C. (2022). Functional graphene oxide for organic pollutants removal from wastewater: a mini review. *Environmental Technology*, *44*, 1–13. https://doi.org/10.1080/09593330.2022.2053754
53. Yu, F., Li, Y., Han, S., & Ma, J. (2016). Adsorptive removal of antibiotics from aqueous solution using carbon materials. *Chemosphere*, *153*, 365–385.
54. Thakur, K., & Kandasubramanian, B. (2019). Graphene and graphene oxide-based composites for removal of organic pollutants: a review. *Journal of Chemical & Engineering Data*, *64*(3), 833–867. https://doi.org/10.1021/acs.jced.8b01057
55. Amanulla, B., Palanisamy, S., Chen, S. M., Chiu, T. W., Velusamy, V., Hall, J. M., ... & Ramaraj, S. K. (2017). Selective colorimetric detection of nitrite in water using chitosan stabilized gold nanoparticles decorated reduced graphene oxide. *Scientific Reports*, *7*(1), 1–9.
56. Yang, A., Yang, P., & Huang, C. P. (2017). Preparation of graphene oxide–chitosan composite and adsorption performance for uranium. *Journal of Radioanalytical and Nuclear Chemistry*, *313*(2), 371–378.
57. Bai, H., Li, C., Wang, X., & Shi, G. (2011). On the gelation of graphene oxide. *The Journal of Physical Chemistry C*, *115*(13), 5545–5551.

58. Dong, F., Li, S., Yan, M., & Li, C. (2014). Preparation and properties of chitosan/nanocrystalline cellulose composite films for food packaging. *Asian Journal of Chemistry*, *26*(17), 5895.
59. Kokkinos, P., Mantzavinos, D., & Venieri, D. (2020). Current trends in the application of nanomaterials for the removal of emerging micropollutants and pathogens from water. *Molecules*, *25*(9), 2016.
60. Cai, Z., Dwivedi, A. D., Lee, W. N., Zhao, X., Liu, W., Sillanpää, M., & Fu, J. (2018). Application of nanotechnologies for removing pharmaceutically active compounds from water: development and future trends. *Environmental Science: Nano*, *5*(1), 27–47.
61. Hummers Jr, W. S., & Offeman, R. E. (1958). Preparation of graphitic oxide. *Journal of the American Chemical Society*, *80*(6), 1339–1339.
62. Dreyer, D. R., Park, S., Bielawski, C. W., & Ruoff, R. S. (2010). The chemistry of graphene oxide. *Chemical Society Reviews*, *39*(1), 228–240.
63. Dimiev, A. M., & Tour, J. M. (2014). Mechanism of graphene oxide formation. *ACS Nano*, *8*(3), 3060–3068.
64. Rao, N., Singh, R., & Bashambu, L. (2021). Carbon-based nanomaterials: synthesis and prospective applications. *Materials Today: Proceedings*, *44*, 608–614.
65. Yu, F., Sun, S., Han, S., Zheng, J., & Ma, J. (2016). Adsorption removal of ciprofloxacin by multi-walled carbon nanotubes with different oxygen contents from aqueous solutions. *Chemical Engineering Journal*, *285*, 588–595.
66. Vuković, G. D., Marinković, A. D., Čolić, M., Ristić, M. Đ., Aleksić, R., Perić-Grujić, A. A., & Uskoković, P. S. (2010). Removal of cadmium from aqueous solutions by oxidized and ethylenediamine-functionalized multi-walled carbon nanotubes. *Chemical Engineering Journal*, *157*(1), 238–248.
67. Huang, Y., Lee, X., Macazo, F. C., Grattieri, M., Cai, R., & Minteer, S. D. (2018). Fast and efficient removal of chromium (VI) anionic species by a reusable chitosan-modified multi-walled carbon nanotube composite. *Chemical Engineering Journal*, *339*, 259–267.
68. Chatterjee, S., Chatterjee, T., & Woo, S. H. (2010). A new type of chitosan hydrogel sorbent generated by anionic surfactant gelation. *Bioresource Technology*, *101*(11), 3853–3858.
69. Hadi, A. G. (2013). Synthesis of chitosan and its use in metal removal. *Chemistry and Materials Research*, *3*(3), 64–73.
70. Domard, A., & Cartier, N. (1992). Glucosamine oligomers: 4. Solid state-crystallization and sustained dissolution. *International Journal of Biological Macromolecules*, *14*(2), 100–106.
71. Muzzarelli, R. A. (2009). Chitins and chitosans for the repair of wounded skin, nerve, cartilage and bone. *Carbohydrate Polymers*, *76*(2), 167–182.
72. Guibal, E., Milot, C., & Tobin, J. M. (1998). Metal-anion sorption by chitosan beads: equilibrium and kinetic studies. *Industrial & Engineering Chemistry Research*, *37*(4), 1454–1463.
73. Deans, J. R., & Dixon, B. G. (1992). Uptake of Pb^{2+} and Cu^{2+} by novel biopolymers. *Water Research*, *26*(4), 469–472.
74. Findon, A., McKay, G., & Blair, H. S. (1993). Transport studies for the sorption of copper ions by chitosan. *Journal of Environmental Science & Health Part A*, *28*(1), 173–185.
75. Bailey, S. E., Olin, T. J., Bricka, R. M., & Adrian, D. D. (1999). A review of potentially low-cost sorbents for heavy metals. *Water Research*, *33*(11), 2469–2479.
76. Hennen, W. J. (1996). *Chitosan: Natural Fat Blocker.* Grove, UT: Woodland Pub.
77. Wang, J., Liu, X., Jin, T., He, H., & Liu, L. (2019). Preparation of nanocellulose and its potential in reinforced composites: a review. *Journal of Biomaterials Science, Polymer Edition*, *30*(11), 919–946.
78. Dai, H., Ou, S., Huang, Y., & Huang, H. (2018). Utilization of pineapple peel for production of nanocellulose and film application. *Cellulose*, *25*(3), 1743–1756.
79. Nasir, M., Hashim, R., Sulaiman, O., & Asim, M. (2017). Nanocellulose: preparation methods and applications, in: J. Mohammad, B. Sami, & K. Abdul (Eds.), *Cellulose-Reinforced Nanofibre Composites: Production, Properties and Applications* (pp. 261–276). Cambridge, MA: Woodhead Publishing.
80. Van Benschoten, J. E., Reed, B. E., Matsumoto, M. R., & McGarvey, P. J. (1994). Metal removal by soil washing for an iron oxide coated sandy soil. *Water Environment Research*, *66*(2), 168–174.
81. Coston, J. A., Fuller, C. C., & Davis, J. A. (1995). Pb^{2+} and Zn^{2+} adsorption by a natural aluminum-and iron-bearing surface coating on an aquifer sand. *Geochimica et Cosmochimica Acta*, *59*(17), 3535–3547.
82. Agrawal, A., & Sahu, K. K. (2006). Kinetic and isotherm studies of cadmium adsorption on manganese nodule residue. *Journal of Hazardous Materials*, *137*(2), 915–924.
83. Henglein, A. (1989). Small-particle research: physicochemical properties of extremely small colloidal metal and semiconductor particles. *Chemical Reviews*, *89*(8), 1861–1873.
84. El-Sayed, M. A. (2001). Some interesting properties of metals confined in time and nanometer space of different shapes. *Accounts of Chemical Research*, *34*(4), 257–264.

85. Deliyanni, E. A., Peleka, E. N., & Matis, K. A. (2009). Modeling the sorption of metal ions from aqueous solution by iron-based adsorbents. *Journal of Hazardous Materials*, *172*(2–3), 550–558.
86. Ghanizadeh, G. H., Ehrampoush, M., & Ghaneian, M. (2010). Application of iron impregnated activated carbon for removal of arsenic from water. *Journal of Environmental Health Science & Engineering*, *7*(2), 145–156.
87. Nassar, N. N. (2012). Iron oxide nanoadsorbents for removal of various pollutants from wastewater: an overview, in: A. Bhatnagar (Ed.), *Application of Adsorbents for Water Pollution Control* (pp. 81–118). Oak Park, IL: Bentham Science Publishers.
88. Lata, S., & Samadder, S. R. (2016). Removal of arsenic from water using nano adsorbents and challenges: a review. *Journal of Environmental Management*, *166*, 387–406.
89. Li, L., Fan, M., Brown, R. C., Van Leeuwen, J., Wang, J., Wang, W., & Zhang, P. (2006). Synthesis, properties, and environmental applications of nanoscale iron-based materials: a review. *Critical Reviews in Environmental Science and Technology*, *36*(5), 405–431.
90. Willis, A. L., Turro, N. J., & O'Brien, S. (2005). Spectroscopic characterization of the surface of iron oxide nanocrystals. *Chemistry of Materials*, *17*(24), 5970–5975.
91. Cushing, B. L., Kolesnichenko, V. L., & O'connor, C. J. (2004). Recent advances in the liquid-phase syntheses of inorganic nanoparticles. *Chemical Reviews*, *104*(9), 3893–3946.
92. Park, J., An, K., Hwang, Y., Park, J. G., Noh, H. J., Kim, J. Y., … & Hyeon, T. (2004). Ultra-large-scale syntheses of monodisperse nanocrystals. *Nature Materials*, *3*(12), 891–895.
93. Wang, X., Zhuang, J., Peng, Q., & Li, Y. (2005). A general strategy for nanocrystal synthesis. *Nature*, *437*(7055), 121–124.
94. Ju-Nam, Y., & Lead, J. R. (2008). Manufactured nanoparticles: an overview of their chemistry, interactions and potential environmental implications. *Science of the Total Environment*, *400*(1–3), 396–414.
95. Saleh, T. A. (2020). Nanomaterials: classification, properties, and environmental toxicities. *Environmental Technology & Innovation*, *20*, 101067.
96. Shop .BSI, bsigroup.com. (2017). PAS 71:2011—Nanoparticles. Vocabulary—BSI British Standards Institute. Available from: https://webstore.ansi.org/standards/bsi/pas712011?srsltid=AfmBOoq3ba0-LrRHKhHzClapiun3XCNkTOVBk7lrrNpHYsRbPlm8H-vk
97. Lozhkomoev, A. S., Pervikov, A. V., Chumaevsky, A. V., Dvilis, E. S., Paygin, V. D., Khasanov, O. L., & Lerner, M. I. (2019). Fabrication of Fe-Cu composites from electroexplosive bimetallic nanoparticles by spark plasma sintering. *Vacuum*, *170*, 108980.
98. Oh, Y., Lee, J., & Lee, M. (2018). Fabrication of Ag-Au bimetallic nanoparticles by laser-induced dewetting of bilayer films. *Applied Surface Science*, *434*, 1293–1299.
99. Abdulhameed, A. S., Mohammad, A. T., & Jawad, A. H. (2019). Application of response surface methodology for enhanced synthesis of chitosan tripolyphosphate/TiO_2 nanocomposite and adsorption of reactive orange 16 dye. *Journal of Cleaner Production*, *232*, 43–56.
100. Saha, S., & Sarkar, P. (2012). Arsenic remediation from drinking water by synthesized nano-alumina dispersed in chitosan-grafted polyacrylamide. *Journal of Hazardous Materials*, *227*, 68–78.
101. Guibal, E. (2004). Interactions of metal ions with chitosan-based sorbents: a review. *Separation and Purification Technology*, *38*(1), 43–74.
102. Pradeep, T. (2009). Noble metal nanoparticles for water purification: a critical review. *Thin Solid Films*, *517*(24), 6441–6478.
103. Pan, B., Pan, B., Zhang, W., Lv, L., Zhang, Q., & Zheng, S. (2009). Development of polymeric and polymer-based hybrid adsorbents for pollutants removal from waters. *Chemical Engineering Journal*, *151*(1–3), 19–29.
104. Mahdavian, A. R., & Mirrahimi, M. A. S. (2010). Efficient separation of heavy metal cations by anchoring polyacrylic acid on superparamagnetic magnetite nanoparticles through surface modification. *Chemical Engineering Journal*, *159*(1–3), 264–271.
105. Zhao, X., Lv, L., Pan, B., Zhang, W., Zhang, S., & Zhang, Q. (2011). Polymer-supported nanocomposites for environmental application: a review. *Chemical Engineering Journal*, *170*(2–3), 381–394.
106. Zhang, Y. U., Yang, M. I. N., Dou, X. M., He, H., & Wang, D. S. (2005). Arsenate adsorption on an Fe-Ce bimetal oxide adsorbent: role of surface properties. *Environmental Science & Technology*, *39*(18), 7246–7253.
107. Kong, S., Wang, Y., Zhan, H., Yuan, S., Yu, M., & Liu, M. (2014). Adsorption/oxidation of arsenic in groundwater by nanoscale Fe-Mn binary oxides loaded on zeolite. *Water Environment Research*, *86*(2), 147–155.
108. Gupta, K., & Ghosh, U. C. (2009). Arsenic removal using hydrous nanostructure iron (III)–titanium (IV) binary mixed oxide from aqueous solution. *Journal of Hazardous Materials*, *161*(2–3), 884–892.

109. Li, Z., Deng, S., Yu, G., Huang, J., & Lim, V. C. (2010). As(V) and As(III) removal from water by a Ce–Ti oxide adsorbent: behavior and mechanism. *Chemical Engineering Journal*, *161*(1–2), 106–113.
110. Ren, Z., Zhang, G., & Chen, J. P. (2011). Adsorptive removal of arsenic from water by an iron–zirconium binary oxide adsorbent. *Journal of Colloid and Interface Science*, *358*(1), 230–237.
111. Gupta, K., Basu, T., & Ghosh, U. C. (2009). Sorption characteristics of arsenic (V) for removal from water using agglomerated nanostructure iron (III)–zirconium (IV) bimetal mixed oxide. *Journal of Chemical & Engineering Data*, *54*(8), 2222–2228.
112. Zhang, G., Ren, Z., Zhang, X., & Chen, J. (2013). Nanostructured iron (III)-copper (II) binary oxide: a novel adsorbent for enhanced arsenic removal from aqueous solutions. *Water Research*, *47*(12), 4022–4031.
113. Zhang, S., Niu, H., Cai, Y., Zhao, X., & Shi, Y. (2010). Arsenite and arsenate adsorption on coprecipitated bimetal oxide magnetic nanomaterials: $MnFe_2O_4$ and $CoFe_2O_4$. *Chemical Engineering Journal*, *158*(3), 599–607.
114. Basu, T., & Ghosh, U. C. (2011). Influence of groundwater occurring ions on the kinetics of As(III) adsorption reaction with synthetic nanostructured Fe(III)–Cr(III) mixed oxide. *Desalination*, *266*(1–3), 25–32.
115. Parsons, J. G., Lopez, M. L., Peralta-Videa, J. R., & Gardea-Torresdey, J. L. (2009). Determination of arsenic (III) and arsenic (V) binding to microwave assisted hydrothermal synthetically prepared Fe_3O_4, Mn_3O_4, and $MnFe_2O_4$ nanoadsorbents. *Microchemical Journal*, *91*(1), 100–106.
116. Ratheesh Chandra, P., Abdussalam, A. K., & Nabeesa, S. (2010). Distribution of bio-accumulated Cd and Cr in two Vigna species and the associated histological variations. *Journal of Stress Physiology & Biochemistry*, *6*(1), 4–12.
117. Ali Khan, M., Govindasamy, R., Ahmad, A., Siddiqui, M. R., Alshareef, S. A., Hakami, A. A. H., & Rafatullah, M. (2021). Carbon based polymeric nanocomposites for dye adsorption: synthesis, characterization, and application. *Polymers*, *13*(3), 419.
118. Beheshti, H., Irani, M., Hosseini, L., Rahimi, A., & Aliabadi, M. (2016). Removal of Cr (VI) from aqueous solutions using chitosan/MWCNT/Fe_3O_4 composite nanofibers-batch and column studies. *Chemical Engineering Journal*, *284*, 557–564.
119. Omanović-Mikličanin, E., Badnjević, A., Kazlagić, A., & Hajlovac, M. (2020). Nanocomposites: a brief review. *Health and Technology*, *10*(1), 51–59.
120. Parker, H. L., Hunt, A. J., Budarin, V. L., Shuttleworth, P. S., Miller, K. L., & Clark, J. H. (2012). The importance of being porous: polysaccharide-derived mesoporous materials for use in dye adsorption. *RSC Advances*, *2*(24), 8992–8997.
121. Schmidt, D., Shah, D., & Giannelis, E. P. (2002). New advances in polymer/layered silicate nanocomposites. *Current Opinion in Solid State and Materials Science*, *6*(3), 205–212.
122. Bharagava, R. N., Saxena, G., & Mulla, S. I. (2020). Introduction to industrial wastes containing organic and inorganic pollutants and bioremediation approaches for environmental management, in: G. Saxena & R. N. Bharagava (Eds.), *Bioremediation of Industrial Waste for Environmental Safety* (pp. 1–18). Singapore: Springer.
123. Saranraj, P., & Sujitha, D. (2013). Microbial bioremediation of chromium in tannery effluent: a review. *International Journal of Microbiology*, *4*(3), 305–320.
124. Tariq, S. R., Shah, M. H., Shaheen, N., Khalique, A., Manzoor, S., & Jaffar, M. (2005). Multivariate analysis of selected metals in tannery effluents and related soil. *Journal of Hazardous Materials*, *122*(1–2), 17–22.
125. Tariq, S. R., Shah, M. H., Shaheen, N., Khalique, A., Manzoor, S., & Jaffar, M. (2006). Multivariate analysis of trace metal levels in tannery effluents in relation to soil and water: a case study from Peshawar, Pakistan. *Journal of Environmental Management*, *79*(1), 20–29.
126. Costa, C. R., Botta, C. M., Espindola, E. L., & Olivi, P. (2008). Electrochemical treatment of tannery wastewater using DSA® electrodes. *Journal of Hazardous Materials*, *153*(1–2), 616–627.
127. Steiert, J. G., & Crawford, R. L. (1985). Microbial degradation of chlorinated phenols. *Trends in Biotechnology*, *3*(12), 300–305.
128. Saxena, G., Purchase, D., & Bharagava, R. N. (2020). Environmental hazards and toxicity profile of organic and inorganic pollutants of tannery wastewater and bioremediation approaches, in: G. Saxena & R. N. Bharagava (Eds.), *Bioremediation of Industrial Waste for Environmental Safety* (pp. 381–398). Singapore: Springer.

129. Saxena, G., Chandra, R., & Bharagava, R. N. (2016). Environmental pollution, toxicity profile and treatment approaches for tannery wastewater and its chemical pollutants. *Reviews of Environmental Contamination and Toxicology, 240*, 31–69.
130. Lyche, J. L., Gutleb, A. C., Bergman, Å., Eriksen, G. S., Murk, A. J., Ropstad, E., … & Skaare, J. U. (2009). Reproductive and developmental toxicity of phthalates. *Journal of Toxicology and Environmental Health, Part B, 12*(4), 225–249.
131. TFL. (2010). TFL Eco Guidelines. Available from: https://studylib.net/doc/10930843/december-2010-tfl-eco-guidelines
132. Dixit, S., Yadav, A., Dwivedi, P. D., & Das, M. (2015). Toxic hazards of leather industry and technologies to combat threat: a review. *Journal of Cleaner Production, 87*, 39–49.
133. He, L., Gielen, G., Bolan, N. S., Zhang, X., Qin, H., Huang, H., & Wang, H. (2015). Contamination and remediation of phthalic acid esters in agricultural soils in China: a review. *Agronomy for Sustainable Development, 35*(2), 519–534.
134. Yadav, A., Mishra, S., Kaithwas, G., Raj, A., & Bharagava, R. N. (2016). Organic pollutants and pathogenic bacteria in tannery wastewater. Organic pollutants and pathogenic bacteria in tannery wastewater and their removal strategies. *Microbes and Environmental Management, 7*, 120–127.
135. Kapdan, I. K., & Oztekin, R. (2003). Decolorization of textile dyestuff Reactive Orange 16 in fed-batch reactor under anaerobic condition. *Enzyme and Microbial Technology, 33*(2–3), 231–235.
136. Yadav, A., Raj, A., Purchase, D., Ferreira, L. F. R., Saratale, G. D., & Bharagava, R. N. (2019). Phytotoxicity, cytotoxicity and genotoxicity evaluation of organic and inorganic pollutants rich tannery wastewater from a common effluent treatment plant (CETP) in Unnao district, India using Vigna radiata and Allium cepa. *Chemosphere, 224*, 324–332.
137. UNEP, I. PAC. (1994). Tanneries and the environment–a technical guide (No. 4, pp. 151–155). Technical Report (2nd Print) Series.
138. Van Groenestijn, J. W., Hazewinkel, J. H. O., Nienoord, M., & Bussmann, P. J. T. (2002). Energy aspects of biological hydrogen production in high rate bioreactors operated in the thermophilic temperature range. *International Journal of Hydrogen Energy, 27*(11–12), 1141–1147.
139. Leghouchi, E., Laib, E., & Guerbet, M. (2009). Evaluation of chromium contamination in water, sediment and vegetation caused by the tannery of Jijel (Algeria): a case study. *Environmental Monitoring and Assessment, 153*(1), 111–117.
140. Hafez, A. I., El-Manharawy, M. S., & Khedr, M. A. (2002). RO membrane removal of unreacted chromium from spent tanning effluent. A pilot-scale study, Part 2. *Desalination, 144*(1–3), 237–242.
141. Reemtsma, T., & Jekel, M. (1997). Dissolved organics in tannery wastewaters and their alteration by a combined anaerobic and aerobic treatment. *Water Research, 31*(5), 1035–1046.
142. Bouwer, E., Durant, N., Wilson, L., Zhang, W., & Cunningham, A. (1994). Degradation of xenobiotic compounds in situ: capabilities and limits. *FEMS Microbiology Reviews, 15*(2–3), 307–317.
143. Qasim, S. R. (1999). *Wastewater Treatment Plants* (2nd Ed.). Lancaster, PA: Technomic Publishing Co., Inc.
144. Herrera-González, A. M., Caldera-Villalobos, M., & Peláez-Cid, A. A. (2019). Adsorption of textile dyes using an activated carbon and crosslinked polyvinyl phosphonic acid composite. *Journal of Environmental Management, 234*, 237–244.
145. Yuan, Y., Ning, X. A., Zhang, Y., Lai, X., Li, D., He, Z., & Chen, X. (2020). Chlorobenzene levels, component distribution, and ambient severity in wastewater from five textile dyeing wastewater treatment plants. *Ecotoxicology and Environmental Safety, 193*, 110257.
146. Hubadillah, S. K., Othman, M. H. D., Tai, Z. S., Jamalludin, M. R., Yusuf, N. K., Ahmad, A., … & Harun, Z. (2020). Novel hydroxyapatite-based bio-ceramic hollow fiber membrane derived from waste cow bone for textile wastewater treatment. *Chemical Engineering Journal, 379*, 122396.
147. Castillo, M., & Barceló, D. (2001). Characterisation of organic pollutants in textile wastewaters and landfill leachate by using toxicity-based fractionation methods followed by liquid and gas chromatography coupled to mass spectrometric detection. *Analytica Chimica Acta, 426*(2), 253–264.
148. Kumar, A., Sharma, G., Naushad, M., Ala'a, H., García-Peñas, A., Mola, G. T., & Stadler, F. J. (2020). Bio-inspired and biomaterials-based hybrid photocatalysts for environmental detoxification: a review. *Chemical Engineering Journal, 382*, 122937.
149. Kishor, R., Bharagava, R. N., & Saxena, G. (2018). Industrial wastewaters: the major sources of dye contamination in the environment, ecotoxicological effects, and bioremediation approaches, in: R. N. Bharagava (Ed.), *Recent Advances in Environmental Management* (pp. 1–25). Boca Raton, FL: CRC Press.

150. Ceretta, M. B., Vieira, Y., Wolski, E. A., Foletto, E. L., & Silvestri, S. (2020). Biological degradation coupled to photocatalysis by ZnO/polypyrrole composite for the treatment of real textile wastewater. *Journal of Water Process Engineering*, *35*, 101230.
151. Garg, N., Garg, A., & Mukherji, S. (2020). Eco-friendly decolorization and degradation of reactive yellow 145 textile dye by Pseudomonas aeruginosa and Thiosphaera pantotropha. *Journal of Environmental Management*, *263*, 110383.
152. Fazal, T., Razzaq, A., Javed, F., Hafeez, A., Rashid, N., Amjad, U. S., & Rehman, F. (2020). Integrating adsorption and photocatalysis: a cost effective strategy for textile wastewater treatment using hybrid biochar-TiO_2 composite. *Journal of Hazardous Materials*, *390*, 121623.
153. Kishor, R., Purchase, D., Saratale, G. D., Saratale, R. G., Ferreira, L. F. R., Bilal, M., ... & Bharagava, R. N. (2021). Ecotoxicological and health concerns of persistent coloring pollutants of textile industry wastewater and treatment approaches for environmental safety. *Journal of Environmental Chemical Engineering*, *9*(2), 105012.
154. Karthik, T., & Gopalakrishnan, D. (2014). Environmental analysis of textile value chain: an overview, in: S. S. Muthu (Ed.), *Roadmap to Sustainable Textiles and Clothing* (pp. 153–188). Singapore: Springer.
155. Torrentó, C., Palau, J., Rodríguez-Fernández, D., Heckel, B., Meyer, A., Domènech, C., & Hunkeler, D. (2017). Carbon and chlorine isotope fractionation patterns associated with different engineered chloroform transformation reactions. *Environmental Science & Technology*, *51*(11), 6174–6184.
156. Ghaly, A. E., Ananthashankar, R., Alhattab, M. V. V. R., & Ramakrishnan, V. V. (2014). Production, characterization and treatment of textile effluents: a critical review. *Journal of Chemical Engineering & Process Technology*, *5*(1), 1–19.
157. Carreón, T. & Herrick, R. L. (2001). Aliphatic hydrocarbons, in: E. Bingham and B. Cohrssen (Eds.), *Patty's Toxicology* (pp. 1–102). New York: John Wiley & Sons, Inc.
158. Ning, X. A., Lin, M. Q., Shen, L. Z., Zhang, J. H., Wang, J. Y., Wang, Y. J., ... & Liu, J. Y. (2014). Levels, composition profiles and risk assessment of polycyclic aromatic hydrocarbons (PAHs) in sludge from ten textile dyeing plants. *Environmental Research*, *132*, 112–118.
159. Agency for Toxic Substances and Disease Registry (ATSDR). (1995). Polycyclic Aromatic Hydrocarbons (PAHs)—ToxFAQs. ATSDR, U.S. Department of Health and Human Services. Available from: https://www.atsdr.cdc.gov/toxfaqs/tfacts69.pdf
160. Roy Choudhury, A. K. (2014). Environmental impacts of the textile industry and its assessment through life cycle assessment, in: S. S. Muthu (Ed.), *Roadmap to Sustainable Textiles and Clothing* (pp. 1–39). Singapore: Springer.
161. Jönsson, C., Posner, S., & Roos, S. (2018). Sustainable chemicals: a model for practical substitution, in: S. S. Muthu (Ed.), *Detox Fashion* (pp. 1–36). Singapore: Springer.
162. Dutta, S., & Singh, S. (2014). Health status of workers of textile and dyeing industries of Pali Industrial Area, (Raj) India. *Global Public Health Conference 2014* (p. 296). Chennai, India: SRM University.
163. Derden, A., & Huybrechts, D. (2013). Brominated flame retardants in textile wastewater: reducing Deca-BDE using best available techniques. *Journal of Cleaner Production*, *53*, 167–175.
164. García-Esquinas, E., Navas-Acien, A., Pérez-Gómez, B., & Artalejo, F. R. (2015). Association of lead and cadmium exposure with frailty in US older adults. *Environmental Research*, *137*, 424–431.
165. Jaishankar, M., Tseten, T., Anbalagan, N., Mathew, B. B., & Beeregowda, K. N. (2014). Toxicity, mechanism and health effects of some heavy metals. *Interdisciplinary Toxicology*, *7*(2), 60.
166. Mondal, M., Halder, G., Oinam, G., Indrama, T., & Tiwari, O. N. (2019). Bioremediation of organic and inorganic pollutants using microalgae, in: A. A. Rastegari, A. N. Yadav, & N. Yadav (Eds.), *New and Future Developments in Microbial Biotechnology and Bioengineering* (pp. 223–235). Amsterdam: Elsevier.
167. Sirtori, C., Zapata, A., Oller, I., Gernjak, W., Agüera, A., & Malato, S. (2009). Decontamination industrial pharmaceutical wastewater by combining solar photo-Fenton and biological treatment. *Water Research*, *43*(3), 661–668.
168. Gadipelly, C., Pérez-González, A., Yadav, G. D., Ortiz, I., Ibáñez, R., Rathod, V. K., & Marathe, K. V. (2014). Pharmaceutical industry wastewater: review of the technologies for water treatment and reuse. *Industrial & Engineering Chemistry Research*, *53*(29), 11571–11592.
169. Deshpande, A. M., Satyanarayan, S., & Ramakant. (2012). Kinetic analysis of an anaerobic fixed-film fixed bed-reactor treating wastewater arising from production of a chemically synthesized pharmaceutical. *Environmental Technology*, *33*(11), 1261–1270.

170. Oktem, Y. A., Ince, O., Donnelly, T., Sallis, P., & Ince, B. K. (2006). Determination of optimum operating conditions of an acidification reactor treating a chemical synthesis-based pharmaceutical wastewater. *Process Biochemistry, 41*(11), 2258–2263.
171. Perez-Vega, S., Ortega-Rivas, E., Salmeron-Ochoa, I., & Sharratt, P. N. (2013). A system view of solvent selection in the pharmaceutical industry: towards a sustainable choice. *Environment, Development and Sustainability, 15*(1), 1–21.
172. Gani, R., Gómez, P. A., Folić, M., Jiménez-González, C., & Constable, D. J. (2008). Solvents in organic synthesis: replacement and multi-step reaction systems. *Computers & Chemical Engineering, 32*(10), 2420–2444.
173. Tripathi, P. K., Rao, N. N., Chauhan, C., Pophali, G. R., Kashyap, S. M., Lokhande, S. K., & Gan, L. (2013). Treatment of refractory nano-filtration reject from a tannery using Pd-catalyzed wet air oxidation. *Journal of Hazardous Materials, 261*, 63–71.
174. Sharma, P., Kumar, S., & Pandey, A. (2021). Bioremediated techniques for remediation of metal pollutants using metagenomics approaches: a review. *Journal of Environmental Chemical Engineering, 9*(4), 105684.
175. Uğurlu, M., Gürses, A., Doğar, Ç., & Yalçın, M. (2008). The removal of lignin and phenol from paper mill effluents by electrocoagulation. *Journal of Environmental Management, 87*(3), 420–428.
176. Lindholm-Lehto, P. C., Knuutinen, J. S., Ahkola, H. S., & Herve, S. H. (2015). Refractory organic pollutants and toxicity in pulp and paper mill wastewaters. *Environmental Science and Pollution Research, 22*(9), 6473–6499.
177. Haq, I., Singh, A., & Kalamdhad, A. S. (2021). Application of biochar for sustainable development in agriculture and environmental remediation, in: I. Haq & A. Kalamdhad (Eds.), *Emerging Treatment Technologies for Waste Management* (pp. 133–153). Singapore: Springer.
178. Bajpai, P. (Ed.). (2018). Biological treatment of pulp and paper mill effluents, in: *Biotechnology for Pulp and Paper Processing* (pp. 313–369). Singapore: Springer.
179. Hooda, R., Bhardwaj, N. K., & Singh, P. (2018). Brevibacillus parabrevis MTCC 12105: a potential bacterium for pulp and paper effluent degradation. *World Journal of Microbiology and Biotechnology, 34*(2), 1–10.
180. Hooda, R., Bhardwaj, N. K., & Singh, P. (2015). Screening and identification of ligninolytic bacteria for the treatment of pulp and paper mill effluent. *Water, Air, & Soil Pollution, 226*(9), 1–11.
181. Gupta, G. K., Liu, H., & Shukla, P. (2019). Pulp and paper industry–based pollutants, their health hazards and environmental risks. *Current Opinion in Environmental Science & Health, 12*, 48–56.
182. Singh, A. K., & Chandra, R. (2019). Pollutants released from the pulp paper industry: aquatic toxicity and their health hazards. *Aquatic Toxicology, 211*, 202–216.
183. Sharma, P., Iqbal, H. M., & Chandra, R. (2022). Evaluation of pollution parameters and toxic elements in wastewater of pulp and paper industries in India: a case study. *Case Studies in Chemical and Environmental Engineering, 5*, 100163.
184. Kringstad, K. P., & Lindström, K. (1984). Spent liquors from pulp bleaching. *Environmental Science & Technology, 18*(8), 236A–248A.
185. Srogi, K. (2008). Levels and congener distributions of PCDDs, PCDFs and dioxin-like PCBs in environmental and human samples: a review. *Environmental Chemistry Letters, 6*(1), 1–28.
186. Leiviskä, T., Nurmesniemi, H., Pöykiö, R., Rämö, J., Kuokkanen, T., & Pellinen, J. (2008). Effect of biological wastewater treatment on the molecular weight distribution of soluble organic compounds and on the reduction of BOD, COD and P in pulp and paper mill effluent. *Water Research, 42*(14), 3952–3960.
187. Leiviskä, T., Sarpola, A., & Tanskanen, J. (2012). Removal of lipophilic extractives from debarking wastewater by adsorption on kaolin or enhanced coagulation with chitosan and kaolin. *Applied Clay Science, 61*, 22–28.
188. Zhang, L., Liu, J., Liu, H., Wan, G., & Zhang, S. (2015). The occurrence and ecological risk assessment of phthalate esters (PAEs) in urban aquatic environments of China. *Ecotoxicology, 24*(5), 967–984.
189. Zhang, Y., Liang, Q., Gao, R., Hou, H., Tan, W., He, X., & Wang, X. (2015). Contamination of phthalate esters (PAEs) in typical wastewater-irrigated agricultural soils in Hebei, North China. *PLoS One, 10*(9), e0137998.
190. Hampannavar, U. S., & Shivayogimath, C. B. (2010). Anaerobic treatment of sugar industry wastewater by upflow anaerobic sludge blanket reactor at ambient temperature. *International Journal of Environmental Sciences, 1*(4), 631.

191. Siddiqui, W. A., & Waseem, M. (2012). A comparative study of sugar mill treated and untreated effluent: a case study. *Oriental Journal of Chemistry*, *28*(4), 1899.
192. Wei, Y. A., & Xu, Y. J. (2004). Eco-friendly management of sugar industry effluents in Guangxi, China. *Sugar Tech*, *6*(4), 285–290.
193. Saranraj, P., & Stella, D. (2012). Effect of bacterial isolates on reduction of physico–chemical characteristics in sugar mill effluent. *International Journal of Pharmaceutical and Biological Archives*, *3*(5), 1077–1084.
194. Sahu, O. (2017). Treatment of sugar processing industry effluent up to remittance limits: suitability of hybrid electrode for electrochemical reactor. *MethodsX*, *4*, 172–185.
195. Salehzadeh, J. (2013). Removal of heavy metals Pb_2, Cu_2, Zn_2, Cd_2, Ni_2, Co_2 and Fe_3 from aqueous solutions by using Xanthium pensylvanicum. *Leonardo Journal of Sciences*, *23*, 97.
196. Renu, M. A., Singh, K., Upadhyaya, S., & Dohare, R. K. (2017). Removal of heavy metals from wastewater using electrocoagulation. *Materials Today: Proceedings*, *4*(9), 10534–10538.
197. Srivastava, S., & Jain, R. (2010). Effect of distillery spent wash on cytomorphological behaviour of sugarcane settlings. *Journal of Environmental Biology*, *31*(5), 809.
198. Damodharan, U., & Reddy, M. V. (2012). Impact of sugar industrial treated effluent on the growth factor in sugarcane—Cuddalore, India. *Journal of Sustainable Bioenergy Systems*, *2*(03), 43.
199. Perovano Filho, N., da Silva, K. F. S., dos Santos, E. C. L., Queissada, D. D., & López, A. M. Q. (2016). Physicochemical monitoring of wastewater from a sugar and ethanol industry after bioaugmentation, with a proposal for reuse. *Acta Scientiarum. Technology*, *38*(4), 383–389.
200. Poddar, P. K., & Sahu, O. (2017). Quality and management of wastewater in sugar industry. *Applied Water Science*, *7*, 461–468.
201. Barakat, M. A. (2011). New trends in removing heavy metals from industrial wastewater. *Arabian Journal of Chemistry*, *4*(4), 361–377.
202. Carolin, C. F., Kumar, P. S., Saravanan, A., Joshiba, G. J., & Naushad, M. (2017). Efficient techniques for the removal of toxic heavy metals from aquatic environment: a review. *Journal of Environmental Chemical Engineering*, *5*(3), 2782–2799.
203. Nogueira, V., Lopes, I., Rocha-Santos, T. A. P., Gonçalves, F., & Pereira, R. (2018). Treatment of real industrial wastewaters through nano-TiO_2 and nano-Fe_2O_3 photocatalysis: case study of mining and kraft pulp mill effluents. *Environmental Technology*, *39*(12), 1586–1596.
204. Singh, P., Srivastava, N., & Singh, P. (2017). Analysis of various iron nanoparticles and compounds in pulp & paper mill waste water treatment. *Integrated Research Advances*, *4*(2), 24–28.
205. Fouda, A., Hassan, S. E. D., Saied, E., & Azab, M. S. (2021). An eco-friendly approach to textile and tannery wastewater treatment using maghemite nanoparticles (γ-Fe_2O_3-NPs) fabricated by Penicillium expansum strain (Kw). *Journal of Environmental Chemical Engineering*, *9*(1), 104693.
206. Fouda, A., Hassan, S. E. D., Saied, E., & Hamza, M. F. (2021). Photocatalytic degradation of real textile and tannery effluent using biosynthesized magnesium oxide nanoparticles (MgO-NPs), heavy metal adsorption, phytotoxicity, and antimicrobial activity. *Journal of Environmental Chemical Engineering*, *9*(4), 105346.
207. Mahmoud, A. S., Mohamed, N. Y., Mostafa, M. K., & Mahmoud, M. S. (2021). Effective chromium adsorption from aqueous solutions and tannery wastewater using bimetallic Fe/Cu nanoparticles: response surface methodology and artificial neural network. *Air, Soil and Water Research*, *14*, 11786221211028162.
208. Mustapha, S., Tijani, J. O., Ndamitso, M. M., Abdulkareem, S. A., Shuaib, D. T., Mohammed, A. K., & Sumaila, A. J. S. R. (2020). The role of kaolin and kaolin/ZnO nanoadsorbents in adsorption studies for tannery wastewater treatment. *Scientific Reports*, *10*(1), 1–22.
209. Ye, J., Li, X., Hong, J., Chen, J., & Fan, Q. (2015). Photocatalytic degradation of phenol over ZnO nanosheets immobilized on montmorillonite. *Materials Science in Semiconductor Processing*, *39*, 17–22.
210. Fouda, A., Hassan, S. E. D., Abdel-Rahman, M. A., Farag, M. M., Shehal-deen, A., Mohamed, A. A., & Azab, M. S. (2021). Catalytic degradation of wastewater from the textile and tannery industries by green synthesized hematite (α-Fe_2O_3) and magnesium oxide (MgO) nanoparticles. *Current Research in Biotechnology*, *3*, 29–41.
211. Ncibi, M. C., & Sillanpää, M. (2015). Optimized removal of antibiotic drugs from aqueous solutions using single, double and multi-walled carbon nanotubes. *Journal of Hazardous Materials*, *298*, 102–110.
212. Álvarez-Torrellas, S., Rodríguez, A., Ovejero, G., & García, J. (2016). Comparative adsorption performance of ibuprofen and tetracycline from aqueous solution by carbonaceous materials. *Chemical Engineering Journal*, *283*, 936–947.
213. Tian, Y., Gao, B., Morales, V. L., Chen, H., Wang, Y., & Li, H. (2013). Removal of sulfamethoxazole and sulfapyridine by carbon nanotubes in fixed-bed columns. *Chemosphere*, *90*(10), 2597–2605.

214. Wei, H., Deng, S., Huang, Q., Nie, Y., Wang, B., Huang, J., & Yu, G. (2013). Regenerable granular carbon nanotubes/alumina hybrid adsorbents for diclofenac sodium and carbamazepine removal from aqueous solution. *Water Research*, *47*(12), 4139–4147.
215. Zhou, Y., He, Y., Xiang, Y., Meng, S., Liu, X., Yu, J., & Luo, L. (2019). Single and simultaneous adsorption of pefloxacin and Cu(II) ions from aqueous solutions by oxidized multiwalled carbon nanotube. *Science of the Total Environment*, *646*, 29–36.
216. Mohammadi, R., Massoumi, B., & Eskandarloo, H. (2015). Preparation and characterization of Sn/Zn/TiO_2 photocatalyst for enhanced amoxicillin trihydrate degradation. *Desalination and Water Treatment*, *53*(7), 1995–2004.
217. Wang, F., Sun, W., Pan, W., & Xu, N. (2015). Adsorption of sulfamethoxazole and 17β-estradiol by carbon nanotubes/$CoFe_2O_4$ composites. *Chemical Engineering Journal*, *274*, 17–29.
218. Wang, Y., Ma, J., Zhu, J., Ye, N., Zhang, X., & Huang, H. (2016). Multi-walled carbon nanotubes with selected properties for dynamic filtration of pharmaceuticals and personal care products. *Water Research*, *92*, 104–112.
219. Yanyan, L., Kurniawan, T. A., Albadarin, A. B., & Walker, G. (2018). Enhanced removal of acetaminophen from synthetic wastewater using multi-walled carbon nanotubes (MWCNTs) chemically modified with NaOH, HNO_3/H_2SO_4, ozone, and/or chitosan. *Journal of Molecular Liquids*, *251*, 369–377.
220. Balarak, D., Mostafapour, F. K., & Joghataei, A. (2017). Application of single-walled carbon nanotubes for removal of aniline from industrial waste water. *Bioscience Biotechnology Research Communications*, *10*(2), 311–318.
221. Babaei, A. A., Lima, E. C., Takdastan, A., Alavi, N., Goudarzi, G., Vosoughi, M., & Shirmardi, M. (2016). Removal of tetracycline antibiotic from contaminated water media by multi-walled carbon nanotubes: operational variables, kinetics, and equilibrium studies. *Water Science and Technology*, *74*(5), 1202–1216.
222. Debnath, B., Majumdar, M., Bhowmik, M., Bhowmik, K. L., Debnath, A., & Roy, D. N. (2020). The effective adsorption of tetracycline onto zirconia nanoparticles synthesized by novel microbial green technology. *Journal of Environmental Management*, *261*, 110235.
223. Muthukumar, H., Shanmugam, M. K., & Gummadi, S. N. (2020). Caffeine degradation in synthetic coffee wastewater using silverferrite nanoparticles fabricated via green route using Amaranthus blitum leaf aqueous extract. *Journal of Water Process Engineering*, *36*, 101382.
224. Abbasi, F., Yaraki, M. T., Farrokhnia, A., & Bamdad, M. (2020). Keratin nanoparticles obtained from human hair for removal of crystal violet from aqueous solution: optimized by Taguchi method. *International Journal of Biological Macromolecules*, *143*, 492–500.
225. Martins, M., Mourato, C., Sanches, S., Noronha, J. P., Crespo, M. B., & Pereira, I. A. (2017). Biogenic platinum and palladium nanoparticles as new catalysts for the removal of pharmaceutical compounds. *Water Research*, *108*, 160–168.
226. Varadavenkatesan, T., Lyubchik, E., Pai, S., Pugazhendhi, A., Vinayagam, R., & Selvaraj, R. (2019). Photocatalytic degradation of Rhodamine B by zinc oxide nanoparticles synthesized using the leaf extract of Cyanometra ramiflora. *Journal of Photochemistry and Photobiology B: Biology*, *199*, 111621.
227. Uddandarao, P., Hingnekar, T. A., Balakrishnan, R. M., & Rene, E. R. (2019). Solar assisted photocatalytic degradation of organic pollutants in the presence of biogenic fluorescent ZnS nanocolloids. *Chemosphere*, *234*, 287–296.
228. Begum, S., & Ahmaruzzaman, M. (2018). Biogenic synthesis of SnO_2/activated carbon nanocomposite and its application as photocatalyst in the degradation of naproxen. *Applied Surface Science*, *449*, 780–789.
229. Ameri, A., Shakibaie, M., Pournamdari, M., Ameri, A., Foroutanfar, A., Doostmohammadi, M., & Forootanfar, H. (2020). Degradation of diclofenac sodium using UV/biogenic selenium nanoparticles/H_2O_2: optimization of process parameters. *Journal of Photochemistry and Photobiology A: Chemistry*, *392*, 112382.
230. Zaheer, Z. (2018). Biogenic synthesis, optical, catalytic, and in vitro antimicrobial potential of Ag-nanoparticles prepared using Palm date fruit extract. *Journal of Photochemistry and Photobiology B: Biology*, *178*, 584–592.
231. Mirsadeghi, S., Zandavar, H., Yousefi, M., Rajabi, H. R., & Pourmortazavi, S. M. (2020). Green-photodegradation of model pharmaceutical contaminations over biogenic Fe_3O_4/Au nanocomposite and antimicrobial activity. *Journal of Environmental Management*, *270*, 110831.
232. Erhardt, C. S., Basegio, T. M., Capela, I., Rodríguez, A. L., Machado, Ê. L., López, D. A. R., ... & Bergmann, C. P. (2021). AOX degradation of the pulp and paper industry bleaching wastewater using nZVI in two different agitation processes. *Environmental Technology & Innovation*, *22*, 101420.

9 Treatment of Fluoride Wastewater Using Clay Minerals

Kawthar Yahya, Noureddine Hamdi, and Yung-Tse Hung

NOMENCLATURE

BDST	Bed depth service time
C	Concentration
CEC	Cation exchange capacity
EBRT	Empty bed residence time
F_2	Fluorine
F	Freundlich
H	Jebel Haidoudi
HCl	Hydrochloric acid
HCO_3^-	Bicarbonate ions
HDPy	Hexadecylpyridinium
HDTMAB	Hexadecyltrimethylammonium bromide
K_1	First-order kinetic constants
K_2	Second-order kinetic constants
K_F	Freundlich constants
K_L	Langmuir constants
L	Langmuir
m	Mass
MK	Sebkhat El Malah-El Mkhacherma
MMT	Montmorillonites
NO^{3-}	Nitrate
O	Octahedral
P	Initial concentration of pollutant
PFO	Pseudo First Order
PSO	Pseudo Second Order
Qads	Quantity adsorbed
qeI	First- and second-order theoretical adsorbed
qeII	Second-order theoretical adsorbed amounts
RO	Reverse osmosis
SO_4^{2-}	Sulfate
T	Contact time
WHO	World Health Organization
ZB	Zemlet El Beidha

DOI: 10.1201/9781003532354-9

9.1 INTRODUCTION

Fluoride is the ionic form of fluorine (F_2), which forms F^- ions in aqueous solution. Fluorine is the most electronegative and reactive chemical element. It has a high tendency to take on a negative charge. Therefore, it almost never occurs in the environment as fluorine, but as fluoride [1]. Fluoride, as a common dissolved component of water, causes either beneficial or toxic health effects on human health depending on its concentration. In fact, the World Health Organization (WHO) has specified the tolerance limit of the fluoride content of drinking water as 1.5 mg/L.

Fluoride is an indispensable element to maintain the human dental and bone well-being of a human at levels between 0.5 and 1.5 mg mg/L [2,3]. However, it constitutes a considerable health risk if its content exceeds the limiting fluoride concentration set by the WHO at 1.5 mg/L; an excessive intake of fluoride for a long period can result in dental fluorosis, skeletal fluorosis, osteoporosis, and brittle bones [3,4].

High fluoride concentration in the environment comes from two different channels: natural sources and anthropogenic activities. Fluoride is released into the environment naturally through weathering and the dissolution of rock minerals. On the other hand, the discharge of wastewater containing fluoride from various industries, e.g. mining, semiconductor fabricating, electroplating, rubber and fertilizer manufacturing, is another cause of fluoride enrichment in the environment [3,5]. Therefore, the removal of fluoride from contaminated water is important to protect the environment and human health.

Fluoride pollution is considered a significant environmental problem in various countries such as America, Africa, Asia, and Europe [6,7]. So far, various potential technologies have been investigated to remove fluoride for remediations of industrial or domestic wastewater. Various reviews of a relatively comprehensive summary of these treatment technologies for removing fluoride were published:

A review of a relatively comprehensive summary of fluoride treatment technologies was published in 2014 by Suneetha et al. [8]. This review article is aimed at providing specific information on attempts made by various researchers in the field of fluoride removal from polluted waters, such as precipitation [9,10], ion exchange [11,12], reverse osmosis (RO) [13,14], electrodialysis [15,16], nanofiltration [17], adsorption [18,19], etc.

Based on Mohapatra et al.'s [4] review paper, fluoride removal has been broadly divided into two sections dealing with membrane and adsorption techniques. Under the membrane techniques RO, nanofiltration, dialysis, and electrodialysis have been discussed. Adsorption, which is a conventional technique, deals with adsorbents. This review article is aimed at providing precise information on efforts made by various researchers in the field of fluoride removal using these techniques.

Also, Singh et al. [20] in a review paper suggest that the defluoridation techniques be generally classified into two, specifically membrane and surface assimilation techniques. In this review article, they emphasized the various techniques like RO, nanofiltration, biopolymers, etc. to solve the problems of higher concentrations of fluoride ions in drinking water.

Based on these review papers, among these approaches, adsorption offers unique advantages due to the use of synthetic and natural materials to achieve efficient fluoride removal. In various reviews, a large number of solid adsorbents such as natural adsorbents, agricultural wastes, industrial wastes, biomass, and nano-adsorbents have been discussed for the removal of fluoride from wastewater [6,8,21–23]. Among these different adsorbents, clay minerals and their derivatives have composed one large family which is considered efficient adsorbents, due to their low cost, wide availability, and excellent adsorption performance. Clay minerals are phyllosilicates in which the individual layers are composed of a tetrahedral (T) and an octahedral (O). Some approaches (e.g. acid washing, thermal treatment) have been applied to modify the surface structure and physicochemical properties of clay minerals, which then enhance the capacity of clay minerals in uptaking fluoride to uptaking various pollutants [24–26].

In fact, clay minerals and their modified derivatives are used by various researchers for the removal of different pollutants from an aqueous environment. Zhu et al. [27] in a review comprehensively introduced the montmorillonite (MMT)-based adsorbents, which can be used for the uptake of most of the chemical contaminants from an aqueous solution. Also, Gu et al. [28] in a review investigated the use of clay-based adsorbents including MMT, attapulgite, halloysite, and vermiculite in heavy metal adsorption.

Therefore, we evaluated the clay minerals' abilities to remove fluoride in wastewater. We emphasize raw clay and modified clay as environmentally safe adsorbents. The main purpose of this review is to help readers choose suitable adsorbents for the removal of target fluoride as well as to facilitate the development of new adsorbents.

9.2 HEALTH EFFECTS OF FLUORIDE

Fluoride present in drinking water is considered an important source for the human body. Fluoride plays an important role in the development of teeth enamel in young children and strengthening the bone matrix. On the contrary, a high ratio of fluoride becomes a very toxic element and is known to cause adverse human and animal health effects due to environmental and occupational exposure [20]. Fluoride has been classified as a priority pollutant by the United States Environmental Protection Agency and the German Research Foundation [20].

Prolonged and excessive intake of fluoride may result in a serious public health problem called fluorosis, which is characterized by dental mottling and skeletal manifestations such as crippling deformities, osteoporosis, and osteosclerosis. Endemic fluorosis is now known to be a global occurrence affecting many millions of people in different countries [29,30].

Dental fluorosis also termed as mottled enamel is characterized by white and yellow glistening patches on the teeth which may eventually turn brown due to hypo-mineralization of tooth enamel caused by excess ingestion of fluoride. Skeletal fluorosis can also damage the fetus if the mother consumes water and food with a high concentration of fluoride during pregnancy/breastfeeding [6]. Other effects include severe pain in the backbone, joints and hip region, stiffness of the backbone, increased density of bones, and paralysis. Non-skeletal manifestations include nervousness, excessive thrust, muscle weakness, stiffness, pain in the muscle and loss of muscle power, acute abdominal pain, diarrhea, and constipation, blood in stools, headache, and very painful skin rashes [6,23]. So, depending on the total intake, fluoride anion F^- is known to have both beneficial and adverse effects on humans, as summarized in Table 9.1.

Therefore, discharges concentrated in fluorides are toxic and require a treatment technique. This is why the removal of excess fluoride in wastewater is a topic of concern.

9.3 TECHNOLOGIES FOR FLUORIDE REMOVAL FROM WASTEWATER

Water purification is a set of techniques that consist in purifying water either to recycle wastewater in the natural environment or to transform natural water into drinking water. Thus, the objective of wastewater treatment is to obtain purified water that meets the discharge standards laid down by the

TABLE 9.1
Health Effects of Fluoride [31]

Concentration of Fluoride (mg/L)	Health Effects
<0.5	Conducive to dental caries
0.5–1.5	Promotes development of strong bones and teeth
1.5–4.0	Promotes dental fluorosis in children
4.0–10.0	Promotes dental and skeletal fluorosis
>10.0	Crippling skeletal fluorosis, possibly cancer

legislation, and which can then be discharged without danger to human health and the environment. The objective in fluoride removal is to treat the contaminated water so as to bring down fluoride concentration to acceptable limits. The current discharge standards for wastewater containing fluoride vary from country to country (Table 9.2).

The presence of fluoride in wastewaters has been identified as one of the acute problems worldwide. Fluoride removal from water has long been an engineering challenge; in fact, much effort by researchers has been put into developing effective defluoridation technologies. The available methods of defluoridation are presented in the following sections.

9.3.1 Conventional Methods of Fluoride Removal

In literature, in order to lessen fluoride concentration to an acceptable level in the water, various technologies have been developed (Figure 9.1).

TABLE 9.2
The Permissible Limit of Fluoride in Discharge Prescribed by Various Countries

Name of Country	Permissible Limit of Fluoride (mg/L)	References
Tunisia	1	[3]
US EPA	4	[32]
China	10	[33]
Japan	5–15	[32]
Poland	25	[32]
India	1.5–10	[32]
Algeria and Taiwan	15	[32]

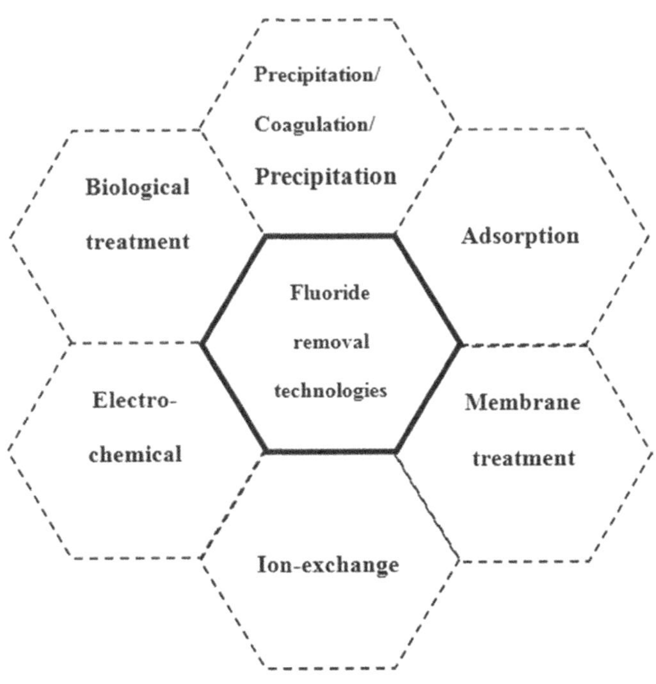

FIGURE 9.1 Fluoride removal technologies.

Ayoob and Gupta [29] have written a comprehensive review on sustainable technologies for the defluoridation of drinking water; different technologies were compared and the strengths and limitations of different methods, used for defluoridation, were critically discussed by the authors:

- The coagulation methods have generally been found effective in defluoridation, but they are unsuccessful in bringing fluoride to desired concentration levels.
- The membrane processes do not require additives but these are relatively expensive to install and operate and prone to fouling, scaling, or membrane degradation.
- The electrochemical techniques, in general, suffer due to the high-cost factor, during installation and maintenance.
- The Nalgonda technique is one of the popular techniques widely used for defluoridation of water in developing countries (e.g. India, Kenya, Senegal, and Tanzania). However, some disadvantages of this technique have also been reported by a few researchers, e.g. high residual aluminum concentration (2–7 mg/L) in the treated water than the set WHO standard: 0.2 mg/L.

Based on various works [8,21,29,34] the comparison of different technologies used in the removal of fluoride is explained in Table 9.3.

TABLE 9.3
Comparison of Fluoride Removal Technologies

		Advantages	Disadvantages	References
		—	Sensitivity of microorganisms to environmental factors and difficulty to control/costly	[35]
	Coagulation	High efficiency Commercially available chemical	Expensive/efficiency depends on pH and presence of co-ions in water/high amount of retained water	[8,35]
	Precipitation	High efficiency Commercially available chemical	Expensive pH adjustment and readjustment are often required due to the elevated residual aluminum concentrations or the formation of aluminum-based precipitates during the defluoridation process	[21]
	Ion exchange	High efficiency	Expensive/requires replacement of media after multiple regenerations/regeneration creates toxic waste, effectiveness highly dependent on pH	[21]
	Adsorption	High accessibility/low cost/simple operation/availability of a wide range of adsorbents	Disadvantages are related to the chosen adsorbent	[21]
Membrane processes	Reverse osmosis	High efficiency	Sensitive to temperature and pH/causes maintenance problems due to clogging, energy consumption is high and a lot of water is wasted/expensive and very complicated	[8,35]
Membrane processes	Nanofiltration	High efficiency/removal of all contaminants	High energy costs/limited retention for salts and univalent ions/membrane fouling/limited lifetime/more expensive than reverse osmosis	[8,35]
Electrochemical techniques	Electro-dialysis	High efficiency High selectivity	Removes all the ions present in the water, including some essential minerals and therefore remineralization is necessary after the treatment/expensive and very complicated	[8]
Electrochemical techniques	Dialysis	—	Reduced efficiency in highly saline waters Expensive technique	[8]
Electrochemical techniques	Electro-coagulation	—	Costly and complicated	[8]

Although these methods have successfully reduced fluoride concentration to acceptable levels, however, according to Table 9.1, it can be seen that most of these methods have high operational and maintenance costs, lack of selectivity for fluoride, undesirable effects on water quality, generation of large volumes of sludge, and complicated procedures involved in the treatment. Therefore, these methods can be effective, but not for a developing country, because of their high cost and especially the operating and maintenance constraints which make their use unattractive.

As a result, research is oriented toward less expensive and environmentally friendly processes, in particular adsorption. Surface adsorption retains a major place in defluoridation research and practice because of its general greater accessibility and lower cost. Thus, even in the past decade, when interest in alternative defluoridation approaches has been increasing rapidly, many researchers have continued to explore the development of low-cost and effective adsorbents and to improve the efficiency of all adsorbents [4,21,22,36].

9.3.2 Adsorption Technology for Defluoridation

Adsorption is a physicochemical wastewater treatment in which dissolved molecules are attached to the surface of an adsorbent by physical/chemical forces. A typical adsorption process has four steps:

1. Addition of feed solution and the adsorbent solid particles;
2. Good mixing to enhance the transfer of solutes from the liquid phase to the surface area of the adsorbent solid particles;
3. Separation of the adsorbent plus adsorbate (solute) from the bulk of solution;
4. Removal of adsorbed solute from the surface of the solid adsorbent using a different solvent. This is called elution of solute and the solvent is called elution solvent;
5. At steady-state adsorption conditions, there is a physical equilibrium between the concentrations of solute in the liquid.

So far, the uptaking fluoride on various materials has been well introduced in a large number of publications [4,21,22,33].

Theoretically, the adsorption of fluoride onto solid particles normally occurs in three steps [37]:

i. Diffusion of fluoride ions to the external surface of the adsorbent from bulk;
ii. Adsorption of fluoride ions onto particle surfaces;
iii. The adsorbed fluoride ions probably exchange with the structural elements inside adsorbent particles.

9.3.2.1 Factors Affecting Adsorption

Generally, adsorption in mode batch depends on ions (adsorbate) in fluid diffusing to the surface of a solid (adsorbent), where they bond with the solid surface [38]. Fluoride removal efficiency always depends on raw water quality profile, i.e. initial fluoride concentration, solution pH, environment temperature, contact time, and adsorbent dose:

- **Contact Time:** The adsorption of fluoride ions by the adsorbent depends on the interactions between the solution and the surface of the adsorbent. The adsorption process can be assumed to be complete when equilibrium is achieved between the fluoride ions and the adsorbent [6,39]. However, specific time was needed to maintain the equilibrium interactions to ensure that the adsorption process was complete [39].
- **pH:** Adsorption of fluoride ions from the wastewater is mainly influenced by the pH of the solution. pH can influence the surface charge of the adsorbent and the degree of ionization of the fluoride ions [39,40]. In a particular pH range, most fluoride ion adsorption was enhanced with pH, increasing to a certain value followed by a reduction when further pH

increases. The dependence of the fluoride ion uptake on pH can be associated with both the surface functional groups on the adsorbent and also the chemistry of the solution [39].

- **Initial Concentration:** Initial concentration of fluoride ions can alter the fluoride removal efficiency through a combination of factors such as the availability of specific surface functional groups and the ability of surface functional groups to bind fluoride ions [41]. The initial concentration of the solution can provide an important driving force to overcome the mass transfer resistance of fluoride ions between the aqueous and solid phases [39].
- **Temperature:** Temperature plays a double role in the fluoride adsorption process. Temperature can impact the physical binding processes of fluoride to the adsorbent. Most adsorption studies were conducted at room temperature in laboratory settings [39].

9.3.2.2 Modeling the Kinetics of an Isotherm of Batch Adsorption

This section intends to give a brief generality to the modeling of adsorption (isotherm and kinetic).

9.3.2.2.1 Kinetic Modeling of Fluoride Adsorption

Adsorption kinetics describes the solute uptake rate (chemical reaction), which in turn governs the residence time of the adsorption process [39]. Parameters of the kinetic constants are identified by linearization followed by both linear and non-linear regression analysis. Many kinetic models of varying complexity have been developed to predict the uptake rate of the adsorptive into the adsorbent.

In recent years, three kinetic models have been widely used in the literature for adsorption processes, namely the pseudo-first-order kinetic model, pseudo-second-order kinetic model, and intra-particle diffusion model.

9.3.2.2.2 Isotherm Modeling of Fluoride Adsorption

Adsorption isotherms are viable for describing the relationship between solute molecules and the adsorbent surfaces, highlighting the distribution of solute molecules between the liquid and solid phases when an adsorption process reaches an equilibrium state.

The analysis of isotherm data by fitting different isotherm models is a significant step to determine the suitable model which can be used for design purposes. Hence, the correlation of equilibrium data using either a theoretical or empirical equation is essential for the interpretation and prediction of the extent of adsorption, in deciding the maximum adsorption capacity of adsorbate for a given adsorbent [39]. Thus, there are several types of isotherm models, out of which Langmuir and Freundlich adsorption isotherm are most widely used and accepted.

Table 9.4 illustrates the corresponding original and linear equations as well as the related plots of the different models (isotherm and kinetic) used in the literature.

9.3.2.3 Adsorbents Used for Defluoridation

Generally, adsorption studies point the most important characteristics which determined adsorbent suitability for practical application: adsorption capacity, selectivity for fluoride ions, regenerability, compatibility, particle and pore size, and finally the overall cost for fluoride removal. Among the above-listed characteristics and process parameters, the adsorbent's selectivity for fluoride ions seems to be the most important adsorbent characteristic since some of the adsorbents showed high efficiency on the test bench but, at the same time, failed under real conditions at water treatment plant due to reducing the effective adsorption capacity caused by adsorbents' active sites occupation by other co-ions present in treated water. Strictly speaking, all solids are adsorbents; however, an effective adsorbent must meet a number of criteria, the most important of which are high adsorption capacity, efficient and selective and easily regenerated fluoride ions, compatibility, particle and pore size, and finally the overall cost for fluoride removal. In the literature, various review papers [4,21,22,36,42,43] provide an overview of various materials as adsorbents used for the effective removal of fluoride from wastewater (Figure 9.2).

TABLE 9.4
Used Models for the Kinetic and Isotherm Studies

Model	Equation	Linear Form	References
Kinetic models			
First order	$\frac{dq_t}{dt} = K_1(q_e - q_t)$	$\ln(q_e - q_t) = \ln q_{eI} - K_1 t$	[6]
Second order	$\frac{dq_t}{dt} = K_2(q_e - q_t)^2$	$\frac{t}{q_t} = \frac{1}{K_2 q_{eII}^2} + \frac{1}{q_{eII}} t$	[6]
Isotherms			
Langmuir	$q_e = \frac{q_{max,L}\ K_l C_e}{1 + K_l C_e}$	$\frac{C_e}{q_e} = \frac{1}{K_L q_{max,L}} + \frac{1}{q_{max,L}} C_e$	[6]
Freundlich	$q_e = K_f C_e^{1/n}$	$\ln q_e = \ln K_F + \frac{1}{n} \ln C_e$	[6]

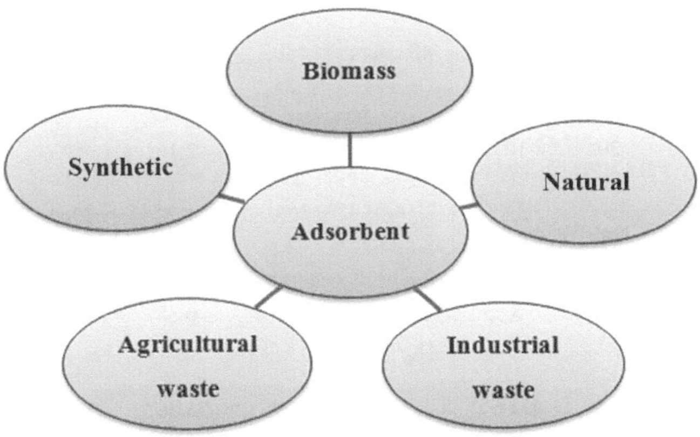

FIGURE 9.2 Different types of adsorbent.

In Table 9.5, the efficiency of some adsorbents has been examined in terms of the quantity adsorbed (Qads) as a function of the initial concentration of pollutant [*P*], contact time (*T*), and pH. For information relating to the methodology and detailed experimental conditions, readers are invited to consult the full articles listed in the references.

9.4 FLUORIDE REMOVAL BY ADSORPTION CLAY MINERALS

9.4.1 Clay Minerals

Clay minerals are mainly phyllosilicates characterized by layered structural units made up of one or two tetrahedral silica sheets around an octahedral aluminum sheet. They have particle sizes of less than 2 μm [28]. Clay minerals are generally categorized into the following groups: MMT, smectite, kaolinite, illite, and chlorite. MMT, kaolinite, and illite are widely used because of their high specific surface area, chemical and mechanical stability, a variety of surface and structural properties, and low cost [61]. Clays have a wide variety of physical properties, such as high specific surface area, the presence of negative charges, chemical and mechanical stability, layered structure, high cation exchange capacity (CEC), etc.

TABLE 9.5
Adsorbents for Fluoride Elimination

Adsorbent	pH	*m* (g)	*C* (mg/L)	*T* (minutes)	*T* (°C)	*Q* (mg/g)	Kinetics	Isotherm	References
Acidic alumina (A)	4.4	4.5	15	90	25	8.4	PSO	Langmuir	[44]
Activated carbon	7	4	5	60	30	1.65	PFO	Langmuir	[45]
Laterite soil	6.8	1	20	—	303	0.8461	PFO	Langmuir	[46]
Magnesium dioxide coated A	7	8	10	180	25	0.16	PSO	Langmuir	[47]
Zirconium-impregnated carbon	4	20	20	360	25	1.95	PSO	Langmuir	[48]
Eichhornia crassipes biomass	5.5	—	2–25	24 hours	25	0.523–1.54	PFO	F	[49]
Biosorbent	—	15	100	24 hours	Room	94%–98%	—	—	[50]
Modified activated alumina	7	10	12	60	20	0.76	—	L and F	[51]
Cow dung carbon @ 500°C	7.53	2	2	40	25	15	PSO	F	[52]
Ceramic adsorbent	4–22	20	10	48 hours	30	2.16	PSO	L and F	[53]
Calcium aluminate material	3–11	3	8.9	1,440	25	4.37	PSO	Langmuir	[54]
Iron-impregnated granular	4	20	10	48 hours	30	2.16	PSO	L and F	[55]
Sawdust raw	6	4	2.5	60	28	1.73	PSO	F	[56]
Copper oxide-coated alumina	4.9	0.4	10	1,440	30	7.22	PSO	Langmuir	[57]
Zirconium-impregnated carbon	3	15	3	180	25	6.38	PSO	F	[58]
Brushite	5.36	1.5	49.06	120	31.96	36.26	PSO	L	[59]
Iron ore	6	5	14.22	120	22	1.72	PFO	F	[60]

For several years, clay has received more attention due to its unique characteristics and properties. In fact, several studies have shown the anti-pollutant role of clays [62]; the use of clay minerals as adsorbents for the adsorption of fluoride has been widely studied by a large number of researchers. From the year 1960 until today, there have been many publications concerning the field of wastewater treatment, in particular the elimination of fluoride by different types of clays. Sometimes, the latter is subjected to modifications of the textural and structural properties in order to optimize their adsorption properties. In the following, we have compiled the past and current research work carried out by various researchers on the adsorption of fluoride by clay and modified clay. We will briefly explain the adsorption efficiency of raw and modified clay in the adsorption of these pollutants in water.

9.4.2 Fluoride Removal by Adsorption of Raw Clay Minerals

Different raw clay is investigated for the effective removal of fluoride from wastewaters (Table 9.6):

Since the year 1967 Bower and Hatcher [63] studied the defluoridation of water by minerals. The equilibrium information fitted well with the Langmuir isotherm model. The adsorption of fluoride at an equilibrium concentration of 16 mg/L for various minerals was kaolinite (190–295 mg/g),

TABLE 9.6
Raw Clay Minerals for Fluoride Elimination Using Adsorption

Adsorbent	pH	*m* (g)	*C* (mg/L)	*T* (minutes)	*T* (°C)	*Q* (mg/g)	Kinetic	Isotherm	References
Montmorillonite	2	70	3	50	298	1.836	—	L	[34]
Montmorillonite	Neutral	—	3.0 mg	50 minutes	30	1.485	—	—	[65]
Clay minerals[a]	3	—	23–2,360	3–4 days	25	69.4–93.4	—	L	[67]
Montmorillonite	6	8	4	180	25	0.263	—	F	[70]
Kaolinite	4.5–6	—	—	120	—	0.556	PSO	F	[68]
Montmorillonite	5	0.1	—	60	—	40	—		[69]
Smectite with kaolin	Acidic	0.1	Low	6 hours	25	1.16	PSO	L	[71]
Smectite with illite	3–4	0.05	100	5 hours	Room	7.04	PSO	L	[72]
Raw kaolinite	3	1		30		0.120	PSO	F L	[73]

[a] The clays used are MK (palygorskite clay, containing kaolinite and a small amount of smectite with quartz as impurity), H (smectic clay associated with kaolinite and calcite) and the ZB (illito-kaolinitic clay).

dehydrated halloysite (1,400 mg/g), hydrated (expanded) halloysite (1,777 mg/g), bentonite, and vermiculite which adsorbed only traces of fluorides.

Chaturvedi et al. [64] investigated the use of China clay for the removal of fluoride from water at different concentrations, temperatures, and pH. Results showed that fluoride removal is favored by low concentration, high temperature, and acidic pH. The alumina constituent of China clay is responsible for the maximum adsorption of fluoride in the pH range of interest.

Batch adsorption studies by Karthikeyan et al. [65] were conducted to determine the effects of contact time and temperature on fluoride removal by MMT clay at neutral pH. The results indicate that the percentage of fluoride removed increased with time and reached an optimum level at 50 minutes. The material with a particle size of 75 μm registered a maximum percentage of fluoride adsorption, compared to the other particle sizes. The preferred mode of isotherm seems to be the Langmuir model. The adsorption mechanism also follows the intraparticle diffusion pattern and surface adsorption.

Tanouayi et al. [66] studied the influence time, solution pH, geo-material dose, and fluoride concentration on the defluoridation by using clay minerals. The adsorption yields of fluoride by these clay minerals at an initial concentration of 10 mg/L and an adsorbent concentration of 10 g/L are 28.2% at pH 6.5 and 36.3% at pH 4.0.

Hamdi and Srasra [67] used three clay soils (MK, H, and ZB) from three regions in Southeast Tunisia for the elimination of fluoride (F) from acidic industrial wastewater and NaF solution. Results showed that the optimum pH for maximum F removal was 3.0. The experimental data fitted well with Langmuir isotherm than the Freundlich isotherm model. The Langmuir adsorption capacity of MK, H, and ZB soils was 73, 64.32, and 54.94 mg/g for NAF solution but the removal capacity of the same materials was 93.45, 84.03, and 69.44 mg/g for wastewater solution. The removal capacity of MK soil was more due to the presence of palygorskite clay and calcite; on the other hand, the adsorption of ZB soil was low due to the presence of an illite-kaolinitic fraction and a little quantity of dolomite.

Nabbou et al. [68] use kaolinite for the elimination of fluoride ions from Saharan groundwater located in the Tindouf region (Algeria). Results of adsorption tests showed that fluoride removal was efficient when the pH varies from 4.5 to 6. Under these conditions, the elimination capacities were 0.442 and 0.448 mg/g, respectively. Thus, findings showed that the pseudo-second-order kinetic model and Freundlich isotherm fit well the experimental data. Thermodynamic calculation indicated the endothermic nature of the sorption process. From this research, it can be deduced that kaolinite is an effective and low-cost adsorbent for the removal of fluoride from groundwater.

Guiza et al. [69] used MMT clay for adsorption of fluoride from aqueous solutions. Results showed that the maximum yield removal of fluoride ions from the aqueous solution was obtained at pH 5.0, the boundary layer resistance to mass transfer was decreased by temperature, and the adsorption equilibrium was obtained in 60 minutes. Under these conditions the maximum removal capacity of the clay was 40 mg/g; the adsorption was spontaneous and endothermic. General Brouers Sotolongo and Brouers Sotolongo are the best models for describing adsorption isotherms and kinetics, respectively.

9.4.3 Fluoride Removal by Adsorption of Modified Clay Minerals

Although the performance of clay minerals is mostly determined by intrinsic properties such as surface area, porosity, pH, etc., the crystal structure and negative charge of natural clays delimit their applications for the removal of fluoride (fluoride could be repelled by negative charges). Modification of clay minerals can achieve a defined property to target a specific type of contaminant. In fact, the modifications of clays through physical grinding, heat and acid treatments, ion exchange, etc. (Figure 9.3) have attracted considerable interest (the limits for the removal of fluoride by raw clay).

9.4.3.1 Fluoride Removal by Organoclay

Organoclays are prepared by incorporating organic molecules, most frequently quaternary ammonium cations, into the clay mineral structure [24]. The resultant organoclay product interacts differently with environmental contaminants as compared to the original natural clay mineral. However, an accumulation of quaternary ammonium cations in the interlayer and on surfaces largely in excess

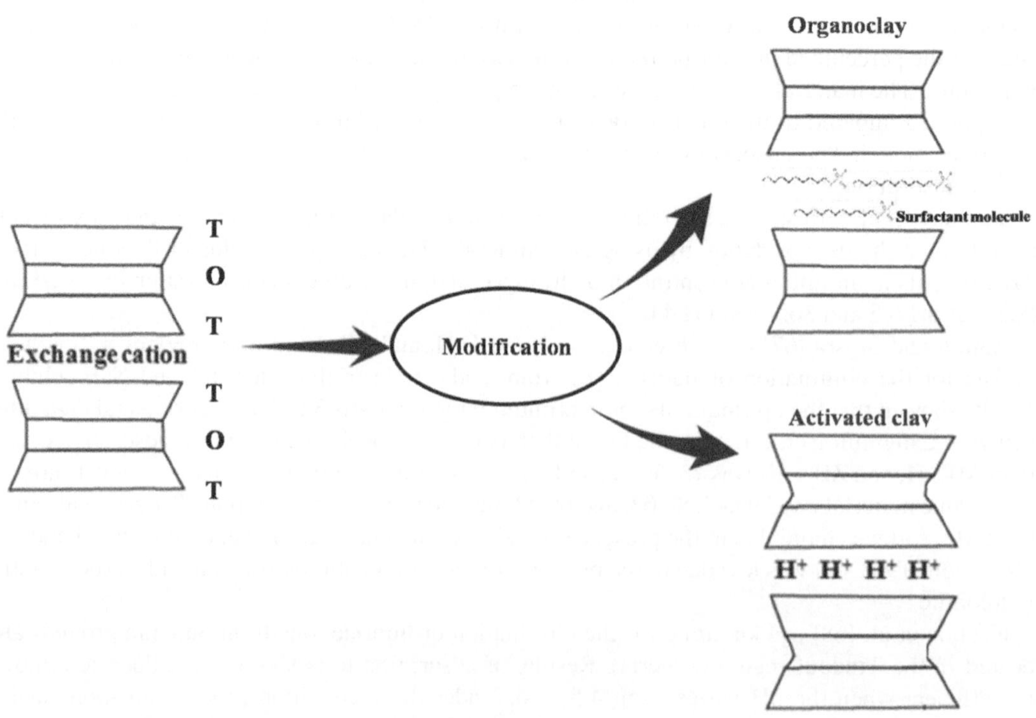

FIGURE 9.3 Structure of raw clay and modified clay.

of the CEC of the clay mineral leads to a build-up of net positive charges. This charge reversal consequently improves the material's affinity to negatively charged contaminants. Smectite or MMT is a better candidate for undertaking an organic modification than kaolinite or palygorskite. In most cases of environmental applications of organoclays, the dominant contaminant removal mechanism has been adsorption via hydrophobic partitioning, electrostatic attraction, and ion exchange. Yahya et al. [74] investigate the performance of organoclay in the adsorption of various hazardous substances such as fluoride. In this chapter, researchers suggest that the organoclay could be considered a cheap and efficient adsorbent for the removal of most of the chemical pollutants from wastewater that could be of socioeconomic and environmental relevance.

Gammoudi et al. [71] prepared various organoclays with smectite and two cationic surfactants (HDPy, hexadecylpyridinium and HDTMAB, hexadecyltrimethylammonium bromide) for fluoride removal. The optimal conditions suitable for fluoride removal consist of acidic pH, little fluoride concentration, and a contact time equal to 6 hours. Under these conditions, organoclays indicate high adsorption capacities for fluoride ions. The results showed that the Langmuir isotherm and pseudo-second-order kinetic model provide the best correlation of the experimental data.

Yahya et al. [72] prepared three organoclays by cation exchange with HDPy at different concentrations (1.0–3.0 cationic exchange capacity (CEC) of clay). The organoclays were tested for the removal of fluoride ions from aqueous solutions. The results indicate that most of the adsorption took place during the first 5 hours and the optimum pH value ranged from 3 to 4. The pseudo-second-order kinetic model and Langmuir isotherm described better the fluoride adsorption. Thus, organoclay (3 CEC) is the best adsorbent and could be considered an efficient and cost-effective adsorbent.

9.4.3.2 Fluoride Removal by Activated Clay

Acid-activated clay mineral is prepared by washing the clay mineral with mineral acids (like sulfuric acid and hydrochloric acid [HCl]) at an intended acid strength. The goal of acid activation of clay minerals is to obtain a material that is of increased surface area, porosity, and surface acidity. During activation, the acid exchanges its protons for the interlayer exchangeable cations and partially dissolves the clay crystalline structure by leaching some of the cations [24].

Ma et al. [75] investigate the adsorption of fluoride from an aqueous solution by granular acid-treated bentonite. The results indicate that the maximum fluoride elimination was obtained at a pH of 4.95 and 40 minutes to attain equilibrium. Kinetics data fitted pseudo-second-order model. Batch adsorption data was better described by Redlich-Peterson and Freundlich isotherm models.

Guiza et al. [76] studied the removal of fluoride using a modified sulfuric acid clay. Results indicate that the rate of fluoride ion removal increased with the mass of the adsorbent and the stirring speed, and decreased with the initial fluoride ion concentration. The optimal adsorption time is 30 minutes for a solution clay concentration of 1.5 g L1, at a minimum stirring speed of 300 rpm. Under optimal conditions, the fluoride removal rate is greater than 90%. Thus, results showed that this coefficient is in the range of 10^{-5} to 10^{-4} m/s. This coefficient varied with the agitation speed, the initial fluoride ion concentration, and the adsorbent mass.

Jianfeng Ma et al. studied the adsorption properties of fluorides by bentonite modified by activation with HCl. The adsorption was carried out according to several parameters: pH, temperature, and contact time. The experimental results indicate that the maximum adsorption capacity is equal to 50.07 mg/g. Thus, they showed that the mechanism of adsorption of these anions on the modified bentonite is an anion/OH exchange reaction.

Tor [70] investigated the influence of contact time, pH, initial fluoride concentration, and adsorbent dosage on the adsorption of fluoride (F) from aqueous solution by using MMT (treated with HCl). Results indicate that the sufficient time for removal equilibrium of (F) ions is 180 minutes. The maximum adsorption of (F) was obtained at pH 6. Under optimal conditions, the fluoride saturation capacity of MMT is 0.263 mg/g at room temperature. The best-fitting adsorption isotherm was the Freundlich isotherm.

Mustakim et al. [77] optimize the conditions of fluoride adsorption by activated clay. As a result of this work, the equilibrium was obtained in 60 minutes with 1.25 g of adsorbent and at a temperature of 45°C with a maximum of 78% fluoride adsorption. The isotherm models of fluoride adsorption mechanism fitted well with the Langmuir and Freundlich isotherm models and followed the pseudo-second-order kinetics model.

Achour and Youcef [78] have assessed the possibilities of groundwater defluoridation by adsorption on raw bentonite and chemical activation. The best findings were obtained with bentonite of Maghnia for a mass ratio acid/bentonite equal to 0.2 and an activation time of 3 hours.

9.4.3.3 Fluoride Removal by Calcined Clay

The exposure of clay minerals to mild to moderate levels of heat treatment can significantly alter the clay surface properties. The alteration in surface properties happens mainly because of the redistribution of hydrated cationic species in the clay interlayers and dehydroxylation of the clay silanol and aluminol groups [24]. As a result, the electronegativity of the clay surface may change, and thus may affect the interaction of the material with environmental contaminants. The thermal treatment of clay minerals can also change the pore size distribution and surface area of clay minerals. A treatment of bentonite at 100°C was able to remove the weakly adsorbed water and volatile compounds from the clay surface and thus increased the surface area. On the other hand, treatment of clay minerals above 500°C can cause dehydroxylation of their surfaces and edges, which consequently reduces their surface area [24].

Hauge et al. [79] studied the defluoridation of drinking water by pottery. A series of pots were prepared from ordinary potter's clay and fired at 500°C–1,000°C. Similarly, small clay bricks were fired, crushed, and sieved. Results showed that the clay fired at 600°C temperature was found to most influence the elimination of fluoride. However, temperatures over 700°C caused a decrease in F-binding, and pottery fired at 900°C and above seemed inadequate to eliminate F from water. Pots fired at 500°C or lesser cracked in water. Finally, the data indicates that clay ware fired at an optimal temperature may be of practical value for partial adsorption removal of drinking water.

Kofa et al. [80] investigate the removal of fluoride from water by adsorption onto fired clay pots. The authors studied the effects of contact time (0–90 minutes), pH (4, 5, 7, 8, and 9), stirring speed (60, 90, 120, and 200 rpm), and ionic strength (0–1,000 mg/L); all experiments were carried out at room temperature (25°C). Findings showed that equilibrium was obtained in 10 minutes whatever the pH. Pseudo-second-order model and pore diffusion model described well the experimental data. The large amount of fluoride removal (1.6 mg/g) was obtained at pH 4–5 and the optimum stirring speed is 120 rpm.

Motora and Tesema [81] studied the elimination of fluoride from aqueous solution by fired clay soil pillared with tea leaf ash components. The authors investigated the effects of pH, adsorbent dose, contact time, initial adsorbate concentration, and temperature. Results showed that the removal of fluoride by these adsorbents was noted to sharply increase from pH 4 to 7 and then reduce. The maximum removal efficiencies at pH 7 for both sorbents were obtained at 75% for fired clay soil and 90% for fired clay soil pillared, and the resulting maximum removal capacity was 0.913 mg/g for fired clay soil and 2.746 mg/g for fired clay soil pillared. Finally, the authors deduced that fired clay soil pillared is a good and cheap sorbent with high potential for the adsorption and removal of fluoride from fluoride-contaminated water and could hence serve as a successful and professional sorbent for the treatment of drinking waters in terms of high removal capacity.

9.4.3.4 Fluoride Removal by Oxide Incorporated Clays

Saturating the interlayer space of a given clay mineral with the desired cation would translate to conferring certain physicochemical properties to the clay mineral product, and these properties are unique to the desired cation. This is extremely important considering the fact that unmodified clay minerals in their natural state have mixed cations occupying the interlayer space. These mixed cations occupying the interlayer space of the clay mineral often make it impossible for the clay mineral to be used for certain purposes for which the surface properties do not allow. Preparing homoionic clay minerals, therefore, involves mixing aqueous clay mineral suspension with excess metallic salts of the desired cation and stirring thoroughly for about 24 hours for the exchange reaction to take place. Thereafter, the suspension is mechanically stirred and centrifuged to produce the homoionic clay mineral [24].

Gitari et al. [82] used Fe(III) modified bentonite clay for fluoride (F) adsorption from water. The optimum Fe(III) loading was done with 100 ppm of Fe^{3+} solution with powdered clay for 30 minutes. The modified bentonite could eliminate 100% of fluoride on the range of pH (2–10) and control the drawback correlated to the release of F. The adsorption capacity of modified bentonite clay was equal to 2.91 mg/g. The experimental result best fitted with the Langmuir isotherm model showing monolayer coverage of the modified bentonite.

Kamble et al. [83] used metal oxide (Lanthanum, magnesium, and manganese) to modify bentonite for defluoridation of water. This work investigated the effect of pH, adsorbent dose, initial fluoride concentration, and the presence of interfering co-ions. Results indicate that the use of 10% La-bentonite showed big fluoride adsorption capacity among Mg-bentonite, Mn-bentonite, and raw bentonite. The adsorption of fluoride in acidic pH solution was important as compared to an alkaline pH solution and maximum fluoride adsorption was attained within 30 minutes. The experimental results fitted well with Langmuir and Freundlich isotherm models. Also, the authors noted co-existing ions can have a positive influence on the elimination of fluoride, but carbonate and bicarbonate anions show a negative effect.

Thakre et al. [84] used magnesium incorporated bentonite for the adsorption of fluoride from an aqueous solution over a wide pH (3–10). The maximum defluoridation capacity of this adsorbent was 2.26 mg/g at 5 mg/L of initial F concentration. Bicarbonate ions very much affected the adsorption of fluoride while other co-anions did not influence the removal of fluoride.

Based on the results presented in Table 9.7, it is evident that the adsorption capacity of modified clays is significantly higher than that of raw clays. The adsorption capacity of fluoride by organoclay is about 2.84 times higher than raw clay [72]. Also, comparing the adsorption capacity of untreated clays and different organoclays, unmodified clay could eliminate only 1.1 mg fluoride ions per gram of clay. While raw clay was modified with surfactants HDPy and HDTMAB in 1.0–3.0 CEC, the adsorption capacities of these adsorbents were greatly improved [71]. Thus Gogoi and Baruah [85] indicate that acid-activated kaolinite clay is effective for defluoridation of water while raw kaolinite has very low defluoridation capacity due to low adsorption.

Kaolinite, a dioctahedral two-layered (silica + alumina) silicate (1:2 type), exhibited very little tendency for fluoride sorption while MMT, a 2:1 type material characterized by octahedral sheet of alumina sandwiched between two tetrahedral sheets of silica, showed significant fluoride sorption. It was investigated that the adsorption of fluoride by kaolinite and bentonite. The results show that bentonite has a higher fluoride adsorption capacity than kaolinite. Also, the sorption capacity of fluoride by MK soil containing palygorskite clay and calcite was superior to other adsorbents (H: smectic clay associated with kaolinite and calcite and the ZB: illite-kaolinitic clay), where the adsorption capacities (Q0) calculated from the Langmuir isotherms were, respectively, 71.94, 66.62, and 55.80 mg/g [91].

Kamble et al. [83] showed that 10% La-bentonite showed higher fluoride uptake capacity for defluoridation of drinking water as compared to Mg-bentonite, Mn-bentonite, and bare bentonite clay. Ten percent La-bentonite had a sorption capacity of 4.24 mg/g and maximum loading capacity was found to be 1.4 mg/g at a sorbent dose of 1.0 g/L.

Some studies evaluate the efficiency and capability of raw clay or modified clay to simultaneously remove various pollutants with fluoride. Yahya et al. investigate the retention and selectivity of phosphate and fluoride from single and industrial solutions using raw clay and organoclay; results showed that the selectivity of fluoride ions is better than phosphate and with coexisting chloride, sulfate ions, and other cations. This good selectivity of fluoride is attributed mainly to steric and charge effects because fluoride is a very small ion that is more strongly hydrated than other

TABLE 9.7
Modified Clays Minerals for Fluoride Elimination

Adsorbent	pH	*m* (g)	*C* (mg/L)	*T* (Minutes)	*T* (°C)	*Q* (mg/g)	Kinetic	Isotherm	References
				Organoclay					
Organoclays	Acidic	0.1	Low	6 hours	25	4.4–18.7	PSO	L	[71]
Organoclays	3–4	0.05	100	5 hours	Room	27.77	PSO	L	[72]
				Activated Clay					
Activated kaolinite	Low	—	3.0	100	25	0.0450	—	—	[85]
Activated montmorillonite	4	3	6	30	—	1.324	—	L	[86]
Activated bentonite	5	1	1.5 g L	30	—	90%	—	—	[76]
Activated kaolinite	3	1	2–10	30		0.134	PFO	F L	[73]
Activated montmorillonite	6	—	—	180	—	0.263	—	—	[70]
Activated clay	—	1.25	—	60	45	78%	—	—	[77]
				Calcined Clay					
Fired clay	4–5	—	—	10	25	1.6 mg	PSO	F	[80]
Fired clay soil pillared	7	2	—	120	25	2.746	—	—	[81]
Fired clay soil	7	2	—	120	25	0.913	—	—	[82]
				Oxide-Incorporated Clays					
Clay mixed hydroxyapatite	6	1.25	5	30	—	18.409	PSO	F	[87]
Magnesium bentonite clay	3–10	3	5	12	25	2.26	PFO	L	[84]
Al^{3+} modified bentonite	2–12	10	60	30	26	5.7	PSO	L F	[88]
Fe^{3+} modified bentonite	2–10	20	10	30	26	2.91	—	L	[82]
Lanthanum bentonite	7	—	—	24 hours	30	4.24	PFO	F L	[83]
Modified attapulgite	7		20–200	48	313 K	41.5	—	—	[89]
Modified montmorillonite	4		1–6	30	28	1.01–1.3	—	—	[86]
				Combined Modification					
Micronized kaolinite	3	1		30	—	0.134	PSO	F L	[73]
Mechanochemical clay	5	2	3.2	60	—	1.87	PSO	F	[90]

monovalent anions because of its high charge density. The selectivity sequence of anion adsorption coexisting on both adsorbents at a pH of about 3.3 has been reported to be: fluoride > phosphate > sulfate > chloride. The presence of co-anions affected the defluoridation capacity in the order of $F^{-} > HCO_3^{-} > Cl^{-}, SO_4^{3-}, NO^{3-}$ [73].

9.4.4 Adsorption of Fluoride Using Fixed-Bed Column

When contemplating future practical applications of adsorbents, forecasting the lifespan of the adsorbent and its scalability is important. According to researchers, full-scale field research is the only accurate way to forecast the long-term performance of water-purifying media. Alternatively, the adsorbent durability can be investigated in laboratory-based fixed-bed column setups. Therefore, various works investigate the fluoride adsorption performance of raw and modified clay in fixed-bed column:

Ma et al. [75] use granular acid-treated bentonite for the adsorption of fluoride from an aqueous solution by column adsorption experiments. The results showed that the capacities of the breakthrough and exhaustion points increased with the reduction of flow rate and the increase of initial fluoride concentration. The experimental data was well fitted with the Thomas model. Exhausted GHB was regenerated by alkali/alum treatment.

Mehari et al. [92] used crushed burnt clay pot as a sorbent in mini-column. The authors showed that the breakthrough volume and time increased with increasing the bed depth of the column. However, the increase in flow rate reduced the treated volume at breakthrough and decreased the service time. The optimum condition for reducing fluoride from 5 to 1.48 mg/L was achieved at 25 cm bed depth and at a 2.5 mL/min flow rate with a breakthrough volume of 7.3 L. The mini-column was scaled up to treat 324 L of water that meets the WHO limit (<1.5 mg/L) and also the constructed household defluoridation unit.

Kalsido et al. [92] evaluate the removal of fluoride (F) from water by bentonite clay in modified and unmodified using a fixed bed column of a borosilicate tube. Results showed that raw bentonite has shown very low F elimination efficiency (47.19%). However, modification of the clay surface with HCl and aluminum oxide increased F adsorption efficiency to 79.77% and 94.38%, respectively. For 5 mg/L initial F concentration, 180 minutes breakthrough time and 10 cm bed depth packed dose of sorbent; a 2.88 mg/g of F removal capacity was noted.

Mehari et al. [94] use the simplified fixed bed design models (bed depth service time [BDST] and empty bed residence time [EBRT]) to investigate the adsorption of fluoride onto crushed burnt clay pot. The results show that the BDST curve was in the form of a straight line, which explains more than 99.6% of the data, thus confirming that it fits the BDST model. For the same operating factors of 50 cm bed depth and 36 cm/h flow rate, using the BDST model 350 L of water could have been defluoridated. But, 324 L was defluoridated from 5 to 1.5 mg/L of fluoride for the case of the pilot. This was 8% higher and consequently not significant. Comparably, when the bed depth data was examined, the finding showed that at big EBRT values, the sorbent depletion rate was similar to batch sorption and thus the EBRT model could be employed to optimize the defluoridation unit design. Hence, both fixed-bed design models (BDST and EBRT) could be successfully used to analyze column performance and design a fluoride removal system based on crushed baked clay pot as the adsorbent medium.

9.5 CONCLUSIONS

Fluoride is a very toxic element and is known to cause adverse health effects on humans and animals due to environmental exposure. In fact, the WHO recommends that the fluoride content in drinking water should be in the range of 1.0–1.5 ppm. Among various technologies employed for defluoridation of wastewater, the adsorption method is relatively simple, economical, and appropriate for fluoride removal, thus offering unique advantages owing to the use of synthetic and natural materials. Although some widely used adsorbent is clay, this work provides an overview of various clay minerals and modified clay minerals as adsorbents used for the effective elimination of fluoride from wastewater. Most of the performance of the clay minerals and modified clay minerals is dependent on various parameters such as pH, initial F concentration, contact time, adsorbent dose, and temperature. The clay modification using acids, calcinations, or surfactants exhibits relatively high absorption ability toward fluoride compared to the raw clay. From the overview, the modification of clay minerals was also remarked for the enhancement of efficiency of fluoride removal from water. It is hoped that it will encourage even more rapid and extensive developments.

GLOSSARY

Defluoridation: Defluoridation is the descending adjustment of the plane of fluoride in water.

Fluoride Wastewater: Fluoride wastewater is water generated after the use of water in various industries, e.g. mining, semiconductor fabricating, electroplating, etc.

Organoclay: Organoclay material is defined as hybrid materials resulting from clay mineral association with surfactant.

World Health Organization: The WHO is a focused agency of the United Nations responsible for international public health.

REFERENCES

1. Lestari RB. Defluoridation of drinking water by hybrid coagulation and filtration process (Doctoral dissertation), Duisburg, Essen, Universität Duisburg-Essen, 2017.
2. Chaudhary M, Maiti A. Defluoridation by highly efficient calcium hydroxide nanorods from synthetic and industrial wastewater. *Colloids and Surfaces A: Physicochemical and Engineering Aspects*. 2019;561:79–88.
3. Arfaoui M, Ouejhani A, Hamrouni B. Physicochemical characterization of a polymeric conductor: application to defluoridation of industrial effluent by electrodialysis. *Turkish Journal of Chemistry*. 2018;42(1):121–131.
4. Mohapatra M, Anand S, Mishra BK, Giles DE, Singh P. Review of fluoride removal from drinking water. *Journal of Environmental Management*. 2009;91(1):67–77.
5. Dobaradaran S, Nabipour I, Mahvi AH, Keshtkar M, Elmi F, Amanollahzade F, et al. Fluoride removal from aqueous solutions using shrimp shell waste as a cheap biosorbent. *Fluoride*. 2014;47(3):253–257.
6. Sivarajasekar N, Paramasivan T, Muthusaravanan S, Muthukumaran P, Sivamani S. Defluoridation of water using adsorbents: a concise review. *Journal of Environment and Biotechnology Research*. 2017;6(1):186–198.
7. Tovar-Gómez R, Moreno-Virgen MR, Dena-Aguilar JA, Hernández-Montoya V, Bonilla-Petriciolet A, Montes-Morán MA. Modeling of fixed-bed adsorption of fluoride on bone char using a hybrid neural network approach. *Chemical Engineering Journal*. 2013;228:1098–1109.
8. Suneetha M, Sundar BS, Ravindhranath K. Studies on defluoridation techniques: a critical review. *International Journal of ChemTech Research*. 2015;8(8):295–309.
9. Turner BD, Binning P, Stipp SLS. Fluoride removal by calcite: evidence for fluorite precipitation and surface adsorption. *Environmental Science and Technology*. 2005;39(24):9561–9568.
10. Larsen MJ, Pearce EIF. Partial defluoridation of drinking water using fluorapatite precipitation. *Caries Research*. 1992;26(1):22–28.
11. Meenakshi S, Viswanathan N. Identification of selective ion-exchange resin for fluoride sorption. *Journal of Colloid and Interface Science*. 2007;308(2):438–450.
12. Apambire WB, Boyle DR, Michel FA. Geochemistry, genesis, and health implications of fluoriferous groundwaters in the upper regions of Ghana. *Environmental Geology*. 1997;33(1):13–24.
13. Shen W, Li Z, Liu Y. Surface chemical functional groups modification of porous carbon. *Recent Patents on Chemical Engineering*. 2008;1(1):27–40.
14. Simons R. Trace element removal from ash dam waters by nanofiltration and diffusion dialysis. *Desalination*. 1993;89(3):325–341.
15. Adhikary SK, Tipnis UK, Harkare WP, Govindan KP. Defluoridation during desalination of brackish water by electrodialysis. *Desalination*. 1989;71(3):301–312.
16. Lahnid S, Tahaikt M, Elaroui K, Idrissi I, Hafsi M, Laaziz I, et al. Economic evaluation of fluoride removal by electrodialysis. *Desalination*. 2008;230(1–3):213–219.
17. Liu J, Xu Z, Li X, Zhang Y, Zhou Y, Wang Z, et al. An improved process to prepare high separation performance PA/PVDF hollow fiber composite nanofiltration membranes. *Separation and Purification Technology*. 2007;58(1):53–60.

18. Raichur AM, Jyoti Basu M. Adsorption of fluoride onto mixed rare earth oxides. *Separation and Purification Technology*. 2001;24(1–2):121–127.
19. Onyango MS, Matsuda H. Fluoride removal from water using adsorption technique. In: T. Nakajima (Ed.), *Advances in Fluorine Science* [Internet]. Amsterdam, the Netherlands: Elsevier; 2006 [cité 4 juill 2022]. pp. 1–48. Disponible sur: https://linkinghub.elsevier.com/retrieve/pii/S187203580602001X
20. Singh J, Singh P, Singh A. Fluoride ions vs removal technologies: a study. *Arabian Journal of Chemistry*. nov 2016;9(6):815–824.
21. Habuda-Stanić M, Ravančić M, Flanagan A. A review on adsorption of fluoride from aqueous solution. *Materials*. 5 sept 2014;7(9):6317–6366.
22. Bhatnagar A, Kumar E, Sillanpää M. Fluoride removal from water by adsorption—a review. *Chemical Engineering Journal*. juill 2011;171(3):811–840.
23. Nagendra Rao CR. Fluoride and environment: a review. In: MJ Bunch, VM Suresh and TV Kumaran, eds., *Proceedings of the Third International Conference on Environment and Health, Chennai, India*, 15–17 December, 2003. Chennai: Department of Geography, University of Madras and Faculty of Environmental Studies, York University. pp. 386–399.
24. Sarkar B, Rusmin R, Ugochukwu UC, Mukhopadhyay R, Manjaiah KM. Modified clay minerals for environmental applications. In: M. Mercurio, B. Sarkar, & A. Langella (Eds.), *Modified Clay and Zeolite Nanocomposite Materials* [Internet]. Amsterdam, the Netherlands:Elsevier; 2019 [cité 22 oct 2021]. pp. 113–127. Disponible sur: https://linkinghub.elsevier.com/retrieve/pii/B9780128146170000037
25. Jlassi K, Krupa I, Chehimi MM. Overview. In: K. Jlassi, M. M. Chehimi, & S. Thomas (eds.), *Clay-Polymer Nanocomposites* [Internet]. Amsterdam, the Netherlands: Elsevier; 2017 [cité 7 mai 2022]. pp. 1–28. Disponible sur: https://linkinghub.elsevier.com/retrieve/pii/B978032346153500001X
26. Guégan R. Organoclay applications and limits in the environment. *Comptes Rendus Chimie*. févr 2019;22(2–3):132–141.
27. Zhu R, Chen Q, Zhou Q, Xi Y, Zhu J, He H. Adsorbents based on montmorillonite for contaminant removal from water: a review. *Applied Clay Science*. avr 2016;123:239–258.
28. Gu S, Kang X, Wang L, Lichtfouse E, Wang C. Clay mineral adsorbents for heavy metal removal from wastewater: a review. *Environmental Chemistry Letters*. juin 2019;17(2):629–654.
29. Ayoob S, Gupta AK. Fluoride in drinking water: a review on the status and stress effects. *Critical Reviews in Environmental Science and Technology*. déc 2006;36(6):433–487.
30. Barbier O, Arreola-Mendoza L, Del Razo LM. Molecular mechanisms of fluoride toxicity. *Chemico-Biological Interactions*. nov 2010;188(2):319–333.
31. Dharmagunawardhane HA, Dissanayake CB. Fluoride problems in Sri Lanka. *Environmental Management and Health*. 1 févr 1993;4(2):9–16.
32. Aoudj S, Drouiche N, Hecini M, Ouslimane T, Palaouane B. Coagulation as a post-treatment method for the defluoridation of photovoltaic cell manufacturing wastewater. *Procedia Engineering*. 2012;33:111–120.
33. Shen F, Chen X, Gao P, Chen G. Electrochemical removal of fluoride ions from industrial wastewater. *Chemical Engineering Science*. févr 2003;58(3–6):987–993.
34. Pillai P, Dharaskar S, Pandian S, Panchal H. Overview of fluoride removal from water using separation techniques. *Environmental Technology & Innovation*. févr 2021;21:101246.
35. Das R, Ali Md E, Hamid SBA, Ramakrishna S, Chowdhury ZZ. Carbon nanotube membranes for water purification: a bright future in water desalination. *Desalination*. mars 2014;336:97–109.
36. Waghmare SS, Arfin T. Fluoride removal by clays, geomaterials, minerals, low cost materials and zeolites by adsorption: a review. *International Journal of Innovative Science, Engineering & Technology (IJISET)*. 2015;4(11):14.
37. Singh NB, Nagpal G, Agrawal S, Rachna. Water purification by using adsorbents: a review. *Environmental Technology & Innovation*. août 2018;11:187–240.
38. Li Y, Zhang P, Du Q, Peng X, Liu T, Wang Z, et al. Adsorption of fluoride from aqueous solution by graphene. *Journal of Colloid and Interface Science*. nov 2011;363(1):348–354.
39. Getachew T, Hussen A, Rao VM. Defluoridation of water by activated carbon prepared from banana (Musa paradisiaca) peel and coffee (Coffea arabica) husk. *International Journal of Environmental Science and Technology*. juin 2015;12(6):1857–1866.
40. Ma W, Ya F, Wang R, Zhao Y. Fluoride removal from drinking water by adsorption using bone char as a biosorbent. *International Journal of Environmental Technology and Management*. 2008;9(1):59.
41. Gupta VK, Ali I, Saini VK. Defluoridation of contaminated water by using low cost adsorbents: A review. *International Journal of Advanced Science and Research*. 2014;1(6):28–32. Retrieved from https://www.allsciencejournal.com/assets/archives/2016/vol1issue6/1-3-18-104.pdf

42. Reddy K, Reddy N, Niharika P, Reddy M, Reddy H, Daneswari V. Defluoridation of water using natural adsorbents. *International Journal of Pedodontic Rehabilitation*. 2017;2(2):51.
43. Mondal P, George S. A review on adsorbents used for defluoridation of drinking water. *Reviews in Environmental Science and Biotechnology*. juin 2015;14(2):195–210.
44. Goswami A, Purkait MK. The defluoridation of water by acidic alumina. *Chemical Engineering Research and Design*. déc 2012;90(12):2316–2324.
45. Ravulapalli S, Kunta R. Defluoridation studies using active carbon derived from the barks of Ficus racemosa plant. *Journal of Fluorine Chemistry*. janv 2017;193:58–66.
46. Sarkar M, Banerjee A, Pramanick PP, Sarkar AR. Use of laterite for the removal of fluoride from contaminated drinking water. *Journal of Colloid and Interface Science*. oct 2006;302(2):432–441.
47. Tripathy S, Raichur A. Abatement of fluoride from water using manganese dioxide-coated activated alumina. *Journal of Hazardous Materials*. 1 oct 2007;153:1043–1051.
48. Sathish RS, Sairam S, Raja VG, Rao GN, Janardhana C. Defluoridation of water using zirconium impregnated coconut fiber carbon. *Separation Science and Technology*. 8 oct 2008;43(14):3676–3694.
49. Sinha S, Pandey K, Mohan D, Singh KP. Removal of fluoride from aqueous solutions by Eichhornia crassipes biomass and its carbonized form. *Industrial & Engineering Chemistry Research*. 1 déc 2003;42(26):6911–6918.
50. Sharma S, Pundhir A. Removal of fluoride from water using bioadsorbents. *Current Research in Microbiology and Biotechnology*. 2014;6:509–512.
51. Rafique T, Chadhar KM, Usmani TH, Memon SQ, Shirin K, Kamaluddin S, et al. Adsorption behavior of fluoride ion on trimetal-oxide adsorbent. *Desalination and Water Treatment*. 6 nov 2015;56(6): 1669–1680.
52. Rajkumar S, Murugesh S, Sivasankar V, Darchen A, Msagati TAM, Chaabane T. Low-cost fluoride adsorbents prepared from a renewable biowaste: syntheses, characterization and modeling studies. *Arabian Journal of Chemistry*. déc 2019;12(8):3004–3017.
53. Chen N, Zhang Z, Feng C, Li M, Zhu D, Chen R, et al. An excellent fluoride sorption behavior of ceramic adsorbent. *Journal of Hazardous Materials*. nov 2010;183(1–3):460–465.
54. Sakhare N, Lunge S, Rayalu S, Bakardjiva S, Subrt J, Devotta S, et al. Defluoridation of water using calcium aluminate material. *Chemical Engineering Journal*. sept 2012;203:406–414.
55. Chen N, Zhang Z, Feng C, Li M, Zhu D, Sugiura N. Studies on fluoride adsorption of iron-impregnated granular ceramics from aqueous solution. *Materials Chemistry and Physics*. janv 2011;125(1–2):293–298.
56. Yadav AK, Abbassi R, Gupta A, Dadashzadeh M. Removal of fluoride from aqueous solution and groundwater by wheat straw, sawdust and activated bagasse carbon of sugarcane. *Ecological Engineering*. mars 2013;52:211–218.
57. Bansiwal A, Pillewan P, Biniwale RB, Rayalu SS. Copper oxide incorporated mesoporous alumina for defluoridation of drinking water. *Microporous and Mesoporous Materials*. avr 2010;129(1–2):54–61.
58. Joshi S, Adhikari M, Pradhananga RR. Adsorption of fluoride ion onto zirconyl-impregnated activated carbon prepared from lapsi seed stone. *Journal of Nepal Chemical Society*. 15 déc 2013;30:13–23.
59. Mourabet M, El Rhilassi A, El Boujaady H, Bennani-Ziatni M, Taitai A. Use of response surface methodology for optimization of fluoride adsorption in an aqueous solution by Brushite. *Arabian Journal of Chemistry*. mai 2017;10:S3292–s3302.
60. Kebede B, Beyene A, Fufa F, Megersa M, Behm M. Experimental evaluation of sorptive removal of fluoride from drinking water using iron ore. *Applied Water Science*. mars 2016;6(1):57–65.
61. Srinivasan R. Advances in application of natural clay and its composites in removal of biological, organic, and inorganic contaminants from drinking water. *Advances in Materials Science and Engineering*. 2011;2011:1–17.
62. Bergaya F, éditeur. *Handbook of Clay Science*. 1. ed., reprint. Amsterdam Heidelberg: Elsevier; 2008. 1224 p. (Developments in Clay Science).
63. Bower CA, Hatcher JT. Adsorption of fluoride by soils and minerals. *Journal of Soil Science*. 1967;3(3)151–154.
64. Chaturvedi AK, Pathak KC, Singh VN. Fluoride removal from water by adsorption on China clay. *Applied Clay Science*. sept 1988;3(4):337–346.
65. Karthikeyan G, Pius A, Alagumuthu G. Fluoride adsorption studies of montmorillonite clay. *Journal of Chemical Technology*. May 2005;12:263–272.
66. Tanouayi G, Gnandi K, Ouro-Sama K, Ahoudi H, Solitoke HD, Badassan TEE, et al. Défluoruration des eaux à l'aide des résidus du traitement des phosphates naturels et des argilites feuilletées. Déchets, sciences et techniques [Internet]. 2017 [cité 13 mars 2021];(N°73). Disponible sur: https://lodel.irevues.inist.fr/dechets-sciences-techniques/index.php?id=3534

67. Hamdi N, Srasra E. Removal of fluoride from acidic wastewater by clay mineral: effect of solid–liquid ratios. *Desalination*. févr 2007;206(1–3):238–244.
68. Nabbou N, Belhachemi M, Boumelik M, Merzougui T, Lahcene D, Harek Y, et al. Removal of fluoride from groundwater using natural clay (kaolinite): optimization of adsorption conditions. *Comptes Rendus Chimie*. févr 2019;22(2–3):105–112.
69. Guiza S, Brouers F, Bagane M. Fluoride removal from aqueous solution by montmorillonite clay: kinetics and equilibrium modeling using new generalized fractal equation. *Environmental Technology & Innovation*. févr 2021;21:101187.
70. Tor A. Removal of fluoride from an aqueous solution by using montmorillonite. *Desalination*. nov 2006;201(1–3):267–276.
71. Gammoudi S, Frini-Srasra N, Srasra E. Preparation, characterization of organosmectites and fluoride ion removal. *International Journal of Mineral Processing*. déc 2013;125:10–17.
72. Yahya K, Msadok I, Mlayah A, Srasra E, Hamdi N. Retention and selectivity of phosphate and fluoride from single and industrial aqueous solutions using purified and surfactant modified Tunisian clay. *Desalination and Water Treatment*. 2021;216:199–210.
73. Meenakshi S, Sundaram CS, Sukumar R. Enhanced fluoride sorption by mechanochemically activated kaolinites. *Journal of Hazardous Materials*. mai 2008;153(1–2):164–172.
74. Yahya K, Hamdi W, Hamdi N. Organoclay nano-adsorbent: preparation, characterization and applications. In W. Oueslati (Ed.), *Nanoclay – Recent Advances, New Perspectives and Applications* (pp. 1–28), IntechOpen, London, 2022.
75. Ma Y, Shi F, Zheng X, Ma J, Gao C. Removal of fluoride from aqueous solution using granular acid-treated bentonite (GHB): batch and column studies. *Journal of Hazardous Materials*. janv 2011;185(2–3):1073–1080.
76. Guiza S, Hajji H, Bagane M. External mass transport process during the adsorption of fluoride from aqueous solution by activated clay. *Comptes Rendus Chimie*. févr 2019;22(2–3):161–168.
77. Mustakim Z, Purwaningtyas FY, Leksono EB, Sarto PA. Optimization condition of fluoride adsorption using activated clay. *IOP Conference Series: Materials Science and Engineering*. 1 avr 2020;821(1):012009.
78. Achour S, Youcef L. Defluoridation of the Algerian north Sahara waters by adsorption onto local bentonites: defluoruration des eaux du Sahara septentrional Algerien par adsorption sur des bentonites locales. *International Journal of Environmental Studies*. avr 2009;66(2):151–165.
79. Hauge S, Österberg R, Bjorvatn K, Selvig KA. Defluoridation of drinking water with pottery: effect of firing temperature. *The European Journal of Oral Sciences*. déc 1994;102(6):329–333.
80. Kofa GP, Gomdje VH, Telegang C, Koungou SN. Removal of fluoride from water by adsorption onto fired clay pots: kinetics and equilibrium studies. *Journal of Applied Chemistry*. 4 juill 2017;2017:1–7.
81. Gebeyehu Motora K, Ayele Tesema D. Removal of fluoride from aqueous solution using fired clay soil pillared with tea leaf ash components. *Indo American Journal of Pharmaceutical Research*, 2017; 7(6):2231–6876.
82. Gitari WM, Ngulube T, Masindi V, Gumbo JR. Defluoridation of groundwater using Fe^{3+}-modified bentonite clay: optimization of adsorption conditions. *Desalination and Water Treatment*. 6 févr 2015;53(6):1578–1590.
83. Kamble SP, Dixit P, Rayalu SS, Labhsetwar NK. Defluoridation of drinking water using chemically modified bentonite clay. *Desalination*. déc 2009;249(2):687–693.
84. Thakre D, Rayalu S, Kawade R, Meshram S, Subrt J, Labhsetwar N. Magnesium incorporated bentonite clay for defluoridation of drinking water. *Journal of Hazardous Materials*. 15 août 2010;180(1–3): 122–130.
85. Gogoi PK, Baruah R. Fluoride removal from water by adsorption on acid activated kaolinite clay. Fluoride adsorption studies of montmorillonite clay. *Indian Journal of Chemical Technology*. September 2008;15:263–272.
86. Ramdani A, Taleb S, Benghalem A, Ghaffour N. Removal of excess fluoride ions from Saharan brackish water by adsorption on natural materials. *Desalination*. janv 2010;250(1):408–413.
87. Dang-I Apea AY, Boansi O, Pedevoah M-M. Reduction of fluorine in water using clay mixed with hydroxyapatite. *International Journal of Applied Science and Technology*. 2015;5:45–55.
88. Vhahangwele M, Mugera GW, Tholiso N. Defluoridation of drinking water using Al^{3+}-modified bentonite clay: optimization of fluoride adsorption conditions. *Toxicological & Environmental Chemistry*. 21 oct 2014;96(9):1294–1309.
89. Zhang J, Xie S, Ho YS. Removal of fluoride ions from aqueous solution using modified attapulgite as adsorbent. *Journal of Hazardous Materials*. 15 juin 2009;165(1–3):218–222.

90. Obijole O, Gitari M, Ndungu P, Samie A. Mechanochemically activated aluminosilicate clay soils and their application for defluoridation and pathogen removal from groundwater. *International Journal of Environmental Research and Public Health*. 22 févr 2019;16(4):654.
91. Hamdi N, Srasra E. Retention of fluoride from industrial acidic wastewater and NaF solution by three Tunisian clayey soils. *Fluoride*. January-March 2009;42(1):39.
92. Mehari BB, Mayabi AO, Kakoi BK. Development of household defluoridation unit based on crushed burnt clay pot as sorbent medium: a case of Keren community, Eritrea. *Environment and Natural Resources Research*. 28 mai 2014;4(3):67.
93. Kalsido AW, Kumar A, Tekola B, Mogessie B, Alemayehu E. Evaluation of bentonite clay in modified and unmodified forms to remove fluoride from water. *Water Science and Technology*. 15 nov 2021;84(10–11):2661–2674.
94. Mehari BB, Mayabi AO, Kakoi B. Investigating simplified fixed bed design models for the adsorption of fluoride onto crushed burnt clay pot. *International Journal of Engineering & Technology*. 27 oct 2014;3(4):512.

10 Fundamentals of Microbial Electrochemical Technology and Innovations in Transforming Industrial Waste Into Energy by MET-Assisted Bioremediation

Akansha Shrivastava, Pragati Singh, Rakesh Kumar Sharma, Yung-Tse Hung, Mu-Hao Sung Wang, and Lawrence K. Wang

NOMENCLATURE

ABC-MFC	Air breathing cathode-based microbial fuel cell
ANMFC	Anaerobic microbial fuel cell
BMFC	Benthic microbial fuel cell
BOD	Biochemical oxygen demand
COD	Chemical oxygen demand
CPCB	Central Pollution Control Board
CW-MFC	Constructed wetland microbial fuel cell
DET	Direct electron transfer
DMP	Dimethyl phthalate
EBC-MFC	Enzyme-based microbial fuel cell
EDTA	Ethylenediamine tetraacetic acid
ETC	Electron transport chain
IPPC	Integrated Pollution Prevention and Control
MET	Mediated electron transfer
MFC	Microbial fuel cell
PEM	Proton exchange membrane
Pt	Platinum
SMFC	Sediment microbial fuel cell
TCMFC	Truncated conical microbial fuel cell
TMDL	Total maximum daily load
TMFC	Thermophilic microbial fuel cell

DOI: 10.1201/9781003532354-10

10.1 FUNDAMENTALS AND ENVIRONMENTAL ENGINEERING SIGNIFICANCE

10.1.1 Fundamentals of Microbial Electrochemical Technologies

10.1.1.1 Key Fundamentals and Working Principles of Microbial Electrochemical Technologies

Microbial Electrochemical Technologies (MET) integrate microbiology, electrochemistry, and material science to convert chemical energy stored in organic materials into electrical power. Some key fundamentals include the following : (a) electroactive microorganisms; (b) electrodes; and (c) bioelectrochemical systems (BES).

1. **Electroactive Microorganisms:** These are specialized bacteria capable of transferring electrons to or from electrodes during their metabolic processes.
2. **Electrodes:** Solid-state electrodes serve as the interface for electron transfer. They can be made from various materials, including carbon-based substances.
3. **Bioelectrochemical Systems (BES):** These systems house the microorganisms and electrodes, facilitating the electrochemical reactions.

The basic working principle includes: (a) electron transfer; (b) electricity generation; and (c) reduction reactions.

1. **Electron Transfer:** Electroactive bacteria oxidize organic compounds, releasing electrons. These electrons are transferred to the anode (negative electrode).
2. **Electricity Generation:** The electrons flow through an external circuit to the cathode (positive electrode), generating an electric current.
3. **Reduction Reactions:** At the cathode, electrons are used to reduce compounds, often resulting in the production of useful chemicals or the removal of contaminants.

10.1.1.2 Applications and Advantages of Microbial Electrochemical Technologies

For wastewater treatment, MET can treat wastewater by breaking down organic pollutants while generating electricity.

For bioremediation, NET can be used to clean up contaminated environments by degrading pollutants.

For energy recovery, MET can recover energy from waste materials, making them a sustainable technology for energy production.

The advantages of MET include the process sustainability and versatility.

1. **Sustainability:** METs offer a green alternative to traditional energy sources by utilizing waste materials.
2. **Versatility:** They can be applied in various fields, from environmental remediation to bioenergy production.

10.1.1.3 Challenges in Implementing Microbial Electrochemical Technologies

Implementing microbial electrochemical technologies (MET) can be quite challenging due to several factors:

1. **Complexity of Biological Systems:** MET relies on the interaction between microorganisms and electrodes. Understanding and controlling these interactions can be difficult due to the complexity and variability of biological systems.
2. **Material Costs and Durability:** The materials used for electrodes and other components in MET systems can be expensive. Additionally, these materials need to be durable

and resistant to biofouling and corrosion, which can increase costs and maintenance requirements.

3. **Scalability:** While MET has shown promise in laboratory settings, scaling up these systems to industrial or municipal levels poses significant challenges. This includes maintaining efficiency and stability over larger volumes and longer periods.
4. **Energy Efficiency:** Achieving a net positive energy balance is crucial for the viability of MET. However, the energy input required for processes like electrode polarization can sometimes outweigh the energy produced, making the system less efficient.
5. **Integration with Existing Infrastructure:** Implementing MET in existing wastewater treatment or energy production facilities requires careful integration. This can involve significant modifications to current infrastructure, which can be costly and time-consuming.
6. **Regulatory and Environmental Concerns:** Ensuring that MET systems comply with environmental regulations and standards is essential. Additionally, the potential environmental impact of these systems, such as the release of by-products, needs to be carefully managed.
7. **Public and Stakeholder Acceptance:** Gaining acceptance from the public and stakeholders is crucial for the successful implementation of MET. This involves educating and convincing them of the benefits and safety of the technology.

These challenges highlight the need for continued research and development to optimize MET systems and make them more viable for widespread use.

10.1.2 Fundamentals of Bioremediation for Industrial Waste Treatment

Bioremediation is a process that uses microorganisms, plants, or microbial enzymes to detoxify and remove pollutants from the environment. It's a sustainable and eco-friendly method for treating industrial waste. The following are some key fundamentals.

10.1.2.1 Mechanisms of Bioremediation

The mechanisms of bioremediation include the basic microbial degradation, phytoremediation, and enzymatic bioremediation, as follows:

1. **Microbial Degradation:** Microorganisms such as bacteria, fungi, and algae break down contaminants into less harmful substances. This can occur through aerobic (with oxygen) or anaerobic (without oxygen) processes.
2. **Phytoremediation:** Plants absorb, accumulate, and detoxify pollutants from soil and water. They can either store contaminants in their tissues or transform them into less toxic forms.
3. **Enzymatic Bioremediation:** Microbial enzymes catalyze the breakdown of pollutants. This method is often used for specific contaminants that are difficult to degrade.

10.1.2.2 Types of Bioremediation

There are two types of bioremediation: (a) in situ bioremediation and (b) ex situ bioremediation.

1. **In Situ Bioremediation:** Treatment occurs at the site of contamination. Techniques include bioventing, biosparging, and phytoremediation.
2. **Ex Situ Bioremediation:** Contaminated material is removed and treated elsewhere. Methods include biopiles, bioreactors, and landfarming.

10.1.2.3 Factors Influencing Bioremediation

The following factors influence bioremediation: (a) Environmental Conditions; (b) Microbial Community; and (c) Contaminant Properties.

1. **Environmental Conditions:** Temperature, pH, oxygen levels, and nutrient availability can significantly impact the efficiency of bioremediation.
2. **Microbial Community:** The presence and activity of specific microorganisms are crucial. Sometimes, bioaugmentation (adding specific strains) or biostimulation (adding nutrients) is necessary to enhance microbial activity.
3. **Contaminant Properties:** The type, concentration, and bioavailability of contaminants affect the bioremediation process. Some pollutants are more readily degraded than others.

10.1.2.4 Bioremediation Applications in Industrial Waste Treatment

Bioremediation has a wide range of applications, particularly in environmental cleanup and restoration. Here are some key areas where bioremediation is effectively used:

1. **Oil Spill Cleanup:** Microorganisms are used to break down and degrade oil in contaminated water and soil. This method has been successfully applied in major oil spill incidents.
2. **Industrial Waste Treatment:** Bioremediation is employed to treat industrial waste sites, including those contaminated with heavy metals, solvents, and other hazardous chemicals.
3. **Groundwater Remediation:** Techniques like biosparging and bioventing are used to inject air or nutrients into the groundwater to stimulate the growth of microorganisms that degrade pollutants.
4. **Soil Remediation:** Contaminated soils, especially those with organic pollutants like pesticides and hydrocarbons, are treated using bioremediation methods such as composting and bioaugmentation.
5. **Wastewater Treatment:** Bioremediation is used in wastewater treatment plants to break down organic matter and reduce pollutants before the water is released back into the environment.
6. **Restoration of Contaminated Sites:** Bioremediation helps restore areas contaminated by hazardous waste, making them safe for human use and ecological recovery.

These applications highlight the versatility and effectiveness of bioremediation in addressing various environmental challenges.

10.1.2.5 Bioremediation Advantages and Challenges

The advantages of bioremediation are associated with costs and process operation. Bioremediation is cost-effective, environmentally friendly, and can be applied to a wide range of contaminants.

Implementing bioremediation comes with several challenges:

1. **Site-Specific Conditions:** The effectiveness of bioremediation can vary significantly depending on the specific conditions of the contaminated site, such as soil type, temperature, pH, and the presence of other contaminants.
2. **Microbial Activity:** The success of bioremediation relies heavily on the activity of microorganisms. Factors such as nutrient availability, oxygen levels, and the presence of toxic substances can inhibit microbial growth and activity.
3. **Ecological Concerns:** The introduction of genetically engineered microorganisms (GEMs) into the environment raises ecological and regulatory concerns. There is a risk of these organisms disrupting local ecosystems.
4. **Incomplete Degradation:** Some pollutants are resistant to microbial degradation, leading to incomplete bioremediation. This can result in the accumulation of intermediate products that may still be harmful.
5. **Monitoring and Control:** Continuous monitoring and control are required to ensure the bioremediation process is proceeding as expected. This can be resource-intensive and costly.

6. **Scalability:** While bioremediation can be effective in laboratory settings, scaling up to field applications presents additional challenges, such as ensuring uniform distribution of microorganisms and nutrients.

Addressing these challenges requires a multidisciplinary approach, combining microbiology, environmental science, and engineering to optimize bioremediation strategies.

Bioremediation represents a promising approach to managing industrial waste sustainably, leveraging natural processes to mitigate environmental pollution.

10.1.3 Microbial Electrochemical Technology-Assisted Bioremediation

Microbial electrochemical technology-assisted bioremediation (MET-assisted bioremediation) is an innovative approach that leverages the capabilities of electrochemically active microorganisms to enhance the bioremediation process.

10.1.3.1 Process Description

A breakdown of how it works is presented below:

1. **Electrodes as Electron Acceptors:** In MET-assisted bioremediation, electrodes are used as inexhaustible electron acceptors. This setup facilitates the degradation of contaminants by providing a continuous supply of electrons, which is crucial for the metabolic processes of certain microorganisms.
2. **Enhanced Microbial Activity:** The electric field generated in the system can stimulate microbial activity, enhancing the degradation of pollutants. This is particularly useful for breaking down complex hydrocarbons and other persistent organic pollutants.
3. **Minimal Chemical Additives:** Unlike traditional methods, MET-assisted bioremediation does not require additional chemicals, making it a greener and more sustainable option. It also tends to produce less sludge and waste.

10.1.3.2 Process Applications

This technology is particularly effective for the remediation of soils and groundwater contaminated with petroleum hydrocarbons, polycyclic aromatic hydrocarbons (PAHs), and other organic pollutants.

Overall, MET-assisted bioremediation offers a promising solution for environmental cleanup, combining the principles of microbiology and electrochemistry to achieve efficient and sustainable remediation.

10.1.4 Fundamentals of Microbial Fuel Cell

A microbial fuel cell (MFC) is an innovative device that converts chemical energy into electrical energy through the action of microorganisms.

10.1.4.1 Process Description

The key elements of the MFC include: (a) microorganisms; (b) electrodes; and (c) anode and cathode.

1. **Microorganisms:** MFCs use bacteria as biocatalysts. These bacteria metabolize organic matter (like waste) and, in the process, release electrons.
2. **Electrodes:** The electrons are captured by electrodes, creating an electric current.
3. **Anode and Cathode:** Typically, MFCs have an anode (where oxidation occurs) and a cathode (where reduction occurs). The bacteria at the anode break down organic matter,

releasing electrons and protons. The electrons travel through an external circuit to the cathode, generating electricity.

10.1.4.2 Process Applications and Benefits

In the area of wastewater treatment, MFCs can treat wastewater while simultaneously generating electricity. MFCs offer a sustainable way to produce renewable energy from organic waste.

In the area of environmental monitoring, MFCs can be used in sensors to monitor environmental conditions.

The benefits of MFCs include: (a) environmental sustainability and (b) dual functionality.

1. **Sustainability:** MFCs also provide a green energy solution by utilizing waste materials; therefore, they are good for environmental sustainability.
2. **Dual Functionality:** MFCs can clean wastewater and produce electricity at the same time.

10.1.5 Environmental Engineering Significance

Industrialization has brought a lot of economic prosperity all over. Booming industrial growth has resulted in an environmental impact closer to the threshold limits of tolerance. Thus, environmental sustainability is now becoming a significant factor. The major negative impact on the environment is the creation of waste. In general industrial waste can be regarded as a non-hazardous and hazardous category [1]. Non-hazardous waste is the waste from an industry that marks any threat to public health or the environment. Hazardous waste is the outcome of industrial activity that poses threat to the environment [2]. Improper disposal of municipal or industrial waste can create unhealthy environmental conditions, resulting in pollution and outbreaks of diseases [3]. Additionally, it generates difficulties for living and non-living things such as sickness, loss of life, ecosystem imbalance, and habitat loss. The major problem associated with industrial waste is global warming due to unrestrained industrial activities [4]. The activity releases a mixture of greenhouse gases into an open environment. Apart from this, the untreated industrial wastes discharged have predominant toxic concentrations. These wastes are then discharged into water bodies, affecting aquaculture, flora, and fauna life cycle. The discharge of untreated industrial wastewater and its usage creates water pollution and soil pollution [5].

The wastewater from industries contains dissolved and suspended substances discharged from different processes such as manufacturing and commercial activities [6]. For example, such industries include mining, laundries, steel/iron, food/beverage, power plants, oil and gas fracking, and others. The major contaminants include chemicals, oils, heavy metals, pesticides, dyes, and other industrial by-products [7]. Phenols and phenolic compounds are also one of the major constituents of toxic pollutants released by drug manufacturing industries. Poorly biodegradable pollutants include sulfides, aniline, petroleum hydrocarbons, nitrobenzene, alkanes, olefins, organochlorines, chloroalkanes, perfluorooctane sulfonate (PFOS), and perfluroοctanoic acid (PFOA) [8–10]. Salt removal from wastewater has become as important as any other pollutant in many countries. The high salinity of wastewater generated by petroleum, food processing, and leather has many adverse effects on life forms [11].

Intake of such amounts of toxic pollutants by flora and fauna can cause detrimental health problems via the food chain (Figure 10.1). They can be carcinogenic and teratogenic. They can cause organ damage, nervous system impairments, and reduced growth and development. The presence of phenols in wastewater causes reflex loss, irregular body temperature, cyanosis, respiratory failure, etc. [12]. Metallic compounds and dumped off oil together form a colloidal matter that acts as a barrier to prevent sunlight from entering and results in decreased dissolved oxygen. Efficient removal of these toxic pollutants is highly necessary for a hale and hearty environment [13].

Industrial waste management deals to reduce the adverse effects of waste on the surroundings and human health. Sustainable management can be defined as "to not use more today than you can replace tomorrow" [14]. The broader context of waste treatment includes commercial sustainability, government policies/regulations, on-site treatment technologies, environmental impacts, etc. [15].

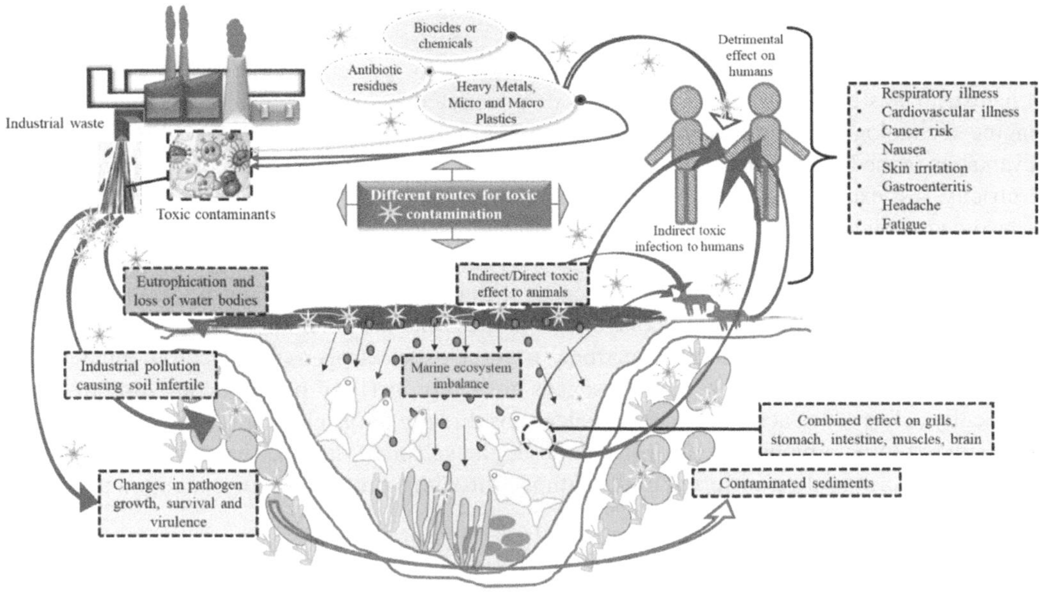

FIGURE 10.1 Contaminants from industrial wastewater and their detrimental effect on environment.

Accordingly, industrial wastewater management deals with the processes or technologies used for the wastewater produced by industries as inevitable by-products. Most industries generate a large amount of wastewater which after being treated is released into the aquatic environment. Currently, the inclination is to reduce waste or to recycle/reuse treated wastewater [16,17]. Waste management itself defines the broad area of collecting, managing, and treating that do not belong to trash. Subsequently, the waste can potentially be utilized as a valuable resource for other value-added products with little or no impact on the environment.

Many countries are establishing limits on the amounts of pollutants that can be safely assimilated into specific water bodies. Some such programs include total maximum daily load (TMDL) under the US Clean Water Act, the Central Pollution Control Board (CPCB) in India, and Integrated Pollution Prevention and Control (IPPC) in Europe, which set MINAS (minimum acceptable standards) for the release of industrial and municipal wastes [18,19]. Numerous treatment plants are also being framed for waste treatment under chemical, biological, and electrochemical processes [20]. With an eye on economic growth and clean water strategy, several industrial developers and manufacturers, nowadays, are embracing technologies to ensure cleaner production and less pollution.

In general, several traditional industrial wastewater treatments include primary, secondary, and final step tertiary process that requires a large amount of energy, high operational cost, and excessive sludge [21]. Due to energy insecurity, and worldwide environmental alarms, there is an immense interest stimulated in sustainable and clean energy sources [22]. Traditional methods focus on removing those toxic pollutants and not on the extraction of energy from industrial waste [23]. Hence, observing all these drawbacks, it is necessary to develop an alternate technology that can be employed or explored to recover precious energy for other sustainable operations. In recent years, anaerobic treatment processes have paved a new way toward bioenergy generation from industrial waste using microbial fuel cell (MFC) technology [24]. These industrial wastes can be aerobically and anaerobically degraded to produce several products such as bioelectricity, biogas, ethanol, and others under waste to bioenergy conversion. The wastewater contains diverse organic compounds that can fuel MFCs. The growth of bioelectrochemically active microbes is enhanced through MFCs during the process. This chapter reports the characteristics of different types of substrates used for their treatment and efficiency and concurrent power generation in MFCs.

10.2 MICROBIAL FUEL CELL

MFC technology has fascinated numerous researchers for power generation and bioremediation of industrial wastewater in recent years. It is demonstrated as one of the most promising and challenging technologies in addressing energy and environmental issues [25,26]. MFCs have many advantages in the bioremediation of wastewater such as the direct conversion of substrate energy to electricity, less excess activated sludge, can be operated at even low temperatures, and may have a widespread application in locations where insufficient electrical infrastructure is a major problem. The principle behind this technology is the conversion of the chemical energy of compounds into electrical energy during the remediation process with the aid of microorganisms as exoelectrogens [27]. It catalyzes two types of MFCs: one is mediatorless and the other one is with the mediator. The application catalyzes the transfer of electrons from the exoelectrogens in the cell to anode using a mediator or without a mediator. Exoelectrogens are the property of bacteria consisting of electrochemically active redox activity to transfer electrons directly to the electrode. Moreover, the system is considered to be a biobattery, as it decomposes the substrate in either batch mode or fed-batch mode. The electrons are transferred to the cathode through a proton exchange membrane (PEM).

10.2.1 Components of MFC

The main components of the MFC system comprise electrodes (anode and cathode), membranes, substrate, temperature, pH, anolyte, catholyte, and others. The following sections provide a brief overview of the cell components.

10.2.1.1 Electrode Materials

The electrode is the most significant component in the electrochemical performance of MFCs. It helps in the rate of electron transfer and adhesion of microorganisms very well. The electrode material (anode or cathode) should have the following properties such as porosity, electrical conductivity, stability, durability, cost-effectiveness, easily accessibility [28]. The anode material enables the electrical conductivity between the anode and substrate. The anaerobic anode compartment is a major part of MFC. Carbonaceous and graphite materials are widely used in anode electrodes [26]. It was also observed that stainless steel modified with polyaniline contributes to a high electron transfer rate. Moreover, anode modification is emerging as a new trend for better power generation. Recently, natural waste material has been developed as a 3D anode electrode for low cost and high performance [29,30]. The pore size of an electrode can be maintained by the carbonization process. Also, the modification readily increases the conductivity, specific surface area, and electrocatalytic activity without blocking the diffusion of substrate.

Several factors are highly dependent upon materials used as an electrode such as electron transfer, adherence of biocatalyst, electrochemical efficiency, etc. The design and material of cathode significantly play an important role in the scale-up of MFCs [31]. The selection of cathode material contains possible cost reduction due to its flexibility in low-cost material and increasing potential. Presently, common types of materials are graphite, carbon (paper, fiber, felt, and cloth) [32], zinc, magnesium, stainless steel, copper, etc. Sometimes to improve power performance modified cathode (e.g. platinum (Pt)) is used to reduce the cathodic reaction activation energy and increase reaction rate as it's an expensive metal and thus limits its practical application.

10.2.1.2 Membranes

In MFCs, a membrane mediates ion transport and separates the anode and cathode chambers, especially by effluents, CO_2, and oxygen. Several polymer electrolyte membranes have been developed for proton exchange during the process. The membranes are classified as hydrocarbon polymer membranes, organic-inorganic hybrid polymer membranes, perfluorinated polymer membranes. Among them, finding suitable and cost-effective separators attracted great attention [33]. Nafion is another most popular membrane due to its highly selective permeability for protons. Though on the high side, it is also very expensive. Therefore, cheaper and more durable substitutes are needed. For this, researchers are trying hard to utilize effective alternatives as separators such as porous cloth, glass fiber, J-cloth, CMI-7000, polymer membranes, AMI-7001, etc. Recently, different types of ceramic membranes (clayware, mfensi clay, terracotta, and earthenware) have been used over conventional membranes [34] as it provides a natural and constant environment for the exoelectrogens and thereby provide an efficient system for energy harvesting.

10.2.1.3 Mediators

Electron transferability in MFC is readily improved if a suitable electron mediator is used. An ideal mediator should be cost-effective, have the ability to cross membrane easily, be non-toxic, be biodegradable toward microbes, have high reaction rate of electrodes, be soluble, and grasp electrons from electron carriers from electron transport chain (ETC) [35]. Electrons shuttle to the anode in three different ways such as exogenous mediators (methylene blue, neutral red, meldola's blue, thionine, 2-hydroxy-1, Fe(III) EDTA, 4-naphthoquinone, etc.) accept the electron from microbes and transfer it to the anode; internal mediators (pyocyanin, flavin, etc.) produced by microbe shuttle the electron to the anode; and direct electron transfer (DET) (cysts, pili, etc.) of an electron from microbe to the anode by nanowires [36]. Due to this cyclic process, the electron accelerates the transfer rate and generates electricity. Unfortunately, the toxicity and instability of synthetic mediators limit their application in MFCs [37]. Perhaps, there is no need for any mediator to transfer the electrons in mediator-less MFCs since there is an electrochemically active microbe that transfers the electron to anode directly from the enzyme in their respiratory chain.

10.3 TYPES OF MFC

Fuel cell design is critical for determining the success of MFCs for bioelectricity and simultaneously wastewater treatment. Different configurations of MFCs can be observed in the course of their performance. The technology offers an environmentally friendly, cost-effective alternative for simultaneous electricity and bioremediation [24]. Wastewater generated from various industries such as the food industry, textile industry, the pharmaceutical industry could be treated while utilizing as a substrate for MFCs. Recently, a few MFC designs have been utilized for wastewater treatment and power generation such as anaerobic microbial fuel cell (ANMFC), sediment microbial fuel cell (SMFC), benthic microbial fuel cell (BMFC), enzyme-based microbial fuel cell (EBC), air-breathing cathode-based microbial fuel cell (ABC-MFC), constructed wetland microbial fuel cell (CW-MFC), and thermophilic microbial fuel cell (TMFC) [38].

10.4 ROLE OF MICROORGANISMS IN MFC-BASED WASTEWATER BIOREMEDIATION SYSTEMS

The exoelectrogenic microorganisms used in MFCs can be broadly classified among those that can mediate DET to the electrodes and those that require exogenous mediators. The complex nature of industrial wastewater for achieving complete degradation and their removal using a single microbial strain has limited feasibility [39]. For that reason, mixed culture or microbial consortia are generally preferred for the accomplishment of the process. During the DET process, electron transfer is possible by the physical contact of the bacteria to the anode without an external mediator. Many mixed microbial communities from activated sludge developing as a biofilm on the electrode have been reported for wastewater treatment in MFCs [40]. Apart from this, many mediators require microorganisms (mediated electron transfer or MET) used in MFCs. The most common electron-producing microbe groups are Actinobacteria, Proteobacteria phyla, Firmicutes, fungi, algae, and yeast which can generate electricity [41]. Furthermore, *Geobacter* spp., *Rhodoferax ferrireducens, Aeromonas*

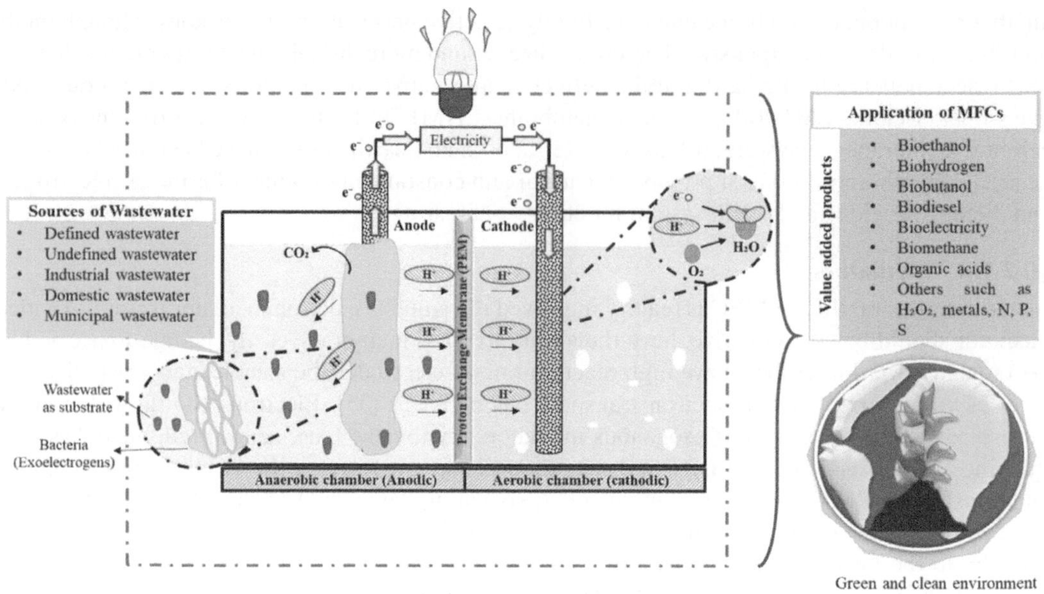

FIGURE 10.2 Microbial fuel cell-assisted wastewater treatment and their possible applications.

hydrophila, Clostridium butyricum, Shewanella spp., *Escherichia coli,* etc. have been reported for the bioremediation system [42]. MFCs remove a diverse group of pollutants and the mechanism involves anodic oxidation where pollutants are degraded as electron donors (Figure 10.2).

Another one is the anodic reduction where pollutants receive electrons at the anode and then the adsorption where they are adsorbed on the electrode or biofilm. Through the electrodynamics process, the electric field generated by MFCs affects the chemistry and distribution of pollutants. The proton is consumed in the cathode through cathode alkalization. The molecular machinery of bacteria helps in transferring electrons to an electrode surface, making MFC technology more interesting [43]. The molecular machinery readily involves the biomolecules, genes, or proteins that chiefly lie between the inner and outer membranes of the bacteria. *Geobacter* spp., *Pelotomaculum thermopropionicum, Methanothermobacter thermautotrophicus,* and *Shewanella* spp. have been studied extensively regarding extracellular electron transfer. They secrete flavin molecules in single-layer biofilms. These molecules combine with outer membrane c-type cytochrome to facilitate the electron transfer adjacent to the electrode. While the bacteria distant from the electrode produce nanowire, i.e. type IV pili, they happen to transport the electrons inside the biofilm and finally to the electrode surface [44]. However, the wastewater treatment also depends on the ability and growth of microbes as the generation of electrons depends on the respiration of microbes which ultimately depends on substrate. The system provides an opportunity to remediate the wastewater through reduction and oxidation reactions to generate electricity. The biofilms adhered to the anode electrode oxidize the organic substrate providing power for an electrochemical reduction of pollutants and recovery procedure at the cathode site [45]. The reduced pollutants are either precipitated in solution or deposited on the surface of the cathode or remain soluble, depending upon the chemistry and properties of reduced pollutants. The reduction turns into an extemporaneous phase at the surface of the cathode which offers the redox potential of a cathode half-cell reaction [46]. It is higher or almost comparable to anode potential which is generated in a fuel cell. For example, the degradation of organic dyes is studied in MFCs to degrade the dyes by promoting the oxidation and reduction process. The energy thus creates a promising and stable environment for the reduction of pollutants. However, the biofilm stability entirely depends on power generation to the cathode, which is required to understand the species interactions to promote electron transformation.

10.5 SUSTAINABLE WASTEWATER MANAGEMENT USING MFCs AND POWER GENERATION

Several types of wastewater have been utilized as substrates in MFC under different designs and operational conditions. Numerous reports have been published on various affecting factors, such as Columbic efficiency, chemical oxygen demand (COD) removal rate, and maximum power density. An inclusive table has been summarized for different industrial wastewater as substrate in MFCs (Table 10.1).

10.5.1 Agro-processing Industry Wastewater

Generally, agro-food processing wastes consist of a large amount of fat, oil, protein, carbohydrate, etc. They are also characterized by the presence of total solids, biochemical oxygen demand (BOD), COD, and pathogens. Since water is commonly used in food processing for various purposes, water pollution seems to be the most serious concern in many agro-industries [67]. Biological processes are widely used for the treatment of diverse groups of wastewaters in MFCs. Substrate composition plays a vital role in affecting power generation in MFCs. In an anodic chamber, anaerobic conditions may result in unwanted side reactions such as methanogenesis and denitrification [68]. Recently, rice mill wastewater was utilized for its degradation and power generation in dual-chambered MFC. The maximum power density of 4.08 W/m^3 with 75% of COD removal was obtained [60].

10.5.2 Beverages Industry Wastewater

Brewery wastewater is generally characterized by the presence of high organics, sugars, starch, and proteins. The organically rich content in the beverage effluent is considered a prominent factor in biofilm adherence to the electrode. The wastewater generated through saccharification cooling, fermentation, and washing causes numerous environmental problems. Many researchers studied the treatment of brewery wastewater in MFCs. The maximum power density observed was 0.35 W/m^3 and 79%–83% COD removal when inoculated with locally isolated microorganisms [48]. In another study, wastewater assessed from the beverage industry was successfully utilized as a fuel in a membrane-less truncated conical microbial fuel cell (TCMFC). The resultant removal efficiency of 84.4% was observed with a 15.03 W/m^3 power yield, showcasing the efficacy of the proposed design of MFC [69].

10.5.3 Dairy Industry Wastewater

Dairy wastewater is abundant in organic matter, fermentable substrate, organic and inorganic pollutants. Real dairy wastewater was investigated using single-chambered MFC and *Shewanella algae* for treatment. The maximum power density of 50 mW/m^2 and 92.21% COD removal was evaluated [70]. The obtained results direct the property of dairy wastewater as a potential substrate for power generation. The utilization of cheese whey was studied using single-chambered MFC inoculated with anaerobic methanogenic sludge which resulted in a maximum power density of 3.26 Wm^3 and COD removal of 76.9% [71]. Cheese whey contains lactose, fats, and proteins which make it a potential substrate for MFC.

10.5.4 Food and Food Processing Industry Wastewater

Food left at the end of the food chain after final consumption is considered food waste. Throughout the world, huge food waste rich in carbohydrate content is produced. This has prompted several researchers to investigate food wastes as potential substrates in MFCs [41]. Wastewater produced by the food industry is a possible risk to the naturally occurring water system. This contains many

TABLE 10.1
Recent Potential Industrial Wastewater as Substrate in MFCs

S. No.	Source of Wastewater	MFC Configuration	Microorganism	Treatment Efficiency	Energy Output	References
1	Beef packing plant (meat wastewater)	Bench-scale tubular	Anaerobic digester sludge	35.9%±9.7%	8.4±0.2 A m^3	[47]
2	Brewery industry wastewater and sludge	Lab scale	*Geobacter* sp. isolated from brewery industry wastewater and sludge	79%–83%	0.35–0.8 W/m^3	[48]
3	Cattle wastewater	Single chamber	Copper (I) oxide/reduced graphene oxide composite (Cu_2O/rGO)	71.50%	1,362 mW/m^2	[49]
4	Crude Oil industrial wastewater	H-shaped dual chambered	Activated sludge	36%	47 mW/m^2	[50]
5	Dairy wastewater	Dual chambered	*Enterococcus faecalis*	53.50%	144 mW/m^2	[51]
6	Dairy wastewater	Air-cathode	*Shewanella oneidensis* and *Clostridium butyricum*	89%	3.5 W/m^3	[52]
7	Electroplating industry wastewater	Dual-chambered anaerobic	Anaerobic mixed culture	87%	260 mW/m^2	[53]
8	Food processing industry	Two chambered	Anaerobic mixed culture	85.86%	272.882 mW/cm^2	[54]
9	Industrial acid mine drainage wastewater	Dual chambered	Anaerobic microorganism	90%	14,000 mW/m^3	[55]
10	Uranium-containing wastewater from mining industry	Dual chambered	Anaerobic sludge	99%	269.5±20.0 mW/m^2	[56]
11	Pharmaceutical industrial wastewater	Paraboloid-shaped graphite	Anaerobic mixed culture	80.55%	2.01 W/m^3	[57]
12	Paper industry wastewater	Dual chambered	*Cellulomonas iranensis* LZ-P1	—	44.05 mW/m^2	[58]
13	Refinery wastewater	Single chamber	Mixed microorganism	87%	113 mW/m^2	[59]
14	Rice mill wastewater	Dual chambered	Anaerobic sludge	72.4%±0.9%	4.08±0.08 W/m^3	[60]
15	Salinomycin-simulated livestock wastewater	Single-chamber air cathode	Gram-positive bacterium *Rhodococcus pyridinivorans*	63.90%	336 mW/m^2	[61]
16	Slaughterhouse wastewater	Photosynthetic	Blue-green algae	99.6%±0.5%	469.6±3.0 mW/m^2	[62]
17	Sugar industry wastewater	Dual chambered	Anaerobic culture	79.8%±1.5%	106±3 mW/m^2	[63]
18	Tannery effluent from leather processing industry	Dual chambered	*Bacillus subtilis, Bacillus flexus, Bacillus endophyticus, Bacillus pumilus,* and *Bacillus licheniformis*	75%	600±5 mW/m^2	[64]
19	Textile dye wastewater	Horizontal sub-surface flow constructed wetland	*Exiguobacterium, Desulfovibrio,* and *Macellibacteroides* DC5	97.32%±1.90%	197.94 mW/m^2	[65]
20	Bagasse-based paper mill wastewater	Dual chambered	*Pre-acclimatized microbes*	85%	73 mW/m^2	[66]

inorganic, organic matter that depletes the oxygen in water streams and poses a detrimental effect on the ecosystem. The feasibility of seafood processing industrial wastewater as substrate was evaluated in air cathode microbial fuel cell (ACMFC) [72]. It was observed that halophilic strains were dominant among sludge, which enhanced the power density of 530±2.4 mW/m^2 with 89% of COD removal. Being rich in micronutrients (phosphorus, nitrogen, etc.) food waste matter obtained from the industrial source is preferred as food by most of the microorganisms. Food processing industries consist of vegetable oil, meat and poultry processing, sugar refining, etc.

10.5.5 Livestock Industry Wastewater

The livestock industry wastewater comprises animal blood, wastewater from cleaning of ambients and animal carcasses, and skin residuals. The quality of slaughter wastewater depends on meat processing activities [73]. Hence without proper treatment, it may deeply contaminate marine life due to high organic content. The use of slaughterhouse and animal carcass wastewater for power generation in MFCs has been reported. Recently, a blue-green algae-based photosynthetic MFC type has been constructed for slaughterhouse wastewater and electricity generation [62]. The results demonstrated a COD removal efficiency of 99.6% with a power density of 469.6 mW/m^2. In another study, wastewater from the beef packing industry was utilized in bench-scale tubular MFC [47]. A maximum current density of 8.4 A/m^3 and removal efficiency of 35.9% was evaluated. An integrated vertical flow constructed wetland MFC was developed for swine wastewater treatment and electricity production [74]. The maximum COD removal of 88% and power density of 4,946 mW/m^3 was observed. Additionally, exoelectrogenic bacteria *Geobacter*, *Comamonas*, and *Desulfuromonas* were detected in the anodic biofilm.

10.5.6 Mining and Allied Industry Wastewater

Acid mine waste is generally caused by the biological oxidation of sulfides to sulfates. It is generally categorized by low pH and metals such as cadmium, copper, arsenic, lead, etc. destroying the environment. The treatment of MFC is suggested not only for electricity generation but also for the removal of these metals and the production of useful products [75]. A study was aimed to evaluate the performance of industrial acid mine wastewater in dual-chambered MFC. The efficiency of heavy metal removal from 42% to 84% and volumetric power density of 14,000 mW/m^3 was evaluated [55].

10.5.7 Paper Recycling Wastewater

Paper industries generate cellulose-rich wastewater used as a substrate for microorganisms. The constant demand in paper production, demand for energy and water is an issue of concern. The main issue that arises during cellulose degradation is the presence of lignin which has an inhibitory effect on microorganisms in the MFC system [14]. The products (vanillin, catechol, and phenolic) thus formed during lignin degradation often hinder the metabolism of microorganisms. Nevertheless, many studies have shown that microorganisms adopt various mechanisms to acclimatize to harmful environments. Many strains explored so far can degrade lignin and organics while contributing to power generation. Some genera are *Pseudomonas*, *Geobacter*, *Fluvicola*, *Desulfuromonas*, *Clostridium*, *Fusibacter*, *Flavobacter*, *Aquimonas*, and Cryomorphaceae. An integrated approach is studied for simultaneous bioelectricity generation and paper wastewater treatment. A system comprising an anaerobic bed biofilm bioreactor and ceramsite-MFC showed 94.5 mW/m^2 with 65.6% of COD removal efficacy [76].

10.5.8 Petrochemical Industry Wastewater

Petrochemical mainly comprises chemicals derived from petroleum, yarn, detergent, plastic, polymers, etc. The extensive use of petrochemical production leads to the contamination of soil, air, and water. The toxic compounds produced belong to a class of phenols, octanols, cyanide, aliphatics, aromatics, and formaldehyde. Many of the toxic compounds lag the disposal criterion and are potent carcinogenics for living beings [77]. Various studies have been carried out for the possible application of MFCs in the treatment of industry effluents. For instance, a dual-chambered MFC was constructed using petroleum and labneh whey wastewater as a substrate for improved electricity generation and treatment efficiency [59]. The resultant output in terms of power density was 832 mW/m^2 along with 63.10% of COD removal. Recently, an integrated study of MFC with a photocatalytic reactor was investigated for simultaneous power generation and treatment efficiency [78]. Dimethyl phthalate (DMP) was employed as a substrate and *Pseudomonas aeruginosa* as a biocatalyst. The employed biocatalyst produced phenazine-1-carboxylic acid in MFC as the metabolite. COD removal was evaluated up to 62% and 0.75–3.84 mW/m^3 of power generation.

10.5.9 Pharmaceutical Industry Wastewater

The end-products of pharmaceutical compounds are the foremost pollutants as their metabolites affect marine ecosystems. The wastewater mainly contains drugs and chemicals, which are difficult to treat. Hormones such as testosterone, estrogen, ethinylestradiol, and estriol among others are also present. During the production of an intermediate of the steroidal drug, hydrocortisone produced largely at the hydrolysis step results in a toxic refining wastewater production [79]. These recalcitrant organic compounds can be treated using microbial processes. The processes are categorized into aerobic and anaerobic based on activated sludge or sequence/membrane batch reactors. Treatment of wastewater mainly depends on pH, organic load, COD, BOD, etc. This remediation process when combined with other microbial remediation technologies such as MFCs gives better degradation and its utilization in a sustainable manner. A dual-chambered MFC was studied for wastewater treatment from non-steroidal sources and synthetic wastewater containing diclofenac sodium [78]. The maximum power density of about 20.5 and 6.5 mW/m^3 was observed along with 93%–78% of COD removal.

10.5.10 Refinery and Distillery Wastewater

Water is generated at various stages of manufacturing, whether cooling in a condenser or fermentation and washing. It releases toxic substances into the topsoil and water bodies, making the environment unhealthy. The dark brown color of the wastewater reduces the penetration of light, soil alkalinity, and the upsurge of heavy metals deep into water bodies [24]. Though there are various physical and chemical treatments available, currently, biological treatment has engrossed the attention of many researchers. A wetland sediment dual-chambered MFC has recently been studied for the removal of cadmium and nickel from distillery wastewater and electricity generation [80]. The maximum power density of 722 mW/m^3 with a 92%–87% of reduction in heavy metals was obtained. Microbial characterization revealed the active involvement of *Ochrobactrum* and *Halomonas* in metal removal efficacy in the cathode chamber, while the anode chamber was dominated by order Clostridiales and Burkholderiales and the genus *Achromobacter* predominantly.

10.5.11 Textile Industry Wastewater

The textile industry produces one of the most complex industrial wastewater. The chemical load includes dyes and chemicals, which thereby increase BOD, alkalinity, and suspended particles. Biological treatment has always emerged as a prominent feasible alternative. The treatment involves a diverse group of bacteria, algae, fungi, and yeast capable of degradation of different pollutants [81]. The dye wastewater degradation efficiency was carried out in three different combinations of MFCs. A single-chambered MFC was constructed for dye textile wastewater and simultaneous power generation. The maximum power density was 123.2 mW/m^3 with 92%–98% of COD removal [82].

10.6 CRITICAL ASPECTS FOR MFC-ENERGY HARVESTING SYSTEMS TO TREAT WASTEWATER AND PRODUCE ENERGY

The ability of MFCs to generate electrical power directly from wastewater treatment is encouraging, leading to profound research in the development. MFCs are emerging as net energy producers in which there is little waste generation. The oxygen reduction reaction is currently a major limiting factor in different MFCs. The issue can be addressed by increasing the ratio of the cathode to anode and using biological catalysts. The overall performance of a biocathode can be enhanced through effective interactions of microorganisms on its surface [83]. In the context of its industrial applicability, the electrode materials used in MFCs must be carefully considered for long-term stability. Further efforts can be made by coupling MFC units to energy harvesting devices such as charge pumps, boost converters, capacitors, etc. From this, the energy can be stored and improved output voltage could be obtained during wastewater treatment. The design of a self-powered energy harvesting system is needed in a manner to utilize only a small fraction of power harvested from MFCs [84]. This will open new opportunities to introduce advanced technology to maximize energy generation at commercial levels.

10.7 ECONOMIC FEASIBILITY OF WASTE MANAGEMENT

MFCs may become a real alternative to conventional methods for the treatment of industrial wastewater. Their integration into the treatment process could reduce the operational cost by diminishing the electricity consumption and reducing the carbon footprint. Recovered by-products possibly increase the economic feasibility of MFCs. Since the recovery of the by-products needs a carbon source in the anode, it is acquainted with wastewater. Moreover, if the projection of electricity price for a particular region is expected to grow, the use of MFCs would be a favorable alternative since the expected monetary savings would also rise. Thus, it makes a useful and eminent tool for future assessment by potential investors in the technology.

10.8 CONCLUSION

MFC technology is a promising technology with great potential for electricity generation and simultaneous applications in the field of wastewater treatment, generation of valuable products, valuable chemicals, recovery of metals, application as biosensors, and as a power source for remote areas. It can generate energy out of the waste without any additional energy. Thus, there are different MFCs and their configurations have been developed for various applications to improve their performance in different aspects. However, several aspects such as power output, material cost, separators, electrogens, and MFC design need to be addressed. Though many pilot-scale MFCs have been developed so far, they are not translated to industrial scale with success due to low power density on scale-up. Future research is needed in this field to make the MFC commercially feasible.

GLOSSARY

Biocatalyst: The use of biological entity or their parts to speed up the biochemical reactions.

Bioremediation: The term broadly refers to any process wherein a biological entity (mainly bacteria, fungi, microalgae, and plants) is employed for removing toxic pollutants from an environment.

Bioremediation for Industrial Waste Treatment: Bioremediation is a process that uses microorganisms, plants, or microbial enzymes to detoxify and remove pollutants from the environment. It's a sustainable and eco-friendly method for treating industrial waste. The mechanisms of bioremediation include the basic microbial degradation, phytoremediation, and enzymatic bioremediation. There are two types of bioremediation: (a) in situ bioremediation and (b) ex situ bioremediation.

Electroactive Microorganisms: These are specialized bacteria capable of transferring electrons to or from electrodes during their metabolic processes.

Electrodes: They are solid-state electrodes serving as the interface for electron transfer. They can be made from various materials, including carbon-based substances.

Enzymatic Bioremediation: Microbial enzymes catalyze the breakdown of pollutants. This method is often used for specific contaminants that are difficult to degrade.

Ex Situ Bioremediation: Contaminated material is removed and treated elsewhere. Methods include biopiles, bioreactors, and landfarming.

Exoelectrogens: The microorganism that has an ability to transfer electrons extracellularly.

Hazardous waste and non-hazardous waste: Hazardous waste is the outcome of industrial activity that poses threat to the environment while non-hazardous waste is the waste from an industry that marks any threat to public health or the environment.

In Situ Bioremediation: Treatment occurs at the site of contamination. Techniques include bioventing, biosparging, and phytoremediation.

Mediators: The chemicals with electrochemical activity which help in exchanging electrons at the reaction sites and then diffuse them to the surface of the electrode.

Microbial Degradation: Microorganisms such as bacteria, fungi, and algae break down contaminants into less harmful substances. This can occur through aerobic (with oxygen) or anaerobic (without oxygen) processes.

Microbial Electrochemical Technologies (MET): MET integrate microbiology, electrochemistry, and material science to convert chemical energy stored in organic materials into electrical power. Some key fundamentals include the following: (a) electroactive microorganisms; (b) electrodes; and (c) bioelectrochemical systems (BES).

Microbial Electrochemical Technology-Assisted Bioremediation: Microbial electrochemical technology-assisted bioremediation (MET-assisted bioremediation) is an innovative approach that leverages the capabilities of electrochemically active microorganisms to enhance the bioremediation process. A breakdown of how it works includes: (a) Electrodes as Electron Acceptors: In MET-assisted bioremediation, electrodes are used as inexhaustible electron acceptors. This setup facilitates the degradation of contaminants by providing a continuous supply of electrons, which is crucial for the metabolic processes of certain microorganisms; (b) Enhanced Microbial Activity: The electric field generated in the system can stimulate microbial activity, enhancing the degradation of pollutants. This is particularly useful for breaking down complex hydrocarbons and other persistent organic pollutants; and (c) Minimal Chemical Additives: Unlike traditional methods, MET-assisted bioremediation does not require additional chemicals, making it a greener and more sustainable option. It also tends to produce less sludge and waste. This technology is particularly effective for the remediation of soils and groundwater contaminated with petroleum hydrocarbons, polycyclic aromatic hydrocarbons (PAHs), and other organic pollutants.

Microbial Fuel Cell (MFC): MFC is an innovative device that converts chemical energy into electrical energy through the action of microorganisms. The key elements of the MFC include: (a) Microorganisms: MFCs use bacteria as biocatalysts. These bacteria metabolize organic matter (like waste) and, in the process, release electrons; (b) Electrodes: The electrons are captured by electrodes, creating an electric current; (c) Anode and Cathode: Typically, MFCs have an anode (where oxidation occurs) and a cathode (where reduction occurs). The bacteria at the anode break down organic matter, releasing electrons and protons.

The electrons travel through an external circuit to the cathode, generating electricity. In the area of wastewater treatment, MFCs can treat wastewater while simultaneously generating electricity. MFCs offer a sustainable way to produce renewable energy from organic waste. In the area of environmental monitoring, MFCs can be used in sensors to monitor environmental conditions. The benefits of MFCs include: (a) environmental sustainability and (b) dual functionality. The technology that converts the chemical energy of compounds into electrical energy during the remediation process with the aid of microorganisms as exoelectrogens. The technology converts the chemical energy of compounds into electrical energy during the remediation process with the aid of microorganisms as exoelectrogens.

Phytoremediation: Plants absorb, accumulate, and detoxify pollutants from soil and water. They can either store contaminants in their tissues or transform them into less toxic forms.

Sustainable management: The application of environmentally friendly practices in a way that will benefit current and future generations.

REFERENCES

1. Sharma J, Goutam J, Dhuriya YK, et al. Bioremediation of industrial pollutants. In: Panpatte DG, Jhala YK (eds) *Microbial Rejuvenation of Polluted Environment: Volume* 2. Singapore: Springer Singapore:1–31.
2. Mezynska M, Brzóska MM. Environmental exposure to cadmium—a risk for health of the general population in industrialized countries and preventive strategies. *Environ Sci Pollut Res* 2018;25:3211–3232.
3. Saxena G, Purchase D, Bharagava RN. Environmental hazards and toxicity profile of organic and inorganic pollutants of tannery wastewater and bioremediation approaches. In: Saxena G, Bharagava RN (eds) *Bioremediation of Industrial Waste for Environmental Safety: Volume I: Industrial Waste and Its Management.* Singapore: Springer Singapore:381–398.
4. Dhanya MS, Kalia A. Bioremediation: an eco-friendly cleanup strategy for polyaromatic hydrocarbons from petroleum industry waste. In: Saxena G, Bharagava RN (eds) *Bioremediation of Industrial Waste for Environmental Safety: Volume I: Industrial Waste and Its Management.* Singapore: Springer Singapore:399–436.
5. Nasrollahi Z, Hashemi M, Bameri S, et al. Environmental pollution, economic growth, population, industrialization, and technology in weak and strong sustainability: using STIRPAT model. *Environ Dev Sustain* 2020;22:1105–1122.
6. Kumar M, Yadav A, Ramanathan AL. Arsenic contamination in environment, ecotoxicological and health effects, and bioremediation strategies for its detoxification. In: Saxena G, Bharagava RN (eds) *Bioremediation of Industrial Waste for Environmental Safety: Volume I: Industrial Waste and Its Management.* Singapore: Springer Singapore:245–264.
7. Devi NL. Persistent organic pollutants (POPs): environmental risks, toxicological effects, and bioremediation for environmental safety and challenges for future research. In: Saxena G, Bharagava RN (eds) *Bioremediation of Industrial Waste for Environmental Safety: Volume I: Industrial Waste and Its Management.* Singapore: Springer Singapore:53–76.
8. Liu Z, Lu Y, Wang P, et al. Pollution pathways and release estimation of perfluorooctane sulfonate (PFOS) and perfluorooctanoic acid (PFOA) in central and eastern China. *Sci Total Environ* 2017;580:1247–1256.
9. Yogalakshmi KN, Singh S. Plastic waste: environmental hazards, its biodegradation, and challenges. In: Saxena G, Bharagava RN (eds) *Bioremediation of Industrial Waste for Environmental Safety: Volume I: Industrial Waste and Its Management.* Singapore: Springer Singapore:99–133.
10. Saratale RG, Rajesh Banu J, Shin H-S, et al. Textile industry wastewaters as major sources of environmental contamination: bioremediation approaches for its degradation and detoxification. In: Saxena G, Bharagava RN (eds) *Bioremediation of Industrial Waste for Environmental Safety: Volume I: Industrial Waste and Its Management.* Singapore: Springer Singapore:135–167.
11. Shankar S, Shikha, Ratnakar A, et al. Environmental contamination, toxicity profile, and bioremediation approaches for detoxification of paper mill wastewater. In: Saxena G, Bharagava RN (eds) *Bioremediation of Industrial Waste for Environmental Safety: Volume I: Industrial Waste and Its Management.* Singapore: Springer Singapore:181–206.

12. Azubuike CC, Chikere CB, Okpokwasili GC. Bioremediation: an eco-friendly sustainable technology for environmental management. In: Saxena G, Bharagava RN (eds) *Bioremediation of Industrial Waste for Environmental Safety: Volume I: Industrial Waste and Its Management*. Singapore: Springer Singapore:19–39.
13. Gaur VK, Sharma P, Sirohi R, et al. Assessing the impact of industrial waste on environment and mitigation strategies: a comprehensive review. *J Hazard Mater* 2020;398:123019.
14. Toczyłowska-Mamińska R. Limits and perspectives of pulp and paper industry wastewater treatment – a review. *Renew Sustain Energy Rev* 2017;78:764–772.
15. Trivedi M, Mathur M, Johri P, et al. Waste management: a paradigm shift. In: Shukla V, Kumar N (eds) *Environmental Concerns and Sustainable Development: Volume 2: Biodiversity, Soil and Waste Management*. Singapore: Springer Singapore:337–363.
16. Moustakas K, Rehan M, Loizidou M, et al. Energy and resource recovery through integrated sustainable waste management. *Appl Energy* 2020;261:114372.
17. Akpan VE, Olukanni DO. Hazardous waste management: an African overview. *Recycling* 2020;5:10–15.
18. Giri S. Water quality prospective in twenty first century: status of water quality in major river basins, contemporary strategies and impediments: a review. *Environ Pollut* 2021;271:116332.
19. Yang B, Usman M, Jahanger A. Do industrialization, economic growth and globalization processes influence the ecological footprint and healthcare expenditures? Fresh insights based on the STIRPAT model for countries with the highest healthcare expenditures. *Sustain Prod Consum* 2021;28:893–910.
20. Woodard R. Waste management in small and medium enterprises (SMEs) – a barrier to developing circular cities. *Waste Manag* 2020;118:369–379.
21. Ranjit P, Jhansi V, Reddy KV. Conventional wastewater treatment processes. In: Maddela NR, García Cruzatty LC, Chakraborty S (eds) *Advances in the Domain of Environmental Biotechnology: Microbiological Developments in Industries, Wastewater Treatment and Agriculture*. Singapore: Springer Singapore:455–479.
22. Kamali M, Persson KM, Costa ME, et al. Sustainability criteria for assessing nanotechnology applicability in industrial wastewater treatment: current status and future outlook. *Environ Int* 2019;125:261–276.
23. Crini G, Lichtfouse E, Wilson LD, et al. Conventional and non-conventional adsorbents for wastewater treatment. *Environ Chem Lett* 2019;17:195–213.
24. Munoz-Cupa C, Hu Y, Xu C, et al. An overview of microbial fuel cell usage in wastewater treatment, resource recovery and energy production. *Sci Total Environ* 2021;754:142429.
25. Wu Q, Jiao S, Ma M, et al. Microbial fuel cell system: a promising technology for pollutant removal and environmental remediation. *Environ Sci Pollut Res* 2020;27:6749–6764.
26. Jatoi AS, Akhter F, Mazari SA, et al. Advanced microbial fuel cell for waste water treatment—a review. *Environ Sci Pollut Res* 2021;28:5005–5019.
27. Sonawane JM, Ezugwu CI, Ghosh PC. Microbial fuel cell-based biological oxygen demand sensors for monitoring wastewater: state-of-the-art and practical applications. *ACS Sensors* 2020;5:2297–2316.
28. Abdallah M, Feroz S, Alani S, et al. Continuous and scalable applications of microbial fuel cells: a critical review. *Rev Environ Sci Bio/Technol* 2019;18:543–578.
29. Cai T, Meng L, Chen G, et al. Application of advanced anodes in microbial fuel cells for power generation: a review. *Chemosphere* 2020;248:125985.
30. Neethu B, Bhowmick GD, Ghangrekar MM. Improving performance of microbial fuel cell by enhanced bacterial-anode interaction using sludge immobilized beads with activated carbon. *Process Saf Environ Prot* 2020;143:285–292.
31. Kaur R, Marwaha A, Chhabra VA, et al. Recent developments on functional nanomaterial-based electrodes for microbial fuel cells. *Renew Sustain Energy Rev* 2020;119: 109551.
32. Narayanasamy S, Jayaprakash J. Application of carbon-polymer based composite electrodes for microbial fuel cells. *Rev Environ Sci Biotechnol* 2020;19:595–620.
33. Shabani M, Younesi H, Pontié M, et al. A critical review on recent proton exchange membranes applied in microbial fuel cells for renewable energy recovery. *J Clean Prod* 2020;264:121446.
34. Dizge N, Unal BO, Arikan EB, et al. Recent progress and developments in membrane materials for microbial electrochemistry technologies: a review. *Bioresour Technol Rep* 2019;8:100308.
35. Kaneko M, Ishihara K, Nakanishi S. Redox-active polymers connecting living microbial cells to an extracellular electrical circuit. *Small* 2020;16:2001849.
36. Agrawal K, Bhardwaj N, Kumar B, et al. Microbial fuel cell: a boon in bioremediation of wastes. In: Shah MP, Rodriguez-Couto S (eds) *Microbial Wastewater Treatment*. Amsterdam:Elsevier:175–194.

37. Heydorn RL, Engel C, Krull R, et al. Strategies for the targeted improvement of anodic electron transfer in microbial fuel cells. *ChemBioEng Rev* 2020;7:4–17.
38. Rahimnejad M, Asghary M, Fallah M. Microbial fuel cell (MFC): an innovative technology for wastewater treatment and power generation. In: Bharagava RN, Saxena G (eds) *Bioremediation of Industrial Waste for Environmental Safety: Volume II: Biological Agents and Methods for Industrial Waste Management*. Singapore: Springer Singapore:215–235.
39. Umar MF, Rafatullah M, Abbas SZ, et al. Advancement in benthic microbial fuel cells toward sustainable bioremediation and renewable energy production. *Int J Environ Res Public Health* 2021;18:3811.
40. Umar MF, Abbas SZ, Mohamad Ibrahim MN, et al. Insights into advancements and electrons transfer mechanisms of electrogens in benthic microbial fuel cells. *Membranes (Basel)* 2020;10:205.
41. Rani N, Sangwan P, Joshi M, et al. Microbes: a key player in industrial wastewater treatment. In: Shah MP, Rodriguez-Couto S (eds) *Microbial Wastewater Treatment*. Amsterdam:Elsevier:83–102.
42. Aiyer KS. How does electron transfer occur in microbial fuel cells? *World J Microbiol Biotechnol* 2020;36:19.
43. Verma P, Daverey A, Kumar A, et al. Microbial fuel cell – a sustainable approach for simultaneous wastewater treatment and energy recovery. *J Water Process Eng* 2021;40:101768.
44. Saratale GD, Saratale RG, Shahid MK, et al. A comprehensive overview on electro-active biofilms, role of exo-electrogens and their microbial niches in microbial fuel cells (MFCs). *Chemosphere* 2017;178:534–547.
45. Nguyen HD, Babel S. Insights on microbial fuel cells for sustainable biological nitrogen removal from wastewater: a review. *Environ Res* 2022;204:112095.
46. Yaqoob AA, Khatoon A, Mohd Setapar SH, et al. Outlook on the role of microbial fuel cells in remediation of environmental pollutants with electricity generation. *Catalysts* 2020;10:819.
47. Li J, Ziara RMM, Li S, et al. Understanding the sustainability niche of continuous flow tubular microbial fuel cells on beef packing wastewater treatment. *J Clean Prod* 2020;257:120555.
48. Negassa LW, Mohiuddin M, Tiruye GA. Treatment of brewery industrial wastewater and generation of sustainable bioelectricity by microbial fuel cell inoculated with locally isolated microorganisms. *J Water Process Eng* 2021;41:102018.
49. Xin S, Shen J, Liu G, et al. High electricity generation and COD removal from cattle wastewater in microbial fuel cells with 3D air cathode employed non-precious Cu_2O/reduced graphene oxide as cathode catalyst. *Energy* 2020;196:117123.
50. Nandy A, Radović JR, Novotnik B, et al. Investigation of crude oil degradation using metal oxide anode-based microbial fuel cell. *Bioresour Technol Rep* 2020;11:100449.
51. Parihar PS, Keshavkant S, Jadhav SK. Electrogenic potential of Enterococcus faecalis DWW1 isolated from the anodic biofilm of a dairy wastewater fed dual chambered microbial fuel cell. *J Water Process Eng* 2022;45:102503.
52. Marassi RJ, López MBG, Queiroz LG, et al. Efficient dairy wastewater treatment and power production using graphite cylinders electrodes as a biofilter in microbial fuel cell. *Biochem Eng J* 2022;178:108283.
53. Karuppiah T, Uthirakrishnan U, Sivakumar SV, et al. Processing of electroplating industry wastewater through dual chambered microbial fuel cells (MFC) for simultaneous treatment of wastewater and green fuel production. *Int J Hydrogen Energy* 2022;47:37569–37576.
54. Sambavi SM, Vishali S, Asthika M, et al. Bioconversion of food waste with cosubstrate to electricity using microbial fuel cell. *J Phys Conf Ser* 2021;1979:12011.
55. Vélez-Pérez LS, Ramirez-Nava J, Hernández-Flores G, et al. Industrial acid mine drainage and municipal wastewater co-treatment by dual-chamber microbial fuel cells. *Int J Hydrogen Energy* 2020;45:13757–13766.
56. Wu X, Lv C, Ye J, et al. Glycine-hydrochloric acid buffer promotes simultaneous U(VI) reduction and bioelectricity generation in dual chamber microbial fuel cell. *J Taiwan Inst Chem Eng* 2021;127:236–247.
57. Rashid T, Sher F, Hazafa A, et al. Design and feasibility study of novel paraboloid graphite based microbial fuel cell for bioelectrogenesis and pharmaceutical wastewater treatment. *J Environ Chem Eng* 2021;9:104502.
58. Xu R, Zhang K, Xie S, et al. Evaluation of electricity production from paper industry wastewater by Cellulomonas iranensis LZ-P1 isolated from giant panda. *J Clean Prod* 2021;278:123576.
59. Mohanakrishna G, Abu-Reesh IM, Pant D. Enhanced bioelectrochemical treatment of petroleum refinery wastewater with Labaneh whey as co-substrate. *Sci Rep* 2020;10:19665.

60. Raychaudhuri A, Behera M. Enhancement of bioelectricity generation by integrating acidogenic compartment into a dual-chambered microbial fuel cell during rice mill wastewater treatment. *Process Biochem* 2021;105:19–26.
61. Cheng P, Shan R, Yuan H-R, et al. Bioelectricity generation from the salinomycin-simulated livestock sewage in a Rhodococcus pyridinivorans inoculated microbial fuel cell. *Process Saf Environ Prot* 2020;138:76–79.
62. Fadhil SH, Ismail ZZ. Bioenergy recovery associated with treatment of actual meat industry wastewater in blue green algal biocathode based microbial fuel cell: experimental and modeling study. *Energy Sources Part A Recover Util Environ Eff* 2022;44(1):1–13.
63. Naina Mohamed S, Thomas N, Tamilmani J, et al. Bioelectricity generation using iron(II) molybdate nanocatalyst coated anode during treatment of sugar wastewater in microbial fuel cell. *Fuel* 2020;277:118119.
64. Naveenkumar M, Senthilkumar K, Sampathkumar V, et al. Bio-energy generation and treatment of tannery effluent using microbial fuel cell. *Chemosphere* 2022;287:132090.
65. Patel D, Bapodra SL, Madamwar D, et al. Electroactive bacterial community augmentation enhances the performance of a pilot scale constructed wetland microbial fuel cell for treatment of textile dye wastewater. *Bioresour Technol* 2021;332:125088.
66. Elakkiya E, Niju S. Bioelectrochemical treatment of real-field bagasse-based paper mill wastewater in dual-chambered microbial fuel cell. 3 Biotech 2021;11:42.
67. Bolognesi S, Cecconet D, Capodaglio AG. Agro-industrial wastewater treatment in microbial fuel cells. In: Abbassi R, Yadav AK, Khan F, et al. (eds) *Integrated Microbial Fuel Cells for Wastewater Treatment*. Amsterdam:Butterworth-Heinemann:93–133.
68. Pandit S, Savla N, Sonawane JM, et al. Agricultural waste and wastewater as feedstock for bioelectricity generation using microbial fuel cells: recent advances. *Fermentation* 2021;7:169. doi:10.3390/fermentation7030169.
69. Nawaz A, Raza W, Gul H, et al. Upscaling feasibility of a graphite-based truncated conical microbial fuel cell for bioelectrogenesis through organic wastewater treatment. *J Colloid Interface Sci* 2020;570:99–108.
70. Choudhury P, Ray RN, Bandyopadhyay TK, et al. Process engineering for stable power recovery from dairy wastewater using microbial fuel cell. *Int J Hydrogen Energy* 2021;46:3171–3182.
71. Antonopoulou G, Ntaikou I, Bebelis S, et al. On the evaluation of filtered and pretreated cheese whey as an electron donor in a single chamber microbial fuel cell. *Biomass Convers Biorefinery* 2021;11:633–643.
72. Pugazhendi A, Al-Mutairi AE, Jamal MT, et al. Treatment of seafood industrial wastewater coupled with electricity production using air cathode microbial fuel cell under saline condition. *Int J Energy Res* 2020;44:12535–12545.
73. Khandaker S, Das S, Hossain MT, et al. Sustainable approach for wastewater treatment using microbial fuel cells and green energy generation – a comprehensive review. *J Mol Liq* 2021;344:117795.
74. Liu F, Sun L, Wan J, et al. Performance of different macrophytes in the decontamination of and electricity generation from swine wastewater via an integrated constructed wetland-microbial fuel cell process. *J Environ Sci* 2020;89:252–263.
75. Kloch M, Toczyłowska-Mamińska R. Toward optimization of wood industry wastewater treatment in microbial fuel cells—mixed wastewaters approach. *Energies* 2020;13:263.
76. Chen F, Zeng S, Luo Z, et al. A novel MBBR–MFC integrated system for high-strength pulp/paper wastewater treatment and bioelectricity generation. *Sep Sci Technol* 2020;55:2490–2499.
77. Jaiswal KK, Kumar V, Vlaskin MS, et al. Microalgae fuel cell for wastewater treatment: recent advances and challenges. *J Water Process Eng* 2020;38:101549.
78. Amari S, Boshrouyeh Ghandashtani M. Non-steroidal anti-inflammatory pharmaceutical wastewater treatment using a two-chambered microbial fuel cell. *Water Environ J* 2020;34:413–419.
79. Bagchi S, Behera M. Pharmaceutical wastewater treatment in microbial fuel cell. In: Abbassi R, Yadav AK, Khan F, et al. (eds) *Integrated Microbial Fuel Cells for Wastewater Treatment*. Amsterdam:Butterworth-Heinemann:135–155.
80. Singh A, Kaushik A. Removal of Cd and Ni with enhanced energy generation using biocathode microbial fuel cell: insights from molecular characterization of biofilm communities. *J Clean Prod* 2021;315:127940.
81. Rathour R, Kalola V, Johnson J, et al. Treatment of various types of wastewaters using microbial fuel cell systems. In: Mohan SV, Varjani S, Pandey A (eds) *Microbial Electrochemical Technology*. Amsterdam:Elsevier:665–692.

82. Logroño W, Pérez M, Urquizo G, et al. Single chamber microbial fuel cell (SCMFC) with a cathodic microalgal biofilm: a preliminary assessment of the generation of bioelectricity and biodegradation of real dye textile wastewater. *Chemosphere* 2017;176:378–388.
83. Gajda I, Greenman J, Ieropoulos IA. Recent advancements in real-world microbial fuel cell applications. *Curr Opin Electrochem* 2018;11:78–83.
84. Afroz AS, Romano D, Inglese F, et al. Towards bio-hybrid energy harvesting in the real-world: pushing the boundaries of technologies and strategies using bio-electrochemical and bio-mechanical processes. *Appl Sci* 2021;11:2220.

11 Fundamentals of Thermochemical Processes and Polyolefin Plastic Waste Treatment for Resource Recycling and Conservation

Anna Matuszewska, Marlena Owczuk, Krzysztof Biernat, Yung-Tse Hung, Mu-Hao Sung Wang, and Lawrence K. Wang

NOMENCLATURE

ABS	Acrylonitrile butadiene styrene
AC	Activated carbon
Al_2O_3	Aluminum(III) oxide
$AlCl_3$	Aluminum trichloride
ATOD	Accelerated thermal oxidative decomposition
C_2H_6	Ethane
C_3H_8	Propane
CE	Circular economy
CH_4	Methane
CNSACC	Cashew nutshell-derived activated carbon catalyst
CO	Carbon oxide
CO_2	Carbon dioxide
CSBR	Conical spouted bed reactor
CtC	Cradle-to-cradle
CtG	Cradle-to-grave
Cu_2O	Copper(I) oxide
DI	Direct injection
DMT	Dimethyl terephthalate
EG	Ethylene glycol
EIA	Energy Information Administration
EN	European Standard
EPDM	Ethylene propylene diene monomer
EPS	Expanded polystyrene
FCC	Fluid catalytic cracking
Fe_2O_3	Iron(III) oxide
$FeCl_3$	Iron(III) chloride
GCMS	Gas chromatography coupled to mass spectrometry
H_2	Hydrogen
HBeta	Zeolite H-beta
HBr	Hydrogen bromide

DOI: 10.1201/9781003532354-11

HDPE	High-density polyethylene
HMOR	Zeolite – hydrogen mordenite zeolite
HUSY	Zeolite – hierarchical H-style ultrastable Y
HZSM-5	Zeolite – hydrogen zeolite socony mobil-5
La_2O_3	Lanthanum(III) oxide
LCA	Life cycle assessment
LDPE	Low-density polyethylene
$M(AlCl_4)_n$	Metal tetrachloroaluminates
MAHs	Monocyclic aromatic hydrocarbons
MgO	Magnesium oxide
Mt	Megatonne
N_2	Nitrogen
NiO	Nickel(II) oxide
PE	Polyethylene
PET	Polyethylene terephthalate
PP	Polypropylene
PS	Polystyrene
PTA	Terephthalic acid
PVC	Polyvinyl chloride
PW	Plastic waste
SAPO	Silicoaluminophosphate catalyst
Sb_2O_3	Diantimony trioxide
SBA-15	Santa Barbara Amorphous-15 – mesoporous silica catalyst
$TiCl_3$	Titanium(III) trichloride
$TiCl_4$	Titanium(IV) tetrachloride
TRL	Technology readiness level
USY	Ultrastable Y zeolite
WCO	Waste cooking oil
WEEE	Waste electrical and electronic equipment
WO_3	Tungsten trioxide
WP	Waste polyolefins
ZnO	Zinc oxide
ZrO_2	Zirconium(IV) oxide

11.1 INTRODUCTION, FUNDAMENTALS AND RESEARCH OBJECTIVES

11.1.1 Thermochemical Chemical Process Fundamentals

Thermochemical processes involve the study of heat energy changes associated with chemical reactions and phase changes, such as melting and boiling. Some key fundamentals are: (a) Heat Energy and Reactions: Thermochemistry focuses on the energy exchange between a system and its surroundings in the form of heat. During a chemical reaction, energy can be either released (exothermic) or absorbed (endothermic); (b) Enthalpy: This is a measure of the total energy of a thermodynamic system. It includes internal energy and the product of pressure and volume. The change in enthalpy during a reaction at constant pressure is a crucial concept in thermochemistry; (c) Calorimetry: This is the measurement of heat flow in a chemical reaction. Calorimeters are devices used to measure the amount of heat involved in a chemical or physical process; (d) Thermochemical Equations: These are chemical equations that include the enthalpy change; and (e) Hess's Law: This principle states that the total enthalpy change of a reaction is the same, no matter how many steps the reaction is carried out in. It allows the calculation of enthalpy changes for complex reactions by breaking them down into simpler steps.

Understanding these fundamentals helps in predicting the energy changes in reactions, which is essential for various applications in chemistry and chemical engineering.

11.1.2 Resource Recycling Fundamentals

Recycling involves converting waste materials into new products to prevent waste, reduce the consumption of fresh raw materials, and minimize energy usage. Here are some key points:

11.1.2.1 Types of Recyclable Materials

In general, the following solid materials are suitable for recycle and reuse:

1. **Paper and Cardboard:** Newspapers, office paper, and packaging.
2. **Plastics:** Bottles, containers, and various packaging materials.
3. **Metals:** Aluminum cans, steel products.
4. **Glass:** Bottles and jars.
5. **Electronics:** Computers, phones, and other electronic devices.

11.1.2.2 Benefits of Recycling

Possible benefits of a recycling program may include the following:

1. **Conserves Natural Resources:** Reduces the need to extract new raw materials like timber, water, and minerals1.
2. **Saves Energy:** Manufacturing products from recycled materials often requires less energy compared to using new materials.
3. **Reduces Pollution:** Decreases greenhouse gas emissions and pollution from manufacturing processes.
4. **Economic Benefits:** Creates jobs in the recycling and manufacturing industries.

11.1.3 Resource Conservation Fundamentals

Conservation focuses on using resources more efficiently and sustainably to ensure their availability for future generations. Key strategies include:

11.1.3.1 Consumption Reduction

Consumption reduction may include energy efficiency, water conservation and material efficiency as follows:

1. **Energy Efficiency:** Using energy-efficient appliances and lighting.
2. **Water Conservation:** Fixing leaks, using water-saving fixtures, and practicing useful water use.
3. **Material Efficiency:** Designing products that use fewer materials and generate less waste.

11.1.3.2 Sustainable Practices

Sustainable practices may include renewable energy, sustainable agriculture, and eco-friendly products as follows:

1. **Renewable Energy:** Utilizing solar, wind, and other renewable energy sources.
2. **Sustainable Agriculture:** Implementing practices that maintain soil health and reduce water usage.
3. **Eco-friendly Products:** Choosing products made from sustainable materials and with minimal environmental impact.

11.1.3.3 Waste Reduction

The following are examples of waste reduction activities which may or may not be applied to the targeted polyolefin plastic solid waste.

1. **Reuse:** Finding new uses for items instead of discarding them.
2. **Composting or Incineration as Examples:** Turning organic waste into valuable compost for gardening and agriculture by composting is one example. Turning organic waste into valuable energy by incineration is another example.
3. **Minimizing Packaging:** Reducing the amount of packaging used in products.

By integrating recycling and conservation practices into our daily lives, we can significantly reduce our environmental footprint and contribute to a more sustainable future.

11.1.4 Polyolefin Plastic Waste Recycle Fundamentals

Polyolefins, such as polyethylene (PE) and polypropylene (PP), are widely used plastics that can be recycled into valuable energy sources and chemicals through various processes. Some key methods are introduced below:

11.1.4.1 Pyrolysis

Pyrolysis involves heating polyolefin waste in the absence of oxygen, breaking down the long polymer chains into smaller hydrocarbons. This pyrolysis process can produce fuels like diesel, gasoline, and kerosene, as well as valuable chemicals such as olefins and aromatics.

11.1.4.2 Gasification

Gasification process converts polyolefin waste into syngas (a mixture of hydrogen and carbon monoxide) by reacting the material at high temperatures with a controlled amount of oxygen or steam. Syngas are the products of the gasification process. Syngas can be used to produce electricity and hydrogen or further processed into liquid fuels and chemicals.

11.1.4.3 Hydrothermal Liquefaction

Hydrothermal liquefaction process uses water at high temperatures and pressures to convert polyolefin waste into liquid hydrocarbons. The resulting products of this hydrothermal liquefaction process include clean fuels and other useful chemicals.

11.1.4.4 Selective Extraction

Selective extraction process involves using solvents to dissolve specific components of polyolefin waste, which can then be separated and purified. This selective extraction process can yield high-purity chemicals and monomers as products that can be reused in the production of new plastics.

11.1.4.5 Expected Benefits of Polyolefins Solid Waste Recycle

In terms of environmental impact, recycling polyolefins reduces plastic pollution and the need for virgin plastic production, contributing to a more sustainable circular economy.

In terms of economic value, the conversion of waste into valuable products can provide economic incentives for recycling and waste management. These methods highlight the potential of polyolefin recycling to not only manage plastic waste but also to create valuable resources, supporting both environmental sustainability and economic growth

11.1.5 Research Needs and Objectives

Access to cheap energy carriers is crucial for the development of many sectors of the economy in all countries of the world. The constantly growing demand for these carriers contributes to the depletion of the world's resources of oil, gas and coal. This growing demand is observed both in developed and developing countries. According to International Energy Agency report [1], world's final consumption of crude oil was 150.1, 169.3 and 153.5 exajoules, while natural gas consumption was 57.2, 69.7 and 68.2 exajoules, in 2010, 2019 and 2020, respectively. The recorded reduction in fuel consumption in 2020 was caused by the global COVID-19 pandemic. The increasing demand for energy carriers shall entail an increase in the prices of these goods. According to Report of U.S. Energy Information Administration (EIA) [2], the price of oil in November 2021 was $81 per barrel, while in the corresponding period of 2020 it was $43. Over the course of 2021, oil prices increased due to the use of global oil reserves at an average of 1.4 million barrels per day. The successive coronavirus waves during the global pandemic situation have contributed greatly to the imbalance in the use and prices of oil resources [3,4]. Despite the slowdown in global oil consumption due to concerns about other COVID-19 variants, the EIA predicts that prices per barrel will remain at a similar level as in 2021.

The growing demand for and rising prices of energy have led to a crisis in the fuel market and the search for substitute fossil fuel carriers. The second reason for replacing fossil fuels with raw materials from renewable sources is the global trend of decarbonizing economies [5]. The aim of decarbonization is to counteract severe climate change, including global warming [6]; and as recommended by the Intergovernmental Panel on Climate Change, to avoid the aggravation of the negative effects of climate change, a temperature rise limit of 1.5°C should be achieved [7].

One way of minimizing the effects of the energy and climate crisis is to, inter alia, move away from a linear economy (a vision called Cradle-to-Grave (CtG)), in which the last stage of a product's life is waste, in favor of a circular economy (CE). The CE treats waste as a valuable raw material and is based on the Cradle-to-Cradle (CtC) model. It is a global strategy for sustainable development, which is assumed to be renovating or regenerative in nature [8]. It replaces the 'end of life' concept with the renovation/regeneration of the use process and moves toward the use of renewable energy. It aims to minimize the amount of generated waste, and products, materials and raw materials should remain in the economic cycle for as long as possible [9].

11.2 CHARACTERISTICS OF PLASTICS

At present, we cannot imagine life without petroleum-based polymers and plastics. Their advantages, such as ease of shaping, non-toxicity, low density, non-conductivity, good mechanical strength or very high corrosion and acid resistance, mean that these materials have become very common in the economy and people's daily life [10]. The constantly growing demand for polymers has led to their continuous increase in production, which, in 2020, amounted to 367 Mt worldwide [11]. For example, in 1950, 1.5 Mt of polymers was produced annually in the world, and in 2020 – almost 370 Mt. It is assumed that in the next decade the amount of plastic waste (PW) will double. The dynamics of development is evidenced by the fact that since 2000 over 55% of the total production of these materials has been produced so far [12] – Figure 11.1.

Plastics have become an indispensable material in many areas of life and economy. In 2020, the largest end-use market of plastic materials was the packaging (40.5%) and building/construction (20.4%) sectors. The third biggest market was the automotive industry sector (8.8%) [11]. This is due to their numerous advantages, such as: low density, easy shaping, high corrosion resistance and good mechanical strength. At present, many groups of polymers are known, but the category of large-volume polymers (mass produced) includes: polyethylene (PE), sometimes differentiated into low-density (LDPE) and high-density (HDPE) material, polypropylene (PP), polyvinyl chloride (PVC), polystyrene (PS), polyethylene terephthalate (PET). These polymers are the main components of plastics, which also contain additives modifying the properties of the final material (e.g., dyes, stabilizers, flame retardants, foaming agents, fillers).

PE is a high-molecular-weight, high-density polymer. It is produced in the process of ethylene polymerization. It is a material with excellent dielectric properties and high resistance to acids,

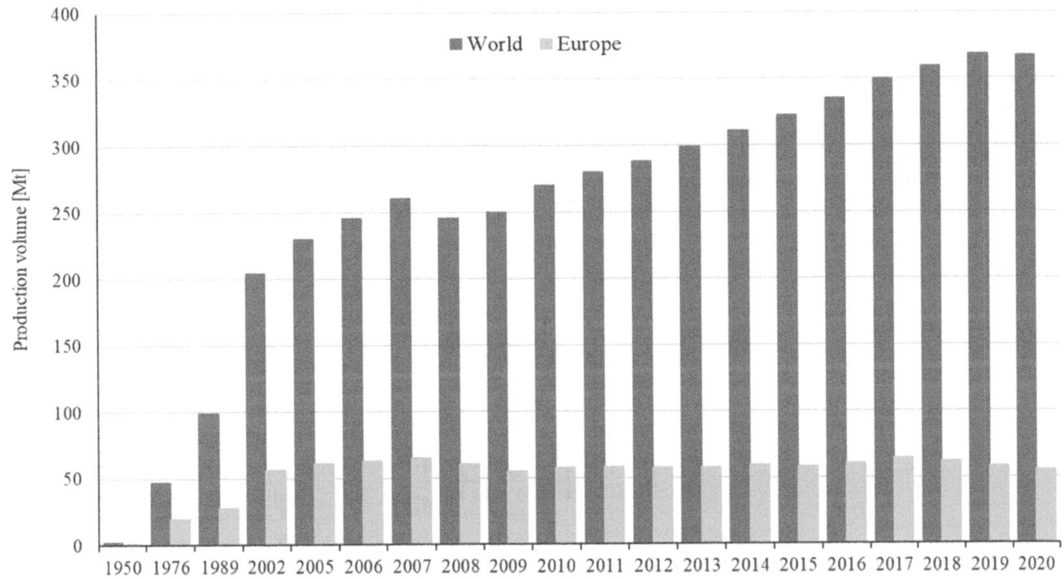

FIGURE 11.1 World and European annual production of plastics from 1950 to 2020. (Own Study Based on Reports from Plastic Europe [13].)

bases, salts and most organic compounds. In aliphatic, aromatic and halogenated hydrocarbons, PE swells, which worsens its properties. The structural formula of PE is as follows:

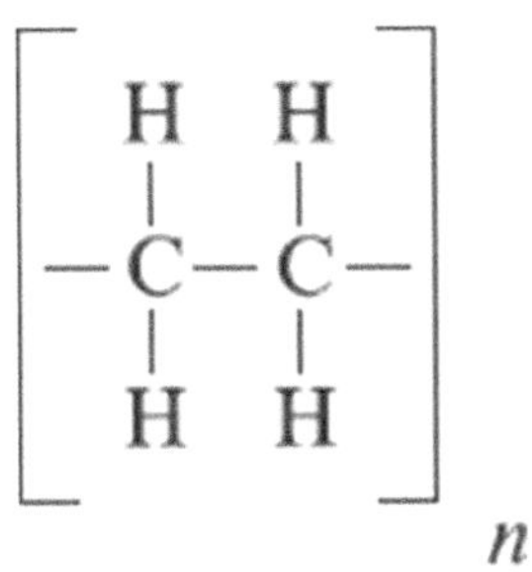

PE is used to produce films and other packaging, pipes for drinking water and sewage, cable sheathing and household items.

PP is a partially crystalline polymer with properties similar to PE. It is obtained by polymerization of propylene. It is resistant to stress corrosion, has a higher mechanical resistance, lower density, higher tensile strength, higher thermal resistance, high brittleness at temperatures below 0°C and greater sensitivity to oxygen than PE. The structural formula of PP is as follows:

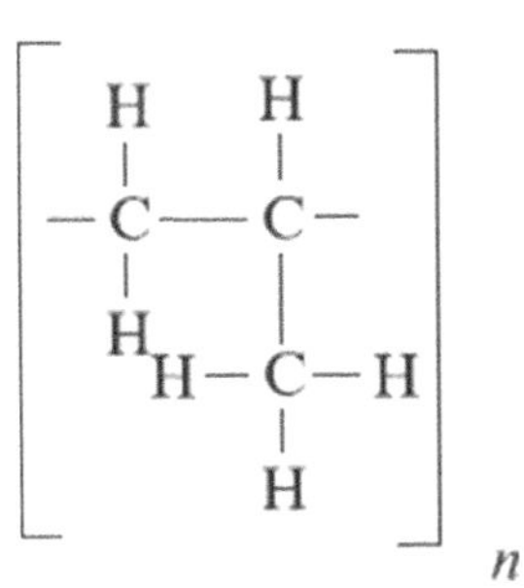

PP is used to manufacture various machine parts, vessels and elements for sterilization with boiling water, pipes for water and aggressive liquids, fibers, and products requiring joining and cooperation with metal elements.

PVC is a compound obtained as a result of polymerization of a monomer – vinyl chloride. It has thermoplastic properties, high mechanical strength, and is resistant to many solvents. At room temperature, it is hard, not very elastic, and becomes brittle at lower temperatures. It is resistant to acids, alkalis, and petrol. It is easy to glue and non-flammable. It is used for pipes, packaging, syringes, gramophone records, and windows. The chemical formula of PVC is as follows:

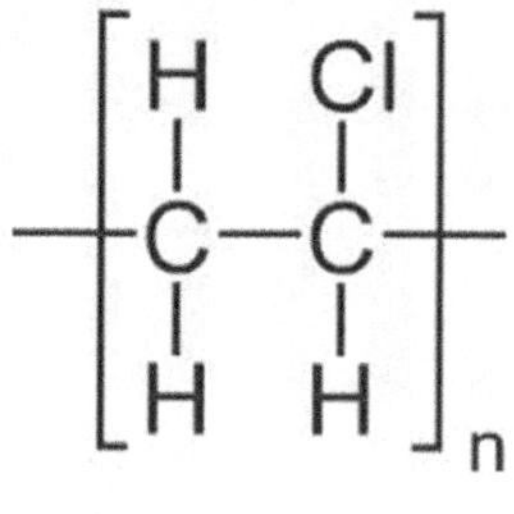

Soft (plasticized) PVC is used to make films, flexible water hoses, anti-corrosion coatings, flooring, and various everyday items.

PS is produced in the process of styrene polymerization, usually derived from the catalytic dehydrogenation of ethylbenzene (about 85% of world production), the Halcon–Oxirane process or the refining of crude oil. The structural formula of PS is as follows:

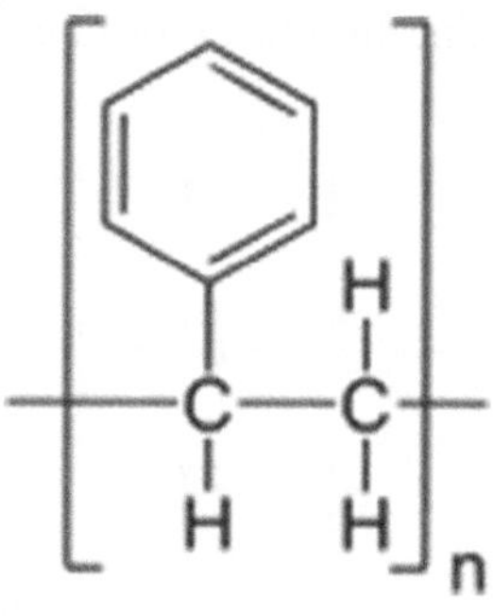

Pure PS is a colorless, hard, brittle thermoplastic (plastic formed by melting), with very limited elasticity. When mixed with additives, PS is the basis for many plastics. It has the advantage over PE and PP of having a lower softening point and lower melt viscosity, which makes it easier to produce small items with complex shapes in injection molding. PS is used to make imitation jewelry, toothbrushes, compact disc boxes, toy parts, etc. However, it has much lower chemical resistance than PE and is therefore not used for the packaging of fatty foods. The most common application for PS is in the production of PS foam.

PET is obtained by reacting terephthalic acid (PTA) or its ester dimethyl terephthalate (DMT) with ethylene glycol (EG) in the presence of a metal catalyst. The structural formula of PET is as follows:

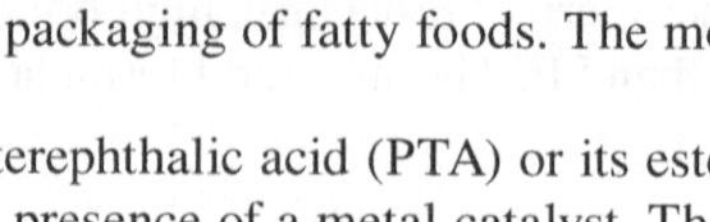

It is characterized by good mechanical properties, a hard surface, good sliding and dielectric properties and chemical resistance to weak acids and alkaline solutions, oils, and fats. It is not resistant to concentrated acid and alkali solutions or prolonged exposure to hot water. PET is used

TABLE 11.1
Selected Properties of Main Polymers [14–19]

Properties	Unit	PE	PP	PS	PVC	PET
Density	g/cm^3	0.89–0.97	0.89–0.92	1.04–1.08	1.19–1.55	1.38–1.41
Melting point	°C	105–135	162–168	240	140–170	242–270
Flash point (auto-ignition)	°C	350–400 (380–440)	350–370 (370–410)	350–390 (480–520)	360–430 (480–550)	(500)
Decomposition temperature	°C	340–400	300	300–400	200–300	340–360
Calorific value	MJ/kg	45	43	40	18	34

to produce fibers, vessels, bottles, packaging, small fittings, and electronic equipment housings. Knitted and woven fabrics (e.g., fleece, dacron, tergal) are produced from PET fibers.

All these plastics belong to the group of thermoplastics (under the influence of heat they soften and can be formed, after cooling they harden). Their basic physical and chemical properties are presented in Table 11.1.

Polymer-based materials are characterized by some disadvantages, such as lower heat resistance, hardness, and mechanical strength in comparison with metals. However, the most important drawbacks of these materials are their non-biodegradable, environmentally unfriendly, and, in most cases, non-recyclable nature [20]. In many cases, the useful life of products made of plastic materials is very short, ranging from 1 day to 3 years [21], this is especially true for packaging. After this time, they become waste, the decomposition time of which is disproportionately long and may last for up to 300–500 years, as is the case with a plastic bottle [22]. New PW is created yearly at a rate of 303 million tonnes (275 million metric tonnes). About 300 million tonnes of PW is produced annually in the world. It is estimated that up to now, 75% of all plastics produced have become waste [23].

PW mainly derives from industrial activities or consumer residue. Postindustrial PW is usually clean, free from organic or other contamination, and often mono-material or its composition is well known. In turn, post-consumer PW most often contains different mixed plastics, the composition of which is unknown, and is contaminated with organic fractions or other materials (paper, metal, textiles) [24]. The largest share (over 61% – 17.8 Mt) is common packaging waste, such as plastic bags, foil, trays for packing and storing food, thermal dishes, cups, and bottles for beverages. The share of individual plastics in the packaging waste stream is as follows: PE > PP > PET >> PS > PVC [25]. Due to the low degradation of the carbon skeleton, PW is a valuable raw material for reprocessing.

11.3 PW MANAGEMENT

The problem with the management of PW is already global, but the situation is improving every year. This is evidenced, inter alia, by the annual data published in the report of the Plastics Europe Foundation, which indicates that, in the European Union in 2020, compared to 2006, recycling of PW increased by almost 118%, energy recovery through incineration increased by 77% and the amount of waste sent to landfills decreased by over 46% [11]. In 2020, over 29 million tonnes of post-consumer PW were collected in European Union countries. Unfortunately, as much as 23.4% of them went to landfills, and only 34.6% was subjected to typical recycling. A total of 42% of collected waste was utilized through energy recovery through incineration (energy recycling) [11]. This form of recycling (in accordance with the principle of waste management hierarchy [26]) should only be used as a last resort for waste that is not suitable for other forms of recycling, such as depolymerization, material or chemical recycling. Sadly, the above data shows that waste incineration is the most common and dominant form of PW management, which unfortunately leads to the complete elimination of plastic material from the economy and makes it impossible to use this material in any other way. In addition, large amounts of harmful and toxic compounds are generated during incineration [27].

The still relatively low level of PW recycling (depolymerization, material, and chemical recycling) results from the complexity of the processing methods, significant energy demand or/and unsatisfactory quality of the obtained products. The best method for managing plastics is their depolymerization to monomers (input ingredients for the production of plastics). However, this method is applicable, with good efficiency, to a few, selected polymers (PET, polymethyl methacrylate, PS) [28–30]. In addition, the raw material must be clean and free from any additives of other materials. Unfortunately, this method cannot be applied to polyolefins, which account for approximately 70%–76% of the waste plastic stream [31].

In the case of polyolefins, the best method for their management is material (mechanical) recycling to re-granulate. This method is considered the most environmentally friendly due to the low value of CO_2 emissions compared to other methods [32], but it is not free from drawbacks. The feedstock for processing must be very well selected and clean, and unfortunately, the material loses quality with each subsequent processing cycle. In addition, the use of different types of plastics on various parts of packaging (e.g., labels, caps, packaging) and the use of additives and fillers for plastics make it difficult to recycle and recover individual groups of materials and reduce the quality of the obtained final products [33,34]. Post-industrial plastics (such as cuttings and trimmings from production, runners from injection molding or granulation residues) are best suited for such recycling because they have a known and homogeneous composition and contain few impurities [35,36].

Regarding post-consumer PW, which is a complex mixture of different plastics and other materials, a sorting process should be carried out to obtain separate components. Selected components, depending on their quality, can be mechanically recycled (for example, into outdoor furniture, flower pots, and elements of terraces), directed to chemical recycling or sent to incineration [37]. However, due to the difficulty of completely separating the individual materials and contaminations of post-consumer waste, chemical recycling methods seem to be the most appropriate for their management [38]. Figure 11.2 shows a schematic diagram of the main directions of PW management.

PW is a valuable raw material for recycling, but the converting method is influenced by the economic efficiency of the process, which depends, among other factors, on the material composition, the size and quality (homogeneity, purity) of the collected waste stream, the availability of sorting and processing technologies, and the market demand for the products obtained as a result of recycling.

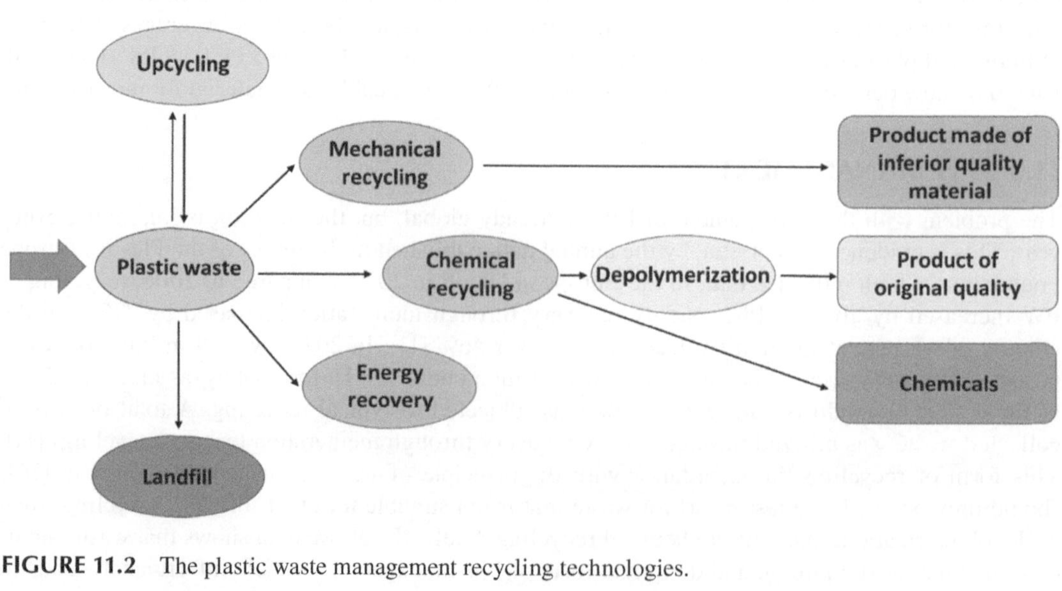

FIGURE 11.2 The plastic waste management recycling technologies.

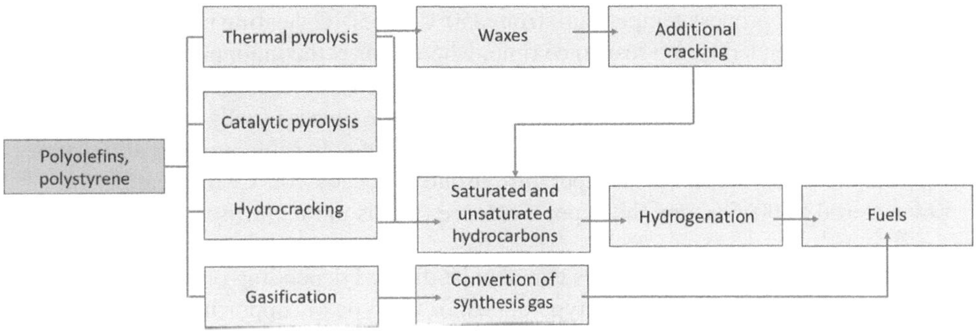

FIGURE 11.3 The main directions of thermochemical conversion of polyolefins and polystyrene to hydrocarbons.

Due to the significant share of polyolefins and PS in the post-consumer waste stream, they are a valuable material for the production of engine fuels [39]. For this reason, many methods are used to convert plastics to hydrocarbons, which could be used as components of liquid fuels. Figure 11.3 shows the main directions of the thermochemical conversion of polyolefins and PS to hydrocarbons, most frequently encountered in the literature.

11.4 PYROLYSIS

One of the methods of converting waste plastics to hydrocarbon fractions is pyrolysis. This enables the transformation of high-molecular compounds, such as waste plastics, into compounds with much shorter carbon chains. It is considered a promising method for the conversion of addition polymers such as PE and PP (polyolefins) [40,41]. It offers the opportunity to recycle this polyolefin stream, which cannot be economically managed in conventional mechanical recycling, thus avoiding their incineration and the generation of hazardous substances such as dioxins and furans [42,43]. The pyrolysis oil (a mixture of waxes and a liquid fraction) obtained as a result of pyrolysis can be used as a substitute for heavy fuel oil or undergo further processing, e.g., in the petrochemical industry [44,45]. Furthermore, as shown in Reference [46], large-scale pyrolysis is economically viable. For example, in the case of an investment in installation that processes 100,000 kg/h of waste plastics, income is generated within the first year and has a positive net present value. The advantage of pyrolysis is that pre-segregated, mixed, and contaminated PW with both organic and inorganic material can be subjected to this process [47,48].

The pyrolysis process is based on the thermochemical decomposition of various materials of organic and synthetic origin at elevated temperatures and in an oxygen-free atmosphere. Typically, the process is carried out at temperatures in the range of 500°C–800°C and, in the case of plastics, it follows four types of mechanism: end-chain scission or depolymerization, random-chain scission, chain stripping, and cross-linking [49]. The main products of the pyrolysis process are pyrolysis oil, gaseous products, and the solid fraction (char). On average, depending on the technology used, 45%–50% of the oil fraction, 35%–40% of the gaseous fraction and 10%–20% of char is obtained in the process [50]. It was generally found that with the increase of the process temperature, the yield of the liquid fraction increases, while the yield of the solid fraction decreases [51]. However, for selected plastics (e.g., for PE over catalyst), it is possible to obtain a liquid fraction in an amount exceeding even 90% [52,53]. The high propensity of plastics to generate significant amounts of oil fraction during pyrolysis is the result of high volatile matter and low ash content in relation to other materials (e.g., biomass). These two parameters are the key factors influencing the pyrolysis oil yield [54].

Due to the process conditions (vapor residence time in the reactor, temperature, raw material heating rate), pyrolysis is divided into three types [55,56]:

a. **Slow Pyrolysis:** process temperature from 350°C to 550°C, heating rate from 1°C to 10°C per minute, extended steam residence time, where char is the main product;
b. **Fast Pyrolysis:** the process temperature is usually from 500°C to 700°C, the raw material heating rate is above 100°C/min, and the vapor residence time is usually within a few seconds, the main product is the liquid fraction, and in the case of polyolefin pyrolysis also waxes;
c. **Flash Pyrolysis:** the process temperature usually exceeds 700°C, the heating rate of the raw material >200°C/s, and the vapor residence time is in the millisecond range.

In addition to the process time, pyrolysis can also be divided depending on other conditions of the process (e.g., the presence of a catalyst, type of gas) or the type of supporting equipment [57].

The composition of pyrolysis oil depends not only on the process conditions but also on the type of waste plastic in the raw material. A higher yield of aromatic hydrocarbons is obtained when PS and PET are pyrolyzed. On the other hand, PP and PE produce oil that is rich in paraffins, olefins, and waxes based on aliphatic hydrocarbons. Wax is an intermediate product, mainly composed of long-chain hydrocarbons (C20+) with a high boiling point. In order to transform it into fuel products, it is necessary to separate it after pyrolysis and subject it to further cracking [55]. The pyrolysis oil also requires further processing (e.g., distillation, refining, hydrogenation) to improve its properties, due to the content of unsaturated hydrocarbons [58–60].

It should also be considered that even well-sorted waste for fuel production (PE, PP and PS) may contain contaminants in the form of other plastics, e.g., PVC. The presence of this plastic will result in the formation of chlorine compounds (including hydrogen chloride) during the recycling process. These compounds will have a corrosive effect on apparatus components and poison the catalyst [61]. Since the proportion of PVC admixture is generally low, in order to eliminate the harmful effect of chlorine, chlorine-binding compounds such as $NaHCO_3$, CaO, $CaCO_3$, Na_2CO_3, and $Ca(OH)_2$ can be fed to the reactor with the feedstock [62].

Plastics, among the additives, may contain flame-retardant compounds. These additives are usually bromine or chlorine derivatives, e.g., chlorinated paraffins, boric acid, polybrominated diphenyl ethers, tetrabromobisphenol, tris(2-chloroethyl)phosphate, hexabromocyclohexane [63]. These compounds are generally not added to packaging, which has the largest share in the PW stream, while they may be included in expanded polystyrene (EPS) waste from construction. The degradation of bromine compounds can lead to the formation of organobromine compounds and bromine hydrogen, whose action is similar to that of chlorine compounds. HBr can be disposed of in a manner analogous to hydrogen chloride. Another method to inactivate bromine compounds is the use of suitable catalysts during pyrolysis, e.g., metal oxides (e.g., ZnO, Fe_2O_3, La_2O_3, Cu_2O, Sb_2O_3, Al_2O_3 [64–66]), calcium and iron-based composites [67,68], zeolites, molecular sieves, red mud, or natural zeolites [24,69].

The reactor also plays an important role in the pyrolysis process [70]. The type of reactor used affects the process control, mass and heat exchange, mixing, liquid and gas phase flows, retention time, etc. There are many ways to classify pyrolysis reactors. According to the method of carrying out the process, there are batch reactors, semi-batch reactors, and reactors with continuous flow. In turn, due to their construction, fixed bed, fluidized bed (bubbling and circulating) and conical spouted bed reactor (CSBR) can be distinguished, among others. Another criterion for the division may be, for example, the method of heat energy supply: to the interior via an inert gas or material, through the external walls via flue gas or electricity or via internal heating elements [71–73]. Examples of studies focused on PW pyrolysis using different reactors are shown in Table 11.2. The main advantages and disadvantages of pyrolysis reactors are presented in Table 11.3.

In general, regardless of the type of reactor that is used, the conventional pyrolysis of polyolefins produces a liquid product of poor quality, which is almost impossible to use as a fuel component. To overcome the problems associated with this type of pyrolysis, new technologies have been developed and are still being developed, that enable better quality liquid products to be obtained from the thermal conversion of waste plastics. In addition to conventional pyrolysis, the process can be carried out using catalysts, in a hydrogen atmosphere, as well as using microwave- and plasma-assisted processes, as shown in Figure 11.4.

TABLE 11.2
Examples of Reactors for Polymer Pyrolysis

Type of Reactor	Type of Feedstock Materials	References
Batch reactor	PS, PE, PP and PET	[24]
	HDPE	[40]
	Plastic medical wastes	[74]
Semi-batch reactor	PS, HDPE, LDPE, PP	[75]
	PE	[76]
	Polyolefin	[77]
Continuous reactor	Plastic waste from landfill	[78]
	HDPE	[79]
Fixed bed reactor	PP, HDPE, LDPE	[80,81]
	HDPE	[82]
Fluidized bed reactor	LDPE	[83]
	PP, PE	[84]
	PE	[85]
CSBR	HDPE, LDPE, PP	[86]
	HDPE, LDPE, PP, PS, PET, PMMA	[87]
	PE	[88]

TABLE 11.3
The Main Advantages and Disadvantages of Pyrolysis Reactors [42,54,71]

Type of Reactor	Advantages	Disadvantages
Batch reactor	• High conversions of feedstock.	• Variability of product from batch to batch; • High labor costs per batch; • Problem with adoption for large-scale production.
Semi-batch reactor	• Possibility of simultaneous addition of reagents and removal of the product; • Flexibility of adding reactants over time.	• High labor costs per batch; • Problem with adoption for large-scale production.
Fixed-bed reactor	• Simple and reliable construction.	• Keeping the size and shape of the raw material particles relatively constant; • Limited accessibility of the catalyst surface area during the reaction; • Low heating rate; • Non-uniform temperature inside the polymeric material.
Fluidized bed reactor	• Better access to the catalyst than in the fixed bed reactor; • Large reaction surface; • Higher degree of cracking than in the fixed bed reactor; • Uniform temperature distribution; • Excellent mixing properties; • Better consistency of the process conditions; • Less frequent loading of raw material and the associated process shutdown; • Easy catalyst regeneration without unloading.	• Small particle size of the raw material; • Problem with separating char from bed material; • It is easy to defluidize bed when melted plastic sticks on it.

(Continued)

TABLE 11.3 (*Continued*)
The Main Advantages and Disadvantages of Pyrolysis Reactors [42,54,71]

Type of Reactor	Advantages	Disadvantages
Rotary kiln reactor	• More efficient in heating raw materials than a fixed bed reactor; • Good mixing of raw material; • Yields uniform pyrolytic products; • Possibility of adjusting the residence time; • Simple maintenance; • Possibility of feeding the raw material with different particle sizes.	• Slow heating rate gives a significant amount of char to the products; • Mainly applicable to conventional pyrolysis.
Stirred tank reactor	• Good temperature control; • Easy to construct and operate; • Uniform heat distribution.	• Tendency to build up char deposits on the walls; • Low conversion rate per volume; • Poor agitation; • Require frequent maintenance.
Conical spouted bed reactor	• Good mixing; • Possibility of using raw material with a large particle size distribution; • High heat transfer; • Protects against defluidization caused by melted plastic; • Relatively simple construction; • Extremely short residence time of the polymers in reactor avoids secondary reactions; • Lower pressure drop in comparison with other fluidized bed reactor.	• Problem with catalyst feeding; • Problem with catalyst entrainment; • Deposition of products on the catalyst; • The reactor requires many pumps (high operating costs).

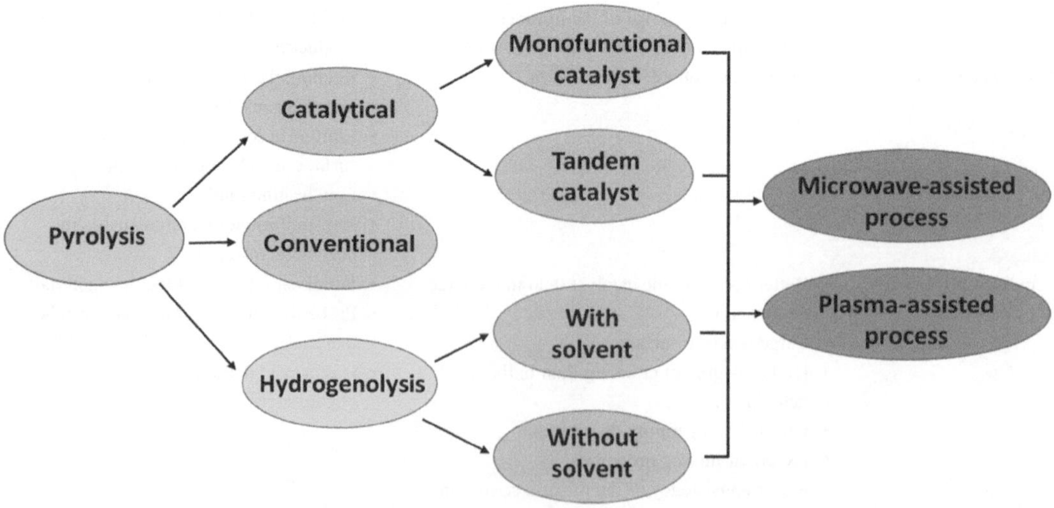

FIGURE 11.4 Pyrolytic methods of plastics conversion.

11.4.1 Catalytic Pyrolysis

A catalyst is a substance whose presence in a mixture of reagents increases the reaction rate. It is involved in a chemical reaction, but is reconstituted during the process. Catalysts make it possible to lower the reaction temperature, speed it up, and increase the selectivity of the product, and, therefore, have been widely used in industry and scientific research. Therefore, catalytic pyrolysis has an advantage over the conventional process and enables a product with higher added value to be obtained. For this reason, catalytic pyrolysis has an advantage over the conventional process and enables the production of a product with higher added value, such as hydrocarbons with a chain length (C8–C16) typical for jet fuel [89].

Catalysis can be carried out homogeneously (catalyst and reagents are in the same phases) or heterogeneously (catalyst and reactants are in different phases). Lewis acids, e.g., $AlCl_3$, $TiCl_3$, $TiCl_4$, and $FeCl_3$, as well as molten metal tetrachloroaluminates $[M(AlCl_4)_n]$ where (M=Li, Na, K, Mg, Ca or Ba; $n=1$ or 2) can be used as homogeneous catalysts in the thermal decomposition of plastics [90,91]. However, due to the easier separation of the catalyst from the product, heterogeneous catalysis has found a much wider application in PW cracking. In the pyrolysis of plastics, nanocrystalline zeolites (HZSM-5 [89,92], HUSY [93,94], HBeta [95], and HMOR [96]) are most often used as heterogeneous catalysts, but are also used as fluid catalytic cracking (FCC) catalysts [97–99] and silica-alumina catalysts [100]. Activated carbon (pure or impregnated with transition metals) can also be used as a catalyst for the process [101]. During the two-stage process of thermal and catalytic cracking, the life of the catalyst can be extended [102].

The use of a catalyst avoids waxes in the liquid product, but can also reduce the liquid fraction in favor of the gaseous fraction. For example, Kassargy et al. [103] investigated the thermal decomposition of PE and PP with and without a catalyst. The process was carried out in a batch reactor at a temperature of 450°C. During pyrolysis without the PE catalyst, the main product was wax, at 80%. With PP, the main product was liquid in the amount of 85.5%. The use of the ultrastable Y zeolite (USY) in the ratio of 1:10 for both polymers resulted in a high yield of liquid products (PE – 71%, PP – 82%). The product of catalytic pyrolysis was a liquid mixture of hydrocarbons in the case of PE – C5–C39 compounds, and PP – C5–C30 compounds.

The use of a catalyst not only determines a better cracking of polymers and a lower reaction temperature but also has an impact on the kind of compounds (naphthene, aromatics, and paraffins) in the product. Dai et al. [104] investigated the influence of various catalysts (ZSM-5, USY, SBA-15 (SiO_2) – supported different metal catalysts) on the composition of HDPE pyrolysis products. They found that some metals (e.g., Cs, Co, Cu) in the metal/SBA-15 catalytic system do not favor the cracking processes, as the product has a high content of C23–C30 hydrocarbons (waxes, from 17% to 36%). Other metals in this system showed a higher cracking capacity of HDPE to C5–C12 hydrocarbons, such as the Al/SBA-15 catalyst (yield approx. 63%). However, aluminum showed a very high selectivity to aromatic compounds (over 45%). Zinc, on the other hand, resulted in a product with a relatively high content of short-chain hydrocarbons (approx. 41% of C5–C12) and low aromatics (4.1%). The zeolite catalysts studied by the authors (ZSM-5, USY), compared to the catalysts based on SBA-15, showed a high cracking efficiency toward C5–C12 hydrocarbons (approx. 90%), but the main components of the oil were aromatic hydrocarbons (over 80%). These hydrocarbons are not desired in such large amounts in the fuel fractions.

The research described by Rahimi and Rostamizadeh [105] showed that the pyrolysis of plastics with the participation of the zeolite 5Fe/B–ZSM-5 catalyst leads to a product containing the highest amount of light hydrocarbons in the C5–C12 range; however, it has a significant predominance of aromatic compounds (76.9%). Similar results were obtained by Akubo et al. [106]. The cause of olefin aromatization may be the strong acidity of the zeolites, which facilitates aromatization during the catalytic cracking of waxes due to Diels–Alder reaction [104]. The number of works on catalytic pyrolysis is very high; Table 11.4 presents a list of selected research works in this area.

TABLE 11.4
Selected Research Works in the Area of Catalytic Pyrolysis of Plastic Wastes

Material	Catalyst	Conditions	Liquid Product	Reference
PE	Ultrastable Y (USY) zeolite	Batch reactor Temperature 450°C Polymer/catalyst ratio 10:1 Reaction time 50 minutes	Yield 71% C5–C39 compounds	[103]
PP	Ultrastable Y (USY) zeolite	Batch reactor Temperature 450°C Polymer/catalyst ratio 10:1 Reaction time 45 minutes	Yield 82% C5–C30 compounds	[103]
LDPE	Zeolite CAT-2	Batch reactor Polymer/catalyst ratio 10:1 Temperature 460°C Nitrogen atmosphere	Yield: • 96% without catalyst (24% C7–C12, 30% C13–C20, 46% C21–C40) • 52% with catalyst (100% C7–C12)	[107]
PP	Zeolite CAT-2	Batch reactor Polymer/catalyst ratio 10:1 Temperature 460°C Nitrogen atmosphere	Yield: • 86% without catalyst (44% C7–C12, 33% C13–C20, 23% C21–C40) • 58% with catalyst (100% C7–C12)	[107]
HDPE, LDPE and PP	Zeolite CAT-2	Batch reactor Polymer/catalyst ratio 10:1 Temperature 460°C Nitrogen atmosphere	Yield (for 34% PP and 66% PE): • 64% without catalyst (36% C7–C12, 32% C13–C20, 32% C21–C40) • 68% with catalyst (100% C7–C12) Yield (for 66% PP and 34% PE): • 84% without catalyst (46% C7–C12, 31% C13–C20, 23% C21–C40) • 63% with catalyst (100% C7–C12)	[107]
HDPE	Y-zeolite impregnated with transition metal (Ni, Fe, Mo, Ga, Ru and Co)	Fixed bed reactor Two-stage process (pyrolysis/catalysis) Temperature both processes 600°C Reaction time 30 minutes	Yield of oil: • without catalyst – 69% • with Y-zeolit – 45% • with metal/Y-zeolit – below 40% Y-zeolit with and without metal causes greater production of aromatic hydrocarbons	[106]
LDPE	FCC catalyst silica alumina	Semi-batch pyrolysis reactor Fluidizing gas – nitrogen Amount of catalyst 5% Temperature 500°C Reaction time 60 minutes	Yield of oil – 93.5% where: • C6–C9 – 20.7% • C10–C15 – 64.7% • C16–C19 – 12.2% • >C20 – 2.4%	[100]

Gopinath et al. [100] for the decomposition of LDPE used a catalyst with a lower acidity (FCC catalyst silica alumina). They obtained light hydrocarbon fractions (C10–C15) with high efficiency (64.7%). GCMS analysis showed that the obtained oil contains similar compounds to diesel fuel (but with a significant number of alkenes), some aromatic hydrocarbons and over 4.5% oxygenated compounds. Blends containing from 20% to 80% pyrolysis oil and diesel fuel were tested in a four-stroke, single-cylinder DI diesel engine. The authors found that a blend containing 20% pyrolytic oil could be used as a diesel fuel without any engine modifications. Recently, works on the possibility of using liquid fractions from the conversion of PW as a fuel component have become very common. These works concern the influence of such a fraction in fuel, e.g., on:

a. Engine work (engine parameters, process of combustion) [108–110];
b. Emissions including life cycle assessment (LCA) [111–116];
c. Combability with fuel components [112,117,118].

The oils obtained in the process as a result of classical and catalytic pyrolysis contain unsaturated hydrocarbons, which are chemically unstable. For this reason, many studies prematurely conclude that the oil fraction of plastics may be a component of conventional fuels. In most of the studies, the obtained fuel fractions were not hydrogenated (thus having poor stability) and/or were not tested for compliance with fuel standards. To use such fractions as fuel components, they must be hydrogenated, and the fuel mixtures are tested for all standard parameters [119,120]. Hydrogenation can be accomplished by:

a. Classical hydrogenation;
b. Hydro-pyrolysis (as well as hydrocracking and hydrogenolysis).

11.4.2 Stabilization of Pyrolysis Oil with Hydrogen

Classical hydrogenation is carried out in a hydrogen atmosphere, under high pressure, and in the presence of the catalyst [121–123]. Due to their high costs, most installations are located on the premises of re-refineries. The process consists of saturation with multiple hydrogen bonds between carbon atoms. In turn, the hydro-pyrolysis of plastics is a thermal decomposition carried out under a hydrogen atmosphere where, apart from the hydrogenation of multiple bonds, a single bond is broken between carbon atoms and hydrogenation with the participation of a metallic catalyst (monofunctional) takes place. A variation in the process is hydrocracking, where a bifunctional metal or acid catalyst is used [124]. These reactions can be performed without or in the presence of solvent as a liquid phase promoting heat and mass exchange. For example, research on HDPE conversion using hydrogenolysis was carried out by Jia et al. [125]. The process was carried out at a temperature of 220°C, under a hydrogen pressure of 0–60 bar, with the Ru/C catalyst, in different solvents. Within an hour of the process, under 30 bar and in the presence of *n*-hexane, approx. 92% efficiency of the liquid fraction was obtained (60.8% jet fuel and 31.6% lubricant-range hydrocarbons). In turn, Rorrer et al. [126] carried out LDPE conversion studies without the presence of a solvent. The process was carried out at a temperature of 200°C, at a pressure of 20 bar of H_2 and in the presence of a Ru/C catalyst. After 16 hours of reaction, the *n*-alkanes were obtained in a yield of up to 45%. Rorrer et al. [127] studied the catalytic depolymerization of PP, in the presence of a Ru/C catalyst, in the absence of solvent. The process was carried out at 200°C–250°C, with a hydrogen pressure of 20–50 bar. The investigated catalyst was found to be effective in the conversion of PP to C5–C32 liquid hydrocarbons with an efficiency of about 68%. As in the case of catalytic pyrolysis, in hydrogenolysis, the type of catalyst affects the reaction efficiency and the yields of individual hydrocarbon fractions [128–130]. It should be mentioned that the hydrogenolysis process of waste plastics can be used to produce not only liquid fuels but also gaseous fuels, such as methane [131].

Due to the high cost of hydrogen, which is the main barrier to its use in the process, other hydrogenation methods are being sought. One of them is the use of tandem catalysis, which uses two or more different catalysts coupled together. Hydrogenolysis conducted under such conditions speeds up polymer conversions compared to a monofunctional process [132]. In tandem catalysis, the metal/acid system catalysts are used. The acid catalyst is responsible for breaking bonds between carbon atoms, while the metal catalyst is responsible for the hydrogenation and minimization of catalyst coking. Therefore, the synergy of catalysts is one of the factors determining the high yields of the liquid fraction. Such studies were carried out by Liu et al. [133], where chain and branched liquid alkanes were obtained from LDPE with a yield of up to 85% in temperature ca. 250°C, in 2 hours. The process was carried out under hydrogen atmosphere (pressure 30 bar), using $Pt/WO_3/ZrO_2$ and HY zeolite as catalysts.

Another example of this process is tandem hydrogenolysis/aromatization, where, during aromatization, hydrogen is generated, which is then used in the hydrogenolysis process [134,135]. This reaction can then be carried out in a single reactor without solvent and hydrogen. For example, tandem hydrogenolysis/aromatization of PE leads to alkylaromatics and alkylnaphthenes [136].

An innovative approach to the conversion of PW and hydration of the liquid fraction is the technology developed by Handerek Technology [137]. The polyolefin thermolysis process is carried out in a reactor, operating on the principle of reactive distillation, filled with Raschig rings. The process is not isothermal; the temperature varies along the axis of the reactor in the range of 360°C–450°C. Gravity moves the plastic down the reactor, where it thermally decomposes. In the final step, the resulting product vapor is hydrogenated at a temperature above 320°C under atmospheric pressure. The reaction gas in both processes is syngas. The mixture of waste plastics (PE and PE) was subjected to liquefaction and hydrogenation. From the 6 kg of waste plastics, 6 L of liquid product (with boiling points below 360°C) and 0.1 kg of solid phase (char) were received; the rest were gases. The liquid fraction from the thermolysis process contained 40.6% saturated compounds, and 59.4% unsaturated compounds (49.1% alkenes, 8.4% alkynes, and 1.9% aromatic compounds). Hydrogenation of this fraction resulted in a significant increase in the content of saturated compounds (to 83.5%) and a decrease in the content of unsaturated compounds (to 16.5%). The solid fraction contains char and mineral impurities that were introduced into the installation together with the raw material, and it may be used as a solid fuel. The gaseous fraction consists of combustible compounds (e.g., methane, ethane) and can be used as a heat carrier to the reactor or as a substrate for the production of syngas. Thus, the process becomes self-sufficient. The scheme of the waste polyolefins thermolysis technology is presented in Figure 11.5.

The authors have shown that this technology is very effective due to the use of a novel reactor solution, operating on the principle of reactive distillation (under non-isothermal conditions). This made it possible to obtain a stable light liquid product, containing no waxes (without using a catalyst). The studies also showed that it is possible to conduct hydrogenation with syngas under atmospheric pressure. Moreover, the authors showed that the use of a mixture of PE:PP:PS in the ratio of 60:30:10 in the thermolysis process makes it possible to obtain a product that, after distillation, can be used as a component of engine fuels [119]. Both diesel oil and gasoline containing 5% v/v of the appropriate fraction met the requirements of the fuel quality standards.

11.4.3 Microwave Pyrolysis

The thermal energy needed for the degradation of polyolefins in the pyrolysis process can be delivered by microwave radiation. Compared to conventional electrical heating, where temperature distribution and heat transfer are carried out by convection, radiation, and conduction, the microwaves pass through the heated particle of the material and are converted into thermal energy inside it [138]. For this reason, microwave heating is faster, runs throughout the entire volume of the heated material and is a non-contact process [139,140]. Materials subjected to microwave pyrolysis are

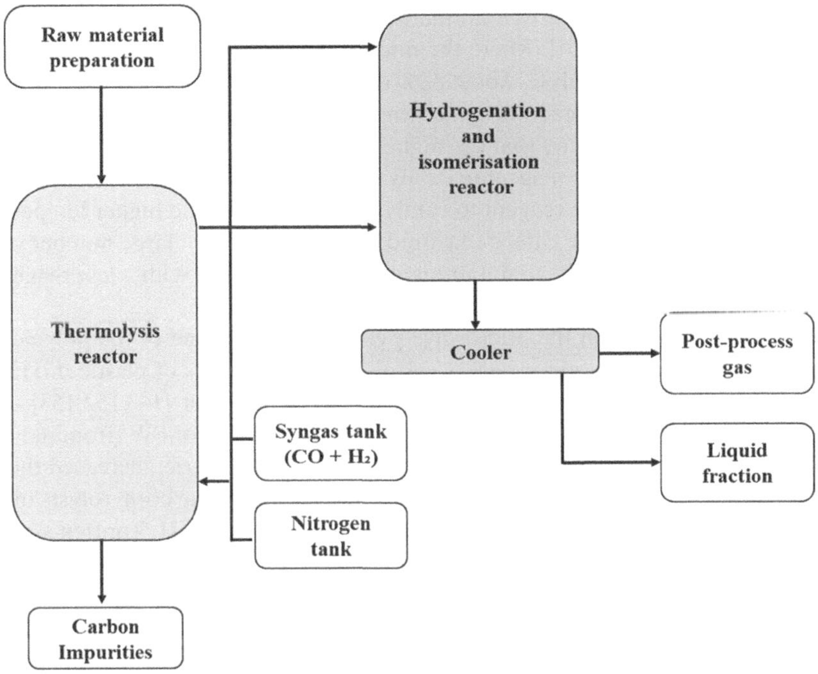

FIGURE 11.5 Scheme of the waste polyolefins thermolysis.

characterized by their differing susceptibility to radiation; therefore, this process is used in the case of materials that absorb radiation. Plastics have low dielectric constants; therefore, a special microwave absorbent (e.g., silicon dioxide, activated carbon, graphene) was added to the microwave pyrolysis process. This absorbent reduces the pretreatment before pyrolysis and the process time, improves the heating efficiency, and increases the efficiency of the process [141,142]. The efficiency of the process can be additionally increased by adding metal (e.g., copper, iron, aluminum) [143,144]. Another way of increasing the process efficiency is to use different types of carbon materials as susceptors. These materials absorb electromagnetic energy and convert it into thermal energy [145]. It should be mentioned that, in this method, polymers are usually introduced without cleaning to facilitate the absorption of microwaves, since dust, moisture, and debris help absorb microwaves.

Due to its many advantages, the microwave-assisted process is the subject of much research. The work available in the literature relates to [146–149]:

a. The general principles of microwave pyrolysis;
b. The influence of parameters on the process efficiency (e.g., microwave power, time and temperature, type of material (absorber, reactor, catalyst));
c. Co-pyrolysis possibilities;
d. Kinetic models of the process;
e. The properties of the obtained products (fractions).

Fan et al. [150] conducted research on the ex situ catalytic upgrading (with MgO) of oil vapors from the microwave pyrolysis process of the LDPE. The ex situ catalytic upgrading of vapors resulted in a lower yield of liquid fraction in comparison with the non-catalytic process. The main product was gasoline fraction, which contained up to 96% of the hydrocarbons. This fraction consists of appropriately 40% aromatic and 60% aliphatic hydrocarbons, while the fraction obtained during the process without a catalyst contained 10% aromatic and 30% aliphatic hydrocarbons. Similar results were obtained by Zhang et al. [151], where the microwave pyrolysis process of LDPE was carried out with the use of the ZSM-5 catalyst. About 88% of the gasoline fraction (containing valuable aromatic hydrocarbons) and the gas fraction containing ethylene, ethane, and hydrogen were obtained. Based on this research, it was found that the highest efficiency of the liquid fraction was obtained at a temperature of 450°C and the reagent to catalyst ratio of 2, while the gas fraction was obtained at a temperature of 375°C and the reagent to catalyst ratio of 4.68. The higher temperature and the higher ratio of the reactants to the catalyst favored the formation of a large number of monocyclic aromatic hydrocarbons. Conversely, maintaining a high temperature with a low reagent-to-catalyst ratio resulted in the formation of polycyclic hydrocarbons.

The above studies focused on the microwave pyrolysis of only one of the polyolefins. In other works, scientists researched the microwave pyrolysis of various types of plastic mixture, including the plastic fraction of waste from electric and electronic equipment [143,152,153]. For example, Aishwarya and Sindhu [154] used the research material of a mixture of PW (from industrial sorting plants and household activities) and carbon black, which, as a dielectric, increased the temperature of the microwave pyrolysis process. There are also some papers on the co-pyrolysis of plastics with biomass mixtures [155–157] or waste oils from cooking [158]. Table 11.5 presents an overview of the work on microwave pyrolysis and the co-pyrolysis of various types of plastic. As the data in the table shows, the process can be carried out for different types of waste plastics and their mixtures with, e.g., biomass. Depending on the type of raw material, process conditions (power, temperature, time), type of catalyst and susceptor, the process efficiency varies. However, according to the research, it is possible to obtain liquid hydrocarbons, which can be used to produce fuel components. The microwave pyrolysis technology is not widespread and still requires a lot of research [143]. Lam et al. [159] showed that the cost of producing oil from the pyrolysis process, which can be a component of or substitute for fossil fuel, is cheaper than diesel oil. For this reason, this technology should be developed and applied on an industrial scale.

TABLE 11.5

Comparison of Research Works on the Conversion of Various Types of Plastic into Hydrocarbon Fractions Using Microwave Pyrolysis and Co-pyrolysis

Type of Plastic/ Material	Type of Catalyst/ Absorber	Process Parameters	Pyrolysis Oil Yield (%)	Oil Composition	References
HDPE	AC, molecular sieves	HDPE to AC ratio from 100:60 to 100:330, heating power combinations (20%–40% of the maximum power of 900 W)	87.8	82.4% C7 and C20	[160]
LLDPE	HZSM-5	Catalyst to feedstock 5%–20%, catalytic temperature 550°C and 500°C	84.1	98.0% gasoline-range hydrocarbons (70.4%–72.3% mono-aromatics)	[161]
LDPE	ZSM-5	Pyrolysis temperature 480°C, catalytic temperature 250°C–500°C, ZSM-5 to plastic ratio from 1:4.68 to 1:1.32, time 10 minutes, power 0.7 kW, N_2 gas	24.4–32.6	>94% aromatics; 74.7%–88.5% C8–C12 MAHs	[151]

(*Continued*)

TABLE 11.5 (*Continued*)
Comparison of Research Works on the Conversion of Various Types of Plastic into Hydrocarbon Fractions Using Microwave Pyrolysis and Co-pyrolysis

Type of Plastic/ Material	Type of Catalyst/ Absorber	Process Parameters	Pyrolysis Oil Yield (%)	Oil Composition	References
HDPE	AC	Pyrolysis temperature 400°C–600°C	27.3–54.9	>90% for C≤21, ~45%–58% normal aliphatics; ~35%–45% aromatics	[162]
LDPE	Silicon carbide, MgO	Pyrolysis and catalytic temperature 350°C–550°C, MgO to plastic ratio from 1:15 to 1:3, plastic to absorber ratio 0.03:1, time 20 minutes, vacuum	24.2–38.5	79.5%–96.0% gasoline fraction; ~15%–50% MAHs	[150]
PS	Carbon, silicon carbide	Time 28 minutes, temperature 574°C, pressure N_2 flux, power 3 kW, PS to absorber ratio 2.21	94.3	40.5%–71.9% styrene, 5.9%–18.4% α-methylstyrene, 6.2%–10.9% toluene	[163]
Waste PS	Tire, carbonaceous char	Power 1.2–6.0 kW, PS to carbon ratio 2.0–0.5, reaction time 13–82 minutes, temperature 364°C–578°C, anoxic	86.5	93.9% single-ring aromatic compounds C6–C10, 66.0% styrene	[164]
Waste: PS, PP, PS+PP	Carbon, AC	Power 0.18–0.9 kW, polymer to an absorbent ratio 10:0.5–10:2, time 10–20 minutes	83.3–86.1	67.6% styrene	[153]
LDPE	NiO, zeolite HY, silicon carbide	LDPE to HY ratio 20:1, 15:1, 10:1, 5:1, pyrolysis temperatures 450°C–600°C, catalysis temperatures 350°C–500°C, HY to NiO 15:1, 10:1, 5:1, 3:1, time 20 minutes, power 1.8 kW, vacuum	56.5	>92% C5–C12 gasoline fraction; 34.6%–46.6% aromatics; 26%–30% isomerized aliphatics	[141]
LDPE	ZSM-5	Pyrolysis and catalytic temperature 375°C, ZSM-5 to plastic ratio 1:10	64.4	97% C8–C16 aromatics	[165]
LDPE, torrefied rice straw	ZSM-5	Temperature 309°C–590°C, ratio catalyst to feedstock from 3.9% to 11.04%	24.5–29.8	~80% gasoline fraction; 26.5%–40.2% long-chain hydrocarbons and cyclic hydrocarbons	[156]

(*Continued*)

TABLE 11.5 (*Continued*)
Comparison of Research Works on the Conversion of Various Types of Plastic into Hydrocarbon Fractions Using Microwave Pyrolysis and Co-pyrolysis

Type of Plastic/ Material	Type of Catalyst/ Absorber	Process Parameters	Pyrolysis Oil Yield (%)	Oil Composition	References
Corn derived plastic bags	Iron powder, carbon powder	Power 1.2–3 kW, temperature (617–808) K, plastic bags to absorber ratio 2.02–2.39	28.5–47.5	Mainly oxygen compounds, (9.6–13.3)% aromatic hydrocarbons, 2.2%–7.1% aliphatic hydrocarbons	[166]
HDPE	AC, molecular sieves	HDPE to AC ratio from 100:60 to 100:330, heating power combinations (20%–40% of the maximum power of 900 W)	87.8	82.4% C7–C20	[160]
LDPE	Chestnut shell (CNSACC)	Carbonization temperature 850°C, catalytic temperature 550°C, CNSACC to LDPE ratio 1.0	—	95.9% aromatics	[167]
PP	Talk, ZSM-5	Pyrolysis temperature 620°C	48.9	73.5% gasoline fraction (45.0% aromatic hydrocarbons and 24.6% isomerized aliphatic)	[79]
Wastes PS	AC	Power 450 W, PS to AC ratio 10:0.5–10:3, temperature 500°C, time 5.5 minutes	80.7–93.0	26.8% benzene derivatives, 23.2% condensed ring aromatics, 8.4% alkenes	[168]
LDPE	AC, zeolite catalyst	Atmospheric pressure, temperature 300°C–550°C, time 45–90 minutes, plastic to absorber ratio 1:1, catalyst to plastic ratio 1:1, N_2 gas	28.1	19.0% hydrocarbons	[169]
Plastic waste from WEEE	Carbon, iron	Power 3 kW, in a nitrogen environment, plastic to absorber ratio 2:1, temperature 386°C–450°C, time 30–60 minutes, N_2 gas	76.6	117.7 mg/mL of styrene and 25.6 mg/mL of xylenes	[152]
HDPE, PP	Chopped tire, carbon	Power 1.2–6 kW, time 33–260 minutes, temperature 702–872 K, plastic to absorber ratio from 0.5:1 to 2.5:1	27.6–83.9	<88.4% aliphatic hydrocarbons, <20.32% aromatic hydrocarbons	[170]

(*Continued*)

TABLE 11.5 (*Continued*)
Comparison of Research Works on the Conversion of Various Types of Plastic into Hydrocarbon Fractions Using Microwave Pyrolysis and Co-pyrolysis

Type of Plastic/ Material	Type of Catalyst/ Absorber	Process Parameters	Pyrolysis Oil Yield (%)	Oil Composition	References
HDPE	Carbon, silicon carbide powder	Temperature 400°C–550°C, absorbent to HDPE ratio 10:1 and 3.3:1, power (3–5) kW, time (60–100) minutes, N_2 gas	34.7–73.4	C8–C35	[171]
PS, PP, LDPE	Iron	Power 500–2500 W, temperature 250°C–800°C, time 19 minutes	28.1–88.7	Aromatic and aliphatic hydrocarbons (C8–C16)	[172]
LDPE, lignin	HZSM-5, MgO, silicon carbide	Power of 750 W, pyrolysis temperature 450°C–600°C, lignin to LDPE ratio (only lignin, 3:1, 2:1, 1:1, 1:2 and only LDPE), MgO to HZSM-5 ratio (only MgO, 1:2, 1:1, 2:1 and only HZSM-5), feedstock to catalyst ratio (only feedstock, 1:2, 1:1 and 2:1), vacuum	22.8–38.5	~(30–85)% aromatic compounds, ~(15–55)% phenolic compounds	[173]
Waste cooking oil (WCO), waste polyolefins (WP)	AC	Power 800 W, WP to WCO ratio 1:2, 1:1.5, 1:1, 1.5:1, 2:1, time 20 minutes, temperature <450°C, N_2 gas, vacuum	24.0–62.0	50% C13–C24 hydrocarbons	[174]
Wheat straw, rice husk, expanded PS, waste PP	Graphite	Power of 600 W, biomass to plastic ratio 1:1, temperature 450°C, N_2 gas	51.5–64.9	50.5%–93.7% aromatic hydrocarbons, <27.1% cycloalkanes and alkanes	[175]

AC, activated carbon; MAHs, monocyclic aromatic hydrocarbons; WEEE, plastic fraction of waste from electric and electronic equipment.

11.4.4 Plasma Pyrolysis

The pyrolysis process can also be plasma assisted [176,177]. In plasma pyrolysis, the necessary energy for the process is provided by the thermal or electric current and electromagnetic radiation [178]. Compared to conventional and microwave pyrolysis, plasma pyrolysis is characterized by a higher energy density and temperature (that is better controlled), better reaction kinetics, greater energy efficiency and lower CO_2 gas emissions to the environment (not the combustion of conventional fuel during the process) [179,180]. Despite its many advantages, this technology also has disadvantages, such as a low technology readiness level (TRL) and high energy demand [124].

Depending on the gas temperature, plasma pyrolysis can be [181,182]:

a. High temperature (thermal), where electrons, ions and particles are in thermodynamic equilibrium and have the same temperature, at approx. $(10^6–10^8)$°C; this method is used to obtain mainly gaseous products;
b. Cold (non-thermal) electrons, ions, and particles are not in thermodynamic equilibrium and have different temperatures of approx. $(10^4–10^6)$°C, while the gas is at room temperature.

This process may be used for biomass [183,184] and municipal waste (including plastic [180,185] and rubber [186] wastes) conversion. Due to the very high energy demand, for economic reasons, this process is mainly used to neutralize hazardous medical waste at a larger scale [187–189]. Despite its drawbacks, this process is being studied. Due to its high temperature, during the processing of plastics by plasma pyrolysis, energy from these materials is mainly recovered in the form of synthesis gas [190,191]. This gas can be used to generate the necessary electricity to power plasma devices, which reduces its consumption [190,192]. The synthesis gas mainly consists of carbon monoxide and hydrogen and may be used for fuel production (such as methanol, diesel oil) [192,193] or hydrogen extraction [180].

Plasma pyrolysis technology is approx. 28%–60% more expensive compared to energy recovery technology by waste incineration [194]. For this reason, the available literature focuses on the search for and optimization of economically and energy-efficient technologies for the plasma pyrolysis of plastics, including the selection of the following systems: plasma generator (torch), reactor, plasma working gas [195]. Mohsenian et al. [196], carried out research on the processing of PW (PP, PE, PVC, acrylonitrile-butadiene-styrene (ABS)) into hydrogen, using the developed twin DC thermal plasma torch. It was found that the concentration of hydrogen in the resulting gaseous product ranged from 0.75% (PE) to 70.5% (ABS), while the concentration of C_xH_y ranged from 21.9% (PVC) to 46.3% (PE). The results are consistent with the results obtained by Mohsenian et al. [197], where a gas containing up to 75% of hydrogen in the process from PW (PP, PE, PVC, ABS) was obtained.

Gabbar et al. [198] carried out research on LDPE conversion using plasma pyrolysis at the temperature of 625°C–860°C. It was shown that the use of direct current thermal plasma circuit in the process allows for a liquid fraction with an efficiency of approx. 57% to be obtained (where the major product was butyl-benzene), and approx. 37% of the gaseous fraction. The process produces up to 6% of tar, and the obtained products are more stable compared to conventional pyrolysis.

Silvarrey [181] used cold plasma technology for the thermal decomposition of HDPE, PP, PET, and their mixtures. The tests were carried out with the use of various catalysts (zeolites, sulfated zirconia, Ni/Al_2O_3, char derived from the pyrolysis of biomass), in nitrogen atmosphere, at the temperature of 500°C–700°C. The cold plasma pyrolysis with thermal and catalytic pyrolysis was compared. It was found that, during thermal pyrolysis HDPE, carried out at the temperature of 500°C, the highest amount of liquid fraction was obtained (in which 92% were mainly C20+ and C12–C20 hydrocarbons). With the increase in the process temperature, as well as with the addition of the catalyst, the amount of the gaseous fraction significantly increased at the expense of the liquid fraction (approx. 40% of the liquid fraction at 700°C and approx. 25% with the catalyst at 600°C). Conducting the cold plasma pyrolysis reaction provided an indirect share of the liquid fraction compared to the thermal and catalytic processes. The plasma process can be applied not only to classical pyrolysis but also to hydrogenolysis. These studies were carried out by Gabbar et al. [199]. The thermal decomposition of HDPE by non-thermal plasma-assisted hydrogenolysis was studied in the presence of 5% catalyst (Pt/C and SAPO), at plasma power 60–90 W and hydrogen partial pressure 0–101 kPa. The results showed that it is possible to obtain light alkanes (mainly CH_4, C_2H_6, and C_3H_8) with 95% selectivity.

The technologies discussed above, used to convert PW to hydrocarbons, have various advantages and disadvantages (Table 11.6), and research is still being conducted to improve them.

TABLE 11.6
The Main Advantages and Disadvantages of Technologies Used to Convert PW to Hydrocarbons

Process	Drawback	Advantage
Conventional pyrolysis	• Very broad product range; • High temperature required; • High content of waxes.	• Lower sensitivity to contamination with other materials.
Microwave-assisted pyrolysis	• Low microwave absorption and conductivity of PW; • Necessity of using susceptors.	• More even heat distribution in liquids; • Heat runs throughout the entire volume of material.
Plasma-assisted pyrolysis	• Low TRL; • Very high energy demand; • High costs.	• High energy density; • Better reaction kinetics and greater energy efficiency than conventional pyrolysis.
Catalytical cracking	• Difficulty in recovering the catalyst after use; • Increased operating costs; • Quick deactivation of the catalyst due to the deposition of carbon matter and poisoning with elements (chlorine, sulfur).	• Reduction in the reaction temperature; • Promotion of decomposition reaction; • Improvement in the product quality.
Hydrogenolysis	• Difficulty in recovering the catalyst after use; • Increased operating costs; • High pressure.	• Very good quality of liquid product; • Reduction in the reaction temperature.

11.5 PATENTS REVIEW

Patents are also an important source of information on the methods and technologies used to convert PW into energy carriers. Patents often illustrate the latest technical and technological solutions within a given field of knowledge. In this review, the authors focused on patents from recent years related to the liquefaction of polyolefins and PS. The number of patents on this subject is quite significant and varies from year to year depending on interest. For example, in 2009, fewer than ten patents were granted related to polyolefin recycling, and, in 2011, 30 were granted [35]. The analysis of patents shows that this topic is still very popular among researchers and inventors. The frequent disadvantages of many solutions are the manual loading of the feed into the reactor or solutions that significantly impede the transition to a continuous process, which is often a condition for industrial implementation. Other disadvantages include poor heat transfer in the waste raw material itself, low conversion efficiencies, the incorporation of multiple steps, and the cost and contamination of consuming catalysts [200].

Often, patents state that the end product is fuel. As is well known, commercial fuels contain only insignificant amounts of unsaturated hydrocarbons, and the thermal decomposition products of polymers contain a lot of alkenes. Without the hydrogenation process, the use of such products as fuels is not possible. Therefore, it would be more correct to write that PW is converted into hydrocarbon fractions. Below, the authors describe a few selected patents.

The patent application [201] describes the multi-stage conversion of PW into hydrocarbon fractions. PW melting and decomposition processes are carried out separately. The molten material is pyrolyzed in an oxygen-free atmosphere in an agitated pyrolysis chamber at 390°C–410°C. The obtained pyrolysis gases are introduced into the contactor with condensation elements. In this contactor, the long-chain gas components are condensed and recycled for further pyrolysis. The condensation of long-chain hydrocarbons takes place on downwardly sloped plates for the run-off of the condensate. The plates include apertures to allow for the upward progression of pyrolysis gases.

The short-chain components leave the contactor in gaseous form (in the temperature ranging from 240°C to 280°C) and are distilled in the distillation column. The thermal decomposition process is carried out without the use of catalysts. This is one of the few patents that has been commercialized. Unfortunately, the patent does not state whether and how the hydrogenation of the multiple bonds of the resulting liquid product is carried out.

Another example of the catalytical conversion of PW waste to fuel is described in the patent application [202]. According to this document, the thermal decomposition of waste plastic is carried out in a temperature in the range from 350°C to 450°C and in a slight negative pressure in the range from 0.0010 psi (0.00007 bar) to 0.030 psi (0.00207 bar). The catalysts used in the process belong to the group of FCC catalysts, Y-zeolite and ZSM-5 zeolite. The thermal decomposition of the waste plastic is realized in the rotary kiln reactor with the helical baffles attached to the inner wall. The obtained gases are then subjected to four-stage sequential cooling to a temperature in the range from −5°C to −15°C. The final products are a gaseous fraction comprising C1 to C4 hydrocarbons and the liquid fraction (fuel). Due to the use of a mixture of the catalyst and materials as the feedstock, it is necessary to separate the catalyst from the char and its regeneration after the process. The use of a catalyst and the operations related to its recovery increase the costs of the entire technology. A comparison of the solutions and process parameters described in the patents is summarized in Table 11.7.

Another example of a patent describing the conversion of PW waste to fuel is by Kumar [200]. The liquefaction process is based on pyrolysis. The process is carried out in one stage in a column reactor. The shredded PW is mixed with additives in a ratio of 10:1. These additives are used to improve heat absorption and can include crude oil, synthetic oil, synthetic sludge, or synthetic heavy oil, which are recovered from the depolymerization plant after distillation. During the process, the temperature is regulated (from 90°C to 550°C) using a temperature controller. It is gradually increased and maintained at a certain level depending on the stage of the process. The product obtained in the final step is fractionated into different hydrocarbon groups by condensation in a condenser comprising a honeycomb structure.

In turn, the patent by Baker [203] concerns the conversion of plastics to hydrocarbons characteristic of diesel fuel. Before the process, the input material is subjected to a pre-treatment, which is fragmentation. The process includes melting the PW material, the pyrolytic decomposition of melted waste material, catalytic conversion of thermolysis products via cracking and reforming to hydrocarbons of appropriate chain length (from C8 to C25), as well as the distillation and condensation of the final product. Catalytic processes are carried out in a catalytic tower analogous to that used in the petrochemical industry – unsaturated bonds are hydrogenated.

Similarly to patent [201], patent application [204] has been commercialized. The described process and apparatus relate to the technology based on the pyrolysis of a mixture of plastic raw materials into petroleum products such as naphtha, distillate (e.g., diesel), and gas oil (e.g., heavy oil and wax). The process is carried out in a reactor comprising several successive and adjacent zones arranged along the horizontal axis of the reactor. These zones are controlled independently. There is a temperature gradient in the reaction zone of the reactor from the lower surface that contacts the raw material to the upper surface, which is provided with an exit port, and may vary widely, for example, from about 50°C to 450°C. The fed raw material moves along the horizontal axis of the reactor that is undergoing pyrolysis. The gaseous products that leave the reactor vessel are condensed and collected for further processing.

A different approach to the degradation of polyolefins is presented in patent [205]. The presented method enables the decomposition of polyolefins without a catalyst into hydrocarbons with chains no longer than those found in diesel fuel (boiling up to 360°C). An additional advantage of the technology is the process of catalytic hydrogenation and isomerization in the synthesis gas atmosphere under atmospheric pressure. The final products are typical fuel fractions. Unlike many other methods, the process is continuously carried out in a moving packed tower flow reactor. Polyolefin waste is mixed with the heated elements of the reactor packing. The surface of the elements is covered

TABLE 11.7
Comparison of Different Patented Processes for the Conversion of PW to Hydrocarbon Fractions

Substrate/Product	Conditions of Thermal Decomposition	Process Description	References
Waste hydrocarbon material such as plastics/fuel	Thermal decomposition process: pyrolysis; Process is carried out without the use of catalysts; Oxygen-free atmosphere (in nitrogen); Process temperature: 390°C–410°C; Agitated (with the use of a double helical agitator with a central screw) pyrolysis chambers.	• Process steps: • Melting the PW material 300°C–320°C; • Pyrolysis of molten material; • Condensation of long-chain hydrocarbons in a contactor and returning them to the pyrolysis chamber; • Distillation of pyrolysis gases (at temperature 260°C–280°C) from the contactor in a distillation column into fuels (temperature of a bottom section of the column 210°C–230°C; the top of the column approx. 100°C); • Heavy hydrocarbons fractions are sent to vacuum distillation.	[201]
Waste plastics PE, PP, PS, PET, EPDM	Continuous process; Thermal decomposition process: catalytic pyrolysis; Carrier gas: nitrogen, argon or the product gas; Process temperature: 350°C–450°C; Slight negative pressure in the range of 0.0010 psi (0.00007 bar) to 0.030 psi (0.00207 bar); Catalysts: belong to the group of FCC catalyst, Y-zeolite and ZSM-5 zeolite; The weight ratio of the feed to the catalyst: from 1:0.153 to 1:1.5.	• Process steps: • Mixing the waste plastic feed and a catalyst in a feed hopper; • Thermal decomposition of waste plastic in the rotary kiln reactor; • Four-stage sequential cooling; • The hydrocarbon vapors must pass through the cyclone in order to separate the residual catalyst and char from their stream; After the process, it is necessary to separate the catalyst from the char and its regeneration.	[201]
Plastic material/diesel fuel	Semi-continuous process; Thermal decomposition processes: pyrolysis (at 390°C–410°C), catalytic conversion (at 220°C or greater on catalytic plates); Pressure in pyrolysis chamber: 1.08 bar; Oxygen-free atmosphere (in nitrogen); Pyrolysis chamber provided with an agitator.	• Process steps: • Melting the PW material; • Pyrolysis chamber is pre-heated to between 170°C and 220°C while the liquefied feedstock fills the chamber to 80% capacity; • Pyrolysis of melted PW at the process temperature; material in the pyrolysis chamber is agitated; • Catalytic conversion (cracking and reforming) of gaseous products from pyrolysis to diesel fuel hydrocarbons; • Distillation and condensation of the final product.	[203]

(*Continued*)

TABLE 11.7 (*Continued*)
Comparison of Different Patented Processes for the Conversion of PW to Hydrocarbon Fractions

Substrate/Product	Conditions of Thermal Decomposition	Process Description	References

Plastic material (PP, PE), light and heavy distillates/fuels	Batch process;
	Thermal decomposition process: pyrolysis;
	Temperature from 90°C to 550°C;
	Oxygen-free atmosphere;
	Atmospheric pressure;
	Whole process time from 18 to 28 hours.
		• The process is one step;
		• PW is grounded into powder form or chips and loaded into the de-polymerization reactor (column thermal cracking reactor) with additives;
		• Melting and gasification of plastic material take place in the same reactor;
		• The reactor works in the temperature range from 90°C to 550°C;
		• Reactor temperature program:
		  • 90°C–140°C moisture removal;
		  • 140°C–220°C purging the reactor and removing oxygen;
		  • 220°C–290°C melting of the plastic material;
		  • 290°C–320°C initial gasification stage;
		  • 350°C for 3 hours – gasification;
		  • Increasing from 350°C to 400°C and holding for 8 hours;
		  • 400°C–440°C pet coke (byproduct) forming process;
		  • Increasing to 480°C–550°C and holding for 4 hours – plastic melt transforms into a gaseous form;
		• Fractionation.
			[200]

Plastic material/petroleum products	Thermal decomposition process: pyrolysis; Oxygen-free atmosphere; Temperature gradient (e.g., from about 50°C to 450°C) in the reaction zone from the lower surface that contacts the raw material to the upper surface provided with an exit port.	• Horizontal reactor, inclined at a certain angle (e.g., 5°–20°); • The reactor consists of several consecutive and adjacent zones (separated by a wall) arranged along the horizontal axis of the reactor; • Reactor vessel can include a variety of mechanical stirrers, e.g., helical screw; • The process involves pyrolysis of a feedstock and in situ reactions that produce solid inert residue, molten fluids and gases inside the reactor vessel; • PW material moves along the horizontal axis of the reactor undergoing pyrolysis; • The raw material and products can be transferred along successive zones of the reactor.	[204]

(*Continued*)

TABLE 11.7 (*Continued*)
Comparison of Different Patented Processes for the Conversion of PW to Hydrocarbon Fractions

Substrate/Product	Conditions of Thermal Decomposition	Process Description	References

Plastic material (PP, PE, PS)	Continuous process;
	Thermal decomposition process: reactive distillation;
	Syngas atmosphere;
	Temperature gradient in the chamber changes along its axis;
	The boiling point of the product vapors controlled by the temperature setting at the reactor outlet.
		• Moving packed tower flow reactor (filled with Raschig rings);
		• Process steps:
		  • Mixing cut plastics with filling and softening of the polymer (240°C–260°C);
		  • PW thermal cracking in a non-isothermal process – temperature of the lower part of the reactor 400°C–440°C, temperature at the outlet in the upper part of the reactor 200°C–360°C; vapors of hydrocarbon with the boiling point the set point at the reactor outlet, condense on the colder rings in the upper parts of the reactor. They stay in reactor and underwent cracking process as long as they reach right temperature;
		  • Catalytic hydrogenation and isomerization (under atmospheric pressure of syngas, temperature 320°C–360°C);
		  • Fractionation.
			[205]

Plastic material	Thermal decomposition process: pyrolysis with catalyst (activated bauxite);
	Oxygen-free atmosphere;
	Process temperature about 450°C;
	Pyrolysis time about 30 minutes;
	The boiling point of the product vapors controlled by the temperature setting in the scrubber (below 350°C).
		• Horizontal cylindrical reactor with heating jacket (tube in tube);
		• The reaction chamber comprises a plurality of stirrers mounted to a horizontal rotating shaft;
		• Process steps:
		  • Catalytic pyrolysis of PW;
		  • Fractionation;
		  • Purification by extraction processes.
			[206]

with a thin layer of plasticized polymer material, and this charge is introduced into the reaction chamber. The temperature in the chamber changes along its axis (highest at the bottom and lowest at the top). The process is carried out in the countercurrent flow of synthesis gas under atmospheric pressure.

In contrast, patent [206] describes a method of pyrolysis of plastics assisted by a catalyst (activated bauxite). In this solution, as by Handerek et al. [205], only hydrocarbon vapors with a certain maximum boiling point can leave the reactor space. The boiling point of the product vapors is controlled by a scrubber that is located at the outlet of the reactor. The scrubber is a packed scrubbing column with packing. In the scrubber, the reaction chamber gas product is brought into contact with a hydrocarbon scrubbing liquid (temperature below 350°C). Hydrocarbons with a boiling point above the scrubbing liquid condense and are returned to the reaction chamber with the liquid. Behind the scrubber, various devices enable the recovery of various components of the hydrocarbon gas product, such as a fractionation column to condense diesel range hydrocarbons, a condenser to condense petrol range hydrocarbons, and a gas compression and cooling device to condense liquid petroleum gas range hydrocarbons. A portion of the hydrocarbons from the fractionation column (diesel oil fraction) is used as the scrubbing liquid. The obtained raw fuels are then subjected to extraction processes to purify them. The patent does not describe whether and how the hydrogenation of the multiple bonds is carried out. The presented purification of products by extraction makes the process quite complex, which probably increases the costs.

The review shows that there are still ongoing works on the improvement in the technology and technological equipment for the recycling of plastics (PP, PE, PS) to fuels. This work is largely focused on the development of a continuous process, which greatly facilitates its implementation in industrial conditions. Much work is still based on simple pyrolysis, which leads to a poor-quality product with no chance of application to fuel channels. After distilling the fuel fraction from such a product, the obtained residue requires further processing. The use of a catalyst in the process makes it possible to obtain hydrocarbon fractions of better quality, but entails the costs of purchasing the catalyst, its separation after the process from the char and regeneration. The best solution seems to be to carry out the process of recycling heavy hydrocarbon vapors to the reactor (nonisothermal process or with the condensation of vapors of long-chain compounds). In this way, relatively light hydrocarbon fractions, boiling up to a certain temperature, are obtained. The light product, unfortunately, contains significant amounts (about 60%) [137] of unsaturated compounds and, despite the declarations of the inventors, according to the authors of this chapter, cannot be used as a fuel. It must be hydrogenated for such purposes.

Regardless of the method used to destroy the polyolefins, whether continuously, semi-continuously or batchwise, and with or without a catalyst, the by-products are the gaseous fraction of light hydrocarbons, which are most often used to heat the process (in some cases as a carrier gas), and the solid fraction (char). Char contains non-volatile substances during the process, such as inorganic compounds (fillers, pigments, catalyst residues), uncracked polymers, carbon as a product of dehydrogenation, and the products of secondary polycondensation reactions [207]. This can be used to produce activated carbon (adsorbents), additives to epoxy resins, fuel briquettes, cement additives, or additives in the production of epoxy composite [208,209]. It can also provide a source of various metal elements that can be recovered as a result of appropriate treatment [210,211]. The high degree of unsaturation of the liquid fraction components is responsible for the unpleasant, pungent smell and instability manifested by the darkening of the product and the formation of resin-like substances. This limits the use of this product as a fuel component. Therefore, further processing is necessary.

In conclusion, the advantages of the technologies presented in the presented patents include:

a. The use of various types of waste polyolefins, including mixed and contaminated;
b. Obtaining hydrocarbons with a small mass distribution and short carbon chains (up to C20–C22 – in the case of catalytic processes, the process with the recycling of heavy

hydrocarbon vapors or the process using reactive distillation), which is possible to use in the production of liquid fuel components;
c. The possibility of running the process continuously (in some cases);
d. Obtaining chemically stable fractions thanks to the application of a cheap hydrogenation process (in the case of one patent).

Despite the fact that, for many years, research was carried out on the conversion of waste polyolefins to fuels, these technologies are still imperfect and have disadvantages, including:

a. The manual loading of the feed into the reactor;
b. Conducting the process in batch mode;
c. Poor-quality product from the conventional pyrolysis process (e.g., high wax content, large dispersion of hydrocarbon masses);
d. No hydrogenation process that chemically stabilizes the product and allows it to be used as a fuel component;
e. High costs related to the use and regeneration of catalysts and product purification;
f. High complexity of some processes.

11.6 INDUSTRIAL EXAMPLES OF POLYOLEFIN WASTE CONVERSION

Despite the many developed and patented technologies used for converting waste polyolefins into hydrocarbons, few have been implemented. This results, among others, from the complexity of these processes, their long reaction times, high energy demand, or the poor quality of the obtained products [212]. A report from 2021, prepared by Closed Loop Partners [213], on plastic recycling technologies in the American and Canadian markets, indicates that only four of the recycling methods analyzed by the company have technologies with TRL of level 8: Plastic Energy, Enerkem, Brightmark and JEPLAN. The technology of the Japanese company JEPLAN is intended for the recycling of PET. Since the authors focused mainly on polyolefin waste in the chapter, this technology was not discussed in more detail.

In turn, Solis and Silveira [38] assessed the TRL of various technologies for the chemical recycling of waste plastics. They analyzed the following technologies: thermal cracking (conventional pyrolysis), plasma pyrolysis, microwave-assisted pyrolysis, catalytic cracking, hydrocracking, conventional gasification, plasma gasification, and pyrolysis with inline reforming. Their research shows that the highest degree of technological readiness is demonstrated by three technologies: pyrolysis, catalytic cracking, and gasification. Other technologies are not yet sufficiently ready and require further work. One of the reasons for the implementation of these technologies was the lower operating temperatures of the installation and their lower sensitivity to the quality of the raw material. High temperatures generate higher installation costs and higher costs to provide the necessary energy for the process. On the other hand, the presence of catalysts (e.g., in hydrocracking) is associated with greater sensitivity to charge contamination, particularly PVC admixtures. As reported by Takada and Bell [214], a high energy consumption, reaching 5%–20% of the calorific value of the polymer waste input, has become the main reason for the failure of the development of pyrolysis technologies on an industrial scale. However, start-ups are currently emerging that are trying to modernize and improve these technologies. Some of the pyrolysis-based industrial installations that are used to convert waste polyolefins to hydrocarbon fractions are still working and some are presented below.

In the technology of British company Plastic Energy [215], the input material is a mixture of PP, PE, and PS. Waste plastics do not require washing or separation into individual types. A Thermal Anaerobic Conversion (TAC Technology) has been patented [201] and its product is TACOIL, which can replace crude oil. The product is distilled into three fractions: raw diesel, light oil, and synthetic, gas components. Naphtha and diesel are sold to refineries, and gaseous products are used to power the installations. From 1 tonne of PW, about 850L of TACOIL is obtained. Plastic Energy has two operational plants, in Seville and Almeria, Spain, which have been operational since 2015 and 2017, each of which can process 5,000 tonnes of plastic annually [216].

In the technology of the Canadian company Enerkem, the PW is first converted into synthesis gas (syngas). The syngas is then transformed (in a catalyzed process) into fuels and chemical products [217]. Since the authors did not focus on waste conversion in the chapter, the technology was not discussed in more detail.

The American company Brightmark has partnered with RES Polyflow, an Ohio-based energy company that has patented [204] the processing of plastics. The companies are building a commercial-scale plastic recycling facility in Ashley, Indiana, with a capacity of 100,000 tonnes of PW per year [218]. The process is expected to be 93% efficient.

A slightly different technology was patented and implemented by BioCellection. The company undergoes conversion of the contaminated PE by an accelerated thermal oxidative decomposition (ATOD) process [219]. The process is performed catalytically in the presence of an oxidant (e.g., oxygen, nitric oxide, nitric acid(V)), at a temperature from 60°C to 200°C and for a period of 30 minutes to 30 hours. The product is a mixture of various high-value chemical compounds and contains, inter alia, succinic, oxalic, glutaric, adipic, pimelic, and azelaic acids, or their esters or salts. The resulting organic acids are intermediate chemicals that are used in the production of materials used, among others, in the clothing, electronics and automotive industries.

In 2018, the international company BASF launched the ChemCycling project. The aim of this project is PW consisting of, e.g., PE, PP, PS by pyrolysis to pyrolysis oil. The oil is then further processed, just like crude oil [220]. In the same year, the Austrian company OMV built a pilot plant at the Schwechat refinery, in which it implemented the proprietary ReOil® technology. This technology enabled the processing of waste plastics such as PE, PP, and PS into a pyrolytic shell. The plant has a processing capacity of up to 100 kg/h, which is equivalent to 100L of synthetic crude oil. Plastics are crushed, mixed with a high boiling solvent and heated in a reactor. The process is a two-stage process – the materials melt at 200°C, and the depolymerization process takes place at temperatures above 390°C. A solvent reduces the viscosity and improves the heat transfer. After distilling the solvent, the obtained product is further processed in the refinery [221,222].

Cynar Plc installed its first full-scale plant in County Laois, Ireland, in 2008. The company implements pyrolysis-based waste processing technology which converts a wide range of end-of-life plastics into hydrocarbons (similar to diesel oil). The plastic (chipped PE, PP, PS) is pyrolyzed at 370°C–420°C, and the pyrolysis gases are condensed in a two-stage condenser to produce a distillate [223]. The distillates are collected in tanks. The raw diesel is further refined in another distillation column, where it is distilled into the main three products: diesel, kerosene, and light oil. The 1-tonne PW is converted into 1,000L of fuels (700L of diesel, 200L of light oil, and 100L of kerosene). The process also produces about 60L of gaseous products, and the remaining char is about 5% [224]. The gaseous product is cleaned and used in the furnaces that heat the pyrolysis chambers.

Alphakat's KDV technology is a catalytic process that enables the conversion of waste plastics to diesel oil. The input to the process may also be waste oils or biomass. Catalytic depolymerization of the fragmented feed occurs in reactor turbines at temperatures from 270°C to 350°C under the influence of a catalyst in the process of oil suspension and circulation. The patented catalyst used in this process has a double function: it catalyzes the reaction, facilitating the formation of valuable products, and works as ion exchanger, retaining the hetero-atoms such as halogens, phosphorus, nitrogen and heavy metals from raw material, and transforming them into inorganic compounds. This process is conducted under low pressure (0.9 bar). The by-products are CO_2, water, light gas, and small amounts of solids [225]. The test facility is located in Eppendorf, Germany. At present, Alphakat is starting a project in Cyprus (Koshi Village Project), under which it will process wood biomass into biofuel [226].

11.7 CONCLUSIONS

The authors have provided an extensive review of the current state of the knowledge in the field of processing plastic liquefaction to fuel fractions as energy carriers. This review shows that thermo-catalytic processing technologies are the most common. The authors did not analyze the methods of biological conversion of plastics, because these methods are at a very early stage of research, and the possibility of their industrial use cannot be determined. As part of the review, the progress in research and development was taken into account, as well as the patent status of selected technologies and examples of industrial use of these technologies. The existing technologies for the processing of waste plastics generally focus on material and energy recovery. The material recycling processes do not cover all types of plastics and, as a result, lead to the degradation of the functional properties of the recycled waste material. This results in the generation of waste that cannot be further recycled. Direct energy recycling also does not cover all groups of PWs and has significant environmental burdens related to the nature of the technological processes. In the field of processing waste plastics into liquid and gaseous energy carriers, the research into pyrolysis processes is well-developed. The analysis of the state of the research into and application of these processes showed their development. However, the research mainly focused on the efficiency and effectiveness of these processes.

On the basis of the review of modern technologies for the liquefaction of waste plastics and their mixtures to liquid fuel fractions, it can be concluded that most existing solutions require further improvement, in terms of both economics and technology. These imperfections result in stagnation of the development of industrial technologies for the chemical recycling of PW. In most of the analyzed studies, no environmental impact assessment of these processes, especially the LCA, was found. Depolymerization processes additionally require hydrogenation from external sources, which brings an additional environmental burden. For this reason, according to the authors, technologies where chemical stabilization of the obtained products is carried out in the process of decomposition should be regarded as promising. Such technologies, despite the need for the refinement of process conditions (pressure, catalyst recovery), include tandem hydrogenolysis/aromatization, where, during aromatization, hydrogen is generated and used in the hydrogenolysis process. The technology with the greatest potential for commercial development, in which the problems of high pressure and catalyst recovery from the reaction mixture have been eliminated, is the thermolysis of plastics, discussed by Matuszewska et al. [137]. In this solution, the saturation of multiple bonds is conducted under atmosphere pressure, with hydrogen from the synthesis gas, using the catalytic effect of carbon monoxide. This technology results in fuel fractions which, when added to conventional fuels, yield products that meet the standards for motor fuels (EN 590 and EN 228). These technologies are part of a new direction called petrosynthesis, as defined and described by Patrick [227]. Petrosynthesis is the latest area of research aimed at leading to a closed production cycle and use of industrial waste. This mainly focuses on carbon dioxide, but the thermolysis of PW could be an element of petrosynthesis processes, leading to the production of saturated hydrocarbon fractions for various applications, forming part of the CE.

ACKNOWLEDGMENT

This book chapter is an expanded and revised version of the original article published in April 2022 (Matuszewska, A., Owczuk, M., Biernat, K. Current trends in waste plastics' liquefaction into fuel fraction: A review. *Energies* 2022, 15, 2719. https://doi.org/10.3390/en15082719).

GLOSSARY

Circular economy: the alternative model of the economy, which is based on closing the product's life cycle, in which resources are used in a rational way, the negative impact of manufactured products on the environment is limited, and waste is used as valuable raw materials for the production of new products.

Hydrocarbons: organic chemical compounds, made up of only carbon and hydrogen.

Hydrocracking: the process of thermal decomposition of chemical compounds into smaller molecules, under high pressure and in the presence of a catalyst and hydrogen.

Hydrogenation: the chemical reaction of unsaturated bond breaking and hydrogen attachment in the presence of a catalyst.

Liquefaction: the process of thermal or chemical conversion of solid wastes into liquid fraction (oil), which can be a source of chemicals or energy carriers.

Plastic wastes: wastes made of plastics, generated by households, as well as manufacturing or industrial processes.

Plastics: materials where the main components are polymers, which also contain additives modifying the properties of the final material (e.g., dyes, stabilizers, flame retardants, foaming agents, fillers).

Polymer: the chemical compound consisting of very large molecules, composed of many repeating subunits called mers.

Polyolefins: polymers containing only carbon and hydrogen, in which there are long carbon chains, which form the basic backbone of the chains of the polymers themselves.

Post-consumer plastic wastes: wastes from households, restaurants, catering establishments, etc., which contain different mixed plastics, with unknown composition, and contaminated with organic fractions or other materials (paper, metal, textiles).

Post-industrial plastic wastes: wastes usually clean, free from organic or other contamination, and often mono-material or its composition is well known, origin from industrial processes.

Pyrolysis: the process of thermal decomposition of solid wastes into liquid, gaseous and carbonized products, under anaerobic atmosphere, with the proportion of each fraction depending on the conditions under which the process is carried out.

Recycling: a method aimed at recovering and reusing waste with the least possible material and energy expenditure.

Thermochemical processes: Thermochemical processes involve the study of heat energy changes associated with chemical reactions and phase changes, such as melting and boiling. Some key fundamentals are: (a) Heat Energy and Reactions: Thermochemistry focuses on the energy exchange between a system and its surroundings in the form of heat. During a chemical reaction, energy can be either released (exothermic) or absorbed (endothermic); (b) Enthalpy: This is a measure of the total energy of a thermodynamic system. It includes internal energy and the product of pressure and volume. The change in enthalpy during a reaction at constant pressure is a crucial concept in thermochemistry; (c) Calorimetry: This is the measurement of heat flow in a chemical reaction. Calorimeters are devices used to measure the amount of heat involved in a chemical or physical process; (d) Thermochemical Equations: These are chemical equations that include the enthalpy change; and (e) Hess's Law: This principle states that the total enthalpy change of a reaction is the same, no matter how many steps the reaction is carried out in. It allows the calculation of enthalpy changes for complex reactions by breaking them down into simpler steps.

REFERENCES

1. World Energy Outlook 2021. Report of International Energy Agency. October 2021. Available online: https://iea.blob.core.windows.net/assets/4ed140c1-c3f3-4fd9-acae-789a4e14a23c/WorldEnergyOutlook2021.pdf (accessed on 10 June 2022).
2. Short–Term Energy Outlook 2021. Report of U.S. Energy Information Administration. December 2021. Available online: https://www.eia.gov/outlooks/steo/pdf/steo_full.pdf (accessed on 10 June 2022).
3. Umar, M., Farid, S., Naeem, M.A. Time–frequency connectedness among clean–energy stocks and fossil fuel markets: Comparison between financial, oil and pandemic crisis. *Energy* 2022, 240, 122702.
4. Tang, C., Aruga, K. Relationships among the fossil fuel and financial markets during the COVID–19 pandemic: Evidence from Bayesian DCC–MGARCH models. *Sustainability* 2022, 14, 51.

5. Strielkowski, W., Civín, L., Tarkhanova, E., Tvaronavičienė, M., Petrenko, Y. Renewable energy in the sustainable development of electrical power sector: A review. *Energies* 2021, 14, 8240.
6. Pisciotta, M., Pilorgé, H., Feldmann, J., Jacobson, R., Davids, J., Swett, S., Sasso, Z., Wilcox, J. Current state of industrial heating and opportunities for decarbonization. *Progress in Energy and Combustion Science* 2022, 91, 100982.
7. Special Report. Global Warming of 1.5°C. 2018. Available online: https://www.ipcc.ch/sr15/ (accessed on 10 June 2022).
8. Shooshtarian, S., Maqsood, T., Caldera, S., Ryley, T. Transformation towards a circular economy in the Australian construction and demolition waste management system. *Sustainable Production and Consumption* 2022, 30, 89–106.
9. Sikdar, S. Circular economy: Is there anything new in this concept? *Clean Technologies and Environmental Policy* 2019, 21, 1173–1175.
10. Dittenber, D.B., GangaRao, H.V.S. Critical review of recent publications on use of natural composites in infrastructure. *Composites Part A: Applied Science and Manufacturing* 2012, 43, 1419–1429.
11. Plastics—The Facts 2021. Report Plastics Europe. The Association of Plastics Manufacturers in Europe. Available online: https://plasticseurope.org/knowledge-hub/plastics-the-facts-2021/ (accessed on 10 June 2022).
12. Plastic Atlas 2019. Heinrich Böll Foundation and Break Free From Plastic, Berlin 2019, Germany. Available online https://www.boell.de/sites/default/files/2020-01/Plastic%20Atlas%202019%202nd%20Edition.pdf (accessed on 01 October 2025).
13. Plastic Europe. Reports from 2000-2022, available online: https://plasticseurope.org/pl/ (accessed on 01 October 2025).
14. Czajka, K. 2005. *Poliolefiny*. WNT, Warszawa, 2005.
15. Pielichowski, J., Puszyński, A. 1998. *Technologia tworzyw sztucznych*. WNT, Warszawa.
16. Migdał, A.R., Kawalec, A., Kędziora, A., Rejewski, P., Śmigiera, E. Odzysk energetyczny materiałów odpadowych z tworzyw sztucznych. *CHEMIK* 2014, 68(12), 1056–1073.
17. German, K. Wybrane zagadnienia pirolizy odpadów tworzyw polimerowych. *Polimery* 2010, 55(5), 351–357.
18. International Chemical Safety Cards (ICSCs). Available online https://chemicalsafety.ilo.org/dyn/icsc/showcard.display?p_lang=pl&p_card_id=1043 (accessed on 01 October 2025).
19. International Chemical Safety Cards (ICSCs). Available online https://chemicalsafety.ilo.org/dyn/icsc/showcard.display?p_lang=pl&p_card_id=1487&p_version=2 (accessed on 01 October 2025).
20. Thakur, V.K., Thakur, M.K. Sustainable Polymers: A Perspective to the Future. In *Handbook of Sustainable Polymers. Structure and Chemistry*; Thakur, V.K., Thakur, M.K., Eds.; Pan Stanford Publishing: Singapore, Singapore, 2016; pp. 1–18.
21. Ncube, L.K., Ude, A.U., Ogunmuyiwa, E.N., Zulkifli, R., Beas, I.N. An overview of plastic waste generation and management in food packaging industries. *Recycling* 2021, 6, 6010012.
22. Jalaluddin, M. Use of plastic waste in civil constructions and innovative decorative material (eco-friendly). *MedCrave Online J. Civ. Eng.* 2017, 3, 00082.
23. Parker, L. Global treaty to regulate plastic pollution gains momentum. National Geografic, June, 2021. Available online: https://www.nationalgeographic.com/environment/article/global-treaty-to-regulate-plastic-pollution-gains-momentum (accessed on 10 June 2022).
24. Manžuch, Z., Akelytė, R., Camboni, M., Carlander, D. Chemical recycling of polymeric materials from waste in the circular economy. RPA Europe Final Report, August 2021. Available online: https://www.echa.europa.eu/documents/10162/1459379/chem_recycling_final_report_en.pdf/887c4182-8327-e197-0bc4-17a5d608de6e?t=1636701265520 (accessed on 10 June 2022).
25. Plastics—The Facts 2020. Report Plastics Europe. The Association of Plastics Manufacturers in Europe. Available online: https://plasticseurope.org/wp-content/uploads/2021/09/Plastics_the_facts-WEB-2020_versionJun21_final.pdf (accessed on 10 June 2022).
26. Menyuka, N.N., Sibanda, M., Bob, U. Perceptions of the challenges and opportunities of utilising organic waste through urban agriculture in the Durban South Basin. *Int. J. Environ. Res. Public Health* 2020, 17, 1158.
27. Jha, K.K., Kannan, T.T.M. Recycling of plastic waste into fuel by pyrolysis—A review. *Mater. Today Proc.* 2021, 37, 3718–3720.
28. Kang, M.J., Yu, H.J., Jegal, J., Kim, H.S., Cha, H.G. Depolymerization of PET into terephthalic acid in neutral media catalyzed by the ZSM-5 acidic catalyst. *Chem. Eng. J.* 2020, 398, 125655.
29. Balema, V.P., Hlova, I.Z., Carnahan, S.L., Seyedi, M., Dolotko, O., Rossini, A.J., Luzinov, I. Depolymerization of polystyrene under ambient conditions. *New J. Chem.* 2021, 45, 2935–2938.

30. dos Santos, P.B., da Silva Ribeiro, H.J., Ferreira, A.C., Ferreira, C.C., Bernar, L.P., da Costa Assunção, F.P., de Castro, D.A.R., Santos, M.C., Duvoisin, S., Borges, L.E.P., Machado, N.T. Process analysis of PMMA-based dental resins residues depolymerization: Optimization of reaction time and temperature. *Energies* 2022, 15, 91.
31. Lahtela, V., Hyvärinen, M., Kärki, T. Composition of plastic fractions in waste streams: Toward more efficient recycling and utilization. *Polymers* 2019, 11, 69.
32. Jubinville, D., Esmizadeh, E., Saikrishnan, S., Tzoganakis, C., Mekonnen, T. A comprehensive review of global production and recycling methods of polyolefin (PO) based products and their post–recycling applications. *Sustain. Mater. Technol.* 2020, 25, e00188.
33. Shen, L., Worrell, E. Plastic Recycling. In *Handbook of Recycling. State-of-the-art for Practitioners, Analysts, and Scientists*; Worrell, E., Reuter, M.A., Eds.; Elsevier: Amsterdam, Boston, Heidelberg, London, New York, Oxford, Paris, San Diego, San Francisco, Sydney, Tokyo, 2014; pp. 179–190.
34. Singh, N., Hui, D., Singh, R., Ahuja, I.P.S., Feo, L., Fraternali, F. Recycling of plastic solid waste: A state of art review and future applications. *Composites Part B: Engineering* 2017, 115, 409–422.
35. Ignatyev, I.A., Thielemans, W., Vander Beke, B. Recycling of polymers: A review. *ChemSusChem* 2014, 7, 1579–1593.
36. Schwabl, D., Bauer, M., Lehner, M. Advancing plastic recycling by wet–mechanical processing of mixed waste fractions. *Processes* 2021, 9, 493.
37. Ragaert, K., Huysveld, S., Vyncke, G., Hubo, S., Veelaert, L., Dewulf, J., Du Bois, E. Design from recycling: A complex mixed plastic waste case study. *Resour. Conserv. Recycl.* 2020, 155, 104646.
38. Solis, M., Silveira, S. Technologies for chemical recycling of household plastics—A technical review and TRL assessment. *Waste Manag.* 2020, 105, 128–138.
39. Li, N., Liu, H.X., Cheng, Z.N., Yan, B.B., Chen, G.Y., Wang, S.B. Conversion of plastic waste into fuels: A critical review. *J. Hazard. Mater.* 2022, 424, 127460.
40. Palos, R., Gutiérrez, A., Vela, F.J., Maña, J.A., Hita, I., Asueta, A., Arnaiz, S., Arandes, J.M., Bilbao, J. Assessing the potential of the recycled plastic slow pyrolysis for the production of streams attractive for refineries. *J. Anal. Appl. Pyrolysis* 2019, 142, 104668.
41. Prajapati, R., Kohli, K., Maity, S.K., Sharma, B.K. Potential chemicals from plastic wastes. *Molecules* 2021, 26, 3175.
42. Gebre, S.H., Sendeku, M.G., Bahri, M. Recent trends in the pyrolysis of non–degradable waste plastics. *ChemistryOpen* 2021, 10, 1202–1226.
43. Ma, H., Cao, Y., Lu, X., Ding, Z., Zhou, W. Review of typical municipal solid waste disposal status and energy technology. *Energy Procedia* 2016, 88, 589–594.
44. Arandes, J.M., Torre, I., Castaño, P., Olazar, M., Bilbao, J. Catalytic cracking of waxes produced by the fast pyrolysis of polyolefins. *Energy Fuels* 2007, 21, 561–569.
45. Rodríguez, E., Gutiérrez, A., Palos, R., Vela, F.J., Arandes, J.M., Bilbao, J. Fuel production by cracking of polyolefins pyrolysis waxes under fluid catalytic cracking (FCC) operating conditions. *Waste Manag.* 2019, 93, 162–172.
46. Fivga, A., Dimitriou, I. Pyrolysis of plastic waste for production of heavy fuel substitute: A techno–economic assessment. *Energy* 2018, 149, 865–874.
47. Fekhar, B., Zsinka, V., Miskolczi, N. Value added hydrocarbons obtained by pyrolysis of contaminated waste plastics in horizontal tubular reactor: In situ upgrading upgrading of the products by chlorine capture. *J. Clean. Prod.* 2019, 241, 118166.
48. Kaminsky, W. Chemical recycling of plastics by fluidized bed pyrolysis. *Fuel Commun.* 2021, 8, 100023.
49. Syamsiro, M., Saptoadi, H., Norsujianto, T., Noviasri, P., Cheng, S., Alimuddin, Z., Yoshikawa, K. Fuel oil production from municipal plastic wastes in sequential pyrolysis and catalytic reforming reactors. *Energy Procedia* 2014, 47, 180–188.
50. Bezergianni, S., Dimitriadis, A., Faussone, G.C., Karonis, D. Alternative diesel from waste plastics. *Energies* 2017, 10, 1750.
51. Nanda, S., Berruti, F. Thermochemical conversion of plastic waste to fuels: A review. *Environ. Chem. Lett.* 2021, 19, 123–148.
52. Marcilla, A., Beltrán, M.I., Navarro, R. Thermal and catalytic pyrolysis of polyethylene over HZSM5 and HUSY zeolites in a batch reactor under dynamic conditions. *Appl. Catal. B Environ.* 2009, 86, 78–86.
53. Fahim, I., Mohsen, O., ElKayaly, D. Production of fuel from plastic: A feasible business. *Polymers* 2021, 13, 915.
54. Sharuddin, S.D.A., Abnisa, F., Daud, W.M.A.W., Aroua, M.K. A review on pyrolysis of plastic wastes. *Energy Convers. Manag.* 2016, 115, 308–326.

55. Papari, S., Bamdad, H., Berruti, F. Pyrolytic conversion of plastic waste to value–added products and fuels: A review. *Materials* 2021, 14, 2586.
56. Papari, S., Hawboldt, K. Development and validation of a process model to describepyrolysis of forestry residues in an auger reactor. *Energy&Fuels* 2017, 31, 10833–10841.
57. Suresh, A., Alagusundaram, A., Kumar, P.S., Vo, D.V.N., Christopher, F.C., Balaji, B., Viswanathan, V., Sankar, S. Microwave pyrolysis of coal, biomass and plastic waste: A review. *Environ. Chem. Lett.* 2021, 19, 3609–3629.
58. Miandad, R., Nizami, A.S., Rehan, M., Barakat, M.A., Khan, M.I., Mustafa, A., Ismail, I.M.I., Murphy, J.D. Influence of temperature and reaction time on the conversion of polystyrene waste to pyrolysis liquid oil. *Waste Manag.* 2016, 58, 250–259.
59. Diaz–Silvarrey, L.S., Zhang, K., Phan, A.N. Monomer recovery through advanced pyrolysis of waste high density polyethylene (HDPE). *Green Chem.* 2018, 20, 1813–1823.
60. Miandad, R., Barakat, M.A., Aburiazaiza, A.S., Rehan, M., Ismail, I.M.I., Nizami, A.S. Effect of plastic waste types on pyrolysis liquid oil. *Int. Biodeterior. Biodegradation* 2017, 119, 239–252.
61. Grabda, M., Oleszek, S., Shibata, E., Nakamura, T. Study on simultaneous recycling of EAF dust and plastic waste containing TBBPA. *J. Hazard. Mater.* 2014, 278, 97–106.
62. Fekhar, B., Gombor, L., Miskolczi, N. Pyrolysis of chlorine contaminated municipal plastic waste: In–situ upgrading of pyrolysis oils by Ni/ZSM–5, Ni/SAPO–11 red mud and $Ca(OH)_2$ containing catalysts. *J. Energy Inst.* 2019, 92, 1270–1283.
63. Hahladakis, J.N., Velis, C.A., Weber, R., Iacovidou, E., Purnell, P. An overview of chemical additives present in plastics: Migration, release, fate and environmental impact during their use, disposal and recycling. *J. Hazard. Mater.* 2018, 344, 179–199.
64. Oleszek, S., Kumagai, S., Grabda, M., Shiota, K., Yoshioka, T., Takaoka, M. Mitigation of bromine–containing products during pyrolysis of polycarbonate-based tetrabromobisphenol A in the presence of copper(I) oxide. *J. Hazard. Mater.* 2021, 409, 124972.
65. Ma, C., Yu, J., Wang, B., Song, Z., Xiang, J., Hu, S., Su, S., Sun, L. Chemical recycling of brominated flame retarded plastics from e–waste for clean fuels production: A review. *Renew. Sustain. Energy Rev.* 2016, 61, 433–450.
66. Terakado, O., Ohhashi, R., Hirasawa, M. Bromine fixation by metal oxide in pyrolysis of printed circuit board containing brominated flame retardant. *J. Anal. Appl. Pyrolysis* 2013, 103, 216–221.
67. Jung, S.H., Kim, S.J., Kim, J.S. Thermal degradation of acrylonitrile–butadiene–styrene (ABS) containing flame retardants using a fluidized bed reactor: The effects of Ca-based additives on halogen removal. *Fuel Process Technol.* 2012, 96, 265–270.
68. Brebu, M., Bhaskar, T., Murai, K., Muto, A., Sakata, Y., Uddin, M.A. Removal of nitrogen, bromine, and chlorine from PP/PE/PS/PVC/ABS–Br pyrolysis liquid products using Fe- and Ca-based catalysts. *Polym. Degrad. Stab.* 2005, 87, 255–230.
69. Charitopoulou, M.A., Kalogiannis, K.G., Lappas, A.A., Achilias, D.S. Novel trends in the thermo–chemical recycling of plastics from WEEE containing brominated flame retardants. *Environ. Sci. Pollut. Res.* 2021, 28, 59190–59213.
70. Al-Salem, S.M., Antelava, A., Constantinou, A., Manos, G., Dutta, A. A review on thermal and catalytic pyrolysis of plastic solid waste (PSW). *J. Environ. Manag.* 2017, 197, 177–198.
71. Wong, S.L., Ngadi, N., Abdullah, T.A.T., Inuwa, I.M. Current state and future prospects of plastic waste as source of fuel: A review. *Renew. Sustain. Energy Rev.* 2015, 50, 1167–1180.
72. Lewandowski, W.M., Januszewicz, K., Kosakowski, W. Efficiency and proportions of waste tyre pyrolysis products depending on the reactor type—A review. *J. Anal. Appl. Pyrolysis* 2019, 140, 25–53.
73. Uddin, M.N., Techato, K., Taweekun, J., Rahman, M.M., Rasul, M.G., Mahlia, T.M.I., Ashrafur, S.M. An overview of recent developments in biomass pyrolysis technologies. *Energies* 2018, 11, 3115.
74. Rasul, S.B., Som, U., Hossain, S., Rahman, W. Liquid fuel oil produced from plastic based medical wastes by thermal cracking. *Sci. Rep.* 2021, 11, 17048.
75. Lee, K.H., Shin, D.H. Characteristics of liquid product from the pyrolysis of waste plastic mixture at low and high temperatures: Influence of lapse time of reaction. *Waste Manag.* 2007, 27, 168–176.
76. Pan, R., Martins, M.F., Debenest, G. Pyrolysis of waste polyethylene in a semi–batch reactor to produce liquid fuel: Optimization of operating conditions. *Energy Convers. Manag.* 2021, 237, 114114.
77. Torres, D., Jiang, Y., Sanchez Monsalve, D.A., Leeke, G.A. Chlorine removal from the pyrolysis of urban polyolefinic waste in a semi–batch reactor. *J. Environ. Chem. Eng.* 2021, 9, 104920.
78. Al-Salem, S.M., Yang, Y., Wang, J., Leeke, G.A. Pyro–oil and wax recovery from reclaimed plastic waste in a continuous auger pyrolysis reactor. *Energies* 2020, 13, 2040.

79. Zhou, N., Dai, L., Lv, Y., Li, H., Deng, W., Guo, F., Chen, P., Lei, H., Ruan, R. Catalytic pyrolysis of plastic wastes in a continuous microwave assisted pyrolysis system for fuel production. *Chem. Eng. J.* 2021, 418, 129412.
80. Tulashie, S.K., Boadu, E.K., Dapaah, S. Plastic waste to fuel via pyrolysis: A key way to solving the severe plastic waste problem in Ghana. *Therm. Sci. Eng. Prog.* 2019, 11, 417–424.
81. Papuga, S.V., Gvero, P.M., Vukić, L.M. Temperature and time influence on the waste plastics pyrolysis in the fixed bed reactor. *Therm. Sci.* 2016, 20, 731–741.
82. Li, H., Mašek, O., Harper, A., Ocone, R. Kinetic study of pyrolysis of high–density polyethylene (HDPE) waste at different bed thickness in a fixed bed reactor. *Can. J. Chem. Eng.* 2021, 99, 1733–1744.
83. Milne, B.J., Behie, L.A., Berruti, F. Recycling of waste plastics by ultrapyrolysis using an internally circulating fluidized bed reactor. *J. Anal. Appl. Pyrolysis* 1999, 51, 157–166.
84. Jung, S.H., Cho, M.H., Kang, B.S., Kim, J.S. Pyrolysis of a fraction of waste polypropylene and polyethylene for the recovery of BTX aromatics using a fluidized bed reactor. *Fuel Process. Technol.* 2010, 91, 277–284.
85. Mastral, J.F., Berrueco, C., Ceamanos, J. Modelling of the pyrolysis of high density polyethylene: Product distribution in a fluidized bed reactor. *J. Anal. Appl. Pyrolysis* 2007, 79, 313–322.
86. Arabiourrutia, M., Elordi, G., Lopez, G., Borsella, E., Bilbao, J., Olazar, M. Characterization of the waxes obtained by the pyrolysis of polyolefin plastics in a conical spouted bed reactor. *J. Anal. Appl. Pyrolysis* 2012, 94, 230–237.
87. Orozco, S., Alvarez, J., Lopez, G., Artetxe, M., Bilbao, J., Olazar, M. Pyrolysis of plastic wastes in a fountain confined conical spouted bed reactor: Determination of stable operating conditions. *Energy Convers. Manag.* 2021, 229, 113768.
88. Artetxe, M., Lopez, G., Elordi, G., Amutio, M., Bilbao, J., Olazar, M. Production of light olefins from polyethylene in a two–step process: Pyrolysis in a conical spouted bed and downstream high-temperature thermal cracking. *Ind. Eng. Chem. Res.* 2012, 51, 13915–13923.
89. Zhang, X., Lei, H., Zhu, L., Qian, M., Yadavalli, G., Wu, J., Chen, S. From plastics to jet fuel range alkanes via combined catalytic conversions. *Fuel* 2017, 188, 28–38.
90. de Marco, I., Caballero, B.M., López, A., Laresgoiti, M.F., Torres, A., Chomón, M.J. Pyrolysis of the rejects of a waste packaging separation and classification plant. *J. Anal. Appl. Pyrolysis* 2009, 85, 384–391.
91. Dwivedi, P., Mishra, P.K., Mondal, M.K., Srivastava, N. Non-biodegradable polymeric waste pyrolysis for energy recovery. *Heliyon* 2019, 5, e02198.
92. Xue, Y., Johnston, P., Bai, X. Effect of catalyst contact mode and gas atmosphere during catalytic pyrolysis of waste plastics. *Energy Convers. Manag.* 2017, 142, 441–451.
93. Colantonio, S., Cafiero, L., De Angelis, D., Ippolito, N.M., Tuffi, R., Cipriotі, S.V. Thermal and catalytic pyrolysis of a synthetic mixture representative of packaging plastics residue. *Front. Chem. Sci. Eng.* 2020, 14, 288–303.
94. Milato, J.V., França, R.J., Rocha, A.S., Calderari, M.R.C.M. Catalytic co-pyrolysis of oil sludge with HDPE to obtain paraffinic products over HUSY zeolites prepared by dealumination and desilication. *J. Anal. Appl. Pyrolysis* 2020, 151, 104928.
95. Zeaiter, J. A process study on the pyrolysis of waste polyethylene. *Fuel* 2014, 133, 276–282.
96. Kasar, P., Sharma, D.K., Ahmaruzzaman, M. Thermal and catalytic decomposition of waste plastics and its co-processing with petroleum residue through pyrolysis process. *J. Clean. Prod.* 2020, 265, 121639.
97. Orozco, S., Artetxe, M., Lopez, G., Suarez, M., Bilbao, J., Olazar, M. Conversion of HDPE into value products by fast pyrolysis using FCC spent catalysts in a fountain confined conical spouted bed reactor. *ChemSusChem* 2021, 14, 4291–4300.
98. de la Puente, G., Klocker, C., Sedran, U. Conversion of waste plastics into fuels: Recycling polyethylene in FCC. *Appl. Catal. B Environ.* 2002, 36, 279–285.
99. Vollmer, I., Jenks, M.J.F., Gonzalez, R.M., Meirer, F., Weckhuysen, B.M. Plastic waste conversion over a refinery waste catalyst. *Angew. Chem. Int. Ed.* 2021, 60, 16101–16108.
100. Gopinath, S., Devan, P.K., Pitchandi, K. Production of pyrolytic oil from ULDP plastics using silica-alumina catalyst and used as fuel for DI diesel engine. *RSC Adv.* 2020, 10, 37266–37279.
101. Kunwar, B., Cheng, H.N., Chandrashekaran, S.R., Sharma, B.K. Plastics to fuel: A review. *Renew. Sustain. Energy Rev.* 2016, 54, 421–428.
102. Aguado, J., Serrano, D.P. Escola, J.M. Fuels from waste plastics by thermal and catalytic processes: A review. *Ind. Eng. Chem. Res.* 2008, 47, 7982–7992.

103. Kassargy, C., Awad, S., Burnens, G., Kahine, K., Tazerout, M. Experimental study of catalytic pyrolysis of polyethylene and polypropylene over USY zeolite and separation to gasoline and diesel-like fuels. *J. Anal. Appl. Pyrolysis* 2017, 127, 31–37.
104. Dai, L., Zhou, N., Lv, Y., Cobb, K., Cheng, Y., Wang, Y., Liu, Y., Chen, P., Zou, R., Lei, H., Ruan, R. Pyrolysis-catalysis for waste polyolefin conversion into low aromatic naphtha. *Energy Convers. Manag.* 2021, 245, 114578.
105. Rahimi, S., Rostamizadeh, M. Novel Fe/B-ZSM-5 nanocatalyst development for catalytic cracking of plastic to valuable products. *J. Taiwan Inst. Chem. Eng.* 2021, 118, 131–139.
106. Akubo, K., Nahil, M.A., Williams, P.T. Aromatic fuel oils produced from the pyrolysis-catalysis of polyethylene plastic with metal-impregnated zeolite catalysts. *J. Energy* 2019, 92, 195–202.
107. Anene, A.F., Fredriksen, S.B., Sætre, K.A., Tokheim, L.A. Experimental study of thermal and catalytic pyrolysis of plastic waste components. *Sustainability* 2018, 10, 3979.
108. Damodharan, D., Sathiyagnanam, A.P., Rana, D., Kumar, B.R., Saravanan, S. Combined influence of injection timing and EGR on combustion, performance and emissions of DI diesel engine fueled with neat waste plastic oil. *Energy Convers. Manag.* 2018, 161, 294–305.
109. Mangesh, V.L., Padmanabhan, S., Tamizhdurai, P., Rameshe, A. Experimental investigation to identify the type of waste plastic pyrolysis oil suitable for conversion to diesel engine fuel. *J. Clean. Prod.* 2020, 246, 119066.
110. Quesada, L., Calero, M., Martín-Lara, M.A., Pérez, A., Blázquez, G. Characterization of fuel produced by pyrolysis of plastic film obtained of municipal solid waste. *Energy* 2019, 186, 115874.
111. Saha, D., Sinha, A., Roy, B. Critical insights into the effects of plastic pyrolysis oil on emission and performance characteristics of CI engine. *Environ. Sci. Pollut. Res.* 2021, 28, 44598–44621.
112. Damodharan, D., Sathiyagnanam, A.P., Kumar, B.R., Ganesh, K.C. Cleaner emissions from a DI diesel engine fueled with waste plastic oil derived from municipal solid waste under the influence of n-pentanol addition, cold EGR, and injection timing. *Environ. Sci. Pollut. Res.* 2018, 25, 13611–13625.
113. Peer, M.S., Peer, M.N. Experimental investigation on engine characteristics fueled with waste HDPE oil and study on NO_x emission variation using thermal imager. *Environ. Sci. Pollut. Res.* 2019, 26, 3436–3446.
114. Kumar, P.S., Sankaranarayanan, G. Investigation on environmental factors of waste plastics into oil and its emulsion to control the emission in DI diesel engine. *Ecotoxicol. Environ. Safety* 2016, 134, 440–444.
115. Khoo, H.H. LCA of plastic waste recovery into recycled materials, energy and fuels in Singapore. *Resour. Conserv. Recycl.* 2019, 145, 67–77.
116. Li, C., Yuan, X., Sun, Z., Suvarna, M., Hu, X., Wang, X., Ok, Y.S. Pyrolysis of waste surgical masks into liquid fuel and its life-cycle assessment. *Bioresour. Technol.* 2022, 346, 126582.
117. Damodharan, D., Sathiyagnanam, A.P., Rana, D., Kumar, B.R., Saravanan, S. Extraction and characterization of waste plastic oil (WPO) with the effect of n-butanol addition on the performance and emissions of a DI diesel engine fueled with WPO/diesel blends. *Energy Convers. Manag.* 2017, 131, 117–126.
118. Kaimal, V.K., Vijayabalan, P. A detailed study of combustion characteristics of a DI diesel engine using waste plastic oil and its blends. *Energy Convers. Manag.* 2015, 105, 951–956.
119. Matuszewska, A., Hańderek, A., Paczuski, M., Biernat, K. Hydrocarbon fractions from thermolysis of waste plastics as components of engine fuels. *Energies* 2021, 14, 7245.
120. Mangesh, V.L., Padmanabhan, S., Tamizhdurai, P., Narayanan, S., Ramesh, A. Combustion and emission analysis of hydrogenated waste polypropylene pyrolysis oil blended with diesel. *J. Hazard. Mater.* 2020, 386, 121453.
121. Scharnagl, F.K., Hertrich, M.F., Ferretti, F., Kreyenschulte, C., Lund, H., Jackstell, R., Beller, M. Hydrogenation of terminal and internal olefins using a biowaste-derived heterogeneous cobalt catalyst. *Sci. Adv.* 2018, 4, eaau1248.
122. Léonard, N.G., Chirik, P.J. Air-stable α-diimine nickel precatalysts for the hydrogenation of hindered, unactivated alkenes. *ACS Catal.* 2018, 8, 342–348.
123. Büschelberger, P., Gärtner, D., Reyes-Rodriguez, E., Kreyenschmidt, F., Koszinowski, K., von Wangelin, A.J., Wolf, R. Alkene metalates as hydrogenation catalysts. *Chemistry* 2017, 23, 3139–3151.
124. Vollmer, I., Jenks, M.J.F., Roelands, M.C.P., White, R.J., van Harmelen, T., de Wild, P., van der Laan, G.P., Meirer, F., Keurentjes, J.T.F., Weckhuysen, B.M. Beyond mechanical recycling: Giving new life to plastic waste. *Angew. Chem. Int. Ed.* 2020, 59, 15402–15423.

125. Jia, C., Xie, S., Zhang, W., Intan, N.N., Sampath, J., Pfaendtner, J., Lin, H. Deconstruction of high-density polyethylene into liquid hydrocarbon fuels and lubricants by hydrogenolysis over Ru catalyst. *Chem. Catal.* 2021, 1, 437–455.
126. Rorrer, J.E., Beckham, G.T., Román-Leshkov, Y. Conversion of polyolefin waste to liquid alkanes with Ru-based catalysts under mild conditions. *JACS Au* 2021, 1, 8–12.
127. Rorrer, J.E., Troyano-Valls, C., Beckham, G.T., Román-Leshkov, Y. Hydrogenolysis of polypropylene and mixed polyolefin plastic waste over Ru/C to produce liquid alkanes. *ACS Sustain. Chem. Eng.* 2021, 9, 11661–11666.
128. Wang, C., Xie, T.J., Kots, P.A., Vance, B.C., Yu, K.W., Kumar, P., Fu, J.Y., Liu, S.B., Tsilomelekis, G., Stach, E.A., Zheng, W.Q., Vlachos, D.G. Polyethylene hydrogenolysis at mild conditions over ruthenium on tungstated zirconia. *JACS Au* 2021, 1, 1422–1434.
129. Celik, G., Kennedy, R.M., Hackler, R.A., Ferrandon, M., Tennakoon, A., Patnaik, S., LaPointe, A.M., Ammal, S.C., Heyden, A., Perras, F.A., Pruski, M., Scott, S.L., Poeppelmeier, K.R., Sadow, A.D., Delferro, M. Upcycling single-use polyethylene into high-quality liquid products. *ACS Central Sci.* 2019, 5, 1795–1803.
130. Nakaji, Y., Tamura, M., Miyaoka, S., Kumagai, S., Tanji, M., Nakagawa, Y., Yoshioka, T., Tomishige, K. Low-temperature catalytic upgrading of waste polyolefinic plastics into liquid fuels and waxes. *Appl. Catal. B Environ.* 2021, 285, 119805.
131. Lee, W.T., Bobbink, F.D., van Muyden, A.P., Lin, K.H., Corminboeuf, C., Zamani, R.R., Dyson, P.J. Catalytic hydrocracking of synthetic polymers into grid-compatible gas streams. *Cell Rep. Phys. Sci.* 2021, 2, 100332.
132. Tennakoon, A., Wu, X., Paterson, A.L., Patnaik, S., Pei, Y., LaPointe, A.M., Ammal, S.C., Hackler, R.A. Heyden, A., Slowing, I.I., Coates, G.W., Delferro, M., Peters, B., Huang, W., Sadow, A.D., Perras, F.A. Catalytic upcycling of high-density polyethylene via a processive mechanism. *Nat. Catal.* 2020, 3, 893–901.
133. Liu, S., Kots, P.A., Vance, B., Danielson, A., Vlachos, D.G. Plastic waste to fuels by hydrocracking at mild conditions. *Sci. Adv.* 2021, 7, eabf8283.
134. Zhang, F., Zeng, M., Yappert, R.D., Sun, J., Lee, Y.H., LaPointe, A.M., Peters, B., Abu-Omar, M.M., Scott, S.L. Polyethylene upcycling to long-chain alkylaromatics by tandem hydrogenolysis/aromatization. *Science* 2020, 370, 437–441.
135. Whajah, B., da Silva Moura, N., Blanchard, J., Wicker, S., Gandar, K., Dorman, J.A., Dooley, K.M. Catalytic depolymerization of waste polyolefins by induction heating: Selective alkane/alkene production. *Ind. Eng. Chem. Res.* 2021, 60, 15141–15150.
136. Hou, Q., Zhen, M., Qian, H., Nie, Y., Bai, X., Xia, T., Rehman, M.L.U., Li, Q., Ju, M. Upcycling and catalytic degradation of plastic wastes. *Cell Rep. Phys. Sci.* 2021, 2, 100514.
137. Matuszewska, A., Hańderek, A., Biernat, K., Bukrejewski, P. Thermolytic conversion of waste polyolefins into fuels fraction with the use of reactive distillation and hydrogenation with the syngas under atmospheric pressure. *Energy Fuels* 2019, 33, 1363–1371.
138. Sun, J., Wang, W., Yue, Q. Review on microwave-matter interaction fundamentals and efficient microwave-associated heating strategies. *Materials* 2016, 9, 231.
139. Chen, P., Xie, Q., Addy, M., Zhou, W., Liu, Y., Wang, Y., Cheng, Y., Li, K., Ruan, R. Utilization of municipal solid and liquid wastes for bioenergy and bioproducts production. *Bioresour. Technol.* 2016, 215, 163–172.
140. Zhang, Y.N., Chen, P., Liu, S.Y., Peng, P., Min, M., Cheng, Y.L., Anderson, E., Zhou, N., Fan, L.L., Liu, C.H., Chen, G., Liu, Y.H., Lei, H.W., Li, B.X., Ruan, R. Effects of feedstock characteristics on microwave-assisted pyrolysis: A review. *Bioresour. Technol.* 2017, 230, 143–151.
141. Ding, K., Liu, S., Huang, Y., Liu, S., Zhou, N., Peng, P., Wang, Y., Chen, P., Ruan, R. Catalytic microwave-assisted pyrolysis of plastic waste over NiO and HY for gasoline-range hydrocarbons production. *Energy Convers. Manag.* 2019, 196, 1316–1325.
142. Jiang, H., Liu, W., Zhang, X., Qiao, J. Chemical recycling of plastics by microwave-assisted high-temperature pyrolysis. *Glob. Chall.* 2020, 4, 1900074.
143. Putra, P.H.M., Rozali, S., Patah, M.F.A., Idris, A. A review of microwave pyrolysis as a sustainable plastic waste management technique. *J. Environ. Manag.* 2022, 303, 114240.
144. Hussain, Z., Khan, K.M., Hussain, K. Microwave–metal interaction pyrolysis of polystyrene. *J. Anal. Appl. Pyrolysis* 2010, 89, 39–43.
145. Bhattacharya, M., Basak, T. Susceptor-assisted enhanced microwave processing of ceramics – A review. *Crit. Rev. Solid State Mater. Sci.* 2017, 42, 433–469.

146. Motasemi, F., Afzal, M.T. A review on the microwave-assisted pyrolysis technique. *Renew. Sustain. Energy Rev.* 2013, 28, 317–330.
147. Kan, T., Strezov, V., Evans, T.J. Lignocellulosic biomass pyrolysis: A review of product properties and effects of pyrolysis parameters. *Renew. Sustain. Energy Rev.* 2016, 57, 1126–1140.
148. Papari, S., Hawboldt, K. A review on the pyrolysis of woody biomass to bio-oil: Focus on kinetic models. *Renew. Sustain. Energy Rev.* 2015, 52, 1580–1595.
149. Martín, M.T., Aguirre, J.L., Baena-González, J., González, S., Pérez-Aparicio, R., Saiz-Rodríguez, L. Influence of specific power on the solid and liquid products obtained in the microwave-assisted pyrolysis of end-of-life tires. *Energies* 2022, 15, 2128.
150. Fan, L.L., Zhang, Y.N., Liu, S.Y., Zhou, N., Chen, P., Liu, Y.H., Wang, Y.P., Peng, P., Cheng, Y.L., Addy, M., Lei, H.W., Ruan, R. Ex-situ catalytic upgrading of vapors from microwave-assisted pyrolysis of low-density polyethylene with MgO. *Energy Convers. Manag.* 2017, 149, 432–441.
151. Zhang, X.S., Lei, H.W., Yadavalli, G., Zhu, L., Wei, Y., Liu, Y.P. Gasoline-range hydrocarbons produced from microwave-induced pyrolysis of low-density polyethylene over ZSM-5. *Fuel* 2015, 144, 33–42.
152. Rosi, L., Bartoli, M., Frediani, M. Microwave assisted pyrolysis of halogenated plastics recovered from waste computers. *Waste Manag.* 2018, 73, 511–522.
153. Rex, P., Masilamani, I.P., Miranda, L.R. Microwave pyrolysis of polystyrene and polypropylene mixtures using different activated carbon from biomass. *J. Energy Inst.* 2020, 93, 1819–1832.
154. Aishwarya, K.N., Sindhu, N. Microwave assisted pyrolysis of plastic waste. *Procedia Technol.* 2016, 25, 990–997.
155. Bu, Q., Chen, K., Xie, W., Liu, Y., Cao, M., Kong, X., Chu, Q., Mao, H. Hydrocarbon rich bio-oil production, thermal behavior analysis and kinetic study of microwave-assisted co-pyrolysis of microwave-torrefied lignin with low density polyethylene. *Bioresour. Technol.* 2019, 291, 121860.
156. Bu, Q., Liu, Y., Liang, J., Morgan, H.M., Yan, L., Xu, F., Mao, H. Microwave-assisted co-pyrolysis of microwave torrefied biomass with waste plastics using ZSM-5 as a catalyst for high quality bio-oil. *J. Anal. Appl. Pyrolysis* 2018, 134, 536–543.
157. Suriapparao, D.V., Vinu, R., Shukla, A., Haldar, S. Effective deoxygenation for the production of liquid biofuels via microwave assisted co-pyrolysis of agro residues and waste plastics combined with catalytic upgradation. *Bioresour. Technol.* 2020, 302, 122775.
158. Zeng, Y., Wang, Y.P., Liu, Y.H., Dai, L.L., Wu, Q.H., Xia, M.L., Zhang, S.M., Ke, L.Y., Zou, R.G., Ruan, R. Microwave catalytic co-pyrolysis of waste cooking oil and low-density polyethylene to produce monocyclic aromatic hydrocarbons: Effect of different catalysts and pyrolysis parameters. *Sci. Total Environ.* 2022, 809, 152182.
159. Lam, S.S., Wan Mahari, W.A., Ok, Y.S., Peng, W.X., Chong, C.T., Ma, N.L., Chase, H.A., Liew, Z., Yusup, S., Kwon, E.E., Tsang, D.C.W. Microwave vacuum pyrolysis of waste plastic and used cooking oil for simultaneous waste reduction and sustainable energy conversion: Recovery of cleaner liquid fuel and techno-economic analysis. *Renew. Sustain. Energy Rev.* 2019, 115, 109359.
160. Jing, X.D., Dong, J.Q., Huang, H.L., Deng, Y.X., Wen, H., Xu, Z.H., Ceylan, S. Interaction between feedstocks, absorbers and catalysts in the microwave pyrolysis process of waste plastics. *J. Clean. Prod.* 2021, 291, 125857.
161. Fan, L.L., Su, Z.Y., Wu, J.B., Xiao, Z.G., Huang, P., Liu, L., Jiang, H.W., Zhou, W.G., Liu, S.Y., Ruan, R. Integrating continuous-stirred microwave pyrolysis with ex-situ catalytic upgrading for linear low-density polyethylene conversion: Effects of parameter conditions. *J. Anal. Appl. Pyrolysis* 2021, 157, 105213.
162. Russell, A.D., Antreou, E.I., Lam, S.S., Ludlow-Palafox, C., Chase, H.A. Microwave-assisted pyrolysis of HDPE using an activated carbon bed. *R. Soc. Chem.* 2012, 2, 6756–6760.
163. Bartoli, M., Rosi, L., Frediani, M., Undri, A., Frediani, P. Depolymerization of polystyrene at reduced pressure through a microwave assisted pyrolysis. *J. Anal. Appl. Pyrolysis* 2015, 113, 281–287.
164. Undri, A., Frediani, M., Rosi, L., Frediani, P. Reverse polymerization of waste polystyrene through microwave assisted pyrolysis. *J. Anal. Appl. Pyrolysis* 2014, 105, 35–42.
165. Zhang, X., Lei, H. Synthesis of high-density jet fuel from plastics via catalytically integral processes. *R. Soc. Chem.* 2016, 6, 6154–6163.
166. Undri, A., Rosi, L., Frediani, M., Frediani, P. Microwave assisted pyrolysis of corn derived plastic bags. *J. Anal. Appl. Pyrolysis* 2014, 108, 86–97.
167. Duan, D., Feng, Z., Dong, X., Chen, X., Zhang, Y., Wan, K., Wang, Y., Wang, Q., Xiao, G., Liu, H., Ruan, R. Improving bio-oil quality from low-density polyethylene pyrolysis: Effects of varying activation and pyrolysis parameters. *Energy* 2021, 232, 121090.

168. Prathiba, R., Shruthi, R., Miranda, L.R. Pyrolysis of polystyrene waste in the presence of activated carbon in conventional and microwave heating using modified thermocouple. *Waste Manag.* 2018, 76, 528–536.
169. Juliastuti, S., Hendrianie, N., Ramadhan, P., Satria, D. Microwave pyrolysis of multilayer plastic waste (LDPE) using zeolite catalyst. *AIP Conf. Proc.* 2017, 1840, 110001.
170. Undri, A., Rosi, L., Frediani, M., Frediani, P. Efficient disposal of waste polyolefins through microwave assisted pyrolysis. *Fuel* 2014, 116, 662–671.
171. Khaghanikavkani, E.M.F.M., Holdem, J., Williamson, A. Microwave pyrolysis of plastic. *J. Chem. Eng. Process Technol.* 2013, 4, 1000150.
172. Arshad, H., Sulaiman, S.A., Hussain, Z., Naz, M.Y., Moni, M.N.Z. Effect of input power and process time on conversion of pure and mixed plastics into fuels through microwave-metal interaction pyrolysis. *Waste Biomass Valor.* 2021, 12, 3443–3457.
173. Fan, L., Chen, P., Zhang, Y., Liu, S., Liu, Y., Wang, Y., Dai, L., Ruan, R. Fast microwave-assisted catalytic co-pyrolysis of lignin and low-density polyethylene with HZSM-5 and MgO for improved bio-oil yield and quality. *Bioresour. Technol.* 2017, 225, 199–205.
174. Mahari, W.A.W., Chong, C.T., Lam, W.H., Anuar, T.N.S.T., Ma, N.L., Ibrahim, M.D., Lam, S.S. Microwave co–pyrolysis of waste polyolefins and waste cooking oil: Influence of N_2 atmosphere versus vacuum environment. *Energy Convers. Manag.* 2018, 171, 1292–1301.
175. Suriapparao, D.V., Yerrayya, A., Nagababu, G., Guduru, R.K., Kumar, T.H. Recovery of renewable aromatic and aliphatic hydrocarbon resources from microwave pyrolysis/co–pyrolysis of agro–residues and plastics wastes. *Bioresour. Technol.* 2020, 318, 124277.
176. Rayaan, M.B.A. Recent advancements of thermochemical conversion of plastic waste to biofuel: A review. *Clean. Eng. Technol.* 2021, 2, 100062.
177. Tavares, R., Ramos, A., Rouboa, A. A theoretical study on municipal solid waste plasma gasification. *Waste Manag.* 2019, 90, 37–45.
178. Sharma, B., Goswami, Y., Sharma, S., Shekhar, S. Inherent roadmap of conversion of plastic waste into energy and its life cycle assessment: A frontrunner compendium. *Renew. Sustain. Energy Rev.* 2021, 146, 111070.
179. Tamosiunas, A., Valatkevicius, P., Valincius, V., Grigaitiene, V. Production of synthesis gas from propane using thermal water vapor plasma. *Int. J. Hydrogen Energy* 2014, 39, 2078–2086.
180. Ruj, B., Ghosh, S. Technological aspects for thermal plasma treatment of municipal solid waste: A review. *Fuel Process. Technol.* 2014, 126, 298–308.
181. Silvarrey, L.S.D. Advanced pyrolysis of plastic waste for chemicals, fuel and materials. Doctoral thesis, School of Engineering Newcastle University, 2019.
182. Fabry, F., Rehmet, C., Rohani, V., Fulcheri, L. Waste gasification by thermal plasma: A review. *Waste Biomass Valor.* 2013, 4, 421–439.
183. Muvhiiwa, R., Kuvarega, A., Llana, E.M., Muleja, A. Study of biochar from pyrolysis and gasification of wood pellets in a nitrogen plasma reactor for design of biomass processes. *J. Environ. Chem. Eng.* 2019, 7, 103391.
184. Fan, Y., Xiong, Y., Zhu, L., Fan, L., Jin, L., Chen, Y., Zhao, W. Comparison of the one-step and two-step plasma-catalytic upgrading of biomass pyrolysis volatiles to bio-fuel. *Chem. Eng. Process. Process Intensif.* 2019, 135, 53–62.
185. Vaidyanathan, A., Mulholland, J., Ryu, J., Smith, M.S., Circeo Jr., L.J. Characterization of fuel gas products from the treatment of solid waste streams with a plasma arc torch. *J. Environ. Manag.* 2007, 82, 77–82.
186. Pak, A.Y., Larionov, K.B., Kolobova, E.N., Slyusarskiy, K.V., Bolatova, J., Yankovsky, S.A., Stoyanovskii, V.O., Vassilyeva, Y.Z., Gubin, V.E. A novel approach of waste tires rubber utilization via ambient air direct current arc discharge plasma. *Fuel Process. Technol.* 2022, 227, 107111.
187. Cai, X., Du, C. Thermal plasma treatment of medical waste. *Plasma Chem. Plasma Process.* 2021, 41, 1–46.
188. Erdogan, A.A., Yilmazoglu, M.Z. Plasma gasification of the medical waste. *Int. J. Hydrogen Energy* 2021, 46, 29108–29125.
189. Zhovtyansky, V.A., Kolesnikova, E.P., Iakymovych, M.V. Plasma-assisted "waste-to-energy" processes. *Probl. At. Sci. Technol. Ser. Plasma Phys.* 2017, 23, 231–236.
190. Paulino, R.F.S., Essiptchouk, A.M., Silveira, J.L. The use of syngas from biomedical waste plasma gasification systems for electricity production in internal combustion: Thermodynamic and economic issues. *Energy* 2020, 199, 117419.

191. Kuo, P.-C., Illathukandy, B., Wu, W., Chang, J.-S. Plasma gasification performances of various raw and torrefied biomass materials using different gasifying agents. *Bioresour. Technol.* 2020, 314, 123740.
192. Punčochář, M., Ruj, B., Chatterje, P.K. Development of process for disposal of plastic waste using plasma pyrolysis technology and option for energy recovery. *Procedia Eng.* 2012, 42, 420–430.
193. Capodaglio, A.G., Bolognesi, S. 2 – Ecofuel feedstocks and their prospects. *Adv. Eco-Fuels Sustain. Enviro.* 2019, 15–51.
194. Yang, L., Wang, H., Wang, H., Wang, D., Wang, Y. Solid waste plasma disposal plant. *J. Electrostat.* 2011, 69, 411–413.
195. Tang, L., Huang, H., Hao, H., Zhao, K. Development of plasma pyrolysis/gasification systems for energy efficient and environmentally sound waste disposal. *J. Electrostat.* 2013, 71, 839–847.
196. Mohsenian, S., Esmaili, M.S., Shokri, B., Ghorbanalilu, M. Physical characteristics of twin DC thermal plasma torch applied to polymer waste treatment. *J. Electrostat.* 2015, 76, 231–237.
197. Mohsenian, S., Esmaili, S.S., Fathi, J., Shokri, B. Hydrogen and carbon black nano-spheres production via thermal plasma pyrolysis of polymers. *Int. J. Hydrogen Energy* 2016, 41, 16656–16663.
198. Gabbar, H.A., Aboughaly, M., Stoute, C.A.B. DC thermal plasma design and utilization for the low density polyethylene to diesel oil pyrolysis reaction. *Energies* 2017, 10, 784.
199. Yao, L., King, J., Wu, D., Chuang, S.S.C., Peng, Z. Non-thermal plasma-assisted hydrogenolysis of polyethylene to light hydrocarbons. *Catal. Commun.* 2021, 150, 106274.
200. Kumar, S.B.V. Process for the conversion of plastic material to fuels. International Patent, 2021, WO 2021171313 A1.
201. McNamara, D., Murray, M. Conversion of waste plastics material to fuel. US Patent Application Publication, 2012, US 2012/0261247 A1.
202. Daggupati, S., Thakur, S., Majhi, S., Mandal, S., Das, A.K., Ajit, V.S. A method for catalytic conversion of waste plastic into liquid fuel. US Patent Application Publication, 2021, US 2021/0348061 A1.
203. Baker, G. Process and plant for conversion of waste material to liquid fuel. European Patent, 2018, EP 1 725 633 B1.
204. Schabel, J., Schwarz, R.A. Grispin, C.W., Gencer, M.A., Hensel, J.D. Process, apparatus, controller and system for producing petroleum products. US Patent Application Publication, 2020, US 2020/0199457 A1.
205. Handerek, A., Kowalczyk, M.P., Kiraga, J., Biernat, K., Matuszewska, A. Method of preparation of hydrocarbon fuels from polyolefin waste materials. US Patent, 2020, US 20200080004.
206. Dooley, B. Production of hydrocarbon fuels from plastics. US Patent, 2018, US 2018/0355256 A1.
207. Jamradloedluk, J., Lertsatitthanakorn, C. Characterization and utilization of char derived from fast pyrolysis of plastic waste. *Procedia Eng.* 2014, 69, 1437–1442.
208. Wijesekara, D.A., Sargent, P., Ennis, C.J., Hughes, D. Prospects of using chars derived from mixed post waste plastic pyrolysis in civil engineering applications. *J. Clean. Prod.* 2021, 317, 128212.
209. Soganciogiu, M., Yel, E., Ahmetli, G. Pyrolysis of waste high density polyethylene (HDPE) and low density polyethylene (LDPE) plastics and production of epoxy composites with their pyrolysis chars. *J. Clean. Prod.* 2017, 165, 369–381.
210. Shen, Y., Chen, X., Ge, X., Chen, M. Chemical pyrolysis of E-waste plastics: Char characterization. *J. Environ. Manag.* 2018, 214, 94–103.
211. Yousef, S., Eimontas, J., Subadra, S.P., Striūgas, N. Functionalization of char derived from pyrolysis of metallised food packaging plastics waste and its application as a filler in fiberglass/epoxy composites. *Process Safety Environ. Protect.* 2021, 147, 723–733.
212. Thiounn, T., Smith, R.C. Advances and approaches for chemical recycling of plastic waste. *J. Polym. Sci.* 2020, 58, 1347–1364.
213. Luu, P., Daly, K., Croke, B., Sherwin, G., Miñana, B. *Transitioning to a Circular System for Plastics.* Assessing Molecular Recycling Technologies in the United States and Canada. Gothenburg, Sweden: Closed Loop Partners, 2021.
214. Takada, H., Bell, L. *Plastic Waste Management Hazards.* Gothenburg, Sweden: International Pollutants Elimination Network (IPEN), June 2021.
215. Plastic Energy Ltd. Plastic energy technology. London: Plastic Energy Ltd, 2022. Available online: https://plasticenergy.com/technology/ (accessed on 27 June 2022).
216. Monreal, C. Chemical recycling in practice. Brightmark LLC, 2021. Available online: https://www.brightmark.com/plastics-renewal/projects/ashley-circularity-center (accessed on 27 June 2022).
217. Enerkem Inc. Our Technology: Proven, Scalable and Sustainable. Available online: https://enerkem.com/solution/technology (accessed on 27 June 2025).

218. Brightmark LLC. Ashley plastics renewal facility. Brightmark LLC, 2019. Available online: https://www.brightmark.com/plastics-renewal/projects/ashley-circularity-center (accessed on 27 June 2022).
219. Yao, J.Y., Wang, Y.W., Muppaneni, T., Shrestha, R., Le Roy, J., Figuly, G.D., Freer, E. Methods for the decomposition of contaminated plastic waste. International Patent, 2019, WO 2019204687 A1.
220. BASF SE. From plastic waste to virgin–grade products. BASF SE, 2020. Available online: https://www.basf.com/us/en/who-we-are/sustainability/we-drive-sustainable-solutions/circular-economy/mass-balance-approach/chemcycling (accessed on 27 June 2022).
221. OMV AG. OMV unveils innovative ReOil® plant to transform end-of-life plastics into circular feedstocks. Available online: https://www.omv.com/en/media/press-releases/2025/250318-omv-unveils-innovative-reoil-plant-to-transform-end-of-life-plastics-into-circular-feedstocks (accessed on 27 June 2025).
222. OMV Refining & Marketing GmbH. OMV Presentation. 2019. Available online: https://www.voeb.at/fileadmin/user_upload/voeeb.at/Intern/2019/Schneider_Chemisches_Recycling.pdf (accessed on 27 June 2022).
223. Ricardo-AEA Ltd. Case Study 3 Cynar Plastics to Diesel. Zero Waste SA and Government of South Australia, 2013. Available online: https://www.greenindustries.sa.gov.au/__media_downloads/165469/Case-Study-3-Cynar-Plastics-to-Fuel-FINAL.pdf (accessed on 28 June 2022).
224. Cynar Plc. Cynar Plc Presentation. Cynar Plc, 2012. Available online: https://wedocs.unep.org/bitstream/handle/20.500.11822/13364/GLOCIIPresentationMurray.pdf?sequence=1&isAllowed=y (accessed on 28 June 2022).
225. Alphakat GmbH. Process Description Alphakat—Update. Alphakat GmbH, December 2010. Available online: https://www.scribd.com/document/118589743/Process-Description-Alphakat-Update-Dec-2010-2 (accessed on 28 June 2022).
226. Alphakat Holdings International Ltd. Koshi Village Project. Alphakat Holdings International Ltd., 2019. Available online: https://alphakatholdings.com/koshi-village-project/ (accessed on 28 June 2022).
227. Patrick Lowe, P. 2021. White paper—Petrosynthesis—The competition of the industrial revolution, November 2021. Industrial Revolution. © 2021 Patrick Lowe, 15 November 2021: Available online: https://zeropetroleum.com/wp-content/uploads/2021/11/White-Paper-Petrosynthesis-15-11-21.pdf (accessed on 12 June 2022).

12 Solvent Extraction Fundamentals and Oil and Grease Determination Using Trichloroethane (Methyl Chloroform)

Lawrence K. Wang, Mu-Hao Sung Wang, Yung-Tse Hung, Hamidi Abdul Aziz, and Nor Habsah Md Sabiani

NOMENCLATURE AND ACRONYM

ACGIH	American Conference of Governmental Industrial Hygienists
AEGL	Acute Exposure Guideline Levels
CFR	Code of Federal Regulations
D	Extract dilution factor
DOT	US Department of Transportation
ERG	Emergency Response Guidebook
ERPG	Emergency Response Planning Guideline
GLP	Good laboratory practice
IARC	International Agency for Research on Cancer
IRIS	Integrated Risk Information System
LEL	Lower explosive limit
LIWT	Lenox Institute of Water Technology
NFPA	National Fire Protection
NIOSH	National Institute for Occupational Safety and Health
NJDEP	New Jersey Department of Environmental Protection
NTP	National Toxicology Program
O&G	Oil and grease
ODS	Ozone-depleting substances
OSHA	Occupational Safety and Health Administration
PAC	Protective Action Criteria
PEL	Permissible exposure limit
PEOSHA	New Jersey Public Employees Occupational Safety and Health Act
ppm	Parts per million parts of air
REL	Recommended exposure limit
STEL	Short-term exposure limit
TCE	Trichloroethane
TLV	Threshold limit value
TOG	Total oil and grease, mg/L
UEL	Upper explosive limit

DOI: 10.1201/9781003532354-12

USEPA	US Environmental Protection Agency
UV	Ultraviolet
V	Volume of sample, L
W	Weight of oil and grease determined from calibration plot, mg

12.1 INTRODUCTION

12.1.1 Fundamentals of Solvent Extraction Analytical Method

Solvent extraction is a widely used analytical method for determining the concentration of oil, grease, and other pollutants in water samples. This method is particularly effective for analyzing non-polar substances that are not easily soluble in water. The key steps in solvent extraction analytical method include:

1. **Sample Collection:** Water samples are collected in clean, contaminant-free containers to avoid any interference;
2. **Sample Preservation:** Samples are often preserved with acid to prevent microbial degradation of the analytes;
3. **Extraction Process:** (a) Solvent Addition: A specific volume of an organic solvent, such as *n*-hexane, is added to the water sample; (b) Mixing: The mixture is vigorously shaken or stirred to ensure thorough contact between the solvent and the water, allowing the non-polar substances to dissolve in the solvent; and (c) Separation: The solvent, now containing the extracted pollutants, is separated from the water phase, typically using a separatory funnel.
4. **Concentration and Analysis:** (a) Evaporation: The solvent is evaporated, leaving behind the extracted pollutants; and (b) Weighing: The residue is weighed to determine the concentration of oil and grease (O&G) in the original water sample.

There are three major applications of the solvent extraction analytical method: (a) Environmental Monitoring: Used to assess the levels of O&G in surface water, groundwater, and wastewater; (b) Industrial Effluent Testing: Ensures compliance with environmental regulations by monitoring the discharge of pollutants from industrial processes; and (c) Drinking Water Quality: Helps in maintaining the quality of drinking water by detecting and quantifying contaminants.

The advantages of solvent extraction analytical method include: (a) Sensitivity: Capable of detecting low concentrations of pollutants; (b) Versatility: Applicable to a wide range of non-polar substances, including hydrocarbons, fats, and oils; and (c) Standardization: Well-established methods like the US Environmental Protection Agency (USEPA) Method 1664 provide standardized procedures for consistent results.

The disadvantages of solvent extraction analytical method are: (a) Solvent Use: Requires the use of organic solvents, which can be hazardous and require proper disposal; (b) Labor Intensive: The process can be time-consuming and requires careful handling to avoid contamination; and (c) Interferences: Other substances in the sample can interfere with the extraction and analysis, potentially affecting accuracy.

12.1.2 O&G Analysis

A knowledge of the concentration of oil and/or grease present in water, wastewater, or liquid sludge is helpful in proper design and operation of treatment systems.

There are six methods [1–9] currently available for quantitative determination of oil, grease, and related matter in water, wastewater, and sludge: (a) liquid-liquid partition-gravimetric method; (b) partition-infrared method; (c) Soxhlet extraction method; (d) extraction method for sludge samples; (e) hydrocarbon method; and (f) solid-phase, partition-gravimetric method.

The partition-gravimetric and Soxhlet methods are very time-consuming and do not provide the needed precision under certain environmental conditions.

The partition-infrared method is designed for samples that might contain volatile hydrocarbons that otherwise would be lost in the solvent removal operations of the gravimetric procedure. When 1 L portions of the sewage were dosed with 14.0 mg of a mixture of No. 2 fuel oil and Wesson oil, the recovery of the added oils was 99% with a standard deviation of 1.4 mg. The method may accurately measure as little as 0.2 mg O&G, but many water quality laboratories are not equipped with the required infrared spectrophotometer.

Selection of an appropriate method for O&G analysis is hinted by APHA-AWWA-WEF [1]. Of the Standard Methods [1], the liquid/liquid partition-gravimetric method (5520B), the partition-infrared method (5520C), the Soxhlet method (5520D), and the solid-phase, partition-gravimetric method (5520G) are for liquid samples. Standard Method 5520C is designed for samples that might contain volatile hydrocarbons that otherwise would be lost in the solvent removal operations of the gravimetric procedure. Standard Method 5520D is the method of choice when relatively polar, heavy petroleum fractions are present, or when the levels of nonvolatile O&Gs may challenge the solubility limit of the solvent. Standard Method 5520C should be chosen for low levels of O&G (less than 10 mg/L) analysis. Standard Method 5520G can be used as an alternative to Standard Method 5520B to reduce solvent volumes and matrix problems [1]. Only Standard Method 5520E is suitable for O&G analysis of sludge or similar materials. It is noted that Standard Method 5520F can be used with 5520B, 5520C, 5520D, or 5520G to obtain a hydrocarbon measurement in addition to, or instead of, the O&G analysis.

With laboratory waste minimization and management in mind [10–12], the authors introduce three spectrophotometric and colorimetric procedures for rapid measurement of organic solvent extractable substances in water and wastewater [13–15]. Almost all the water quality laboratories in the USA are equipped with a filter photometer, a visible spectrophotometer, and/or a UV/visible spectrophotometer. Any one of the aforementioned instrumentations can be used in the newly developed rapid procedures. Furthermore, more than one organic solvent is recommended for adoption in order to provide an added flexibility [13–15].

It should be noted that the definition of O&G is based on the procedure used. The O&G may include hydrocarbons, vegetable oils, animal fats, waxes, soaps, greases, and other related substances. The source of the oil and/or grease, and the presence of extractable non-oily matter will influence the substance measured and interpretation of results. An *Unknown Oil* is defined as one for which a representative sample of the oil or grease is not available for the preparation of a standard. The O&G in a mixed sewage or an unidentified oil slick on a surface water are typical examples of unknown oil. A *Known Oil* is defined as a sample of oil and/or grease that represents the only material of that type used or produced in the processes or represented by the process water or wastewater.

12.1.3 Solvent Selection Based on Environmental Impacts, Solvent Hazards, and Waste Disposal Technologies

It is necessary to review the history of Standard Methods, in turn, to understand the importance of solvent selection. It is the authors' view that the solvent selection for an analytical method development should be mainly decided by the seriousness of solvent hazards and the convenience of waste disposal. Environmental protection, such as minimization of ozone-depleting substance (ODS) discharge, can be done by a Good Laboratory Practice (GLP), such as solvent recycle, or proper laboratory waste management [1,16–19].

The 12th edition of the Standard Methods for the Examination of Water and Wastewater (i.e. Standard Methods) originally adopted the use of petroleum ether as the solvent for natural and treated waters and *n*-hexane for polluted waters. The 13th edition Standard Methods adopted trichlorotrifluoroethane as an optional solvent for all O&G sample types. In the 14th to 17th editions of Standard Methods, only trichlorotrifluoroethane was specified for O&G analysis. Since trichlorotrifluoroethane (commercially known as Freon 113, or CFC 113) is one of the major ODS which destroys the Earth's precious ozone protection layer [20], an alternative solvent (80% *n*-hexane and 20% methyl-tert-butyl ether) was included for gravimetric methods in the 19th edition Standard Methods. In the 20th edition Standard Methods, trichlorotrifluoroethane was dropped from all gravimetric analytical procedures (retained for Standard Method 5520C, an infrared method), and replaced with *n*-hexane. In the latest 23rd edition Standard Methods [1], *n*-hexane is used for 5520B and trichlorotrifluoroethane is used for 5520C for O&G analysis.

In the authors' research, several organic solvents were investigated. Three organic solvents (trichlorotrifluoroethane, trichloromethane, and trichloroethane [TCE]) were finally chosen for extraction and analysis of O&G. Accordingly three analytical methods have been developed and reported separately [13–15] for the convenience of the readers.

It is the authors' opinion that chemist's health (health hazard, permissible exposure limit [PEL], recommended exposure limit [REL], threshold limit value [TLV], short-term exposure limit [STEL], toxicity, etc.) and safety (flammability hazard, reactivity hazard, etc.) are as important as the solvent's environmental impacts (such as ozone layer protection, hazardous waste disposal) [10,12,21–24].

In each Lenox Institute of Water Technology (LIWT) research report [13–15], the principle, sampling, sample storage, apparatus, reagents, the analytical procedures for preparation of calibration curves of known oil reference standards, the analytical procedures for preparation of calibration curves of unknown oil reference standards, the procedures for sample analysis, precision and accuracy, typical example of calibration curve, risk management, solvent selection, hazardous solvent disposal, and glossary are presented.

Each selected solvent in this research is compared with *n*-hexane which is the standard solvent adopted by the Standard Methods [1] for O&G analysis. Trichloromethane (chloroform, $CHCl_3$) was also selected as one of the three solvents in this O&G research series because chloroform is an organic solvent adopted by the Standard Methods [1] for examination of phenol (Standard Method 5530C), anionic surfactant (Standard Method 5540C), etc. and by Wang et al. [25–27] for analysis of anionic and cationic surfactants.

12.1.4 Dedication

This publication is dedicated to the author's beloved colleague, Aimee E. Thayer (BS, MS) of the Lenox Institute of Water Technology, Massachusetts, USA.

12.1.5 Summary and Recommendation

Current Standard Methods 5520B, 5520D, 5520E, and 5520G for O&G analysis adopt *n*-hexane for solvent extraction. The authors compare the selected solvent (1,1,1-trichloroethane) with *n*-hexane for O&G analysis and conclude that 1,1,1-trichloroethane is safer than *n*-hexane considering the safety of both the laboratory workers and the laboratory, although 1,1,1-trichloroethane is an ODS. Solvent recycle and/or proper laboratory waste management will eliminate the opportunity of ODS release to the environment.

12.2 PRINCIPLE

The liquid sample containing O&G is acidified to a low pH (equal to or less than pH 2) and extracted with an organic solvent (TCE). Under controlled conditions, the O&G are extractable into a selected organic solvent, in which the intensity of yellow color is proportional to the O&G concentration. At the recommended wavelength and light path, the O&G in the liquid sample can be qualitatively determined. The O&G is determined by comparison of the absorbance of the sample extract with oil standards.

12.3 SAMPLING AND SAMPLE STORAGE

A representative sample of at least 250 mL volume should be collected in a clean glass sample bottle, glass or Teflon stoppered. If the sample is known to have less than 0.1 mg/L of O&G, 1 L volume of the representative sample should be collected in a large glass bootie, glass, or Teflon stoppered.

If the sample volume cannot be readily measured in the field, the sample level in the sampling bottle should be marked for later determination of sample volume in the laboratory.

Losses of O&G may occur on sampling equipment; therefore, the collection of a composite sample for analysis is not practical. Individual portions collected at prescribed time intervals must be analyzed separately to obtain an average O&G concentration over an extended period of time.

Always collect a separate sample for an O&G determination and do not subdivide the sample in the laboratory.

If O&G analysis is to be done immediately, no sample preservation is required. Otherwise, the liquid sample should be preserved by the addition of 5 mL of 1:1 sulfuric acid or 1:1 hydrochloric acid per every liter of sample. For example, for exact 300 mL sample in a glass sample bottle, 1.5 mL of 1:1 sulfuric acid will be sufficient for sample preservation if the O&G analysis is to be delayed for more than a few hours. The readers are referred to the Standard Methods [1] for more technical information for O&G sample collection, preservation, and storage.

In sampling sludge, precaution must be taken to obtain a true representative sample. Sludge samples should be preserved with 1 mL concentrated hydrochloric acid per 80 g of sludge sample if O&G analysis cannot be made immediately.

Never preserve any samples with organic solvent or other non-recommended chemicals.

12.4 INTERFERENCE

The selected organic solvent (TCE) has the ability to dissolve not only O&G but also other organic substances.

Heavier residuals of petroleum may contain a significant portion of substances insoluble in the recommended organic solvents.

For reduction of interferences caused by synthetic surfactant or other known chemicals, special O&G standards should be prepared by the same solvent extraction procedure in the presence of synthetic surfactant, etc.

12.5 TCE SPECTROPHOTOMETRIC METHOD

12.5.1 Apparatus and Instrumentation

1. **Separatory Funnel:** One liter capacity with inert Teflon stopcock (or equivalent). Note: Do not use glass stopcock lubricated with grease.
2. **Analytical Equipment:** One of the following is required: (a) Spectrophotometer: For use at 400–500 nm, providing light paths of 1, 5, and 10 cm; or (b) Filter photometer, providing light paths of 1 cm, 5 cm, or longer and equipped with a color filter exhibiting maximum transmittance near 400–500 nm.
3. **Cells:** Quartz, 1, 5, and 10 cm light path length.
4. **Syringes or Micropipettes:** 10, 25, 50, 100 μL capacity (optional).
5. **Filter Paper:** Whatman No. 40 or equivalent, 11 cm diameter.
6. Glass wool.
7. **Glass or Teflon Graduate Cylinders:** 10, 50, 500, 1,000 mL capacity.
8. **Volumetric Flasks:** 100-mL capacity, at least two; 50 mL capacity, at least seven.
9. **pH Paper:** For measuring pH of water sample at pH 2 or lower.
10. **Graduated Glass Pipettes:** For measuring oil standards in the 0–30 mL range.

11. **Glass Bottles:** 50, 500, and 1,000 mL capacity, glass or Teflon stoppered.
12. Glass or Teflon filter funnel cone to hold the filter paper with 11 cm diameter.

12.5.2 Reagents

1. 1,1,1-Trichloroethane or methyl chloroform), reagent grade. The solvent should leave no measurable residue on evaporation; distill if necessary. Do not use any plastic tubing to transfer solvent between containers.
2. **Strong Acid:** One of the following is required: (a) Sulfuric acid, H_2SO_4, 1+1: Mix equal volumes of concentrated sulfuric acid and distilled water; (b) Hydrochloric acid, HC1, 1+1: Mix equal volumes of concentrated hydrochloric acid and distilled water.
3. Sodium sulfate, anhydrous crystal.
4. *N*-hexadecane, reagent grade.
5. Iso-octane, reagent grade.
6. Benzene, reagent grade.

12.5.3 Procedure for Preparation of Calibration Curves of Known Oil Reference Standard

1. Prepare a *Known Oil Reference Standard Stock Solution* if the oil and/or grease used or manufactured in the process is known. Accurately weigh slightly over 0.25 g (i.e. about 0.5 mL) of known oil directly into a 100 mL tared, glass stoppered volumetric flask. Add 80 mL TCE and dissolve the oil. (If, as in the case of heavy fuel oil, all the oil does not go into solution, let stand overnight. The next day filter through a filter paper into another 100 mL volumetric flask). Dilute with TCE to the 100 mL mark. Treat analysis as if all oil had gone into solution. Calculate the volume of TCE required and dilute the freshly prepared 100-mL oil-in-TCE solution to a concentration that 1 mL of oil-TCE mixture gives 2.500 mg oil.
2. Use syringes or micropipettes to transfer 0.2, 0.4, 1, 2, 4, and 8 mL of the *Known Oil Reference Standard Stock Solution* to six 50-mL volumetric flasks separately, and dilute each flask to the 50-mL mark with pure TCE. There are now six *Known Oil Reference Standard Working Solutions* with oil concentrations of 0.5, 1.0, 2.5, 5.0, 10.0, and 20.0 mg, respectively.
3. Select appropriate wavelength, and cell light path length for calibration curves preparation.

The following table serves as an approximate guide for the wavelength range of 400–500 nm:

Light Path Length, cm	Oil and Grease Range, mg
1	4–40
5	<0.4–8
10	<0.1–4

Transfer the six *Known Oil Reference Standard Working Solutions*, one by one, to moisture-free clean cells. Use pure TCE to zero the spectrophotometer (or a filter photometer) at a selected wavelength, and take absorbance readings of these Standard Working Solutions, one by one, at this wavelength. At least four wavelengths (400, 425, 450, and 500 nm) are recommended to be scanned for calibration curves preparation.

1. Prepare at least four calibration curves (absorbance versus mg oil) at wavelengths of 400, 425, 450, and 500 nm.
2. For reduction of interferences caused by synthetic surfactant or other known chemicals, special O&G standard working solutions should be prepared by the solvent extraction procedure in the presence of synthetic surfactant, etc.

12.5.4 Procedure for Preparation of Calibration Curves of Unknown Oil Reference Standard

1. **Prepare an Unknown Oil Reference Primary Standard:** Pipette 30.0 mL *n*-hexadecane, 30.0 mL iso-octane, and 20.0 mL benzene into a 100 mL glass stoppered volumetric flask to prevent evaporation. Assume the specific gravity of this mixture to be 0.769 and maintain the integrity of the mixture by keeping stoppered except when withdrawing aliquots. 10 microliter (μL) = 7.69 mg of oil; or 1 mL = 769 mg of oil.
2. **Prepare an Unknown Oil Reference Secondary Standard:** Micropipette 0.650195 mL (or 650.195 μL) of Unknown Oil Reference Primary Standard into a 100 mL glass stoppered volumetric flask, and dilute it with TCE to the 100 mL mark. 1 mL of Secondary Standard gives 5.00 mg oil. The Secondary Standard remains valid within 1 week.
3. Use syringes or micropipettes to transfer 0.1, 0.2, 0.5, 1, 2, and 4 mL of the *Unknown Oil Reference Secondary Standard* to six 50-mL volumetric flasks separately, and dilute each flask to the 50-mL mark with pure TCE. There are now six *Unknown Oil Reference Standard Working Solutions* with oil concentrations of 0.5, 1.0, 2.5, 5.0, 10.0, and 20 mg, respectively.
4. Select appropriate wavelength, and cell light path length for calibration curves preparation using the same guide as in the last section. Transfer the six *Unknown Oil Reference Standard Working Solutions*, one by one, to moisture-free clean cells. Use pure TCE to zero the spectrophotometer (or a filter photometer) at a selected wavelength, and take absorbance readings of these Standard Working Solutions, one by one, at this wavelength. At least four wavelengths (400, 425, 450, and 500 nm) are recommended to be scanned for calibration curves preparation.
5. Prepare at least four calibration curves (absorbance versus mg oil) at wavelengths of 400, 425, 450, and 500 nm.
6. For reduction of interferences caused by synthetic surfactant or other known chemicals, special O&G standard working solutions should be prepared by the solvent extraction procedure in the presence of synthetic surfactants, etc.

12.5.5 Procedure for Sample Analysis

1. Measure the exact sample volume on-site, or mark the sample bottle at the water meniscus for later determination of sample volume. If the sample was not acidified at the time of sample collection, add 5 mL/L of 1:1 sulfuric acid (or 1:1 hydrochloric acid) to the sample bottle. After mixing the sample, check the pH by touching the pH-sensitive paper to the cap to ensure that the pH is 2 or lower. Add more acid if necessary.
2. Pour the sample into a separatory funnel. Record the sample volume.
3. Add 50 mL TCE solvent to the sample bottle, stopper it, and rotate the sample bottle to rinse the sides. Transfer the TCE solvent into the separatory funnel containing the known volume of sample. Extract by shaking vigorously for 2.5 minutes. Allow the water layer and TCE solvent layer to separate. The TCE solvent layer will stay below the water layer and will be in yellow color if the TCE now contains extractable O&G.
4. Insert a sufficient amount of glass wool in the stem of the separatory funnel.

5. Filter the TCE solvent layer into a 100 mL volumetric flask through the glass wool first, then through a filter funnel cone containing a TCE-moistened Whatman No. 40 filter. Note: An emulsion that fails to dissipate can be broken by pouring about 1 g sodium sulfate into the filter paper cone and draining the emulsion through the salt. Additional 1 g portions can be added to the cone as required.
6. Clean the cell thoroughly with TCE and be sure there is no moisture in the cell. Transfer the oil sample's TCE extract into the cell. Measure the extracts' absorbance with a spectrophotometer or a filter photometer at a recommended wavelength in the range of 400–500 nm. For each extract's absorbance measurement, the pure TCE in an identical light path cell shall be used to zero the instrument.
7. Use a calibration plot of absorbance versus mg oil prepared from the standards to determine the mg oil (W) in the sample solution. If the absorbance from one calibration curve exceeds 0.8 for a sample, select another wavelength, or another shorter path length, or dilute sample as required.
8. Calculation:

$$\text{TOG} = (W \times D)/V$$

where TOG = total O&G concentration in mg/L; W = weight of O&G determined from calibration plot, in mg; D = extract dilution factor, if used; and V = volume of sample, either measured exactly on-site or determined by refilling sample bottle to mark line and correcting for acid addition if necessary, in L.

12.5.6 Precision and Accuracy

The O&G concentration as low as 0.1 mg/L can be quantitatively measured with a cell of 5 cm light path length. For the determination of O&G concentrations below 0.1 mg/L, the use of a cell of 10 cm light path at optimum wavelength is required.

If a filter photometer or field spectrophotometer (such as HACH Spectrophotometer DL-4) is used, the standard deviation for O&G measurement is 1 mg/L.

12.6 TYPICAL EXAMPLE OF CALIBRATION CURVE OF KNOWN OIL REFERENCE STANDARD

The following are the general analytical conditions:

1. Instrument = Bausch & Lomb Spectronic 710 or equivalent
2. Cells = Quartz
3. Sample Size (V) = 250 mL
4. Organic Solvent Volume = 50 mL
5. Blank (Zeroing the Instrument) = Pure Organic Solvent
6. Dilution Factor = 1 (Note: No Dilution)
7. Known Oil = Kendall SAE 30, Dual Action Heavy Duty Motor Oil, Part # 531-7113, M.L-L-46152.
8. Calibration Curve for Trichloroethane Extraction Spectrophotometric Method (10 cm Light Path Length) – See Figure 12.1.

12.7 SOLVENT HEALTH HAZARD AND WASTE DISPOSAL

Tables 12.1 and 12.2 indicate the physical properties and the exposure limits, respectively, of 1,1,1-Trichloroethane (methyl chloroform) (Source: OSHA).

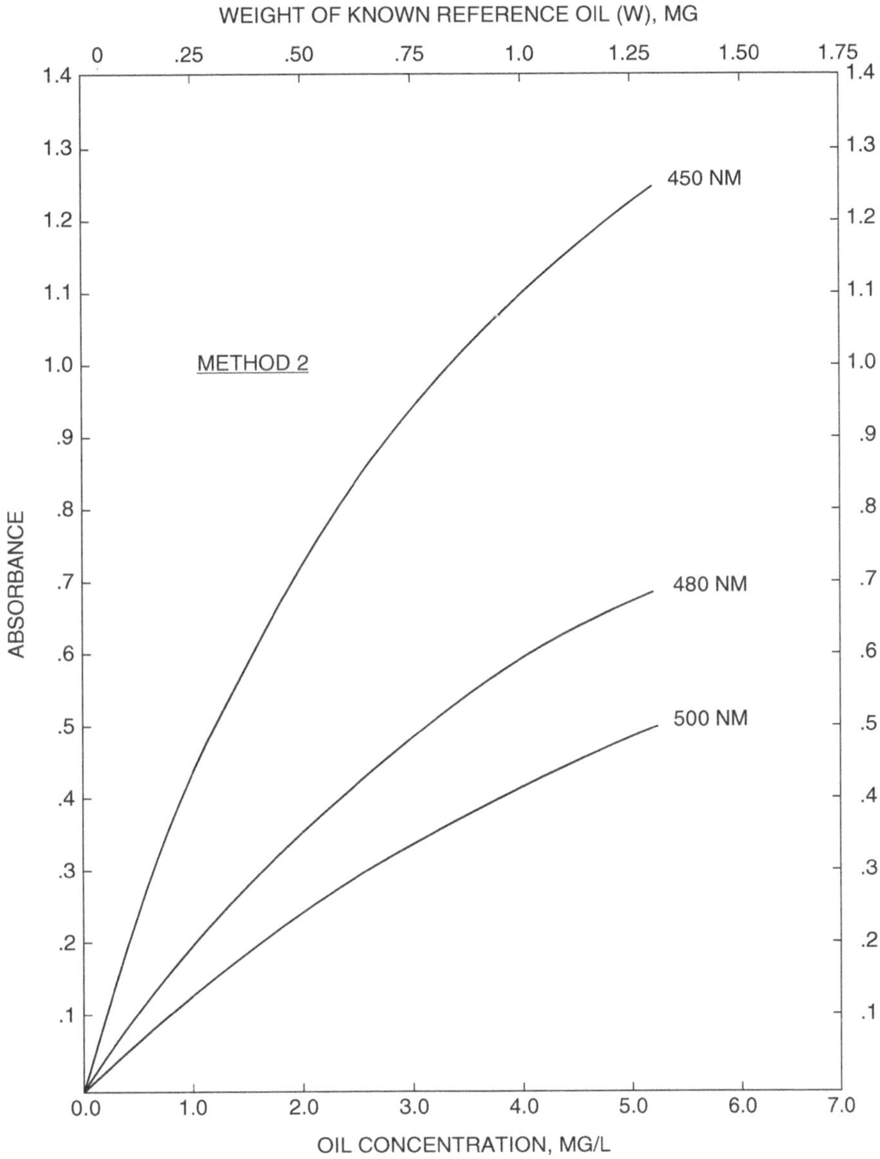

FIGURE 12.1 Calibration curves for trichloroethane extraction spectrophotometric method (10 cm light path length).

1,1,1-Trichloroethane (methyl chloroform) is a colorless organic liquid, with an ether-like odor, which is used as a laboratory or industrial solvent, spot cleaner, or degreaser, and a raw material in making other chemicals. Its chemical numbers are: CAS No. = 71-55-6, DOT No. = UN 2832. Its NFPA hazard rating are: health = 2 (moderate); flammability = 1 (slight); reactivity = 0 (minimal). Additional hazard information are: hazard class = 6 (poison); OSHA legal airborne PEL = 350 ppm averaged over an 8-hour work shift; NIOSH airborne REL = 350 ppm, which should not be exceeded during any 15-minute work period; and ACGIH TLV = 350 ppm averaged over an 8-hour work shift, and 450 ppm as a STEL. It is not classifiable as to its potential to cause cancer. It may cause mutations or genetic changes. There is limited evidence that methyl chloroform may damage the

TABLE 12.1
Physical Properties of 1,1,1-Trichloroethane (Methyl Chloroform)

METHYL CHLOROFORM (1,1,1-TRICHLOROETHANE)†

Chemical Identification			
CAS #	71-55-6		
Formula	$C_2H_3Cl_3$		
Synonyms	chlorothene; 1,1,1-trichloroethane; 1,1,1-trichloroethane (stabilized)		
Physical Properties			
Physical description	Colorless liquid with a sweet, chloroform-like odor.		
Boiling point	165°F	Molecular weight	133.4
Freezing point/melting point	-26.5°F	Vapor pressure	100 mmHg
Flash point	>200°F	Vapor density	4.6
Specific gravity	1.31	Ionization potential	11.00 eV
Lower explosive limit (LEL)	7.5%	Upper explosive limit (UEL)	12.5%
NFPA health rating	2	NFPA fire rating	1
NFPA reactivity rating	0	NFPA special instruction	
Vapor hazard ratio (VHR)			
Historical exceedance percentage			

TABLE 12.2
Exposure Limits of 1,1,1-Trichloroethane (Methyl Chloroform)

Exposure Limits							
OSHA PEL 8-hour TWA (ST) STEL (C) Ceiling Peak		**NIOSH REL** Up to 10-hour TWA (ST) STEL (C) Ceiling		**ACGIH TLV©** 8-hour TWA (ST) STEL (C) Ceiling		**CAL/OSHA PEL** 8-hour TWA (ST) STEL (C) Ceiling Peak	
PEL-TWA	350 ppm (1900 mg/m³)	**REL-TWA**		**TLV-TWA**	350 ppm [1992]	**PEL-TWA**	350 ppm (1900 mg/m³)
PEL-STEL		**REL-STEL**		**TLV-STEL**	450 ppm [1992]	**PEL-STEL**	450 ppm (2450 mg/m³)
PEL-C		**REL-C**	350 ppm (1900 mg/m³) [15 minutes]	**TLV-C**		**PEL-C**	800 ppm
Skin notation	N	**Skin notation**	N	**Skin notation**	N	**Skin notation**	N
Notes: See 29 CFR 1910.1000 Table Z-1.		**Notes:**		**Notes:** BEI®		**Notes:**	
Health factors: See NIH-NLM PubChem.		**IDLH**	700 ppm				
Carcinogenic classifications: IARC-3, TLV-A4, EPA-II		**Notes:**					
AIHA emergency response planning guidelines - ERPG-1/ERPG-2/ERPG-3: 350 ppm/700 ppm/3500 ppm							

developing fetus in animals. Its prolonged or repeated contact can cause a skin rash, dryness, and redness. Its exposure to very high levels may cause irregular heartbeat.

n-Hexane is currently adopted by the Standard Methods for the determination of O&G in water and wastewater [1]. *n*-Hexane is a colorless liquid (normal hexane, $C_{16}H_{14}$), with a gasoline-like odor. It is used in laboratories and as a solvent to remove vegetable oils from crops. It is also found in gasoline and rubber cement. Its chemical numbers are: CAS No. = 110–54-3, RTK substance No. = 1340; DOT No. = UN 1208. Its NJDOH hazard rating is: health = 2 (moderate); NFPA hazard rating are: flammability = 3 (serious); reactivity = 0 (minimal). It is flammable. Poisonous gases are produced in fire. Its containers may explode in fire. Additional hazard information are: hazard class = 3 (flammable); OSHA legal airborne PEL = 500 ppm averaged over an 8-hour work shift; NIOSH airborne REL = 50 ppm averaged over a 10-hour work shift; and ACGIH TLV = 50 ppm averaged over an 8-hour work shift. The above exposure limits are for air levels only. *n*-Hexane can affect a chemist when inhaled and by passing through the skin. It can cause reproductive damage, so it must be handled with extreme caution. Contact can irritate and burn the skin and eyes. Prolonged or repeated contact can cause a skin rash, dryness, and redness. Inhaling *n*-hexane can irritate the nose, throat, and lungs. Exposure can cause headache, nausea, vomiting, dizziness, lightheadedness, and passing out. Higher levels can cause coma and death. *n*-Hexane may damage the nervous system causing numbness, tingling, blurred vision, "pins and needles", and weakness in the hands and feet. *n*-Hexane is a very flammable liquid and a dangerous fire hazard [16].

n-Hexane is adopted by the Standard Methods [1] for O&G analysis. The authors compare the selected solvent with *n*-hexane and conclude that the TCE is better than *n*-hexane considering the safety of both laboratory workers and the laboratory although TCE is an ODS. Solvent recycle and/or proper laboratory waste management will eliminate the opportunity of ODS release into the environment.

Recycling and reclamation have limited potential in small water quality laboratories, due to the fact that the waste volumes generated are generally too small for economical waste management and reagent purity requirements are too high. However, organic solvents often can be distilled and recovered for reuse. Figure 12.2 shows the setup of distillate recovery apparatus (Source: APHA-AWWA-WEF) [1]. Alternatively, a water quality laboratory may use commercially available solvent recovery equipment.

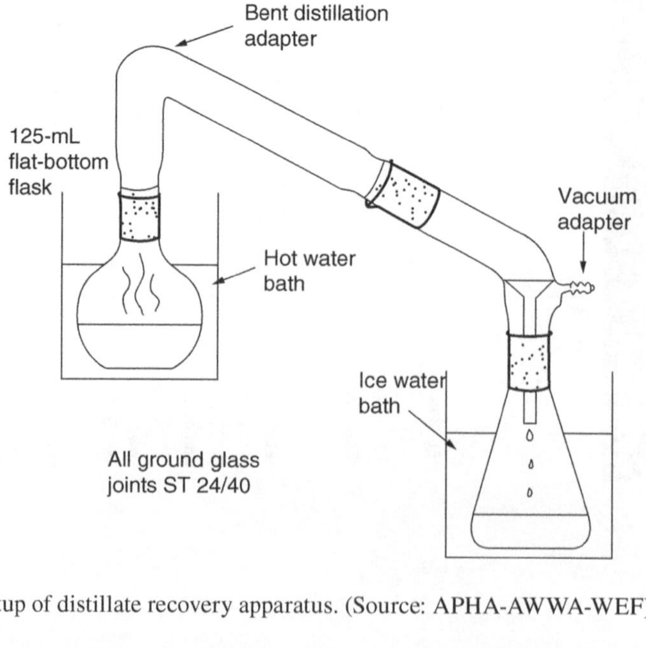

FIGURE 12.2 A setup of distillate recovery apparatus. (Source: APHA-AWWA-WEF) [1].

The laboratory hazardous wastes may be sent off site for further treatment and disposal. Extreme care must be taken in selecting a reputable waste handler and disposal firm because liability does not disappear when the waste leaves the generator's facility. A good waste management firm will assist laboratories in packaging and manifesting "lab packs" 19–208-L (5–55 gal) drums containing several smaller containers of wastes. A laboratory manager must ensure that the laboratory receives a copy of the completed manifest and certificate of treatment and/or disposal [1,10,17–19].

GLOSSARY

1,1,2-Trichloro-1,2,2-trifluoroethane (Freon 113): It is a colorless liquid (CFC 113, $C_2Cl_3F_3$), with a faint, sweet, or ether-like odor at high concentrations. It is used as a refrigerant, heat transfer medium, solvent, and chemical intermediate. Its chemical numbers are: CAS No.=76-13-1, RTK substance No.=1904. Its NJDOH hazard rating are: health=2 (moderate); flammability=0 (minimal); reactivity=0 (minimal). It does not burn, although poisonous gases may be produced in fire. Additional hazard information are: hazard class=6 (poison); OSHA legal airborne PEL=1,000ppm averaged over an 8-hour work shift; NIOSH airborne REL=1,000ppm averaged over a 10-hour work shift and 1,250ppm, not to be exceeded during any 15-minute work period; and ACGIH TLV=1,000ppm averaged over an 8-hour work shift, and 1,250ppm as a STEL (short-term exposure limit). Contact can irritate and burn the skin and cause frostbite. It is not classifiable as to its potential to cause cancer. There is no evidence that it may affect reproduction. Its prolonged or repeated exposure may cause bronchitis to develop with coughing, phlegm, and/or shortness of breath.

ACGIH: American Conference of Governmental Industrial Hygienists.

Action level: A concentration designated in 29 CFR Part 1910 for a specific substance, calculated as an 8-hour time-weighted average, which initiates certain required activities such as exposure monitoring and medical surveillance.

Acute Exposure Guideline Levels (AEGLs): AEGLs are established by the USEPA. They describe the risk to humans resulting from once-in-a lifetime, or rare, exposure to airborne chemicals.

Allergen: A substance which produces an allergic reaction upon contact with body tissues.

Assistant Secretary: The Assistant Secretary of Labor for Occupational Safety and Health, U.S. Department of Labor, or designee.

American Conference of Governmental Industrial Hygienists (ACGIH): It is an organization which publishes guidelines called TLVs for exposure to workplace chemicals.

Boiling point: It is the temperature at which a substance can change its physical state from a liquid to a gas.

Carcinogen: A carcinogen is a substance that causes cancer.

CAS number: It is unique, identifying number, assigned by the Chemical Abstracts.

CFR 29 Section 1910 Subpart Z: The section of OSHA regulations addressing Toxic and Hazardous Substances. Contains the Air Contaminants Standard (or PELs) in Section 1910.1000.

Chloroform: It is a colorless organic liquid (trichloromethane, $CHCl_3$), with a pleasant, sweet odor, which is used as a laboratory or industrial solvent, and to make refrigerants, resins, and plastics, but it is no longer used as an anesthetic. Its chemical numbers are: CAS No.=67-66-3, DOT No.=UN 1888. Its NJDOH hazard rating are: health=3 (serious); flammability=0 (minimal); reactivity=0 (minimal). Additional hazard information are: hazard class=6.1; OSHA legal airborne PEL=50ppm; NIOSH airborne REL=2ppm, which should not be exceeded in any 60-minute work period; and ACGIH TLV=10ppm averaged over an 8-hour work shift. It is a carcinogen, and its poisonous gases are produced in fire. It can affect workers when inhaled and may be absorbed through the skin. It may damage the developing fetus. Its exposure can cause headache, dizziness, lightheadedness, and passing out. High exposure can cause the heart to beat irregularly or to stop.

Code of Federal Regulations (CFR): It is the Code of Federal Regulations, which are the regulations of the US government. The collection of rules and regulations was originally published in the Federal Register by various governmental departments and agencies. OSHA regulations are found in 29 CFR; USEPA regulations in 40 CFR; and US Department of Transportation regulations in 49 CFR.

Combustible liquid: Any liquid having a flashpoint at or above 100°F, but below 200°F, except any mixture having components with flashpoints of 200°F, or higher, the total volume of which makes up 99% or more of the total volume of the mixture.

Combustible substance: A combustible substance is a solid, liquid, or gas that will burn.

Compressed gas: (a) A gas or mixture of gases having, in a container, an absolute pressure exceeding 40 psi at 70°F; or (b) A gas or mixture of gases having, in a container, an absolute pressure exceeding 104 psi at 130°F regardless of the pressure at 70°F; or (c) A liquid having a vapor pressure exceeding 40 psi at 100°F as determined by ASTM D-323–72.

Corrosive (OSHA): A chemical that causes visible destruction of, or irreversible alterations in, living tissue by chemical action at the site of contact.

Corrosive (USEPA): A waste is corrosive if it has either of the following properties: (a) It is aqueous and has a pH less than or equal to 2 or greater than or equal to 12.5; (b) It is a liquid and corrodes steel (SAE 1020) at a rate greater than 6.35 mm (0.250 inch) per year at a test temperature of 55°C (130°F).

Corrosive substance: A corrosive substance is a gas, liquid, or solid that causes destruction of human skin or severe corrosion of containers.

Critical temperature: It is the temperature above which a gas cannot be liquefied, regardless of the pressure applied.

Designated area: An area that may be used for work with "select carcinogens", reproductive toxins or substances which have a high degree of acute toxicity. A designated area may be the entire laboratory, an area of a laboratory, or a device such as a laboratory hood.

Emergency Response Guidebook (ERG): ERG is a guide for emergency responders for transportation emergencies involving hazardous substances.

Emergency Response Planning Guideline (ERPG): ERPG values provide estimates of concentration ranges where one reasonably might anticipate observing adverse effects.

Emergency: Any occurrence such as, but not limited to, equipment failure, rupture of containers, or failure of control equipment which results in an uncontrolled release of a hazardous chemical into the workplace.

Explosive: A chemical that causes a sudden, almost instantaneous release of pressure, gas, and heat when subjected to sudden shock, pressure, or high temperature.

Fetus: A fetus is an unborn human or animal.

Flammable substance: A flammable substance is a solid, liquid, vapor, or gas that will ignite easily and burn rapidly.

Flammable: A chemical that falls into one of the following categories: (a) Aerosol flammable means an aerosol that, when tested by the method described in 16 CFR 1500.45, yields a flame protection exceeding 18 inches at full valve opening, or a flashback at any degree of valve opening; (b) Gas flammable means: (b1) A gas that, at ambient temperature and pressure, forms a flammable mixture with air at a concentration of 13% by volume or less; or (b2) A gas that, at ambient temperature and pressure, forms a range of flammable mixtures with air wider than 12% by volume, regardless of the lower limit. (c) Liquid flammable means any liquid having a flashpoint below 100°F, except any mixture having components with flashpoints of 100°F or higher, the total of which makes up 99% or more of the total volume of the mixture. (d) Solid flammable means a solid, other than blasting agent or explosive as defined in § 1910.109(a), that is liable to cause fire through friction, absorption of moisture, spontaneous chemical change, or retained heat from manufacturing or processing, or which can be ignited readily and when ignited burns so vigorously and persistently as to create a serious hazard. A chemical shall be considered to be a flammable solid if, when tested by the method described in 16 CFR 1500.44, ignites and burns with a self-sustained flame at a rate of one-tenth of an inch per second along its major axis.

Flashpoint: The minimum temperature at which a liquid or solid gives off a vapor in sufficient concentration to ignite with air when tested by one of the following methods: (a) Tagliabue Closed Tester. (b) Pensky-10 Martens Closed Tester. (c) Setaflash Closed Tester. Organic peroxides, which undergo auto accelerating thermal decomposition, are excluded from any of the flashpoint determination methods specified above.

Hazard warning: Any words, pictures, symbols, or combination thereof appearing on a label or other appropriate form of warning which convey the hazard(s) of the chemical(s) in the container(s).

Hazardous chemical: A chemical for which there is statistically significant evidence based on at least one study conducted in accordance with established scientific principles that acute or chronic health effects may occur in exposed employees.

Health hazard: Includes chemicals that are carcinogens, toxic or highly toxic agents, reproductive toxins, irritants, corrosives, sensitizers, hepatotoxins, nephrotoxins, neurotoxins, agents which act on the hematopoietic system, and agents which damage the lungs, skin, eyes, or mucous membranes.

Highly toxic: A chemical falling within any of the following categories: (a) A chemical that has a median lethal dose (LD50) of 50 mg or less per kilogram of body weight when administered orally to albino rats weighing between 200 and 300 g each; (b) A chemical that has a median lethal dose (LD50) of 200 mg or less per kilogram of body weight when administered by continuous contact for 24 hours (or less if death occurs within 24 hours) with the bare skin of albino rabbits weighing between 2 and 3 kg each; (c) A chemical that has a median lethal concentration (LC50) in air of 200 ppm by volume or less of gas or vapor, or 2 mg/L or less of mist, fume, or dust, when administered by continuous inhalation for 1 hour (or less if death occurs within 1 hour) to albino rats weighing between 200 and 300 g each.

Integrated Risk Information System (IRIS): It is the Integrated Risk Information System database on human health effects that may result from exposure to various chemicals, maintained by federal USEPA.

International Agency for Research on Cancer (IARC): It is an international scientific group involved in cancer research.

Ionization potential: It is the amount of energy needed to remove an electron from an atom or molecule. It is measured in electron volts.

Known oil: A *Known Oil* is defined as a sample of oil and/or grease that represents the only material of that type used or produced in the processes or represented by the process water or wastewater.

Laboratory scale: Work with substances in which the containers used for reactions, transfers, and other handling of substances are designed to be easily and safely manipulated by one person. "Laboratory scale" excludes those workplaces whose function is to produce commercial quantities of materials.

Laboratory use of hazardous chemicals: Handling or use of such chemicals in which all of the following conditions are met: (a) Chemical manipulations are carried out on a "laboratory scale"; (b) Multiple chemical procedures or chemicals are used; (c) The procedures involved are not part of a production process, nor in any way simulate a production process; and (d) "Protective laboratory practices and equipment" are available and in common use to minimize the potential for employee exposure to hazardous chemicals.

Laboratory: A facility where the "laboratory use of hazardous chemicals" occurs. It is a workplace where relatively small quantities of hazardous chemicals are used on a non-production basis.

Laboratory-type hood: A device located in a laboratory, enclosure on five sides with a movable sash or fixed, partially enclosed on the remaining side; constructed and maintained to draw air from the laboratory and to prevent or minimize the escape of air contaminants into the laboratory; and allows 11 chemical manipulations to be conducted in the enclosure without insertion of any portion of the employee's body other than the hands and arms.

Lower Explosive Limit (LEL): It is the lowest concentration of a combustible substance (gas or vapor) in the air capable of continuing an explosion.

Median Lethal Concentration (LC50): The concentration of a material in air that on the basis of laboratory tests (respiratory route) is expected to kill 50% of a group of test animals as a single exposure in a specified time period.

Median Lethal Dose (LD50): The single dose of a substance that causes the death of 50% of an animal population from exposure to the substance by any route other than inhalation.

Medical consultation: A consultation that takes place between an employee and a licensed physician for the purpose of determining what medical examinations or procedures, if any, are appropriate in cases where a significant exposure to a hazardous chemical may have taken place.

Methyl chloroform: It is a colorless organic liquid (1,1,1-trichloroethane, or methyltrichloromethane, $C_2H_3Cl_3$), with an ether-like odor, which is used as a laboratory or industrial solvent, spot cleaner, or degreaser, and a raw material in making other chemicals. Its chemical numbers are: CAS No.=71-55-6, DOT No.=UN 2832. Its NFPA hazard rating are: health=2 (moderate); flammability=1 (slight); reactivity=0 (minimal). Additional hazard information are: hazard class=6 (poison); OSHA legal airborne PEL=350 ppm averaged over an 8-hour work shift; NIOSH airborne REL=350 ppm, which should not be exceeded during any 15-minute work period; and ACGIH TLV=350 ppm averaged over an 8-hour work shift, and 450 ppm as a STEL. It is not classifiable as to its potential to cause cancer. It may cause mutations or genetic changes. There is limited evidence that methyl chloroform may damage the developing fetus in animals. Its prolonged or repeated contact can cause a skin rash, dryness, and redness. Its exposure to very high levels may cause irregular heartbeat.

mg/m^3: It means milligrams of a chemical in a cubic meter of air. It is a measure of concentration (weight/volume).

MSDS: It is Material Safety Data Sheet. See Safety Data Sheet.

Mutagen: A mutagen is a substance that causes mutations.

Mutation: A mutation is a change in the genetic material in a body cell. Mutations can lead to birth defects, miscarriages, or cancer.

National Fire Protection Association (NFPA): It is an organization that classifies substances according to their fire and explosion hazard.

National Institute for Occupational Safety and Health (NIOSH): NIOSH is an organization that tests equipment, evaluates and approves respirators, conducts studies of workplace hazards, and proposes standards to OSHA.

National Toxicology Program (NTP): It is a US national program that tests chemicals and reviews evidence for cancer.

NFPA: National Fire Protection Association.

***n*-Hexane :** It is a colorless liquid (normal hexane, $C_{16}H_{14}$), with a gasoline-like odor. Is used in laboratories and as a solvent to remove vegetable oils from crops. It is also found in gasoline and rubber cement. Its chemical numbers are: CAS No.=110–54-3, RTK substance No.=1340; DOT No.=UN 1208. Its NJDOH hazard rating is: health=2 (moderate); NFPA hazard rating are: flammability=3 (serious); reactivity=0 (minimal). It is flammable. Poisonous gases are produced in fire. Its containers may explode in fire. Additional hazard information are: hazard class=3 (flammable); OSHA legal airborne PEL=500 ppm averaged over an 8-hour work shift; NIOSH airborne REL=50 ppm averaged over a 10-hour work shift; and ACGIH TLV=50 ppm averaged over an 8-hour work shift. The above exposure limits are for air levels only. *n*-Hexane can affect a chemist when inhaled and by passing through the skin. It can cause reproductive damage, so it may be handled with extreme caution. Contact can irritate and burn the skin and eyes. Prolonged or repeated contact can cause a skin rash, dryness, and redness. Inhaling n-hexane can irritate the nose, throat, and lungs. Exposure can cause headache, nausea, vomiting, dizziness, lightheadedness, and passing out. Higher levels can cause coma and death. *n*-Hexane may damage the nervous system causing numbness, tingling, blurred vision, "pins and needles", and weakness in the hands and feet. *n*-Hexane is a very flammable liquid and a dangerous fire hazard.

NJDEP: It is the New Jersey Department of Environmental Protection, USA.

Occupational Safety and Health Administration (OSHA): It is the US federal agency that adopts and enforces health and safety standards.

Oil and grease (O&G): The O&G definition is based on the procedure used. The O&G may include hydrocarbons, vegetable oils, animal fats, waxes, soaps, greases, and other related substances. The source of the oil and/or grease, and the presence of extractable non-oily matter will influence the substance measured and interpretation of results.

Organic peroxides: An organic compound that contains the bivalent -O-O- structure and which may be considered to be a structural derivative of hydrogen peroxide where one or both of the hydrogen atoms has been replaced by an organic radical.

OSHA: Occupational Safety and Health Administration.

Oxidizer: A chemical other than a blasting agent or explosive as defined in §1910.109(a), that initiates or promotes combustion in other materials, thereby causing fire either of itself or though the release of oxygen or other gases.

PEOSHA: It is the New Jersey Public Employees Occupational Safety and Health Act, which adopts and enforces health and safety standards in public workplaces.

Permeated: It is the movement of chemicals through protective materials.

Permissible Exposure Limit (PEL): The maximum air contaminant concentration a worker can be exposed to on a repeated basis without developing adverse effects. Published and enforced by OSHA as a legal standard.

Physical hazard: A chemical for which there is scientifically valid evidence that it is a combustible liquid, a compressed gas, explosive, flammable, an organic peroxide, an oxidizer pyrophoric, unstable, or water reactive.

ppm: It means parts of a substance per million parts of air. It is a measure of concentration by volume in air.

Principal investigator: An individual with primary responsibility for a program or project in a specific laboratory.

Protective Action Criteria (PAC): They are values established by the US Department of Energy and are based on AEGLs and ERPGs. They are used for emergency planning of chemical release events.

Protective laboratory practices and equipment: Those laboratory procedures, practices, and equipment accepted by laboratory health and safety experts as effective, or that the employer can show to be effective, in minimizing the potential for employee exposure to hazardous chemicals.

Reactive (EPA): A waste is reactive if it has any of the following properties: (a) It is normally unstable and readily undergoes violent change without detonating; (b) It reacts violently with water; (c) It forms potentially explosive mixtures with water; (d) When mixed with water, it generates toxic gases, vapors, or fumes in a quantity sufficient to present a danger to human health or the environment; (e) It is a cyanide or sulfide bearing waste which, when exposed to pH conditions between 2 and 12.5, can generate toxic gases, vapors, or fumes in a quantity sufficient to present a danger to human health or the environment; (f) It is capable of detonation or explosive reaction if it is subjected to a strong initiating source or if heated under confinement; (g) It is readily capable of detonation or explosive decomposition or reaction at standard temperature and pressure; (h) It is identified by the DOT as a forbidden explosive, Class A explosive, or Class B explosive.

Reactive substance: A reactive substance is a solid, liquid, or gas that releases energy under certain conditions.

Reproductive toxins: Chemicals that affect the reproductive capabilities including chromosomal damage (mutations) and effects on fetuses (teratogenesis).

Safety Data Sheet (SDS): Written or printed material concerning a hazardous chemical which is prepared in accordance with the requirements of the Hazard Communication Standard.

Select carcinogen: Any substance that meets one of the following criteria: (a) It is regulated by OSHA as a carcinogen; or (b) It is listed under the category, "known to be carcinogens" in the Annual Report on Carcinogens published by the National Toxicology Program (NTP) (latest edition); or (c) It is listed under Group 1 ("carcinogenic to humans") by the International Agency for Research on Cancer Monographs (IARC) (latest editions); or (d) It is listed in either Group 2A or 2B by IARC or under the category, "reasonably anticipated to be carcinogens" by NTP, and causes statistically significant tumor incidence in experimental animals in accordance with any of the following criteria: (a) After inhalation exposure of 6–7 hours per day, 5 days per week, for a significant portion of a lifetime to dosages of less than 10 mg/m^3; (b) After repeated skin application of less than 300 (mg/kg of body weight) per week; or (c) After oral dosages of less than 50 mg/kg of body weight per day.

Sensitizer: A chemical that causes a substantial proportion of exposed people or animals to develop an allergic reaction in normal tissue after repeated exposure to the chemical.

Short Term Exposure Limit (STEL): STEL is usually a 15-minute exposure that should not be exceeded at any time during a workday.

Teratogen: A teratogen is a substance that causes birth defects by damaging the fetus.

Threshold Limit Value (TLV): Represents the air concentrations of chemical substances to which it is believed that workers may be exposed daily without adverse effect. Published by the ACGIH.

Unknown oil: An *Unknown Oil* is defined as one for which a representative sample of the oil or grease is not available for preparation of a standard. The O&G in a mixed sewage or an unidentified oil slick on a surface water are typical examples of unknown oil.

Unstable (reactive) (OSHA): A chemical which is the pure state, or as produced or transported, will vigorously polymerize, decompose, condense, or become self-reactive under conditions of shocks, pressure, or temperature.

Upper Explosive Limit (UEL): UEL is the highest concentration in air above which there is too much fuel (gas or vapor) to begin a reaction or explosion.

USDOT: It is the US Department of Transportation, the federal agency that regulates the transportation of chemicals.

USEPA: It is the US Environmental Protection Agency, the federal agency responsible for regulating environmental hazards and drinking water, and other environmental affairs.

Vapor density: It is the ratio of the weight of a given volume of one gas to the weight of another (usually air), at the same temperature and pressure.

Vapor pressure: The vapor pressure is a force exerted by the vapor in equilibrium with the solid or liquid phase of the same substance. The higher the vapor pressure the higher concentration of the substance in air.

Water reactive: A chemical that reacts with water to release a gas that is either flammable or presents a health hazard.

REFERENCES

1. Baird, RB, Eaton, AD & Rice, EW (2017). *Standard Methods for the Examination of Water and Wastewater*, 23rd Edition. American Public Health Association, Washington, DC, USA.
2. USEPA (1979). Methods for chemical analysis of water and wastes. Report No. EPA-600/4-79-020, US. Environmental Protection Agency, Washington, DC, USA, p. 229–235.

3. API (1958). *Manual in Disposal of Refinery Wastes, Volume* IV, *Method 733–58*. American Petroleum Institute, Washington, DC, USA.
4. Hatfield, WD & Symons, GE (1945). The determination of grease in sewage, *Sewage Works J.* 17:16.
5. Kirschman, HD & Pomeroy, R (1949). Determination of oil in oil field waste waters, *Anal. Chem.* 21:793.
6. Gilcreas, FW, Sanderoon, WW & Elmer, RP (1953). TWO new methods for the determination of grease in sewage, *Sewage Ind. Wastes* 25:1379.
7. Ullmann, WW & Sanderoon, WW (1959). A further study of methods for the determination of grease in sewage, *Sewage Ind. Wastes* 31:8.
8. Chanin, G, Chow, EH, Alexander, RB & Powers, JF (1967). A safe solvent for oil and grease analysis, *J. Water Pollution Cont. Fed.* 39:1892.
9. Gruenfeld, M (1973). Extraction of dispersed oils from water for quantitative analysis by infrared spectrophotometry, *Environ. Sci. Technol.* 7:636.
10. Krofta, M & Wang, LK (1985). Hazardous waste management in institutions and colleges. PB86-184180/AS, US Department of Commerce, National Technical Information Service, Alexandria, VA, USA.
11. Snider, EH (1992). Waste minimization. In: *Handbook of Industrial Waste Treatment*, Wang, LK and Wang, MHS (editors). Marcel Dekker, New York, USA, pp. 1–60.
12. Wang, MHS & Wang, LK (2025). COD determination and cleaning solution preparation using potassium permanganate for hazardous waste minimization. In: *Control of Heavy Metals in the Environment, Volume 1*, Wang, LK, Wang, MHS, Hung, YT, and Chen, JP (editors). CRC Press, Boca Raton, FL, USA, pp. 335–344.
13. Wang, LK & Wang, MHS (2025). Oil and grease determination, solvent risks, and waste disposal when using trichlorotrifluoroethane for extraction. In: *Industrial Water and Waste Treatment: Technology Developments*, Wang, LK, Wang, MHS and Hung, YT (editors). CRC Press, Boca Raton, FL, USA.
14. Wang, LK, Wang, MHS, Hung, YT, Aziz, HA & Sabiani, NH (2026). Oil and grease determination, solvent risks, and waste disposal when using trichloroethane (methyl chloroform) for extraction. In: *Industrial Water and Waste Treatment: Fundamentals,* Wang, LK, Wang, MHS and Hung, YT (editors). CRC Press, Boca Raton, FL, USA. pp. 399–419.
15. Wang, LK, Wang, MHS, Omar, FM & Aziz, HA (2025). Oil and grease determination, solvent risks, and waste disposal when using trichloromethane (chloroform) for extraction. In: *Industrial Water and Waste Treatment: Research Developments*, Wang, LK, Wang, MHS and Hung, YT (editors). CRC Press, Boca Raton, FL, USA.
16. NJDOH (2022). *Hazardous Substance Fact Sheet*. New Jersey Department of Health, Trenton, NJ, USA.
17. Old Dominion University (2022). *Chemical Hygiene Plan*. Old Dominion University, Norfolk, VA, USA.
18. Rutgers University (2022). *OSHA Laboratory Standard & Chemical Hygiene Guide*. Rutgers University, New Brunswick, NJ, USA.
19. The City College of New York (2018). *Chemical Hygiene Plan*. The City College of New York, New York, USA, August 16.
20. Wang, MHS, Wang, LK & Shammas, NK (2020). Glossary of climate change, global warming and ozone layer protection. In: *Handbook of environment and Waste Management: Acid Rain and Greenhouse Gas Pollution Control*, Hung, YT, Wang, LK and Shammas, NK (editors). World Scientific Singapore, Singapore, pp. 689–749.
21. Wang, MHS, Wang, LK & DeMichele, E (2018). BOD determination, cleaning solution preparation, and waste disposal in laboratories. In: *Handbook of Advanced Industrial and Hazardous Wastes Management*, Wang, LK, Wang, MHS, Hung, YT, Shammas, NK, and Chen, JP (editors). CRC Press, Boca Raton, FL, USA, pp. 797–808.
22. Wang, MHS, Wang, LK & DeMichele, E (2018). Principles, procedures, and heavy metal management of dichromate reflux method for COD determination in laboratories. In: *Handbook of Advanced Industrial and Hazardous Wastes Management*, Wang, LK, Wang, MHS, Hung, YT, Shammas, NK, and Chen, JP (editors). CRC Press, Boca Raton, FL, USA. pp. 809–818.
23. Wang, LK, Wang, MHS, Shammas, NK, Renak, V, Hung, YT & Amuda, OS (2021). Cationic surfactant analysis with good laboratory practice and waste management. In: *Integrated Natural Resources Research*, Wang, LK, Wang, MHS, and Hung, Y-T (editors). Springer Nature, Cham, Switzerland. pp. 481–502.
24. Wang, LK, Wang, MHS & Thayer, AE (2021). A spectrophotometrical method for determination of dissolved proteins in water and wastewater. In: *Environmental Flotation Engineering*, Wang, LK, Wang, MHS, Shammas, NK, and Aulenbach, DB (editors). Springer Nature, Cham, Switzerland, pp. 371–398.

25. Wang, LK (1975). Modified methylene blue method for estimating MBAS concentration. *J. Am. Water Works Assoc.* 67(1):19–21. https://www.jstor.org/stable/41267738. Accessed 17 May 2022.
26. Wang, LK, Shuster, WW & Panzardi, PJ (1975). Evaluation of two methylene blue methods for analyzing MBAS concentrations in aqueous solutions. *J. Am. Water Works Assoc.* 67(4):182–184. https://www.jstor.org/stable/41268195. Accessed 17 May 2022.
27. Wang, LK & Langley, D (1975). Determining cationic surfactant concentration. *Ind. Eng. Chem. Prod. Res. Dev.* 14(3):210–212; American Chemical Society. https://doi.org/10.1021/i360055a601

13 Colorimetric Technology Fundamentals and A Modified Standard Colorimetric Method for Determination of Lignin and Tannin in Industrial Waters

Mu-Hao Sung Wang and Lawrence K. Wang

ACRONYM

APHA	American Public Health Association
AWWA	American Water Works Association
LIWT	Lenox Institute of Water Technology
QC	Quality control
WEF	Water Environment Federation

13.1 INTRODUCTION

13.1.1 Fundamentals of Colorimetric Analytical Method

Colorimetric analysis is an analytical technique used to determine the concentration of a substance in a sample by measuring the intensity of color produced during a chemical reaction. This method relies on the principle that the amount of light absorbed by a colored solution is proportional to the concentration of the analyte. Applications of the colorimetric analysis include: (a) Clinical Diagnostics: Used for measuring concentrations of glucose, cholesterol, and other biomarkers in blood and urine samples; (b) Environmental Monitoring: Detection of pollutants like tannins, lignins, nitrates, phosphates, and heavy metals in water samples; (c) Food and Beverage Industry: Analysis of food additives, pollutants, impurities, and nutritional content; and (d) Research Laboratories: Quantification of various substances in biochemical, physicochemical, and chemical research.

There are two key principles: (a) Beer-Lambert Law: This law states that the absorbance of light by a solution is directly proportional to the concentration of the absorbing species and the path length of the light through the solution; and (b) Chromogens: Substances that develop color when they react with specific reagents. The intensity of the color is measured to determine the concentration of the analyte.

The advantages of colorimetric analytical method can be summarized here: (a) simplicity: easy to perform with minimal training required; (b) cost-effective: less expensive compared to other analytical techniques; (c) rapid analytical results: provides quick results, which is crucial in fields like clinical diagnostics; (d) high sensitivity: capable of detecting low concentrations of analytes; (e) quantitative analysis: allows for precise measurement of analyte concentrations; and (f) versatility: applicable to a wide range of analytes, including ions, small molecules, proteins, and nucleic acids.

DOI: 10.1201/9781003532354-13

There are many disadvantages and limitations when using the colorimetric analytical method.

1. **Limited to Colored Compounds:** Cannot analyze colorless compounds.
2. **Interferences:** Other substances with similar colors can interfere with the results and cause inaccuracies.
3. **Sample Preparation:** Requires preparation and handling of samples, reagents, and standard solutions and sometimes requires more sample volume.
4. **Sensitivity:** Lower sensitivity compared to some advanced analytical techniques.
5. **Wavelength Limitations:** Limited to the visible light spectrum, not suitable for UV or IR analysis.
6. **Accuracy:** May not be as accurate for complex mixtures due to overlapping absorption spectra.

Colorimetric analysis is a valuable tool in various fields due to its simplicity, cost-effectiveness, and rapid results. However, it is limited to colored compounds and can be affected by interferences and sample preparation requirements. Despite these limitations, it remains a widely used method for quantitative analysis in environmental monitoring, food testing, clinical diagnostics, and research.

13.1.2 Environmental Engineering Significance

Lignin is a plant constituent that often is discharged as a waste during the manufacture of paper pulp [1,2]. Another plant constituent, tannin, may enter the water supply through the process of vegetable matter degradation or through the wastes of the tanning industry [3,4]. Tannin also is applied in the so-called internal treatment of boiler waters, where it reduces scale formation by causing the production of a more easily handled sludge [5–8].

13.1.3 Summary

A modified Standard Method is developed for elimination of the interferences caused by calcium ions which are present in many industrial water or wastewater, such as cooling tower water, boiler wastewater, lime-treated paper, pulp wastewater, etc. According to the "Standard Methods for the Examination of Water and Wastewater", Method 5550 Tannin and Lignin [1], 1 mL Folin phenol reagent and 10 mL carbonate-tartrate reagent are added to 50 mL of clear sample (above 20°C) and mixed well. After waiting 30 minutes for color development, a spectrophotometric or photometric reading of the sample is to be made and lignin/tannin concentration determined from a calibrated curve previously prepared. When industrial water or wastewater contains high concentrations of calcium ions, precipitates ($CaCO_3$) are formed at the same time of the color development, therefore causing interferences at the time of obtaining instrument readings. This publication introduces a modified Standard Method which adds the extra analytical steps of: (a) filtration of the test water sample with 0.45 μm pore-size membrane filter immediately after the completion of the color development period, so accurate instrument readings can be obtained; (b) preparation of calibration curves when using instrument; and (c) preparation of Standard Nessler tubes for color comparison in case an instrument is not available.

Since the analytical procedures introduced in this publication are meant to be a modified "Standard Method", the authors include the exact words in original "Standard Methods 5550 Tannin and Lignin" [1], but add the authors' recommended extra analytical procedures for the convenience of the readers.

13.1.4 Chemical Reagent Addition, Chemical Reactions, and Color Development

Add in rapid succession 1 mL Folin phenol reagent and 10 mL carbonate-tartrate reagent. Allow 30 minutes for chemical reactions and color development.

13.2 PRINCIPLE OF THE STANDARD COLORIMETRIC METHOD

Both lignin and tannin contain aromatic hydroxyl groups that react with Folin phenol reagent (tungstophosphoric and molybdophosphoric acids) to form a blue color suitable for estimation of concentrations up to at least 9 mg/L. However, the reaction is not specific for lignin or tannin, nor for compounds containing aromatic hydroxyl groups, inasmuch as many other reducing materials, both organic and inorganic, respond similarly [1].

13.3 APPLICABILITY OF THE STANDARD COLORIMETRIC METHOD

This method is generally suitable for the analysis of any organic chemical that will react with Folin phenol reagent to form measurable blue color at the concentration of interest. However, many compounds are reactive and each yields a different molar extinction coefficient (color intensity). Hence, the analyst must demonstrate conclusively the absence of interfering substances.

13.4 INTERFERENCES OF THE STANDARD COLORIMETRIC METHOD

Any substance able to reduce Folin phenol reagent will produce a false-positive response. Organic chemicals known to interfere include hydroxylated aromatics, proteins, humic substances, nucleic acid bases, fructose, and amines. Inorganic substances known to interfere include iron(II), manganese, nitrite cyanide, bisulfite, sulfite, sulfide, hydrazine, and hydroxylamine hydrochloride. Both 2 mg ferrous iron/L and 125 mg sodium sulfite/L individually produce a color equivalent to 1 mg tannic acid/L [1].

13.4.1 Minimum Detectable Concentrations

Approximately 0.025 mg/L for phenol and tannic acid and 0.1 mg/L for lignin with a 1-cm-path-length spectrophotometer.

13.4.2 Quality Control (QC)

The QC practices considered to be an integral part of each method are summarized in Table 5020:1 of the Standard Methods [1].

13.5 ANALYTICAL APPARATUS

One of the following colorimetric equipment is required:

13.5.1 Spectrophotometer

A spectrophotometer is used at 700 nm. A light path of 1 cm or longer yields satisfactory results.

13.5.2 Filter Photometer

A filter photometer is provided with a red filter exhibiting maximum transmittance in the wavelength range of 600–700 nm. Sensitivity improves with increasing wavelength. A light path of 1 cm or longer yields satisfactory results.

13.5.3 Nessler Tubes

They are matched, 100-mL, tall form, marked at 50-mL volume [1].

13.6 ANALYTICAL REAGENTS

13.6.1 Folin Phenol Reagent

Transfer 100 g sodium tungstate, Na_2WO_4 $2H_2O$, and 25 g sodium molybdate, $Na_2MoO_4 \bullet 2H_2O$, together with 700 mL distilled water, to a 2,000-mL flat-bottom boiling flask. Add 50 mL 85% H_3PO_4 and 100 mL cone HCl. Connect to a reflux condenser and boil gently for 10 hours. Add 150 g Li_2SO_4, 50 mL distilled water, and a few drops of liquid bromine. Boil without condenser for 15 minutes to remove excess bromine. Cool to 25°C, dilute to 1 L, and filter. Store finished reagent, which should have no greenish tint, in a tightly stoppered bottle to protect against reduction by air-borne dust and organic materials.

Alternatively, purchase commercially prepared Folin phenol reagent and use it before the recommended expiration date.

13.6.2 Carbonate-Tartrate Reagent

Dissolve 200 g Na_2CO_3 and 12 g sodium tartrate, $Na_2C_4H_4O_6$ $2H_2O$, in 750 mL hot distilled water, cool to 20°C, and dilute to 1 L.

13.6.3 Stock Solution

The nature of the substance present in the sample dictates the choice of chemical used to prepare the standard because each substance produces a different color intensity. Weigh 1.000 g tannic acid, tannin, lignin, or other compound being used for boiler water treatment or known to be a contaminant of the water sample. Dissolve in distilled water and dilute to 1,000 mL. If the identity of the compound in the water sample is not known, use phenol and report results as "substances reducing Folin phenol reagent" in mg phenol/L. Interpret such results with caution.

Note that tannin and lignin are not individual chemical species of known molecular weight and structure; rather, they are substances containing a spectrum of chemicals of different molecular weights.

Their chemical properties depend on the source and method of isolation. If a particular substance is being added to the water, use it to prepare the stock solution.

13.6.4 Standard Solutions Preparation

Dilute the stock solution with distilled water to desired range. Prepare a minimum of three standards bracketing the expected sample concentration range [1].

13.7 ANALYTICAL PROCEDURE

13.7.1 Sample Storage and Preparation

Bring 50-mL portions of clear sample and standards to a temperature above 20°C and maintain within a ±2°C range.

13.7.2 Chemical Reagent Addition, Chemical Reactions, and Color Development

Add in rapid succession 1 mL Folin phenol reagent and 10 mL carbonate-tartrate reagent. Allow 30 minutes for chemical reactions and color development.

13.7.3 Filtration after Color Development

Filter the water sample from Step 13.7.2 with 0.45 μm pore-size membrane filter right after the completion of the color development period.

13.7.4 Standard Solution Calibration Curves and/or Standard Solution Nessler Tubes Preparation

13.7.4.1 Calibration Curves Preparation

Follow the analytical procedures in Sections 13.7.1–13.7.3 using the Standard Solutions prepared in Section 13.6.4 and using spectrophotometer or photometer for plotting Standard Solution Calibration Curves for lignin or tannin, depending on the standard solution and the target contaminant. Use the following guide for instrumental measurement at a wavelength of 700 nm:

13.7.4.1.1 Tannic Acid in 61-mL Final Volume

1. Light path 1 cm if tannic acid is in the range of 50–600 μg
2. Light path 5 cm if tannic acid is in the range of 10–150 μg

13.7.4.1.2 Lignin in 61-mL Final Volume

1. Light path 1 cm if lignin is in the range of 100–1,500 μg
2. Light path 5 cm if lignin is in the range of 30–400 μg

13.7.4.2 Standard Nessler Tubes Preparation

Follow the analytical procedures in Sections 13.7.1–13.7.3 using the Standard Solutions prepared in Section 13.6.4 and using Nessler Tubes for the preparation of a set of Standard Nessler Tubes with known lignin or tannin concentrations.

13.7.5 Final Lignin or Tannin Concentration Determination

13.7.5.1 Spectrophotometer or Photometer Application

Make spectrophotometric readings or photometric readings against a reagent blank prepared at the same time, and determine the concentration in accordance with the calibration curves prepared in Procedure Section 13.7.4.1. Report results in mg/L of the compound known to be present or as "substances reducing Folin phenol reagent" in mg phenol/L.

When using this Modified Standard Method for regulatory compliance, use spectrophotometric or photometric detection.

13.7.5.2 Nessler Tube Application

Compare visually against the lignin or tannin Standard Nessler Tubes simultaneously prepared in Section 13.7.4.2, match the Standard Nessler Tubes, and determine the lignin or tannin concentration. Report results in mg/L of the compound known to be present or as "substances reducing Folin phenol reagent" in mg phenol/L.

13.8 PRECISION AND BIAS

In a single laboratory analyzing seven replicates for phenol at 0.1 mg/L the precision was ±7% and recovery was 107% [1].

REFERENCES

1. Baird, R.B., A.D. Eaton, and E.W. Rice (2017). *Standard Methods for the Examination of Water and Wastewater.* 23rd edition, Century Celebration; APHA/AWWA/WEF, American Public Health Association, Washington, DC, USA.
2. Wang, M.H.S. (1972). Separation of lignin from aqueous solution by adsorptive bubble separation processes. PhD Thesis, Rutgers University, New Brunswick, NJ, USA. 241 pages.
3. Wang, M.H.S., M.L. Granstrom, T.E. Wilson, and L.K. Wang (1974). Removal of lignin from water by precipitate flotation. *J. Environ. Eng. Div. ASCE* 100:629–640.
4. Wang, M.H.S., and L.K. Wang (2026). Cooling tower and boiler water management. In: *Industrial Water and Waste Treatment: Pollution Control*, Wang, L.K., Wang, M.H.S. and Hung, Y.T. (editors). CRC Press, Boca Raton, FL, USA.
5. Wang, L.K., and M.H.S. Wang (2025). Cooling tower and boiler water treatment s. In: *Industrial Water and Waste Treatment: Process Developments*, Wang, L.K., Wang, M.H.S. and Hung, Y.T. (editors). CRC Press, Boca Raton, FL, USA.
6. Berk, A.A., and W.C. Schroeder (1942). Determination of tannin substances in boiler water. *Ind. Eng. Chem. Anal. Ed.* 14:456.
7. Kloster, M.B. (1974). Determination of tannin and lignin. *J. Am. Water Works Assoc.* 66:44.
8. Box, J.D. (1983). Investigation of the Folin-Ciocalteu phenol reagent for the determination of polyphenolic substances in natural waters. *Water Res.* 17:511; discussion by SJ. Randtke & R.A. Larson. 1984. *Water Res.* 18:1597.

14 Ion Exchange Engineering Fundamentals

Mu-Hao Sung Wang and Lawrence K. Wang

14.1 INTRODUCTION TO IX PROCESS

14.1.1 PROCESS DESCRIPTION AND APPLICATIONS

Ion exchange (IX) is a chemical process where ions are exchanged between a solution and an IX material, typically a resin. It is a reversible chemical process in which ions from an insoluble permanent solid medium (ion exchanger, or IX resin) are exchanged for ions in a solution or fluid mixture surrounding the insoluble medium. The direction of the exchange depends on the selective attraction of the ion exchanger resin for the certain ions present and the concentration of the ions in the solution [1]. This process is widely used for water purification, separation, and decontamination. The IX material can be either cationic (exchanging positive ions) or anionic (exchanging negative ions), depending on the application.

IX process has a broad range of environmental, municipal, and industrial applications across various industries: (a) water treatment: used for water softening, deionization, and removal of contaminants like iron, manganese, calcium, magnesium, heavy metals, and nitrates; (b) food and beverage: purification of ingredients and removal of unwanted ions to improve product quality; (c) pharmaceuticals: purification of drugs and removal of impurities; (d) power generation: production of high-purity water for boilers and cooling systems; (e) metal finishing: recovery and purification of metals from plating solutions; and (f) nuclear industry: separation and recovery of valuable radioactive isotopes [1–4].

14.1.2 IX PROCESS ADVANTAGES AND DISADVANTAGES

Advantages of the IX process include the following: (a) high efficiency: IX is capable of removing a wide range of contaminants to very low levels; (b) selectivity: IX can be tailored to selectively remove specific ions; (c) regenerability: IX resins can be regenerated and reused multiple times; and (d) environmental benefits: IX reduces the need for chemical additives and minimizes waste.

IX process has many limitations and disadvantages, so it must be selected properly for specific applications. Apparent disadvantages include: (a) cost factor: initial IX process setup and IX resin regeneration can be expensive; (b) process complexity: IX requires careful monitoring and maintenance to ensure optimal performance; (c) IX waste disposal: spent IX resins and regeneration waste need proper disposal to avoid environmental pollution; and (d) limited capacity: IX resins can become saturated and require periodic regeneration.

Iron and manganese removal can be accomplished by IX in the same way as calcium and magnesium (hardness) removal in water. Iron and manganese are exchanged with sodium on the IX resin. The resin can then be backwashed to remove the trapped iron and manganese and then can be regenerated for reuse. Softening with potassium chloride regeneration, however, may be more effective than sodium chloride in removing manganese. Some vendors recommend using special softener salt that contains additives (e.g., food-grade acids) that remove accumulated iron from the resin during regeneration. Not all water softeners are able to remove iron and manganese from water. Be sure to check the manufacturer's specifications. The raw water should not come in contact with any

DOI: 10.1201/9781003532354-14

oxidizing agent like air or chlorine before entering the softener. If oxidized iron and manganese are present in raw water, filtration should be used for the removal.

14.1.3 Types of IX Resins

IX resins are classified based on the type of ions they exchange and their functional groups. Here are the main types:

1. **Strong Acid Cation (SAC) Resins:** (a) functional groups are sulfonic acid groups; and (b) applications include water softening, demineralization, and removal of hardness ions like calcium and magnesium;
2. **Weak Acid Cation (WAC) Resins:** (a) functional groups are carboxylic acid groups; and (b) applications include dealkalization and softening of streams with high salinity levels;
3. **Strong Base Anion (SBA) Resins:** (a) functional groups are quaternary ammonium groups; and (b) applications include removal of silica, uranium, and nitrates; used in demineralization;
4. **Weak Base Anion (WBA) Resins:** (a) functional groups are primary, secondary, and tertiary amino groups; and (b) applications include effective demineralization where removal of SiO_2 and CO_2 is not required; also used for acid absorption;
5. **Specialty Resins:** (a) types include chelating resins (e.g., iminodiacetic acid, thiourea-based resins); and (b) applications include heavy metal removal and selective ion recovery.

The above IX resins are widely used in various industries for water treatment, purification, and separation processes.

14.2 REGENERATION OF IX RESINS

14.2.1 Regeneration Process Description

Regeneration of IX resins is a crucial chemical engineering process that restores the IX capacity of resins, allowing them to be reused multiple times. Here's a detailed overview of how it works:

The following is an operational procedure of the IX and resin regeneration:

1. **Exhaustion:** During the IX service cycle, the IX resin becomes saturated with unwanted ions from the treated solution. This state is known as exhaustion.
2. **Backwashing:** The resin bed is backwashed with water to remove any suspended solids and to reclassify the resin beads.
3. **Regenerant Introduction:** A regenerant solution is introduced to the resin bed. The type of regenerant depends on the resin: (a) cation exchange resins: typically regenerated with a strong acid like hydrochloric acid (HCl) or sulfuric acid (H_2SO_4); and (b) anion exchange resins: typically regenerated with a strong base like sodium hydroxide (NaOH).
4. **Ion Exchange:** The regenerant solution displaces the unwanted ions from the resin beads, replacing them with the ions from the regenerant. For example, in a cation exchange resin, H^+ ions from the acid replace the captured cations (e.g., Ca^{2+}, Mg^{2+}).
5. **Rinsing:** The resin bed is rinsed with water to remove any excess regenerant and displaced ions.

14.2.2 Types of Regeneration

There are two types of regeneration:

1. **Co-flow Regeneration:** The regenerant solution flows in the same direction as the service flow. This method is simpler but less efficient, often requiring more regenerant and resulting in higher leakage of contaminants.

2. **Counterflow Regeneration (CFR):** The regenerant solution flows in the opposite direction to the service flow. This method is more efficient, using less regenerant and reducing contaminant leakage, but requires more complex equipment.

14.2.3 Advantages and Disadvantages of IX Resin Regeneration

The advantages of IX resin regeneration include: (a) high efficiency: Regeneration allows for the reuse of resins, reducing operational costs; and (b) high flexibility: Different regenerants can be used depending on the specific application.

In comparison with other separation processes, the IX and its regeneration have many disadvantages, such as: (a) Chemical Handling: Requires careful handling and disposal of strong acids and bases; and (b) Complexity: CFR systems can be more complex and costly to implement.

14.3 MIXED-BED IX SYSTEMS

Mixed-bed IX systems are advanced water treatment units that combine both cation and anion exchange resins in a single vessel. This configuration allows for highly efficient removal of dissolved ions, resulting in exceptionally pure water.

The following are the process configuration, process function, and process objective of a mixed-bed IX process system.

1. **Mixed-Resin Bed Process Configuration:** The IX process configuration may contain a mixture of SAC and SBA resins. These types of resins are uniformly mixed to ensure thorough IX.
2. **IX Process Function:** As water passes through the mixed-bed IX process system, cations (e.g., Ca^{2+}, Mg^{2+}) are exchanged with H^+ ions from the SAC resin, and anions (e.g., Cl^-, SO_4^{2-}) are exchanged with OH^- ions from the SBA resin.
3. **Polishing Process Objective:** This mixed-bed setup provides a polishing effect, removing trace amounts of total dissolved solids (TDS) that might remain after initial treatment steps like reverse osmosis (RO) or two-bed deionization.

The applications of a mixed-bed IX process system include: (a) high-purity water production: used in industries requiring ultra-pure water, such as electronics manufacturing, pharmaceuticals, and power generation; (b) laboratory use: essential for producing reagent-grade water for analytical and research purposes; and (c) final polishing: often used as the final step in water treatment processes to ensure the highest water quality.

The advantages of a mixed-bed IX process system include: (a) high purity: capable of producing water with very low conductivity and minimal ionic contamination; (b) compact design: combines two processes in one vessel, saving space and simplifying system design; and (c) efficiency: provides a high level of ion removal in a single pass.

The disadvantages of a mixed-bed IX process system include: (a) regeneration complexity: regenerating mixed-bed resins is more complex than separate cation and anion beds, often requiring specialized equipment; (b) cost: higher initial and operational costs due to the need for precise control and maintenance; and (c) operational sensitivity: requires careful monitoring to maintain optimal performance and prevent resin fouling.

14.4 COMPARISON OF MIXED-BED AND TWO-BED DEIONIZATION SYSTEMS

Although mixed bed and two-bed deionization systems are used for producing high-purity water, they have distinct differences in design, operation, and performance.

14.4.1 Mixed-Bed Deionization Systems

14.4.1.1 Design and Function

The following are the design features and functions of the mixed-bed deionization systems: (a) single vessel: contains a mixture of cation and anion exchange resins in one vessel; and (b) polishing effect: provides a final polishing step, removing trace ions to achieve ultra-pure water.

14.4.1.2 Performance, Regeneration, and Applications

The performance of a mixed-bed IX process system is excellent because: (a) the process system produces high purity of process effluent water with very low conductivity and minimal ionic contamination; and (b) the process system achieves high level of ion removal in a single pass.

Regeneration of a mixed-bed IX process system is difficult because: (a) its complex process system requires separation of resins for regeneration, often needing specialized equipment; and (b) its operational costs are higher (in comparison with a two-bed IX process system) due to the complexity of regeneration.

Special applications of the mixed-bed IX process systems that may ignore their complex process configuration and the high operational costs are: (a) ultra-pure water production: ideal for industries requiring extremely high-purity water, such as electronics and pharmaceuticals; and (b) final effluent polishing requirement: often used as the final step in water treatment processes.

14.4.2 Two-Bed Deionization Systems

14.4.2.1 Design and Function

The following are the design features and functions of the two-bed deionization systems: (a) separate vessels: utilizes two separate vessels, one for cation exchange resin and one for anion exchange resin; and (b) sequential process: water passes through the cation bed first, then the anion bed.

14.4.2.2 Performance, Regeneration, and Applications

The performance of a two-bed IX process system is moderate because: (a) it produces moderate-purity water and is not as pure as mixed-bed systems; and (b) it is effective for general deionization needs, but may require additional polishing for ultra-pure water.

Regeneration of a two-bed IX process system is easy because: (a) it is a simpler process, for easier regeneration as each resin type is handled separately; and (b) its operational costs are lower compared to mixed-bed systems.

Applications of a two-bed IX process system include: (a) general deionization: it is suitable for applications where (moderately) high-purity water is needed, but ultra-pure water is not critical; and (b) water pretreatment: it is often used as a pretreatment step before mixed-bed systems or other water polishing processes.

As a summary, the mixed-bed IX systems are best for applications requiring ultra-pure water, but with higher complexity and cost in regeneration. The two-bed IX systems are suitable for general high-purity water needs, with simpler and more cost-effective regeneration.

When deciding between mixed-bed and two-bed IX systems, several key factors should be considered to ensure the chosen system meets your specific needs:

14.5 RESIN FOULING PREVENTION AND TROUBLESHOOTING

14.5.1 Resin Fouling Prevention

Resin fouling is a common issue in IX systems, where unwanted materials accumulate on the resin beads, reducing their efficiency. Preventing resin fouling involves a combination of proper pretreatment, regular cleaning, optimized regeneration, diligent monitoring, and proper storage.

Implementing these strategies can significantly extend the life of your IX resins and maintain system performance. Here are some key strategies to prevent resin fouling:

1. **Proper Pretreatment:** (a) filtration: use pre-filters to remove suspended solids and particulates before they reach the IX resin; and (b) chemical treatment: add chemicals to the feed water to prevent the precipitation of insoluble salts that can cause fouling.
2. **Regular Cleaning:** (a) backwashing: periodically backwash the resin bed to remove accumulated solids and reclassify the resin beads; and (b) chemical cleaning: use specific cleaning agents to remove fouling substances. For example, citric acid can be used to remove iron fouling, while sodium hypochlorite can help with organic fouling.
3. **Optimized Regeneration:** (a) increased regenerant quantities: use sufficient quantities of regenerant to ensure complete removal of fouling ions; (b) frequent regeneration: increase the frequency of regeneration cycles to prevent foulants from gaining a permanent hold; and (c) elevated regeneration temperature: higher temperatures can enhance the effectiveness of the regeneration process.
4. **Monitoring and Maintenance:** (a) regular sampling: take regular samples of the resin to monitor its condition and detect early signs of fouling; and (b) performance tracking: keep track of the system's performance metrics, such as pressure drop and ion leakage, to identify potential fouling issues early.
5. **Proper Storage:** (a) dry storage: store resins in a dry environment to prevent microbial growth and contamination; and (b) controlled storage environment: maintain appropriate temperature and humidity levels to preserve resin integrity.

14.5.2 Troubleshooting Resin Fouling Issues

Resin fouling can significantly impact the performance of IX systems. By identifying the type of fouling, evaluating resin condition, implementing appropriate cleaning procedures, optimizing regeneration, and taking preventive measures, an environmental engineer can effectively troubleshoot and mitigate resin fouling issues. Here are some steps and strategies to troubleshoot and address resin fouling:

1. **Identify the Type of Fouling:** (a) total suspended solids (TSS) fouling: look for signs of TSS particulate matter like silica, iron, or manganese. These can cause physical blockages in the resin bed; (b) organic fouling: check for organic substances, oils, and greases that can coat the resin beads; (c) microbial growth: inspect for bacteria and algae, which can form biofilms on the resin; and (d) chemical fouling: identify any chemical precipitates or insoluble salts that may have formed on the resin.
2. **Evaluate Resin Condition:** (a) sampling: Take resin samples and analyze them for signs of fouling. This can include visual inspection, microscopic examination, and chemical analysis; and (b) performance metrics: Monitor system performance indicators such as pressure drop, flow rate, and effluent quality to detect any deviations from normal operation.
3. **Implement Cleaning Procedures:** (a) backwashing: perform a thorough backwash to remove suspended solids and reclassify the resin bed; (b) chemical cleaning: use appropriate cleaning agents based on the type of fouling: (b1) acids: for removing metal fouling (e.g., citric acid for iron); (b2) bases: for removing organic fouling (e.g., sodium hydroxide); and (b3) surfactants: for oil and grease fouling, ensuring the surfactant itself does not foul the resin.
4. **Optimize Regeneration:** (a) regenerant quantity: ensure sufficient regenerant is used to fully regenerate the resin; (b) regeneration frequency: increase the frequency of regeneration cycles to prevent foulants from accumulating; and (c) temperature control: use elevated temperatures during regeneration to enhance the effectiveness of the process.

5. **Preventive Measures:** (a) pretreatment: implement proper pretreatment steps such as filtration and chemical dosing to remove potential foulants before they reach the resin; (2) regular maintenance: schedule regular maintenance and cleaning to keep the system in optimal condition; and (c) monitoring: continuously monitor system performance and resin condition to detect and address fouling early.

By identifying the type of fouling, evaluating resin condition, implementing appropriate cleaning procedures, optimizing regeneration, and taking preventive measures, an environmental engineer can effectively troubleshoot and mitigate resin fouling issues.

14.6 IX TECHNOLOGIES AND TERMINOLOGIES

Acidity: It is an expression of the concentration of hydrogen ions present in a solution.

Adsorption: It is a physicochemical process or reaction in which the charged particles are attached to the chemically active groups on the surface and in the pores of an adsorbent, or ion exchanger.

Alkalinity: It is an expression of the total basic anions (hydroxyl groups) present in a solution. It also represents the bicarbonate, carbonate, and, occasionally, the borate, silicate, and phosphate salts which will react with water to produce the hydroxyl groups.

Aluminum Fouling (Ion Exchange): Insoluble aluminum hydroxide, resulting from clarification or precipitation softening, coats the IX resin in a sodium zeolite softener. It is removed by cleaning with either acid or caustic during a normal regeneration.

Anion: It is a negatively charged ion in water.

Anion Exchange: (a) It is a proven physicochemical treatment process in which water passes through a bed of anionic synthetic resin and negatively charged contaminants in the water are exchanged with more innocuous negatively charged ions, typically chloride, on the resin's surface; (b) Anion exchange is useful for the greater than 99% removal of negatively charged contaminants including arsenic, chromium-6, cyanide, nitrate, perchlorate, per- and polyfluoroalkyl substances (PFAS), sulfate, and uranium although treatment capacities for different contaminants vary depending on the properties of the resin used and characteristics of the influent water; (c) Since vendors manufacture different resins, including those designed to selectively remove specific contaminant ions, laboratory or pilot plant demonstration experiments must be conducted to determine the resin's technical and economical feasibilities for removing the targeted pollutants; (d) Since the resin cost is high, when the capacity of the resin is exhausted, it normally is regenerated using a saturated solution, usually of sodium chloride (also known as brine) to restore resin to its initial condition for reuse. The spent regenerant brine is a concentrated solution of the removed contaminants and will be high in dissolved solids and excess regenerant ions (e.g., sodium, chloride). This waste stream will require disposal or discharge. Anion exchange treatment also can lower the pH of the treated water and, therefore, may require post-treatment corrosion control; (e) An alternative to resin regeneration is to dispose of the exhausted resin and replace it with fresh resin, and this alternative is often employed when selective resins are used to remove perchlorate or PFAS, or other extremely toxic substances. When replacement with fresh resin is used as an alternative to regeneration, the spent resin, loaded with removed contaminants, will require disposal. In some cases, disposal of the resin may require a special hazardous waste handling permit.

Anion Interchange: The displacement of one negatively charged particle by another on an anion exchange material which can be a synthetic anion exchange resin or a natural material, such as soil or zeolites.

Anionic: To have a negative ionic charge.

Ash: The residual mineral content of a substance, such as sludge or resin after incineration at 800°C.

ASTM: American Society of Testing and Materials

Attrition: It is the rubbing or mechanical wear of one particle (such as one ion exchanger grain) against another in a resin bed that will affect the size of resin particles.

Backwash: (a) The upward flow of water through a resin bed or filtration bed (i.e., in at the bottom of the ion exchange (IX) bed or filtration bed, out at the top) to clean and reclassify the bed after exhaustion; (b) The operational step of an IX bed or a filtration bed.

Backwash in Ion Exchange Regeneration Cycle: Backwash is the first step of ion exchange (IX) regeneration cycle for the removal of accumulated material and reclassifies the IX resin bed. Clean water (such as filtered water) flows from the underdrain distributor up through the resin bed and out the service distributor to waste. The upward flow lifts and expands the resin, allowing for removal of particulate material and resin fines and the classification of the resin. IX resin classification brings the smaller beads to the top of the unit while the larger beads settle to the bottom. Backwashing should continue for a minimum of 10 minutes or until effluent from the backwash outlet is clear. The backwash flow should be sufficient to expand the resin bed volume by 50% or more, depending on the available free-board. Backwash flow rates usually vary between 4–8 (ambient temperature) and 12–15 (hot service) gpm/ft^2 of bed area, or are controlled according to the manufacturer's recommendation. The backwash wastewater containing total suspended solids(TSS) can be returned to the clarifier or filter influent for treatment.

Backwash-Sedimentation-Drainage Procedure (BS&D Procedure) (Ion Exchange): A backwash-sedimentation-drainage procedure for resin volume measurement where an ion exchange (IX) resin bed is initially backwashed, then allowed to settle and then drained of water. The resultant IX resin bed's height is measured for volume and flow calculations.

Base: It is the hydroxyl form of an ion or a compound that can neutralize an acid.

Batch Operation: The utilization of chemicals, adsorbents, or ion exchange (IX) resins to treat a solution in a container where the removal of pollutants/contaminants is accomplished by (a) sequencing batch operation of influent filling, chemical addition, mixing/flocculation of the solution, clarification and effluent discharge, or (b) sequencing batch operation of influent filling, IX resin (or an adsorbent) addition, mixing of the solution and resin, settling of resin, and effluent discharge, or (c) sequencing batch operation of influent filling, chemical and/or activated sludge addition, aeration, sludge settling, and effluent discharge.

Bed: An ion exchange resin contained in a column, or a filtration media contained in a column.

Bed Depth (Ion Exchange): The height of the resinous material in the ion exchange column after the exchanger has been properly conditioned for effective operation.

Bed Expansion: The effect produced during backwashing: The media (ion exchange resin particles or filtration media, or adsorbent media) become separated and rise in the column. The expansion of the bed due to the increase in the space between media particles may be controlled by regulating backwash flow.

Bed Volumes per Hour (BV/H): BV/H is the flow rate expressed in relation to the volume of ion exchange (IX) resin in the IX column.

Bicarbonate Alkalinity: The presence in a solution of hydroxyl (OH^-) ions resulting from the hydrolysis of carbonates or bicarbonates. When these salts react with water a strong base and a weak acid are produced, and the solution is alkaline.

Boiler Water Treatment: (a) Scale and deposit buildup in boilers and the formation of insoluble soap curds in washing operations have created a large demand for softened water. Sodium zeolite ion exchange (IX) softeners are able to satisfy this demand economically and are widely used in the preparation of water for low and medium pressure boilers, laundries, and chemical processes. The treated water has a very low scaling tendency because zeolite softening reduces the hardness level of most water supplies to less than 2 mg/L; (b) Sodium zeolite softeners do not function efficiently on turbid waters with an influent turbidity in excess of 1.0 NTU and the IX resin can be fouled by heavy metal contaminants, iron ions, and aluminum ions; (c) If excess iron or manganese is present in the water supply, and/or aluminum coagulants are used ahead of zeolite softeners, proper equipment operation and close control of sedimentation or dissolved air flotation (DAF) clarifier pH are essential to good softener performance; (d) Do not apply chlorination (or any strong oxidation) prior to zeolite; and (e) Sometimes softening alone is insufficient for most high-pressure boiler feedwaters and for many process streams, especially those used in the manufacture of electronics equipment. In addition to the removal of hardness, these processes require the removal of all dissolved solids, such as sodium, silica, alkalinity, and the mineral anions (Cl^-, SO_4^{2-}, NO_3^-) by demineralization.

Breakthrough: It is the first appearance in the solution flowing from an ion exchange (IX) unit of un-adsorbed ions similar to those that are depleting the activity of the IX resin bed, so it is an indication that regeneration of the resin is necessary.

Breakthrough (Adsorption): The first appearance in the solution flowing from an adsorption unit of unadsorbed pollutants similar to those that are depleting the activity of the adsorption bed. Breakthrough is an indication that regeneration of the adsorbent is necessary.

Breakthrough (Ion Exchange): The first appearance in the solution flowing from an ion exchange unit of unadsorbed ions similar to those that are depleting the activity of the resin bed. Breakthrough is an indication that regeneration of the resin is necessary.

Breakthrough Curve: In a continuous ion exchange process controlled at a constant flow rate, the unwanted ion concentrations left in the treated water are periodically or continuously monitored at different operational times and these data plotted as a two-dimensional curve (ion concentrations vs. time; or ion concentrations vs. bed volumes) in order to determine the ion leakage and the maximum permissible ion leakage requiring the production cycle to be shut down.

Brine: (a) Brine is any solution with an extremely high concentration of salts such as sodium chloride, sodium sulfate, etc.: (b) It is a salt solution, generally sodium chloride in a saturated solution.

Brining in Ion Exchange Regeneration Cycle: Brining is the second step of ion exchange (IX) resin regeneration after the first backwash step. The brine stream enters the unit through the regenerant distributor and flows down through the resin bed at a slow rate (usually between 0.5 and 1 gpm/ft^2 of IX resin). Brine flow is collected through the underdrain and sent to waste. The slow flow rate increases contact between the brine and resin. To achieve optimum efficiency from the brine, the solution strength should be 10% during brine introduction.

Brining System Design (Ion Exchange): (a) The brining system consists of salt dissolving/brine measuring equipment, and dilution control equipment to provide the desired regenerant strength. The dissolving/measuring equipment is designed to provide the correct amount of concentrated brine (approximately 26% NaCl) for each regeneration, without allowing any undissolved salt into the resin. The concentrated brine can be diluted to the optimum regenerant strength (8%–10% NaCl), or be pumped from the concentrated salt tank and mixed with dilution water to provide the desired regenerant strength (8%–10% NaCl); (b) Regenerant brine is applied after backwash. The brine stream enters the unit through the regenerant distributor and flows down through the resin bed at a slow rate (usually between 0.5 and 1 gpm/ft^2 of resin). Brine flow is collected through the underdrain and sent to waste. The slow flow rate increases contact between the brine and resin. To achieve optimum efficiency from the brine, the solution strength should be 10% during brine introduction.

Buffer: It is a solution that is capable of opposing small changes in chemical composition, particularly in pH.

Calcium (Ca): It is a metallic element often found in water, usually as dissolved calcium carbonate ($CaCO_3$). It is soluble in water causing hardness and subsequent scaling.

Capacity (Ion Exchange): (a) It is defined as the amount of exchangeable ions a unit quantity of resin will remove from a solution. It is usually expressed in kilograins per cubic foot as calcium carbonate or milliequivalents per milliliter; (b) It is the portion of the total capacity utilized in practical ion exchange operation.

Capacity, Salt Splitting (Ion Exchange): The portion of total capacity to split neutral salt.

Capacity, Total (Adsorption): It is the ultimate adsorption capacity of the adsorbent.

Capacity, Total (Ion Exchange): It is the ultimate exchange capacity of the resin.

Carbonaceous Exchangers: They are the ion exchange materials of limited capacity prepared by the sulfonation of coal, lignite, peat, etc.

Carbonaceous Zeolite: It is a sulfonated coal cation exchange medium which is used for hydrogen cycle operation, allowing for the reduction of alkalinity as well as hardness.

Cation: It is a positively charged ion or cationic ion.

Cation Exchange: (a) It is a proven physicochemical treatment process in which water passes through a bed of cationic synthetic resin and positively charged contaminants in the water are exchanged with more innocuous positively charged ions, typically sodium, on the resin's surface; (b) Cation exchange is useful for water softening by removing greater than 99% of hardness ions such as calcium and magnesium as well as other positively charged contaminants, such as barium, radium, and strontium; (c) Since cation exchange treatment capacities for different contaminants vary depending on the properties of the resin used and characteristics of the influent water, and since a number of vendors manufacture different resins, including those designed to selectively remove specific contaminant ions; it is recommended that laboratory or pilot plant demonstration experiments be conducted to determine the resin's technical and economical feasibilities for removing the targeted pollutants; (d) Normally when the capacity of the resin is exhausted, it is regenerated to restore it to its initial condition. The regeneration process uses a saturated solution, usually of sodium chloride (also known as brine); (e) The spent regenerant brine is a concentrated solution of the removed contaminants and also will be high in dissolved solids and excess regenerant ions (e.g., sodium, chloride). This waste stream will require disposal or discharge.

Cation Exchange Capacity (CEC): (a) It is the sum total of exchangeable cations that a sediment or soil can adsorb. (b) CEC is expressed in milliequivalents of negative charge per 100 g (mEq/100 g) of soil or sediment.

Cationic: To have a positive ionic charge.

Channeling: Cleavage and furrowing of the granular media (ion exchange resins, filtration media, granular activated carbons [GACs], activated alumina, green sands, etc.) bed due to faulty operational procedures, in which the solution being treated follows the path of least resistance, runs through these furrows, and fails to contact active groups in other parts of the media bed.

Chemical Stability (Ion Exchange): It is the resistance to chemical change which ion exchange resins must possess despite contact with aggressive solutions.

Chromated Copper Arsenate (CCA) and Its Decontamination: CCA is a water-soluble wood preservative chemical containing inorganic pesticide, chromium arsenate $Cr(AsO_4)$ and copper arsenate $Cu_3(AsO_4)_2$ for treating the construction woods and making them resistant to attack by termites and fungi that cause decay. The wood is dipped in a solution of CCA and subjected to vacuum pressure to force penetration of CCA into the wood. CCA-treated wood is also referred to as pressure-treated wood and is used widely in outdoor structures such as decks, playground equipment, picnic tables, garden-bed borders and docks. Since CCA can be leached from CCA-treated wood by rainwater and CCA-treated wood weathering, the soil and storm runoff water adjacent to CCA-treated wood structures have been shown to be contaminated by toxic arsenic (V), chromium (III), and copper (II). Both cationic chromium ions and copper ions can be removed by cation exchanger (Z-H or Z-Na), while toxic arsenic in anionic arsenate form can be removed by anion exchanger (Z-OH or Z-Cl). The following are the chemical reactions for environmental decontamination assuming Z-Na and Z-Cl are used as the first-stage cation exchanger and the second-stage anion exchanger, respectively:

Decontamination Chemical Reactions for Removal of Copper, Chromium Cations in the First Stage:

$$Cr(AsO_4) + 3Z\text{-}Na \Rightarrow Z_3\text{-}Cr + Na_3(AsO_4)$$

$$Cu_3(AsO_4)_2 + 6Z\text{-}Na \Rightarrow 3Z_2\text{-}Cu + 2Na_3(AsO_4)$$

Decontamination Chemical Reaction for Removal of Arsenate in the Second Stage:

$$Na_3(AsO_4) + 3Z\text{-}Cl \Rightarrow Z_3\text{-}(AsO_4) + 3NaCl$$

Co-current Operation (Ion Exchange) and Sodium Leakage: (a) The co-current operation is the basis for most conventional ion exchange (IX) system designs, in which the regenerant flow is in the same direction as the service flow, down through the IX resin bed. During the regeneration, the contaminants are displaced through the resin bed; (b) At the end of the regeneration, some ions, predominately sodium ions, remain in the bottom of the resin bed. The sodium ions that remained in the bed during regeneration diffuse into the decationized water before it leaves the IX unit, called sodium leakage.

Co-flow Regeneration: The regenerant solution flows in the same direction as the service flow. This method is simpler but less efficient, often requiring more regenerant and resulting in higher leakage of contaminants.

Colloidal Substance or Colloid: Colloids are composed of extremely small size particles which are dispersed throughout another substance and cannot be removed by normal filtration but can be removed by membrane filtration with semipermeable membranes.

Column Operation (Ion Exchange): Conventional utilization of ion exchange resins in columns through which pass, either upflow or downflow, the solution to be treated.

Community Water System (CWS): It is a public water system that serves at least 15 service connections used by year-round residents or regularly serves at least 25 year-round residents. Examples of a community water system include, but are not limited to, cities, villages, nursing homes, and mobile home parks.

Condensate Polishers: Ion exchange resins being used to remove or exchange ions as well as to filter condensate for reuse in the steam cycle are called condensate polishers.

Condensate Polishing (Ion Exchange): (a) Condensate polishing is one of the important ion exchange (IX) applications. For boiler water treatment, IX has been used to process boiler makeup water and returned condensate, removing corrosion products that could cause harmful deposits in boilers. For condensate polishing, the condensate polishes filter out the iron and copper particulates and IX removes soluble contaminants entering the boiler water system through condenser, pump seal leaks, or carry-over of boiler water into the steam; (b) Most paper mill condensate polishers operate at temperatures approaching 200°F, precluding the use of anion resin. Cation resin, which is stable up to temperatures of over 270°F, is used for deep bed condensate polishing in the paper mills. The deep bed cation resin condensate polisher is regenerated with 15 lb/ft^3 of sodium chloride per resin. The high service flow rate for a deep bed condensate polisher (20–50 gpm/ft^2 of IX resin surface area) is permissible because the level of soluble ions in the condensate can be usually very low; and (c) In systems requiring total dissolved solids (TDS) and total suspended soilds (TSS) removal, an IX mixed-bed condensate polisher may be used, if the temperature of the condensate can be below 140°F, which is the maximum continuous operating temperature for the anion resin. The service flow rate through the IX unit should be reduced to about 20 gpm/ft^2.

Conductivity: (a) It is the reciprocal of resistivity, for measuring the ability to conduct an electric current or measuring ionic purity because ionized impurities increase the conductivity of water. (b) To measure it, current is passed between two electrodes a fixed distance apart; (c) Conductivity is normally expressed as microSiemens/cm, identical to micromhos/cm; (d) It is the ability of electric current to flow through water as a measure of its ion content in mhos or micromhos.

Counterflow and Mixed-Bed Exchanger Deionization: (a) The deionization process is a modified standard demineralization process to increase the purity of the treated water; (b) It includes: (b1) counterflow cation exchangers and (b2) mixed-bed exchangers; (c) Counterflow and mixed-bed systems produce a purer water than conventional cation-anion demineralizers, but require more sophisticated equipment and have a higher initial cost. The more complicated regeneration sequences require closer operator attention than standard systems. This is especially true for a mixed-bed unit.

Counterflow Cation Exchangers: (a) In a counterflow regenerated cation exchanger, the regenerant flows in the opposite direction of the service flow. If the service flow is downward through the ion exchange (IX) bed, the regenerant acid flow is up through the IX bed. As a result, the most highly regenerated resin is located where the service water leaves the IX unit. The highly regenerated IX resin removes the low level of contaminants that have escaped removal in the top of the bed. This results in higher water purity than co-current designs can produce; (b) To maximize contact between the acid and resin and to keep the most highly regenerated resin from mixing with the rest of the bed, the resin bed must stay compressed during the regenerant introduction. This compression is usually achieved in one of two ways: (b1) a blocking flow of water or air is used; and (b2) the acid flow is split, and acid is introduced at both the top and the bottom of the resin bed.

Counterflow Regeneration (CFR): The regenerant solution flows in the opposite direction to the service flow. This method is more efficient, using less regenerant and reducing contaminant leakage, but requires more complex equipment.

Cross-linkage: It is the degree of binding of a monomer or set of monomers to form an insoluble tri-dimensional resin matrix.

Cycle (Ion Exchange): It is a complete course of ion exchange operation. For instance, a complete cycle of cation exchange may involve: exhaustion of regenerated bed, backwash, regeneration, and rinse to remove excess regenerant.

Dealkalization: (a) The dealkalization process (instead of expensive demineralization process) is used for improving the boiler or similar process operating conditions by removal of hardness and the reduction of alkalinity, but not the removal of the other solids; (b) There are three kinds of dealkalization processes: (b1) Sodium zeolite and hydrogen zeolite (split stream) dealkalization, (b2) Sodium zeolite and chloride-anion dealkalization, and (b3) weak acid cation (WAC) dealkalization; (c) Ion exchange dealkalization systems produce hardness-free, low-alkalinity water at a reasonable cost, and with a high degree of reliability. They are well suited for processing feedwater for medium-pressure boilers, and for process water for the beverage industry.

Deashing (Ion Exchange): (a) The removal from solution of inorganic salts by means of adsorption by ion exchange resins of both the cations and the anions that comprise the salts; (b) Same as deionization.

Decarbonation Process or Degassing Process: (a) It is a physicochemical process (decarbonator or degasser) which uses a countercurrent stream of air for converting unstable carbonic acid to carbon dioxide gas and water and then stripping carbon dioxide gas from water. Decarbonation is typically used in a sodium zeolite and hydrogen zeolite (split stream) dealkalization system; (b) Decarbonators and degassers are economically beneficial to many demineralization systems, because they reduce the amount of caustic required for regeneration.

Decarbonators Equipment Design: Water from a cation exchanger is broken into small droplets by sprays and trays or packing in a decarbonator; The water then flows through a stream of air flowing in the opposite direction. Carbonic acid present in the cation effluent dissociates into carbon dioxide and water. The carbon dioxide is stripped from the water by the air, reducing the load to the anion exchangers. Typical forced draft decarbonators are capable of removing carbon dioxide down to 10–15 ppm. However, water effluent from a decarbonator is saturated with oxygen.

Decationization: Strong acid cation (SAC) exchangers or resins in the hydrogen cycle for removal of all cations in water, so the process is called "decationization". SAC resins remove nearly all raw water cations, replacing them with hydrogen ions.

Deionization: It is a technical l term similar to deashing, meaning the removal of all charged constituents or ionizable salts (both inorganic and organic) from solution.

Demineralization (Distillation): Demineralization of water is accomplished by evaporation of water for formation of steams, and then by condensation of steams for formation of demineralized water containing no inorganic mineral salts.

Demineralization (Ion Exchange): (a) In the ion exchange (IX) demineralization process, strong acid cation (SAC) resin in the hydrogen form is used for converting dissolved salts into their corresponding acids, and strong base anion (SBA) resin in the hydroxide form is used for removing these acids. IX demineralization produces water similar in quality to distillation at a lower cost for most fresh waters; (b) A demineralizer system consists of one or more IX resin columns, which include a SAC unit and a SBA unit. The cation resin exchanges hydrogen for the raw water cations as shown by the following reactions:

Demineralization (Ion Exchange) Chemical Reactions Using Strong Acid Cation (SAC) Resin in the Hydrogen Form:

$$CaSO_4 + 2Z\text{-}H \Rightarrow Z_2\text{-}Ca + H_2SO_4$$

$$Ca(HCO_3)_2 + 2Z\text{-}H \Rightarrow Z_2\text{-}Ca + 2H_2CO_3$$

$$CaCl_2 + 2Z\text{-}H \Rightarrow Z_2\text{-}Ca + 2HCl$$

$$Ca(NO_3)_2 + 2Z\text{-}H \Rightarrow Z_2\text{-}Ca + 2HNO_3$$

$$MgSO_4 + 2Z\text{-}H \Rightarrow Z_2\text{-}Mg + H_2SO_4$$

$$Mg(HCO_3)_2 + 2Z\text{-}H \Rightarrow Z_2\text{-}Mg + 2H_2CO_3$$

$$MgCl_2 + 2Z\text{-}H \Rightarrow Z_2\text{-}Mg + 2HCl$$

$$Mg(NO_3)_2 + 2Z\text{-}H \Rightarrow Z_2\text{-}Mg + 2HNO_3$$

$$2NaNO_3 + 2Z\text{-}H \Rightarrow 2Z\text{-}Na + 2HNO_3$$

$$Na_2SO_4 + 2Z\text{-}H \Rightarrow 2Z\text{-}Na + H_2SO_4$$

$$Na(HCO_3) + Z\text{-}H \Rightarrow Z\text{-}Na + H_2CO_3$$

$$NaCl + Z\text{-}H \Rightarrow Z\text{-}Na + HCl$$

(c) To complete the demineralization process, water from the cation unit is passed through a strong base anion (SBA) exchange resin in the hydroxide form. The resin exchanges hydrogen ions for both highly ionized mineral ions and the more weakly ionized carbonic and silicic acids, as shown below:

Demineralization (Ion Exchange) Chemical Reactions Using Strong Base Anion (SBA) Resin in the Hydroxide Form:

$$H_2SO_4 + 2Z\text{-}OH \Rightarrow Z_2\text{-}SO_4 + 2H_2O$$

$$H_2SiO_3 + Z\text{-}OH \Rightarrow Z\text{-}HSiO_3 + H_2O$$

$$H_2CO_3 + Z\text{-}OH \Rightarrow Z\text{-}HCO_3 + H_2O$$

$$HNO_3 + Z\text{-}OH \Rightarrow Z\text{-}NO_3 + H_2O$$

Demineralization: Demineralization of water is the total removal of all inorganic salts by ion exchange, distillation (evaporation/condensation), or membrane filtration (MF).

Demineralization (Ion Exchange): It is the physicochemical process (same as deionization), for removal of minerals and mineral salts using ion exchange.

Demineralization (Membrane Filtration): Demineralization of water is accomplished by membrane filtration (MF) process for total removal of all inorganic mineral salts from water.

Demineralization: Strong acid cation (SAC) exchangers or resins derive their functionality from sulfonic acid groups (HSO_3^-). When used in demineralization, SAC resins remove nearly all raw water cations, replacing them with hydrogen ions. When strong base anion (SBA) resins in the hydroxide form, they remove all commonly encountered anions. The process for total removal of nearly all raw water cations and anions from raw water is called demineralization.

Density (Ion Exchange): The weight of a given volume of exchange material, backwashed, and in place in the column.

Diffusion (Ion Exchange): It is usually referred to ion exchange (IX) resins as the diffusion of ions through the IX resin beads.

Direct Acid Injection and Decarbonation: In this ion exchange (IX) process, an acid injection system is used to convert alkalinity in water to carbonic acid which dissociates to form carbon dioxide and water. The carbon dioxide gas is removed in a decarbonator. Since an acid overfeed or a breakdown in the pH control system can produce acidic feedwater, which corrodes the iron surfaces of process equipment and boilers, correct pH monitoring and caustic feed after decarbonation are required.

Displacement (Slow Rinse) in Ion Exchange Regeneration Cycle: Displacement (slow rinse) is the third step of ion exchange (IX) resin regeneration after the first backwash step and the second brining step for ensuring proper contact of the regenerant with the bottom of the IX resin bed, and for moving the brine completely through the IX bed. Following the regenerant brining, a slow flow of water continues through the regenerant distribution system. This water flow displaces the regenerant through the bed at the desired flow rate. The flow rate for the displacement water is usually the same rate used for the dilution of the concentrated brine. The duration of the displacement step should be sufficient to allow for approximately one resin bed volume of water to pass through the unit.

Dissociation: (a) It is an ionization process in which an electrolyte or a salt upon is dissolved in water to form ions of cation and anion; (b) It is a process in which a chemical combination (such as sodium chloride) breaks up into similar ionic constituents (such as sodium cation and chloride anion).

Dissolved Solids, Total (TDS): It is the total amount of nonvolatile matter dissolved in a water sample, and is usually expressed in parts-per-million by weight.

Downflow: Conventional direction of solutions to be processed in a granular media (ion exchange, granular activated carbon, sand, etc.) column operation, i.e., in at the top, out at the bottom of the column.

Dry Solids (DS): It is the matter, usually expressed in weight percent, remaining after liquid removal.

Efficiency: (a) The effectiveness of the operational performance of a process, such as ion exchange (IX), adsorption, etc. (b) For IX process, the efficiency in the adsorption of ions is expressed as the quantity of regenerant required to effect the removal of a specified unit weight of adsorbed material, e.g., pounds of acid per kilograin of salt removed.

Effluent: The solution that emerges from a process unit such as an ion exchange column or any other process reactor.

Electrolyte: (a) It is a chemical compound that dissociates or ionizes in water to produce a solution which will conduct an electric current; an acid, base or salt; (b) It is a substance that when dissolved in a suitable solvent or when fused becomes an ionic conductor.

Elution: It is a chemical process that involves the stripping of adsorbed ions from an ion exchange material by the use of solutions containing other ions in relatively high concentrations.

Equilibrium Reactions (Ion Exchange): It is the interaction of ionizable compounds in which the products obtained tend to revert to the substances from which they were formed until a balance is reached in which both reactants and products are present in definite ratios.

Equivalent Weight: The molecular weight of any element or radical expressed as grams, pounds, etc., divided by the valence.

Exchange Capacity: This is the weight of ions that can be retained per unit volume (or sometimes per unit weight) of the exchange material concerned.

Exchange Sites: The reactive groups in an ion exchange resin.

Exchange Velocity: The rate with which one ion is displaced from an exchanger in favor of another.

Exempt Water System: It means a water system that is exempt from the State of Federal drinking water regulations; All of the following conditions must be met: (a) It consists only of distribution and storage facilities and does not have any collection and treatment facilities; (b) It obtains all of its water from, but is not owned or operated by a public water system; (c) It does not sell water to any person; and (d) It is not a carrier which conveys passengers in interstate commerce (e.g., airline, railroad, bus line, boat line, etc.).

Exhaustion (Adsorption): It is the state in which the adsorbent is no longer capable of useful adsorption.

Exhaustion (Ion Exchange): It is the state in which the ion exchange (IX) resin is no longer capable of useful IX; the depletion of the exchanger's supply of available ions. The exhaustion point is determined arbitrarily in terms of: (a) a value in parts per million of ions in the effluent solution; (b) the reduction in quality of the effluent water determined by a conductivity bridge which measures the electrical resistance of the water; (c) After final rinse of the available IX resins and start-up the IX process unit for operation, the IX process unit (such as a softener) produces a very low, nearly constant level of effluent contaminant ions (such as residual hardness) until the IX resin nears exhaustion. At exhaustion, the effluent contaminant ions (such as hardness) increases sharply, the IX resins are no longer available for further removal of the contaminant ions, then regeneration is required.

Fast Rinse in Ion Exchange Regeneration Cycle: Fast rinse is the fourth step of ion exchange (IX) resin regeneration after the first backwash step, the second brining step and the third displacement (slow rinse) step for ensuring high effluent quality during the service run. After completion of the third displacement rinse step, clean water (such as the filtered water) is introduced through the inlet distributor at a high flow rate between 1.5 and 2 gpm/ft^2 of IX resin bed, until the effluent chlorides or conductivity are near influent levels. If the softener has been in standby following a regeneration, a second fast rinse, known as a service rinse, can be used to remove any contaminant ion (such as hardness) that has entered the water during standby.

Fines: (a) Extremely small particles of ion exchange materials in an IX process; (b) Extremely small particles of granular activated carbons (GAC) in a GAC adsorption process.

Flow Rate (Ion Exchange): The volume of solution passing through a given quantity of resin within a given time. Usually expressed in terms of gallons per minute per cubic foot of resin, as milliliters per minute per milliliter of resin, or as gallons per square foot of resin per minute.

FMA: The free mineral acidity, or sum of the mineral acids.

Free Mineral Acidity (FMA) (Ion Exchange): (a) FMA is a measure of the total concentration of the strong acids in the cation effluent. In a typical service run, the FMA content is stable most of the time, during an ion exchange (IX) process run; (b) If cation exchange were 100% efficient, the FMA from the exchanger would be equal to the theoretical mineral acidity (TMA) of the water. The FMA is usually slightly lower than the TMA because a small amount of sodium leaks through the cation exchanger; (c) As a cation exchange unit nears exhaustion, FMA in the effluent drops sharply, indicating that the exchanger should be removed from service. At this time the IX resin should be regenerated with an acid solution, which returns the exchange sites to the hydrogen form. Sulfuric acid or hydrochloric acid can be used for regeneration; (d) In order to prevent the calcium sulfate fouling, the sulfuric acid is usually applied at a high flow rate (1 gpm/ft^2 of resin) and an initial concentration of 2% or less. The acid concentration is gradually increased to 6%–8% to complete regeneration; (e) In order to prevent the hydrochloric acid corrosion, special materials of construction in the regenerant system are needed; and (f) An excess of regenerant (sulfuric or hydrochloric acid) is required up to three times the theoretical dose.

Free-board: The space provided above the granular media (resin, GAC, etc.) bed in a process column to allow for expansion of the bed during backwashing.

Gel: Ion exchange resins that are made up of firm gel structure in a solid bead form allowing for the diffusion of ions through the gel.

Grain: A unit of weight; 0.0648 g.

Grains per Gallon: It is an expression of concentration of material in a solution, generally in terms of calcium carbonate; a unit of water hardness. One grain (as calcium carbonate) per gallon is equivalent to 17.1 ppm. One grain per US gallon is equivalent to 0.017118061 kg/m^3.

Gram-Milliequivalents: The equivalent weight in grams, divided by 1,000.

Greensand: It is a naturally occurring material that has a lower exchange capacity than the synthetic exchanger material, but greensand has greater physical stability making it more suitable for practical municipal and industrial applications.

Hardness: (a) The scale-forming and lather-inhibiting qualities which water, high in calcium and magnesium ions, possesses; (b) Temporary hardness, caused by the presence of magnesium or calcium bicarbonate, is so called because it may be removed by boiling the water to convert the bicarbonates to the insoluble carbonates; (c) Calcium sulfate, magnesium sulfate, and the chlorides of these two metals cause permanent hardness.

Hardness as Calcium Carbonate: (a) It is the expression ascribed to the value obtained when the hardness forming salts are calculated in terms of equivalent quantities of calcium carbonate; (b) It is a convenient method of reducing all salts to a common basis for comparison.

Hardness Leakage (Ion Exchange): Water from a properly operated zeolite softener is nearly free from detectable hardness. Some small amounts of hardness, known as leakage, are present in the treated water. The level of hardness leakage is dependent on the hardness and sodium level in the influent water and the amount of salt used for regeneration.

Hardness Precipitate Fouling (Ion Exchange): Hardness precipitates that carry through a filter from a precipitation softener or form after filtration by post-precipitation can foul resins used for sodium zeolite softening. These hardness precipitates can be dissolved and removed with acid.

Hardness-Free Water or Soft Water: The product water from a softening process is termed hardness-free water or soft water after the scale-forming hardness ions, such as calcium and magnesium from raw water. The hardness-free water, or soft water, can then be used for low to medium pressure boiler feedwater, reverse osmosis system makeup, some chemical processes, and commercial applications, such as laundries.

Headloss: (a) It is the reduction in liquid pressure associated with the passage of a liquid through a bed of granular material (such as ion exchanger, granular activated, carbon, activated alumina, or filtration media), or through a pipe or a reactor; (b) It is a measure of the resistance of a process reactor (such as an ion exchange reactor, a GAC reactor, or a biological reactor), or a pipe to the flow of the liquid passing through it.

Hot Sodium Zeolite Softening: Sodium zeolite softeners can be used to remove residual hardness in the effluent from a hot process lime or lime-soda softening process system. The hot process effluent flows through filters and then through an ion exchange (IX) bed of strong acid cation (SAC) resin in the sodium form. The equipment and operation of a hot zeolite softener is identical to that of an ambient temperature softener, except that (a) the valves, piping, controllers, and instrumentation must be suitable for the high temperature (220°F–250°F). StandardSAC resin can be used at temperatures of up to 270°F, but for a longer service life a premium gel or macroreticular resin is recommended; (b) The IX system should be designed to eliminate flow surges in the hot lime unit; (c) Backwash water storage tanks may be used in the hot lime unit; and (d) Extended slow rinses may be used for the zeolite in lieu of a standard fast rinse.

Hydraulic Classification: The rearrangement of granular media particles (such as resins, granular activated carbons, sands, etc.) in a granular media process unit. As the backwash water flows up through the granular media bed, the particles are placed in a mobile condition wherein the larger particles settle and the smaller particles rise to the top of the granular media bed.

Hydrogen Cycle (Ion Exchange): A complete course of cation exchange operation in which the adsorbent is employed in the hydrogen or free acid form.

Hydroxyl: The term used to describe the anionic radical (OH^-) which is responsible for the alkalinity of a solution.

Hydroxymethyl Furfural (HMF): 5 (hydroxymethyl)-2-furaldehyde, a precursor of the coloring matter from the decomposition of glucose and also thereby assisting in the color development in sugars. HMF is produced during contact with strong acid cation (SAC) resins in H^+ form at elevated temperatures.

Influent: The liquid which enters a process unit (such as an ion exchange unit, or any other physicochemical or biological process reactor).

Ion: (a) Ion is an atom or molecule with a net electric charge due to the loss or gain of one or more electrons; (b) It is an atom that is electrically unstable because it has more or fewer electrons than protons, therefore it is an electrically charged particle; (c) A positive ion is called a cation, and a negative ion is called an anion; (d) It is any particle of less than colloidal size possessing either a positive or a negative electric charge.

Ion Exchange: (a) Ion exchange (IX) is a reversible chemical process in which ions from an insoluble permanent solid medium (the ion exchanger, or IX resin) are exchanged for ions in a solution or fluid-mixture surrounding the insoluble medium. The direction of the exchange depends on the selective attraction of the ion exchanger resin for the certain ions present and the concentration of the ions in the solution; (b) there are three types of IX process: cation exchange, anion exchange, and mixed-bed exchange (consisting of both cation exchange and anion exchange); (c) IX is the basis for clay peptization, which can improve a poorly performing clay. In water muds, cation exchange is an ongoing and dynamic process. Cation exchange on clay minerals occurs in muds and during mud testing in which the methylene blue dye is a cationic dye that goes onto clays in the mud sample being tested to show the extent of cation exchange capacity. In muds, various positive ions (for example, Ca^{+2}) on clay surfaces can be replaced by other positive ions (for example, Na^+) that occur in higher concentration or have a higher affinity for the substrate.

Ion Exchange (IX) Inlet Distribution System: It is usually located at the top of the tank, providing even distribution of influent water and consisting of a central header/hub with distributing laterals/radials or simple baffle plates, which direct the flow of water evenly over the resin bed.

Ion Exchange (IX) Free-Board Space: The volume between the inlet distributor and the top of the resin bed is called the free-board space. The free-board allows for the expansion of the resin during the backwash portion of the regeneration without loss of resin. It should be a minimum of 50% of the resin volume (80% preferred).

Ion Exchange (IX) Minimum Bed Depth: A minimum bed depth of 24 inch is recommended for all IX process systems.

Ion Exchange (IX) Process Advantages: The IX advantages in water and wastewater treatment include: (a) Highly efficient removal of dissolved ions and ionic contaminants; (b) Versatile applications to a wide range of water and wastewater treatment; and (c) Targeted applications for specific ion removal needs.

Ion Exchange (IX) Process Equipment: The IX process equipment used for sodium zeolite softening consists of a softener exchange vessel, control valves and piping, and a system for brining, or regenerating, the resin. Usually, the softener tank is a vertical steel pressure vessel with dished heads. The inlet distribution system, free-board space, regenerant distributor, ion exchange resin bed depth, and resin-retaining underdrain collection system must be designed properly.

Ion Exchange (IX) Process Limitations: The IX process limitations in water and wastewater treatment may include: (a) Limited IX resin capacity and lifespan, requiring periodic replacement; (b) High IX operation and maintenance (O&M) costs for large-scale water and wastewater treatment or for treatment of contaminated water; and (c) Potential secondary pollution caused by the regenerant chemicals used in the IX process.

Ion Exchange (IX) Regenerant Distributor: The regenerant distributor is usually a header-lateral system that evenly distributes the regenerant brine during regeneration. The location of the distributor, 6 inch above the top of the resin bed, prevents the dilution of regenerant by water in the free-board space. It also reduces water and time requirements for displacement and fast rinse. The regenerant distributor should be secured to the tank structure to prevent breakage and subsequent channeling of the regenerant.

Ion Exchange (IX) Resin Degradation: Ion exchange (IX) resin fouling is reversible, but IX resin degradation is irreversible. There are three major types of resin degradation: (a) Oxidant degradation: Oxidizing agents, such as chlorine, ozone, chlorine dioxide, air, oxygen, etc. degrade both cation and anion IX resins. (a1) Oxidants attack the divinylbenzene cross-links in a cation resin, reducing the overall strength of the resin bead. As the attack continues, the cation resin begins to lose its exchange capacity, spherical shape, rigidity, causing bed compaction, bed channeling, and pressure drop across the resin bed; (a2) Anion resin is not directly affected in water chlorination or ozonation, or oxidation, because the chlorine is consumed by the cation resin. However, downstream strong base anion (SBA) resins are fouled by certain degradation products (such as organic disinfectant by-products, DBP) from oxidized cation resin. The oxidants should be removed prior to IX with activated carbon filtration or sodium sulfite; (a3) Oxygen-saturated water, such as that formed following forced draft decarbonation, accelerates the destruction of strong base exchange sites that occurs naturally over time. It also accelerates degradation due to organic fouling; (b) Thermal degradation: Thermal degradation occurs if the anion resin becomes overheated during the service or regeneration cycle. This is especially true for acrylic resins, which have temperature limitations as low as 100°F, and Type II SBA resins, which have a temperature limit of 105°F when in the hydroxide form; and (c) **Organic fouling and degradation: (c1)** Organic fouling and degradation is very common and expensive. Groundwaters contains only low levels of natural organic compounds (NOC). Surface waters can contain high concentrations of NOC and synthetic organic compounds (SOC). NOC include tannins, tannic acid, humic acid, fulvic acid, etc. Both NOC and SOC can complex heavy metals and iron to form organometallic compounds (OMC), and block the strong base sites on a resin. This blockage causes long final rinses and reduces salt splitting capacity. As the NOC, SOC, and OMC continue to remain on the resin, the functionality of the site changes from strong base to weak base, and finally to an irreversible nonactive IX site. Organic fouling of anion resin is evidenced by the color of the effluent from the anion unit during regeneration, which ranges from tea-colored to dark brown. During operation, the treated water has higher conductivity and a lower pH; (c2) Prevention of organic fouling and degradation can be accomplished or significantly reduced by application of prechlorination, clarification, granular activated carbon (GAC) filtration, special IX process system (such as macroporous and weak base resin ahead of strong base resin. The weak base or macroporous resin absorbs the organic material and is eluted during regeneration), and specialty resins (such as, acrylic and other specialty resins that are less susceptible to organic fouling and degradation).

Ion Exchange (IX) Resin Fouling: IX resin can become fouled with contaminants that hinder the IX process by one or more of the following cases: (a) Iron and manganese fouling: Iron may exist in water as a ferrous or ferric inorganic salt or as a sequestered organic complex. Soluble ferrous iron exchanges in IX resin; Insoluble ferric iron coats cation resin, preventing exchange, thus requiring an acid or a strong reducing agent to remove this insoluble ferric iron. Organically bound iron passes through a cation unit and fouls the anion resin. It must be removed along with the organic material. Manganese, present in some well waters, fouls a resin in the same manner as iron; (b) Aluminum fouling: Insoluble aluminum hydroxide, resulting from clarification or precipitation softening, coats the IX resin in a sodium zeolite softener. It is removed by cleaning with either acid or caustic during a normal regeneration; (c) Hardness precipitate fouling: Hardness precipitates that carry through a filter from a precipitation softener or form after filtration by post-precipitation can foul resins used for sodium zeolite softening. These hardness precipitates can be removed with acid; (d) Sulfate precipitation fouling: (d1) Calcium sulfate precipitation can occur in a strong acid cation (SAC) unit operated in the hydrogen cycle and foul the IX resin, if sulfuric acid is used as the regenerant, and the sulfuric acid concentration is too high or the sulfuric acid flow rate is too low. After calcium sulfate has formed, it is very difficult to redissolve; therefore, resin fouled by calcium sulfate is usually discarded. Mild cases of calcium sulfate fouling may be reversed with a prolonged soak in hydrochloric acid; (d2) Barium sulfate is even less soluble than calcium sulfate. If a water source contains measurable amounts of barium, hydrochloric acid regeneration should be considered; (e) Oil fouling: Oil coats resin and blocks the passage of ions to and from exchange sites. A surfactant can be used to remove oil. Oil-fouled anion resins should be cleaned with nonionic surfactants; (f) Microbiological fouling: It can occur in the IX resin beds, especially IX beds that are idled for a long time. Microbiological fouling can lead to severe plugging of the resin bed, and even mechanical damage due to an excessive pressure drop across the fouled resin. A constant flow of recirculating water through the IX bed may minimize the fouling problem. Severe conditions may require the application of suitable disinfectants or surfactants; (g) Silica fouling: Silica fouling is removed by a prolonged soak in warm (120°F) caustic soda. Silica fouling can occur in: (g1) an exhausted strong base anion (SBA) resins if the regenerant temperature is too low, or (g2) a weak base resin if the effluent caustic from the SBA unit used to regenerate the weak base unit contains too much silica; or (g3) a weak base resin at low pH levels that polymerization of the silica can occur.

Ion Exchange (IX) Resin: (a) An IX resin has a plastic matrix and contains ionizable functional groups. These functional groups consist of both positively charged cation elements and negatively charged anion elements. Only one of the ionic species is mobile. The other ionic group is attached to the bead structure; (b) For instance, a strong acid cation (SAC) exchange resin bead consisting of immobile anionic (SO_3^-) radicals and mobile sodium cations (Na^+). IX occurs when raw water ions diffuse into the bead structure and exchange for the mobile sodium cations (Na^+) and ions displaced from the bead diffuse back into the water solution.

Ion Exchange (IX) Underdrain System: It is located at the bottom of the IX vessel, retains IX resin in the tank, evenly collects the service flow, and evenly distributes the backwash flow. Uneven collection of water in service or uneven distribution of the backwash water can result in channeling, resin fouling, or resin loss. There are two primary types of underdrain system: (a) subfill system which consists of multiple layers of support media (such as graded gravel or anthracite) which support the resin, and a collector incorporating drilled pipes or subfill strainer; and (b) A resin-retaining collector system, such as a screened lateral or profile wire strainer, is more expensive than a subfill system but protects against resin loss

Ion Exchange Capacity: It is the ability of an ion exchange material to absorb ions, and is usually expressed in kilograins per cubic foot or milliequivalents per milliliter.

Ion Exchange Membrane (Electrodialysis Membrane): It is an ion exchange resin cast in flat sheet form and used in electrodialysis (ED) and electrodialysis reversal (EDR) process systems.

Ion Exchange Regenerant: It is a chemical solution used to restore an exhausted bed of ion exchange resins to the fully ionic (regenerated) form necessary for the desired ion exchange to again take place effectively.

Ion Exchange Resin: It is bead-like material that is used in deionizers for removing ions from water.

Ion Exchange Softener (Ion Exchange Water Softener): It is a water treatment unit consisting of ion exchange resins for removal of hardness (calcium and magnesium ions) from water.

Ion Exchange Total Capacity: It is the maximum theoretical quantity of ions that an ion exchange resin can hold or the total number of exchange sites.

Ion Exchanger: It is a permanent insoluble material (either a synthetic resin or a natural material) containing ions that will exchange reversible with other ions in a surrounding solution.

Ion Leakage: This is the concentration of unwanted ion left in the ion exchange process treated liquid.

Ionization: (a) It is a chemical process by which an atom or a molecule acquires a negative or positive charge by gaining or losing electrons, often in conjunction with other chemical changes; (b) It is a process or phenomenon in which the dissociation of molecules into charged particles occurs.

Ionization Constant or Dissociation Constant: (a) It is an expression in absolute units of the extent of dissociation into ions of a chemical compound in solution; (b) It is an equilibrium constant for a chemical reaction known as dissociation in the context of acid-base reactions; (c) It uses the symbol K to describe the equilibrium between ions and molecules that are not ionized in a solution or liquid.

Iron and Manganese Fouling (Ion Exchange): Iron may exist in water as a ferrous or ferric inorganic salt or as a sequestered organic complex. Soluble ferrous iron exchanges in IX resin; Insoluble ferric iron coats cation resin, preventing exchange, thus requiring an acid or a strong reducing agent to remove this insoluble ferric iron. Organically bound iron passes through a cation unit and fouls the anion resin. It must be removed along with the organic material. Manganese, present in some well waters, fouls a resin in the same manner as iron.

Leakage (Ion Exchange): It is the phenomenon in which some of the influent ions are not adsorbed or exchanged and appear in the effluent when a solution is passed through an under-regenerated exchange resin bed.

Macroporous: Resins or adsorbents that have a rigid solid porous network in which there exists a true pore structure even after drying. The pores are larger than atomic distances and are not part of the solid structure.

Macroreticular, or Macroporous Anion Resin: It is an improved polystyrene-divinylbenzene-based anion exchanger having discrete pores within a highly cross-linked polystyrene-divinylbenzene matrix and being more resistant to organic fouling due to their more porous structure.

Media-Retaining Underdrain System (Ion Exchange Media or Filtration Media): A media-retaining collector system, such as a screened lateral or profile wire strainer, is more expensive than a subfill underdrain system but protects against media loss.

Microbiological Fouling (Ion Exchange): It can occur in the ion exchange (IX) resin beds, especially IX beds that are idled for a long time. Microbiological fouling can lead to severe plugging of the resin bed, and even mechanical damage due to an excessive pressure drop across the fouled resin. A constant flow of recirculating water through the IX bed may minimize the fouling problem. Severe conditions may require the application of suitable disinfectants or surfactants.

Mixed-Bed Deionization System: (a) It is a single vessel ion exchange process system containing a mixture of cation and anion exchange resins in one vessel for the purpose of deionization; (b) it provides a final polishing step, removing trace ions to achieve ultra-pure water with very low conductivity and minimal ionic contamination.

Mixed-Bed Exchangers: (a) A mixed-bed exchanger has both cation and anion resin mixed together in a single ion exchange (IX) unit. As service water flows through the resin bed, the IX process is repeated many times, "polishing" the water to a very high purity. During regeneration, the resin is separated into distinct cation and anion fractions due to their density difference. The resin is separated by backwashing, with the lighter anion resin settling on top of the cation resin. Regenerant acid is introduced through the bottom distributor, and caustic is introduced through distributors above the resin bed. The regenerant streams meet at the boundary between the cation and anion resin and discharge through a collector located at the resin interface. Following regenerant introduction and displacement rinse, air and water are used to mix the resins. Then the resins are rinsed, and the unit is ready for service; (b) It is an IX process consisting of both cation exchange (using cation exchange resin) and anion exchange (using anion exchange resin) in different "cation IX/anion IX" ratios for removal of both cations and anions from water in one mixed-bed IX reactor at the same time; (c) The most common mixed-bed IX process unit has 40% cation exchange resin and 60% anion exchange resin in one single IX reactor for deionization

Monomer: A single reactive molecule capable of combining with another different monomer to form a polymer. Where two or more different monomers combine the resulting polymer is called a polymer.

Negative Charge: The electrical potential which an atom acquires when it gains one or more electrons; a characteristic of an anion.

Nontransient Noncommunity Water System (NTNC): It is a public water system that is not a community water system and that regularly serves at least 25 of the same persons at least 6 months/year. Examples of a nontransient noncommunity water system include, but are not limited to, schools, day care centers, factories, and other places of employment.

Oil Fouling (Ion Exchange): Oil coats resin and blocks the passage of ions to and from exchange sites. A surfactant can be used to remove oil. Oil-fouled anion resins should be cleaned with non-ionic surfactants.

Organic Fouling and Degradation (Ion Exchange): (a) Organic fouling and degradation is very common and expensive. Groundwaters contains only low levels of natural organic compounds (NOC). Surface waters can contain high concentrations of NOC and synthetic organic compounds (SOC). NOC include tannins, tannic acid, humic acid, fulvic acid, etc. Both NOC and SOC can complex heavy metals and iron to form organometallic compounds (OMC), and block the strong base sites on a resin. This blockage causes long final rinses and reduces salt splitting capacity. As the NOC, SOC, and OMC continue to remain on the resin, the functionality of the site changes from strong base to weak base, and finally to an irreversible nonactive IX site. Organic fouling of anion resin is evidenced by the color of the effluent from the anion unit during regeneration, which ranges from tea-colored to dark brown. During operation, the treated water has higher conductivity and a lower pH; (b) Prevention of organic fouling and degradation can be accomplished or significantly reduced by application of pre-chlorination, clarification, granular activated carbon (GAC) filtration, special IX process system (such as macroporous and weak base resin ahead of strong base resin. The weak base or macroporous resin absorbs the organic material and is eluted during regeneration), and specialty resins (such as, acrylic and other specialty resins that are less susceptible to organic fouling and degradation).

Oxidant Degradation (Ion Exchange): Oxidizing agents, such as chlorine, ozone, chlorine dioxide, air, oxygen, etc. degrade both cation and anion ion exchange (IX) resins. (a) Oxidants attack the divinylbenzene cross-links in a cation resin, reducing the overall strength of the resin bead. As the attack continues, the cation resin begins to lose its exchange capacity, spherical shape, rigidity, causing bed compaction, bed channeling, and pressure drop across the resin bed; (b) Anion resin is not directly affected in water chlorination or ozonation, or oxidation, because the chlorine is consumed by the cation resin. However, downstream strong base anion (SBA) resins are fouled by certain degradation products (such as organic disinfectant by-products, DBP) from oxidized cation resin. The oxidants should be removed prior to IX with activated carbon filtration or sodium sulfite; (c) Oxygen-saturated water, such as that formed following forced draft decarbonation, accelerates the destruction of strong base exchange sites that occurs naturally over time. It also accelerates degradation due to organic fouling.

Permanent Hardness: Calcium sulfate, magnesium sulfate, and the chlorides of these two metals are called permanent hardness because it may not be removed by boiling the water.

pH: (a) An expression of the acidity of a solution; the negative logarithm of the hydrogen ion concentration (pH 1 very acidic; pH 14, very basic); (b) The measurement of the acidity of a solution where 1 is very acidic, 7 is neutral, and 14 is very basic.

Physical Stability (Ion Exchange): The quality which an ion exchange resin must possess to resist changes that might be caused by attrition, high temperatures, and other physical conditions.

Point-of-Entry Treatment Device (POE): It is a treatment device applied to the drinking water entering a house or building for the purpose of reducing the contaminants in the drinking water distributed through all, or a portion of, the house or building.

Point-of-Use Treatment Device (POU): It is a treatment device applied to a single tap used for the purpose of reducing contaminants in drinking water at that one tap.

Polymer: Where two or more different monomers combine the resulting polymer is called a polymer.

Polymeric Adsorbent: It is a synthetic resin possessing the ability to attract and to hold charged particles, and is an adsorbent used in polymeric adsorption process.

Polystyrene-Divinylbenzene-Based Anion Exchanger: It is manufactured by copolymerization of styrene cross-linked with divinylbenzene. It is very stable and has high exchange capacities for removing anions including silicic and carbonic acids.

Porosity: An expression of the degree of permeability in granular media (such as ion exchange resins, or granular activated carbons) to liquids and large organic molecules. Gel resins, even when referred to as highly porous, have a negligible porosity in comparison to the macropores inherent in the macroporous resins or adsorbents.

Positive Charge: The electrical potential acquired by an atom which has lost one or more electrons; a characteristic of a cation.

Public Water System (PWS): (a) A public water system provides piped water for human consumption through at least 15 service connections or serves at least 25 people at least 60 days a year; or (b) It is any water supply system serving an agricultural labor camp, as defined in a Federal or the State environmental regulation.

Quaternary Ammonium: A specific basic group $[-N(CH_3)_3^+]$ on which depends the exchange activity of certain anion exchange resins. Any of numerous strong bases and their salts derived from ammonium by replacement of the hydrogen atoms with organic radicals.

Raw Water: It is the untreated water from wells or from surface sources.

Regeneration Cycle (Ion Exchange Softener): (a) When an ion exchange (IX) softener is exhausted, the IX resin must be regenerated. Monitoring of the effluent hardness reveals resin exhaustion. When hardness increases, the unit is exhausted. Many facilities regenerate softeners before exhaustion, based on a predetermined time period or number of gallons processed. Most softening systems consist of more than one softener which are often operated so that one softener is in regeneration cycle or standby while the other units are in service cycle. The regeneration cycle of a sodium zeolite softener consists of four steps: (a) backwash, (b) regeneration (brining), (c) displacement (slow rinse), and (d) fast rinse.

Regenerant: The solution used to restore the activity of an ion exchanger. Acids are employed to restore a cation exchanger to its hydrogen form; brine solutions may be used to convert the cation exchanger to the sodium form. The anion exchanger may be rejuvenated by treatment with an alkaline solution.

Regenerant Reuse (Ion Exchange): Due to the high cost of caustic soda and the increasing problems of waste disposal, many ion exchange (IX) systems are now equipped with a caustic reclaim feature. The reclaim system uses a portion of the spent caustic from the previous regeneration at the beginning of the next regeneration cycle. The reused caustic is followed by fresh caustic to complete the regeneration. The new caustic is then reclaimed for use in the next regeneration. Typically, sulfuric acid is not reclaimed, because it is lower in cost and calcium sulfate precipitation is a potential problem.

Regeneration: (a) Restoration of the activity of an ion exchanger by replacing the ions adsorbed from the treated solution by ions that were adsorbed initially on the resin; (b) Restoration of the activity of granular activated carbon (GAC) or nonionic polymeric adsorbent resin by physicochemical or thermal processes, so they can be reused.

Regeneration (Ion Exchange): When exhausted ion exchange (IX) resin is regenerated, a high concentration of sodium ions is applied to the resin to replace calcium and magnesium. During regeneration, a large excess of IX regenerant (approximately 3 times the amount of calcium and magnesium in the resin) is used. The eluted hardness is removed from the softening unit in the waste brine and by rinsing. The duration of a service cycle depends on the rate of softener flow, the hardness level in the water, and the amount of salt used for regeneration. The resin is treated with a 10% sodium chloride solution, and regeneration proceeds according to the following equations:

Regeneration (Ion Exchange) Chemical Reactions:

$$\text{Z-Ca} + 2\text{NaCl (concentrated)} \Rightarrow \mathbf{Na_2\text{-}Z} + \mathbf{CaCl_2}$$

$$\text{Z-Mg} + 2\text{NaCl (concentrated)} \Rightarrow \mathbf{Na_2\text{-}Z} + \mathbf{MgCl_2}$$

Resistivity: A measure of specific resistance to the flow of electricity. In water, it is an accurate measure of ionic purity.

Rinse (Ion Exchange): It is a physicochemical operation which follows regeneration for flushing out of excess regenerant solution.

Salt Splitting (Ion Exchange): It is the conversion of salts to their corresponding acids or bases by passage through ion exchange resins containing strongly acidic or strongly basic functional groups.

Selectivity: (a) It is the difference in attraction of one ion over another by an ion exchange (IX) resin. IX media have a greater affinity for certain ionic species than for others and are preferentially sorbed; (b) The difference in attraction of one ion over another by an IX resin.

Service Cycle (Ion Exchange Softener): In the service cycle, influent water enters the ion exchange (IX) softener through the inlet distribution system and flows through the resin bed at almost constant flow rate between 6 and 12 gpm/ft^2 (Note: some special designs utilize a deep resin bed to permit operation at 15–20 gpm/ft^2) of resin surface area. The hardness ions diffuse into the resin and exchange with sodium ions, which return to the bulk water. Soft water effluent is collected in the underdrain system and discharged.

Service Rinse in Ion Exchange Regeneration Cycle: If the softener has been in standby following a regeneration, a second fast rinse, known as a service rinse, can be used to remove any contaminant ions (such as hardness) that has entered the water during standby.

Silica Breakthrough Monitoring and Removal in Ion Exchange (IX) Process: (a) Demineralization using strong base anion (SBA) resins removes silica as well as other dissolved solids. IX effluent silica and conductivity are important parameters to monitor during an IX demineralizer service run. Both silica and conductivity are low at the end of the fast rinse. When silica breakthrough occurs at the end of a service run, the treated water silica level increases sharply. Often, the conductivity of the water decreases momentarily, then rises rapidly; (b) When the end of a demineralizer run is detected, the unit must be removed from service immediately. If the demineralizer is allowed to remain in service past the breakpoint, the level of silica in the treated water can rise above that of the influent water, due to the concentrating of silica that takes place in the anion resin during the service; (c) strong base anion (SBA) exchangers are regenerated with a 4% sodium hydroxide (NaOH) solution. As with cation regeneration, the relatively high concentration of hydroxide drives the regeneration reaction. To improve the removal of silica from the resin bed, the regenerant caustic is usually heated to 120°F or to the temperature specified by the resin manufacturer. Silica removal is also enhanced by a resin bed preheat step before the introduction of warm caustic.

Silica Fouling (Ion Exchange): Silica fouling is removed by a prolonged soak in warm (120°F) caustic soda. Silica fouling can occur in: (a) an exhausted strong base anion (SBA) resins if the regenerant temperature is too low, or (b) a weak base resin if the effluent caustic from the SBA unit used to regenerate the weak base unit contains too much silica; or (c) a weak base resin at low pH levels that polymerization of the silica can occur.

Siliceous Gel Zeolite: A synthetic, inorganic exchanger produced by the aqueous reaction of alkali with aluminum salts.

Sodium Zeolite and Hydrogen Zeolite (Split Stream) Dealkalization: In a split stream dealkalization unit, a portion of the raw water flows through a sodium zeolite softener. The remainder flows through a hydrogen-form strong acid cation (SAC) unit (hydrogen zeolite). The effluent from the sodium zeolite is combined with the hydrogen zeolite effluent. The effluent from the hydrogen zeolite unit contains carbonic acid, produced from the raw water alkalinity, and free mineral acids. When the two streams are combined, free mineral acidity (FMA) in the hydrogen zeolite effluent converts sodium carbonate and bicarbonate alkalinity in the sodium zeolite effluent to carbonic acid (note: Unstable carbonic acid H_2CO_3 forms carbon dioxide gas and water) as shown below:

$$H_2CO_3 \Rightarrow CO_2 \text{ (gas)} + H_2O$$

Sodium Zeolite and Chloride Anion Dealkalization: (a) Strong base anion (SBA) **resin in the chloride form (Z-Cl) can be used to reduce the alkalinity of a water. Water flows through a**

zeolite softener and then an SBA anion unit, which replaces the carbonate, bicarbonate, sulfate, and nitrate ions with chloride ions; (b) The chloride anion dealkalizer reduces alkalinity by approximately 90% but does not reduce total solids. ; (c) The zeolite softener is regenerated as previously described. In addition, the anion resin is also regenerated with a sodium chloride brine that returns the resin to the chloride form **(Z-Cl).**

Sodium Zeolite Softener Operation: The operation include two basic cycles: (a) the service cycle, which produces soft water for use, and (b) the regeneration cycle, which restores resin capacity at exhaustion.

Sodium Zeolite Softening: (a) It is a physicochemical (PC) process involving the use of zeolite ion exchange (IX) bed containing strong acid cation (SAC) resin in sodium form for removal of scale-forming hardness ions, such as calcium and magnesium from water; The product water is the hardness-free water, or soft **water;** (b) In the PC-IX process, the hardness ions (calcium and magnesium ions) are exchanged with the sodium, and the sodium ions diffuse into the bulk water solution when water passes through the SAC resin bed (sodium form).

Softening Process (Ion Exchange): In the softening process, a strong acid cation (SAC) (SAC) exchange resin bead consisting of immobile anionic (SO_3^-) radicals and mobile sodium cations (Na^+) is used. Ion exchange occurs when raw water ions diffuse into the bead structure and exchange for the mobile sodium cations (Na^+) and ions displaced from the bead diffuse back into the water solution.

Specialty Resins: (a) Types include chelating resins (e.g., iminodiacetic acid, thiourea-based resins); and (b) applications include heavy metal removal and selective ion recovery.

Sphericity: It is a physical character relating to the spherical nature and whole.

Strong Acid Cation Exchanger or Resin: (a) SAC resins can neutralize strong bases and convert neutral salts into their corresponding acids. strong acid cation (SAC) exchangers function well at all pH ranges. These resins have found a wide range of applications. They can be used in the sodium cycle (sodium as the mobile ion) for softening, and in the hydrogen cycle for decationization; (b) In the softening process, a SAC exchange resin bead consisting of immobile anionic (SO_3^-) radicals and mobile sodium cations (Na^+) is used. Ion exchange (IX) occurs when raw water ions diffuse into the bead structure and exchange for the mobile sodium cations (Na^+) and ions displaced from the bead diffuse back into the water solution; (c) The SAC exchange reaction is reversible. When its capacity is exhausted, the SAC resin can be regenerated with an excess of appropriate mineral compound; and (d) The IX regeneration is less efficient at the higher regenerant levels. The softener operating costs increase as the regenerant level increases. A 150% increase in regenerant salt provides only a 67% increase in operating capacity. The effect of regenerant salt level on SAC resin softening capacity is reported in below: (d1) 6 lb/ft^3 salt for 18,000 g/ft^3 capacity; (d2) 8 lb/ft^3 salt for 20,000 g/ft^3 capacity; (d3) 10 lb/ft^3 salt for 24,000 g/ft^3 capacity; and (d4) 15 lb/ft^3 salt for 30,000 g/ft^3 capacity.

Strong Base Anion Resin: Strong base anion (SBA) resins can neutralize strong acids and convert neutral salts into their corresponding bases. These resins are utilized in most softening and full demineralization applications; (b) SBA resins derive their functionality from quaternary ammonium functional groups. There are two types of quaternary ammonium groups: Type I exchanger sites have three methyl groups [R-N $(CH_3)_3$)], while in a Type II exchanger, one of the methyl groups (CH_3) is replaced with an ethanol group. The Type I resin has a greater stability than the Type II resin and is able to remove more of the weakly ionized acids. Type II resins provide a greater regeneration efficiency and a greater capacity for the same amount of regenerant chemical used; (c) As with the cation resins, these reactions are reversible, allowing for the regeneration of the resin with a strong alkali, such as caustic soda, to return the resin to the hydroxide form; (d) When in the hydroxide form, SBA resins remove all commonly encountered anions as shown below:

Strong Base Anion (SBA) Resin Chemical Reactions:

$$H_2SO_4 + 2Z\text{-}OH \Rightarrow Z_2\text{-}SO_4 + 2H_2O$$

$$H_2SiO_3 + Z\text{-}OH \Rightarrow Z\text{-}HSiO_3 + H_2O$$

$$H_2CO_3 + Z\text{-}OH \Rightarrow Z\text{-}HCO_3 + H_2O$$

$$HCl + Z\text{-}OH \Rightarrow Z\text{-}Cl + H_2O$$

Strong Electrolyte Resin: The equivalent of strongly acidic or strongly basic resins and capable of splitting neutral salts. They're good conductors of electric current in a solution.

Subfill Underdrain System (Ion Exchange or Filtration): It consists of multiple layers of support media (such as ion exchange resins, graded gravel or anthracite) which support the median, and a collector incorporating drilled pipes or subfill strainer. As long as the support layers remain intact, the media in the bed will remain in place. If the supporting media becomes disturbed, usually due to improper backwash, the media can move through the disrupted layers and exit the vessel.

Sulfate Precipitation Fouling (Ion Exchange): (a) Calcium sulfate precipitation can occur in a strong acid cation (SAC) unit operated in the hydrogen cycle and foul the IX resin, if sulfuric acid is used as the regenerant, and the sulfuric acid concentration is too high or the sulfuric acid flow rate is too low. After calcium sulfate has formed, it is very difficult to redissolve; therefore, resin fouled by calcium sulfate is usually discarded. Mild cases of calcium sulfate fouling may be reversed with a prolonged soak in hydrochloric acid; (b) Barium sulfate is even less soluble than calcium sulfate. If a water source contains measurable amounts of barium, hydrochloric acid regeneration should be considered.

Sulfonic: It is a chemical specific acidic group (SO_3H) on which depends the exchange activity of certain cation exchange resins

Superficial Linear Velocity: It is the flow of water per unit of area, or gpm/ft^2 or L/min/m^2.

Swelling: The expansion of an ion exchange bed which occurs when the reactive groups on the resin are converted into certain forms.

Temporary Hardness: It is caused by the presence of magnesium or calcium bicarbonate because it may be removed by boiling the water to convert the bicarbonates to the insoluble carbonates.

Thermal Degradation (Ion Exchange): Thermal degradation occurs if the anion resin becomes overheated during the service or regeneration cycle. This is especially true for acrylic resins, which have temperature limitations as low as 100°F, and Type II strong base anion (SBA) resins, which have a temperature limit of 105°F when in the hydroxide form.

Throughput (Ion Exchange): It is the amount of solution treated prior to the exhaustion of the ion exchange resin.

Throughput Volume (Ion Exchange): It is the amount of solution passed through an ion exchange (IX) bed before exhaustion of the IX resin.

Total Dissolved Solids (TDS): It is the total amount of nonvolatile matter dissolved in a water sample, and is usually expressed in parts-per-million by weight.

Total Ionized Solids: (a) It is the concentration of dissolved ions in solution, expressed in concentration units of NaCl (sodium chloride); (b). It determines the operating life of ion exchange resins, and is calculated from measurements of specific resistance.

Total Organic Carbon (TOC): (a) It is a measure of the level of organic impurities (in terms of total carbon content) in water; (b) TOC determines the operating life of activated carbon beds.

Transient Noncommunity Water System (TNC): It is a noncommunity public water system that serves at least 25 persons per day for at least 60 days/year. Examples of a transient noncommunity water system may include, but are not limited to, campgrounds, churches, restaurants, and rest areas.

Two-Bed Deionization System (Two-Bed Ion Exchange System): (a) a two-bed deionization system utilizes two separate vessels, one for cation exchange resin and one for anion exchange resin; and is operated as a sequential process that water passes through the cation bed first, then the anion bed; (b) the performance of a two-bed ion exchange process system is moderate because it produces moderate-purity water, and not as pure as mixed-bed systems; and it is effective for general deionization needs, not for production of ultra-pure water.

Upflow: The operation of a granular media (such as ion exchange, granular activated carbon, sand filtration, biological filtration, etc.) unit in which water or wastewater is passed in at the bottom and out at the top of the process reactor.

Vacuum Degasser Equipment Design: In a vacuum degasser, water droplets are introduced into a packed column that is operated under a vacuum. Carbon dioxide is removed from the water due to its decreased partial pressure in a vacuum. A vacuum degasser usually reduces carbon dioxide to less than 2 ppm and also removes most of the oxygen from the water. However, vacuum degassers are more expensive to purchase and operate than forced draft decarbonators.

Valance: (a) It is a measurement of the number of atoms or ions of hydrogen it takes to combine with or be replaced by an element or radical; (b) It is the number of positive or negative charges of an ion; (c) It is the combining power of an element, as measured by the number of hydrogen atoms it can displace or combine with.

Void Volume: The space between particles of ion exchange resins or other granular media bed.

Weak Acid Cation Dealkalization: (a) It is a dealkalization process which uses weak acid cation (WAC) resins. WAC resins are similar in operation to strong acid cation (SAC) resins, but only exchange for cations that are associated with alkalinity; (b) During the initial portion of a WAC service run (the first 40%–60%) some cations associated with mineral anions exchange, producing small amounts of mineral acids in the effluent. As the service cycle progresses, alkalinity appears in the effluent. When the alkalinity in the effluent exceeds 10% of the influent alkalinity, the unit is removed from service and regenerated with a 0.5% sulfuric acid solution. The concentration of regenerant acid should be kept below 0.5%–0.7%, to prevent calcium sulfate precipitation in the resin. WAC resin exchange is very efficient. Therefore, the amount of acid required is virtually equal (chemically) to the amount of cations removed during the service cycle: (c) The carbonic acid (H_2CO_3) formed is removed by a decarbonator or degasser as in a split stream system. Assuming Z represents the IX resin the chemical reactions of WAC dealkalization are presented below:

Weak Acid Cation (WAC) Dealkalization Chemical Reactions:

$$Ca(HCO_3)_2 + 2Z\text{-}H \Rightarrow Z_2\text{-}Ca + 2H_2CO_3$$

$$Mg(HCO_3)_2 + 2Z\text{-}H \Rightarrow Z_2\text{-}Mg + 2H_2CO_3$$

$$Na(HCO_3) + Z\text{-}H \Rightarrow Z\text{-}Na + H_2CO_3$$

$$H_2CO_3 \text{ (unstable)} \Rightarrow CO_2 + H_2O$$

Weak Acid Cation Exchange Resins: (a) Weak acid cation (WAC) resins derive their exchange activity from a carboxylic group (-COOH). When operated in the hydrogen form, WAC resins remove cations that are associated with alkalinity, producing carbonic acid; (b) These reactions are reversible and permit the return of the exhausted WAC resin to the regenerated form. WAC resins are not able to remove all of the cations in most water supplies. Their primary value is their high regeneration efficiency in comparison with strong acid cation (SAC) resins. This high efficiency reduces the amount of acid required to regenerate the resin, thereby reducing the waste acid and minimizing disposal problems; (c) WAC resins are used primarily for softening and dealkalization of high-hardness, high-alkalinity waters, frequently in conjunction with SAC sodium cycle polishing systems. In full demineralization systems, the use of WAC and SAC resins in combination provides the economy of the more efficient WAC resin along with the full exchange capabilities of the SAC resin; and (d) WAC resins are able to neutralize strong bases, and these WAC resins are used for dealkalization, partial demineralization, or (in combination with strong resins) full demineralization.

Weak Acid Cation Exchange Chemical Reactions:

$$Ca(HCO_3)_2 + 2Z(COO)H \Rightarrow [Z(COO)]_2\text{-}Ca + 2H_2CO_3$$

$$Mg(HCO_3)_2 + 2Z(COO)H \Rightarrow [Z(COO)]_2\text{-}Mg + 2H_2CO_3$$

$$Na(HCO_3) + Z(COO)H \Leftrightarrow [Z(COO)]\text{-}Na + H_2CO_3$$

Weak Base Anion Resin: (a) WBA resins are able to neutralize strong acids and these WBA resins are used for dealkalization, partial demineralization, or (in combination with strong resins) full demineralization; and (b) WBA resin functionality originates in primary (R-NH_2), secondary (R-NHR′), or tertiary (R-NR'_2) amine groups. WBA resins readily remove sulfuric, nitric, and hydrochloric acids, as represented by the following reactions:

Weak Base Anion (WBA) Resin Chemical Reactions:

$$H_2SO_4 + 2Z\text{-}OH \Rightarrow Z_2\text{-}SO_4 + 2H_2O$$

$$HCl + Z\text{-}OH \Rightarrow Z\text{-}Cl + H_2O$$

$$HNO_3 + Z\text{-}OH \Rightarrow Z\text{-}NO_3 + H_2O$$

Weak Electrolyte: It is an electrolyte that gives a low percentage yield of ions when dissolved in water.

Zeolite Softening or Sodium Zeolite Softening: The removal of hardness from water by a zeolite softening process is described by the following reaction:

Zeolite Softening or Sodium Zeolite Softening Chemical Reactions:

$$CaSO_4 + Z\text{-}Na_2 \Rightarrow Z\text{-}Ca + Na_2SO_4$$

$$CaCl_2 + Z\text{-}Na_2 \Rightarrow Z\text{-}Ca + 2NaCl$$

$$Ca(HCO_3)_2 + Z\text{-}Na_2 \Rightarrow Z\text{-}Ca + 2Na(HCO_3)$$

$$MgSO_4 + Z\text{-}Na_2 \Rightarrow Z\text{-}Mg + Na_2SO_4$$

$$MgCl_2 + Z\text{-}Na_2 \Rightarrow Z\text{-}Mg + 2NaCl$$

$$Mg(HCO_3)_2 + Z\text{-}Na_2 \Rightarrow Z\text{-}Mg + 2Na(HCO_3)$$

Zeolite: (a) It is a synthetic aluminosilicate material which is used as the exchangers in the first ion exchange water softeners. Although aluminosilicate materials are rarely used today, the term "zeolite softener" is commonly used to describe any cation exchange process mainly for domestic and commercial water purification and softening. They have also been used as catalysts and sorbents.

REFERENCES

1. Wang, LK and Shammas, NK (2016). *Water Engineering: Hydraulics, Distribution and Treatment*, Wiley, New York, USA, 806 pages.
2. Wang, MHS and Wang, LK (2026). Fundamentals and innovations of iron and manganese control. In: Industrial *Water and Waste Treatment: Fundamentals*, Wang, LK, Wang, MHS and Hung, YT (editors), CRC Press, Boca Raton, FL, USA. pp. 183–209.
3. Wang, MHS and Wang, LK (2026). Fundamentals and innovations of calcium and magnesium control. In: Industrial *Water and Waste Treatment: Fundamentals*, Wang, LK, Wang, MHS and Hung, YT (editors), CRC Press, Boca Raton, FL, USA. pp. 211–221.
4. Wang, MHS and Wang, LK (2025). Cooling tower and boiler water management, In: Industrial *Water and Waste Treatment: Pollution Control*, Wang, LK, Wang, MHS and Hung, YT (editors), CRC Press, Boca Raton, FL, USA.

15 Sequencing Batch Reactors Fundamentals and Innovations

Mu-Hao Sung Wang, Lawrence K. Wang, and Kenneth A. Mikkelson

DEDICATION

This publication is dedicated to Dr. Kenneth A. Mikkelson who contributed significantly in the field of sequencing batch reactor technology and wrote the preliminary draft of this chapter before his passing. Dr. Mikkelson (1941–2021) came from a family of Norwegian descent and passed away on May 15, 2021, at age 79. He earned a Bachelor's degree in Chemistry from Missouri State, a Masters in Environmental Engineering, and a Ph.D. in Environmental Health Science. Both higher degrees were from Kansas University. His advanced education led to a career in environmental health engineering. It kept him busy traveling the world, teaching, lecturing, designing, selling, and writing about wastewater treatment (WWT) systems. His family followed his career from Kansas to Alabama, Illinois, and finally Madison, WI. This became a permanent home as his career matured. The choice to reside in Madison was a considerable commitment for Dr. Mikkelson since he was working for Aqua-Aerobic Systems, Inc. in Rockford, IL. He commuted 150 miles a day for years so his family could live in Madison where they would be comfortable. It was just another example of how much Ken was capable of giving. He truly had an inexhaustible capacity to provide for and care for his family and the society. To say he was a devoted family man and a devoted environmental engineer is an understatement. Dr. Mikkelson would go to any lengths to build something for the society and those he loved. His most important academic contribution is a book entitled, "AquaSBR Design Manual" [1], which is widely viewed and cited by professional engineers, government officials, and university scholars. Everyone loved Dr. Mikkelson, and will forever remember him and his accomplishment in the field of environmental health engineering.

ACRONYMS

BIO-SBR	Biological sequencing batch reactor
BNR	Biological nutrient removal
BOD	Biochemical oxygen demand
CBOD	Carbonaceous biochemical oxygen demand
COD	Chemical oxygen demand
DAF	Dissolved air flotation
DOB	Depth of blanket
ENR	Enhanced nutrient removal
EPS	Extracellular polymeric substances
F/M	Food-to-microorganism ratio
HRT	Hydraulic retention time
LIWT	Lenox Institute of Water Technology
MCRT	Mean cell residence time
ML	Mixed liquor
MLSS	Mixed-liquor suspended solids
MLVSS	Mixed-liquor volatile suspended solids

DOI: 10.1201/9781003532354-15

MPI	Meat and poultry industry
NH_3-N	Ammonia-nitrogen
NMPAN	Niche Meat Processor Assistance Network
NO_2-N	Nitrite-nitrogen
NO_3-N	Nitrate-nitrogen
NPDES	National Pollutant Discharge Elimination System
O&G	Oil and grease
O&M	Operation and maintenance
ORP	Oxidation reduction potential
OUR	Oxygen-uptake rate
PC-BIO-SBR	Physicochemical and biological sequencing batch reactor
PC-SBR	Physicochemical sequencing batch reactor
PLC	Programmable logic controller
PO_4-P	Phosphate-Phosphorus
POTW	Publicly owned treatment works
RUWWTP	Riviera Utilities Wastewater Treatment Plant
SBR	Sequencing batch reactor
SOUR	Specific oxygen-uptake rate
SSV	Settled-sludge volume
SVI	Sludge volume index
TKN	Total Kjeldahl nitrogen
TOC	Total organic carbon
TP	Total phosphorus
TSS	Total suspended solids
USEPA	US Environmental Protection Agency
UV	Ultraviolet
VSS	Volatile suspended solids
WAS	Waste-activated sludge
WWT	Wastewater treatment
WWTP	Wastewater treatment plant
XRF	X-ray fluorescence

15.1 INTRODUCTION

15.1.1 Sequencing Batch Reactor Systems Overview

There are many types of sequencing batch reactor (SBR) process systems: (a) physicochemical sequencing batch reactor (PC-SBR); (b) biological sequencing batch reactor (BIO-SBR, or simply SBR), and (c) combined physicochemical-biological sequencing batch reactor (PC-BIO-SBR) [2–6].

A physicochemical sequencing batch reactor (PC-SBR) is a variation of the water treatment process or independent physicochemical wastewater treatment (WWT) process. As a fill and draw or batch process, all physicochemical treatment phases, such as chemical feeding, rapid mixing, slow flocculation, clarification (sedimentation or flotation), and clarifier effluent discharge, occur in a single tank. This differs from the conventional flow-through physicochemical process system in that PC-SBRs do not require separate process tanks for chemical feeding, rapid mixing, slow flocculation, and clarification [2,3]. In some special cases, the final disinfection and corrosion control steps can also be incorporated into the PC-SBR's single tank system, although it is not recommended. Small PC-SBR systems have been adopted by the industry [3] for treating the wastewaters containing toxic substances, and large industrial and municipal PC-SBR systems are gaining momentum [6].

Similarly, a biological sequencing batch reactor (BIO-SBR) is a variation of the conventional biological activated sludge process. As a fill and draw or batch process, all biological treatment phases occur in a single SBR tank. This differs from the conventional continuous flow-through activated sludge process in that a SBR does not require separate tanks for equalization, anoxic, aeration, and clarification. SBR systems contain either two or more reactor tanks that are operated in parallel or one equalization tank and one reactor tank.

15.1.2 Summary

The F.B. Purnell Sausage Co. wastewater treatment plant (WWTP) is a wastewater total treatment system consisting of: (a) preliminary treatment by screening; (b) primary treatment by dissolved air flotation (DAF); (c) secondary treatment by SBR; (d) equalization of SBR effluents; (e) tertiary filtration by sand filtration; (f) ultraviolet (UV) disinfection; and (g) final WWTP effluent discharge to a receiving stream. The WWTP sludge treatment train includes: (a) sludge from SBR; (b) aerobic digestion; (c) mechanical dewatering or drying beds; and (d) final disposal of dried sludge by landfill. The Purnell's process is both technically and economically feasible for total waste treatment aiming at direct effluent discharge to a receiving stream. The AquaSBR is a conventional SBR system that has proven to be an efficient, flexible, and reliable means of meat processing WWT.

A new AquaNereda® Aerobic Granular Sludge System at the Riviera Utilities Wastewater Treatment Plant (RUWWTP) at Wolf Creek, Foley, AL, USA, was installed and started in January 2020 serving about 20,000 population of the City of Foley. The advanced SBR (AquaNereda) provides WWT for an average flow of 3.5 MDG and peak flow of 6 MGD. It consists of three aerobic granular sludge reactors (AquaNereda) operating similar to a continuous flow system with a FILL/DRAW phase that alternates between reactors. Downstream polishing is performed by AquaDisk® Pile cloth media filters to produce Class B reuse-quality water. Although not currently required by NPDES permit to achieve total nitrogen (TN) limit, the inherent biological nutrient removal (BNR) properties of Aerobic Granular Sludge provide TN removal now and for future NPDES permit limits. The new SBR technology was highly appealing as an innovative approach to secondary treatment for a number of reasons, including: (a) The innate BNR capabilities provided by the layered microbial community of the granules enable the system to meet future TN and total phosphorus (TP) limits; (b) The high operating MLSS in excess of 8 g/L significantly reduces settling time minimizing system footprint by up to 75%; (c) Up to a 50% decrease in energy use due to reduced equipment demand, smart aeration control, and improved substrate and oxygen utilization with granular sludge; (d) The robust nature of the granules provides superior protection against plant upsets and toxic shocks compared to conventional activated sludge; (e) The RUWWTP is now capable of providing BNR for future total nitrogen (TN) and total phosphorus (TP) limits. Downstream polishing by AquaDisk pile cloth media filters is capable of achieving biochemical oxygen demand (BOD) and total suspended solids (TSS) concentrations of <5 mg/L. The RUWWTP has consistently met effluent requirements since startup in January 2020. It has been proven in full scale that the combination of newly developed AquaNereda technology and AquaDisk pile cloth media filters is reliable for municipal WWT up to 6 MGD flow.

15.2 CONVENTIONAL SEQUENCING BATCH REACTOR SYSTEMS AND MEAT PROCESSING WASTEWATER TREATMENT

15.2.1 Conventional SBR Process Systems

There are many types of SBR systems, including continuous influent/time-based, non-continuous influent/time-based, volume-based, an intermittent cycle system (a SBR that utilizes jet aeration), and various other system modifications based on different manufacturer designs [7]. The type of SBR system used depends on site and wastewater characteristics as well as the needs of the area or community installing the unit. Package SBRs are manufactured to treat wastewater flow rates between 0.01 and 5.0 MGD (million gallons per day). A package SBR WWTP is a complete treatment system that includes a WWT train and a sludge treatment train. The WWT train processes include: (a) preliminary WWT (screening, grinding), (b) primary WWT, such as dissolved air flotation (DAF) for oil and grease (O&G) removal, if necessary, (c) SBR; (d) equalization, (e) tertiary filtration, and (f) disinfection. The sludges wasted from SBR are treated separately by a sludge treatment train including: (a) sludge thickening; (b) sludge dewatering; (c) sludge digestion; and (d) sludge disposal, or beneficial reuse [1–11].

It is important to note that (a) an equalization tank after a batch SBR is typically needed before the wastewater disinfection in order to store large volumes of SBR effluent. If the SBR effluent flow is not equalized, a sizable tertiary filter may be necessary to accommodate the large flow of water entering the final disinfection system; (b) SBR systems typically have no primary or secondary clarifiers as settling takes place in the SBR; (c) a primary flotation clarification using DAF is needed only if the wastewater O&G is very high; and (d) a SBR system does not require a return activated sludge (RAS) system, as both aeration and settling occur in the same SBR tank. This prevents any sludge from being lost during the react step and eliminates the need to return sludge from the clarifier to the aeration chamber.

There are normally five phases in the SBR treatment cycle: fill, react, settle, draw (decant), and idle. The length of time that each phase occurs is controlled by a programmable logic controller (PLC), which allows the system to be controlled from remote locations [1–11]. The react phase can be expanded (to anoxic, aerobic, etc.) for biological nutrient removal (BNR), when necessary. Figure 15.1 shows this sequence of operation in an SBR.

In the FILL PHASE, shown in Figure 15.1, raw wastewater enters the SBR basin, where it is mixed with settled biomass from the previous cycle. Some aeration may occur during this phase, as shown in Figure 15.1.

Then, in the REACT PHASE (Figure 15.1), the basin's mixed liquor (ML) is initially mixed without aeration, creating an anoxic condition for denitrification, then the mixed liquor is aerated, allowing aerobic bio-oxidation and nitrification to occur.

During the SETTLING PHASE (Figure 15.1), aeration and mixing are stopped and the solids in the mixed liquor are allowed to settle. The treated wastewater is then discharged from the basin in the DRAW PHASE. In the final IDLE PHASE (not shown in Figure 15.1), the basin is idle as it waits for the start of the next cycle. During this time, parts of the solids are removed from the basin and disposed of as waste sludge.

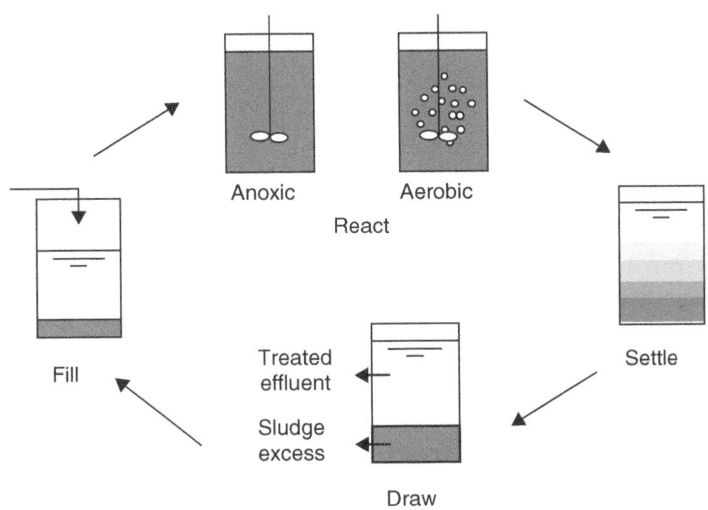

FIGURE 15.1 The operational phases of a typical biological sequencing batch reactor [7].

As stated previously, SBRs are typically used at wastewater flowrates of 5 MGD or less. As these small WWT systems have a relatively small footprint, SBRs are useful for areas where the available land is limited. In addition, the WWT cycles within the SBR system can be easily modified for BNR in the future, if BNR becomes necessary. This makes SBRs extremely flexible to adapt to regulatory changes for effluent parameters such as BNR. SBRs are also very cost-effective if WWT beyond biological treatment is required, such as additional tertiary filtration. Some additional advantages of SBRs are listed below: (a) Equalization, primary clarification (in most cases), biological treatment, and secondary clarification can be achieved in a single reactor vessel; (b) Operating flexibility and control; (c) Minimal footprint; (d) Potential capital cost savings by eliminating clarifiers and other equipment.

There are some minor disadvantages for operating a SBR system, however, in accordance with the US Environmental Protection Agency (USEPA) [11]: (a) A higher level of sophistication is required (compared to conventional systems), especially for larger systems, of timing units and controls; (b) Higher level of maintenance (compared to conventional systems) associated with more sophisticated controls, automated switches, and automated valves; (c) Potential of discharging floating or settled sludge during the DRAW or decant phase with some SBR configurations; (d) Potential plugging of aeration devices during selected operating cycles, depending on the aeration system used by the manufacturer; and (e) Potential requirement for equalization after the SBR, depending on the downstream processes.

The wastewater influent parameters typically include design flow, maximum daily flow BOD_5, TSS, pH, alkalinity, wastewater temperature, total Kjeldahl nitrogen (TKN), ammonia-nitrogen (NH_3-N), and TP. The state regulatory agency should be contacted to determine the specific effluent requirements of the proposed WWTP. These effluent discharge parameters will be dictated by the state in the National Pollutant Discharge Elimination System (NPDES) permit [7,12]. The parameters typically permitted for municipal systems are flow rate, BOD_5, TSS, and fecal coliform. In addition, many states are moving toward requiring nutrient removal. Therefore, TN, TKN, NH_3-N, or TP may also be required. It is imperative to establish effluent requirements because they will impact the operating sequence of the SBR. For example, if there is a BNR requirement and NH_3-N and/or TKN is required, then nitrification will be necessary. If there is a TN limit, then both nitrification and denitrification will be necessary [7,12]. Once the influent and effluent characteristics of the system are determined, the engineer will typically consult SBR manufacturers for a recommended design [1].

The WWT performance of SBRs is typically comparable to conventional activated sludge process systems and depends on system design and site-specific criteria. Depending on the mode of SBR operation, SBRs can achieve good BOD and nutrient removal. The BOD removal efficiency of a typical SBR process system is generally 85%–95%. SBR manufacturers [11] will typically provide a process guarantee to produce an effluent of less than: (a) 10 mg/L BOD; (b) 10 mg/L TSS; (c) 5–8 mg/L TN; and (d) 1–2 mg/L TP.

The heart of a SBR system is the process controls, automatic valves, and automatic switches, these systems may require more operation and maintenance (O&M) than a conventional activated sludge process system. An increased level of sophistication usually equates to more items that can fail or require more O&M. The level of sophistication may be very advanced in larger SBR WWTPs requiring a higher level of O&M on the automatic valves and switches. The use of higher grade of certified WWT operators for operation of SBRs is recommended. Some useful troubleshooting charts prepared by the University of Florida TREEO Centers are listed in Appendix 15.A (University of Florida [13]) for the convenience of the readers.

15.2.2 Meat Processing Wastewater Treatment Case History

15.2.2.1 A Review of Meat and Poultry Processing WWT

Niche Meat Processor Assistance Network (NMPAN) is advising meat processing facilities in the meat and poultry industry (MPI) of the following: (a) A typical septic tank and leaching field system will not work well for most meat processing plants because of the high levels of BOD, TSS, and

FOG in the wastewater; (b) If it is possible to connect to publicly owned treatment works (POTW), this is often the small facility's best bet. If the small facility is located within reach of these services, in the long run, it will probably be worth it to pay the initial connection fees and monthly sewer costs rather than build and manage a small facility's own WWT system. It is necessary to determine the wastewater flow and strength and contact the local POTW to find out the connection fees, the estimated monthly charges with the effluent sewer discharge service, and the estimated maintenance fees for emergence service only. Having a POTW emergence service connection is important regardless of the situation; (c) Constructed wetlands are a WWT option that NMPAN recommends highly. Brushy Prairie Packing, a very small, red meat and poultry slaughter facility at Gunthrop Farms in Indiana, is quite pleased with their constructed wetland WWT plant; (d) All of the meat processing facilities, even connecting to a POTW sewer system, require some form of pre-treatment. Pre-treatment can range from management practices (sweeping up the kill floor and processing floor before wash down to capture any solids, capturing blood on the kill floor so it doesn't go down the drain, and similar practices), to simple filter systems, to more advanced pre-treatment options, such as DAF. Blood, fat, and tissue are all nutrient-rich materials. A facility's WWT will have to remove these nutrients before that wastewater can be discharged. Pretreatment pollutant removal is equal to cost saving; (e) Wastewater pretreatment by DAF (or equivalent) and secondary treatment by waste stabilization ponds (anaerobic pond, facultative pond, and aerobic maturation ponds) is recommended for large meat processing facilities if there will be a cost-saving in comparison with facility effluent discharge to a POTW sewer system, or if the facility effluent will meet the government standards for direct discharge to a receiving water; and (f) Anaerobic digesters are an attractive technology for large meat processing facilities because it treats the facility waste *and* generate electricity in the process, but it is cost-effective for a small meat processing facility [14].

The US government environmental laws governing the meat and poultry processing WWT can be found from the literature [15–18].

15.2.2.2 History of WWT at F.B. Purnell Sausage Co.

Like many food processors, the F.B. Purnell Sausage Co. employed a flow-through activated sludge system followed by a polishing lagoon for treatment of process and other wastewater. Due to widely varying flows and highly variable organic loading rates, the flow-through system often was unable to produce the required effluent quality. Under such hydraulic and organic loading conditions, Purnell Sausage Co. was not able to meet its effluent discharge permit. In many states, the government has set guidelines whereby the best technology available must be used to meet the goal of zero discharge of pollutants. While the State of Kentucky, USA, has not yet imposed such strict discharge limitations, Purnell wanted to ensure that, with the installation of a new WWT system, the company would have the ability to meet more stringent discharge limitations in the future.

15.2.2.3 Selection of a Treatment System

After investigating several WWT systems, Purnell's consulting engineer, A.W. Goodman & Associates of New Albany, Indiana, decided on a three-step process. The WWT system was installed with DAF for removal of blood and for grease recovery; and a SBR system for BOD, suspended solids, and nitrification. In anticipation of future, more restrictive discharge requirements, Goodman & Associates also recommended the installation of a tertiary sand filter.

The SBR process was chosen for several reasons, including: (a) It requires a minimum amount of land for installation; (b) Capital costs are significantly less than a flow-through activated sludge system; (c) It is controlled by microprocessor and requires considerably less operator attention than a comparably sized flow-through activated sludge system; and (d) Its flexibility enables it to be operated for today's discharge requirements and future, more stringent requirements usually without the addition of equipment or additional reactors.

The SBR is an activated sludge process in which all treatment steps, including clarification, take place in a single reactor. In a flow-through activated sludge system, the waste flows continuously through a series of basins composed of a minimum of aeration and secondary clarification (Figure 15.1). Discharge is continuous.

In the SBR process, the influent also feeds continuously, but in batches to each SBR reactor. For example, in a two-reactor system, 100% of the influent flows to each reactor 50% of the time, while one reactor receives 100% of the flow the other reactor continues through other treatment phases. Discharge is intermittent.

The SBR system operates on the very simple concept of introducing a quantity of waste to a reactor, providing an adequate time period for the treatment of the waste, and subsequently discharging a volume of effluent plus waste sludge that is essentially equal to the original volume of waste introduced to the reactor.

This "Fill and Draw" principle of operation will, by virtue of its very nature, involve the basic steps of Fill, React, Settle, and Draw (Decant). The SBR systems are designed, within the overall Fill and Draw mode, to include seven individual phases of operation.

The seven individual phases of operation are based upon process considerations related to effluent quality requirements of the specific WWT application. The time duration of each particular phase has been established based on specific waste characteristics and effluent objectives. As a general principle, the total time duration in a dual-reactor system allotted for *all* Fill phases of operation is *equal* to the total time duration designated for the React, Settle, Decant and Idle, plus Waste Sludge phases. This condition permits a mode of operation featuring continuous flow to the "system" and the introduction of 100% of the flow to each reactor for 50% of the time. For example, flow is typically entering the treatment system for 24 hours of each day and each reactor in the dual-reactor system will receive 100% of the flow, for a total of 12 hours. As reactor No. 1 receives flow for a specific Fill period, reactor No. 2 completes the React, Settle, Decant, Idle, and Waste Sludge phases of operation on a simultaneous basis. During the next cycle of operation, the two reactors will essentially exchange modes of operation and reactor No. 2 will complete a Fill period while reactor No. 1 will complete the React, Settle, Decant, Idle, and Waste Sludge phases of operation.

If the SBR system is designed as a single reactor, continuous flow system, which means that flow is entering the SBR reactor during most of the treatment cycle, it may be operated at one cycle per day (24 hours). In most single-reactor industrial facilities, flow only enters the system during times when the plant is in operation. The React, Settle, and Decant phases of the SBR cycle may be programmed to occur during the period when processing operations are shut down and no flow is entering the SBR. This enables the reactor contents to react, settle, and decant under the same conditions as would be found in a multiple reactor SBR. The single reactor system may also be designed to include an influent storage tank to precede the SBR. The storage tank would serve as a means to hold flow during the React, Settle, Decant, and Waste Sludge phases of operation.

Inherent to the description of each individual phase of operation is the fact that each AquaSBR reactor is equipped with mechanical mixing and aeration systems that are operated on an independent basis. This capability to mix only, without introducing oxygen to the reactor, or mix and supply oxygen simultaneously, provides flexibility. By introducing or withholding the oxygen supply at appropriate time intervals, the operator is able to manipulate the reactor environment, and readily create an anoxic or aerobic condition.

At the beginning of the process cycle, the SBR reactor is at its minimum liquid level and influent wastewater is allowed to enter the reactor.

Depending on the effluent required, the SBR will either be mixed in the absence of oxygen (Figure 15.1, anoxic) or mixed and aerated (Figure 15.1, aerobic), or may be operated with alternate periods of mixing and mixing/aeration to achieve nitrification and denitrification. Mixing is provided by means of a floating mechanical downdraft mixer; while aeration can be provided by any of several means including membrane or coarse bubble diffusers, or mechanical surface aerators.

In most SBRs, when the microprocessor (using preset timer values) or level sensors signal that the reactor has reached the maximum water level, influent flow will be cut off and the basin will enter the React phase of treatment (Figure 15.1, React). The influent flow will either be diverted to a second SBR where the treatment phases described above will begin, or to an influent storage tank (if a single reactor system).

Following the React phase, all equipment will shut off and the reactor contents will be allowed to settle (Figure 15.1, Settle). Settling in an SBR occurs under perfectly quiescent conditions, since no flow is entering or leaving the reactor, and no equipment is in operation.

When the sludge has settled for a predetermined time period, the floating decanter weir will open and draw clear supernatant from 8 to 10 inches below the surface of the reactor (Figure 15.1, Draw for Decant/Waste Sludge). The weir will close when the reactor reaches its designed low water level. During the Decant phase or following, sludge will be wasted by means of a submersible solid handling pump and discharged to a holding tank or aerobic digester (Figure 15.1, Draw for Decant/Waste Sludge).

When the Decant and Waste Sludge phases are complete, the reactor will enter the Idle phase (not shown in Figure 15.1) until the Fill phases are complete in reactor two, and the influent wastewater is once again diverted back to reactor one.

15.2.2.4 Process Description and Case History of the SBR System at F.B. Purnell Sausage Co.

The F.B. Purnell Sausage Co. wastewater treatment plant (WWTP) is a wastewater total treatment system consisting of: (a) preliminary treatment by screening; (b) primary treatment by DAF; (c) secondary treatment by SBR; (d) equalization of SBR effluents; (e) tertiary filtration by sand filtration; (f) UV disinfection, and (g) final WWTP effluent discharge to a receiving stream. The WWTP sludge treatment train includes: (a) sludge from SBR; (b) aerobic digestion; (c) mechanical dewatering or drying beds, and (d) final disposal of dried sludge by landfill.

Based on the above WWT train and the sludge treatment train, the process sequences are presented here. Following screening, the sausage plant's wastewater passes through DAF for removal of oil, grease, and blood. From there, it flows to the SBR reactor. Following treatment in the SBR, the effluent flows to an equalization basin and then to a sand filter. After it passes through ultraviolet (UV) disinfection, the final effluent is discharged to a stream. Sludge is pumped from the SBR basin to an aerobic digester and then to mechanical dewatering or drying beds. The dried sludge is then disposed of in a local landfill.

The SBR system at Purnell Sausage utilizes one 77 ft. × 77 ft. square reactor with a sidewater depth of 14.4 ft. The reactor is equipped with a coarse bubble diffused aeration system, 25 HP positive displacement blowers, one 20 HP AquaDDM mixer, and one gravity AquaDecanter.

The SBR was designed to treat 150,000 gallons/day (Monday–Friday) of meat processing wastewater with BOD_5 of 1,000 mg/L, TSS of 250 mg/L, and NH_3-N of 7 mg/L. On the weekends, the SBR treats about 5,000 gallons/day of combined washdown water, sanitary waste, and process wastewater. The cycle structure remains the same as it is on weekdays.

At the time of the process system design, the wastewater entering the SBR is nutrient deficient so, instead of operating the SBR for nitrification, plant operators *add* nutrients in the form of ammonium hydroxide and monosodium phosphate.

The SBR at Purnell is operated at one 24-hour cycle/day. Influent wastewater enters the reactor during most of the cycle, except during the Settle and Decant/Waste Sludge phases which have been set up to occur during the hours when the meat processing operation is shut down and no wastewater is being discharged to the SBR.

Even though wastewater influent T-BOD_5 average data near the design of 1,000 mg/L and sometimes exceeds the design value of 1,000 mg/L, the C-BOD_5 in the final WWTP effluent always falls far below the required effluent limitation of 10 mg/L. Post-DAF data (preceding SBR) for TSS are better than the system's design of 250 mg/L; The final WWTP effluent TSS is always much lower than the required TSS level of 10 mg/L.

15.3 INNOVATIVE SEQUENCING BATCH REACTOR SYSTEMS AND MUNICIPAL WASTEWATER TREATMENT15.3.1 INNOVATIVE SBR PROCESS SYSTEMS

One of the recent advances in environmental biotechnology is the development of granular sludge technology [19–24]. The AquaNereda® Aerobic Granular Sludge Technology is one of these innovative biological WWT systems that provides advanced treatment using the unique features of aerobic granular biomass.

The aerobic granular biomass is comprised of compact granules which consist of layered microbial communities and provide superior settling compared to conventional activated sludge. Within a single tank, the process creates proper conditions to develop and reliably maintain a stable granule population without the need for a supplemental carrier. The layered aerobic and anaerobic zones within the granule allow for simultaneous processes to take place in the granular biomass, including enhanced biological phosphorus reduction and simultaneous nitrification and denitrification.

The AquaNereda® is another form of biological SBR that adopts the aerobic granular biomass for enhanced bio-oxidation, nitrification, and denitrification. Based on the unique characteristics of granular biomass, the AquaNereda® Aerobic Granular Sludge System uses an optimized batch cycle structure. There are three main phases of the cycle to meet advanced WWT objectives. The duration of the phases will be based on the specific waste characteristics, the flow, and the effluent objectives. Figure 15.2 shows the three main phases of the cycle and the granular biomass used in the new SBR system. The outside layer of the granular biomass is in aerobic zone suitable for bio-oxidation and nitrification. The inside layer of the biomass is in anoxic zone suitable for denitrification. The center core of the biomass (not shown in Figure 15.2) is in anaerobic zone sometimes. The three main phases of the process cycle are: (a) FILL/DRAW PHASE; (b) REACT PHASE; and (c) SETTLE PHASE, which are described below.

The first simultaneous FILL/DRAW PHASE (Figure 15.2) involves the following process operation and consequences: (a) simultaneous filling influent wastewater flow and drawing the AquaNereda SBR effluent from the reactor; (b) the substrate in the wastewater influent is readily available as a carbon source for supplemental denitrification biochemical reactions; (c) anoxic and anaerobic conditions are present in this phase suitable for denitrification; (d) biomass conditioning

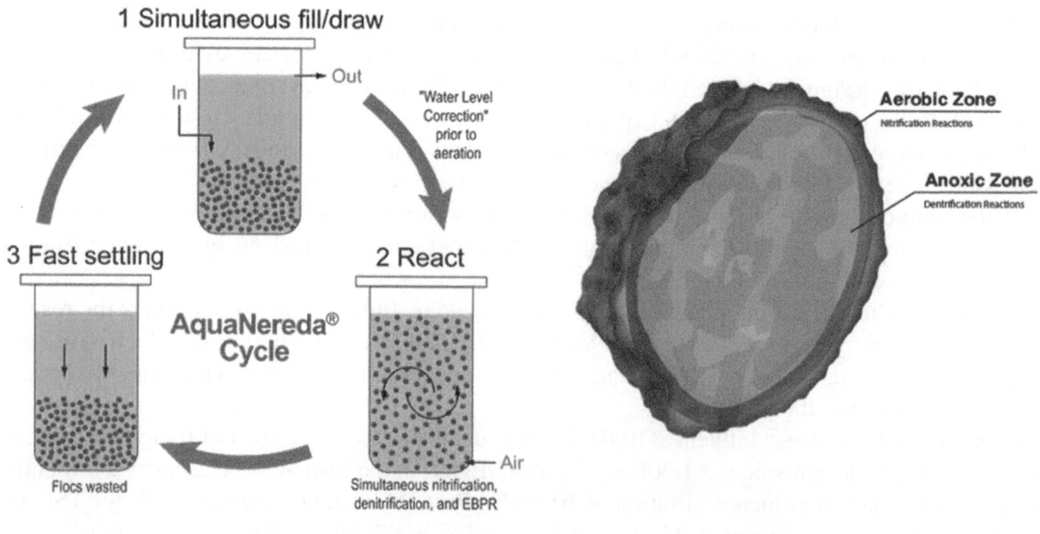

FIGURE 15.2 Three operating phases and aerobic granular sludge of AquaNereda SBR system.

is accomplished in this phase; (e) phosphorus release to promote enhanced biological phosphorus removal; and (f) the treated SBR effluent having low BOD, nitrogen (N), and phosphorus (P) is discharged.

The second REACT PHASE (Figure 15.2) is responsible for major biochemical reactions and involves the following process operation and consequences: (a) wastewater influent flow is terminated; (b) the biomass having both aerobic zone and anoxic zone is subjected to aerobic condition (bio-oxidation and nitrification) and anoxic biochemical conditions (for denitrification), respectively; (c) accordingly simultaneous bio-oxidation, nitrification, and denitrification occur; (d) nitrate-nitrogen is transported by diffusion between outer aerated layers and inner anoxic layers of the granule eliminating the need for pumping large recycle flows in the WWTP; (e) luxury uptake of phosphorus is promoted; (f) automated control of the SBR process allows energy savings and process optimization.

The third and final phase of AquaNereda operation is SETTLE PHASE (Figure 15.2) in which: (a) the wastewater influent does not enter the SBR reactor; (b) the granular biomass settles from the treated wastewater rapidly; (c) excess sludge is wasted in order to maintain the desired amount of biomass; and (d) the SBR system is ready for the beginning of a new process cycle.

15.3.1 Municipal Wastewater Treatment Case Histories

The new SBR system using granular activated sludge represents a step change in the WWT industry. Aerobic granulation of activated sludge was successfully achieved by Li et al. [24] in a full-scale SBR with 50,000 m^3/day for treating a town's wastewater in 2014. After operation for 337 days, in this full-scale SBR, aerobic granules with an average SVI30=47.1 mL/g, diameter of 0.5 mm, and settling velocity of 42 m/h were obtained. Three kinds of sludge had a high content of protein in extracellular polymeric substances (EPS). X-Ray fluorescence (XRF) analysis revealed that metal ions and some inorganics from raw wastewater precipitated in sludge acted as core to enhance granulation. Raw wastewater characteristics had a positive effect on the granule formation, but the SBR mode operating with periodic feast-famine, shorter settling time, and no return sludge pump played a crucial role in aerobic sludge granulation [24].

A new AquaNereda® Aerobic Granular Sludge System at the Riviera Utilities RUWWTP at Wolf Creek, Foley, AL, USA, was installed and started in January 2020 serving about 20,000 population of the City of Foley [25]. The advanced SBR (AquaNereda) provides WWT for an average flow of 3.5 MDG and peak flow of 6 MGD. It consists of three aerobic granular sludge reactors operating similar to a continuous flow system with a FILL/DRAW phase that alternates between reactors. Downstream polishing is performed by AquaDisk® Pile Cloth Media Filters to produce Class B reuse-quality water. Although not currently required by NPDES permit to achieve TN limit, the inherent BNR properties of Aerobic Granular Sludge provide TN removal now and for future NPDES permit limits. The new SBR technology was highly appealing as an innovative approach to secondary treatment for a number of reasons, including: (a) The innate BNR capabilities provided by the layered microbial community of the granules enable the system to meet future TN and TP limits; (b) The high operating MLSS in excess of 8 g/L significantly reduces settling time minimizing system footprint by up to 75%; (c) Up to a 50% decrease in energy use due to reduced equipment demand, smart aeration control, and improved substrate and oxygen utilization with granular sludge; (d) The robust nature of the granules provides superior protection against plant upsets and toxic shocks compared to conventional activated sludge; (e) The RUWWTP is now capable of providing BNR for future TN and TP limits. Downstream polishing by AquaDisk pile cloth media filters is capable of achieving BOD and TSS concentrations of <5 mg/L. The RUWWTP has consistently met effluent requirements since its startup in January 2020 [25].

15.4 CONCLUSIONS

15.4.1 Conventional Sequencing Batch Reactors and Meat Processing Wastewater Treatment

15.4.1.1 Conventional SBRs

The SBR typically eliminates the need for separate primary and secondary clarifiers in most municipal systems, which reduces operations and maintenance requirements. In addition, RAS pumps are not required. In conventional BNR systems, anoxic basins, anoxic zone mixers, toxic basins, toxic basin aeration equipment, and internal MLSS nitrate-nitrogen recirculation pumps may be necessary. With the SBR, this can be accomplished in one reactor using aeration/mixing equipment, which will minimize O&M requirements otherwise needed for clarifiers and pumps. Significant operating flexibility is associated with SBR systems. An SBR can be set up to simulate any conventional activated sludge process, including BNR systems. For example, holding times in the Aerated React mode of an SBR can be varied to achieve simulation of a contact stabilization system with a typical hydraulic retention time (HRT) of 3.5–7 hours or, on the other end of the spectrum, an extended aeration treatment system with a typical HRT of 18–36 hours. For a BNR plant, the aerated react mode (oxic conditions) and the mixed react modes (anoxic conditions) can be alternated to achieve nitrification and denitrification. The mixed fill mode and mixed react mode can be used to achieve denitrification using anoxic conditions. In addition, these modes can ultimately be used to achieve an anaerobic condition where phosphorus removal can occur. Conventional activated sludge systems typically require additional tank volume to achieve such flexibility. SBRs operate in time rather than in space and the number of cycles per day can be varied to control desired effluent limits, offering additional flexibility with an SBR.

15.4.1.2 Meat Processing Wastewater Treatment

The F.B. Purnell Sausage Co. WWTP is a wastewater total treatment system consisting of: (a) preliminary treatment by screening; (b) primary treatment by DAF; (c) secondary treatment by SBR; (d) equalization of SBR effluents; (e) tertiary filtration by sand filtration; (f) UV disinfection; and (g) final WWTP effluent discharge to a receiving stream. The WWTP sludge treatment train includes: (a) sludge from SBR; (b) aerobic digestion; (c) mechanical dewatering or drying beds, and (d) final disposal of dried sludge by landfill. The Purnell's process is both technically and economically feasible for total waste treatment aiming at direct effluent discharge to a receiving stream.

The AquaSBR is a system that has proven to be an efficient, flexible, and reliable means of WWT for a variety of industrial and municipal wastewaters – including food processing wastewater. To date, nearly 125 AquaSBR systems are in operation or under construction throughout the USA, Puerto Rico, Mexico, Netherlands, Middle East, and Germany. These systems are treating a wide range of waste, including municipal, food processing, pharmaceutical, landfill leachate, and more. The AquaSBR system is a cost-effective and viable means of WWT.

15.4.2 Innovative Sequencing Batch Reactors and Municipal Wastewater Treatment

15.4.2.1 Innovative SBRs

The unique process features of the AquaNereda technology translate into a flexible and compact process that offers energy efficiency and significantly lower chemical consumption culminating in a low life-cycle cost. System features and advantages of AquaNereda technology are: (a) Optimal biological treatment is accomplished in one effective aeration step; (b) Settling properties at SVI values of 30–50 mL/g allow MLSS concentrations of 8,000 mg/L or greater; (c) 25% of the footprint compared to conventional activated sludge systems; (d) Energy savings up to 50% compared to activated sludge processes; (e) No secondary clarifiers, selectors, separate compartments, or return sludge pumping stations; (f) Proven enhanced nutrient removal (ENR), (g) Robust structure of granule withstands fluctuations in chemical spikes, load, salt, pH, and toxic shocks; (h) Significant reduction of chemicals for nutrient removal due to the layered structure and biopolymer backbone of the granule; (i) Ease of operation with fully automated controls; and (j) Lowest life-cycle cost. Typical applications may include: (a) Retrofit of existing tanks to increase treatment capacity; (b) Upgrade of existing treatment systems to meet BNR requirements; (c) New construction plants; and (d) Municipal and industrial applications.

15.4.2.2 Municipal WWT

It has been proven in full scale that the combination of newly developed AquaNereda technology and AquaDisk pile cloth media filters is reliable for municipal WWT up to 6 MGD flow.

GLOSSARY

Sequencing batch reactor (SBR), conventional: The conventional SBR is a fill-and-draw activated sludge system for wastewater treatment (WWT). In this system, wastewater is added to a single "batch" reactor, treated to remove undesirable components, and then discharged. In a typical SBR operation, influent wastewater generally passes through screens and grit removal prior to the SBR. The wastewater then enters a partially filled reactor, containing biomass, which is acclimated to the wastewater constituents during preceding cycles. Once the reactor is full, it behaves like a conventional activated sludge system, but without a continuous influent or effluent flow. The aeration and mixing is discontinued after the biological reactions are complete, the biomass settles, and the treated supernatant is removed. Excess biomass is wasted at any time during the cycle. Frequent wasting results in holding the mass ratio of influent substrate to biomass nearly constant from cycle to cycle.

Sequencing batch reactor, innovative (AquaNereda®): It is another form of biological sequencing batch reactor (BIO-SBR) that adopts the aerobic granular biomass for enhanced bio-oxidation, nitrification, and denitrification. The outside layer of the granular biomass is in aerobic zone suitable for bio-oxidation and nitrification. The inside layer of the biomass is in anoxic zone suitable for denitrification. The center core of the biomass is in anaerobic zone sometimes. The three main phases of the process cycle are: (a) FILL/DRAW PHASE: It involves the following process operation and consequences: (a1) simultaneous filling influent wastewater flow and drawing the AquaNereda SBR effluent from the reactor; (a2) the substrate in the wastewater influent is readily available as a carbon source for supplemental denitrification biochemical reactions; (a3) anoxic and anaerobic conditions are present in this phase suitable for denitrification; (a4) biomass conditioning is accomplished in this phase; (a5) phosphorus release to promote enhanced biological phosphorus removal; and (a6) the treated SBR effluent having low BOD, nitrogen (N), and phosphorus (P) is discharged; (b) REACT PHASE: It is responsible for major biochemical reactions and involves the following process operation and consequences: (b1) wastewater influent flow is terminated; (b2) the biomass having both aerobic zone and anoxic zone is subjected to aerobic conditions (fobio-oxidation and nitrification) and anoxic biochemical conditions (for denitrification), respectively; (b3) accordingly simultaneous bio-oxidation, nitrification, and denitrification occur; (b4) nitrate-nitrogen is transported by diffusion between outer aerated layers and inner anoxic layers of the granule eliminating the need for pumping large recycle flows in the WWTP; (b5) luxury uptake of phosphorus is promoted; and (b6) automated control of the SBR process allows energy savings and process optimization; (c) SETTLE PHASE: (c1) the wastewater influent does not enter the SBR reactor; (c2) the granular biomass settles from the treated wastewater rapidly; (c3) excess sludge is wasted in order to maintain the desired amount of biomass; and (c4) the SBR system is ready for the beginning of a new process cycle.

APPENDIX 15.A: SEQUENCING BATCH REACTOR TROUBLESHOOTING CHARTS [13]

SEQUENCING BATCH REACTOR TROUBLESHOOTING CHART

PROBLEM OR OBSERVATION	CONDITION	PROCESS CONTROL ANALYSIS	POSSIBLE CAUSES	CONTROL ACTION
I. Loss of solids from reactor due to a high blanket	Poor sludge settling velocity and compaction	SSV_X, SSV_5, SVI, diluted SSV_X, microscopic examination, NH_3 - N, COD, D.O., SOUR	• Glutting (old sludge)	• Decrease MCRT.
			• Classic bulking (young sludge)	• Increase MCRT.
			• Filamentous bulking	• Identify conditions contributing to filamentous growth and correct. See comments in narrative below.
			• Slime bulking	• Add nutrients.
			• Foam Trapping	• Optimize pretreatment removal of oil and grease.
			• Highly nitrified or oxidized sludge	• Increase anoxic cycle, reduce aerobic cycle.
			• Toxicity	• Isolate or split flow, identify source of toxic influent and eliminate, increase aeration cycle, increase MCRT.
			• High organic loading	• Short-term, increase aerobic cycle; long-term, increase MCRT.
II. Rapidly settling blanket leaving particulate. Difficulty in maintaining waste concentration	Rapid sludge settling velocity and compaction	SSV_X, SSV_5, SVI, F/M, SOUR	• Low F/M ratio	• Increase F/M ratio by decreasing MLVSS.
III. Turbid or cloudy effluent, disinfection problems	A. High effluent BOD or TS	MLSS, MLVSS, D.O., pH, temperature, Influent COD or TOC, Influent NH_3 –N, D.O., SOUR	• Low MLSS or MLVSS	• Increase MLSS/MLVSS.
			• Low D.O., temperature or pH	• Increase aeration cycle in fill react, increase MLSS, add alkalinity.
			• High organic loading	• If long-term, increase MLSS/MLVSS and aeration cycle.
			• High nitrogenous loading	• If long-term, increase MLSS/MLVSS and aeration cycle.
			• Toxicity	• Isolate or split flow, identify source of toxic influent and eliminate, increase aeration cycle, increase MCRT.
	B. High effluent NH_3 – N (Incomplete nitrification)	Influent and process NH_3 – N, influent and process alkalinity, pH, temperature, SOUR, D.O.	• Influent NH_3-N overload	• Increase aerobic cycle.
			• Low D.O.	• Increase aerobic cycle.
			• Low temperature	• Increase aerobic cycle.
			• Inadequate aerobic retention time	• Increase aerobic cycle.
			• Low pH or alkalinity	• Add alkalinity.
			• Low MLVSS (nitrifiers)	• Increase MLVSS.
			• Toxicity	• Isolate or split flow, identify source of toxic influent and eliminate, increase aeration cycle, increase MCRT.

APPENDIX 15.B

SEQUENCING BATCH REACTOR TROUBLESHOOTING CHART (CONTINUED)

PROBLEM OR OBSERVATION	CONDITION	PROCESS CONTROL ANALYSIS	POSSIBLE CAUSES	CONTROL ACTION
IV. High-effluent TSS	Individual particle washout	Effluent and recycle TSS or turbidity, F/M, microscopic exam, SOUR	• Pin floc – low F/M,	• Increase waste cycle, decrease MLSS.
			• Pin floc – denitrification	• Increase waste cycle, decease MLSS, increase anoxic cycle.
			• Pin floc – solids recycle	• Optimize solids handling.
			• Straggler floc – high F/M	• Decrease waste cycle, increase MLSS, increase aeration cycle.
			• Straggler floc – filamentous	• Identify filamentous organism (see filamentous control above).
			• Straggler floc – hydraulic	• See mechanical troubleshooting section.
			• Individual bacterial cells in effluent	• Decrease waste cycle, raise MLSS, increase aeration cycle, if toxicity, remove source of toxic influent.
V. High-effluent NO_3 - N	High effluent NO_3 – N	NO_3 – N, pH, TOC or COD	• Lack of or inadequate anoxic conditions	• Increase anoxic cycle (may require decreasing oxic cycle).
			• Lack of or inadequate carbon source	• Add carbon (methanol or acetic acid).
			• Low pH, temperature or MCRT	• Add alkalinity, increase MCRT.
VI. Difficulty in maintaining chlorine residual	Chlorine (Cl_2)residual fluctuation, no chlorine residual	Cl_2 residual, supernatant NH_3-N., NO_2-N, turbidity or TSS	• Incomplete nitrification/denitrification resulting in high NO_2-N in supernatant.	• High NO_2-N in supernatant will result in increased demand. Optimize nitrification and denitrification processes.
			• High TSS in supernatant	• High TSS in supernatant will result in increased demand. See Problems I, III, IV.
			• Reducing agents in supernatant	• Reducing agents such as H_2S, Fe, Mn in supernatant. Investigate source and eliminate. Increase chlorine feed rate to overcome demand.
VII. High fecal coliform values	Sufficient chlorine (Cl_2)residual, but high fecal coliform values	Supernatant TSS, free and total Cl_2 residual, supernatant NH_3-N, theoretical and actual CCC detention time	• Excessive TSS in supernatant	• High TSS in supernatant can result in "blinding" of disinfection process. See Problems I, III, IV.
			• Short circuiting of chlorine contact chamber (CCC)	• Calculate the theoretical CCC detention time. Conduct dye testing to determine actual detention time.
			• Chloro-organic compounds	• If there is no NH_3-N in effluent but organic nitrogen is present, then false residual (DPD)may be present due to formation of chloro-organic compounds. Use free chlorine to establish residual not total chlorine. Reduce aeration cycle to de-optimize nitrification rate.

Source: University of Florida TREEO Center's *Sequencing Batch Reactor Operations and Troubleshooting Manual*.

APPENDIX 15.C

Sequencing Batch Reactor Troubleshooting Chart (continued)				
PROBLEM OR OBSERVATION	**CONDITION**	**PROCESS CONTROL ANALYSIS**	**POSSIBLE CAUSES**	**CONTROL ACTION**
VIII. Foam	Excessive foam or scum on surface of SBR, flow EQ tank or chlorine contact chamber	Microbiological examination, NO_3-N, C-N-P ratio, SRT, oils and grease, D.O.	• Excessive filamentous bacteria.	• The presence of hydrophobic filamentous bacteria may lead to excessive scum and foam. See section I.5.
			• Denitrification	• Denitrification can result in sludge and foam on surface of SBR.
			• Nutrient deficiency	• Foam may also indicate a possible nutrient deficiency. This type of foam may be due to bacteria producing a natural polymer when subjected to nutrient deficient conditions for an excessive period of time.
			• SRT	• Both too low and too high an SRT can cause foam problems.
			• Fats, oil or grease	• Fats, oils grease and other non-degraded surface active organics can cause foam problems.
			• Overaeration	• Excessive (D.O. > 4.0 mg/L) may cause foaming.

Source: University of Florida TREEO Center's *Sequencing Batch Reactor Operations and Troubleshooting Manual.*

REFERENCES

1. Mikkelson, KA (1995). *AquaSBR Design Manual*, Aqua-Aerobic Systems, Rockford, IL, USA.
2. Wang, LK, Kurylko, L, and Wang, MHS (1994). Sequencing Batch Liquid Treatment. US Patent No. 5354458. US Patent and Trademarks Office, Washington, DC, USA.
3. Wang, LK, Krougzek, JV, and Kounitson, U (1995). Case Studies of Cleaner Production and Site Remediation. UNIDO-Registry No. DTT-5-4-95. United Nations Industrial development Organization (UNIDO), Vienna, Austria, 136 pages, April.
4. Wang, LK and Li, Y (2009). Sequencing batch reactors. In: *Biological Treatment Processes*, Wang, LK, Shammas, NK, and Hung, YT (editors), Humana Press, Totowa, NJ, USA, pp. 459–512.
5. Shammas, NK and Wang, LK (2009). SBR systems for biological nutrient removal. In: *Advanced Biological Treatment Processes*, Wang, LK, Pereira, NC, Hung, YT, and Shammas, NK (editors), Humana Press, Totowa, NJ, USA, pp. 157–184.
6. Wang, LK, Wang, MHS, Wang, P, and Clesceri, NL (2021). Biological and physicochemical sequencing batch reactors using sedimentation or flotation. In: *Environmental Flotation Engineering*, Wang, LK, Wang, MHS, Shammas, NK, and Aulenbach, DB (editors), Springer Nature, Cham, Switzerland, pp. 397–417.
7. NYRWA (2022). *SBR -- Sequencing Batch Reactor Systems for Wastewater Treatment*, New York Rural Water Association, New York, NY, USA.
8. Norcross, KL (1992). Sequencing batch reactors: An overview. *Water Science and Technology* 26(9–11):2523–2526.
9. USEPA (1986). Sequencing Batch Reactors. EPA/625/8–86/011. US Environmental Protection Agency, Washington, DC, USA.
10. USEPA (2000). Wastewater Technology Fact Sheet: Package Plants. EPA 832-F-00–016. US Environmental Protection Agency, Washington, DC, USA.
11. USEPA (1999). Sequencing Batch Reactors. EPA932-F-99-073. US Environmental Protection Agency, Washington, DC, USA.
12. NEIWPC (2005). *Sequencing Batch Reactor Design and Operational Considerations*, New England Interstate Water Pollution Control Commission, Lowell, MA, USA; mail@neiwpcc.org, www.neiwpcc.org.
13. University of Florida (2015). University of Florida TREEO Centers Sequencing Batch Reactors (SBR) Operations and Troubleshooting Manual, University of Florida, USA.
14. Oregon State University Extension (2015). Wastewater Treatment for Meat Processors. Niche Meat Processor Assistance Network: www.nichemeatprocessing.org., Oregon State University, USA, https://www.nichemeatprocessing.org/wastewater-treatment-for-meat-processors/
15. USEPA (2004). *Technical Development Document for the Final Effluent Limitations Guidelines and Standards for the Meat and Poultry Products Point Source Category*, US Environmental Protection Agency, Washington, DC, USA.
16. USEPA (2004). *Economic and Environmental Benefits Analysis of the Final Meat and Poultry Products Rule*. February, 821-R-04-010. US Environmental Protection Agency, Washington, DC, USA.

17. USEPA (2023). *Meat and Poultry Products Effluent Guidelines: Slides from Industry Update*, US Environmental Protection Agency, Washington, DC, USA.
18. USEPA (2023). *Stormwater Fact Sheet. Sector U: Food and Kindred Products Facilities*, US Environmental Protection Agency, Washington, DC, USA.
19. Liu, Y-Q, Kong, Y, Tay, J-H, and Zhu, J (2011). Enhancement of start-up of pilot-scale granular SBR fed with real wastewater. *Separation and Purification Technology* 82(1):190–196.
20. Verawaty, M, Pijuan, M, Yuan, Z, and Bond, PL (2012). Determining the mechanisms for aerobic granulation from mixed seed of floccular and crushed granules in activated sludge wastewater treatment. *Water Research* 46(3):761–771.
21. Su, KZ, Ni, BJ, and Yu, HQ (2013). Modeling and optimization of granulation process of activated sludge in sequencing batch reactors. *Biotechnology and Bioengineering* 110(5):1312–1322.
22. Tay, JH, Liu, Y, Tay, STL, and Hung, YT (2009). Aerobic granulation technology. In: *Advanced Biological Treatment Processes*, Wang, LK, Shammas, NK, and Hung, YT (editors), Humana Press, Totowa, NJ, USA, pp. 109–128.
23. Adav, SS, Lee, D-J, Show, K-Y, and Tay, J-H (2008). Aerobic granular sludge: Recent advances. *Biotechnology Advances* 26(5):411–423.
24. Li, J, Ding, LB, Cai, A, Hung, GX, and Horn, H (2014). Aerobic sludge granulation in a full-scale sequencing batch reactor. *Biomed Research International* 2014:268789. doi:10.1155/2014/268789
25. TPO (2020). Alabama WWTP Upgrades to State-of-the-Art AquaNereda Aerobic Granular Sludge Technology. Treatment Plant Operators (TPO) Magazine, November. https://www.tpomag.com/

16 Biowaste and Value-Added Products

Ahmed Albahnasawi, Motasem Y. D. Alazaiza, Ercan Gürbulak, Murat Eyvaz, Ebubekir Yüksel, and Yung-Tse Hung

NOMENCLATURE

AD Anaerobic digestion
CO_2 Carbon dioxide
CSTR Continuously stirred tank reactor
DF Dark fermentation
FBB Fibrous bed bioreactor
H_2 Hydrogen gas
MSW Municipal solid wastes
NH_4^+ Ammonium
NO_2 Nitrite
NO_3 Nitrate
RDF Refuse derived fuel
SA Succinic acid
SSF Synchronous saccharification and fermentation
TOP Torrefied pellets
VFA Volatile fatty acids

16.1 INTRODUCTION

Due to its harmful and detrimental effects on the ecosystem and human health, environmental pollution brought on by anthropogenic activities and increased industrialization is a significant issue in the modern day [1]. In order to fulfill consumer demand, there is an equivalent increase in the market for these product lines as the need for functional food rises globally [2]. In the end, a major problem currently facing the world is the generation of large-scale organic waste from these various sources, including the food, agricultural, municipal, urban, starch industries, etc. Therefore, by reducing waste and minimizing reliance on fossil fuels, their bioconversion into value-added goods might turn this issue into an opportunity [3]. Currently, it is anticipated that biomass will make up 10%–14% of the world's energy generation [4,5]. Urban pollutants, industrial leftovers, and agricultural trash all include valuable organic elements that can be turned into high-end products [6]. Given the enormous volume of organic waste produced annually, roughly 13×10^9 tons, its utilization in the upcoming years is probably going to be advantageous [7,8].

Compared to conventional treatment, bioconversion produces relevant bioproducts [9]. To address the problem of ecological deterioration, this has increased the value of organic wastes. Biorefineries have historically been a top priority for using organic waste as a sustainable strategy for converting biomass [3]. Previous research on the transformation of biomass focused mainly

DOI: 10.1201/9781003532354-16

on bioenergy, such as bioethanol and biodiesel, due to the dearth of fossil fuels [10]. However, the development of genetic engineering, which began in the early 20th century, was the key innovation to the improvement in waste bioconversion and depending on the type of organic wastes, numerous intended bio-based products can be created from various biotransformation processes [11].

Three phases can be used to summarize the full biotransformation approach. The genetic modification phase of a microbe comes after the first selection of the organic waste biomass. Utilizing suitable pre-treatments based on a precise analysis of the organic waste's content might increase the efficacy of biomass conversion [12]. Scientists can employ gene editing to increase productivity by using metabolic engineering to help them understand the metabolic processes of microorganisms [13]. The second is midstream regulation, which increases the productivity of the entire bioprocess. A common technique for improving product concentration from the conversion of organic waste biomass is fermentation optimization. Additionally, important variables in enhancing the rate of organic waste conversion are the fermenter's configuration and the transformation parameters. Finally, it is helpful to remove products from the conversion system using an effective downstream technique. The circular economy concept is implemented by resource recovery from these organic streams, which may help to ease resource shortages [14].

Integration of the circular economic approach, which transitions to the use of renewable energy sources, eliminates the use of toxic synthetic compounds that prevent reuse and focuses on appropriate material utilization planning, to replace the concept of "end of life" with "reclamation" [15]. Therefore, novel biorefinery approaches must be developed while taking into account the fundamental components of the financial system, society, and ecosystem to adhere to the circular economy concept, which would ideally permit the re-emergence of generated bio-based food products into the food chain [16]. The current chapter supplies information on the classification of biomass. Moreover, this chapter supplies a discussion on existing biowaste management techniques. In addition, the assessment also goes over possible uses for trash produced by several streams that have been bio-converted. It also emphasizes the sector's existing possibilities and technological obstacles related to the valorization of organic waste.

16.2 ORGANIC WASTE GENERATION

Overpopulated nations and cities supply more goods and services to the populace while taking part in international trade and investments. As a result, they are challenged by the growing amount of garbage they must treat and dispose of [17]. Approximately 46% of the world's solid wastes are now organic wastes. According to statistics from the Global Waste Management Reviewers, the amount of organic waste generated globally in 2016 was up to 2.01 billion tons [18]. Up to 43% of the world's trash is produced in Europe, East, and Middle Asia. In contrast, the Central East and North African regions produce the least trash, accounting for 15% of the world's production of organic waste [19].

This organic waste is produced in such massive quantities, which suggests that its effects will be positive in the years to come. Waste biomass has become the preferred organic substrate for biotransformation into energy and chemicals as biomass uses have advanced [8]. A significant amount of the organic waste found in agricultural, industrial, and urban leftovers can be successfully converted into high-value resources [20]. The concentration of organic waste with various compositions is indicated by industries including food production, agriculture, starch, and animal rearing [21]. The industry and kind of trash have a significant impact on the regulations governing garbage disposal. The percentage of the most significant organic waste kinds in the total amount of waste produced globally is shown in Figure 16.1.

Organic waste can be produced by nearly all biomass production streams, and the origin of the biomass is mostly related to the waste content [22,23]. To ensure efficient use, biodegradable garbage is categorized according.

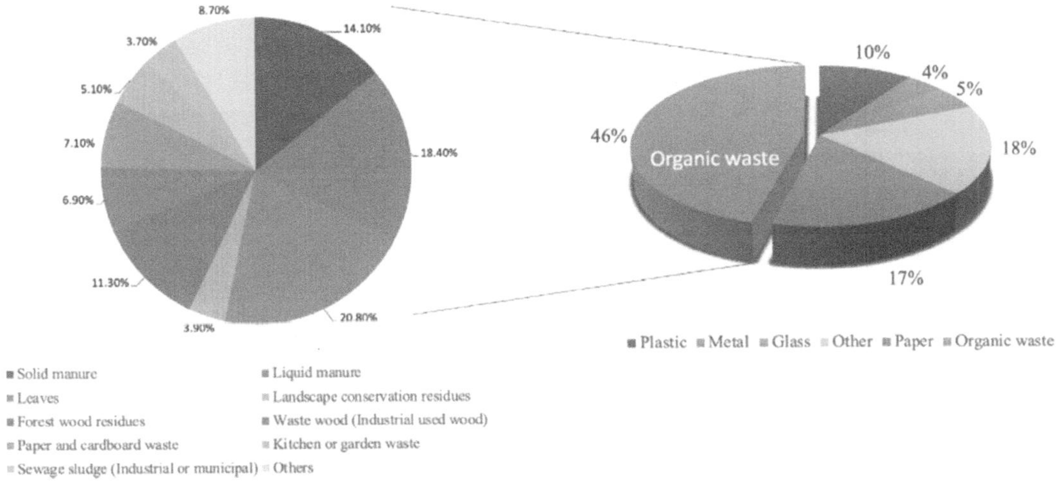

FIGURE 16.1 Organic waste generation at global level [24].

16.3 BIOWASTE MANAGEMENT STRATEGIES

Because it helps with sustainable agriculture, crop production, and fertility restoration, organic matter is a crucial part of the soil. It is acknowledged as a valuable source of organic material with many benefits [25]. However, when organic waste is disposed of in landfills, it generates greenhouse gases and may have negative health effects [26]. In order to transform this essential resource into a soil amendment, it is crucial to adopt efficient collection, transit, and governance techniques [27]. Utilizing a variety of technologies, including biological processes like anaerobic digestion (AD) and various composting procedures, organic waste can be transformed into several useful product lines [6]. In the sections that follow, these techniques will be briefly explored.

16.3.1 Thermal Conversion

This method thermally treats the organic matrix already present in the municipal solid wastes (MSW) to produce either gas, heat energy, or fuel oil [28]. The heat treatment choice is suitable for trash that has a lower moisture content (dry waste) and a higher percentage of organic stuff that is still alive. A flammable material (RDF) with a greater heat value is frequently subjected to thermal treatment technology [29]. By palletizing and/or grating the leftover trash, non-combustible and recyclable components from MSW are passively combined to create RDF [30]. Incineration is one of the main thermal treatment technologies that is used extensively. In this method, controlled combustion is used to raise the temperature of the waste wastes. Additionally, techniques for thermal conversion like pyrolysis and gasification are still in the preliminary stages of development. Due to improper facility design, a lack of adequate MSW data and characterization, and poor feedstock quality, neither is suited for commercial applications for large-scale manufacturing [31]. A couple of gasification/pyrolysis-based plants are in operation worldwide for the treatment of MSW. These facilities are in operation to process MSW along with other waste types like biomass, biomedical waste, industrial waste, etc. [32]. The operating temperature and external factors are where these three thermal treatment procedures diverge most. The reaction's quality of the final product and beneficial intermediate products depend on both factors. Like this, the design of the process and the feedstock material will have the biggest impact on the operational temperature of the process. In underdeveloped nations, MSW supplementation is not very well developed for incineration [33]. As a result, raw MSW is used directly as feedstock. Figure 16.2 illustrates the process of thermal conversation of biomass.

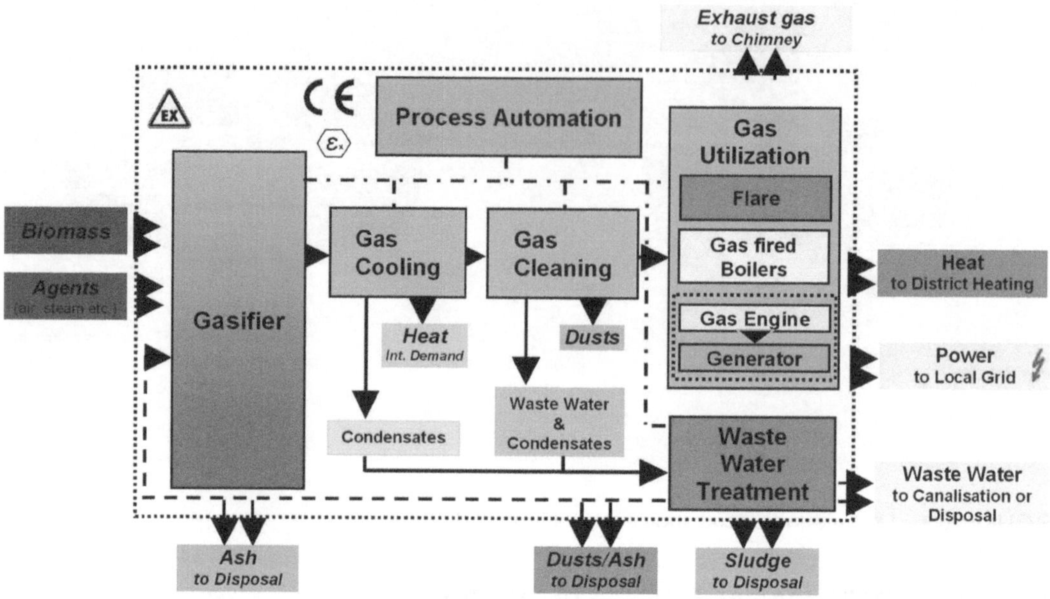

FIGURE 16.2 Thermal conversation of biomass.

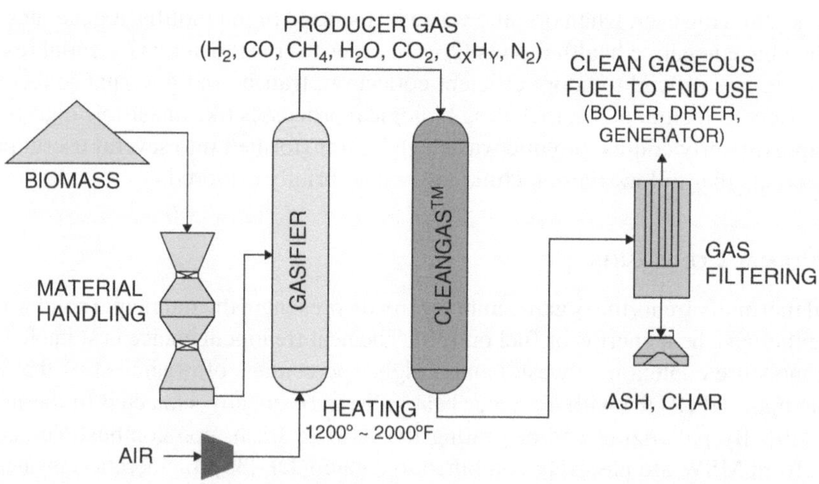

FIGURE 16.3 Gasification process for biomass.

16.3.2 Gasification

In this method, the organic molecules are converted into syngas under exacting temperature and oxygen conditions [34]. The primary outcome of the gasification process is the gas, which is also used to fuel the combustion process for the generation of energy. Other byproducts of the process include liquid fuels and chemical feedstock [35]. The procedure is depicted schematically in Figure 16.3. Specific types of MSW and homogenized solid fuel flow like coal, wood, etc., are the areas of concern in most of the documented reports. Although gasification is currently seen as an important potential alternative for energy recovery from MSW, it is most commonly used in the coal industry [34]. When Panepinto and colleagues [36] looked into how many plants used the gasification approach, they discovered that 100 plants were using the technique worldwide to treat MSW. Eighty-five gasification facilities were operating in Japan as of 2007 [37]. While other nations including Germany, the United States, Norway, the United Kingdom, Iceland, and Italy employ gasification to treat MSW on a commercial or pilot scale, leveraging it as a means to enhance energy recovery and reduce environmental impact, its adoption remains limited in many regions due to economic, regulatory, and technological barriers [36]. When dealing with the same amount of MSW, gasification technology produces significantly less CO_2 than incineration [38]. According to Defra [39], modern gasification facilities have additions that effectively reduce the likelihood of land and water contamination. Asia is now thought to be the region most appropriate for gasification technology after Africa, Europe, and the United States due to the recent tremendous rise in this technology there [40]. Gasification and pyrolysis are thought to be the best methods for treating MSW in terms of energy recovery and environmental impact [41]. When compared to incineration, gasification and pyrolysis technologies can reduce waste volume by 95% and require less extensive flue gas cleaning [35]. Both systems are more effective in terms of energy recovery efficiency and environmental contamination when compared to other waste-to-energy methods. However, due to inadequate gas cleaning techniques, gasifier efficiency, particle size, variability in high moisture content, and MSW composition, they must be developed internationally on a large scale, particularly in developing nations for the recovery of MSW energy [42].

16.3.3 Pyrolysis

Another innovative method that uses heat to treat MSW is pyrolysis. It operates at temperatures between 400°C and 800°C without air or oxygen [43]. The rate of heating, residence time, process temperature, particle size, and content of the waste all play major roles in determining how much char, oil, and pyrolysis gas are created during the process [44,45]. Wax, tar, and pyrolysis oil are the end products of the pyrolysis reaction at temperatures below 500°C–550°C, but the principal end products of the reaction at temperatures above 700°C are the gases created by the reaction [46,47]. The following types of feedstocks, such as wood waste, plastic, electronic waste, electrical waste, tires, etc., can be used to produce items of high quality. Reports on the kind of trash for pyrolysis have been offered by several investigations. These studies are focused more on the pyrolysis process than the potentially lucrative commercial applications of the pyrolysis products. Pyrolysis has recently drawn attention, especially for the reprocessing of fight tires to recover oil, gas, wire, and carbon black [45]. Additionally, pyrolysis has been used on a large scale commercially for energy recovery using MSW. Since 1987, a factory in Germany's Burgau region has been successfully employing pyrolysis to generate power from 110 tons of MSW per day [45]. Figure 16.4 shows the biomass treatment by pyrolysis.

16.3.4 Torrefaction

The process of thermally upgrading biomass to produce a more homogeneous product that is then palletized to produce a denser energy product known as briquettes or torrefied pellets (TOPs), which have qualities similar to coal [48]. More thermochemical transformations can be facilitated by this kind of energy [49]. The term "trivial pyrolysis process" is another name for this technology (torrefaction). This thermochemical method operates in the 200°C–300°C temperature range with minimal heating and an inert environment [50]. The process entails chipping biomass to enable impurity screening and effective drying before sizing and drying to a moisture content threshold of 20% (Figure 16.5). Torrefied biomass can produce briquettes that retain up to 96% of their chemical energy, are hydrophobic by nature, and are resistant to biodegradation [51]. As a result, it can replace charcoal or coal in residential heating, co-firing, gasification, and power production systems [52].

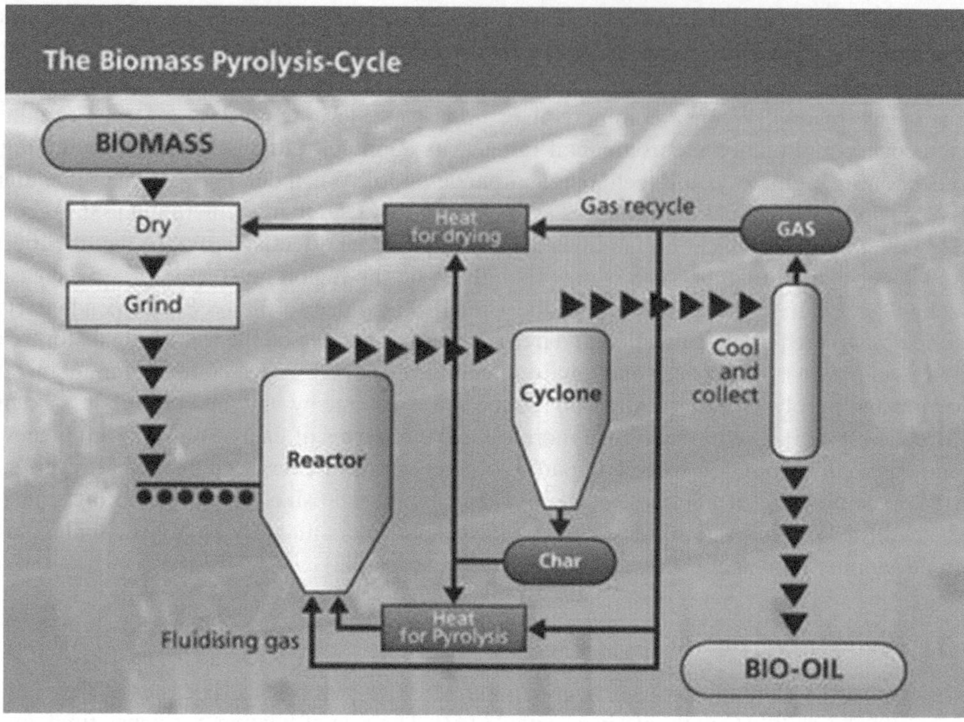

FIGURE 16.4 Biomass treatment by pyrolysis.

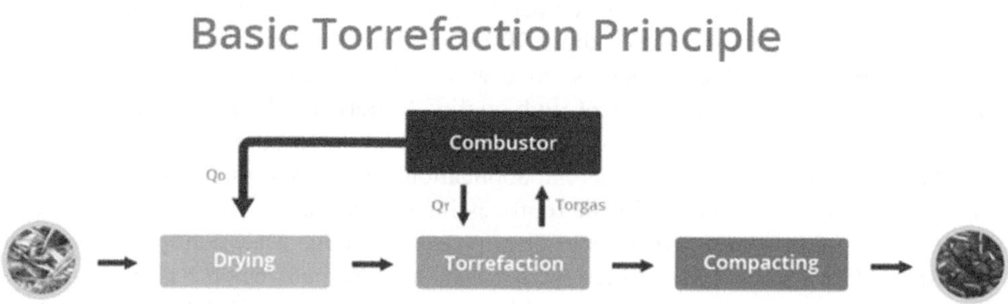

FIGURE 16.5 Biomass treatment by torrefaction.

16.3.5 Anaerobic Digestion

AD is the controlled oxidative breakdown of organic material. The disintegration process occurs in specially designed reactors in the case of engineered AD frameworks [53]. To increase biogas generation and waste decay rates, it is crucial to closely monitor environmental factors such as water content, heat, and pH levels. The method is used to denature lignocellulose, food scraps, live-stock manure, wastewater sludge, etc. This process in particular produces biogas, which contains trace amounts of volatile organic compounds, methane, CO_2, H_2S, and other gases [54]. The AD consists of four steps, the first of which is hydrolysis, in which organic polymers are broken down into simpler structural elements like monosaccharides and amino acids. In stage 2, which is acidification, organic polysaccharide subunits are then broken down, producing volatile fatty acids (VFAs), acetate, hydrogen gas, carbon dioxide (CO_2), and ammonia [55]. The third stage, known as "acidogenesis," sees the breakdown of long-chain fatty acids to produce acetate, CO_2, and H_2.

Methanogenic bacteria convert H_2 and acetic acid into methane gas and CO_2 in the process known as methanogenesis. Variations in the balance of different microbial species or other parameters (such as temperature or pH) might have an impact on the AD process. These four phases can take place in a single reactor or a multi-system. For instance, temperature changes cause harmful chemicals to build up, which lowers bacteria's metabolic activity and finally leads to the failure of the digesting reactor. Therefore, careful monitoring and management of these processes are necessary because these rectors are also vulnerable to pollution from plastic, silt, and chemicals [56]. Pre-treatments like physical (separation, crushing, hydrothermolysis), chemical (acidic or alkaline), and biological (microbial) are extensively researched in AD to increase methane output [57]. Pre-treatments using chemicals and biological processes effectively remove hemicellulose and make cellulose more processable for microbial species [58]. According to certain research, sisal fiber waste produced more methane when the particle size was reduced from 100 to 2 mm in AD. Process optimization can improve the economics of AD; for instance, by using a co-digestion method of urban sludge, the net cost was reduced [59]. It was shown that AD is preferred with a yearly output of 50,000 MT biowaste in the Murphy and Power study. As a result, recent AD research has focused mostly on enhancing workflows and raising the value of finished goods [60]. Figure 16.6 illustrates how AD works.

16.3.6 Composting Process

The controlled biological decomposition of garbage under aerobic or anaerobic conditions is referred to as composting. Because anaerobic procedures are harder to regulate and require more specialized technology to be effective, they are not used in developing nations [61]. Aerobic composting

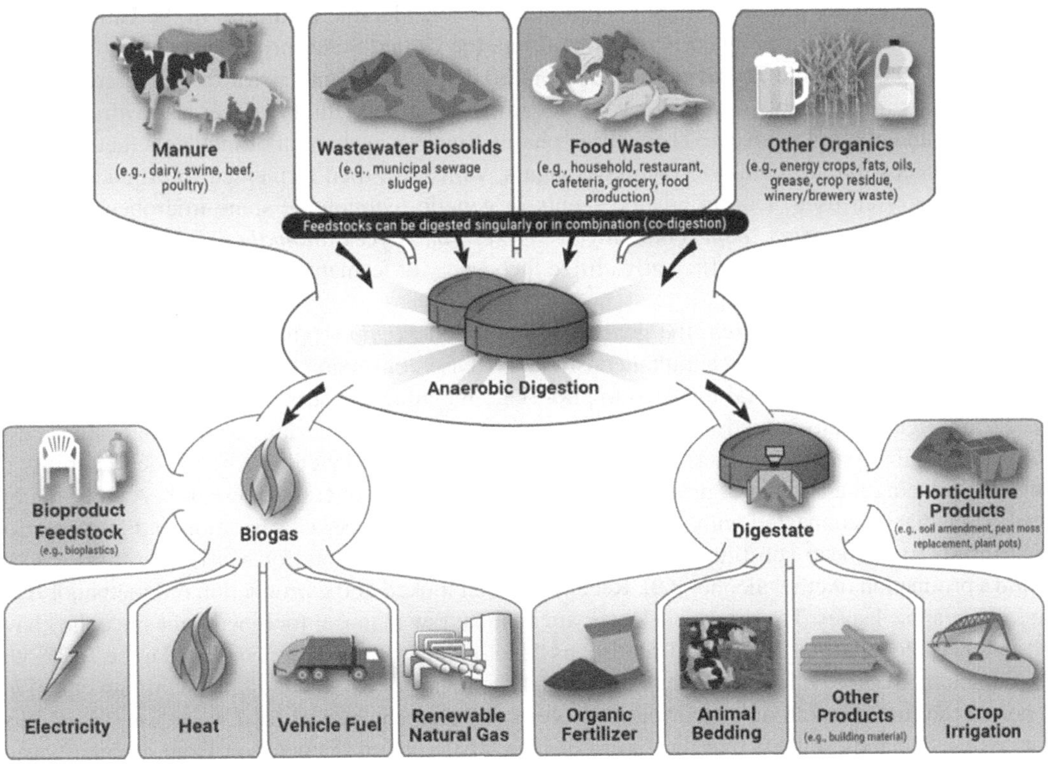

FIGURE 16.6 How AD is work (https://africalive.net/article/converting-biowaste-to-biogas-could-power-cleaner-sustainable-earth-future/).

happens in three steps. At temperatures above 60°C, the first step of degradation involves the breakdown of carbohydrates and amino acids into less complex substances like CO_2 and water. Mesophilic bacteria break down organic material during the transformation stage. The process by which nitrite (NO_2) is changed into nitrate (NO_3) and then ammonium (NH_4^+) is known as "synthesis" or "maturation," which occurs as the compost is brought under control. The full composting process can take anywhere from 3 to 6 months to finish, depending on some variables. Composting can have a faster rate of organic disintegration than AD because of the diversity of bacterial species and faster bacterial growth rates. Composting requires aeration, which needs to be considerably enhanced. However, no single aeration technique can effectively control emissions while still supplying oxygen [62]. Multi-stage inoculation can enhance the composting process, but it also makes the technique more complex. Pre-treating the substrate might also hasten the composting process. In a recent study, pretreatment with H_2O_2 caused composting of sewage sludge to mature in 20 days [63]. However, these pre-treatment techniques are not proper for composting because of the energy and chemical inputs. More research is needed to find the paths of improvement and interactions of the provided microbes with the existing microbiota to increase AD efficiency or regulate emissions in composting.

16.4 BIOCONVERSION APPLICATIONS

16.4.1 Bioethanol

An energy crisis is brought on by a considerable increase in the consumption of fossil fuels and their eventual depletion. This increases the need for energy-producing methods. Attention has been drawn to the generation of liquids like ethanol and biodiesel as well as the production of hydrogen utilizing biowaste [64]. Bioethanol is produced in considerable quantities in the United States, Brazil, and China. When using corn as raw material, the United States produced roughly 39.5 109 L of ethanol in 2009, while Brazil produced about 30 109 L of ethanol when using sugarcane [65]. The production of ethanol from organic wastes involves three primary processes: hydrolysis, fermentation, and ethanol recovery. These three phases function as a chain of incidents that affect the stages in succession. To obtain fermentable sugars, which are then fermented to produce ethanol, hydrolysis is often used. Due to advancements in genetic technology, some microbes, including *Saccharomyces cerevisiae, Aspergillus niger, Wickerhamomyces anomalus, Trichoderma reesei, Zymomonas mobilis*, etc., can efficiently utilize the variety of fermentable sugars obtained from the hydrolysis process [66].

According to recent research, the genetically modified *Pichia* strain can consume both hexose and pentose and ferment them simultaneously to increase conversion efficiency [67]. Starch is the most well-known raw material for a lab to pilot-scale manufacture ethanol among a variety of biowaste, and lignocellulosic feedstocks [68]. Genetically modified bacterial isolates can also successfully address the issue of low yield during the fermentation stage by enhancing the degradability of sugars. Reducing the number of processing stages is another aim of current research. A cutting-edge method primarily connected with recent research is synchronous saccharification and fermentation (SSF) [68]. As a cheap substitute for gasoline, ethanol fuel accounts for nearly two-thirds of the world's production of ethyl alcohol [69]. Recent research looked at the production of bioethanol from waste pineapple leaves. The pineapple leaves are an ideal raw material for bioethanol since they have 60%–80% holocellulose. In the ideal conditions, bioethanol production was reported to be 7.12% v/v. In a single bioreactor without the use of hazardous chemicals or harsh conditions, Dhiman et al. [70] reported the first research on concurrent hydrolysis and fermentation of food waste into bioethanol. The rising demands suggest that the future of large-scale ethanol generation from organic wastes is promising. The process enhancements greatly reduce reaction time and cost, enabling the use of biodegradable material in upcoming industrial ethanol fermentation applications [70].

16.4.2 Biohydrogen

Since water vapor is the only byproduct of burning pure hydrogen, it is the cleanest and least polluting fuel currently on the market [71]. In comparison to chemical production, biological hydrogen production offers several benefits, such as the fermentation of organic materials or the bio-photolysis of water by algae, among others. The biological processes that produce biohydrogen can be divided into those that require light (such as photo-fermentation and bio-photolysis) and those that don't (such as dark fermentation (DF) and using a microbial electrolysis cell) [72]. Over the years, a novel system, hybrid technique, has also been researched.

According to a study by Yang et al. [73], combining dark- and photo-fermentation with a pre-treatment maize cob increased the output of biohydrogen. The first step showed that the maximal biohydrogen output and rate for DF were 120.3 mL H_2/g corn cob and 150 mL H_2/L/h, respectively. Another common form of bio-photolysis is the production of hydrogen by biological processes using water and sunshine, which has its own set of benefits and drawbacks. The cost of raw materials, which can be addressed by using waste materials for the same, is one of the key issues with the generation of bio-hydrogen via DF or photo fermentation. Cost, availability, carbohydrate content, and biodegradability are the main factors considered when choosing waste materials for the generation of biohydrogen. The substrate used, the price, the pH, the kind of production process, etc., are all variables that affect the generation of biohydrogen [74].

Simple sugars (glucose and sucrose), waste from the agricultural industry having starch, waste from the pulp and paper industry or food industry, wastewater from the food industry, cheese whey, or waste sludge from waste treatment plants are a few examples of waste materials that could be used as a starter or raw ingredients to produce hydrogen. Various other waste materials have also been tested over time [75].

In one of their research, Pawar et al. [76] showed that wheat straw hydrolysate could produce biohydrogen when *Caldicellulosiruptor saccharolyticus* was used in a continuously stirred tank reactor (CSTR), and its effluents could then be anaerobically digested to produce methane (CH_4). Instead of the ideal described scenarios where they are connected, the process was a two-step process where the reactors executing DF and AD were uncoupled. Although there is room for improvement, the two-step process showed a respectable efficiency in sugar conversion.

The wheat hydrolysate was successfully converted to hydrogen by *C. saccharolyticus* at a high production rate of about 5.2 L H_2/L/day. The maximum energy output of the entire process was estimated to be 10.9 kJ/g of straw, or around 57% of the total energy, with the sugar fraction in the wheat straw accounting for 67% of this output. The study found that a simplified and economically viable version of the technology might improve a biorefinery; in addition to producing bioenergy, renewable biomass is also utilized to create additional value-added goods [76]. Producing biohydrogen sequentially through photo- and dark-fermentation is another strategy under investigation because it would break the production limit for organic acids. The best effluent quality in terms of chemical oxygen requirement is another benefit of greater organic acid use (COD).

Like all other techniques, employing biological wastes for biohydrogen production has several downsides, including poor rates and yields of H_2 creation [75,77]. Higher numbers of reactors are needed to manufacture biohydrogen due to lower production yields. The aforementioned problems can be resolved by choosing and using the right organisms or microbial consortium effectively, developing more skillful processing setups, optimizing environmental conditions, increasing light utilization efficiency, and creating successful photo-bioreactors [71].

16.4.3 Enzymes and Chemicals

Enzymes are biological catalysts that enable the completion of biochemical reactions, and as a result, play a crucial part in many facets of life and its fundamental processes [78]. Plants, animals, and microorganisms are all natural providers of enzymes. However, the rising costs of enzymes have made it possible for researchers to concentrate on and investigate the many ways of producing enzymes, as well as to improve the currently used methods for producing enzymes affordably and sustainably [79]. Microbial sources have been preferred over plant and animal sources due to technological advantages such as quicker fermentation times, larger yields, and economic feasibility [80]. As biotechnology improves, several bioprocesses are appearing for the use of different industrial wastes to produce enzymes. The agro-industry is one of the key sectors that contributes to this.

Wastes produced during the industrial processing of either agricultural or animal products, or products obtained because of agricultural activities, are referred to as agro-industrial wastes. These wastes can take the form of stalks, leaves, straw, lint, pulp, grains, and many other things. Fibers, proteins, minerals, and carbohydrates are typically present in these wastes. Lignocellulosic materials are the main components of these wastes (consisting of hemicelluloses, cellulose, and lignin). Because of their variety in the trash, some people like to use them [1]. Numerous research has investigated the impact of pretreatment techniques on the concentration of the product generated. Pre-treatments, which often use lignocellulosic precursors, are used in the procedures. In a recent study, four different enzymes' yields were examined using various agricultural fragments as growth feedstocks and new species of Bacillus [82]. To figure out their impact on enzyme production many pre-treatment techniques, including acid, alkali, microwave, and ultrasound, were applied. Higher quantities of enzymes like amylase and protease were produced using chemical pre-treatment techniques [83].

Additionally, researchers have confirmed the use of ultrasonication as a potential pre-treatment method to increase the enzyme output from olive pomace using solid-state fermentation [84]. *Aspergillus niger* was used to report a threefold increase in xylanase activity and a 1.2-fold increase in cellulase activity. Maximum cellulase (38 U/g of the solid substrate) and xylanase (28 U/g of the solid substrate) activity was seen when SSF was compared to crude and exhausted olive pomace. Like this, when liquid fraction was added during SSF, cellulase, and xylanase activities were 46 ± 3 and 100 ± 15 U/g, respectively, as opposed to the prior 21 and 71 U/g.

Additionally, intensive research and testing are done on *Rhizopus microsporus var. oligosporus* to improve a solid-state fermentation process using a variety of agro-industrial wastes [85]. The study included four different substrates: wheat bran, soybean meal, sugarcane bagasse, and wheat flour (Type 2). Out of these, wheat bran was found to be the best carbon source for amylase synthesis (392.5 U/g). Another substrate made from wheat, type 2 wheat flour, produced amylase after wheat bran but at a rate that was reported to be 2.7 times lower.

The most affordable and easily accessible source of carbohydrates for the augmentation and later addition for enzyme production, according to numerous research, is a lignocellulosic waste. By using agricultural wastes, bacteria and fungi can be employed to produce enzymes. Enzyme production may be increased economically with a pretreatment that allows high rates of saccharification. To find the best way to create enzymes, a thorough study of the numerous studies must be done.

Several compounds have also been discovered as byproducts of industrial bioprocesses in addition to enzymes. As a result, their use has been researched. Abghari and Chen's [86] work was one of them and focused on the improved production of citric acid and lipids by engineering *Yarrowia lipolytica* at various phases of the processes, metabolic pathways of lipid synthesis, and up/downregulation of genes. In addition to site-specific gene knockout techniques, they used a single-step double gene knock-in technique to increase the synthesis of lipids and citric acid in the strain grown under nitrogen constraint [86]. Another study by Chong et al. [87] concentrated on the generation of high-efficiency succinic acid (SA) by glycerol utilizing an engineered *Y. lipolytica* in situ fibrous bed bioreactor (FBB). The study successfully increased succinic acid production by *Y. lipolytica* and reported this. In the past, it has been shown that *Y. lipolytica* has a high tolerance for glycerol in the formation of organic acids, which influences other parameters affecting SA production. The highest succinic acid titer (198.2 g/L) ever recorded was reported in this study that included FBB, which also claimed productivity of 1.46 g/L/h. The use of *Y. lipolytica* to create SA under low pH circumstances and the immobilization of agricultural residues including wheat straw, corn stalk, and sugarcane bagasse have both been studied in other studies. The best immobilization substance was said to be sugarcane bagasse. Under the optimized settings, a final SA concentration of 53.6 g/L and an average yield of 1.45 g/L/h with a final yield of 0.45 g/g were reported [87].

16.4.4 Biogas

The most efficient way to produce biofuels is by the production of biogas, which is a mixture of methane, CO_2, H_2S, moisture, and siloxanes. An excellent substitute is the anaerobic conversion of food waste to biogas [88]. However, this waste transformation process requires a preparatory treatment as well for yield augmentation. Previously, the focus was on AD in urban garbage, but lignocellulosic waste might also be used to produce biogas-rich biomethane [89]. The three steps in the formation of biogas are hydrolysis, acidogenesis/acetogenesis, and methane generation [88]. It has been demonstrated that the structure and diversity of the bacterial population significantly affect the yield of biogas and methane as well as operating efficiencies [90]. According to Deepanraj et al. [91], substrate pre-treatment significantly affects the production of biogas from food waste. In another study, biogas yield from food scraps was boosted by 19% by applying pre-treatment procedures by co-digestion with – grease waste [92]. According to Zhang et al. [93] investigation into the AD of food waste and cow manure, co-digestion boosts overall methane outputs by 41.1% in batch mode and by 55.2% in semi-continuous mode.

Studies are performed to better use and improve the process of waste product valorization for some additional compounds that are created as byproducts of the organic waste industry. Lactic acid, acetic acid, citric acid, L-ascorbic acid, polyhydroxyalkanoates, perillic acid, fumaric acid, and xylitol are a few examples of compounds created using bioconversion technology.

16.5 PROSPECTS OF ORGANIC WASTE USE FOR BIOREFINERIES

The concept of biorefinery, which primarily focuses on the reclamation of organic wastes (abundant sources of the renewable and inexpensive substrate) and their effective fermentation via microbes, has been seen too often to have biohydrogen as a key component over time. Biorefining is the economically practical conversion of biomass into a wide range of commodities, such as energy, materials, chemicals, food, and feed. For instance, the carbohydrates from which furfural is generated are seen as possible bio-based platform chemicals, having the ability to be further converted into a variety of goods like bioplastics, medications, and agrochemicals.

Due to the demand for resources by humanity, biorefining is seen to be expanding internationally. It typically uses biotechnological procedures and methods to generate different biochemicals. Both sustainability and the cascade principles are typically at the center of them. The following are a few examples of the various biorefinery methods that have been discussed in the literature [80,94–96]:

1. Increased lipid production through yeast's conversion of glucose to lipids using dry algae as the substrate
2. Using xylose and cellobiose to produce fatty acids
3. Using rice straw as the substrate, laccase production (to oxidize phenolic chemicals)
4. Bioethanol production using glycerine from a biodiesel plant as the starting material
5. Using agricultural residues as the raw substrates, the production of hydrolytic enzymes (xylanase, chitinase, proteases, and amylases)
6. Using the organic components of MSW as substrates, the production of biohydrogen, methane, and enzymes

Regarding the employment and depreciation of biorefineries, many technological and nontechnical gaps and impediments have been identified. Most of the technical challenges associated with the use of energy crops are related to the production costs and issues with the storage of biomass along with its harvesting. When crops must be harvested in a limited amount of time, this poses a fundamental problem. Transport costs appear to be of the utmost importance for determining the total cost of biomass feedstock, particularly in small-scale biorefineries. Biomass production on a local or regional scale could have significant economic benefits.

16.6 TECHNICAL CHALLENGES, CURRENT AND FUTURE PERSPECTIVES FOR SUSTAINABLE WASTE MANAGEMENT

Water, soil, and air contamination have resulted from the expansion of waste leftovers and their poor management, which pose serious risks. The techniques of disposing of biological waste that are addressed in this article are composting and AD. However, they require larger energy inputs and may spread germs and odors [96,97].

Therefore, waste valorization is a fascinating area that is currently attracting growing interest in many nations. As a result, research teams are concentrating not only on methods of valorization but also on the creation of numerous environmentally friendly alternatives. Organic wastes are often employed for commercial advantage. Existing research on the utilization of biowaste involves three fundamental steps: examining the composition of particular biowaste, separating useful elements from raw wastes through pre-treatment, and fermenting this content with microorganisms to create an end product [98]. Compared to other bio-based goods like bio-fertilizers, enzymes, organic acids, etc., the success of producing biofuel from biomass has increased dramatically. Increasing productivity and yield and streamlining the biowaste transformation process are the two most crucial issues that these biotechnology advancements must address. Additionally, the refinement of downstream processes for product purification may be the key to the future of renewable fuel conversion from organic leftovers [99].

A significant advancement in the use of biomass is the conversion of waste biomass into advantageous industrial chemicals [100]. The new trend is still in its prime, with plenty of opportunity for fresh disputes. Another efficient method to speed up the bioconversion process is to use genetically modified strains that can simultaneously ingest many organic waste-derived precursors and convert them to chemical molecules [101]. Technically practical to produce organic acids and enzymes from waste biomass in a laboratory size, which is the first step in putting bioconversion into practice on a pilot scale is needed. Process development and improvement are also needed. A crucial factor in industrial applications is product quality. Before items like enzymes and antibiotics are utilized commercially, increased efficiency and constant quality must be proven in a pilot-scale setting [67].

The overexpression of fungus, bacteria, or yeast genes in a host organism for the manufacture of desired fuels or compounds is referred to as metabolic science in the literature. Examples of the aforementioned biotechnology include studies carried out by Almeida et al. [99] and Blankschien et al. [101] Understanding and exploiting the metabolic pathways of bacteria can benefit from studies on transformation inhibitors. To prove its commercial viability, the biological conversion process needs rigorous optimization based on several factors, including the choice of feedstocks, pre-treatment techniques, and byproduct output. Process optimization has also been thoroughly studied in the current transformation of biomass wastes. In the cumulative process of industrial biotechnology, enzymes or bacterial strains are essential components. Enzyme immobilization has been widely used in the production of biochemicals and biofuels. The effectiveness of solid-state fermentation for the production of enzyme products may be improved by the use of immobilized catalysts and other bioprocess optimization approaches [24]. Despite significant advancements, the process of biomass conversion still confronts significant obstacles in large-scale applications. Future research must concentrate on developing low biocatalysts for the industrial manufacturing of value-added product lines. A circular bioeconomy, based on recovery, reuse, and reprocessing, must be built by moving away from a systemic linear approach.

With a focus on waste reduction, resource efficiency, and employment development, the circular economy has attracted much attention in industrial economics. Industrial waste has a significant potential to produce renewable energy within the circular economy concept, which includes waste management, reuse, and recycling. By creating integrated and multipurpose elements for the exploitation of biomass waste to create appealing intermediates and final products, the biorefinery concept may be a crucial part of enhancing the circular economy. Such biotechnological procedures are created based on economic viability, which equates to natural sustainability, as shown by a decreased carbon footprint.

16.7 CONCLUSION

The chapter offers a thorough examination of the organic waste valorization produced by various businesses. Due to its composition and low cost, organic waste is a highly invested source of fuel in the production of useful bioproducts. For their use, it is helpful on both an economic and an environmental scale to enhance the idea of value addition to the trash by producing an intended product with desired properties. However, the majority of biowaste usage has only been studied in the laboratory; commercial applications have not yet been fully used. Therefore, bio-valorization of surplus waste and side streams into diverse value-added products greatly encourages the development of a circular bioeconomy that is sustainable.

GLOSSARY

Anaerobic Digestion: a process through which bacteria break down organic matter such as animal manure, wastewater biosolids, and food wastes – in the absence of oxygen.

Bioconversion: also known as biotransformation, is the conversion of organic materials, such as plant or animal waste, into usable products or energy sources by biological processes or agents, such as certain microorganisms.

Biorefineries: a refinery that converts biomass to energy and other beneficial byproducts (such as chemicals).

Circular Economy: a model of production and consumption, which involves sharing, leasing, reusing, repairing, refurbishing, and recycling existing materials and products as long as possible.

Fermentation: a metabolic process that produces chemical changes in organic substrates through the action of enzymes.

Fossil Fuels: a hydrocarbon-containing material formed naturally in the earth's crust from the remains of dead plants and animals that are extracted and burned as a fuel.

Gasification: a process that converts biomass- or fossil fuel-based carbonaceous materials into gases, including the largest fractions: nitrogen (N_2), carbon monoxide (CO), hydrogen (H_2), and carbon dioxide (CO_2).

Incineration: a waste treatment process that involves the combustion of substances contained in waste materials.

Industrial Waste: waste generated by manufacturing or industrial processes.

Organic Waste: includes any organic matter in waste which can be broken down into carbon dioxide, water, methane. or simple organic molecules by microorganisms and other living things by composting, aerobic digestion, anaerobic digestion, or similar processes.

Photobioreactor: refers to any cultivation system designed for growing photoautotrophic organisms using artificial light sources or solar light to facilitate photosynthesis.

Pyrolysis: process of the thermal decomposition of materials at elevated temperatures in an inert atmosphere.

Torrefaction: of biomass, e.g., wood or grain, is a mild form of pyrolysis at temperatures typically between 200°C and 320°C. Torrefaction changes biomass properties to provide a better fuel quality for combustion and gasification applications. Torrefaction produces a relatively dry product, which reduces or eliminates its potential for organic decomposition.

Valorization: the process of reusing, recycling, or composting waste materials and converting them into more useful products including materials, chemicals, fuels, or other sources of energy.

Waste Management: includes the processes and actions required to manage waste from its inception to its final disposal. This includes the collection, transport, treatment, and disposal of waste, together with monitoring and regulation of the waste management process and waste-related laws, technologies, economic mechanisms.

REFERENCES

1. C. S. Shivamathi et al., Process optimization and characterization of pectin derived from underexploited pineapple peel biowaste as a value-added product, *Food Hydrocoll.*, vol. 123, no. August 2021, p. 107141, 2022.
2. G. Sailer et al., Improving the energetic utilization of household food waste: Impact of temperature and atmosphere during storage, *Waste Manag.*, vol. 144, no. January, pp. 366–375, 2022.
3. S. A. Siddiqui et al., Black soldier fly larvae (BSFL) and their affinity for organic waste processing, *Waste Manag.*, vol. 140, no. January, pp. 1–13, 2022.
4. I. Anastopoulos et al., Agricultural biomass/waste as adsorbents for toxic metal decontamination of aqueous solutions, *J. Mol. L*iq., vol. 295, p. 111684, 2019.
5. A. Yuvaraj, R. Thangaraj, B. Ravindran, S. W. Chang, and N. Karmegam, Centrality of cattle solid wastes in vermicomposting technology – A cleaner resource recovery and biowaste recycling option for agricultural and environmental sustainability, *Environ. Pollut.*, vol. 268, p. 115688, 2021.
6. R. W. I. B. Priyadarshana, P. E. Kaliyadasa, S. R. W. M. C. J. K. Ranawana, and K. G. C. Senarathna, Biowaste management: Banana fiber utilization for product development, *J. Nat. Fibers*, vol. 19, no. 4, pp. 1461–1471, 2022.
7. C. R. Chilakamarry et al., Advances in solid-state fermentation for bioconversion of agricultural wastes to value-added products: Opportunities and challenges, *Bioresour. Technol.*, vol. 343, no. August 2021, p. 126065, 2022.
8. M. Chopping, Z. Wang, C. Schaaf, M. A. Bull, and R. R. Duchesne, Forest aboveground biomass in the southwestern United States from a MISR multi-angle index, 2000–2015, *Remote Sens. Environ.*, vol. 275, no. January, p. 112964, 2022.
9. W. J. Elliot and H. Rhee, Impacts of forest biomass operations on forest hydrologic and soil erosion processes, *Trees For. People*, vol. 7, no. January, p. 100186, 2022.
10. A. T. Ubando, A. J. R. Del Rosario, W. H. Chen, and A. B. Culaba, A state-of-the-art review of biowaste biorefinery, *Environ. Pollut.*, vol. 269, p. 116149, 2021.
11. N. Howaniec and A. Smoliński, Biowaste utilization in the process of co-gasification with bituminous coal and lignite, *Energy*, vol. 118, pp. 18–23, 2017.
12. L. Dai et al., Toward green production of xylooligosaccharides and glucose from sorghum straw biowaste by sequential acidic and enzymatic hydrolysis, *Ind. Crops Prod.*, vol. 179, no. 159, p. 114662, 2022.
13. S. Yasir et al., Microalgae biomass as a sustainable source for biofuel, biochemical and biobased value-added products: An integrated biorefinery concept, *Fuel*, vol. 307, no. August 2021, p. 121782, 2022.
14. R. F. Al-Thani and B. T. Yasseen, Perspectives of future water sources in Qatar by phytoremediation: Biodiversity at ponds and modern approach, *Int. J. Phytoremediation*, vol. 23, no. 8, pp. 866–889, 2021.
15. G. K. Sharma et al., Circular economy fertilization: Phycoremediated algal biomass as biofertilizers for sustainable crop production, *J. Environ. M*anage., vol. 287, no. February, p. 112295, 2021.
16. J. Ullmann and D. Grimm, Algae and their potential for a future bioeconomy, landless food production, and the socio-economic impact of an algae industry, *Org. Agric.*, vol. 11, no. 2, pp. 261–267, 2021.
17. S. Krishnamoorthy and P. Manickam, Phycoremediation of industrial wastewater: Challenges and prospects, In: V. Kumar, G. Saxena, and M. P. Shah (eds.), *Bioremediation for Environmental Sustainability*. Elsevier, Amsterdam, pp. 99–123, Jan. 2021.
18. S. Jafarinejad, *Petroleum Waste Treatment and Pollution Control*. Butterworth-Heinemann, pp. 1–362, Oct. 2016.
19. N. Mahesh et al., A review on mitigation of emerging contaminants in an aqueous environment using microbial bio-machines as sustainable tools: Progress and limitations, *J. Water Process Eng.*, vol. 47, no. February, p. 102712, 2022.
20. M. K. Nguyen et al., Evaluate the role of biochar during the organic waste composting process: A critical review, *Chemosphere*, vol. 299, no. January, p. 134488, 2022.
21. H. Wang, L. Shan, Q. Lv, S. Cai, G. Quan, and J. Yan, Production of hierarchically porous carbon from natural biomass waste for efficient organic contaminants adsorption, *J. Clean. Prod.*, vol. 263, p. 121352, 2020.
22. M. H. de Sousa et al., Valorizing municipal organic waste to produce biodiesel, biogas, organic fertilizer, and value-added chemicals: An integrated biorefinery approach, *Biomass Convers. Biorefinery*, vol. 12, no. 3, pp. 827–841, 2022.
23. F. Lü, Z. Wang, H. Zhang, L. Shao, and P. He, Anaerobic digestion of organic waste: Recovery of value-added and inhibitory compounds from liquid fraction of digestate, *Bioresour. Technol.*, vol. 333, no. April, p. 125196, 2021.

24. H. J. Choi, Agricultural biowaste, rice bran, as carbon source to enhance biomass and lipid production: Analysis with various growth rate models, *Water Sci. Technol.*, vol. 82, no. 6, pp. 1120–1130, 2020.
25. S. K. S. Patel, D. Das, S. C. Kim, B. K. Cho, V. C. Kalia, and J. K. Lee, Integrating strategies for sustainable conversion of waste biomass into dark-fermentative hydrogen and value-added products, *Renew. Sustain. Energy Rev.*, vol. 150, no. July, p. 111491, 2021.
26. K. W. Chew, S. R. Chia, W. Y. Chia, W. Y. Cheah, H. S. H. Munawaroh, and W. J. Ong, Abatement of hazardous materials and biomass waste via pyrolysis and co-pyrolysis for environmental sustainability and circular economy, *Environ. Pollut.*, vol. 278, p. 116836, 2021.
27. N. Esercizio et al., Fermentation of biodegradable organic waste by the family thermotogaceae, *Resources*, vol. 10, no. 4, pp. 1–26, 2021.
28. H. I. Abdel-Shafy and M. S. M. Mansour, Solid waste issue: Sources, composition, disposal, recycling, and valorization, *Egypt. J. Pet.*, vol. 27, no. 4, pp. 1275–1290, Dec. 2018.
29. M. Azam, S. S. Jahromy, W. Raza, C. Jordan, M. Harasek, and F. Winter, Comparison of the combustion characteristics and kinetic study of coal, municipal solid waste, and refuse-derived fuel: Model-fitting methods, *Energy Sci. Eng.*, vol. 7, no. 6, pp. 2646–2657, Dec. 2019.
30. H. Rezaei, F. Y. Panah, C. J. Lim, and S. Sokhansanj, Pelletization of refuse-derived fuel with varying compositions of plastic, paper, organic and wood, *Sustainablity*, vol. 12, no. 11, p. 4645, Jun. 2020.
31. L. Appels et al., Anaerobic digestion in global bio-energy production: Potential and research challenges, *Renew. Sustain. Energy Rev.*, vol. 15, no. 9, pp. 4295–4301, Dec. 2011.
32. G. Ionescu, E. C. Rada, M. Ragazzi, C. Mărculescu, A. Badea, and T. Apostol, Integrated municipal solid waste scenario model using advanced pretreatment and waste to energy processes, *Energy Convers. Manag.*, vol. 76, pp. 1083–1092, Dec. 2013.
33. Y. Nie, Development and prospects of municipal solid waste (MSW) incineration in China, *Front. Environ. Sci. Eng. China*, vol. 2, no. 1, pp. 1–7, Mar. 2008.
34. H. A. Arafat and K. Jijakli, Modeling and comparative assessment of municipal solid waste gasification for energy production, *Waste Manag.*, vol. 33, no. 8, pp. 1704–1713, Aug. 2013.
35. H. Y. Yap and J. D. Nixon, A multi-criteria analysis of options for energy recovery from municipal solid waste in India and the UK, *Waste Manag.*, vol. 46, pp. 265–277, Dec. 2015.
36. D. Panepinto, V. Tedesco, E. Brizio, and G. Genon, Environmental performances and energy efficiency for MSW gasification treatment, *Waste Biomass Valorization*, vol. 6, no. 1, pp. 123–135, Feb. 2015.
37. V. S. Sikarwar et al., An overview of advances in biomass gasification, *Energy Environ. Sci.*, vol. 9, no. 10, pp. 2939–2977, Oct. 2016.
38. J. D. Murphy, E. McKeogh, and G. Kiely, Technical/economic/environmental analysis of biogas utilisation, *Appl. Energy*, vol. 77, no. 4, pp. 407–427, Apr. 2004.
39. A. Kumar and S. R. Samadder, A review on technological options of waste to energy for effective management of municipal solid waste, *Waste Manag.*, vol. 69, pp. 407–422, Nov. 2017.
40. O. K. M. Ouda, S. A. Raza, A. S. Nizami, M. Rehan, R. Al-Waked, and N. E. Korres, Waste to energy potential: A case study of Saudi Arabia, *Renew. Sustain. Energy Rev.*, vol. 61, pp. 328–340, Aug. 2016.
41. A. U. Zaman, Comparative study of municipal solid waste treatment technologies using life cycle assessment method, *Int. J. Environ. Sci. Technol.*, vol. 7, no. 2, pp. 225–234, Mar. 2010.
42. F. C. Luz et al., Techno-economic analysis of municipal solid waste gasification for electricity generation in Brazil, *Energy Convers. Manag.*, vol. 103, pp. 321–337, Oct. 2015.
43. D. Czajczyńska et al., Potential of pyrolysis processes in the waste management sector, *Therm. Sci. Eng. Prog.*, vol. 3, pp. 171–197, Sep. 2017.
44. K. A. Kalyani and K. K. Pandey, Waste to energy status in India: A short review, *Renew. Sustain. Energy Rev.*, vol. 31, pp. 113–120, Mar. 2014.
45. L. Lombardi, E. Carnevale, and A. Corti, A review of technologies and performances of thermal treatment systems for energy recovery from waste, *Waste Manag.*, vol. 37, pp. 26–44, Mar. 2015.
46. M. Gholizadeh et al., Progress of the development of reactors for pyrolysis of municipal waste, *Sustain. Energy Fuels*, vol. 4, no. 12, pp. 5885–5915, Nov. 2020.
47. S. Kern, M. Halwachs, G. Kampichler, C. Pfeifer, T. Pröll, and H. Hofbauer, Rotary kiln pyrolysis of straw and fermentation residues in a 3 MW pilot plant – Influence of pyrolysis temperature on pyrolysis product performance, *J. Anal. Appl.* Pyrolysis, vol. 97, pp. 1–10, Sep. 2012.
48. B. Batidzirai, A. P. R. Mignot, W. B. Schakel, H. M. Junginger, and A. P. C. Faaij, Biomass torrefaction technology: Techno-economic status and future prospects, *Energy*, vol. 62, pp. 196–214, Dec. 2013.
49. W. Yan, J. T. Hastings, T. C. Acharjee, C. J. Coronella, and V. R. Vásquez, Mass and energy balances of wet torrefaction of lignocellulosic biomass, *Energy Fuels*, vol. 24, no. 9, pp. 4738–4742, Sep. 2010.
50. D. Medic, M. Darr, B. Potter, and A. Shah, Effect of torrefaction process parameters on biomass feedstock upgrading, *Am. Soc. Agric. Biol. Eng. Annu. Int. Meet.* 2010, vol. 5, pp. 1–8, 2010.

51. W. H. Chen and P. C. Kuo, A study on torrefaction of various biomass materials and its impact on lignocellulosic structure simulated by a thermogravimetry, *Energy*, vol. 35, no. 6, pp. 2580–2586, Jun. 2010.
52. M. Phanphanich and S. Mani, Impact of torrefaction on the grindability and fuel characteristics of forest biomass, *Bioresour. Technol.*, vol. 102, no. 2, pp. 1246–1253, Jan. 2011.
53. L. Bi and M. Haight, Anaerobic digestion and community development: A case study from Hainan province, China, *Environ. Dev. Sustain.*, vol. 9, no. 4, pp. 501–521, Jul. 2006.
54. Q. Zhang, J. Xu, X. Wang, T. Zhu, J. Liu, and S. Qin, Performance of full-scale aerobic composting and anaerobic digestion on the changes of antibiotic resistance genes in dairy manure, *Bioresour. Technol.*, vol. 342, p. 125898, Dec. 2021.
55. M. Cucina, P. De Nisi, L. Trombino, F. Tambone, and F. Adani, Degradation of bioplastics in organic waste by mesophilic anaerobic digestion, composting and soil incubation, *Waste Manag.*, vol. 134, pp. 67–77, Oct. 2021.
56. R. E. Speece. *Anaerobic Biotechnology and Odor/Corrosion Control for Municipalities and Industries.* Nashville, TN: Archa Press, 2008. [Online]. Available: https://scholar.google.com/scholar_lookup?title=Anaerobic%20biotechnology%20and%20odor%20Fcorrosion%20control%20for%20municipalities%20and%20industries&author=R.E.%20Speece&publication_year=2008. (Accessed: 27-Jun-2022).
57. Y. Zheng, J. Zhao, F. Xu, and Y. Li, Pretreatment of lignocellulosic biomass for enhanced biogas production, *Prog. Energy Combust. Sci.*, vol. 42, no. 1, pp. 35–53, Jun. 2014.
58. K. Weerasai et al., Sequential acid and alkaline pretreatment of rice straw for bioethanol fermentation, *BioResources*, vol. 9, pp. 5988–6001, 2014. [Online]. Available: https://ojs.cnr.ncsu.edu/index.php/BioRes/article/view/BioRes_09_4_5988_Weerasai_Sequential_Acid_Rice_Straw. (Accessed: 27-Jun-2022).
59. L. N. Liew, J. Shi, and Y. Li, Enhancing the solid-state anaerobic digestion of fallen leaves through simultaneous alkaline treatment, *Bioresour. Technol.*, vol. 102, no. 19, pp. 8828–8834, Oct. 2011.
60. J. D. Murphy and N. M. Power, A technical, economic and environmental comparison of composting and anaerobic digestion of biodegradable municipal waste, *J. Environ. Sci. Health Part A*, vol. 41, no. 5, pp. 865–879, May 2007. https://dx.doi.org/10.1080/10934520600614488
61. R. Assandri, C. Canetta, G. Viganò, E. Buscarini, A. Scartabellati, and A. Montanelli, Laboratory markers included in the corona score can identify false negative results on COVID-19 RT-PCR in the emergency room, *Biochem. Medica*, vol. 30, no. 3, pp. 1–3, Oct. 2020.
62. J. P. Sheets, L. Yang, X. Ge, Z. Wang, and Y. Li, Beyond land application: Emerging technologies for the treatment and reuse of anaerobically digested agricultural and food waste, *Waste Manag.*, vol. 44, pp. 94–115, Oct. 2015.
63. H. D. Chang and C. Y. Chen, Composting of biosolids enhanced by a combined pretreatment with hydrogen peroxide and triton X-100, *Waste Biomass Valorization*, vol. 6, no. 1, pp. 45–51, Feb. 2015.
64. A. Kang and T. S. Lee, Converting sugars to biofuels: Ethanol and beyond, *Bioengineering*, vol. 2, no. 4, pp. 184–203, Oct. 2015.
65. L. Bhatia, S. Johri, and R. Ahmad, An economic and ecological perspective of ethanol production from renewable agro waste: A review, *AMB Express*, vol. 2, no. 1, p. 65, Dec. 2012.
66. G. Koçar and N. Civaş, An overview of biofuels from energy crops: Current status and future prospects, *Renew. Sustain. Energy Rev.*, vol. 28, pp. 900–916, Dec. 2013.
67. R. C. Kuhad, R. Gupta, Y. P. Khasa, A. Singh, and Y. H. P. Zhang, Bioethanol production from pentose sugars: Current status and future prospects, *Renew. Sustain. Energy Rev.*, vol. 15, no. 9, pp. 4950–4962, Dec. 2011.
68. X. Yang et al., Production of bioethanol and biodiesel using instant noodle waste, *Bioprocess Biosyst. Eng.*, vol. 37, no. 8, pp. 1627–1635, Feb. 2014.
69. X. Yang, S. J. Lee, H. Y. Yoo, H. S. Choi, C. Park, and S. W. Kim, Biorefinery of instant noodle waste to biofuels, *Bioresour. Technol.*, vol. 159, pp. 17–23, May 2014.
70. S. S. Dhiman et al., Simultaneous hydrolysis and fermentation of unprocessed food waste into ethanol using thermophilic anaerobic bacteria, *Bioresour. Technol.*, vol. 244, pp. 733–740, Nov. 2017.
71. V. Singh Yadav, R. Vinoth, and D. Yadav, Bio-hydrogen production from waste materials: A review, *MATEC Web Conf.*, vol. 192, p. 02020, Aug. 2018.
72. W. Nabgan et al., Catalytic biohydrogen production from organic waste materials: A literature review and bibliometric analysis, *Int. J. Hydrogen Energy*, vol. 46, no. 60, pp. 30903–30925, Sep. 2021.
73. H. Yang, L. Guo, and F. Liu, Enhanced bio-hydrogen production from corncob by a two-step process: Dark- and photo-fermentation, *Bioresour. Technol.*, vol. 101, no. 6, pp. 2049–2052, Mar. 2010.

74. S. Dahiya, A. N. Kumar, J. Shanthi Sravan, S. Chatterjee, O. Sarkar, and S. V. Mohan, Food waste biorefinery: Sustainable strategy for circular bioeconomy, *Bioresour. Technol.*, vol. 248, pp. 2–12, Jan. 2018.
75. S. Sung, Bio-hydrogen production from renewable organic wastes, Technical Report, Golden Field Office, Golden, CO, Apr. 2004.
76. S. S. Pawar, V. N. Nkemka, A. A. Zeidan, M. Murto, and E. W. J. Van Niel, Biohydrogen production from wheat straw hydrolysate using Caldicellulosiruptor saccharolyticus followed by biogas production in a two-step uncoupled process, *Int. J. Hydrogen Energy*, vol. 38, no. 22, pp. 9121–9130, Jul. 2013.
77. K. Yan, R. Li, Z. Yang, X. Li, Y. Wang, and G. Wu, Biomass waste-derived porous carbon efficient for simultaneous removal of chlortetracycline and hexavalent chromium, iScience, vol. 24, no. 5, p. 102421, 2021.
78. I. N. Muliarta and I. K. Darmawan, Processing household organic waste into eco-enzyme as an effort to realize zero waste, *Agriwar J.*, vol. 1, no. 1, pp. 6–11, 2021. [Online]. Available: https://ejournal.warmadewa.ac.id/index.php/agriwar/article/view/3658. (Accessed: 27-Jun-2022).
79. P. S. Panesar, R. Kaur, G. Singla, and R. S. Sangwan, Bio-processing of agro-industrial wastes for production of food-grade enzymes: Progress and prospects, *Appl. Food Biotechnol.*, vol. 3, no. 4, pp. 208–227, Oct. 2016.
80. S. Zhang, J. Wang, and H. Jiang, Microbial production of value-added bioproducts and enzymes from molasses, a by-product of sugar industry, *Food Chem.*, vol. 346, p. 128860, Jun. 2021.
81. R. Liguori, A. Amore, and V. Faraco, Waste valorization by biotechnological conversion into added value products, *Appl. Microbiol.* Biotechnol., vol. 97, no. 14, pp. 6129–6147, Jul. 2013.
82. R. Ravindran, S. S. Hassan, G. A. Williams, and A. K. Jaiswal, A review on bioconversion of agro-industrial wastes to industrially important enzymes, *Bioengineering*, vol. 5, no. 4, p. 93, Oct. 2018.
83. R. Singh, M. Kumar, A. Mittal, and P. K. Mehta, Microbial enzymes: Industrial progress in 21st century, *3 Biotech*, vol. 6, no. 2, pp. 1–15, Dec. 2016.
84. P. Leite, J. M. Salgado, A. Venâncio, J. M. Domínguez, and I. Belo, Ultrasounds pretreatment of olive pomace to improve xylanase and cellulase production by solid-state fermentation, *Bioresour. Technol.*, vol. 214, pp. 737–746, Aug. 2016.
85. E. G. Fernández Núñez et al., Artificial intelligence approach for high level production of amylase using Rhizopus microsporus var. oligosporus and different agro-industrial wastes, *J. Chem. Technol.* Biotechnol., vol. 92, no. 3, pp. 684–692, Mar. 2017.
86. A. Abghari and S. Chen, Engineering Yarrowia lipolytica for enhanced production of lipid and citric acid, *Fermentation*, vol. 3, no. 3, p. 34, Jul. 2017.
87. P. S. Chong et al., Enhancement of batch biohydrogen production from prehydrolysate of acid treated oil palm empty fruit bunch, *Int. J. Hydrogen Energy*, vol. 38, no. 22, pp. 9592–9599, Jul. 2013.
88. B. Schink, Energetics of syntrophic cooperation in methanogenic degradation, *Microbiol. Mol. Biol.* Rev., vol. 61, no. 2, pp. 262–280, Jun. 1997.
89. K. Braber, Anaerobic digestion of municipal solid waste: A modern waste disposal option on the verge of breakthrough, *Biomass Bioenergy*, vol. 9, no. 1–5, pp. 365–376, Jan. 1995.
90. P. Weiland, Biogas production: Current state and perspectives, *Appl. Microbiol.* Biotechnol., vol. 85, no. 4, pp. 849–860, Sep. 2010.
91. B. Deepanraj, V. Sivasubramanian, and S. Jayaraj, Effect of substrate pretreatment on biogas production through anaerobic digestion of food waste, *Int. J. Hydrogen Energy*, vol. 42, no. 42, pp. 26522–26528, Oct. 2017.
92. L. J. Wu, T. Kobayashi, H. Kuramochi, Y. Y. Li, K. Q. Xu, and Y. Lv, High loading anaerobic co-digestion of food waste and grease trap waste: Determination of the limit and lipid/long chain fatty acid conversion, *Chem. Eng. J.*, vol. 338, pp. 422–431, Apr. 2018.
93. L. Qian et al., Nanoscale zero-valent iron supported by biochars produced at different temperatures: Synthesis mechanism and effect on Cr(VI) removal, *Environ. Pollut.*, vol. 223, pp. 153–160, 2017.
94. A. S. Gong, C. H. Bolster, M. Benavides, and S. L. Walker, Extraction and analysis of extracellular polymeric substances: Comparison of methods and extracellular polymeric substance levels in Salmonella pullorum SA 1685, *Environ. Eng. Sci.*, vol. 26, no. 10, pp. 1523–1532, Oct. 2009.
95. A. P. Trzcinski, E. Hernandez, and C. Webb, A novel process for enhancing oil production in algae biorefineries through bioconversion of solid by-products, *Bioresour. Technol.*, vol. 116, pp. 295–301, Jul. 2012.
96. I. C. Lee et al., Comparative toxicity and biodistribution of copper nanoparticles and cupric ions in rats, *Int. J. Nanomed.*, vol. 11, pp. 2883–2900, Jun. 2016.
97. L. Lin, F. Xu, X. Ge, and Y. Li, Improving the sustainability of organic waste management practices in the food-energy-water nexus: A comparative review of anaerobic digestion and composting, *Renew. Sustain. Energy Rev.*, vol. 89, pp. 151–167, Jun. 2018.

98. G. S. Dhillon, S. K. Brar, M. Verma, and R. D. Tyagi, Apple pomace ultrafiltration sludge – A novel substrate for fungal bioproduction of citric acid: Optimisation studies, *Food Chem.*, vol. 128, no. 4, pp. 864–871, Oct. 2011.
99. J. R. M. Almeida, L. C. L. Fávaro, and B. F. Quirino, Biodiesel biorefinery: Opportunities and challenges for microbial production of fuels and chemicals from glycerol waste, *Biotechnol. Biofuels*, vol. 5, no. 1, pp. 1–16, Jul. 2012.
100. S. Mazumdar, J. Lee, and M. K. Oh, Microbial production of 2,3 butanediol from seaweed hydrolysate using metabolically engineered Escherichia coli, *Bioresour. Technol.*, vol. 136, pp. 329–336, May 2013.
101. M. D. Blankschien, J. M. Clomburg, and R. Gonzalez, Metabolic engineering of Escherichia coli for the production of succinate from glycerol, *Metab. Eng.*, vol. 12, no. 5, pp. 409–419, Sep. 2010.

Index

Note: **Bold** page numbers refer to tables; *italic* page numbers refer to figures and page numbers followed by "n" denote endnotes.

T

U

V

W

Y

Z